Practical Methods of Optimization

Practical Methods of Optimization

Second Edition

R. Fletcher
Department of Mathmatics and Computer Science
University of Dundee, Scotland, UK

A Wiley-Interscience Publication

JOHN WILEY & SONS

Chichester · New York · Brisbane · Toronto · Singapore

Telephone (+44) 1243 779777

Email (for orders and customer service enquiries): cs-books@wiley.co.uk
Visit our Home Page on www.wileyeurope.com or www.wiley.co.uk

Reprinted March 1991, July 1993, January 1995, January 1996, January 1999,
Reprinted in paperback May 2000, October 2001, June 2003, September 2004,
January 2006, July 2006, September 2007, June 2008

Other Wiley Editorial Offices

John Wiley & Sons Inc., 111 River Street, Hoboken, NJ 07030, USA

Jossey-Bass, 989 Market Street, San Francisco, CA 94103-1741, USA

Wiley-VCH Verlag GmbH, Boschstr. 12, D-69469 Weinheim, Germany

John Wiley & Sons Australia Ltd, 33 Park Road, Milton, Queensland 4064, Australia

John Wiley & Sons (Asia) Pte Ltd, 2 Clementi Loop #02-01, Jin Xing Distripark, Singapore 129809

John Wiley & Sons Canada Ltd, 22 Worcester Road, Etobicoke, Ontario, Canada M9W 1L1

British Library Cataloguing in Publication Data

A catalogue record for this book is available from the British Library

ISBN: 13: 978 0 471 91547 8 (HB)
ISBN: 13: 978 0 471 49463 8 (PB)

Contents

Preface

The subject of optimization is a fascinating blend of heuristics and rigour, of theory and experiment. It can be studied as a branch of pure mathematics, yet has applications in almost every branch of science and technology. This book aims to present those aspects of optimization methods which are currently of foremost importance in solving real life problems. I strongly believe that it is not possible to do this without a background of practical experience into how methods behave, and I have tried to keep practicality as my central theme. Thus basic methods are described in conjunction with those heuristics which can be valuable in making the methods perform more reliably and efficiently. In fact I have gone so far as to present comparative numerical studies, to give the feel for what is possible, and to show the importance (and difficulty) of assessing such evidence. Yet one cannot exclude the role of theoretical studies in optimization, and the scientist will always be in a better position to use numerical techniques effectively if he understands some of the basic theoretical background. I have tried to present such theory as shows how methods are derived, or gives insight into how they perform, whilst avoiding theory for theory's sake.

Some people will approach this book looking for a suitable text for undergraduate and postgraduate classes. I have used this material (or a selection from it) at both levels, in introductory engineering courses, in Honours mathematics lectures, and in lecturing to M.Sc. and Ph.D. students. In an attempt to cater for this diversity, I have used a Jekyll and Hyde style in the book, in which the more straightforward material is presented in simple terms, whilst some of the more difficult theoretical material is nonetheless presented rigorously, but can be avoided if need be. I have also tried to present worked examples for most of the basic methods. One observation of my own which I pass on for what it is worth is that the students gain far more from a course if they can be provided with computer subroutines for a few of the standard methods, with which they can perform simple experiments for themselves, to see for example how badly the steepest descent method handles Rosenbrock's problem, and so on.

In addition to the worked examples, each chapter is terminated by a set of questions which aim to not only illustrate but also extend the material in the text. Many of the questions I have used in tutorial classes or examination papers. The reader may find a calculator (and possibly a programmable calculator) helpful in some cases. A few of the questions are taken from the Dundee Numerical Analysis M.Sc. examination, and are open book questions in the nature of a one day mini research project.

The second edition of the book combines the material in Volumes 1 and 2 of the first edition. Thus unconstrained optimization is the subject of Part 1 and covers the basic theoretical background and standard techniques such as line search methods, Newton and quasi-Newton methods and conjugate direction methods. A feature not common in the literature is a comprehensive treatment of restricted step or trust region methods, which have very strong theoretical properties and are now preferred in a number of situations. The very important field of nonlinear equations and nonlinear least squares (for data fitting applications) is also treated thoroughly. Part 2 covers constrained optimization which overall has a greater degree of complexity on account of the presence of the constraints. I have covered the theory of constrained optimization in a general (albeit standard) way, looking at the effect of first and second order perturbations at the solution. Some books prefer to emphasize the part played by convex analysis and duality in optimization problems. I also describe these features (in what I hope is a straightforward way) but give them lesser priority on account of their lack of generality.

Most finite dimensional problems of a continuous nature have been included in the book but I have generally kept away from problems of a discrete or combinatorial nature since they have an entirely different character and the choice of method can be very specialized. In this case the nearest thing to a general purpose method is the branch and bound method, and since this is a transformation to a sequence of continuous problems of the type covered in this volume, I have included a straightforward description of the technique. A feature of this book which I think is lacking in the literature is a treatment of non-differentiable optimization which is reasonably comprehensive and covers both theoretical and practical aspects adequately. I hope that the final chapter meets this need. The subject of geometric programming is also included in the book because I think that it is potentially valuable, and again I hope that this treatment will turn out to be more straightforward and appealing than others in the literature. The subject of nonlinear programming is covered in some detail but there are difficulties in that this is a very active research area. To some extent therefore the presentation mirrors my assessment and prejudice as to how things will turn out, in the absence of a generally agreed point of view. However, I have also tried to present various alternative approaches and their merits and demerits. Linear constraint programming, on the other hand, is now well developed and here the difficulty is that there are two distinct points of view. One is the traditional approach in which algorithms are presented as generalizations of early linear programming methods which carry out pivoting in a tableau. The other is a more recent approach in terms of active set strategies:

I regard this as more intuitive and flexible and have therefore emphasized it, although both methods are presented and their relationship is explored.

This second edition has given me the opportunity to improve the presentation of some parts of the book and to introduce new developments and a certain amount of new material. In Part 1 the description of line searches is improved and some new results are included. The variational properties of the BFGS and DFP methods are now described in some detail. More simple proofs of the properties of trust region methods are given. Recent developments in hybrid methods for nonlinear least squares are described. A thorough treatment of the Dennis–Moré theorem characterizing superlinear convergence in nonlinear systems is given and its significance is discussed. In Part 2 the treatment of linear programming has been extended considerably and includes new methods for stable updating of LU factors and the reliable treatment of degeneracy. Also, important recent developments in polynomial time algorithms are described and discussed, including ellipsoid algorithms and Karmarkar's method. The treatment of quadratic programming now includes a description of range space and dual active set methods. For general linear constraint programming some new theorems are given, including convergence proofs for a trust region method. The chapter on nonlinear programming now includes an extra section giving a direct treatment of the L_1 exact penalty function not requiring any convex analysis. New developments in sequential quadratic programming (SQP) are described, particularly for the case that only the reduced Hessian matrix is used. A completely new section on network programming is given relating numerical linear algebraic and graph theoretic concepts and showing their application in various types of optimization problem. For non-smooth optimization, Osborne's concept of structure functionals is used to unify the treatment of regularity for second order conditions and to show the equivalence to nonlinear programming. It is also used to demonstrate the second order convergence of a non-smooth SQP algorithm. The Maratos effect and the use of second order corrections are described. Finally a new section giving optimality conditions for constrained composite non-smooth optimization is included. A considerable number of new exercises is also given.

It is a great pleasure to me to acknowledge those many people who have influenced my thinking and contributed to my often inadequate knowledge. Amongst many I must single out the assistance and encouragement given to me by Professor M. J. D. Powell, my former colleague at AERE Harwell, and one whose contributions to the subject are unsurpassed. I am also indebted to Professor A. R. Mitchell and other members of the University of Dundee for providing the stimulating and yet relaxed environment in which this book was prepared. I also wish to thank Professor D. S. Jones for his interest and encouragement in publishing the book, and Drs M. P. Jackson, G. A. Watson and R. S. Womersley for their constructive advice on the contents. I gratefully acknowledge those various people who have taken the trouble to write or otherwise inform me of errors, misconceptions, etc., in text. Whilst this new edition has given me the opportunity to correct previous errors, it has also inevitably enabled me to introduce many new ones for which I apologise in advance. I am also grateful for the invaluable secretarial help that I have received over the years in preparing various drafts of this book.

Last, but foremost, I wish to dedicate this book to my parents and family as some small acknowledgement of their unfailing love and affection.

Dundee, December 1986 R. Fletcher

Table of Notation

$\mathbf{A}$	matrix (Jacobian matrix, matrix of constraint normals)
$\mathbf{I}$	unit matrix
$\mathbf{L}, \mathbf{U}$	lower or upper triangular matrices respectively
$\mathbf{P}, \mathbf{Q}$	permutation matrix or orthogonal matrix
$\mathbf{a}$	vector (usually a column vector)
$\mathbf{a}_i\ i = 1, 2, \ldots$	set of vectors (columns of $\mathbf{A}$)
$\mathbf{e}_i\ i = 1, 2, \ldots$	coordinate vectors (columns of $\mathbf{I}$)
$\mathbf{e}$	vector of ones $(1, 1, \ldots, 1)^{\mathrm{T}}$
$\mathbf{A}^{\mathrm{T}}, \mathbf{a}^{\mathrm{T}}$	transpose
$\mathbb{R}^n$	n-dimensional space
$\mathbf{x}$	variables in an optimization problem
$\mathbf{x}^{(k)}\ k = 1, 2, \ldots$	iterates in an iterative method
$\mathbf{x}^*$	local minimizer or local solution
$\mathbf{s}, \mathbf{s}^{(k)}$	search direction (on iteration k)
$\alpha, \alpha^{(k)}$	step length (on iteration k)
$\boldsymbol{\delta}, \boldsymbol{\delta}^{(k)}$	correction to $\mathbf{x}^{(k)}$
$f(\mathbf{x})$	objective function
∇	first derivative operator (elements $\partial/\partial x_i$)
$\mathbf{g}(\mathbf{x}) = \nabla f(\mathbf{x})$	gradient vector
$f^*, \mathbf{g}^*, \ldots$	$f(\mathbf{x}^*), \mathbf{g}(\mathbf{x}^*), \ldots$
$f^{(k)}, \mathbf{g}^{(k)}, \ldots$	$f(\mathbf{x}^{(k)}), \mathbf{g}(\mathbf{x}^{(k)}), \ldots$
∇^2	second derivative operator (elements $\partial^2/\partial x_i \partial x_j$)
$\mathbf{G}(\mathbf{x}) = \nabla^2 f(\mathbf{x})$	Hessian matrix (second derivative matrix)
$\mathbb{C}^k$	set of k times continuously differentiable functions
$l(\mathbf{x})$	linear function (1.2.8)
$q(\mathbf{x})$	quadratic function (1.2.11)
$[a, b]$	closed interval
(a, b)	open interval
$\|\cdot\|$	norm of a vector or matrix
$\square$	end of proof
$\exists, \forall$	'there exists', 'for all'
$\Rightarrow, \Leftrightarrow$	'implies', 'equivalent to'
$O(.), o(.)$	'big O' and 'little o' notation (Hardy, 1960) (let $h \to 0$: then $a = O(h)$ iff $\exists$ a constant c such that $\lvert a \rvert \leqslant ch$, and $a = o(h)$ iff $a/h \to 0$)
$\subset$	set inclusion
$\in$	element in set

$\emptyset$	empty set
$\triangleq$	equal by definition
$\mathscr{L}$	Lagrangian function
σ, ρ, τ	fixed parameters in algorithms
$\mathbf{r}$	residual vector, artificial variables vector
$\mathbf{c}_i(\mathbf{x})\ i = 1, 2, \ldots$	constraint functions
E, I	set of equality and inequality constraints respectively (I is set of integer variables in Chapter 13)
$\mathscr{A}$	set of active constraints
$\mathbf{c}(\mathbf{x})$	vector of constraint functions (usually for equality constraints only)
$\mathbf{a}_i(\mathbf{x}) = \nabla \mathbf{c}_i(\mathbf{x})$	constraint gradient vector (normal vector)
x_i^+, x_i^-	max $(x, 0)$ and max $(-x, 0)$ respectively
$\mathbf{A}_B$	basis matrix in linear programming
B, N	sets of basic and nonbasic variables respectively
$\hat{\mathbf{A}}, \hat{\mathbf{c}}, \hat{\mathbf{b}}$	tableau matrix, reduced costs, basic variable values
$\mathbf{x}_c.\ \mathbf{x}_+$	current and next point (Section 8.7)
$\lambda_i\ i = 1, 2, \ldots$	Lagrange multipliers
$\boldsymbol{\lambda}$	vector of Lagrange multipliers (usually for equality constraints only)
I*	$\mathscr{A}^* \cap \mathrm{I}$ (set of active inequality constraints at $\mathbf{x}^*$)
$\mathscr{F}, F, \mathscr{G}, G$	feasible direction sets for optimality conditions
K	convex set
$\mathbf{Y}, \mathbf{Z}$	left inverse and null space matrices in generalized elimination
:=	assignment operator
$q^{(k)}(\boldsymbol{\delta})$	model quadratic function obtained by Taylor series expansion about $\mathbf{x}^{(k)}$
$\mathbf{l}^{(k)}(\boldsymbol{\delta})$	linear model about $\mathbf{x}^{(k)}$ to vector of constraint functions
$\lvert\mathscr{A}\rvert$	number of elements in active set
ν, σ	weighting parameters in penalty functions
$\dot{\mathbf{x}}$	$\mathrm{d}x/\mathrm{d}\theta$ for some trajectory $\mathbf{x}(\theta)$
∇_x, ∇_λ	partial derivative operators with respect to x, λ respectively
$\blacktriangledown = \begin{pmatrix} \nabla_x \\ \nabla_\lambda \end{pmatrix}$	combined vector of partial derivatives
$\mathbf{W} = \nabla_x^2 \mathscr{L}(\mathbf{x}, \boldsymbol{\lambda})$	Hessian of Lagrangian function
$\mathbf{W}^*, \mathbf{W}^{(k)}, \ldots$	$\mathbf{W}(\mathbf{x}^*, \boldsymbol{\lambda}^*), \mathbf{W}(\mathbf{x}^{(k)}, \boldsymbol{\lambda}^{(k)}), \ldots$
$\mathbf{M}$	approximation to reduced Hessian matrix ($\mathbf{Z}^T\mathbf{G}\mathbf{Z}$ in Chapter 11, or $\mathbf{Z}^T\mathbf{W}\mathbf{Z}$ in Chapter 12)
$\phi(\mathbf{x})$	penalty function (possibly with additional parameters in argument list)
$\psi^{(k)}(\boldsymbol{\delta})$	model approximating function about $\mathbf{x}^{(k)}$ to $\phi(\mathbf{x}^{(k)} + \boldsymbol{\delta})$
[x]	greatest integer not larger than x
G, T	graph, tree
$h(\mathbf{c})$	convex non-smooth function
$\partial h(\mathbf{c})$	subdifferential (set of all subgradients at $\mathbf{c}$)
$h^*, h^{(k)}, \ldots$	$h(\mathbf{c}^*), h(\mathbf{c}^{(k)}), \ldots$ (may be $h(\mathbf{c}(\mathbf{x}^*))$ etc. when $\mathbf{c} = \mathbf{c}(\mathbf{x})$)
$\mathbf{D}^*$	basis vectors for subdifferential (14.2.30)
$\partial h - \boldsymbol{\lambda}$	the set $\{\mathbf{u}: \mathbf{u} = \boldsymbol{\gamma} - \boldsymbol{\lambda}, \boldsymbol{\gamma} \in \partial h\}$
$\partial h \backslash \boldsymbol{\lambda}$	the set $\{\mathbf{u}: \mathbf{u} \in \partial h,\ \mathbf{u} \neq \boldsymbol{\lambda}\}$

PART 1

UNCONSTRAINED OPTIMIZATION

Chapter 1
Introduction

1.1 HISTORY AND APPLICATIONS

Optimization might be defined as the science of determining the 'best' solutions to certain mathematically defined problems, which are often models of physical reality. It involves the study of optimality criteria for problems, the determination of algorithmic methods of solution, the study of the structure of such methods, and computer experimentation with methods both under trial conditions and on real life problems. There is an extremely diverse range of practical applications. Yet the subject can be studied (not here) as a branch of pure mathematics.

Before 1940 relatively little was known about methods for numerical optimization of functions of many variables. There had been some least squares calculations carried out, and steepest descent type methods had been applied in some physics problems. The Newton method in many variables was known, and more sophisticated methods were being attempted such as the self-consistent field method for variational problems in theoretical chemistry. Nonetheless anything of any complexity demanded armies of assistants operating desk calculating machines. There is no doubt therefore that the advent of the computer was paramount in the development of optimization methods and indeed in the whole of numerical analysis. The 1940s and 1950s saw the introduction and development of the very important branch of the subject known as linear programming. (The term 'programming' by the way is synonymous with 'optimization' and was originally used to mean optimization in the sense of optimal planning.) All these methods however had a fairly restricted range of application, and again in the post-war period the development of 'hill-climbing' methods took place—methods of wide applicability which did not rely on any special structure in the problem. The latter methods were at first very crude and inefficient, but the subject was again revolutionized in 1959 with the publication of a report by W. C. Davidon which led to the introduction of variable metric methods. My friend and colleague M. J. D. Powell describes a meeting he attended in 1961 in which the speakers were telling of the difficulty of minimizing functions of ten variables, whereas he had just programmed a method based on Davidon's ideas which had solved problems of 100 variables in a short time. Since that time the development of the subject has proceeded apace and has included methods for a wide variety of problems. This book describes these developments in what is hoped will be a systematic and comprehensive way.

The applicability of optimization methods is widespread, reaching into almost every activity in which numerical information is processed (Science, Engineering, Mathematics, Economics, Commerce, etc.). To provide a comprehensive account of all these applications would therefore be unrealistic, but a selection might include:

(a) chemical reactor design;
(b) aero-engine or aero-frame design;
(c) structural design—buildings, bridges, etc.;
(d) commerce—resource allocation, scheduling, blending;

and applications to other branches of numerical analysis:

(e) data fitting;
(f) variational principles in p.d.e.s;
(g) nonlinear equations in o.d.e.s; and
(h) penalty functions.

More such applications can be found in the proceedings of a conference on 'Optimization in Action' (Dixon, 1976), and many more of course in the specialized technical literature. However to give some idea of what is involved consider the optimum design of a distillation column, which can be modelled in an idealized way as in Figure 1.1.1. The aim of such a column is to separate out a more volatile component from a mixture of components in the input

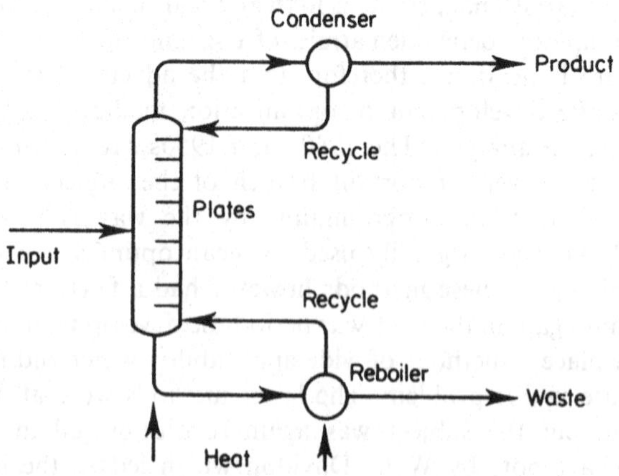

Figure 1.1.1 A model distillation column

stream. An *objective function* to be optimized might therefore be the quantity of the product or the profit from operating the system. The variables would be the rate of flow in the input, the heat rates applied and on each plate the liquid and vapour compositions of each component, and the temperature and vapour pressure. The variables are subject to restrictions or interrelations of many kinds, which are referred to as *constraints*. For instance compositions and flows must be non-negative ($x_i \geqslant 0$) and temperatures must not exceed certain upper bounds ($T_i \leqslant T_{\max}$). Relationships such as the unit sum of percentage compositions must be included explicitly ($\sum_i x_i = 1$). More complicated constraints state how components interact physically, for instance vapour and liquid compositions are related by $v_i = l_i \phi(T_i)$, where $\phi(T_i)$ is a given but highly nonlinear function of temperature. A more difficult situation arises if the number of plates in the column is allowed to vary, and this is an example of an *integer variable* which can take on only integer values.

This book however is not concerned with applications, except insofar as they indicate the different types of optimization problem which arise. It is possible to categorize these into a relatively small number of standard problems and to state algorithms for each one. The user's task is to discover into what category his problem fits, and then to call up the appropriate optimization subroutine on a computer. This subroutine will specify to the user how the problem data is to be presented, for example nonlinear functions usually have to be programmed in a user-written subroutine in a certain standard format. It is also as well to remember that in practice the solution of an optimization problem is not the only information that the user might need. He will often be interested in the *sensitivity* of the solution to changes in the parameters, especially so if the mathematical model is not a close approximation to reality, or if he cannot build his design to the same accuracy as the solution. He may indeed be interested in the variation of the solution obtained by varying some parameters over wide ranges, and it is often possible to provide this information without re-solving the problem numerous times.

This book therefore is concerned with some of the various standard optimization problems which can arise. In fact the material is divided into Part 1 and Part 2. Part 1 is devoted to the subject of *unconstrained optimization*, in which the optimum value is sought of an objective function of many variables, without any constraints. This problem is important in its own right and also as a major tool in solving some constrained problems. Also many of the ideas carry over into constrained optimization. The special case of sums of squares functions, which arise in data fitting problems, is also considered. This also includes the solution of sets of simultaneous nonlinear equations, which is an important problem in its own right, but which is often solved by optimization methods. Part 2 is devoted to *constrained optimization* in which the additional complication arises of the various types of constraint referred to above. An overview of constrained optimization is given in Section 7.

In this book a selection has had to be made amongst the extensive literature about optimization methods. I have been concerned to present *practical* methods (and associated theory) which have been implemented and for which a body of satisfactory numerical experience exists. I am equally concerned about reliability of algorithms and whether there is proof or good reason to think that convergence to a solution will occur at a reasonably rapid rate. However, I shall also be trying to point out which new ideas in the subject I feel are significant and which might lead to future developments. Many people may read this book seeking a particular algorithm which best solves their specific problem. Such advice is not easy to give, especially in that the decision is not as clear-cut as it may seem. There are many special cases which should be taken into account, for instance the relative ease of computing the function and its derivatives. Similarly, considerations of how best to pose the problem in the first instance are relevant to the choice of method. Finally, and of most importance, the decision is subject to the availability of computer subroutines or packages which implement the methods. However some program libraries now give a decision tree in the documentation to help the user choose his method. Whilst these are valuable, they should only be used as a rough guide, and never as a substitute for common sense or the advice of a specialist in optimization techniques.

1.2 MATHEMATICAL BACKGROUND

The book relies heavily on the concepts and techniques of matrix algebra and numerical linear algebra, which are not set out here (see Broyden, 1975, for example), although brief explanations are given in passing in certain cases. A *vector* is represented by a lower case bold letter (e.g. **a**) and usually refers to a column vector. A *matrix* is referred to by a bold upper case letter (e.g. **B**). That is

$$\mathbf{a} = \begin{pmatrix} a_1 \\ a_2 \\ \vdots \\ a_n \end{pmatrix}; \qquad \mathbf{B} = \begin{bmatrix} B_{11} & B_{12} & \cdots & B_{1s} \\ B_{21} & B_{22} & \cdots & B_{2s} \\ \vdots & \vdots & & \vdots \\ B_{r1} & B_{r2} & \cdots & B_{rs} \end{bmatrix}.$$

Sometimes b_{ij} is used for elements of **B** in place of B_{ij}. Transposition is referred to by superscript T so that $\mathbf{a}^{\mathrm{T}}$ is a row vector and $\mathbf{a}^{\mathrm{T}}\mathbf{z}$ for instance is the *scalar product* $\mathbf{a}^{\mathrm{T}}\mathbf{z} = \mathbf{z}^{\mathrm{T}}\mathbf{a} = \sum_i a_i z_i$.

The ideas of vector spaces are also used, although often only in a simple minded way. A *point* $\mathbf{x}$ in n-dimensional space ($\mathbb{R}^n$) is the vector $(x_1, x_2, \ldots, x_n)^{\mathrm{T}}$, where x_1 is the component in the first coordinate direction, and so on. Most of the methods to be described are *iterative methods* which generate a *sequence* of points, $\mathbf{x}^{(1)}$, $\mathbf{x}^{(2)}$, $\mathbf{x}^{(3)}$,... say, or $\{\mathbf{x}^{(k)}\}$ (the superscripts denoting iteration number), hopefully converging to a fixed point $\mathbf{x}^*$ which is the solution of the problem (see Figure 1.2.2). The idea of a *line* is important, and is the set of points

$$\mathbf{x}(=\mathbf{x}(\alpha)) = \mathbf{x}' + \alpha\mathbf{s} \tag{1.2.1}$$

for all α (sometimes for all $\alpha \geqslant 0$; this is strictly a half-line), in which $\mathbf{x}'$ is a fixed

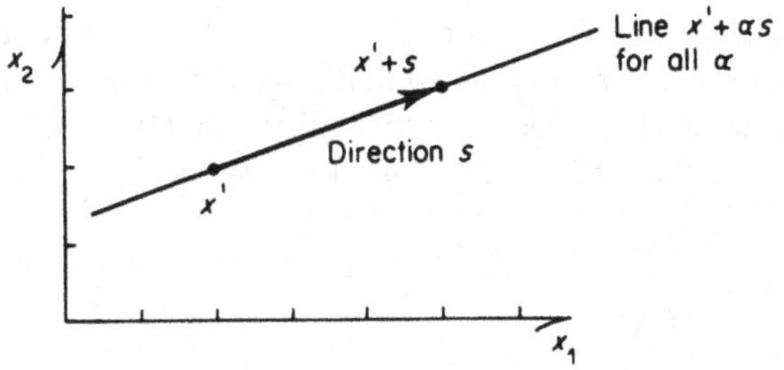

Figure 1.2.1 A line in two dimensions

point along the line (corresponding to $\alpha = 0$), and $\mathbf{s}$ is the *direction* of the line. For instance in Figure 1.2.1 $\mathbf{x}'$ is the point $\begin{pmatrix} 2 \\ 2 \end{pmatrix}$ and $\mathbf{s}$ the direction $\begin{pmatrix} 3 \\ 1 \end{pmatrix}$. The vector $\mathbf{s}$ is indicated by the arrow. Sometimes it is convenient to *normalize* $\mathbf{s}$ so that for instance $\mathbf{s}^T\mathbf{s} = \sum_i s_i^2 = 1$; this does not change the line, but only the value of α associated with any point.

The calculus of any *function* of many variables, $f(\mathbf{x})$ say, is clearly important. Some pictorial intuition for two variable problems is often gained by drawing *contours* (surfaces along which $f(\mathbf{x})$ is constant). A well-known test function for optimization methods is Rosenbrock's function

$$f(\mathbf{x}) = 100(x_2 - x_1^2)^2 + (1 - x_1)^2 \tag{1.2.2}$$

the contours for which are shown in Figure 1.2.2. Some other contours are

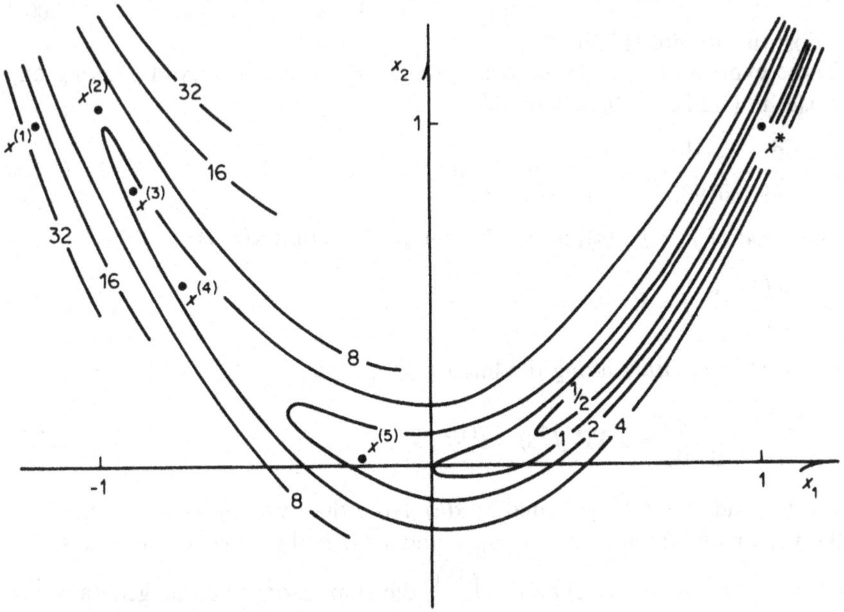

Figure 1.2.2 Contours for Rosenbrock's function, equation (1.2.2)

illustrated in Figure 6.2.2 in Chapter 6. In general it will be assumed that the problem functions which arise are *smooth*, that is continuous and continuously (Fréchet) differentiable ($\mathbb{C}^1$). Therefore for a function $f(\mathbf{x})$ at any point $\mathbf{x}$ there is a *vector of first partial derivatives*, or *gradient vector*

$$\begin{pmatrix} \partial f/\partial x_1 \\ \partial f/\partial x_2 \\ \vdots \\ \partial f/\partial x_n \end{pmatrix}_{\mathbf{x}} = \nabla f(\mathbf{x}) \tag{1.2.3}$$

where ∇ denotes the gradient operator $(\partial/\partial x_1, \ldots, \partial/\partial x_n)^{\mathrm{T}}$. If $f(\mathbf{x})$ is twice continuously differentiable ($\mathbb{C}^2$) then there exists a *matrix of second partial derivatives* or *Hessian matrix*, written $\nabla^2 f(\mathbf{x})$, for which the i, jth element is $\partial^2 f/(\partial x_i \partial x_j)$. This matrix is square and symmetric. Since any column (the jth, say) is $\nabla(\partial f/\partial x_j)$, the matrix can strictly be written as $\nabla(\nabla f^{\mathrm{T}})$. For example, in (1.2.2)

$$\begin{aligned} \nabla f(\mathbf{x}) &= \begin{pmatrix} -400x_1(x_2 - x_1^2) - 2(1 - x_1) \\ 200(x_2 - x_1^2) \end{pmatrix} \\ \nabla^2 f(\mathbf{x}) &= \begin{bmatrix} 1200x_1^2 - 400x_2 + 2 & -400x_1 \\ -400x_1 & 200 \end{bmatrix} \end{aligned} \tag{1.2.4}$$

and this illustrates that ∇f and $\nabla^2 f$ will in general depend upon $\mathbf{x}$, and vary from point to point. Thus at $\mathbf{x}' = \begin{pmatrix} 0 \\ 0 \end{pmatrix}$, $\nabla f(\mathbf{x}') = \begin{pmatrix} -2 \\ 0 \end{pmatrix}$ and $\nabla^2 f(\mathbf{x}') = \begin{bmatrix} 2 & 0 \\ 0 & 200 \end{bmatrix}$ by substitution into (1.2.4).

These expressions can be used to determine the derivatives of f along any line $\mathbf{x}(\alpha)$ in (1.2.1). By the chain rule

$$\frac{\mathrm{d}}{\mathrm{d}\alpha} = \sum_i \frac{\mathrm{d}}{\mathrm{d}\alpha} x_i(\alpha) \frac{\partial}{\partial x_i} = \sum_i s_i \frac{\partial}{\partial x_i} = \mathbf{s}^{\mathrm{T}} \nabla \tag{1.2.5}$$

so the *slope* of $f(= f(\mathbf{x}(\alpha))$ along the line at any point $\mathbf{x}(\alpha)$ is

$$\frac{\mathrm{d}f}{\mathrm{d}\alpha} = \mathbf{s}^{\mathrm{T}} \nabla f = \nabla f^{\mathrm{T}} \mathbf{s}. \tag{1.2.6}$$

Likewise the *curvature* along the line is

$$\frac{\mathrm{d}^2 f}{\mathrm{d}\alpha^2} = \frac{\mathrm{d}}{\mathrm{d}\alpha} \frac{\mathrm{d}f}{\mathrm{d}\alpha} = \mathbf{s}^{\mathrm{T}} \nabla(\nabla f^{\mathrm{T}} \mathbf{s}) = \mathbf{s}^{\mathrm{T}} \nabla^2 f \mathbf{s} \tag{1.2.7}$$

where ∇f and $\nabla^2 f$ are evaluated at $\mathbf{x}(\alpha)$. Note that, writing $\mathbf{G} = \nabla^2 f$, then $\mathbf{Gs}$ is the vector for which $(\mathbf{Gs})_i = \sum_j G_{ij} s_j$, and $\mathbf{s}^{\mathrm{T}}\mathbf{Gs}$ is the scalar product of $\mathbf{s}$ and $\mathbf{Gs}$. For example, for (1.2.2) at $\mathbf{x}' = \begin{pmatrix} 0 \\ 0 \end{pmatrix}$, the slope along the line generated by

$\mathbf{s} = \begin{pmatrix} 1 \\ 0 \end{pmatrix}$ (the x_1-axis in Figure 1.2.2) is $\mathbf{s}^T\nabla f = -2$ and the curvature is $\mathbf{s}^T\mathbf{G}\mathbf{s} = 2$ $\left(\text{since } \mathbf{G}\mathbf{s} = \begin{pmatrix} 2 \\ 0 \end{pmatrix}\right)$.

These definitions of slope and curvature depend on the size of $\mathbf{s}$, and this ambiguity can be resolved by requiring that $\|\mathbf{s}\| = 1$. (Note: the *norm* $\|\mathbf{s}\|$ is just a measure of the size of $\mathbf{s}$; one common norm is the L_2 norm $\|\mathbf{s}\|_2 = \sqrt{(\mathbf{s}^T\mathbf{s})}$.) Denoting $\nabla f(\mathbf{x}')$ by $\mathbf{g}'$, then $\pm\mathbf{g}'/\|\mathbf{g}'\|_2$ are the directions of greatest and least slope at $\mathbf{x}'$, over all directions for which $\|\mathbf{s}\|_2 = 1$, and are orthogonal to the contour and tangent plane of $f(\mathbf{x})$ at $\mathbf{x}'$ (see Figure 1.2.3 and Question 1.4).

Special cases of many variable functions include the general *linear function* which can be written

$$l(\mathbf{x}) = \sum_{i=1}^{n} a_i x_i + b = \mathbf{a}^T\mathbf{x} + b \tag{1.2.8}$$

where $\mathbf{a}$ and b are constant. (Strictly this should be described as an *affine* function on account of the existence of the constant b. However, the use of *linear* to describe a function whose graph is a line (or a hyperplane) is common in optimization. I do not intent to depart from this usage, but apologise to the erudite reader.) If the coordinate vector

$$\mathbf{e}_i = \begin{pmatrix} 0 \\ 0 \\ \vdots \\ 1 \\ \vdots \\ 0 \end{pmatrix} \leftarrow i\text{th position} \tag{1.2.9}$$

is defined, then the identity $\nabla x_i = \mathbf{e}_i$ gives

$$\nabla\mathbf{x}^T = \nabla(x_1, x_2, \ldots, x_n) = [\mathbf{e}_1, \mathbf{e}_2, \ldots, \mathbf{e}_n] = \mathbf{I} \tag{1.2.10}$$

since the vectors $\mathbf{e}_i$ are the columns of the *unit matrix* $\mathbf{I}$. Thus for (1.2.8), $\nabla l = \mathbf{a}$

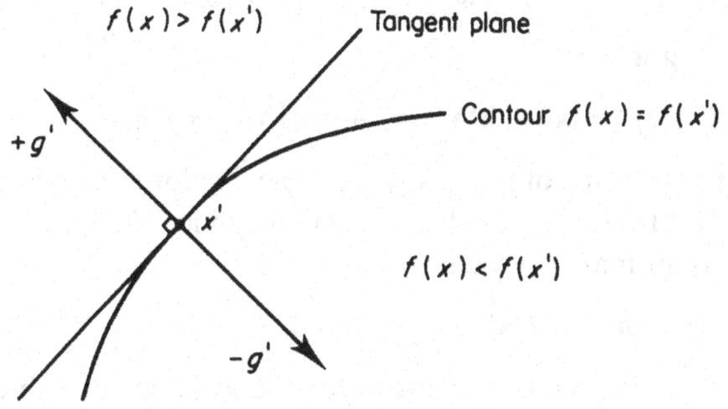

Figure 1.2.3 Properties of the gradient vector

is a constant vector, and $\nabla^2 l = \mathbf{0}$ is the *zero matrix*. A general *quadratic function* can be written

$$q(\mathbf{x}) = \tfrac{1}{2}\mathbf{x}^{\mathrm{T}}\mathbf{G}\mathbf{x} + \mathbf{b}^{\mathrm{T}}\mathbf{x} + c \tag{1.2.11}$$

where $\mathbf{G}, \mathbf{b}$, and c are constant and $\mathbf{G}$ is symmetric, or as

$$q(\mathbf{x}) = \tfrac{1}{2}(\mathbf{x} - \mathbf{x}')^{\mathrm{T}}\mathbf{G}(\mathbf{x} - \mathbf{x}') + c' \tag{1.2.12}$$

where $\mathbf{G}\mathbf{x}' = -\mathbf{b}$ and $c' = c - \tfrac{1}{2}\mathbf{x}'^{\mathrm{T}}\mathbf{G}\mathbf{x}'$. From the rule for differentiating a product, it can be verified that

$$\nabla(\mathbf{u}^{\mathrm{T}}\mathbf{v}) = (\nabla\mathbf{u}^{\mathrm{T}})\mathbf{v} + (\nabla\mathbf{v}^{\mathrm{T}})\mathbf{u} \tag{1.2.13}$$

if $\mathbf{u}$ and $\mathbf{v}$ depend upon $\mathbf{x}$. It therefore follows from (1.2.11) (using $\mathbf{u} = \mathbf{x}$, $\mathbf{v} = \mathbf{G}\mathbf{x}$) that

$$\nabla q(\mathbf{x}) = \tfrac{1}{2}(\mathbf{G} + \mathbf{G}^{\mathrm{T}})\mathbf{x} + \mathbf{b} = \mathbf{G}\mathbf{x} + \mathbf{b} \tag{1.2.14}$$

using the symmetry of $\mathbf{G}$. Likewise $\nabla^2 q = \mathbf{G}$ can be established. Thus $q(\mathbf{x})$ has a constant Hessian matrix $\mathbf{G}$ and its gradient is a linear function of $\mathbf{x}$. A consequence of (1.2.14) is that if $\mathbf{x}'$ and $\mathbf{x}''$ are two given points and if $\mathbf{g}' = \nabla q(\mathbf{x}')$ and $\mathbf{g}'' = \nabla q(\mathbf{x}'')$ then

$$\mathbf{g}'' - \mathbf{g}' = \mathbf{G}(\mathbf{x}'' - \mathbf{x}') \tag{1.2.15}$$

that is the Hessian matrix maps differences in position into differences in gradient. This result is used widely.

An indispensable technique for handling more general smooth functions of many variables is the *Taylor series*. For functions of one variable the infinite series is

$$f(\alpha) = f(0) + \alpha f'(0) + \tfrac{1}{2}\alpha^2 f''(0) + \cdots \tag{1.2.16}$$

although the series may be truncated after the term in α^p, replacing $f^{(p)}(0)$ by $f^{(p)}(\xi)$ where $\xi \in [0, \alpha]$. An integral form of the remainder can also be used. Now let $f(\alpha) = f(\mathbf{x}(\alpha))$ be the value of a function of many variables along the line $\mathbf{x}(\alpha)$ (see (1.2.1)). Then using (1.2.6) and (1.2.7) in (1.2.16)

$$f(\mathbf{x}' + \alpha\mathbf{s}) = f(\mathbf{x}') + \alpha\mathbf{s}^{\mathrm{T}}\nabla f(\mathbf{x}') + \tfrac{1}{2}\alpha^2\mathbf{s}^{\mathrm{T}}[\nabla^2 f(\mathbf{x}')]\mathbf{s} + \cdots \tag{1.2.17}$$

or by writing $\mathbf{h} = \alpha\mathbf{s}$

$$f(\mathbf{x}' + \mathbf{h}) = f(\mathbf{x}') + \mathbf{h}^{\mathrm{T}}\nabla f(\mathbf{x}') + \tfrac{1}{2}\mathbf{h}^{\mathrm{T}}[\nabla^2 f(\mathbf{x}')]\mathbf{h} + \cdots \tag{1.2.18}$$

These are two forms of the many variable Taylor series. Furthermore, consider applying (1.2.18) to the function $\partial f(\mathbf{x})/\partial x_i$. Since $\nabla(\partial f(\mathbf{x})/\partial x_i)$ is the ith column of the Hessian matrix $\nabla^2 f$, it follows that

$$\nabla f(\mathbf{x}' + \mathbf{h}) = \nabla f(\mathbf{x}') + [\nabla^2 f(\mathbf{x}')]\mathbf{h} + \cdots \tag{1.2.19}$$

which is a Taylor series expansion for the gradient of f. Neglecting the higher terms in the limit $\mathbf{h} \to \mathbf{0}$, then this reduces to (1.2.15) showing that a general function behaves like a quadratic function in a sufficiently small neighbourhood of $\mathbf{x}'$.

It is hoped that a grasp of simple mathematical concepts such as these will enable the reader to follow most of the developments in the book. In certain places more complicated mathematics is used without detailed explanation. This is usually in an attempt to establish important results rigorously; however they often can be skipped over without losing the thread of the explanation. A summary of the notations used in the book is given immediately following the Preface.

QUESTIONS FOR CHAPTER 1

1.1. Obtain expressions for the gradient vector and Hessian matrix for the functions of n variables:
 (i) $\mathbf{a}^T\mathbf{x}$: $\mathbf{a}$ constant;
 (ii) $\mathbf{x}^T\mathbf{A}\mathbf{x}$: $\mathbf{A}$ unsymmetric and constant;
 (iii) $\frac{1}{2}\mathbf{x}^T\mathbf{A}\mathbf{x} + \mathbf{b}^T\mathbf{x}$: $\mathbf{A}$ symmetric, $\mathbf{A}$, $\mathbf{b}$ constant;
 (iv) $\mathbf{f}^T\mathbf{f}$: $\mathbf{f}$ is an m-vector depending on $\mathbf{x}$ and $\nabla\mathbf{f}^T$ is denoted by $\mathbf{A}$ which is not constant.

1.2. Write down the Taylor expansion for the gradient $\mathbf{g}(\mathbf{x}' + \boldsymbol{\delta})$ about $\mathbf{x}'$, neglecting terms of order $\|\boldsymbol{\delta}\|^2$. Hence show that if $f(\mathbf{x})$ is a quadratic function with Hessian $\mathbf{G}$, then $\boldsymbol{\gamma} = \mathbf{G}\boldsymbol{\delta}$, where $\boldsymbol{\delta}$ is the difference between any two points and $\boldsymbol{\gamma}$ is the corresponding difference in gradients.

1.3. Write down the Taylor expansion for the m-vector $\mathbf{f}(\mathbf{x})$ about $\mathbf{x}'$, where $\nabla\mathbf{f}^T$ is denoted by $\mathbf{A}$.

1.4. At a point $\mathbf{x}'$ for which $\mathbf{g}' \neq \mathbf{0}$, show that the direction vector $\mathbf{s} = \mathbf{g}'/\|\mathbf{g}'\|_2$ has the greatest slope, over all vectors for which $\mathbf{s}^T\mathbf{s} = 1$. (The *steepest ascent* vector.)

1.5. At a point $\mathbf{x}'$ for which $\mathbf{g}' \neq \mathbf{0}$, show that the direction vectors $\pm\mathbf{g}'$ are orthogonal to the contour and the tangent plane surface at $\mathbf{x}'$.

1.6. If $\mathbf{x}(\alpha)$ is any twice differentiable arc, if $f(\mathbf{x}(\alpha))$ is regarded as $f(\alpha)$, and if $\mathrm{d}\mathbf{x}(\alpha_0)/\mathrm{d}\alpha = \mathbf{s}$ and $\mathrm{d}^2\mathbf{x}(\alpha_0)/\mathrm{d}\alpha^2 = \mathbf{t}$, use the chain rule to obtain expressions for $\mathrm{d}f(\alpha_0)/\mathrm{d}\alpha$ and $\mathrm{d}^2f(\alpha_0)/\mathrm{d}\alpha^2$ in terms of $\mathbf{s}, \mathbf{t}$ and the derivatives of $f(\mathbf{x})$ evaluated at $\mathbf{x}(\alpha_0)$.

(Some other questions which partly refer to the material of Section 1.2 are given at the end of Chapter 2.)

Chapter 2
Structure of Methods

2.1 CONDITIONS FOR LOCAL MINIMA

In the following chapters the problem of finding a local solution to the problem

$$\text{minimize } f(\mathbf{x}), \qquad \mathbf{x} \in \mathbb{R}^n \tag{2.1.1}$$

is considered. $f(\mathbf{x})$ is referred to as the *objective function*, and the minimizing point or *minimizer* is denoted by $\mathbf{x}^*$. Optimization problems also exist which are *maximization problems* and these can be cast in the form of (2.1.1) through the simple transformation

$$\max_{\mathbf{x}} f(\mathbf{x}) = -\min_{\mathbf{x}} -f(\mathbf{x}). \tag{2.1.2}$$

Usually one approaches an optimization problem presupposing that $\mathbf{x}^*$ exists, is unique, and is located by the method to be used. Whilst this is often the case, it is important to realize that there are a number of ways in which this ideal situation may fail to hold, and some of these are described, together with some simple examples. First of all $\mathbf{x}^*$ might not exist when f is unbounded below ($f = x^3$), or even (unusually) when f is bounded below ($f = e^{-x}$). If $\mathbf{x}^*$ does exist it may not be unique ($f = \max(-x-1, 0, x-1)$ or $f = \cos x$). Of particular importance is that it is only generally practicable to locate a *local minimizer* and this may not be a *global minimizer*. An example in which local minima exist that are not global minimizers of $f(\mathbf{x})$ is $f = x^2 - \cos x$ (see also Figures 2.1.1 and 6.2.2). Local minimizers can exist and be of interest even if $f(\mathbf{x})$ is unbounded below ($f = x^3 - 3x$). There is a considerable literature on finding global minima, see Dixon and Szego (1975) for example, and also Section 6.2. However, the difficulties in guaranteeing to find global solutions are considerable. The only simple advice in practice (not guaranteed to work) is to solve the problem from a number of different starting points and take the best local solution that is obtained. Even the definition of a local minimum has its pitfalls. A convenient working definition is that $f(\mathbf{x}) \geqslant f(\mathbf{x}^*)$ for all $\mathbf{x}$ sufficiently close to $\mathbf{x}^*$ (i.e. for all $\mathbf{x}$ in some neighbourhood of $\mathbf{x}^*$). This nonetheless allows situations that

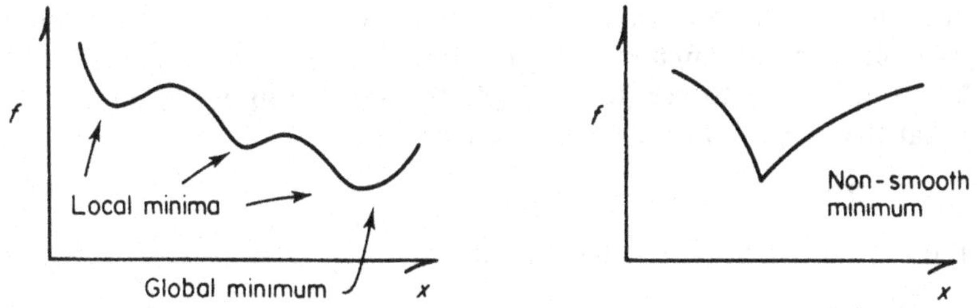

Figure 2.1.1 Types of minima

are hardly typical of a local minimizer, for example, $x = 0$ is a local minimizer of $f = \min(1 + x, 0, 1 - x)$ and also a global maximizer! This situation is eliminated by the definition of a *strict local minimizer* in which $f(\mathbf{x}) > f(\mathbf{x}^*)$ for all $\mathbf{x} \neq \mathbf{x}^*$ sufficiently close to $\mathbf{x}^*$. A stronger definition is that of an *isolated local minimizer* in which $\mathbf{x}^*$ is the only local minimizer in a neighbourhood of $\mathbf{x}^*$. For the function $f = x^2(1 + x^2 + \sin(1/x))$ (with $f(0) = 0$), $x = 0$ is a strict local minimizer but is not isolated. This example is fairly pathological, however, and stronger statements can be made for special classes of function, e.g. quadratic functions (see Question 2.19), smooth functions (Theorem 2.1.1), or convex functions (see Section 9.4).

Other difficulties are caused when $f(\mathbf{x})$ is a *non-smooth* function because non-smooth minima (see Figure 2.1.1) do not satisfy the same conditions as smooth minima. However, non-smooth optimization is an important practical study, and is considered in some detail in Chapter 14. Subsequently, however, in Part 1 it is assumed that first and also second derivatives exist and are continuous, so as to eliminate these possibilities. These derivatives are referred to by $\mathbf{g}(\mathbf{x}) = \nabla f(\mathbf{x})$ and $\mathbf{G}(\mathbf{x}) = \nabla^2 f(\mathbf{x})$ respectively. The notation $f^* = f(\mathbf{x}^*)$, $\mathbf{g}^* = \mathbf{g}(\mathbf{x}^*)$, etc, is used for quantities derived from $\mathbf{x}^*$, likewise $f^{(k)} = f(\mathbf{x}^{(k)})$, $\mathbf{g}^{(k)} = \mathbf{g}(\mathbf{x}^{(k)})$ and so on.

The main aim of this section is to state and discuss some simple conditions which hold at a local minimizer $\mathbf{x}^*$. These arise from the observation that along any line $\mathbf{x}(\alpha) = \mathbf{x}^* + \alpha\mathbf{s}$ through $\mathbf{x}^*$, then $f(= f(\mathbf{x}(\alpha)))$ has both *zero slope* and *non-negative curvature* at $\mathbf{x}^*$. This is illustrated in Figure 2.1.2 and is the usual

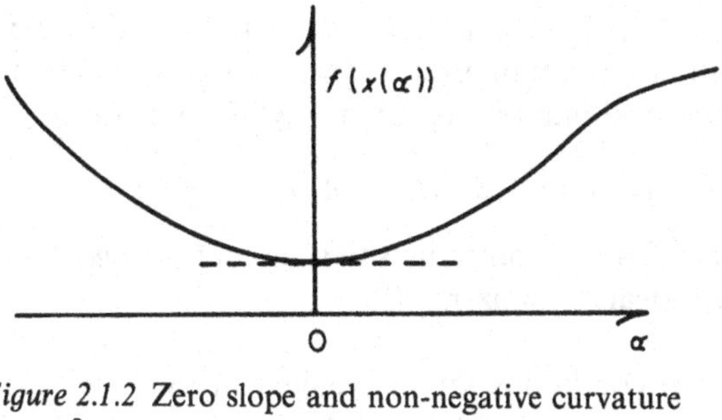

Figure 2.1.2 Zero slope and non-negative curvature at $\alpha = 0$

condition derived from a Taylor series for a local minimum of a function of one variable. From (1.2.6) and (1.2.7) it follows for all $\mathbf{s}$ that both $\mathbf{s}^T\mathbf{g}^* = 0$ and $\mathbf{s}^T\mathbf{G}^*\mathbf{s} \geqslant 0$. It is easy to see (for example by considering $\mathbf{s} = \mathbf{e}_1, \mathbf{s} = \mathbf{e}_2$, etc. in turn) that the first condition is equivalent to

$$\mathbf{g}^* = \mathbf{0} \tag{2.1.3}$$

and the second condition can be written as

$$\mathbf{s}^T\mathbf{G}^*\mathbf{s} \geqslant 0, \qquad \forall \mathbf{s}. \tag{2.1.4}$$

Because these conditions are implied by $\mathbf{x}^*$ being a local minimizer, they are *necessary conditions* for a local solution. Condition (2.1.3) is referred to as a *first order necessary condition* since it is based on first order variations in f and therefore first derivatives. Condition (2.1.4) is a *second order necessary condition*, and is the condition that $\mathbf{G}^*$ is a *positive semi-definite matrix*, by definition of this property.

It is also possible to derive *sufficient conditions* (those which imply that $\mathbf{x}^*$ is a local minimizer) as follows.

Theorem 2.1.1

Sufficient conditions for a strict and isolated local minimizer $\mathbf{x}^*$ *are that* (2.1.3) *holds and that* $\mathbf{G}^*$ *is positive definite, that is*

$$\mathbf{s}^T\mathbf{G}^*\mathbf{s} > 0 \qquad \forall \mathbf{s} \neq \mathbf{0}. \tag{2.1.5}$$

Proof

Consider any point $\mathbf{x}^* + \boldsymbol{\delta}$, $\boldsymbol{\delta} \neq \mathbf{0}$. A Taylor series about $\mathbf{x}^*$ and (2.1.3) imply that

$$f(\mathbf{x}^* + \boldsymbol{\delta}) = f^* + \tfrac{1}{2}\boldsymbol{\delta}^T\mathbf{G}^*\boldsymbol{\delta} + o(\boldsymbol{\delta}^T\boldsymbol{\delta})$$

(see the list of notation for a definition of $o(\cdot)$). Now (2.1.5) implies that there exists an $a > 0$ such that $\boldsymbol{\delta}^T\mathbf{G}^*\boldsymbol{\delta} \geqslant a\boldsymbol{\delta}^T\boldsymbol{\delta}$ (a is the smallest eigenvalue of $\mathbf{G}^*$) and hence that

$$f(\mathbf{x}^* + \boldsymbol{\delta}) \geqslant f^* + (\tfrac{1}{2}a + o(1))\boldsymbol{\delta}^T\boldsymbol{\delta}.$$

As $\boldsymbol{\delta} \to \mathbf{0}$, $o(1) \to 0$ and $a > 0$ is fixed, so it follows that $f(\mathbf{x}^* + \boldsymbol{\delta}) > f^*$, and hence $\mathbf{x}^*$ is a strict local minimizer. Now consider a sequence of points $\mathbf{x}^* + \boldsymbol{\delta}$, $\boldsymbol{\delta} \to \mathbf{0}$, which are local minimizers. By a Taylor series for $\mathbf{g}$,

$$\boldsymbol{\delta}^T\mathbf{g}(\mathbf{x}^* + \boldsymbol{\delta}) = \boldsymbol{\delta}^T\mathbf{g}^* + \boldsymbol{\delta}^T\mathbf{G}^*\boldsymbol{\delta} + o(\|\boldsymbol{\delta}\|^2).$$

But $\boldsymbol{\delta}^T\mathbf{G}^*\boldsymbol{\delta} \geqslant a\boldsymbol{\delta}^T\boldsymbol{\delta}$ which contradicts the fact that $\mathbf{g}^* = \mathbf{g}(\mathbf{x}^* + \boldsymbol{\delta}) = \mathbf{0}$. Thus $\mathbf{x}^*$ is also an isolated local minimizer. □

These sufficient conditions are convenient in that they are readily checked numerically. For instance, if $f(\mathbf{x})$ is given by (1.2.2), then at $\mathbf{x}^* = (1, 1)^T$, $\mathbf{g}^* = (0, 0)^T$

and $\mathbf{G}^* = \begin{bmatrix} 802 & -400 \\ -400 & 200 \end{bmatrix}$ which is positive definite (see below), so that it follows from Theorem 2.1.1 that $\mathbf{x}^*$ is an isolated local minimizer. (In fact since $f^* = 0$ and $f(\mathbf{x}) \geq 0$ it is clear that $\mathbf{x}^*$ is also a global minimizer.) The necessary conditions (2.1.3) and (2.1.4), and the sufficient conditions of Theorem 2.1.1 are almost necessary and sufficient, and there is only a 'gap' in the case of zero curvature. Examples which satisfy the necessary but not the sufficient conditions are $f(x) = x^3$ and $f(x) = x^4$. $x^* = 0$ is a local minimizer of the second function but not the first.

The notion of a positive definite matrix $\mathbf{G}^*$ may be unfamiliar to some readers, and the definition (2.1.5) does not help in that it cannot be checked numerically. However there are several different equivalent definitions which can be checked, namely

(i) all eigenvalues of $\mathbf{G}^* > 0$,
(ii) $\mathbf{LL}^T$ (Choleski) factors of $\mathbf{G}^*$ exist with $l_{ii} > 0$,
(iii) $\mathbf{LDL}^T$ factors exist with $l_{ii} = 1$ and $d_{ii} > 0$,
(iv) all pivots > 0 in Gaussian elimination without pivoting, and
(v) all principal minors of $\mathbf{G}^* > 0$.

The matrices $\mathbf{L}$ and $\mathbf{D}$ are lower triangular and diagonal, respectively. For small $n(\leq 3)$, condition (v) is most readily checked $\left(\text{for } \mathbf{G}^* \text{ above we have } \det(802) > 0 \text{ and } \det\begin{bmatrix} 802 & -400 \\ -400 & 200 \end{bmatrix} > 0\right)$, but in general conditions (ii) or (iii) are the most

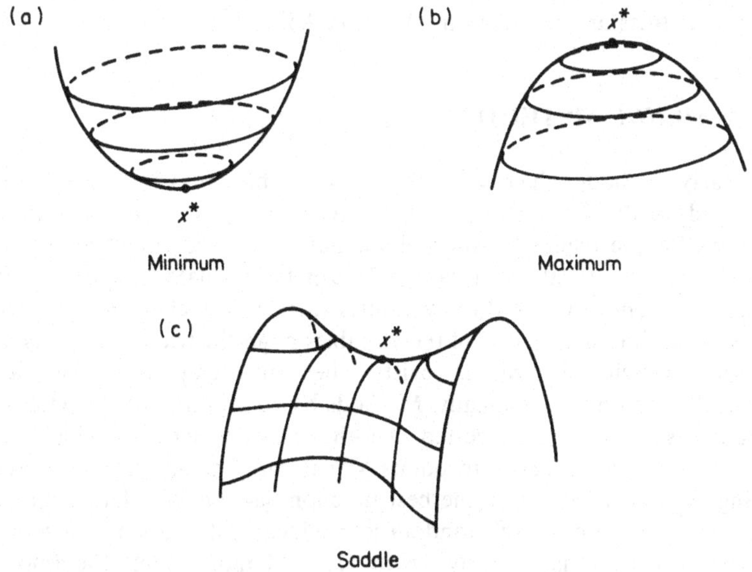

Figure 2.1.3 Types of stationary point

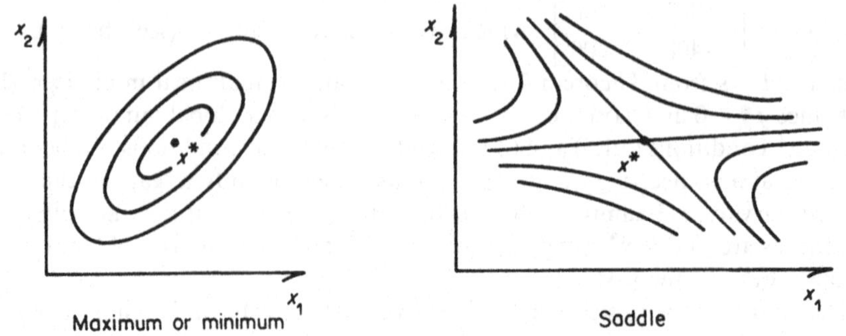

Figure 2.1.4 Contours for stationary points

efficient and they also enable linear equations with coefficient matrix $\mathbf{G}$ to be solved subsequently (see Section 3.1).

In fact many minimization methods are based only upon trying to locate a point $\mathbf{x}^*$ such that $\mathbf{g}(\mathbf{x}^*) = \mathbf{0}$. This may not be a local minimizer and in general is referred to as a *stationary point.* Different types of stationary point are illustrated in Figure 2.1.3 and their contours in Figure 2.1.4. Note that in Figure 2.1.3, whilst all the graphs have zero slope at $\mathbf{x}^*$, for (a) there is positive curvature in every direction, for (b) negative curvature in every direction, whereas for (c) there is negative curvature across the saddle and positive curvature along the saddle. Thus usually a minimizer corresponds to a positive definite Hessian matrix, a maximizer to a negative definite matrix, and a saddle point to an indefinite matrix (that is one in which the eigenvalues have both positive and negative sign). Numerical methods for finding stationary points which are not minimizers are occasionally of interest (see Sinclair and Fletcher, 1974) and some possibilities are described in Question 4.5.

2.2 *AD HOC* METHODS

Many early methods which were suggested for minimization were based on rough and ready ideas without very much theoretical background. It is instructive for the reader to think about how he or she would go about the problem, given that values of $f(\mathbf{x})$ only can be evaluated for any $\mathbf{x}$. If the problem is in only two or three variables then it is likely that some sort of repeated bisection in each one of the variables might be tried so as to establish a region in which the minimum exists. Then an attempt could be made to contract this region systematically. Methods based on this sort of sectioning in n-dimensions have been suggested. Another possible method which readily springs to mind is to generate points $\mathbf{x}^{(k)}$ at random in some fixed region, selecting the one which gives the best function value over a large number of trials. Many variations on this random search idea exist. A good review of these two types of method is given by Swann (1972). Unfortunately the amount of effort required to implement these methods goes up rapidly (typically as 2^n)

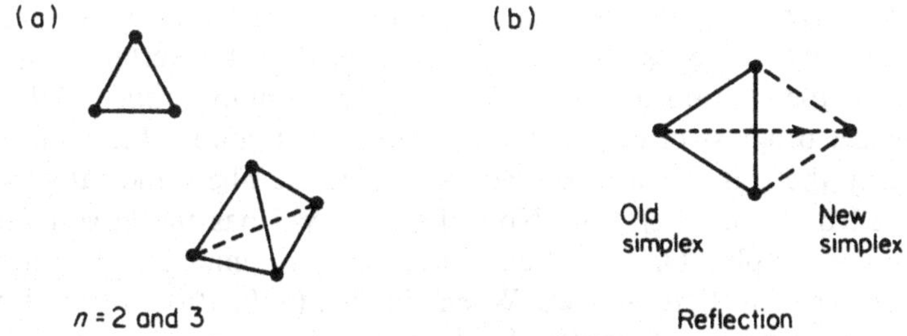

Figure 2.2.1 Regular simplexes

which caused their authors to coin the phrase the 'curse of dimensionality'. In fact this curse is largely self-inflicted, and for better methods the rate of increase of effort with n is rarely worse than a multiple of n^2. (Note that effort is being measured in terms of the number of times the function needs to be evaluated. In fact the housekeeping effort for some methods can go up like $\sim n^3$, but this is not usually the major cost.)

The most successful of the methods which merely compare function values is that known as the *simplex method* (Spendley *et al.*, 1962). This is not to be confused with the simplex method of linear programming, although its name is derived from the same geometrical concept. A (regular) simplex is a set of $n+1$ (equidistant) points in $\mathbb{R}^n$, such as the (equilateral) triangle for $n=2$ or tetrahedron for $n=3$ (see Figure 2.2.1(a)). The current information kept in the method is the coordinates of the $n+1$ points and their corresponding function values.

On the first iteration of the simplex method the vertex at which the function value is largest is determined. This vertex is then reflected in the centroid of the other n vertices, thus forming a new simplex (see Figure 2.2.1(b)). The function value at this new vertex is evaluated and the process repeated. On iterations after the first it might be that the newest vertex still has the largest function value in the new simplex, and to reflect this vertex would cause oscillation. Hence the largest function value other than that at the newest vertex is subsequently used to decide which vertex to reflect. Ultimately this iteration will fail to make further progress, so an additional rule has to be introduced. When a certain vertex $\mathbf{x}'$ has been in the current simplex for more than a fixed

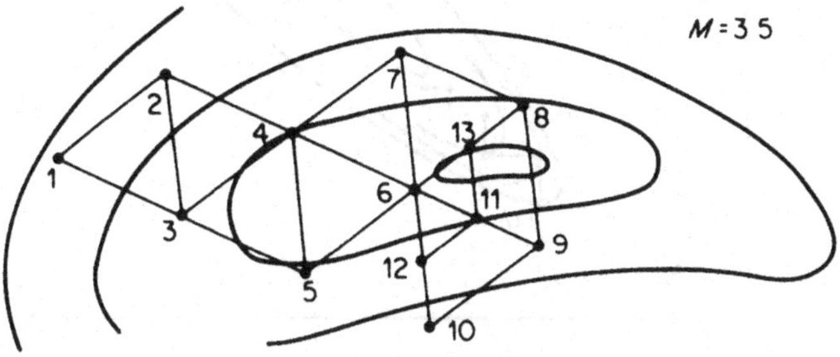

Figure 2.2.2 The simplex method in two variables

number M iterations, then the simplex should be contracted by replacing the other vertices by new ones half way along the edge to the vertex $\mathbf{x}'$. Spendley *et al.* suggest the relationship $M = 1.65n + 0.05n^2$ from experimental data.

The typical progress of the iteration is shown in Figure 2.2.2. Vertices 1, 2, and 3 form the initial simplex, and increasing numbers indicate the new vertices added at each iteration. Note that vertex 7 has the largest function value for the simplex (4, 6, 7) but is not reflected immediately since it is the newest vertex in that simplex. When simplex (6, 9, 10) is reached, vertex 6 has been in the current simplex for four iterations, and since $M = 3.5$, the simplex is contracted at this stage to the new simplex (6, 11, 12), and the iteration continues anew from this simplex. A modified simplex method which allows irregular simplexes has been suggested by Nelder and Mead (1965), and distortions of the simplex are performed automatically in an attempt to take into account the local geometry of the function.

Simplex methods do not compare well with some of the methods to be described in later chapters but may be useful when the function values suffer from substantial noise. An example of this is in the real time control of a chemical plant, when repeated evaluation of the objective function for the same parameters might only give agreement to say 1 per cent.

Another simple method which readily suggests itself is the *alternating variables method*, in which on iteration k ($k = 1, 2, \ldots, n$), the variable x_k alone is changed in an attempt to reduce the objective function value, and the other variables are kept fixed. After iteration n, when all the variables have been changed, then the whole cycle is repeated until convergence occurs. One possibility is to make a change in each coordinate direction which reduces the objective function as much as possible. Unfortunately in practice the alternating variable method is usually very inefficient and unreliable. The progress of the iteration is characterized by the oscillatory behaviour shown in Figure 2.2.3. Furthermore a problem

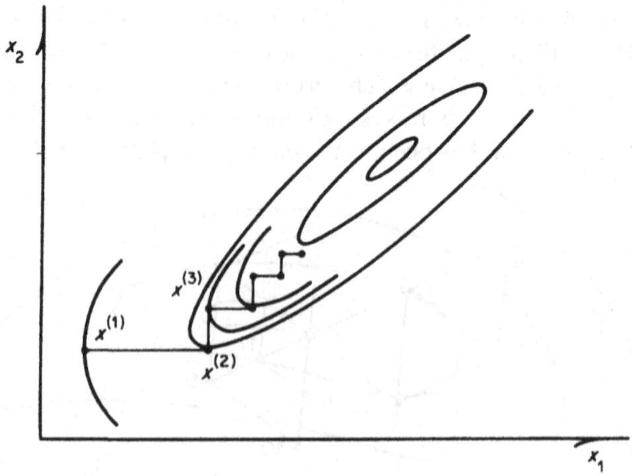

Figure 2.2.3 The method of alternating variables

has been constructed (Powell, 1973) for which the method fails to converge to a stationary point.

It is tempting to dismiss the method by saying that it has no theoretical background and cannot therefore be expected to work well. Perhaps it is more accurate however to say that the theory which supports the method is deficient. It ignores the possibility of correlation between the variables, which causes the search parallel to x_2 to destroy the property that $\mathbf{x}^{(2)}$ is the minimizer in the direction parallel to x_1. This illustrates that it is important to study which are the most relevant theoretical features in the design of minimization methods, and this is taken up in the next section.

Ad hoc modifications have been proposed to the alternating variables method, based on the observation that the points at the beginning and end of a cycle determine a line along which more substantial progress might be made (along the valley in Figure 2.2.3, for example). Thus if provision is made for searching along this line, a more efficient method might result. This idea is incorporated into many early methods (see again Swann, 1972), the most efficient of which are probably the Hooke and Jeeves and the DSC methods. However the conjugate direction methods described in Section 4.2 also generate lines by joining up points in a similar way, yet have a stronger theoretical background, and have been developed to be more efficient. Thus the early *ad hoc* methods are gradually falling out of use.

2.3 USEFUL ALGORITHMIC PROPERTIES

In this section a closer study is made of the structural elements of an iterative method for unconstrained optimization, and in particular of the desirable features of such an algorithm. The typical behaviour of an algorithm which is regarded as acceptable is that the iterates $\mathbf{x}^{(k)}$ move steadily towards the neighbourhood of a local minimizer $\mathbf{x}^*$ (see Figure 1.2.2, for example), and then converge rapidly to the point $\mathbf{x}^*$ itself, the iteration being terminated when some user-supplied convergence test becomes satisfied. In certain cases it may be possible to prove that some elements of this idealized behaviour must occur. For instance it might be possible to show that any accumulation point (limit point of a subsequence) of the sequence $\{\mathbf{x}^{(k)}\}$ is a local minimum point. This is not quite as strong as proving that the sequence converges, but in practice would be quite an acceptable result. In fact it is often only possible to show that the gradient vector $\nabla f(\mathbf{x}^{(k)})$ tends to zero, which (if $\{\mathbf{x}^{(k)}\}$ converges) corresponds to convergence to a stationary point. This might appear to be less than satisfactory. However, there are often other features of an algorithm, such as the fact that $f(\mathbf{x}^{(k)})$ is always reduced on every iteration, which usually imply that the stationary point turns out to be a local minimizer, except in rather rare circumstances. Proofs of this type are referred to as global convergence proofs in that they do not require $\mathbf{x}^{(1)}$ to be close to $\mathbf{x}^*$.

Another aspect of acceptable behaviour arises when the neighbourhood of a local minimum point $\mathbf{x}^*$ is reached. If the error

$$\mathbf{h}^{(k)} = \mathbf{x}^{(k)} - \mathbf{x}^* \tag{2.3.1}$$

is defined, and if $\mathbf{h}^{(k)} \to \mathbf{0}$ (convergence), then it may be possible to give *local convergence* results about how rapidly the iterates converge in a neighbourhood of $\mathbf{x}^*$. If the errors behave according to $\|\mathbf{h}^{(k+1)}\|/\|\mathbf{h}^{(k)}\|^p \to a$ where $a > 0$ then the *order of convergence* is defined to be pth order. The most important cases are $p = 1$ (*first order* or *linear convergence*) and $p = 2$ (*second order* or *quadratic convergence*). It is also possible to state these definitions in terms of bounds on the ratios, when the limit above does not exist. Thus first order convergence would correspond to

$$\|\mathbf{h}^{(k+1)}\|/\|\mathbf{h}^{(k)}\| \leqslant a \quad \text{or} \quad \mathbf{h}^{(k+1)} = O(\|\mathbf{h}^{(k)}\|)$$

and second-order convergence to

$$\|\mathbf{h}^{(k+1)}\|/\|\mathbf{h}^{(k)}\|^2 \leqslant a \quad \text{or} \quad \mathbf{h}^{(k+1)} = O(\|\mathbf{h}^{(k)}\|^2).$$

Linear convergence is only satisfactory in practice if the *rate constant* a is small, say $a \leqslant \frac{1}{4}$ would be acceptable. However, many methods look for something better than linear convergence in which the local rate constant tends to zero. This is known as *superlinear convergence* defined by

$$\|\mathbf{h}^{(k+1)}\|/\|\mathbf{h}^{(k)}\| \to 0 \quad \text{or} \quad \mathbf{h}^{(k+1)} = o(\|\mathbf{h}^{(k)}\|).$$

Some fundamental results regarding superlinear convergence are given in Section 6.2.

Nonetheless, it must be appreciated that the existence of convergence and order of convergence results for any algorithm is not a guarantee of good performance in practice. Not only do the results themselves fall short of a guarantee of acceptable behaviour, but also they neglect computer round-off error which can be crucial. Often the results impose certain restrictions on $f(\mathbf{x})$ which it may not be easy to verify, and in some cases (for example when $f(\mathbf{x})$ is assumed to be a *convex* function) these conditions may not be satisfied in practice. Thus the development of an optimization method also relies on *experimentation*. That is to say, the algorithm is shown to have acceptable behaviour on a variety of *test functions* which should be chosen to represent the different features which might arise in general (insofar as this is possible). Clearly experimentation can never give a guarantee of good performance in the sense of a mathematical proof. My experience however is that well-chosen experimental testing is often the most reliable indication of good performance. The ideal of course is a good selection of experimental testing backed up by convergence and order of convergence proofs.

In considering algorithms, some important general features which occur in currently successful methods for unconstrained optimization are described in the rest of this section. A method is usually based on a *model*, that is some convenient approximation to the objective function, which enables a prediction of the location of a local minimizer to be made. Most successful has been the use of *quadratic models* and this is described in more detail in the next section. A method is also based on some *prototype algorithm* which describes the broad strategy of the approach, but without all the small details that might be required in a computer program. The prototype algorithm is mainly concerned with how to use the model prediction in such a way as to obtain satisfactory convergence properties. There are two main prototypes that have been used: most recent is the *restricted step* or *trust region* approach, described in Chapter 5. However the more traditional approach, and one which is still used successfully, is that of a *line search algorithm.* Such a prototype algorithm is the subject of the rest of this chapter, and different realizations of line search algorithms are considered in Chapters 3, 4, and 6.

Even in early methods (Section 2.2) the idea occurs of searching along coordinate directions or in more general directions. This is an example of a line search method. The user is required to supply an initial estimate $\mathbf{x}^{(1)}$ and the basic structure of the kth iteration is

(a) determine a direction of search $\mathbf{s}^{(k)}$.
(b) find $\alpha^{(k)}$ to minimize $f(\mathbf{x}^{(k)} + \alpha\mathbf{s}^{(k)})$ with respect to α. (2.3.2)
(c) set $\mathbf{x}^{(k+1)} = \mathbf{x}^{(k)} + \alpha^{(k)}\mathbf{s}^{(k)}$.

Different methods correspond to different ways of choosing $\mathbf{s}^{(k)}$ in step (a), based on information available in the model. Step (b) is the *line search subproblem* and is carried out by repeatedly sampling $f(\mathbf{x})$ and possibly its derivatives for different points $\mathbf{x} = \mathbf{x}^{(k)} + \alpha\mathbf{s}^{(k)}$ along the line. More details of this are given in Section 2.5 and 2.6. Step (b) is idealized in that the exact minimizing value of α is required (an *exact line search*) and this cannot be implemented in practice in a finite number of operations. (Essentially the nonlinear equation $\mathrm{d}f/\mathrm{d}\alpha = 0$ must be solved.) Also it may be that the minimizing value of $\alpha^{(k)}$ might not exist ($\alpha^{(k)} = \infty$). Nonetheless the idea is conceptually useful, and occurs in some idealized proofs of convergence. In this respect it is convenient to point out the consequential property that the slope $\mathrm{d}f/\mathrm{d}\alpha$ at $\alpha^{(k)}$ must be zero, which from (1.2.6) gives

$$\nabla f^{(k+1)\mathrm{T}}\mathbf{s}^{(k)} = 0 \tag{2.3.3}$$

(see Figure 2.3.1). In practice it is necessary to terminate the line search when certain conditions for an approximate minimum along the line are satisfied. Since it is not efficient to determine line search minima to a high accuracy when

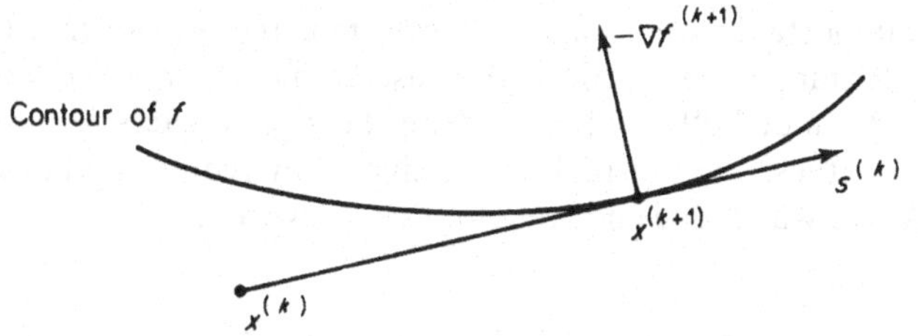

Figure 2.3.1 The exact line search

$\mathbf{x}^{(k+1)}$ is remote from $\mathbf{x}^*$, these conditions are chosen so as to be readily satisfied. In these circumstances the algorithm is said to employ an *inexact* or *approximate* line search. The motivation for the choice of these conditions is given in Section 2.5. Algorithms for an inexact line search aimed at satisfying these conditions are discussed in Section 2.6.

Associated with these ideas is the concept of a *descent method.* This is a line search method in which the direction of search $\mathbf{s}^{(k)}$ satisfies the *descent property*

$$\mathbf{s}^{(k)\mathrm{T}}\mathbf{g}^{(k)} < 0. \tag{2.3.4}$$

From (1.2.6) this ensures that the slope $\mathrm{d}f/\mathrm{d}\alpha$ (2.3.2(b)) is always negative at $\alpha = 0$ unless $\mathbf{x}^{(k)}$ is a stationary point. This condition guarantees that the function can be reduced in the line search for some $\alpha^{(k)} > 0$. In certain cases, by a suitable choice of line search conditions, it is possible to incorporate the descent property into a convergence proof. Although this proof may not always be realistic, in practice the descent property is closely associated with global convergence. This point and the proof of convergence are taken up in more detail in Section 2.5.

A simple line search descent method is the *steepest descent method* in which $\mathbf{s}^{(k)} = -\mathbf{g}^{(k)}$ for all k. That is to say, the method searches in the steepest descent direction, along which the objective function decreases most rapidly local to $\mathbf{x}^{(k)}$. Unfortunately, although this consideration is appealing, in practice the method usually exhibits oscillatory behaviour similar to that described for the alternating variables method in Section 2.2. In fact for the example illustrated by Figure 2.2.3 the same sequence $\{\mathbf{x}^{(k)}\}$ is generated. Furthermore although there is a theoretical proof of convergence (see Theorem 2.5.1) the method usually terminates far from the solution owing to round-off effects. In practice, therefore, the method is usually both inefficient and unreliable. In fact the local convergence result for steepest descent (see Question 2.11) does predict the possibility of an arbitrarily slow rate of linear convergence, as observed.

This inadequacy of the steepest descent method can be put down to a failure in the model situation, perhaps because the steepest descent property along the line holds only at $\alpha = 0$ and not for all α. A type of model which in practice does usually give rise to methods with a rapid rate of convergence is the *quadratic model.* One possible reason is that this model is to some extent associated with the property of second order convergence. Certainly descent methods based on a quadratic model have proved to be very powerful in practice. Some possible reasons for this are discussed further in Section 2.5.

Another important feature of any algorithm is the convergence test which is required to terminate the iteration. Most useful to the user would be a guarantee that $f^{(k)} - f^* \leqslant \varepsilon$ or $|x_i^{(k)} - x_i^*| \leqslant \varepsilon_i$, where the ε parameters are user-supplied. Unfortunately these are not practicable since they require a knowledge of the solution. A test which does not require this knowledge is

$$\|\mathbf{g}^{(k)}\| \leqslant \varepsilon \tag{2.3.5}$$

and this has been used on occasions. However it has the disadvantage that it is not easy for the user to know what magnitude to choose for ε. Also it does not have certain important invariance properties (see Section 3.3), and can work badly for ill-conditioned problems such as penalty functions. When the algorithm can be expected to converge rapidly, tests based on

$$|x_i^{(k)} - x_i^{(k+1)}| \leqslant \varepsilon_i \qquad \forall i \tag{2.3.6}$$

or

$$f^{(k)} - f^{(k+1)} \leqslant \varepsilon \tag{2.3.7}$$

usually work well. These are both convenient for the user, but (2.3.6) requires a vector $\boldsymbol{\varepsilon}$ and is invariant only to scaling. (Alternatively a test of the form $\|\mathbf{x}^{(k)} - \mathbf{x}^{(k+1)}\| \leqslant \varepsilon$ requires only a scalar ε, but is not invariant and requires the user to scale his variables appropriately.) Yet another test is on the predicted change in f

$$\tfrac{1}{2}\mathbf{g}^{(k)\mathrm{T}}\mathbf{H}^{(k)}\mathbf{g}^{(k)} \leqslant \varepsilon$$

where $\mathbf{H} = \mathbf{G}^{-1}$ or an approximation to it (see Chapter 3). I have found that tests based on (2.3.7) work well for the methods of Chapter 3. For algorithms which converge less rapidly (for example the conjugate gradient methods of Section 4.1) a test based on (2.3.5) is more appropriate, especially if some attention to scaling is required anyway for the method to work well. An additional consideration, which applies if second derivatives are available, is that the test should prevent termination at a saddle point. Also useful in general is the ability to terminate when a user-supplied maximum iteration count is reached. In addition, some consideration has to be given to the effects of round-off near the solution, and to terminate when it is judged that these effects are preventing further progress. It is difficult to be certain what strategy is best in this respect.

Finally, it can be seen that many of the considerations in this section are only relevant when the method is able to compute the gradient vector $\mathbf{g}^{(k)}$ at any point $\mathbf{x}^{(k)}$. It also turns out in practice that these first derivative methods are much more reliable than no-derivative methods, in which $\mathbf{g}^{(k)}$ is not available. Indeed the best no-derivative methods seem to be those which estimate derivatives by the finite difference approximations

$$g_i(\mathbf{x}) \approx (f(\mathbf{x} + h\mathbf{e}_i) - f(\mathbf{x}))/h \tag{2.3.8}$$

(forward differences) or

$$g_i(\mathbf{x}) \approx \tfrac{1}{2}(f(\mathbf{x} + h\mathbf{e}_i) - f(\mathbf{x} - h\mathbf{e}_i))/h \tag{2.3.9}$$

(central differences), where $\mathbf{e}_i$ is defined in (1.2.9). Apropos of this, for first derivative methods it is all too easy for the user to make errors in programming formulae for derivatives. It is therefore very wise to check derivative formulae systematically by using (2.3.8) or (2.3.9). Some enlightened libraries now associate a subroutine to do this with any minimization subroutine requiring derivatives. Second derivatives can be checked by differences in first derivatives in a similar way.

2.4 QUADRATIC MODELS

A feature of many methods for unconstrained minimization is that they are derived in such a way that they will work well, or even exactly, if applied to a quadratic function (with positive definite Hessian **G** unless otherwise specified). Yet the methods can be applied iteratively to minimize general functions. (Indeed there would be no sense in applying the methods in practice to minimize a quadratic function, since this can be achieved simply by solving a linear system of equations.) Such methods are said to be derived from a *quadratic model*. There is no doubt that this approach has proved very fruitful in practice and that there is a correlation between this and the efficiency and rapid local convergence of the method. Some tentative reasons for this are the following.

(i) A quadratic function is one of the simplest smooth functions with a well determined minimum, so is easy to manipulate.

(ii) A general function expanded about a local minimizer $\mathbf{x}^*$ is approximated well by a quadratic function. Thus methods based on quadratic models should have a rapid ultimate rate of convergence.

(iii) Even remote from the minimum it seems preferable to use quadratic information to a large extent rather than to reject it. Quadratic information is more effective than linear information (steepest descent) in predicting directions along which substantial progress can be made. This reason is presumably related to the fact that a Taylor series of $f(\mathbf{x})$ about an arbitrary point $\mathbf{x}^{(k)}$, taken to quadratic terms, will agree with $f(\mathbf{x})$ to a given accuracy over a much greater neighbourhood of $\mathbf{x}^{(k)}$ than will the series taken to linear terms.

(iv) Methods based on quadratic models can be made invariant under a linear transformation of variables to a large extent. The relevance of this property is discussed in more detail in Section 3.3.

Almost all the methods described in the following chapters of Part 1 are based on quadratic models. If both first and second derivatives are available to the method, then it is clear how to set up the quadratic model and this gives rise to the classical *Newton method* in Chapter 3. A similar quadratic model is also used in the *restricted step methods* of Chapter 5. Even when second derivatives are not available they can be estimated in various ways and the resulting quadratic model can then be used in the algorithm. In this category are the *quasi-Newton methods* of Chapter 3 which estimate second derivatives in a most subtle and efficient way. Also there are the *conjugate direction methods* of Chapter 4 which use a quadratic model is a less direct way. Finally there are methods described in Chapter 6 which use a sums of squares structure in $f(\mathbf{x})$ to make a readily calculated approximation to the Hessian matrix. Even when no derivatives are available to the minimization method, then the estimation of first derivatives by finite differences, combined with a quasi-Newton method, has provided the currently most efficient method.

There are therefore two particular situations in which a quadratic model might be said to be used. One occurs when the method is a *Newton-like method*

(Newton, quasi-Newton, etc.) which uses or approximates the Hessian matrix in Newton's method. The motivation for this approach is a fundamental result common to any equation solving process that superlinear convergence can be obtained if and only if the step is asymptotically equal to that of the Newton–Raphson method (the Dennis–Moré theorem, Section 6.2). Another situation in which a quadratic model is used arises when a method is derived which has the property known as *quadratic termination*. The definition of quadratic termination is that the method will locate the minimizing point $\mathbf{x}^*$ of a quadratic function in a known finite number of iterations, yet can be applied iteratively to minimize non-quadratic functions. Many good algorithms possess this property, although successful methods do exist (some Newton-like methods) which do not terminate for quadratic functions. In particular it is not possible to state either that superlinear convergence implies quadratic termination or the converse, although there do exist methods having both properties.

A particular way of obtaining quadratic termination is to invoke the concept of the *conjugacy* of a set of non-zero vectors $\mathbf{s}^{(1)}, \mathbf{s}^{(2)}, \ldots, \mathbf{s}^{(n)}$ to a given positive definite matrix $\mathbf{G}$. This is the property that

$$\mathbf{s}^{(i)\mathrm{T}}\mathbf{G}\mathbf{s}^{(j)} = 0 \qquad \forall i \neq j \tag{2.4.1}$$

A *conjugate direction method* is one which generates such directions when applied to a quadratic function with Hessian $\mathbf{G}$. In this case termination can be obtained if exact line searches are made, as the following theorem shows.

Theorem 2.4.1

A conjugate direction method terminates for a quadratic function in at most n exact line searches, and each $\mathbf{x}^{(k+1)}$ *is the minimizer in the subspace generated by* $\mathbf{x}^{(1)}$ *and the directions* $\mathbf{s}^{(1)}, \mathbf{s}^{(2)}, \ldots, \mathbf{s}^{(k)}$ *(that is the set of points* $\{\mathbf{x} \mid \mathbf{x} = \mathbf{x}^{(1)} + \sum_{j=1}^{k} \alpha_j \mathbf{s}^{(j)}\ \forall \alpha_j\}$*).*

Proof

Because $\mathbf{G}$ is positive definite, and because conjugate directions are independent (see Question 2.16), both results follow by establishing for all $k \leqslant n$ that

$$\mathbf{g}^{(k+1)\mathrm{T}}\mathbf{s}^{(j)} = 0 \qquad j = 1, 2, \ldots, k \tag{2.4.2}$$

(zero slope along any line in the subspace). Now

$$\mathbf{g}^{(k+1)\mathrm{T}}\mathbf{s}^{(j)} = \mathbf{g}^{(j+1)\mathrm{T}}\mathbf{s}^{(j)} + \sum_{i=j+1}^{k} \boldsymbol{\gamma}^{(i)\mathrm{T}}\mathbf{s}^{(j)}$$

where $\boldsymbol{\gamma}^{(i)} = \mathbf{g}^{(i+1)} - \mathbf{g}^{(i)}$, so by (2.3.3), (1.2.15), (2.3.2) and (2.4.1) it follows that (2.4.2) holds. □

This proof, although simple, does not make the basic structure of conjugacy entirely clear. To do this let the minimizer be written as

$$\mathbf{x}^* = \mathbf{x}^{(1)} + \sum_{i=1}^{n} \alpha_i^* \mathbf{s}^{(i)}$$

and a general point as

$$\mathbf{x} = \mathbf{x}^{(1)} + \sum \alpha_i \mathbf{s}^{(i)}.$$

Then the expression (1.2.12) for a quadratic function (with $\mathbf{x}' = \mathbf{x}^*$ and ignoring c') can be written in terms of the variables $\boldsymbol{\alpha}$ as

$$q(\boldsymbol{\alpha}) = \tfrac{1}{2}(\mathbf{x} - \mathbf{x}^*)^{\mathrm{T}}\mathbf{G}(\mathbf{x} - \mathbf{x}^*) = \tfrac{1}{2}(\boldsymbol{\alpha} - \boldsymbol{\alpha}^*)\mathbf{S}^{\mathrm{T}}\mathbf{G}\mathbf{S}(\boldsymbol{\alpha} - \boldsymbol{\alpha}^*)$$

where $\mathbf{S}$ is the matrix with columns $\mathbf{s}^{(1)}, \mathbf{s}^{(2)}, \ldots, \mathbf{s}^{(n)}$. If the vectors $\mathbf{s}^{(i)}$ are conjugate then this reduces to

$$q(\boldsymbol{\alpha}) = \tfrac{1}{2} \sum_{i=1}^{n} (\alpha_i - \alpha_i^*)^2 d_i$$

where $d_i = \mathbf{s}^{(i)\mathrm{T}}\mathbf{G}\mathbf{s}^{(i)}$, and $q(\boldsymbol{\alpha})$ can then be minimized by choosing $\alpha_1 = \alpha_1^*$, $\alpha_2 = \alpha_2^*$, etc. This is equivalent to making the exact line searches in $\mathbf{x}$-space. Thus conjugacy implies a diagonalizing transformation $\mathbf{S}^{\mathrm{T}}\mathbf{G}\mathbf{S}$ of $\mathbf{G}$ to a new coordinate system ($\boldsymbol{\alpha}$) in which the variables are decoupled. A conjugate direction method is then the alternating variables method applied in this new coordinate system.

All conjugate direction methods rely on this theorem, the differences between methods being in the ingenuity by which directions are generated without explicit knowledge of the Hessian, and in the techniques needed to make the methods efficient for general functions. Of particular importance is the *conjugate gradient method* which uses these ideas in conjunction with the steepest descent method. For no-derivative problems, some other conjugate direction methods have been developed. These methods are discussed in more detail in Sections 4.1 and 4.2.

2.5 DESCENT METHODS AND STABILITY

Line search descent methods have frequently been used as a means of introducing a degree of reliability into optimization software and this idea is followed up in this section. In the early days one common strategy was to choose the step $\alpha^{(k)}$ close to the value given by an exact line search. This is motivated by early theory which shows that the steepest descent method with an exact line search is globally convergent to a stationary point (Curry (1944); there is also an unbelievably distant reference to a paper of Cauchy in 1847!). However accurate line searches are expensive to carry out, and there is also the nuisance that the exact minimizer may not exist. Other researchers weakened the line search tolerance considerably and used the descent property merely to force a decrease $f^{(k+1)} < f^{(k)}$ in the objective function on each iteration. This usually turned out to be more efficient. However merely requiring a decrease in f does not ensure global convergence so there were doubts about the stability of this more efficient approach. This fact, and the proliferation of numerous different codes for the line search, led to the development of a generally accepted set of conditions for terminating the line search which would allow low accuracy line searches whilst forcing global convergence. These conditions and the consequent global theorem are the subject of this section. Algorithms which enable a step length to be determined which satisfies these conditions are considered in Section 2.6.

The requirement $f^{(k+1)} < f^{(k)}$ that the objective function decreases on each iteration is unsatisfactory because it allows negligible reductions in f relative say to the optimum reduction that could be obtained in an exact line search. For example, if $\bar{\alpha}^{(k)}$ denotes the least positive value of α for which $f(\mathbf{x}^{(k)} + \alpha\mathbf{s}^{(k)}) = f(\mathbf{x}^{(k)})$, then negligible reductions can occur either if $\alpha^{(k)} \to \bar{\alpha}^{(k)}$ or if $\alpha^{(k)} \to 0$ as $k \to \infty$ (see Question 3.4). This clarifies the aim of the line search as being to find a step $\alpha^{(k)}$ which gives a significant reduction in f on each iteration and which is not close to the extremes of the interval $[0, \bar{\alpha}^{(k)}]$. The resulting conditions must be such that an *acceptable point* (that is a value of α which satisfies the conditions) always exists and can be determined in a finite number of steps. It is also important in regard to proving superlinear convergence of some methods that the conditions do not exclude the minimizing value of α when $f(\mathbf{x}^{(k)} + \alpha\mathbf{s}^{(k)})$ is a quadratic with positive curvature.

Two conditions on $\alpha^{(k)}$ which together meet these requirements are given by Goldstein (1965). In stating these conditions, $f(\mathbf{x}^{(k)} + \alpha\mathbf{s}^{(k)})$ is written as $f(\alpha)$, so that $f(0)$ corresponds to $f^{(k)}$, and the descent condition (2.3.4) becomes $f'(0) < 0$. The Goldstein conditions are

$$f(\alpha) \leqslant f(0) + \alpha\rho f'(0) \tag{2.5.1}$$

to exclude the right-hand extreme of $[0, \bar{\alpha}^{(k)}]$, and

$$f(\alpha) \geqslant f(0) + \alpha(1-\rho) f'(0) \tag{2.5.2}$$

to exclude the left-hand extreme, where $\rho \in (0, \frac{1}{2})$ is a fixed parameter. If $\alpha^{(k)}$ satisfies (2.5.1) then it follows that the resulting reduction in $f(\mathbf{x})$ satisfies

$$f^{(k)} - f^{(k+1)} \geqslant -\rho \mathbf{g}^{(k)\mathrm{T}} \boldsymbol{\delta}^{(k)} \tag{2.5.3}$$

where $\boldsymbol{\delta}^{(k)} = \alpha^{(k)}\mathbf{s}^{(k)} = \mathbf{x}^{(k+1)} - \mathbf{x}^{(k)}$. A similar bound can be derived from (2.5.2). The Goldstein conditions are illustrated in Figure 2.5.1 for $\rho = \frac{1}{4}$, although the resulting algorithm is usually not too sensitive to this choice. A global convergence proof for a line search descent method based on these conditions can be given (see the previous edition) which is similar to Theorem 2.5.1 below. The requirement that $\rho < \frac{1}{2}$ allows the property that the minimizing value of a quadratic function is acceptable. However when $f(\alpha)$ is non-quadratic, the second condition (2.5.2) may exclude the minimizing point of $f(\alpha)$ as Figure 2.5.1 also illustrates.

For this reason, (2.5.2) may be replaced by a different test on the slopes

$$f'(\alpha) \geqslant \sigma f'(0) \qquad \sigma \in (\rho, 1) \tag{2.5.4}$$

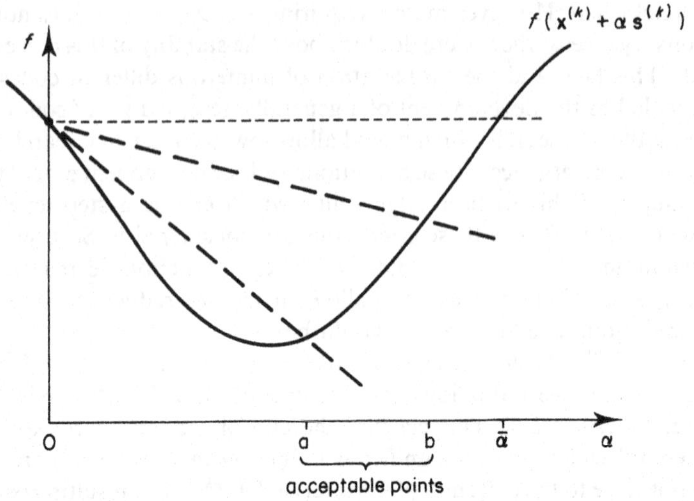

Figure 2.5.1 Goldstein conditions

due to Wolfe (1968b), which also arises in more complicated theorems given by Powell (1976). Here σ is another fixed parameter. This inequality implies that $\mathbf{x}^{(k+1)}$ satisfies the condition

$$\mathbf{g}^{(k+1)\mathrm{T}}\boldsymbol{\delta}^{(k)} \geqslant \sigma \mathbf{g}^{(k)\mathrm{T}}\boldsymbol{\delta}^{(k)}. \tag{2.5.5}$$

Because $\sigma < 1$, this test can be used to exclude the left-hand extreme of $[0, \bar{\alpha}]$, whilst (2.5.1) is used to exclude the right-hand extreme as before. I shall refer to conditions (2.5.1) and (2.5.4) as the Wolfe–Powell conditions, and they are illustrated in Figure 2.5.2 for $\sigma = \frac{1}{2}$ and $\rho = \frac{1}{4}$. The restriction that $\sigma > \rho$ ensures

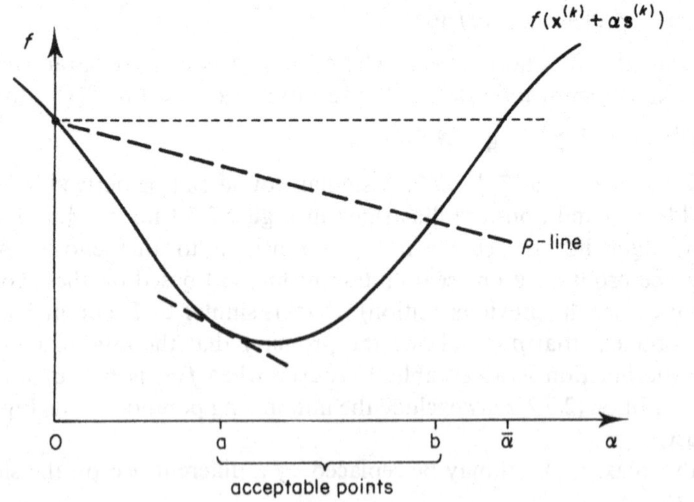

Figure 2.5.2 Wolfe–Powell conditions

that acceptable points exist and can be located in a finite number of steps. In fact in some cases when $f \to -\infty$, acceptable points may not exist and a line search algorithm will continue to reduce f indefinitely. Otherwise the graph of f must intersect the *ρ-line* defined by taking equality in (2.5.1). Let $\acute{\alpha}$ be the least value of $\alpha > 0$ at which this intersection occurs: clearly a sequence of intersection points converging to zero cannot occur by continuity of f' at $\alpha = 0$. Then the following lemma holds

Lemma 2.5.1

If $\acute{\alpha}$ exists and $\sigma \geqslant \rho$ then there exists an interval of acceptable points for both (2.5.1) *and* (2.5.4) *in* $(0, \acute{\alpha})$.

Proof

Consider any point $\varepsilon \in (0, \acute{\alpha})$. By definition of $\acute{\alpha}$, $f(\varepsilon) < f(0) + \varepsilon\rho f'(0)$ and $f(\acute{\alpha}) = f(0) + \acute{\alpha}\rho f'(0)$. By the mean value theorem there exists $\alpha \in (\varepsilon, \acute{\alpha})$ such that

$$f'(\alpha) = (f(\acute{\alpha}) - f(\varepsilon))/(\acute{\alpha} - \varepsilon) > \rho f'(0) \geqslant \sigma f'(0)$$

after using $\sigma \geqslant \rho$ and $f'(0) < 0$. Thus α satisfies both (2.5.1) and (2.5.4) strictly, and so by continuity of f and f' there exists an interval of acceptable points. □

This result is the best possible in that if $\sigma < \rho$, Al-Baali and Fletcher (1986) show how to construct a $\mathbb{C}^1$ function for which $\acute{\alpha}$ exists, having no acceptable points (see Question 2.20).

In practice a more stringent two-sided test on the slope

$$|f'(\alpha)| \leqslant -\sigma f'(0) \tag{2.5.6}$$

is preferred in place of (2.5.4). An acceptable point in (2.5.1) and (2.5.4) may not be close to the value of a given by an exact line search, as $\alpha = b$ in Figure 2.5.2 illustrates. With (2.5.6) on the other hand it is possible by reducing σ to restrict the set of acceptable points to an arbitrarily small neighbourhood of a local minimizer of $f(\alpha)$. It is possible to guarantee the existence of acceptable points when using the two-sided test in place of (2.5.4).

Lemma 2.5.2

If $\acute{\alpha}$ exists and $\sigma \geqslant \rho$ then there exists an interval of acceptable points for (2.5.1) *and* (2.5.6) *in* $(0, \acute{\alpha})$.

Proof

Extending Lemma 2.5.1, let ε be sufficiently small so that $f(0) > f(\varepsilon) > f(\acute{\alpha})$. It then follows that the point α which exists by virtue of the mean value theorem satisfies $0 > f'(\alpha) > \sigma f'(0)$. Thus α satisfies (2.5.1) and (2.5.6) strictly and the lemma follows as before. □

In practice values of $\sigma = 0.9$ for a weak (not restrictive) line search or $\sigma = 0.1$ for a fairly accurate line search have both been used satisfactorily in different circumstances. Smaller values of σ require substantially more effort to find an acceptable point with very little return, so are rarely used, except in testing theoretical hypotheses about an exact line search. As for ρ, a value of $\rho = 0.01$ would be typical, although this choice is not very significant as it is usually (2.5.6) which limits the range of acceptable points. It is possible to construct cases in which (2.5.1) excludes the only local minimum point of $f(\alpha)$ but this is unlikely to occur and there is the definite advantage that the otherwise troublesome case is excluded in which $f(\alpha)$ decreases to a constant value as $\alpha \to \infty$. The line search subproblem of finding a step that satisfies both (2.5.1) and (2.5.6) is considered further in the next section.

It is now appropriate to present the global convergence result which motivated the development of these tests. It is given here for the Wolfe–Powell tests (and hence *a fortiori* for tests (2.5.1) and (2.5.6)), although a similar result holds for the Goldstein tests. However it is necessary first of all to place some restriction on the choice of search direction $\mathbf{s}^{(k)}$ since it is also possible to make negligible reductions in $f(\mathbf{x})$ if the directions $\mathbf{s}^{(k)}$ are chosen so as to be increasingly close to being orthogonal to the steepest descent vector. This possibility can be excluded if $\mathbf{s}^{(k)}$ satisfies an *angle criterion* that the angle $\theta^{(k)}$ between $\mathbf{s}^{(k)}$ and $-\mathbf{g}^{(k)}$ is uniformly bounded away from orthogonality, that is if

$$\theta^{(k)} \leqslant \frac{\pi}{2} - \mu \qquad \forall k \tag{2.5.7}$$

where $\mu > 0$ is independent of k, and $\theta^{(k)} \in [0, \pi/2)$ is defined by

$$\begin{aligned} \cos\theta^{(k)} &= -\mathbf{g}^{(k)T}\mathbf{s}^{(k)}/(\|\mathbf{g}^{(k)}\|_2 \|\mathbf{s}^{(k)}\|_2) \\ &= -\mathbf{g}^{(k)T}\boldsymbol{\delta}^{(k)}/(\|\mathbf{g}^{(k)}\|_2 \|\boldsymbol{\delta}^{(k)}\|_2). \end{aligned} \tag{2.5.8}$$

With this restriction it is possible to state a global convergence theorem.

Theorem 2.5.1 (Global convergence of descent methods)

For a descent method with an inexact line search based on (2.5.1) *and* (2.5.4), *if* ∇f *exists and is uniformly continuous on the level set* $\{\mathbf{x}: f(\mathbf{x}) < f^{(1)}\}$, *and if* (2.5.7) *holds then either* $\mathbf{g}^{(k)} = \mathbf{0}$ *for some* k, *or* $f^{(k)} \to -\infty$, *or* $\mathbf{g}^{(k)} \to \mathbf{0}$.

Proof

Assume that $\mathbf{g}^{(k)} \neq \mathbf{0}$ for all k (and hence $\boldsymbol{\delta}^{(k)} \neq \mathbf{0}$) and that $f^{(k)}$ is bounded below; it follows that $f^{(k)} - f^{(k+1)} \to 0$ and hence from (2.5.1) and (2.5.3) that $-\mathbf{g}^{(k)T}\boldsymbol{\delta}^{(k)} \to 0$. Assume $\mathbf{g}^{(k)} \to \mathbf{0}$ does not hold. Then $\exists$ an $\varepsilon > 0$ and a subsequence

$\mathscr{S}$ such that $\|\mathbf{g}^{(k)}\|_2 \geqslant \varepsilon$ for $k \in \mathscr{S}$, and it follows from (2.5.8) and (2.5.7) that $\|\boldsymbol{\delta}^{(k)}\| \to 0$ for $k \in \mathscr{S}$. It is possible to rearrange (2.5.5) to give

$$-\mathbf{g}^{(k)\mathrm{T}}\boldsymbol{\delta}^{(k)} \leqslant \frac{(\mathbf{g}^{(k+1)} - \mathbf{g}^{(k)})^{\mathrm{T}}\boldsymbol{\delta}^{(k)}}{1-\sigma} \leqslant \frac{\|\mathbf{g}^{(k+1)} - \mathbf{g}^{(k)}\| \, \|\boldsymbol{\delta}^{(k)}\|}{1-\sigma}. \tag{2.5.9}$$

The uniform continuity assumption implies $\mathbf{g}^{(k+1)} - \mathbf{g}^{(k)} = o(1)$ for $k \in \mathscr{S}$ so it follows from (2.5.9), (2.5.8) and $\|\mathbf{g}^{(k)}\| \geqslant \varepsilon$ that $\cos\theta^{(k)} \leqslant o(1)$ for $k \in \mathscr{S}$ which contradicts (2.5.7). Thus $\mathbf{g}^{(k)} \to \mathbf{0}$. □

The possibility that $f^{(k)} \to -\infty$ can usually be eliminated by a priori knowledge about $f(\mathbf{x})$. It then follows from the theorem that the termination test (2.3.5) will always be satisfied for some k sufficiently large, and that the iteration cannot converge to a non-stationary point. It also implies that any accumulation point of the sequence $\{\mathbf{x}^{(k)}\}$ is stationary. However, convergence to a local minimizer is not guaranteed, although this usually occurs since f is always reduced by the line search.

It is important to consider whether the angle criterion (2.5.7) is likely to be satisfied by practical algorithms. Of course the steepest descent method satisfies the criterion with $\theta^{(k)} = 0$ but unfortunately is not good in practice. It is shown in Chapter 3 that Newton-like algorithms define $\mathbf{s}^{(k)}$ by

$$\mathbf{s}^{(k)} = -\mathbf{H}^{(k)}\mathbf{g}^{(k)} \tag{2.5.10}$$

where $\mathbf{H}^{(k)}$ is a symmetric matrix. A *sufficient condition for the descent property* is that $\mathbf{H}^{(k)}$ is positive definite (because $\mathbf{g}^{(k)\mathrm{T}}\mathbf{s}^{(k)} = -\mathbf{g}^{(k)\mathrm{T}}\mathbf{H}^{(k)}\mathbf{g}^{(k)} < 0$ when $\mathbf{g}^{(k)} \neq \mathbf{0}$). In this case a *sufficient condition for the angle criterion* is that the spectral condition number $\kappa^{(k)}$ of $\mathbf{H}^{(k)}$ (the ratio λ_1/λ_n of largest to smallest eigenvalues of $\mathbf{H}^{(k)}$) is bounded above, independently of k. (This result is established from (2.5.8) using the inequalities $\|\mathbf{Hg}\|_2 \leqslant \lambda_1 \|\mathbf{g}\|_2$, $\mathbf{g}^{\mathrm{T}}\mathbf{Hg} \geqslant \lambda_n \mathbf{g}^{\mathrm{T}}\mathbf{g}$ and $\sin x \leqslant x$ to get

$$\theta^{(k)} \leqslant \frac{\pi}{2} - \kappa^{(k)-1} \tag{2.5.11}$$

as required.) It then follows that Theorem 2.5.1 holds. In fact for most good algorithms it has not been possible to show that the angle criterion (2.5.7) is satisfied. Because of the convergence theorem, some modified algorithms have been suggested in which $\mathbf{s}^{(k)}$ is adjusted so as to satisfy (2.5.7). This can be done for example by adding some multiple of $-\mathbf{g}^{(k)}$ to $\mathbf{s}^{(k)}$, or by modifying $\mathbf{H}^{(k)}$ so that $\kappa^{(k)}$ is bounded. However, this can be unprofitable in that an algorithm with a superlinear convergence property can degrade to being linearly convergent when the modification operates. It may therefore be unwise to employ such *ad hoc* modifications unless there is good reason to think that the algorithm will otherwise fail.

The alternative is to try to improve the global convergence theorem itself, and this can be done if the weaker aim is considered of proving $\liminf \|\mathbf{g}^{(k)}\| = 0$

(that is $\mathbf{g}^{(k)} \to \mathbf{0}$ on a subsequence). This is contradicted if it is assumed that $\|\mathbf{g}^{(k)}\| \geqslant \varepsilon > 0$ for all k. It then follows from (2.5.3) and (2.5.8) that

$$f^{(k)} - f^{(k+1)} \geqslant \rho\varepsilon \|\boldsymbol{\delta}^{(k)}\|_2 \cos\theta^{(k)}. \tag{2.5.12}$$

If the stronger (but reasonable) assumption of Lipschitz continuity of $\mathbf{g}$ is made ($\|\mathbf{g}^{(k+1)} - \mathbf{g}^{(k)}\|_2 \leqslant \lambda \|\mathbf{x}^{(k+1)} - \mathbf{x}^{(k)}\|_2$ on the level set) then it follows from (2.5.9) that $\|\boldsymbol{\delta}^{(k)}\|_2 \geqslant (1-\sigma)\varepsilon\lambda^{-1}\cos\theta^{(k)}$. Incorporating this result with (2.5.12) gives

$$f^{(k)} - f^{(k+1)} \geqslant (1-\sigma)\rho\varepsilon^2\lambda^{-1}\cos^2\theta^{(k)}. \tag{2.5.13}$$

Summing over all k and assuming that f is bounded below, it follows that $\sum_k \cos^2\theta^{(k)}$ is bounded. Since a contradiction is being sought, it is assumed that the search directions satisfy the property that

$$\sum_k \cos^2\theta^{(k)} = \infty. \tag{2.5.14}$$

Then $\|\mathbf{g}^{(k)}\| \geqslant \varepsilon$ is contradicted and $\liminf \|\mathbf{g}^{(k)}\| = 0$ follows. It can be seen that (2.5.14) is a much weaker condition than $\cos\theta^{(k)} \geqslant \mu > 0$, and to my knowledge it is the weakest condition that has been used to prove global convergence theorems. Proofs that some standard methods satisfy (2.5.14) have been given by Powell (1976) (by implication) and Al-Baali (1985).

Finally the interaction of the line search tests with the property of superlinear convergence of an algorithm is considered. Using the notation $\mathbf{h}^{(k)} = \mathbf{x}^{(k)} - \mathbf{x}^*$ and $\boldsymbol{\delta}^{(k)} = \mathbf{x}^{(k+1)} - \mathbf{x}^{(k)}$ as in previous sections, then the following result can be given.

Lemma 2.5.3

If $\mathbf{x}^*$ is a local minimizer of $f(\mathbf{x}) \in \mathbb{C}^2$, if $\mathbf{G}^*$ is positive definite, and if $\mathbf{x}^{(k)} \to \mathbf{x}^*$ superlinearly, then

$$\frac{f^{(k)} - f^{(k+1)}}{-\mathbf{g}^{(k)\mathrm{T}}\boldsymbol{\delta}^{(k)}} \to \tfrac{1}{2} \quad \text{and} \quad \frac{\mathbf{g}^{(k+1)\mathrm{T}}\boldsymbol{\delta}^{(k)}}{\mathbf{g}^{(k)\mathrm{T}}\boldsymbol{\delta}^{(k)}} \to 0.$$

Proof

Using the above assumptions together with Taylor series, continuity and the results of Lemma 6.2.1 and Theorem 6.2.3, the following expressions can be established.

$$\begin{aligned}
f^{(k)} - f^* &= \tfrac{1}{2}\mathbf{h}^{(k)\mathrm{T}}\mathbf{G}^*\mathbf{h}^{(k)} + o(\|\mathbf{h}^{(k)}\|^2)\\
f^{(k+1)} - f^* &= o(\|\mathbf{h}^{(k)}\|^2)\\
-\mathbf{g}^{(k)\mathrm{T}}\boldsymbol{\delta}^{(k)} &= \mathbf{h}^{(k)\mathrm{T}}\mathbf{G}^*\mathbf{h}^{(k)} + o(\|\mathbf{h}^{(k)}\|^2)\\
\mathbf{g}^{(k+1)\mathrm{T}}\boldsymbol{\delta}^{(k)} &= o(\|\mathbf{h}^{(k)}\|^2)
\end{aligned}$$

and

$$\mathbf{h}^{(k)\mathrm{T}}\mathbf{G}^*\mathbf{h}^{(k)} \geqslant (\lambda + o(1))\|\mathbf{h}^{(k)}\|_2^2,$$

where $\lambda > 0$ is the least eigenvalue of $\mathbf{G}^*$. The lemma follows directly from these expressions. □

It follows from the lemma that tests such as (2.5.3) and (2.5.5) are satisfied for all k sufficiently large under these conditions. Consider therefore any line search algorithm which is locally superlinearly convergent when $\alpha^{(k)} = 1$ for all k. If the initial estimate of the step in the line search algorithm is taken asymptotically as $\alpha_1 = 1$ then it follows ultimately that this value is accepted as $\alpha^{(k)}$. Hence $\alpha^{(k)} = 1$ for all k sufficiently large, and the superlinear convergence property is preserved. This result is true for algorithms based on any of the tests (2.5.1), (2.5.2), (2.5.4) and (2.5.6).

2.6 ALGORITHMS FOR THE LINE SEARCH SUBPROBLEM

Numerous line search algorithms have been proposed over the years, and a good choice is important since it can have a considerable effect on the performance of the method in which it is embedded. The availability or not of first derivatives is a primary consideration: if derivatives are not available then there is not much theory available to act as a guide to how the line search should be terminated. There are therefore a wide range of possibilities and some of these are considered towards the end of this section. However, the case in which first derivatives are available is of major importance. In this case it is widely accepted that the line search should attempt to satisfy the conditions (2.5.1) and (2.5.6) for the reasons set out in Section 2.5. Therefore this section describes a line search algorithm which satisfies these conditions by making a finite (usually small) number of evaluations of $f(\alpha)$ and $f'(\alpha)$. (The notation $f(\alpha) = f(\mathbf{x}^{(k)} + \alpha\mathbf{s}^{(k)})$ as in Section 2.5 is used and the descent condition $f'(0) < 0$ is assumed to hold.)

A line search algorithm is an iterative method which generates a sequence of estimates $\{\alpha_i\}$. The sequence terminates when an iterate is located which satisfies some standard conditions for an acceptable point. There are two distinct parts to any line search algorithm. First comes the *bracketing phase* which searches to find a *bracket*, that is a non-trivial interval, $[a_i, b_i]$ say, which is known to contain an interval of acceptable points. This is followed by the *sectioning phase* in which the bracket is *sectioned* (i.e. divided) so as to generate a sequence of brackets $[a_j, b_j]$ whose length tends to zero. This forms the basis of a finite termination proof for the line search algorithm. In addition, since it is preferable to find an acceptable point which is close to a local minimizer of $f(\alpha)$, some form of *interpolation* is also desirable. This involves fitting usually a quadratic or cubic polynomial in α to known data, and choosing the next iterate α_{j+1} so as to minimize the polynomial, possibly subject to some sectioning restrictions.

One particular line search algorithm is now described for the case that first derivatives are available. Whilst it is not necessarily uniquely best, it is the one which I currently prefer for general unconstrained minimization, and it is closely similar to other line search algorithms that are also currently used. Some results which substantiate the properties of the algorithm are also proved. It has already been pointed out in Lemma 2.5.2 *either* that there exist acceptable points, *or* that the graph of $f(\alpha)$ never intersects the ρ-line. The latter possibility is avoided by assuming that the user is able to supply a lower bound $\bar{f}$ on $f(\alpha)$. (More precisely it is assumed that the user is prepared to accept any value of $f(\alpha)$ for which $f(\alpha) \leqslant \bar{f}$.) For example in a nonlinear least squares problem $\bar{f}=0$ would be appropriate. A consequence of this assumption is that the line search can be restricted to the interval $(0, \mu]$ where

$$\mu = (\bar{f} - f(0))/(\rho f'(0)) \tag{2.6.1}$$

is the point at which the ρ-line intersects the line $f=\bar{f}$.

In the bracketing phase the iterates α_i move out to the right in increasingly large jumps until either $f \leqslant \bar{f}$ is detected or a bracket on an interval of acceptable points is located. Initially $\alpha_0 = 0$, α_1 is given $(0 < \alpha_1 \leqslant \mu)$, and this phase of the algorithm can be described as follows:

<u>for</u> $i := 1, 2, \ldots$ <u>do</u>
<u>begin</u> evaluate $f(\alpha_i)$;
  <u>if</u> $f(\alpha_i) \leqslant \bar{f}$ <u>then</u> terminate;
  <u>if</u> $f(\alpha_i) > f(0) + \alpha_i f'(0)$ <u>or</u> $f(\alpha_i) \geqslant f(\alpha_{i-1})$
    <u>then begin</u> $a_i := \alpha_{i-1}$; $b_i := \alpha_i$; terminate B <u>end</u>;
  evaluate $f'(\alpha_i)$; (2.6.2)
  <u>if</u> $|f'(\alpha_i)| \leqslant -\sigma f'(0)$ <u>then</u> terminate;
  <u>if</u> $f'(\alpha_i) \geqslant 0$
    <u>then begin</u> $a_i := \alpha_i$; $b_i := \alpha_{i-1}$; terminate B <u>end</u>;
  <u>if</u> $\mu \leqslant 2\alpha_i - \alpha_{i-1}$
    <u>then</u> $\alpha_{i+1} := \mu$
    <u>else</u> choose $\alpha_{i+1} \in [2\alpha_i - \alpha_{i-1}, \min(\mu, \alpha_i + \tau_1(\alpha_i - \alpha_{i-1}))]$
<u>end</u>.

In this algorithm $\tau_1 > 1$ is a preset factor by which the size of the jumps is increased, typically $\tau_1 = 9$. The choice of α_{i+1} in the next last line of (2.6.2) can be made in any way, but a sensible choice would be to minimize in the given interval a cubic polynomial interpolating $f(\alpha_i)$, $f'(\alpha_i)$, $f(\alpha_{i-1})$ and $f'(\alpha_{i-1})$. Because μ exists and by virtue of (2.6.1), algorithm (2.6.2) clearly must terminate. In (2.6.2) 'terminate B' is used to indicate termination of just the bracketing phase, whereas 'terminate' alone indicates termination of the line search with an acceptable point α_i, or a point for which $f(\alpha_i) \leqslant \bar{f}$. If 'terminate B' occurs then a bracket $[a_i, b_i]$ is known (it is convenient to allow either $a_i < b_i$ or $b_i < a_i$ in this notation for a bracket) which satisfies the following properties.

(i) a_i is the current best trial point (least f) that also satisfies (2.5.1).
(ii) $f'(a_i)$ has been evaluated and satisfies

$$(b_i - a_i)f'(a_i) < 0 \text{ but not (2.5.6).} \tag{2.6.3}$$

(iii) b_i satisfies either $f(b_i) > f(0) + b_i \rho f'(0)$ or $f(b_i) \geqslant f(a_i)$ or both.

For such a bracket the following result holds.

Lemma 2.6.1

If $\sigma \geqslant \rho$, a bracket which satisfies (2.6.3) *contains an interval of acceptable points for* (2.5.1) *and* (2.5.6).

Proof

If $b_i > a_i$, consider a line, L say, through $(a_i, f(a_i))$ having slope $\rho f'(0)$ (i.e. parallel to the ρ-line). Let $\hat{\alpha}_i$ be the point in (a_i, b_i) closest to a_i at which the graph of $f(\alpha)$ intersects L. Existence of $\hat{\alpha}_i$ comes from (ii) and (iii) and continuity. Use of the mean value theorem as in Lemma 2.5.1 then implies the required result. If $b_i < a_i$ then a line through $(a_i, f(a_i))$ having zero slope is considered and a similar argument is used. □

This lemma shows that the bracketing phase has achieved its aim of bracketing an interval of acceptable points. Next comes the sectioning phase which generates a sequence of brackets $[a_j, b_j]$ for $j = i, i+1, \ldots$ whose lengths tend to zero. Each iteration picks a new trial point α_j in $[a_j, b_j]$ and the next bracket is either $[a_j, \alpha_j]$, $[\alpha_j, a_j]$ or $[\alpha_j, b_j]$, the choice being made so that properties (2.6.3) are preserved. The sectioning phase terminates when the current trial point α_j is found to be acceptable in (2.5.1) and (2.5.6). The algorithm can be described as follows

<u>for</u> $j := i, i+1, \ldots$ <u>do</u>

<u>begin</u> choose $\alpha_j \in [a_j + \tau_2(b_j - a_j), b_j - \tau_3(b_j - a_j)]$;
evaluate $f(\alpha_j)$;

<u>if</u> $f(\alpha_j) > f(0) + \rho\alpha_j f'(0)$ <u>or</u> $f(\alpha_j) \geqslant f(a_j)$

<u>then</u> <u>begin</u> $a_{j+1} := a_j$; $b_{j+1} := \alpha_j$ <u>end</u>

<u>else</u> <u>begin</u> evaluate $f'(\alpha_j)$; (2.6.4)

<u>if</u> $|f'(\alpha_j)| \leqslant -\sigma f'(0)$ <u>then</u> terminate;

$a_{j+1} := \alpha_j$;

<u>if</u> $(b_j - a_j)f'(\alpha_j) \geqslant 0$ <u>then</u> $b_{j+1} := a_j$ <u>else</u> $b_{j+1} := b_j$

<u>end</u>.

<u>end</u>.

In this algorithm τ_2 and τ_3 are preset factors $(0<\tau_2<\tau_3\leqslant\frac{1}{2})$ which restrict α_j from being arbitrarily close to the extremes of the interval $[a_j, b_j]$. It then follows that

$$|b_{j+1}-a_{j+1}|\leqslant(1-\tau_2)|b_j-a_j| \tag{2.6.5}$$

and this guarantees convergence of the interval lengths to zero. Typical values are $\tau_2=\frac{1}{10}$ ($\tau_2\leqslant\sigma$ is advisable) and $\tau_3=\frac{1}{2}$, although the algorithm is insensitive to the precise values that are used. The choice of α_j in the second line of (2.6.4) can be made in any way, but a sensible choice would be to minimize in the given interval a quadratic or cubic polynomial which interpolates $f(a_j)$, $f'(a_j)$, $f(b_j)$, and $f'(b_j)$ if it is known. When (2.6.4) terminates then α_j is the required acceptable point and becomes the step length $\alpha^{(k)}$ to be used in (2.3.2). The convergence properties of the sectioning scheme are given in the following result (Al-Baali and Fletcher, 1986).

Theorem 2.6.1

If $\sigma\geqslant\rho$ and the initial bracket $[a_i, b_i]$ satisfies (2.6.3), *then the sectioning scheme* (2.6.4) *has the following properties.*

Either (*a*) *the scheme terminates with an α_j which is an acceptable point in* (2.5.1) *and* (2.5.6)

or (*b*) *the scheme fails to terminate and there exists a point c such that for j sufficiently large $a_j\uparrow c$ and $b_j\downarrow c$ monotonically (not strictly) and c is an acceptable point in* (2.5.1) *and* (2.5.6).

Moreover if $\sigma>\rho$ then (*b*) *cannot occur and the scheme must terminate.*

Proof

It follows from (2.6.5) that $|a_j-b_j|\to 0$ and from (2.6.4) that $[a_{j+1}, b_{j+1}]\subset[a_j, b_j]$. Therefore $\exists$ a limit point c such that $a_j\to c$, $b_j\to c$ and hence $\alpha_j\to c$. Since a_j satisfies (2.5.1) but not (2.5.6) it follows that c satisfies (2.5.1) and that

$$|f'(c)|\geqslant-\sigma f'(0) \tag{2.6.6}$$

Let there exist an infinite subsequence of brackets for which $b_j<a_j$. It follows from (2.6.3) (iii) that $f(b_j)-f(a_j)\geqslant 0$ and hence by the mean value theorem and the existence of c that $f'(c)\leqslant 0$. But from (2.6.3) (ii), $f'(a_j)(b_j-a_j)<0$ implies that $f'(c)\geqslant 0$ in the limit. These inequalities contradict (2.6.6), proving that $a_j\uparrow c$ and $b_j\downarrow c$ for j sufficiently large. Consider therefore the case that $b_j>a_j$. It follows from (2.6.3) (iii) and a_j satisfying (2.5.1) that

$$f(b_j)-f(a_j)\geqslant(b_j-a_j)\rho f'(0),$$

so the mean value theorem and the existence of c imply that $f'(c)\geqslant\rho f'(0)$. But $f'(a_i)(b_i-a_i)<0$ implies that $f'(c)\leqslant 0$. If $\sigma>\rho$, these inequalities contradict (2.6.6), showing that case (b) cannot arise, in which case the algorithm must terminate as in (a). If $\sigma=\rho$ the possibility of non-termination exists, with a limit point c for which (2.5.1) holds and $f'(c)=\rho f'(0)$. □

An example in which case (b) arises is described by Al-Baali and Fletcher (1986). This reference also gives a similar theorem for more simple algorithm aimed at satisfying the Wolfe–Powell conditions. For practical purposes, however, Theorem 2.6.1 indicates that $\sigma > \rho$ should be selected, in which case the sectioning algorithm (2.6.4) is guaranteed to terminate with an acceptable point. This algorithm is not the only one which has been used, for instance a different one is given by Moré and Sorensen (1982). This differs from (2.6.4) in that α_j is allowed to become arbitrarily close to a_j, but a bisection step is used if the bracket length is not reduced by a factor of $\frac{1}{2}$ after any two iterations. This is similar to an idea of Brent (1973a) in the context of solving a nonlinear equation.

The use of polynomial extrapolation or interpolation in conjunction with (2.6.2) and (2.6.4) is recommended. Given an interval [0, 1], and data values f_0, f'_0 and f_1 ($f_0 = f(0)$ etc.) then the unique quadratic that interpolates these values is defined by

$$q(z) = f_0 + f'_0 z + (f_1 - f_0 - f'_0)z^2.$$

If in addition f'_1 is given then the corresponding (Hermite) interpolating cubic is

$$c(z) = f_0 + f'_0 z + \eta z^2 + \xi z^3$$

where

$$\eta = 3(f_1 - f_0) - 2f'_0 - f'_1$$

and

$$\xi = f'_0 + f'_1 - 2(f_1 - f_0).$$

In the schemes (2.6.2) and (2.6.4) it is required to work with an interval $[a, b]$ rather than [0, 1], and $a > b$ is allowed. To do this the transformation

$$\alpha = a + z(b - a)$$

is used which maps [0, 1] into $[a, b]$. The chain rule gives $\mathrm{d}f/\mathrm{d}z = \mathrm{d}f/\mathrm{d}\alpha \cdot \mathrm{d}\alpha/\mathrm{d}z = (b - a)\mathrm{d}f/\mathrm{d}\alpha$ which relates the derivatives f'_0 and f'_1 above to known values of $\mathrm{d}f/\mathrm{d}\alpha$ obtained in the line search. To find the minimizers of $q(z)$ or $c(z)$ in a given interval, it is necessary to examine not only the stationary values, but also the values and derivatives at the ends of the interval. This requires some simple calculus.

An example is now given of the use of this line search, illustrating the bracketing, sectioning and interpolation processes. Consider using the function (1.2.2) and let $\mathbf{x}^{(k)} = \mathbf{0}$ and $\mathbf{s}^{(k)} = (1, 0)^{\mathrm{T}}$. Then $f(\alpha) = 100\alpha^4 + (1 - \alpha)^2$ and $f'(\alpha) = 400\alpha^3 - 2(1 - \alpha)$. The parameters $\sigma = 0.1$, $\rho = 0.01$, $\tau_1 = 9$, $\tau_2 = 0.1$ and $\tau_3 = 0.5$ are used. The progress of the line search is set out in Table 2.6.1. The first part shows what happens if the initial guess is $\alpha_1 = 0.1$. The initial interval does not give a bracket, so the bracketing algorithm requires that the next iterate is chosen in [0.2, 1]. Mapping [0, 0.1] on to [0, 1] in z-space, the resulting cubic fit is $c(z) = 1 - 0.2z + 0.02z^3$ and the minimum value of $c(z)$ in [2, 10] is at $z = 2$. Thus $\alpha_2 = 0.2$ is the next iterate. This iterate give a bracket [0.2, 0.1] which satisfies (2.6.3). The sectioning algorithm comes

Table 2.6.1 The line search using first derivatives

Starting from $\alpha_1 = 0.1$

α	0	0.1	0.2	0.160948
$f(\alpha)$	1	0.82	0.8	0.771111
$f'(\alpha)$	-2	-1.4	1.6	-0.010423

Starting from $\alpha_1 = 1$

α	0	1	0.1	0.19	0.160922
$f(\alpha)$	1	100	0.82	0.786421	0.771112
$f'(\alpha)$	-2	—	-1.4	1.1236	-0.011269

into play and seeks a new iterate in $[0.19, 0.15]$. Mapping $[0.2, 0.1]$ on to $[0, 1]$ gives a cubic $c(z) = 0.8 - 0.16z + 0.24z^2 - 0.06z^3$. The minimizer of this cubic in $[0.1, 0.5]$ is locally unconstrained at $z = 0.390524$, and this gives a new iterate $\alpha_3 = 0.160948$. This point is found to be acceptable and the line search is terminated. Alternatively let the line search start with $\alpha_1 = 1$. Then the initial value of $f(1) > f(0)$ so this point immediately gives a bracket and it is not necessary to evaluate $f'(1)$ (this saves unnecessary gradient evaluations). The sectioning phase is therefore entered. Making a quadratic interpolation gives $q(\alpha) = 1 - 0.2\alpha + 99.2\alpha^2$, which is minimized in $[0.1, 0.5]$ by $\alpha_2 = 0.1$. Thus $[0.1, 1]$ is the new bracket that satisfies (2.6.3) and the next iterate is chosen as a point in $[0.19, 0.55]$. This could be done either by interpolating a quadratic at $\alpha = 0.1$ and $\alpha = 1$, or interpolating a cubic at $\alpha = 0$ and $\alpha = 0.1$ (my current algorithm would do the latter). In both cases here, $\alpha_3 = 0.19$ is the minimizing point and $[0.19, 0.1]$ becomes the new bracket. A cubic interpolation in this bracket leads finally to an iterate $\alpha_4 = 0.160922$ which is acceptable.

This description of the algorithm neglects the effect of round-off errors, and these can cause difficulties when $\mathbf{x}^{(k)}$ is close to $\mathbf{x}^*$. In this case, although $f'(0) < 0$, it may happen that $f(\alpha) \geqslant f(0)$ for all $\alpha > 0$, due to round-off error. This situation can also arise if the user has made errors in his formulae for derivatives. Therefore I have found it advisable also to terminate after line 3 of (2.6.4) if $(a_j - \alpha_j)f'(a_j) \leqslant \varepsilon$ where ε is a tolerance on f such as in (2.3.7). This causes an exit from the minimization method with an indication that no progress can be made in the line search.

The initial choice of α_1 also merits some study. If an estimate $\Delta f > 0$ is available of the likely reduction in f to be achieved by the line search, then it is possible to interpolate a quadratic to $f'(0)$ and Δf, giving the estimate

$$\alpha_1 = -2\Delta f / f'(0). \tag{2.6.7}$$

On the first iteration of the minimization method Δf must be user supplied, but subsequently $\Delta f = \max(f^{(k-1)} - f^{(k)}, 10\varepsilon)$ has been found to work well, that is the reduction from the previous iteration, suitably safeguarded. For Newton-like methods, however, the choice $\alpha^{(k)} = 1$ is significant in giving rapid ultimate convergence by virtue of the Dennis–More theorem (Theorem 6.2.3). In this case

$$\alpha_1 = \min(1, -2\Delta f/f'(0)) \tag{2.6.8}$$

is used. This is failsafe in that ultimately the choice $\alpha_1 = 1$ is always made by a method which is converging superlinearly. This is proved as in Lemma 2.5.3.

The rest of this section is devoted to the line search for a no-derivative method. In this case there is no theory analogous to that in Section 2.5 which can act as a guide. Since even the precise aim of the line search is unclear, there is no general agreement on how it should be carried out. Also the lack of derivative information makes the search more difficult. One possibility is to emulate the line search of (2.6.2) and (2.6.4), but using difference approximations to derivatives, and this might be the most suitable for a no-derivative quasi-Newton method.

However this approach would not be suitable for a higher accuracy line search suitable say for a conjugate direction method. There are various other methods that have been used for a no-derivative line search, and their main features are similar to those described earlier in this section. Any three α-values $\alpha_1 < \alpha_2 < \alpha_3$ form a *bracket* (on the minimizing α-value) when $f_2 \leqslant \min(f_1, f_3)$ (see Figure 2.6.1 for example). Once located, the bracket is contracted, either by sectioning or interpolation or some combination of the two. A simple, purely sectioning approach is the *golden section search* in which the ratio of $\alpha_3 - \alpha_2$: $\alpha_2 - \alpha_1$ is kept fixed at $\tau:1$ (or $1:\tau$), where $\tau > 1$. When an additional point α_4 is inserted to contract the bracket (see Figure 2.6.1), it is required that both potential new brackets $(\alpha_1, \alpha_2, \alpha_4)$ or $(\alpha_2, \alpha_4, \alpha_3)$ have intervals in the same ratio. Since by symmetry $\alpha_3 - \alpha_4 = \alpha_2 - \alpha_1$ in the figure, it follows that τ satisfies $1 = \tau^2/(\tau + 1)$ and is therefore the positive root of $\tau^2 - \tau - 1 = 0$, that is $\tau = (1 + \sqrt{5})/2 \simeq 1.618$. The method is to evaluate f at the new point and choose the new smaller interval which maintains the bracket. In the figure this would be $(\alpha_1, \alpha_2, \alpha_4)$. The process is then repeated until some convergence test is satisfied. The application of golden section search to find the minimizing point of $f(\alpha) = 1 - \alpha\exp(-\alpha^2)$ to within one decimal place is shown in Table 2.6.2. The initial bracket is (0, 0.618, 1) and the final bracket (0.674, 0.708, 0.764). Similar ideas occur in the *Fibonnaci section search* (Kiefer, 1957) which is optimal in a certain (not very relevant) sense.

Unfortunately sectioning alone is not very efficient (see Question 2.12) and it is more effective to use interpolation in some way. For instance given a bracket $(\alpha_1, \alpha_2, \alpha_3)$, a quadratic could be fitted to f_1, f_2, f_3 and its minimizer chosen as the next trial point. This often works well but can fail unless supplemented by

Table 2.6.2 An example of golden section search

α	0	0.618	1	0.382	0.764	0.854	0.708	0.674
$f(\alpha)$	1	0.578	0.632	0.670	0.574	0.588	0.571	0.572

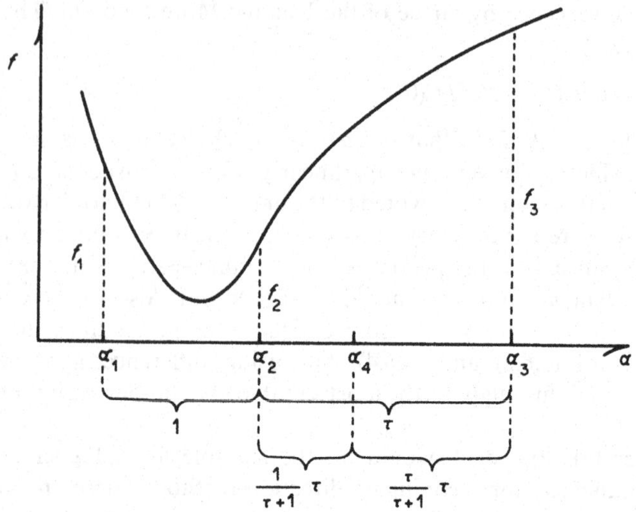

Figure 2.6.1 Golden section search

some sectioning ideas. This is illustrated by Figure 2.6.1 in which a quadratic interpolation gives $\alpha = \alpha_2$ as the new trial point. As yet however there is no general agreement on how best to combine sectioning and interpolation in the no-derivative line search.

QUESTIONS FOR CHAPTER 2

2.1. Obtain expressions for all first and second derivatives of the function of two variables

$$f(\mathbf{x}) = 100(x_2 - x_1^2)^2 + (1 - x_1)^2.$$

Verify that the minimizer $\mathbf{x}^* = (1,1)^T$ satisfies $\mathbf{g}^* = \mathbf{0}$ and $\mathbf{G}^*$ positive definite. Show that $\mathbf{G}(\mathbf{x})$ is singular if and only if $\mathbf{x}$ satisfies the condition

$$x_2 - x_1^2 = 0.005.$$

Hence show that $\mathbf{G}$ is positive definite for all $\mathbf{x}$ such that $f(\mathbf{x}) < 0.0025$.

2.2. Obtain expressions for all first and second derivatives of the function of two variables

$$f(\mathbf{x}) = x_1^4 + x_1 x_2 + (1 + x_2)^2.$$

Evaluate these derivatives at $\mathbf{x} = \mathbf{0}$ and show that $\mathbf{G}(\mathbf{0})$ is not positive definite. Verify that a local minimizer is $\mathbf{x}^* = (0.6959, -1.3479)^T$ to four decimal places

2.3. Find the stationary points of the function

$$f(\mathbf{x}) = 2x_1^3 - 3x_1^2 - 6x_1 x_2 (x_1 - x_2 - 1).$$

Which of these points are local minima, which are local maxima, and which are neither?

2.4. Show that the function $f(\mathbf{x}) = (x_2 - x_1^2)^2 + x_1^5$ has only one stationary point which is neither a local maximum nor a local minimum.

2.5. Find the stationary point of the function

$$f(\mathbf{x}) = 2x_1^2 + x_2^2 - 2x_1x_2 + 2x_1^3 + x_1^4$$

which is a *global* minimizer.

2.6. Investigate the stationary points of the function

$$f(\mathbf{x}) = x_1^2x_2^2 - 4x_1^2x_2 + 4x_1^2 + 2x_1x_2^2 + x_2^2 - 8x_1x_2 + 8x_1 - 4x_2.$$

2.7. (a) Sketch contours of the function

$$f(\mathbf{x}) = x_1^2 + 4x_2^2 - 4x_1 - 8x_2.$$

Deduce the value $\mathbf{x}^*$ which minimizes f.

(b) Show that if the method of steepest descent is started from $\mathbf{x}^{(1)} = (0, 0)^T$ it cannot converge to $\mathbf{x}^*$ in a finite number of steps. Are there any values of $\mathbf{x}^{(1)}$ for which there will be convergence in a finite number of steps?

2.8. The method of steepest descent applied to the function

$$f(\mathbf{x}) = 2x_1^2 - 2x_1x_2 + x_2^2 + 2x_1 - 2x_2$$

generates a sequence $\{\mathbf{x}^{(k)}\}$. If $\mathbf{x}^{(2k+1)} = (0, 1 - 1/5^k)^T$, show that $\mathbf{x}^{(2k+3)} = (0, 1 - 1/5^{k+1})^T$. Sketch in the $x_1 - x_2$ plane the successive steps taken by the method if $\mathbf{x}^{(1)} = \mathbf{0}$ is chosen, and deduce the minimizer of $f(\mathbf{x})$.

2.9. For the function

$$f(\mathbf{x}) = x_1^2 + 2x_2^2 + 4x_1 + 4x_2$$

prove by induction that the method of steepest descent applied with an initial guess $\mathbf{x}^{(1)} = \mathbf{0}$ generates the sequence $\{\mathbf{x}^{(k)}\}$ where

$$\mathbf{x}^{(k+1)} = \left(\frac{2}{3^k} - 2, (-\tfrac{1}{3})^k - 1\right)^T$$

Hence deduce the minimizer of $f(\mathbf{x})$.

2.10. Consider applying the method of steepest descent to a positive definite quadratic function with Hessian $\mathbf{G}$. Suppose that $\mathbf{x}^{(1)}$ $(\neq \mathbf{x}^*)$ can be expressed as

$$\mathbf{x}^{(1)} = \mathbf{x}^* + \mu\mathbf{s}$$

where $\mathbf{s}$ is an eigenvector of $\mathbf{G}$ with eigenvalue λ. Show that $\mathbf{g}^{(1)} = \mu\lambda\mathbf{s}$ and that, if an exact line search in the direction of steepest descent is assumed, then the method terminates in one iteration. Hence show that the method terminates in one iteration for any $\mathbf{x}^{(1)}$ if $\mathbf{G}$ is a multiple of a unit matrix. If

$\mathbf{x}^{(1)}$ cannot be expressed as above, then it can be expressed as

$$\mathbf{x}^{(1)} = \mathbf{x}^* + \sum_{i=1}^{m} \mu_i \mathbf{s}_i$$

where $m > 1$, and for all i, $\mu_i \neq 0$ and the $\mathbf{s}_i$ are eigenvectors of $\mathbf{G}$ corresponding to distinct eigenvalues λ_i. In this case show that the method does not terminate in one iteration.

2.11. For the steepest descent method with exact line searches, show that the search direction $\mathbf{s}^{(k+1)}$ is orthogonal to $\mathbf{s}^{(k)}$ for all k. Apply the method to the function $f(\mathbf{x}) = 10x_1^2 + x_2^2$ from $\mathbf{x}^{(1)} = (1/10, 1)^{\mathrm{T}}$. It can be shown that in the worst case the method has linear convergence at a rate $(\lambda_1 - \lambda_n)/(\lambda_1 + \lambda_n)$, where λ_1 and λ_n are the largest and smallest eigenvalues of $\mathbf{G}^*$ (Akaike, 1959). Verify numerically that this rate is achieved in this example. Notice that if $\mathbf{G}^*$ is sufficiently ill-conditioned, this rate can be arbitrarily close to unity.

2.12. For the problem of Table 2.6.2, show that a single quadratic interpolation at α_1, α_2, and α_3 gives an estimate of the minimum α-value which is within the final bracket achieved by golden section search, although without the same guaranteed accuracy.

2.13. Find the minimum of the function $1 - \alpha \exp(-\alpha)$ by golden section search to an uncertainty in α of within 0.1. Verify that $\alpha_1 = 0$, $\alpha_2 = 1.236$, $\alpha_3 = 2$ gives an initial bracket, and work to three decimal places. Also try the effect of a single quadratic interpolation at these points.

2.14. Find the range of acceptable α-values for the Goldstein conditions, the Wolfe–Powell conditions, and the conditions (2.5.1) and (2.5.6) in regard to the function of the previous question for $\sigma = \rho = \frac{1}{10}$ and $\sigma = \rho = \frac{1}{4}$. In the latter case, apply the line search of (2.6.2) and (2.6.4) with initial values $\alpha = 0$ and 1, and verify that an acceptable point is readily obtained.

2.15. Apply the line search methods of the previous two questions to the function $1 - (5\alpha^2 - 6\alpha + 5)^{-1}$ from the same initial α-values. Verify that golden section search obtains a good initial approximation, but is relatively slow in reducing the bracket.

2.16. Show that if the non-zero vectors $\mathbf{s}^{(1)}, \mathbf{s}^{(2)}, \ldots, \mathbf{s}^{(n)}$ are conjugate to a positive definite matrix $\mathbf{G}$, then the vectors are independent.

2.17. If $\mathbf{S}$ is a matrix with conjugate columns $\mathbf{s}^{(1)}, \mathbf{s}^{(2)}, \ldots, \mathbf{s}^{(n)}$ such that $\mathbf{S}^{\mathrm{T}}\mathbf{G}\mathbf{S} = \mathbf{I}$, show that the same is true of $\bar{\mathbf{S}}$ if and only if $\bar{\mathbf{S}} = \mathbf{S}\mathbf{Q}$, where $\mathbf{Q}$ is orthogonal.

2.18. Show that the vectors $\mathbf{s}^{(i)}$, $i = 1, 2, \ldots, n$, defined by $s_j^{(i)} = \mathit{if}\ j \leqslant i\ \mathit{then}\ j\ \mathit{else}$ 0 are conjugate with respect to the tridiagonal matrix $\mathbf{G}$ defined by $G_{ii} = 2$ and $G_{i,i+1} = G_{i+1,i} = -1\ \forall i$.

2.19. For a quadratic function (1.2.11) show that
(a) a minimizer exists iff $\mathbf{G}$ is positive semi-definite and $\mathbf{b}$ is in the range of $\mathbf{G}$,
(b) the minimizer is unique iff $\mathbf{G}$ is positive definite, and
(c) if $\mathbf{G}$ is positive semi-definite then every stationary point is a global minimizer.

2.20. Consider the $\mathbb{C}^1$ quadratic spline

$$f(\alpha)=\begin{cases}\frac{1}{2}\left(\dfrac{1-\sigma^+}{t}\right)\alpha^2-\alpha & \text{if } \alpha\leqslant t\\ \frac{1}{2}(\sigma^+-1)t-\sigma^+\alpha & \text{if } \alpha\geqslant t\end{cases}$$

where σ^+ is marginally greater than σ and $\sigma<\rho<\frac{1}{2}$. Show that its graph intersects the ρ-line at $\alpha=1$, but that there are no points that are simultaneously acceptable in both (2.5.1) and (2.5.4).

Chapter 3

Newton-like Methods

3.1 NEWTON'S METHOD

In Section 2.4 a number of reasons are put forward for deriving a method on the basis of a quadratic model. The most straightforward way of doing this gives rise to what is usually known as *Newton's method*. The quadratic model is obtained from a truncated Taylor series expansion of $f(\mathbf{x})$ about $\mathbf{x}^{(k)}$, which can be written

$$f(\mathbf{x}^{(k)}+\boldsymbol{\delta})\approx q^{(k)}(\boldsymbol{\delta})=f^{(k)}+\mathbf{g}^{(k)\mathrm{T}}\boldsymbol{\delta}+\tfrac{1}{2}\boldsymbol{\delta}^{\mathrm{T}}\mathbf{G}^{(k)}\boldsymbol{\delta} \tag{3.1.1}$$

where $\boldsymbol{\delta}=\mathbf{x}-\mathbf{x}^{(k)}$, and $q^{(k)}(\boldsymbol{\delta})$ is the resulting quadratic approximation for iteration k. Then the iterate $\mathbf{x}^{(k+1)}$ in Newton's method is simply taken to be $\mathbf{x}^{(k)}+\boldsymbol{\delta}^{(k)}$, where the correction $\boldsymbol{\delta}^{(k)}$ minimizes $q^{(k)}(\boldsymbol{\delta})$. The method requires zero, first and second derivatives of f to be available at any point, so that the coefficients $f^{(k)}, \mathbf{g}^{(k)}$, and $\mathbf{G}^{(k)}$ which define $q^{(k)}(\boldsymbol{\delta})$ are available. Also since $q^{(k)}(\boldsymbol{\delta})$ only has a unique minimizer if $\mathbf{G}^{(k)}$ is *positive definite*, the method is only well defined in these circumstances. In this case $\boldsymbol{\delta}^{(k)}$ is defined by the condition that $\nabla q^{(k)}(\boldsymbol{\delta}^{(k)})=\mathbf{0}$ (Section 2.1), so by (1.2.14) the kth iteration of Newton's method can be written

$$\begin{aligned}&\text{(a) solve } \mathbf{G}^{(k)}\boldsymbol{\delta}=-\mathbf{g}^{(k)} \quad \text{for } \boldsymbol{\delta}=\boldsymbol{\delta}^{(k)},\\ &\text{(b) set } \mathbf{x}^{(k+1)}=\mathbf{x}^{(k)}+\boldsymbol{\delta}^{(k)}.\end{aligned} \tag{3.1.2}$$

Step (a) involves the solution of an $n\times n$ system of linear equations. This is most conveniently solved by factorizing $\mathbf{G}^{(k)}=\mathbf{LDL}^{\mathrm{T}}$ (see Section 2.1), which also enables the positive definite condition to be checked. This requires $\frac{1}{6}n^3+O(n^2)$ multiplications per iteration.

An example of the application of the method to the function

$$f(\mathbf{x})=x_1^4+x_1x_2+(1+x_2)^2 \tag{3.1.3}$$

from $\mathbf{x}^{(1)}=(0.75,-1.25)^{\mathrm{T}}$, working to about ten significant digits, is given in

Table 3.1.1 Newton's method

k	1	2	3	4	5
$x_1^{(k)}$	0.75	0.7	0.6959107807	0.6958843872	0.6958843861
$x_2^{(k)}$	−1.25	−1.35	−1.347955390	−1.347942194	−1.347942193
$f^{(k)}$	−0.55859375	−0.5824	−0.5824451726	−0.5824451744	−0.5824451744
$g_1^{(k)}$	0.4375	0.022	0.0001401889	0.0000000058	0
$g_2^{(k)}$	0.25	0	0	0	0
$h_1^{(k)}$	0.0541156139	0.0041156139	0.0000263946	0.0000000011	0
$h_2^{(k)}$	0.097942193	−0.002057807	−0.0000131973	−0.0000000006	0

Table 3.1.1 (the error $\mathbf{h}^{(k)}$ is defined in (2.3.1)). The matrix $\mathbf{G}^{(k)}$ is positive definite for all k in this example. This fact, and the convergence of the method at a rapid rate which the example shows, are typical of what happens when the initial point $\mathbf{x}^{(1)}$ is close to a local minimizing point $\mathbf{x}^*$. In fact this behaviour can be established rigorously as follows.

Theorem 3.1.1

Assume that $f \in \mathbb{C}^2$ and that the Hessian matrix satisfies a Lipschitz condition $\|\mathbf{G}(\mathbf{x}) - \mathbf{G}(\mathbf{y})\| \leqslant \lambda \|\mathbf{x} - \mathbf{y}\|$ in a neighbourhood of a local minimizer $\mathbf{x}^*$. If $\mathbf{x}^{(k)}$ is sufficiently close to $\mathbf{x}^*$ for some k, and if $\mathbf{G}^*$ is positive definite, then Newton's method is well defined for all k, and converges at second order.

Proof

The continuity properties of f ensure that the Taylor series (1.2.19) for $\mathbf{g}(\mathbf{x}^{(k)} + \mathbf{h})$ about $\mathbf{x}^{(k)}$ can be written as

$$\mathbf{g}(\mathbf{x}^{(k)} + \mathbf{h}) = \mathbf{g}^{(k)} + \mathbf{G}^{(k)}\mathbf{h} + O(\|\mathbf{h}\|^2)$$

and $\mathbf{h} = -\mathbf{h}^{(k)}$ gives

$$\mathbf{0} = \mathbf{g}^* = \mathbf{g}^{(k)} - \mathbf{G}^{(k)}\mathbf{h}^{(k)} + O(\|\mathbf{h}^{(k)}\|^2).$$

Let $\mathbf{x}^{(k)}$ be in a neighbourhood of $\mathbf{x}^*$ for which $\mathbf{G}^{(k)}$ is positive definite and $\mathbf{G}^{(k)^{-1}}$ is bounded above. Such a neighbourhood exists by continuity of $\mathbf{G}(\mathbf{x})$. Then the kth iteration exists, and multiplying through by $\mathbf{G}^{(k)^{-1}}$ gives

$$\mathbf{0} = -\boldsymbol{\delta}^{(k)} - \mathbf{h}^{(k)} + O(\|\mathbf{h}^{(k)}\|^2) = -\mathbf{h}^{(k+1)} + O(\|\mathbf{h}^{(k)}\|^2)$$

by definition of $\mathbf{h}^{(k+1)}$. Hence by definition of $O(\cdot)$ there exists a constant $c > 0$ such that

$$\|\mathbf{h}^{(k+1)}\| \leqslant c\|\mathbf{h}^{(k)}\|^2. \tag{3.1.4}$$

If $\mathbf{x}^{(k)}$ is in a closer neighbourhood for which $\|\mathbf{h}\| \leqslant \alpha/c$, where $0 < \alpha < 1$, then it follows that $\|\mathbf{h}^{(k+1)}\| \leqslant \alpha\|\mathbf{h}^{(k)}\|$. Thus $\mathbf{x}^{(k+1)}$ is in the neighbourhood, and by induction the iteration is well defined for all k and $\|\mathbf{h}^{(k)}\| \to \mathbf{0}$. Thus the iteration converges and the order is shown to be second order by (3.1.4). □

An illustration of the second order convergence is provided by Table 3.1.1. If $\|\mathbf{h}^{(k+1)}\|_2/\|\mathbf{h}^{(k)}\|_2^2$ is computed for $k = 2$ and 3 then a value of close to 1.4 is observed in both cases. Higher precision would be required to sample the ratio for larger values of k. In passing, it is pointed out that a variation of Newton's method exists in which the matrix $\mathbf{G}^{(k)}$ is not evaluated on every iteration, but the factors from a previous iteration are used in its place. This saves effort in carrying out the iteration, but slows down the overall rate of convergence. In general if the Hessian is evaluated every m iterations then the ultimate order is reduced to $(m+1)^{1/m}$. In practice my impression is that no significant advantage has been demonstrated for such variations.

The basic Newton method as it stands is not suitable for a general purpose algorithm since $\mathbf{G}^{(k)}$ may not be positive definite when $\mathbf{x}^{(k)}$ is remote from the solution. Furthermore even if $\mathbf{G}^{(k)}$ is positive definite then convergence may not occur, in fact $\{f^{(k)}\}$ may not even decrease. The latter possibility can be eliminated by *Newton's method with line search* in which the Newton correction is used to generate a direction of search

$$\mathbf{s}^{(k)} = -\mathbf{G}^{(k)^{-1}}\mathbf{g}^{(k)} \tag{3.1.5}$$

(by solving $\mathbf{G}^{(k)}\mathbf{s}^{(k)} = -\mathbf{g}^{(k)}$). This is then used in a line search algorithm (2.3.2). In fact if $\mathbf{G}^{(k)}$ and hence $\mathbf{G}^{(k)^{-1}}$ are positive definite then $\mathbf{s}^{(k)}$ can be expressed by (2.5.10) and therefore is a descent direction. The discussion in Sections 2.5 and 2.6 about when convergence occurs, and what type of line search should be used, is then relevant. [In fact it is possible to interpret Newton's method with line search in a different way (Fletcher, 1978) so that its domain of definition is extended to include certain cases where $\mathbf{G}^{(k)}$ is singular or has one negative eigenvalue. However the counter-example which is outlined in the next paragraph is still a fundamental difficulty.]

The main difficulty therefore in modifying Newton's method arises when $\mathbf{G}^{(k)}$ is not positive definite. In most cases it is still possible to compute $\mathbf{s}^{(k)}$ from (3.1.5) and to search along $\pm\mathbf{s}^{(k)}$, the sign being chosen to ensure a descent direction. However the stationary point of the approximating quadratic function $q^{(k)}(\boldsymbol{\delta})$ is then not a minimizing point, and the relevance of searching in such a direction is questionable. In fact the function (3.1.3) provides a simple example (due to Powell) that Newton's method with line search can fail. If $\mathbf{x}^{(1)} = (0,0)^T$ is chosen, then it can be verified that

$$\mathbf{g}^{(1)} = \begin{pmatrix} 0 \\ 2 \end{pmatrix}, \quad \mathbf{G}^{(1)} = \begin{bmatrix} 0 & 1 \\ 1 & 2 \end{bmatrix}, \quad \mathbf{s}^{(1)} = \begin{pmatrix} -2 \\ 0 \end{pmatrix}.$$

Since a search along $\pm\mathbf{s}^{(1)}$ changes only the x_1 component of $\mathbf{x}^{(1)}$, reference to (3.1.3) shows that $x_1 = 0$ is the minimizing value in the search, and the algorithm fails to make progress. The difficulty arises because $\mathbf{s}^{(1)T}\mathbf{g}^{(1)} = 0$, and the directions $\pm\mathbf{s}^{(1)}$ are not downhill. Yet the function has a well-determined minimizer (see Table 3.1.1), and there is no difficulty in reducing the function, for instance by a search along the steepest descent direction. Another example of failure occurs whenever $\mathbf{x}^{(1)}$ is a saddle point of $f(\mathbf{x})$. Then $\mathbf{s}^{(1)}$ itself is zero and no search direction is defined. Yet $\mathbf{x}^{(1)}$ is not a minimizer, and for instance a search direction which is an eigenvector of $\mathbf{G}^{(1)}$ with negative eigenvalue gives a direction along which f can be reduced.

Clearly then some additional modification is required in order to determine a method of general applicability. In order to make a critical appraisal of the various modifications which have been suggested, the desirable features of an algorithm, as set out in Section 2.3, should be taken into account. One possibility (Goldstein and Price, 1967) is to revert to the steepest descent direction $\mathbf{s}^{(k)} = -\mathbf{g}^{(k)}$ whenever $\mathbf{G}^{(k)}$ is not positive definite. If this is done in conjunction with the angle criterion (2.5.7), then it is easy to see that convergence to a stationary point can be proved (Theorem 2.5.1). However if this modification operates successively for a number of iterations, then it is likely that the slow oscillatory behaviour of the steepest descent method would be observed (Section 2.3). This feature arises because the information in the model quadratic function (3.1.1) is ignored.

An alternative approch is to modify the Newton search direction (3.1.5) by giving it a bias towards the steepest descent vector $-\mathbf{g}^{(k)}$. This is most conveniently achieved by adding a multiple of the unit matrix to $\mathbf{G}^{(k)}$ and solving the system

$$(\mathbf{G}^{(k)} + \nu\mathbf{I})\mathbf{s}^{(k)} = -\mathbf{g}^{(k)}. \tag{3.1.6}$$

If $\mathbf{G}^{(k)}$ is close to being positive definite then it may only be necessary to add a small multiple of $\mathbf{I}$ so as to give a good direction of search. In a sense therefore this type of method uses the quadratic information to a large extent even when $\mathbf{G}^{(k)}$ is not positive definite, and as a consequence is more effective than the Goldstein and Price (1967) modification. The idea of modifying matrices by the addition of a multiple of the unit matrix dates back many years (Levenberg, 1944; Marquardt, 1963), although the first application to Newton's method is by Goldfeld *et al.* (1966). Because of its importance, the subject is treated in much more detail in Section 5.2 and it is shown that it is necessary to choose ν in (3.1.6) so that the modified matrix $\mathbf{G}^{(k)} + \nu\mathbf{I}$ is positive definite. The property of convergence to a stationary point is again not difficult to establish. Even these methods however do not make the best use of the available quadratic information, especially in the vicinity of a saddle point (for example set $\mathbf{g}^{(k)} = \mathbf{0}$ in (3.1.6)). It is possible to regard this type of method in a different way as being a *restricted step method*, in which the model quadratic function (3.1.1) is minimized subject to a restriction on the length of step. A model algorithm of this nature is described in Section 5.1 and it is shown that global convergence to a stationary point *which satisfies the second order conditions* (2.1.4) can be established, and the order of convergence is usually second order. The possibility of implementing this algorithm by using a modified matrix $\mathbf{G}^{(k)} + \nu\mathbf{I}$ is discussed in Section 5.2. Some numerical results are given in Table 5.2.1 which show that the method is effective in practice. This approach is currently regarded as the best for unconstrained minimization when second derivatives are readily calculated.

Some authors (Murray, 1972; Hebden, 1973) have developed different modifications in which a positive definite matrix $\mathbf{G}^{(k)} + \mathbf{D}$ ($\mathbf{D}$ diagonal) is used to determine the search direction in (3.1.6). The modification occurs as the matrix is being factorized, so that if it is recognized that the $\mathbf{LL}^{\mathrm{T}}$ factors do not exist, then an *ad hoc* modification is made which enables the factorization to proceed. An advantage of this approach is that the extra effort required to calculate the factors when $\mathbf{G}^{(k)}$ is indefinite is negligible. The method is implemented in the NAG library by a subroutine developed by Gill, Murray and Picken (1972). Unfortunately these methods do not cater adequately for the case in which $\mathbf{x}^{(k)}$ is close to a saddle point. Also it is in the nature of

Murray's (1972) method that quite a substantial modification to the Hessian can take place. To some extent this seems to have an adverse effect on the performance of the method, as the comparative figures in Table 5.2.1 illustrate.

Another possibility for modifying Newton's method is to compute a *negative curvature* descent direction of search when $\mathbf{G}^{(k)}$ is indefinite (that is a direction which satisfies $\mathbf{s}^{(k)T}\mathbf{G}^{(k)}\mathbf{s}^{(k)} < 0$ and $\mathbf{s}^{(k)T}\mathbf{g}^{(k)} \leqslant 0$). This is first suggested by Fiacco and McCormick (1968) who use $\mathbf{LDL}^T$ factors of $\mathbf{G}^{(k)}$, and solve $\mathbf{L}^T\mathbf{s}^{(k)} = \sigma\mathbf{a}$, where $a_i = if\, d_{ii} < 0$ *then* 1 *else* 0. It is easily verified that $\mathbf{s}^{(k)T}\mathbf{G}^{(k)}\mathbf{s}^{(k)} < 0$, and $\sigma = \pm 1$ is chosen so that $\mathbf{s}^{(k)T}\mathbf{g}^{(k)} \leqslant 0$. Unfortunately if $\mathbf{G}^{(k)}$ is indefinite then computation of the $\mathbf{LDL}^T$ factorization is unstable, and it may not even exist. However a similar way of computing negative curvature directions from a stable factorization is given by Fletcher and Freeman (1977). Although the idea of using negative curvature directions is in some ways attractive, Fletcher and Freeman find that it is not satisfactory to use such directions on successive iterations and that to alternate positive curvature and negative curvature searches gives better results in practice. Other negative curvature search methods have been given by McCormick (1977) and Goldfarb (1980), both of whom show that any accumulation points of the algorithm satisfy the first and second order necessary conditions of Section 2.1. Clearly then negative curvature search methods are interesting and merit further research to determine the best practical approach.

Both methods which modify $\mathbf{G}^{(k)}$ (as in (3.1.6)) and negative curvature search methods have one less attractive aspect in that they lose certain invariance properties (see Section 3.3) associated with Newton's method and Newton's method with line search. However, Fletcher (1978) shows that it is inevitable that this must happen to some extent. What is required is that the non-invariant aspects of the algorithm do not operate too frequently and in such a way that they degrade the efficiency of the method.

3.2 QUASI-NEWTON METHODS

The main disadvantage of Newton's method, even when modified to ensure global convergence, is that the user must supply formulae from which the second derivative matrix $\mathbf{G}$ can be evaluated. This is often a major disincentive to its use. However methods closely related to Newton's method can be derived when only first derivatives are available. The most obvious is a *finite difference Newton method* in which increments h_i in each coordinate direction $\mathbf{e}_i$ are taken so as to estimate $\mathbf{G}^{(k)}$ by differences in gradient vectors. That is to say, the matrix $\bar{\mathbf{G}}$ whose ith column is $(\mathbf{g}(\mathbf{x}^{(k)} + h_i\mathbf{e}_i) - \mathbf{g}^{(k)})/h_i$ is evaluated. Then $\bar{\mathbf{G}}$ is made symmetric by taking $\frac{1}{2}(\bar{\mathbf{G}} + \bar{\mathbf{G}}^T)$ and the resulting matrix used to replace $\mathbf{G}^{(k)}$ in Newton's method. Disadvantages of the method include the following. The matrix $\frac{1}{2}(\bar{\mathbf{G}} + \bar{\mathbf{G}}^T)$ may not be positive definite (requiring modification techniques), n gradient evaluations are required to estimate $\mathbf{G}^{(k)}$, and a set of linear equations must be solved at each iteration. Nonetheless the method can be useful, especially on large sparse problems when the amount of differencing can be reduced (see Curtis *et al.*, 1974). Also in these circumstances the option of re-using factors of previous approximate $\mathbf{G}$ matrices might be more advantageous.

The above disadvantages are all avoided in the much more important class of *quasi-Newton methods*, the introduction of which greatly increased the range of problems which could be solved. This type of method is like Newton's method with line search, except that $\mathbf{G}^{(k)-1}$ is *approximated* by a symmetric positive definite matrix $\mathbf{H}^{(k)}$, which is corrected or *updated* from iteration to iteration. Thus the kth iteration has the basic structure

$$\begin{aligned}&\text{(a) set } \mathbf{s}^{(k)} = -\mathbf{H}^{(k)}\mathbf{g}^{(k)},\\ &\text{(b) line search along } \mathbf{s}^{(k)} \text{ giving } \mathbf{x}^{(k+1)} = \mathbf{x}^{(k)} + \alpha^{(k)}\mathbf{s}^{(k)},\\ &\text{(c) update } \mathbf{H}^{(k)} \text{ giving } \mathbf{H}^{(k+1)}.\end{aligned} \tag{3.2.1}$$

The initial matrix $\mathbf{H}^{(1)}$ can be any positive definite matrix, although in the absence of any better estimate, the choice $\mathbf{H}^{(1)} = \mathbf{I}$ is often made. Potential advantages of the method (as against Newton's method) are:

(i) only first derivatives required (second);
(ii) $\mathbf{H}^{(k)}$ positive definite implies the descent property ($\mathbf{G}^{(k)}$ may be indefinite); and
(iii) $O(n^2)$ multiplications per iteration (n^3).

Property (iii) can be achieved by maintaining an approximation to $\mathbf{G}^{-1}$ rather than $\mathbf{G}$ so that the solution of a system of equations at each iteration is avoided (also see (3.2.16)). The storage requirement is for $\frac{1}{2}n^2 + O(n)$ locations, which is the same as for Newton's method. Another advantage over some other methods, for instance some conjugate gradient methods, is that there is no need to restart the iteration periodically. In fact some quasi-Newton methods do not have property (ii), but these are not the most important. Methods which do maintain property (ii) are sometimes called *variable metric methods* (see Section 3.3).

Much of the interest lies in the *updating formula* which enables $\mathbf{H}^{(k+1)}$ to be calculated from $\mathbf{H}^{(k)}$. This represents an attempt to augment $\mathbf{H}^{(k)}$ with second derivative information gained on the kth iteration. It is desirable that repeated updating changes the arbitrary matrix $\mathbf{H}^{(1)}$ into a close approximation to $\mathbf{G}^{(k)-1}$. One way of doing this is the following. If differences

$$\boldsymbol{\delta}^{(k)} = \alpha^{(k)}\mathbf{s}^{(k)} = \mathbf{x}^{(k+1)} - \mathbf{x}^{(k)} \tag{3.2.2}$$

$$\boldsymbol{\gamma}^{(k)} = \mathbf{g}^{(k+1)} - \mathbf{g}^{(k)} \tag{3.2.3}$$

are defined, then the Taylor series (1.2.19) gives

$$\boldsymbol{\gamma}^{(k)} = \mathbf{G}^{(k)}\boldsymbol{\delta}^{(k)} + o(\|\boldsymbol{\delta}^{(k)}\|) \tag{3.2.4}$$

The higher order terms are zero for a quadratic function and it is convenient to neglect them. Since $\boldsymbol{\delta}^{(k)}$ and $\boldsymbol{\gamma}^{(k)}$ can only be calculated after the line search

(3.2.1(b)) is completed, the matrix $\mathbf{H}^{(k)}$ does not usually relate them correctly (in the sense of $\mathbf{H}^{(k)}\boldsymbol{\gamma}^{(k)} \approx \boldsymbol{\delta}^{(k)}$ since $\mathbf{H} \approx \mathbf{G}^{-1}$). Thus $\mathbf{H}^{(k+1)}$ is chosen so that it does correctly relate these differences, that is so that

$$\mathbf{H}^{(k+1)}\boldsymbol{\gamma}^{(k)} = \boldsymbol{\delta}^{(k)}. \tag{3.2.5}$$

This is sometimes called the *quasi-Newton condition.* Various possible ways exist of achieving this condition. In what follows the aim is to find something which is simple and involves only a small amount of computation, and yet which is effective.

Perhaps the simplest possibility is to have

$$\mathbf{H}^{(k+1)} = \mathbf{H}^{(k)} + \mathbf{E}^{(k)} = \mathbf{H}^{(k)} + a\mathbf{u}\mathbf{u}^{\mathrm{T}} \tag{3.2.6}$$

in which a symmetric rank one matrix $\mathbf{E}^{(k)} = a\mathbf{u}\mathbf{u}^{\mathrm{T}}$ is added into $\mathbf{H}^{(k)}$. Note that $E_{ij}^{(k)} = au_iu_j$ so that $\mathbf{E}^{(k)}$ can be calculated by $n^2 + n$ multiplications only. Now if (3.2.5) is to be satisfied, it follows that

$$\mathbf{H}^{(k)}\boldsymbol{\gamma}^{(k)} + a\mathbf{u}\mathbf{u}^{\mathrm{T}}\boldsymbol{\gamma}^{(k)} = \boldsymbol{\delta}^{(k)}$$

and hence that $\mathbf{u}$ is proportional to $\boldsymbol{\delta}^{(k)} - \mathbf{H}^{(k)}\boldsymbol{\gamma}^{(k)}$. Since any change of length can be taken up in a, $\mathbf{u} = \boldsymbol{\delta}^{(k)} - \mathbf{H}^{(k)}\boldsymbol{\gamma}^{(k)}$ is set, in which case $a\mathbf{u}^{\mathrm{T}}\boldsymbol{\gamma}^{(k)} = 1$ must hold, which defines a. Thus the *rank one formula* is given by

$$\mathbf{H}^{(k+1)} = \mathbf{H} + \frac{(\boldsymbol{\delta} - \mathbf{H}\boldsymbol{\gamma})(\boldsymbol{\delta} - \mathbf{H}\boldsymbol{\gamma})^{\mathrm{T}}}{(\boldsymbol{\delta} - \mathbf{H}\boldsymbol{\gamma})^{\mathrm{T}}\boldsymbol{\gamma}}. \tag{3.2.7}$$

Note that superscript (k) has been suppressed on the right-hand side: this convention is often adopted. The formula was suggested independently by Broyden (1967), Davidon (1968), Fiacco and McCormick (1968), Murtagh and Sargent (1969), and Wolfe (1968a).

Table 3.2.1 illustrates the rank one method (i.e. the rank one formula within a quasi-Newton method), applied to minimizing the quadratic function $f(\mathbf{x}) = 10x_1^2 + x_2^2$ (see also Question 2.11). The quantities are tabulated in the order in which they are evaluated and exact line searches are used. The table also illustrates that the method terminates for quadratic functions with $\mathbf{H}^{(k+1)} = \mathbf{G}^{-1}$, and this justifies the way in which the updating formula is motivated. Indeed this result can be established in most cases as follows.

Theorem 3.2.1

If it is well defined, and if $\boldsymbol{\delta}^{(1)}, \boldsymbol{\delta}^{(2)}, \ldots, \boldsymbol{\delta}^{(n)}$ are independent, then the rank one method terminates on a quadratic function in at most $n+1$ searches, with $\mathbf{H}^{(n+1)} = \mathbf{G}^{-1}$.

Proof

In all these quadratic termination proofs the result is used that

$$\boldsymbol{\gamma}^{(k)} = \mathbf{G}\boldsymbol{\delta}^{(k)} \tag{3.2.8}$$

Table 3.2.1 The rank one method

k	$\mathbf{x}^{(k)}$	$\mathbf{g}^{(k)}$	$\boldsymbol{\delta}^{(k-1)}$	$\boldsymbol{\gamma}^{(k-1)}$	$\mathbf{u}^{(k-1)}$	$\mathbf{H}^{(k)}$		$\mathbf{s}^{(k)}$	$\alpha^{(k)}$
1	$\frac{1}{10}$	2	—	—	—	1	0	-2	$\frac{1}{11}$
	1	2	—	—	—	0	1	-2	
2	$-\frac{9}{110}$	$-\frac{18}{11}$	$-\frac{2}{11}$	$-\frac{40}{11}$	$\frac{38}{11}$	$\frac{21}{382}$	$-\frac{19}{382}$	$\frac{360}{2101}$	$\frac{191}{400}$
	$\frac{9}{11}$	$\frac{18}{11}$	$-\frac{2}{11}$	$-\frac{4}{11}$	$\frac{2}{11}$	$-\frac{19}{382}$	$\frac{381}{382}$	$-\frac{3600}{2100}$	
3	0	0	$\frac{9}{110}$	$\frac{18}{11}$	$-\frac{1881}{21010}$	$\frac{1}{20}$	0		
	0	0	$-\frac{9}{11}$	$\frac{18}{11}$	$\frac{1881}{2101}$	0	$\frac{1}{2}$		

For the DFP method:

$$\mathbf{H}^{(2)} = \begin{bmatrix} \frac{123}{2222} & -\frac{119}{2222} \\ -\frac{119}{2222} & \frac{2301}{2222} \end{bmatrix} \quad \mathbf{s}^{(2)} = \begin{bmatrix} \frac{18}{101} \\ -\frac{180}{101} \end{bmatrix} \quad \alpha^{(2)} = \frac{101}{220}.$$

For the BFGS method:

$$\mathbf{H}^{(2)} = \begin{bmatrix} \frac{15}{242} & -\frac{29}{242} \\ -\frac{29}{242} & \frac{411}{242} \end{bmatrix} \quad \mathbf{s}^{(2)} = \begin{bmatrix} \frac{36}{121} \\ -\frac{360}{121} \end{bmatrix} \quad \alpha^{(2)} = \frac{11}{40}.$$

Otherwise these methods generate the same values given by the rank one method.

for a quadratic function with positive definite Hessian $\mathbf{G}$ (see (1.2.15)). First the *hereditary property* that

$$\mathbf{H}^{(i)}\boldsymbol{\gamma}^{(j)} = \boldsymbol{\delta}^{(j)}, \quad j = 1, 2, \ldots, i-1 \tag{3.2.9}$$

is established by induction. It is true for $i = 2$ directly from (3.2.7). Let it be true for any $i \geqslant 2$. For any $j < i$ the symmetry of $\mathbf{H}^{(i)}$ and (3.2.9) give

$$(\boldsymbol{\delta}^{(i)} - \mathbf{H}^{(i)}\boldsymbol{\gamma}^{(i)})^{\mathrm{T}}\boldsymbol{\gamma}^{(j)} = \boldsymbol{\delta}^{(i)\mathrm{T}}\boldsymbol{\gamma}^{(j)} - \boldsymbol{\gamma}^{(i)\mathrm{T}}\boldsymbol{\delta}^{(j)} = 0$$

by (3.2.8). Thus by (3.2.7)

$$\mathbf{H}^{(i+1)}\boldsymbol{\gamma}^{(j)} = \mathbf{H}^{(i)}\boldsymbol{\gamma}^{(j)} + \mathbf{0} = \boldsymbol{\delta}^{(j)}$$

by (3.2.9). Also $\mathbf{H}^{(i+1)}\boldsymbol{\gamma}^{(i)} = \boldsymbol{\delta}^{(i)}$ directly from (3.2.7) so the induction is established with $i+1$ replacing i. Then by (3.2.5) and (3.2.8)

$$\boldsymbol{\delta}^{(j)} = \mathbf{H}^{(n+1)}\boldsymbol{\gamma}^{(j)} = \mathbf{H}^{(n+1)}\mathbf{G}\boldsymbol{\delta}^{(j)}, \quad j = 1, 2, \ldots, n$$

and since the $\boldsymbol{\delta}^{(j)}$ are independent it follows that $\mathbf{H}^{(n+1)}\mathbf{G} = \mathbf{I}$, that is $\mathbf{H}^{(n+1)} = \mathbf{G}^{-1}$. Therefore iteration $n+1$ is a Newton iteration, so termination occurs then, if not before. □

A feature of this proof is that it does not require exact line searches, and indeed does not even require $\mathbf{s}^{(k)} = -\mathbf{H}^{(k)}\mathbf{g}^{(k)}$ to be used except on iteration $n+1$. Table 3.2.1 however is computed with $\mathbf{s}^{(k)} = -\mathbf{H}^{(k)}\mathbf{g}^{(k)}$ and with exact searches, and this accounts for the fact that termination occurs after two iterations, and not three as the theorem allows. Another feature is that condition (3.2.9) is true not only for $j = i-1$ (the quasi-Newton condition), but also for $j < i-1$. Thus when f is quadratic the update also preserves the conditions from earlier iterations (the hereditary property).

Curiously enough, the rank one formula was not the first updating formula to be discovered. This is perhaps as well since it suffers from two serious disadvantages, even when applied to a quadratic function. It does not in general maintain the positive definite property in $\mathbf{H}^{(k)}$, and also the denominator in (3.2.7) may become zero so that the algorithm is no longer defined. Moreover numerical difficulties arise when the denominator is small. Therefore a practical algorithm requires *ad hoc* safeguards to circumvent these difficulties, which is not entirely satisfactory.

A more flexible formula is obtained by allowing the correction to be of rank two, and such a formula can always be written

$$\mathbf{H}^{(k+1)} = \mathbf{H}^{(k)} + a\mathbf{u}\mathbf{u}^{\mathrm{T}} + b\mathbf{v}\mathbf{v}^{\mathrm{T}}.$$

Again the quasi-Newton condition (3.2.5) must be satisfied, giving

$$\boldsymbol{\delta}^{(k)} = \mathbf{H}^{(k)}\boldsymbol{\gamma}^{(k)} + a\mathbf{u}\mathbf{u}^{\mathrm{T}}\boldsymbol{\gamma}^{(k)} + b\mathbf{v}\mathbf{v}^{\mathrm{T}}\boldsymbol{\gamma}^{(k)}. \tag{3.2.10}$$

Now $\mathbf{u}$ and $\mathbf{v}$ are no longer determined uniquely. However, an obvious choice in (3.2.10) is to try $\mathbf{u} = \boldsymbol{\delta}^{(k)}$ and $\mathbf{v} = \mathbf{H}^{(k)}\boldsymbol{\gamma}^{(k)}$. Then $a\mathbf{u}^{\mathrm{T}}\boldsymbol{\gamma}^{(k)} = 1$ and $b\mathbf{v}^{\mathrm{T}}\boldsymbol{\gamma}^{(k)} = -1$

determine a and b. Thus

$$\mathbf{H}_{\mathrm{DFP}}^{(k+1)} = \mathbf{H} + \frac{\boldsymbol{\delta}\boldsymbol{\delta}^{\mathrm{T}}}{\boldsymbol{\delta}^{\mathrm{T}}\boldsymbol{\gamma}} - \frac{\mathbf{H}\boldsymbol{\gamma}\boldsymbol{\gamma}^{\mathrm{T}}\mathbf{H}}{\boldsymbol{\gamma}^{\mathrm{T}}\mathbf{H}\boldsymbol{\gamma}} \tag{3.2.11}$$

is obtained after dropping the superscript (k) and using the symmetry of $\mathbf{H}$. This formula was first suggested as part of a method due to Davidon (1959), and later presented as described here by Fletcher and Powell (1963), thus becoming known as the DFP formula (and method), hence the subscript DFP in (3.2.11). An illustration of the DFP method applied to the simple quadratic function $f = 10x_1^2 + x_2^2$ is also given in Table 3.2.1.

The method was found to work well in practice, and has been used widely. It proved to be much more efficient than the steepest descent method, and also somewhat more efficient than the conjugate gradient methods introduced in 1964 onwards (compare the results in Sections 3.5 and 4.1). Early implementations attempted to carry out fairly accurate line searches. The coming of low accuracy line searches in 1970 however showed the DFP formula in a less satisfactory light than other formulae which were being introduced, and currently the DFP method is no longer preferred (see Section 3.5). However the method has a number of important properties as follows:

(1) *for quadratic functions* (with exact line searches)
- (i) terminates in at most n iterations with $\mathbf{H}^{(n+1)} = \mathbf{G}^{-1}$;
- (ii) previous quasi-Newton conditions are preserved (cf. (3.2.9));
- (iii) generates conjugate directions, and conjugate gradients when $\mathbf{H}^{(1)} = \mathbf{I}$;

(2) *for general functions*
- (iv) preserves positive definite $\mathbf{H}^{(k)}$ matrices – hence the descent property holds;
- (v) requires $3n^2 + O(n)$ multiplications per iteration;
- (vi) superlinear order of convergence;
- (vii) global convergence for strictly convex functions (with exact line searches).

These properties are proved or referenced in a more general setting in Section 3.4. It is convenient here however to justify property (iv).

Theorem 3.2.2

If $\boldsymbol{\delta}^{(k)\mathrm{T}}\boldsymbol{\gamma}^{(k)} > 0$ for all k, then the DFP formula preserves positive definite matrices $\mathbf{H}^{(k)}$.

Proof

The proof is inductive and shows that $\mathbf{z}^{\mathrm{T}}\mathbf{H}^{(k)}\mathbf{z} > 0$ for all $\mathbf{z} \neq \mathbf{0}$. The result is true for $\mathbf{H}^{(1)}$ by choice. Assume it is true for some $k \geqslant 1$. Writing $\mathbf{H}^{(k)} = \mathbf{L}\mathbf{L}^{\mathrm{T}}$, since Choleski factors exist, and if $\mathbf{a} = \mathbf{L}^{\mathrm{T}}\mathbf{z}$ and $\mathbf{b} = \mathbf{L}^{\mathrm{T}}\boldsymbol{\gamma}$, then

$$\mathbf{z}^{\mathrm{T}}\left(\mathbf{H} - \frac{\mathbf{H}\boldsymbol{\gamma}\boldsymbol{\gamma}^{\mathrm{T}}\mathbf{H}}{\boldsymbol{\gamma}^{\mathrm{T}}\mathbf{H}\boldsymbol{\gamma}}\right)\mathbf{z} = \mathbf{a}^{\mathrm{T}}\mathbf{a} - \frac{(\mathbf{a}^{\mathrm{T}}\mathbf{b})^2}{\mathbf{b}^{\mathrm{T}}\mathbf{b}} \geqslant 0$$

by Cauchy's inequality. Since $\mathbf{z} \neq \mathbf{0}$, equality holds only if $\mathbf{a} \propto \mathbf{b}$, that is if $\mathbf{z} \propto \boldsymbol{\gamma}$. But since $\boldsymbol{\delta}^{\mathrm{T}}\boldsymbol{\gamma} > 0$,

$$\mathbf{z}^{\mathrm{T}}\left(\frac{\boldsymbol{\delta}\boldsymbol{\delta}^{\mathrm{T}}}{\boldsymbol{\delta}^{\mathrm{T}}\boldsymbol{\gamma}}\right)\mathbf{z} \geqslant 0$$

with strict inequality if $\mathbf{z} \propto \boldsymbol{\gamma}$. The induction is then established with $k+1$ replacing k, by expanding $\mathbf{z}^{\mathrm{T}}\mathbf{H}^{(k+1)}\mathbf{z}$ using (3.2.11), and using the above inequalities. □

It is important to establish that the condition $\boldsymbol{\delta}^{(k)\mathrm{T}}\boldsymbol{\gamma}^{(k)} > 0$ is realistic and can always be achieved. For a quadratic function, using (3.2.8), $\boldsymbol{\delta}^{(k)\mathrm{T}}\boldsymbol{\gamma}^{(k)} = \boldsymbol{\delta}^{(k)\mathrm{T}}\mathbf{G}\boldsymbol{\delta}^{(k)} > 0$ since $\mathbf{G}$ is positive definite, and it can be seen that $\boldsymbol{\delta}^{(k)\mathrm{T}}\boldsymbol{\gamma}^{(k)}$ is naturally associated with the curvature of the objective function along $\mathbf{s}^{(k)}$. For a general function $\boldsymbol{\delta}^{(k)\mathrm{T}}\boldsymbol{\gamma}^{(k)} = \mathbf{g}^{(k+1)\mathrm{T}}\boldsymbol{\delta}^{(k)} - \mathbf{g}^{(k)\mathrm{T}}\boldsymbol{\delta}^{(k)}$, and $\mathbf{g}^{(k)\mathrm{T}}\boldsymbol{\delta}^{(k)} < 0$ holds by the descent property. If an exact line search is used, (2.3.3) implies that $\mathbf{g}^{(k+1)\mathrm{T}}\boldsymbol{\delta}^{(k)} = 0$ and so $\boldsymbol{\delta}^{(k)\mathrm{T}}\boldsymbol{\gamma}^{(k)} > 0$ again follows. However even for approximate line searches, as long as condition (2.5.5) is imposed, it also follows that $\boldsymbol{\delta}^{(k)\mathrm{T}}\boldsymbol{\gamma}^{(k)} > 0$. Essentially all that is required is to take the differences over an interval for which the curvature estimate is positive.

Yet another important formula was suggested by Broyden (1970), Fletcher (1970a), Goldfarb (1970), and Shanno (1970), and is known as the *BFGS formula*

$$\mathbf{H}_{\mathrm{BFGS}}^{(k+1)} = \mathbf{H} + \left(1 + \frac{\boldsymbol{\gamma}^{\mathrm{T}}\mathbf{H}\boldsymbol{\gamma}}{\boldsymbol{\delta}^{\mathrm{T}}\boldsymbol{\gamma}}\right)\frac{\boldsymbol{\delta}\boldsymbol{\delta}^{\mathrm{T}}}{\boldsymbol{\delta}^{\mathrm{T}}\boldsymbol{\gamma}} - \left(\frac{\boldsymbol{\delta}\boldsymbol{\gamma}^{\mathrm{T}}\mathbf{H} + \mathbf{H}\boldsymbol{\gamma}\boldsymbol{\delta}^{\mathrm{T}}}{\boldsymbol{\delta}^{\mathrm{T}}\boldsymbol{\gamma}}\right). \tag{3.2.12}$$

It can be motivated in the following way. If $\mathbf{H}^{-1}$ is denoted by $\mathbf{B}$ (so that $\mathbf{B}^{(k)} = \mathbf{H}^{(k)-1}$ approximates $\mathbf{G}$), then it can be verified (by establishing that $\mathbf{B}_{\mathrm{BFGS}}^{(k+1)}\mathbf{H}_{\mathrm{BFGS}}^{(k+1)} = \mathbf{I}$) that

$$\mathbf{B}_{\mathrm{BFGS}}^{(k+1)} = \mathbf{B} + \frac{\boldsymbol{\gamma}\boldsymbol{\gamma}^{\mathrm{T}}}{\boldsymbol{\gamma}^{\mathrm{T}}\boldsymbol{\delta}} - \frac{\mathbf{B}\boldsymbol{\delta}\boldsymbol{\delta}^{\mathrm{T}}\mathbf{B}}{\boldsymbol{\delta}^{\mathrm{T}}\mathbf{B}\boldsymbol{\delta}}. \tag{3.2.13}$$

This resembles the DFP formula (3.2.11) but with the interchanges $\mathbf{B} \leftrightarrow \mathbf{H}$ and $\boldsymbol{\gamma} \leftrightarrow \boldsymbol{\delta}$ having been made. Formulae related in this way are said to be *dual* or *complementary*. Similarly it follows from (3.2.12) that

$$\mathbf{B}_{\mathrm{DFP}}^{(k+1)} = \mathbf{B} + \left(1 + \frac{\boldsymbol{\delta}^{\mathrm{T}}\mathbf{B}\boldsymbol{\delta}}{\boldsymbol{\gamma}^{\mathrm{T}}\boldsymbol{\delta}}\right)\frac{\boldsymbol{\gamma}\boldsymbol{\gamma}^{\mathrm{T}}}{\boldsymbol{\gamma}^{\mathrm{T}}\boldsymbol{\delta}} - \left(\frac{\boldsymbol{\gamma}\boldsymbol{\delta}^{\mathrm{T}}\mathbf{B} + \mathbf{B}\boldsymbol{\delta}\boldsymbol{\gamma}^{\mathrm{T}}}{\boldsymbol{\gamma}^{\mathrm{T}}\boldsymbol{\delta}}\right). \tag{3.2.14}$$

In fact given any formula for $\mathbf{H}^{(k+1)}$ the dual formula for $\mathbf{B}_{\mathrm{D}}^{(k+1)}$ follows by interchanging as above, and then a dual formula for $\mathbf{H}_{\mathrm{D}}^{(k+1)}$ can be obtained using the Sherman–Morrison formula (see Question 3.13). Of course taking the dual operation twice restores the original formula. Also the quasi-Newton condition (3.2.5) is preserved by the duality operation. It can be established that the rank one formula is *self-dual*, and another self-dual formula which preserves positive definite $\mathbf{H}$ matrices is given by Hoshino (1972) (see Question 3.14).

The BFGS formula in particular has been found to work well in practice, perhaps even better than the DFP formula. A simple illustration of the BFGS method is again given in Table 3.2.1. It has usually been implemented in conjunction with low accuracy line searches. Some computational experiments are described in Section 3.5, where the current opinion that the BFGS method is the 'best' quasi-Newton method is also discussed. As to the theoretical side, properties (i)–(vii) for the DFP method also hold for the BFGS method. In addition, global convergence of the BFGS method with inexact line searches which satisfy (2.5.3) and (2.5.5) has been proved, a result which has not yet been shown for the DFP method. These results are proved or referenced in a more general setting in Section 3.4, except property (iv) which follows by applying Theorem 3.2.2 to the dual formula (3.2.13).

The formulae which have been introduced so far by no means exhaust all the possibilities. In fact a one-parameter (ϕ) family of rank two formulae can be generated by taking

$$\mathbf{H}_{\phi}^{(k+1)} = (1-\phi)\mathbf{H}_{\mathrm{DFP}}^{(k+1)} + \phi\mathbf{H}_{\mathrm{BFGS}}^{(k+1)}. \tag{3.2.15}$$

This family (the *Broyden family*) includes of course the DFP ($\phi = 0$) and BFGS ($\phi = 1$) formulae, and also the rank one formula (3.2.7) and the Hoshino formula ($\phi = 1/(1 \mp \boldsymbol{\gamma}^{\mathrm{T}}\mathbf{H}\boldsymbol{\gamma}/\boldsymbol{\delta}^{\mathrm{T}}\boldsymbol{\gamma})$). The Broyden family is important in that many of the properties of the DFP and BFGS formulae are common to the whole family. In addition it can be developed naturally from the quasi-Newton condition and certain invariance considerations which are introduced in the next section. Many of the theoretical properties of the Broyden family can be unified, and this is done in Section 3.4. Some experiments which help to indicate which is the best formula to choose in a practical algorithm are set out and discussed in Section 3.5. Finally some different or more recent areas of research into quasi-Newton methods are described briefly in Section 3.6.

Before finishing this section however two other important aspects of a quasi-Newton method are described. One is the application to no-derivative problems, in which *finite difference approximations* to first derivatives are used to estimate elements of the vector ∇f (see (2.3.8) and (2.3.9)). This idea was tried by Stewart (1967) who suggested a modification of the DFP method, and later by Gill and Murray (1972) in the context of the Broyden family. One obvious question which arises is how best to carry out the line search (Section 2.6). Other considerations include how the differencing interval h is to be chosen and when to use forward differences (2.3.8) (for efficiency) or central differences (2.3.9) (for higher accuracy). Assuming that reasonable decisions of this type are made, then in practice the methods currently seem to be superior to the conjugate direction methods of Section 4.2.

Finally the question of how best to *represent* the approximating matrix $\mathbf{H}^{(k)}$ or $\mathbf{B}^{(k)} = \mathbf{H}^{(k)^{-1}}$ is discussed. In early work $\mathbf{H}^{(k)} \approx \mathbf{G}^{-1}$ was chosen since this avoids solving the system of equations

$$\mathbf{B}^{(k)}\mathbf{s}^{(k)} = -\mathbf{g}^{(k)} \tag{3.2.16}$$

which would arise if $\mathbf{B}^{(k)}$ were selected. A more recent idea, pioneered by Gill and Murray (1972), is to represent **B** by the factors $\mathbf{LDL}^T$ where **L** is unit lower triangular and **D** is diagonal and positive definite. This method is viable since it is possible to update the factors in $O(n^2)$ multiplications (that is to compute new factors $\mathbf{L}^{(k+1)}$ and $\mathbf{D}^{(k+1)}$ from $\mathbf{L}^{(k)}$ and $\mathbf{D}^{(k)}$ so that $\mathbf{B}^{(k+1)} = \mathbf{H}^{(k+1)^{-1}} = \mathbf{L}^{(k+1)}\mathbf{D}^{(k+1)}\mathbf{L}^{(k+1)^T}$). Then (3.2.16) can be solved merely by a forward and a backward substitution. The rank two change in $\mathbf{B}^{(k)}$ can be achieved by writing it as the sum of two rank one changes, and using the updating methods of Fletcher and Powell (1974) or Gill and Murray (1978a). However for the BFGS method there is an even more effective update described by Powell (1985e). The advantage of using factorizations is claimed to be that round-off errors are controlled better, and that it is easier to recognize and correct an indefinite matrix (by making $D_i > 0$). Thus these representations have been used in many recent implementations of quasi-Newton methods. However some recent research (see Grandinetti, 1979, for instance) claims that even on ill-conditioned problems no practical advantage in using factorizations can be detected. Further research can be expected here.

3.3 INVARIANCE, METRICS AND VARIATIONAL PROPERTIES

The possibilities for quasi-Newton formulae are endless, yet none have been found which improve significantly on the DFP and particularly the BFGS formulae. In this section some theoretical results are derived which help to explain this fact. The formulae have some invariance properties related to affine transformations which are described and their significance is discussed. Likewise there are some relevant variational properties which are considered later in the section.

It is instructive to consider the effect on any Newton-like method of making an *affine transformation of variables*

$$\mathbf{y} = \mathbf{A}\mathbf{x} + \mathbf{a} \tag{3.3.1}$$

in which **A** is non-singular. The transformation is one-to-one and the inverse transformation is $\mathbf{x} = \mathbf{A}^{-1}(\mathbf{y} - \mathbf{a})$. Thus $f(\mathbf{x})$ can be regarded as being computed either from **x** ($f_x(\mathbf{x})$ say) or from **y** ($f_y(\mathbf{y}) = f_x(\mathbf{A}^{-1}(\mathbf{y} - \mathbf{a}))$ say). The chain rule gives

$$\frac{\partial}{\partial x_i} = \sum_k \frac{\partial y_k}{\partial x_i}\frac{\partial}{\partial y_k} = \sum_k A^T_{ik}\frac{\partial}{\partial y_k}$$

so that $\nabla_x = \mathbf{A}^T\nabla_y$. Operating on f gives $\nabla_x f = \mathbf{A}^T\nabla_y f$ or

$$\mathbf{g}_x = \mathbf{A}^T\mathbf{g}_y \tag{3.3.2}$$

(using an obvious notation $\mathbf{g}_x = \nabla_x f$, etc.). Likewise operating on $\mathbf{g}_x$ gives $\nabla_x\mathbf{g}_x^T = \mathbf{A}^T\nabla_y\mathbf{g}_y^T\mathbf{A}$ or

$$\mathbf{G}_x = \mathbf{A}^T\mathbf{G}_y\mathbf{A}. \tag{3.3.3}$$

Thus (3.3.2) and (3.3.3) relate derivatives in the $\mathbf{x}$ and $\mathbf{y}$ coordinate systems. Consider any Newton or Newton-like method in which $\mathbf{H}$ is equal to or approximates $\mathbf{G}^{-1}$. Then the following theorem holds.

Theorem 3.3.1

If $\mathbf{H}^{(k)}$ *transforms according to* (3.3.3), *that is if*

$$\mathbf{H}_x^{(k)} = \mathbf{A}^{-1}\mathbf{H}_y^{(k)}\mathbf{A}^{-\mathrm{T}}, \quad \forall k, \tag{3.3.4}$$

then a Newton-like method with fixed step $\alpha^{(k)}$ *is invariant under the transformation* (3.3.1).

Proof

The proof is inductive and shows that the sequences $\{\mathbf{x}^{(k)}\}$ and $\{\mathbf{y}^{(k)}\}$ satisfy (3.3.1) for all k. The result is true for $k=1$ by choice of $\mathbf{y}^{(1)}$. Assume $\mathbf{y}^{(k)} = \mathbf{A}\mathbf{x}^{(k)} + \mathbf{a}$ for some $k \geqslant 1$, and let the Newton-like method be applied in both the $\mathbf{x}^{(k)}$ and $\mathbf{y}^{(k)}$ variables. Then

$$\mathbf{x}^{(k+1)} = \mathbf{x}^{(k)} - \alpha^{(k)}\mathbf{H}_x^{(k)}\mathbf{g}_x^{(k)}$$

and

$$\begin{aligned}\mathbf{y}^{(k+1)} &= \mathbf{y}^{(k)} - \alpha^{(k)}\mathbf{H}_y^{(k)}\mathbf{g}_y^{(k)}\\ &= \mathbf{A}\mathbf{x}^{(k)} + \mathbf{a} - \alpha^{(k)}\mathbf{A}\mathbf{H}_x^{(k)}\mathbf{A}^{\mathrm{T}}\mathbf{A}^{-\mathrm{T}}\mathbf{g}_x^{(k)}\\ &= \mathbf{A}\mathbf{x}^{(k+1)} + \mathbf{a}\end{aligned}$$

which completes the induction with $k+1$ replacing k. □

Corollary

A method is also invariant if $\alpha^{(k)}$ *is determined by tests on* $f^{(k)}$, $\mathbf{g}^{(k)\mathrm{T}}\mathbf{s}^{(k)}$ *or other invariant scalars.*

Proof

f is invariant by definition. $\mathbf{g}_x^{\mathrm{T}}\mathbf{s}_x = -\mathbf{g}_x^{\mathrm{T}}\mathbf{H}_x\mathbf{g}_x = -\mathbf{g}_y^{\mathrm{T}}\mathbf{A}\mathbf{A}^{-1}\mathbf{H}_y\mathbf{A}^{-\mathrm{T}}\mathbf{A}^{\mathrm{T}}\mathbf{g}_y = \mathbf{g}_y^{\mathrm{T}}\mathbf{s}_y$ and so is invariant. □

It follows directly that Newton's method and Newton's method with the line search of (2.6.2) and (2.6.4) are invariant. However the steepest descent method is not invariant since $\mathbf{H} = \mathbf{I}$ does not transform correctly. Likewise modified Newton methods are not usually invariant because $\mathbf{G} + \nu\mathbf{I}$ does not transform correctly when $\nu > 0$. For quasi-Newton methods, if $\mathbf{H}^{(1)} = \mathbf{I}$ is chosen, then invariance also does not hold. However, assume that $\mathbf{H}^{(1)}$ is chosen so as to transform correctly, for example by $\mathbf{H}^{(1)} = \mathbf{G}(\mathbf{x}^{(1)})^{-1}$, or $\mathbf{H}^{(1)} = [2\mathbf{A}^{(1)}\mathbf{A}^{(1)\mathrm{T}}]^{-1}$ as in (6.1.8) (not the same $\mathbf{A}$ as in (3.3.1)). Then to prove invariance it is necessary to show that the updating formula preserves the transformation property (3.3.4). Consider the DFP formula in the **x**-coordinates

$$\mathbf{H}_x^{(k+1)} = \mathbf{H}_x + \frac{\boldsymbol{\delta}_x\boldsymbol{\delta}_x^{\mathrm{T}}}{\boldsymbol{\delta}_x^{\mathrm{T}}\boldsymbol{\gamma}_x} - \frac{\mathbf{H}_x\boldsymbol{\gamma}_x\boldsymbol{\gamma}_x^{\mathrm{T}}\mathbf{H}_x}{\boldsymbol{\gamma}_x^{\mathrm{T}}\mathbf{H}_x\boldsymbol{\gamma}_x}.$$

Pre- and post-multiplying by $\mathbf{A}$ and $\mathbf{A}^{\mathrm{T}}$, and using $\mathbf{A}\boldsymbol{\delta}_x = \boldsymbol{\delta}_y$ (from (3.3.1)) and $\boldsymbol{\gamma}_x = \mathbf{A}^{\mathrm{T}}\boldsymbol{\gamma}_y$ from (3.3.2), it follows that $\mathbf{A}\mathbf{H}_x^{(k)}\mathbf{A}^{\mathrm{T}} = \mathbf{H}_y^{(k)}$ implies $\mathbf{A}\mathbf{H}_x^{(k+1)}\mathbf{A}^{\mathrm{T}} = \mathbf{H}_y^{(k+1)}$. Thus the DFP formula does preserve (3.3.4). A similar result follows for the BFGS formula, and in fact for any formula in which the correction is a sum of rank one terms constructed from vectors $\boldsymbol{\delta}$ and $\mathbf{H}\boldsymbol{\gamma}$, multiplied by invariant scalars.

It is important to discuss the *significance* of the invariance property. It suggests that an invariant algorithm is not easily upset by a problem in which $\mathbf{G}$ is ill-conditioned, because one can implicity transform to $\mathbf{G} = \mathbf{I}$ without changing the methods. The steepest decent method, which is not invariant, performs badly when $\mathbf{G}$ is ill-conditioned. In quasi-Newton methods the situation is less clear when $\mathbf{H}^{(1)} = \mathbf{I}$ is chosen. However after n iterations $\mathbf{H}^{(n+1)} \approx \mathbf{G}^{(n+1)^{-1}}$ and the method then becomes close to one in which invariance is preserved. Of course if $\mathbf{H}^{(n+1)}$ were replaced by $\mathbf{G}^{(n+1)^{-1}}$ then the method would subsequently be invariant. This is only true for those quasi-Newton formulae which maintain property (3.3.4), and suggests that those formulae which do not maintain (3.3.4) may have difficulty in solving problems in which $\mathbf{G}$ is ill-conditioned.

When using methods which are not invariant, it can be advantageous to find a linear transformation which improves the conditioning of the problem. To transform to coordinates in which $\mathbf{G}_y = \mathbf{I}$ requires from (3.3.3) that $\mathbf{G}_x = \mathbf{A}^{\mathrm{T}}\mathbf{A}$; for instance one could use $\mathbf{A}^{\mathrm{T}} = \mathbf{L}$, the Choleski factor of $\mathbf{G}_x$. Unfortunately this just leads to Newton's method in circumstances where $\mathbf{G}$ is positive definite and so does not help much. (Except that it indicates that the Newton step is equivalent to a unit steepest descent step in the transformed coordinates.) However, it is often possible to improve the conditioning by the lesser aim of *scaling* the variables. To achieve this $\mathbf{A}$ is chosen to be a diagonal matrix (and $\mathbf{a} = \mathbf{0}$) so that $\mathbf{A}^2$ estimates $\mathbf{G}_x$. Then $\mathbf{G}_y = \mathbf{A}^{-\mathrm{T}}\mathbf{G}_x\mathbf{A}^{-1}$ is required to be close to $\mathbf{I}$ in some sense; perhaps in that $G_{ii} = 1$ for all i. It is not usually necessary to perform the scaling explicity, but rather to replace $\mathbf{I}$ in the methods by a suitable diagonal matrix. Thus the steepest descent or quasi-Newton methods can be improved by choosing $\mathbf{H}^{(1)} = \mathbf{A}^{-2}$ rather than $\mathbf{I}$. Similarly the modified Newton method can be improved by using $\mathbf{G} + \nu\mathbf{A}^2$ rather than $\mathbf{G} + \nu\mathbf{I}$. It may be possible to choose $\mathbf{A}$ in such a way that the resulting method is *invariant under scaling* although not under a more general linear transformation. This is usually an improvement.

The behaviour of certain *metrics* is also related to this discussion. For instance Euclidean distance

$$\|\mathbf{y}'' - \mathbf{y}'\|_2 = \|\mathbf{x}'' - \mathbf{x}'\|_2$$

is *not* preserved in a general transformation of the form (3.3.1). ($\mathbf{y}''$ and $\mathbf{y}'$ are the images of distinct points $\mathbf{x}''$ and $\mathbf{x}'$ under the transformation.) However in terms of the metric

$$\|\mathbf{h}\|_G = \sqrt{(\mathbf{h}^T\mathbf{G}\mathbf{h})} \tag{3.3.5}$$

where $\mathbf{G}(=\nabla^2 f)$ is positive definite, then

$$\|\mathbf{y}'' - \mathbf{y}'\|_{\mathbf{G}_y} = \|\mathbf{x}'' - \mathbf{x}'\|_{\mathbf{G}_x}$$

so distance measured in terms of this metric is preserved in the transformation. This also relates to the invariance of methods. One possible way of defining the steepest descent direction is to let $\mathbf{x}_r$ minimize $f(\mathbf{x})$ on a ball of centre $\mathbf{x}'$ and radius r. Then $\lim_{r\to 0}(\mathbf{x}_r - \mathbf{x}')/\|\mathbf{x}_r - \mathbf{x}'\|_2$ is the normalized steepest descent direction at $\mathbf{x}'$. Since this depends on $\|\cdot\|_2$ it is not invariant. Likewise it can be shown that the Newton direction (normalized to $\mathbf{s}^T\mathbf{G}\mathbf{s} = 1$) is defined when $\|\cdot\|_\mathbf{G}$ is used. Since $\|\cdot\|_\mathbf{G}$ is invariant, so is the resulting direction. Also it shows that Newton's method takes a step in the steepest descent direction with respect to $\|\cdot\|_\mathbf{G}$. Likewise a quasi-Newton method takes a step in the steepest descent direction with respect to $\|\cdot\|_\mathbf{B}$, where $\mathbf{B} = \mathbf{H}^{-1}$. Hence the term *variable metric method* is used for quasi-Newton methods in which $\mathbf{B}^{(k)}$ is positive definite and changes from iteration to iteration.

Another interesting result for the BFGS and DFP formulae is that they possess a so-called *variational property*. When updating $\mathbf{H}^{(k)}$ in a quasi-Newton method it is desirable that $\mathbf{H}^{(k+1)}$ should be symmetric and satisfy the quasi-Newton condition (3.2.5), but there is still a lot of freedom in the choice of $\mathbf{H}^{(k+1)}$. It can be argued that this freedom should be taken up by making $\mathbf{H}^{(k+1)}$ as close as possible to $\mathbf{H}^{(k)}$ in some sense, whilst satisfying (3.2.5). The reason for this is that if some second order information (hereditary properties etc.) has been built up in $\mathbf{H}^{(k)}$ then it is undesirable to make arbitrary large changes to $\mathbf{H}^{(k)}$ which might corrupt this information. Thus the idea is suggested of deriving an updating formula which minimizes some measure of the correction to $\mathbf{H}^{(k)}$, subject to symmetry and to (3.2.5) holding. To get an invariant formula out of this process it is necessary to use a measure with an invariance property. A measure which allows this and yet is convenient to handle is the weighted Euclidean norm $\|\cdot\|_\mathbf{W}$ defined by

$$\|\mathbf{A}\|_\mathbf{W}^2 \triangleq \|\mathbf{W}^{1/2}\mathbf{A}\mathbf{W}^{1/2}\|_E^2 = \operatorname{trace} \mathbf{W}\mathbf{A}^T\mathbf{W}\mathbf{A} \tag{3.3.6}$$

where $\mathbf{W}$ is a symmetric positive definite matrix. The BFGS and DFP formulae can be deduced directly as a consequence of using such a measure. This result is due to Goldfarb (1970) consequent on ideas of Greenstadt (1970).

Theorem 3.3.2

If the BFGS formula (3.2.12) *is written as* $\mathbf{H}^{(k+1)} = \mathbf{H} + \mathbf{E}$ *where* $\mathbf{H}$ *is symmetric, then* $\mathbf{E}$ *solves the variational problem*

$$\begin{aligned} &\underset{\mathbf{E}}{\text{minimize}}\ \|\mathbf{E}\|_\mathbf{W} \\ &\text{subject to } \mathbf{E}^T = \mathbf{E} \quad \text{and} \quad \mathbf{E}\gamma = \eta \end{aligned} \tag{3.3.7}$$

where $\boldsymbol{\eta} = \boldsymbol{\delta} - \mathbf{H}\boldsymbol{\gamma}$ *and where* $\mathbf{W}$ *also satisfies* $\mathbf{W}\boldsymbol{\delta} = \boldsymbol{\gamma}$. *(The constraints in* (3.3.7) *are symmetry and the quasi-Newton condition respectively.)*

Proof

The problem is a convex programming problem so it is sufficient to find $\mathbf{E}$ to satisfy first order conditions. After squaring the norm, a suitable Lagrangian function is

$$\mathscr{L} = \tfrac{1}{4}\text{trace}\,\mathbf{W}\mathbf{E}^{\mathrm{T}}\mathbf{W}\mathbf{E} + \text{trace}\,\boldsymbol{\Lambda}^{\mathrm{T}}(\mathbf{E}^{\mathrm{T}} - \mathbf{E}) - \boldsymbol{\lambda}^{\mathrm{T}}\mathbf{W}(\mathbf{E}\boldsymbol{\gamma} - \boldsymbol{\eta})$$

where $\boldsymbol{\Lambda}$ is a Lagrange multiplier matrix for the constraint $\mathbf{E}^{\mathrm{T}} = \mathbf{E}$ (use of the trace is just a convenient way of summing $\Lambda_{ij}(E^{\mathrm{T}}_{ij} - E_{ij})$ over all i, j) and $\boldsymbol{\lambda}^{\mathrm{T}}\mathbf{W}$ is a vector of Lagrange multipliers for the constraint $\mathbf{E}\boldsymbol{\gamma} = \boldsymbol{\eta}$. $\mathscr{L}$ must be stationary with respect to $\mathbf{E}, \boldsymbol{\Lambda}$ and $\boldsymbol{\lambda}$. Setting derivatives with respect to $\boldsymbol{\Lambda}$ and $\boldsymbol{\lambda}$ to zero just gives the constraints in (3.3.7). For derivatives with respect to $\mathbf{E}$, $\partial\mathbf{E}/\partial E_{ij} = \mathbf{e}_i\mathbf{e}_j^{\mathrm{T}}$ so in this case

$$\partial\mathscr{L}/\partial E_{ij} = \tfrac{1}{4}(\text{trace}\,\mathbf{W}\mathbf{e}_j\mathbf{e}_i^{\mathrm{T}}\mathbf{W}\mathbf{E} + \text{trace}\,\mathbf{W}\mathbf{E}^{\mathrm{T}}\mathbf{W}\mathbf{e}_i\mathbf{e}_j^{\mathrm{T}}$$
$$+ \text{trace}\,\boldsymbol{\Lambda}(\mathbf{e}_j\mathbf{e}_i^{\mathrm{T}} - \mathbf{e}_i\mathbf{e}_j^{\mathrm{T}}) - \boldsymbol{\lambda}^{\mathrm{T}}\mathbf{W}\mathbf{e}_i\mathbf{e}_j^{\mathrm{T}}\boldsymbol{\gamma} = 0$$

or, using symmetry and invariance of the trace to cyclic permutations,

$$\tfrac{1}{2}[\mathbf{W}\mathbf{E}\mathbf{W}]_{ij} + \Lambda_{ij} - \Lambda_{ji} = [\mathbf{W}\boldsymbol{\lambda}\boldsymbol{\gamma}^{\mathrm{T}}]_{ij}.$$

Transposing and adding eliminates $\boldsymbol{\Lambda}$ to give

$$\mathbf{W}\mathbf{E}\mathbf{W} = \mathbf{W}\boldsymbol{\lambda}\boldsymbol{\gamma}^{\mathrm{T}} + \boldsymbol{\gamma}\boldsymbol{\lambda}^{\mathrm{T}}\mathbf{W},$$

and using $\mathbf{W}\boldsymbol{\delta} = \boldsymbol{\gamma}$ and the nonsingularity of $\mathbf{W}$ it follows that

$$\mathbf{E} = \boldsymbol{\gamma}\boldsymbol{\delta}^{\mathrm{T}} + \boldsymbol{\delta}\boldsymbol{\lambda}^{\mathrm{T}} \tag{3.3.8}$$

Thus the result that the correction is of rank two is seen to arise naturally out of the analysis. It is now possible to solve for $\boldsymbol{\lambda}$ using the constraints of (3.3.7). Substituting (3.3.8) into $\mathbf{E}\boldsymbol{\gamma} = \boldsymbol{\eta}$ and rearranging gives

$$\boldsymbol{\lambda} = (\boldsymbol{\eta} - \boldsymbol{\delta}\boldsymbol{\lambda}^{\mathrm{T}}\boldsymbol{\gamma})/\boldsymbol{\delta}^{\mathrm{T}}\boldsymbol{\gamma}. \tag{3.3.9}$$

Postmultiplying by $\boldsymbol{\gamma}^{\mathrm{T}}$ gives $\boldsymbol{\lambda}^{\mathrm{T}}\boldsymbol{\gamma} = \tfrac{1}{2}\boldsymbol{\gamma}^{\mathrm{T}}\boldsymbol{\eta}/\boldsymbol{\delta}^{\mathrm{T}}\boldsymbol{\gamma}$ so (3.3.9) becomes

$$\boldsymbol{\lambda} = (\boldsymbol{\eta} - \tfrac{1}{2}\boldsymbol{\delta}\boldsymbol{\gamma}^{\mathrm{T}}\boldsymbol{\eta}/\boldsymbol{\delta}^{\mathrm{T}}\boldsymbol{\gamma})/\boldsymbol{\delta}^{\mathrm{T}}\boldsymbol{\gamma} = (\mathbf{H}\boldsymbol{\gamma} - \tfrac{1}{2}\boldsymbol{\delta}(\boldsymbol{\gamma}^{\mathrm{T}}\mathbf{H}\boldsymbol{\gamma}/\boldsymbol{\delta}^{\mathrm{T}}\boldsymbol{\gamma}))/\boldsymbol{\delta}^{\mathrm{T}}\boldsymbol{\gamma}.$$

Substituting this into (3.3.8) gives the correction defined by the BFGS formula (3.2.12). □

Corollary

If the DFP formula (3.2.14) *is written as* $\mathbf{B}^{(k+1)} = \mathbf{B} + \mathbf{E}$ *where* $\mathbf{B}$ *is symmetric, then* $\mathbf{E}$ *solves the variational problem*

$$\underset{\mathbf{E}}{\text{minimize}}\ \|\mathbf{E}\|_{\mathbf{W}}$$
$$\text{subject to } \mathbf{E}^{\mathrm{T}} = \mathbf{E} \quad \text{and} \quad \mathbf{E}\boldsymbol{\delta} = \boldsymbol{\xi}$$

where $\boldsymbol{\xi} = \boldsymbol{\gamma} - \mathbf{B}\boldsymbol{\delta}$ *and where* $\mathbf{W}$ *satisfies* $\mathbf{W}\boldsymbol{\gamma} = \boldsymbol{\delta}$.

Proof

This is the complementary result to (3.3.7) obtained by interchanging $\mathbf{B}\leftrightarrow\mathbf{H}$ and $\boldsymbol{\delta}\leftrightarrow\boldsymbol{\gamma}$ in the theorem. □

Thus it can be seen that the BFGS formula is one in which the correction to the inverse Hessian approximation **H** is minimal, whereas in the DFP method it is the correction to **B**. The correct invariance properties in each case are obtained by imposing an appropriate condition on **W** in the definition of the norm. Observe that it is not necessary to know **W** directly, but only that it satisfies $\mathbf{W}\boldsymbol{\delta}=\boldsymbol{\gamma}$ say in the case of the BFGS formula. There are various matrices which could be used for **W**, for instance any Broyden family update $\mathbf{B}_{\phi}^{(k+1)}$ or the averaged Hessian matrix

$$\int_0^1 \mathbf{G}(\mathbf{x}^{(k)}+\theta\boldsymbol{\delta}^{(k)})\,\mathrm{d}\theta$$

when these matrices are positive definite. The minimum value of $\|\mathbf{E}\|_{\mathbf{W}}$ itself can also be obtained (see Question 3.17) and has been used for example by Al-Baali and Fletcher (1985) as a way of estimating errors in Hessian approximations. There are many other interesting applications of variational methods in deriving updating formulae: more are mentioned in Section 3.6. Another interesting idea examined in Question 3.18 is just to follow the above theory but without the symmetry constraint. An unsymmetric rank 1 formula is obtained that is always well defined and has termination and conjugate direction properties for quadratic functions.

3.4 THE BROYDEN FAMILY

As demonstrated in Section 3.2, there is only one symmetric rank one formula which satisfies the quasi-Newton condition (3.2.5). Consider therefore symmetric rank two correction formulae (assuming $n\geqslant 2$) which for reasons of invariance are constructed from vectors $\boldsymbol{\delta}$ and $\mathbf{H}\boldsymbol{\gamma}$. The relevance of using rank two formula is suggested by the variational result in (3.3.8). The general formula is

$$\mathbf{H}^{(k+1)}=\mathbf{H}+a\boldsymbol{\delta}\boldsymbol{\delta}^{\mathrm{T}}+b(\mathbf{H}\boldsymbol{\gamma}\boldsymbol{\delta}^{\mathrm{T}}+\boldsymbol{\delta}\boldsymbol{\gamma}^{\mathrm{T}}\mathbf{H})+c\mathbf{H}\boldsymbol{\gamma}\boldsymbol{\gamma}^{\mathrm{T}}\mathbf{H} \tag{3.4.1}$$

after suppressing superscript (k) on the right, and using the symmetry of **H**. If condition (3.2.5) is to hold, then equating terms in $\boldsymbol{\delta}$ and $\mathbf{H}\boldsymbol{\gamma}$ gives the equations

$$\begin{aligned}1&=a\boldsymbol{\delta}^{\mathrm{T}}\boldsymbol{\gamma}+b\boldsymbol{\gamma}^{\mathrm{T}}\mathbf{H}\boldsymbol{\gamma}\\0&=1+b\boldsymbol{\delta}^{\mathrm{T}}\boldsymbol{\gamma}+c\boldsymbol{\gamma}^{\mathrm{T}}\mathbf{H}\boldsymbol{\gamma}.\end{aligned} \tag{3.4.2}$$

These two equations relate the three unknown parameters a, b, and c, leaving one degree of freedom. To isolate this conveniently $b=-\phi/\boldsymbol{\delta}^{\mathrm{T}}\boldsymbol{\gamma}$ is set, in which case after eliminating a, b, and c from (3.4.2) and (3.4.1) it follows that

$$\mathbf{H}_\phi^{(k+1)} = \mathbf{H} + \frac{\boldsymbol{\delta}\boldsymbol{\delta}^{\mathrm{T}}}{\boldsymbol{\delta}^{\mathrm{T}}\boldsymbol{\gamma}} - \frac{\mathbf{H}\boldsymbol{\gamma}\boldsymbol{\gamma}^{\mathrm{T}}\mathbf{H}}{\boldsymbol{\gamma}^{\mathrm{T}}\mathbf{H}\boldsymbol{\gamma}} + \phi\mathbf{v}\mathbf{v}^{\mathrm{T}} \tag{3.4.3}$$

$$= \mathbf{H}_{\mathrm{DFP}}^{(k+1)} + \phi\mathbf{v}\mathbf{v}^{\mathrm{T}} \tag{3.4.4}$$

where

$$\mathbf{v} = (\boldsymbol{\gamma}^{\mathrm{T}}\mathbf{H}\boldsymbol{\gamma})^{1/2}\left(\frac{\boldsymbol{\delta}}{\boldsymbol{\delta}^{\mathrm{T}}\boldsymbol{\gamma}} - \frac{\mathbf{H}\boldsymbol{\gamma}}{\boldsymbol{\gamma}^{\mathrm{T}}\mathbf{H}\boldsymbol{\gamma}}\right). \tag{3.4.5}$$

An alternative rearrangement which emphasizes the rank two correction property is

$$\mathbf{H}_\phi^{(k+1)} = \mathbf{H} + [\boldsymbol{\delta}, \mathbf{H}\boldsymbol{\gamma}]\begin{bmatrix} (1 + \phi\boldsymbol{\gamma}^{\mathrm{T}}\mathbf{H}\boldsymbol{\gamma}/\boldsymbol{\delta}^{\mathrm{T}}\boldsymbol{\gamma})/\boldsymbol{\delta}^{\mathrm{T}}\boldsymbol{\gamma} & -\phi/\boldsymbol{\delta}^{\mathrm{T}}\boldsymbol{\gamma} \\ -\phi/\boldsymbol{\delta}^{\mathrm{T}}\boldsymbol{\gamma} & (\phi - 1)/\boldsymbol{\gamma}^{\mathrm{T}}\mathbf{H}\boldsymbol{\gamma} \end{bmatrix}[\boldsymbol{\delta}, \mathbf{H}\boldsymbol{\gamma}]^{\mathrm{T}}. \tag{3.4.6}$$

It is not difficult to see that these formulae correspond to (3.2.15), so that the *Broyden family* of formulae (Broyden, 1967), generated by the parameter ϕ, has been established by this approach.

It is mentioned in passing that more general formulae have been suggested, for which some of the results of this section also hold. Huang (1970) gives a three-parameter family in which **H** is allowed to become unsymmetric, and in which the quasi-Newton condition is replaced by

$$\mathbf{H}^{(k+1)}\boldsymbol{\gamma}^{(k)} = \rho^{(k)}\boldsymbol{\delta}^{(k)}. \tag{3.4.7}$$

A special case is the two-parameter *symmetric Huang family*, in which **H** is symmetric, but (3.4.7) holds, and this is of some interest, for instance $\rho^{(k)} \neq 1$ might be a useful choice on non-quadratic functions (Biggs, 1973). This idea also features indirectly in some other methods in Section 3.6. However the Broyden family (that is $\rho^{(k)} = 1$ for all k) is by far the most important.

The expression (3.4.4) is important since it shows that all formulae $\mathbf{H}_\phi^{(k+1)}$ in the Broyden family differ by a multiple of the rank one matrix $\mathbf{v}\mathbf{v}^{\mathrm{T}}$. Because the eigenvalues of $\mathbf{H} + \phi\mathbf{v}\mathbf{v}^{\mathrm{T}}$ interpolate those of **H** from above if $\phi > 0$ and from below if $\phi < 0$ (Wilkinson, 1965), it follows from Theorem 3.2.2 that $\mathbf{H}_\phi^{(k+1)}$ maintains positive definiteness for any $\phi \geqslant 0$. In fact this is true for any $\phi > \bar{\phi}$, where $\bar{\phi}$ is the value which causes $\mathbf{H}_{\bar{\phi}}^{(k+1)}$ to be singular, and is given (see Question 3.14, and equate the denominator of θ to zero) by

$$\bar{\phi} = \frac{1}{(1 - \boldsymbol{\gamma}^{\mathrm{T}}\mathbf{H}\boldsymbol{\gamma}\boldsymbol{\delta}^{\mathrm{T}}\mathbf{B}\boldsymbol{\delta}/\boldsymbol{\delta}^{\mathrm{T}}\boldsymbol{\gamma}^2)}. \tag{3.4.8}$$

If $\mathbf{H}^{(k)}$ is positive definite it follows by Cauchy's inequality that $\bar{\phi} < 0$. If $\phi < \bar{\phi}$ is chosen, then $\mathbf{H}^{(k+1)}$ is indefinite. When exact line searches are made, a further property of $\bar{\phi}$ emerges. From the definition $\mathbf{s}^{(k+1)} = -\mathbf{H}^{(k+1)}\mathbf{g}^{(k+1)}$, together with (3.4.3) and (2.3.3) it follows that

$$\mathbf{s}^{(k+1)} = \mathbf{v}\left\{\frac{\gamma^{\mathrm{T}}\mathbf{s}}{\gamma^{\mathrm{T}}\mathbf{H}\gamma^{1/2}} - \phi\mathbf{v}^{\mathrm{T}}\mathbf{g}\right\} \tag{3.4.9}$$

after a little manipulation, and suppressing the superscript (k). It is clear from this that a so-called *degenerate value* of ϕ exists which causes $\mathbf{s}^{(k+1)} = \mathbf{0}$, so that the algorithm fails. In fact this is the value $\phi = \bar{\phi}$ in (3.4.8), since if $\mathbf{g}^{(k+1)} \neq \mathbf{0}$ then $\mathbf{s}^{(k+1)}$ can only be zero if $\mathbf{H}^{(k+1)}$ is singular.

Equation (3.4.9) has another fundamental implication. It shows that when exact line searches are used, then the effect of different choices $\phi^{(k)}$ is just to change the *length* of $\mathbf{s}^{(k+1)}$ and not its *direction*. Thus it can be expected that a method using any updating formula from the Broyden family (a *Broyden method*, say) will be *independent* of $\phi^{(k)}$ to some extent. In fact it is shown in Theorem 3.4.3 that if exact line searches are used then the sequence $\{\mathbf{x}^{(k)}\}$ is independent of the choice of the sequence $\{\phi^{(k)}\}$, even for general functions. In many cases therefore it is possible to state results which hold for any Broyden method. A number of such results are now proved. The possibility that there exist degenerate values of $\phi^{(k)}$ which cause the algorithm to fail must however be borne in mind. It is assumed that these values are avoided, so that the algorithm is well defined with $\mathbf{H}^{(k)}$ existing, and that $\mathbf{g}^{(k)} \neq \mathbf{0}$ implies $\mathbf{s}^{(k)} \neq \mathbf{0}$, $\mathbf{g}^{(k)\mathrm{T}}\mathbf{s}^{(k)} \neq 0$ and hence $\alpha^{(k)} \neq 0$. These conditions follow immediately if $\phi^{(k)} > \bar{\phi}^{(k)}$ because $\mathbf{H}^{(k)}$ is then positive definite for all k, and this is by far the most important case. However the above assumptions do enable the results to be stated in slightly more general terms. The degenerate value $\bar{\phi}$ above is not the only degenerate value and this question is considered further by Fletcher and Sinclair (1981). The first two theorems therefore relate to any Broyden method with exact line searches, applied to a positive definite quadratic function with Hessian $\mathbf{G}$, and avoiding the use of any degenerate values. The first theorem establishes a hereditary result (3.4.10), a conjugacy property (3.4.11), and the existence of an iteration m ($\leqslant n$) after termination occurs with $\mathbf{g}^{(m+1)} = \mathbf{0}$.

Theorem 3.4.1

A Broyden method with exact line searches terminates after $m \leqslant n$ iterations on a quadratic function, and the following hold for all $i = 1, 2, \ldots, m$.

$$\mathbf{H}^{(i+1)}\gamma^{(j)} = \delta^{(j)}, \quad j = 1, 2, \ldots, i \tag{3.4.10}$$

$$\mathbf{s}^{(i)\mathrm{T}}\mathbf{G}\mathbf{s}^{(j)} = 0, \quad j = 1, 2, \ldots, i-1. \tag{3.4.11}$$

Also if $m = n$, then $\mathbf{H}^{(n+1)} = \mathbf{G}^{-1}$.

Proof

Trivial for $m = 0$. For any m such that $1 \leqslant m \leqslant n$, equations (3.4.10) and (3.4.11) are established by induction. For $i = 1$, (3.4.10) holds by substitution in (3.4.3) and (3.4.11) holds vacuously. Let them be true for some i, $1 \leqslant i < m$. Since $\mathbf{g}^{(i+1)} \neq \mathbf{0}$, the iteration is well defined with $\mathbf{s}^{(i+1)} \neq \mathbf{0}$ and $\alpha^{(i+1)} \neq 0$. For any $j \leqslant i$

$$\mathbf{s}^{(i+1)\mathrm{T}}\mathbf{G}\mathbf{s}^{(j)} = \frac{-\mathbf{g}^{(i+1)\mathrm{T}}\mathbf{H}^{(i+1)}\boldsymbol{\gamma}^{(j)}}{\alpha^{(j)}} = \frac{-\mathbf{g}^{(i+1)\mathrm{T}}\boldsymbol{\delta}^{(j)}}{\alpha^{(j)}} = 0$$

by (2.4.2). Hence (3.4.11) is true with $i+1$ replacing i. Also by substitution in (3.4.3), $\mathbf{H}^{(i+2)}\boldsymbol{\gamma}^{(i+1)} = \boldsymbol{\delta}^{(i+1)}$, and for any $j \leqslant i$, $\mathbf{H}^{(i+2)}\boldsymbol{\gamma}^{(j)} = \mathbf{H}^{(i+1)}\boldsymbol{\gamma}^{(j)}$ *plus* terms multiplied by the scalars $\boldsymbol{\gamma}^{(i+1)\mathrm{T}}\mathbf{H}^{(i+1)}\boldsymbol{\gamma}^{(j)}$ or $\boldsymbol{\delta}^{(i+1)\mathrm{T}}\boldsymbol{\gamma}^{(j)}$. Using (3.4.11) with i replaced by $i+1$, and (3.4.10), both scalars are zero, and so $\mathbf{H}^{(i+2)}\boldsymbol{\gamma}^{(j)} = \mathbf{H}^{(i+1)}\boldsymbol{\gamma}^{(j)} = \boldsymbol{\delta}^{(j)}$ for both $j \leqslant i$ and $j = i+1$. This is (3.4.10) with i replaced by $i+1$, so the induction is complete. Since the vectors $\mathbf{s}^{(i)}$, $i = 1, 2, \ldots, m$, are conjugate, so $m \leqslant n$ follows by Theorem 2.4.1. If $m = n$, then (3.4.11) and $\alpha^{(i)} \neq 0$ imply that the vectors $\boldsymbol{\delta}^{(i)}$. $i = 1, 2, \ldots, n$, are independent, so that (3.4.10) implies $\mathbf{H}^{(n+1)} = \mathbf{G}^{-1}$ as proved in Theorem 3.2.1. □

The next theorem applies when $\mathbf{H}^{(1)} = \mathbf{I}$ and establishes both a precise value for m and equivalence with the Fletcher–Reeves conjugate gradient method (and hence with other conjugate gradient methods). The result of Theorem 4.1.1. is assumed. Quantities relevant to the conjugate gradient algorithm are subscripted CG.

Theorem 3.4.2

If $\mathbf{H}^{(1)} = \mathbf{I}$ and $\mathbf{x}^{(1)} = \mathbf{x}^{(1)}_{\mathrm{CG}}$, then the Broyden method with exact line searches is equivalent to the Fletcher–Reeves conjugate gradient method when applied to a quadratic function, and m is the number of linearly independent vectors in the sequence

$$\mathbf{g}^{(1)}, \mathbf{G}\mathbf{g}^{(1)}, \mathbf{G}^2\mathbf{g}^{(1)}, \ldots \tag{3.4.12}$$

Proof

Trivial for $m = 0$. For $m \geqslant 1$ the following are established inductively. For any $i: 1 \leqslant i \leqslant m$

$$\{\mathbf{g}^{(1)}, \mathbf{g}^{(2)}, \ldots, \mathbf{g}^{(i)}\} \text{ is an orthogonal basis for a linear space, } S_i \text{ say} \tag{3.4.13}$$

$$\{\mathbf{s}^{(1)}_{\mathrm{CG}}, \mathbf{s}^{(2)}_{\mathrm{CG}}, \ldots, \mathbf{s}^{(i)}_{\mathrm{CG}}\} \text{ is a linearly independent basis for } S_i \tag{3.4.14}$$

$$\{\mathbf{g}^{(1)}, \mathbf{G}\mathbf{g}^{(1)}, \ldots, \mathbf{G}^{i-1}\mathbf{g}^{(1)}\} \text{ is a linearly independent basis for } S_i \tag{3.4.15}$$

$$\mathbf{H}^{(i)}\mathbf{u} = \mathbf{u} \quad \forall \mathbf{u} \in S_i^{\perp} = \{\mathbf{v} : \mathbf{v}^{\mathrm{T}}\mathbf{w} = 0 \ \ \forall \mathbf{w} \in S_i\} \tag{3.4.16}$$

$$\mathbf{s}^{(i)} \propto \mathbf{s}^{(i)}_{\mathrm{CG}}; \qquad \mathbf{x}^{(i+1)} = \mathbf{x}^{(i+1)}_{\mathrm{CG}}; \qquad \mathbf{g}^{(i+1)} = \mathbf{g}^{(i+1)}_{\mathrm{CG}}. \tag{3.4.17}$$

These conditions are readily established when $i = 1$. Assume they are true for any $i: 1 \leqslant i < m$. Now (4.1.5), (2.4.2), and (3.4.14) imply that $\mathbf{g}^{(i+1)} \in S_i^{\perp}$. Since $\mathbf{g}^{(i+1)} \neq \mathbf{0}$ so (3.4.13) is true for $i+1$ replacing i, and also iteration $i+1$ of both algorithms is well defined with $\alpha^{(i+1)} \neq 0$. Since by (4.1.3), $\mathbf{s}_{\text{CG}}^{(i+1)}$ has a non-zero component of $\mathbf{g}^{(i+1)}$, it follows that (3.4.14) is true with $i+1$ replacing i. Now consider $\boldsymbol{\delta}^{(i)}$ expanded in terms of the basis (3.4.15). Since $\boldsymbol{\delta}^{(i)} \in S_i$ and $\notin S_{i-1}$ it follows that the component of $\mathbf{G}^{i-1}\mathbf{g}^{(1)}$ is non-zero. Then by (3.2.8), $\mathbf{g}^{(i+1)} = \mathbf{G}\boldsymbol{\delta}^{(i)} + \mathbf{g}^{(i)}$ implies that $\mathbf{G}^i\mathbf{g}^{(1)} \in S_{i+1}$ and has a non-zero component of $\mathbf{g}^{(i+1)}$. Thus (3.4.15) is established with $i+1$ replacing i. Next, let $\mathbf{u} \in \mathbf{S}_{i+1}^{\perp}$ so that $\mathbf{H}^{(i+1)}\mathbf{u} = \mathbf{H}^{(i)}\mathbf{u}$ *plus* terms multiplied by the scalars $\boldsymbol{\delta}^{(i)\text{T}}\mathbf{u}$ and $\boldsymbol{\gamma}^{(i)\text{T}}\mathbf{H}^{(i)}\mathbf{u}$. Clearly $\boldsymbol{\delta}^{(i)} \in S_{i+1}$. Also $\mathbf{H}^{(i)}\boldsymbol{\gamma}^{(i)} = \mathbf{H}^{(i)}\mathbf{g}^{(i+1)} + \mathbf{s}^{(i)} = \mathbf{g}^{(i+1)} + \mathbf{s}^{(i)}$ by (3.4.16) since $\mathbf{g}^{(i+1)} \in S_i^{\perp}$, so $\mathbf{H}^{(i)}\boldsymbol{\gamma}^{(i)} \in S_{i+1}$. Since $\mathbf{u} \in S_{i+1}^{\perp}$, both scalars are zero. Hence $\mathbf{H}^{(i+1)}\mathbf{u} = \mathbf{H}^{(i)}\mathbf{u} = \mathbf{u}$ by (3.4.16), so (3.4.16) is established with $i+1$ replacing i. Then by (3.4.9), $\mathbf{s}^{(i+1)} \propto \mathbf{v}^{(i)}$ which is a linear combination of $\boldsymbol{\delta}^{(i)}$ and $\mathbf{H}^{(i)}\boldsymbol{\gamma}^{(i)}$. As above, $\boldsymbol{\delta}^{(i)}$ and $\mathbf{H}^{(i)}\boldsymbol{\gamma}^{(i)} \in S_{i+1}$ so $\mathbf{s}^{(i+1)} \in S_{i+1}$. Since $\mathbf{s}_{\text{CG}}^{(i+1)} \in S_{i+1}$, it follows from (3.4.17) and the conjugacy conditions (3.4.11) for $\mathbf{s}^{(i+1)}$ that $\mathbf{s}^{(i+1)} \propto \mathbf{s}_{\text{CG}}^{(i+1)}$. By assumption $\mathbf{s}^{(i+1)} \neq \mathbf{0}$, so by the exact line search and (3.4.17) it follows that $\mathbf{x}^{(i+2)} = \mathbf{x}_{\text{CG}}^{(i+2)}$ and so (3.4.17) is established with $i+1$ replacing i, completing the induction.

Finally by definition of m, $\mathbf{g}^{(m+1)} = \mathbf{G}\boldsymbol{\delta}^{(m)} + \mathbf{g}^{(m)} = \mathbf{0}$ so it follows that the set $\{\mathbf{g}^{(1)}, \mathbf{G}\mathbf{g}^{(1)}, \ldots, \mathbf{G}^m\mathbf{g}^{(1)}\}$ is linearly dependent. But by (3.4.15) with $i = m$, the set $\{\mathbf{g}^{(1)}, \mathbf{G}\mathbf{g}^{(1)}, \ldots, \mathbf{G}^{m-1}\mathbf{g}^{(1)}\}$ is linearly independent. Hence (3.4.12) follows. □

Corollary

If $\mathbf{H}^{(1)}$ is any positive definite matrix then the property

$$\mathbf{g}^{(i)\text{T}}\mathbf{H}^{(1)}\mathbf{g}^{(j)} = 0, \qquad j = 1, 2, \ldots, i-1 \tag{3.4.18}$$

holds for all $i = 1, 2, \ldots, m$, and m is number of independent vectors in the sequence

$$\mathbf{g}^{(1)}, \mathbf{G}\mathbf{H}^{(1)}\mathbf{g}^{(1)}, (\mathbf{G}\mathbf{H}^{(1)})^2\mathbf{g}^{(1)}, \ldots \tag{3.4.19}$$

Proof

These results are obtained by making a transformation of variables from $\mathbf{H}_x^{(1)} = \mathbf{H}^{(1)}$ to $\mathbf{H}_y^{(1)} = \mathbf{I}$ by writing $\mathbf{H}_x^{(1)} = \mathbf{A}^{-1}\mathbf{A}^{-\text{T}}$. Then the results (3.4.13) and (3.4.12) in the y-variables, and the invariance result of Theorem 3.3.1 and its corollary, imply that (3.4.18) and (3.4.19) hold. □

In fact the independence of the Broyden method to changes in ϕ goes far beyond quadratic functions, as a surprising theorem due to Dixon (1972) shows. It is stated in the following form (Powell, 1972a) which also illustrates a special property of the BFGS formula.

Theorem 3.4.3

When applied to any $\mathbb{C}^1$ function, a Broyden method with exact line searches has the property that for all $k \geqslant 1$

$$\mathbf{x}^{(k+1)} \textit{ and } \mathbf{H}_{\text{BFGS}}^{(k+1)} \textit{ are independent of } \phi^{(1)}, \phi^{(2)}, \ldots, \phi^{(k-1)} \tag{3.4.20}$$

assuming that multiple local minima in the line search are resolved consistently, degenerate values of ϕ are avoided, and the algorithm is well-defined. [*By* $\mathbf{H}_{\text{BFGS}}^{(k+1)}$ *is meant the result of applying the BFGS formula to* $\mathbf{H}^{(k)}$, *whereas* $\mathbf{H}^{(k+1)}$ *implies using a Broyden family update with parameter* $\phi^{(k)}$.]

Proof

The result is clearly true for $k=1$. By induction, let it be true for any $k \geqslant 1$. Then by (3.4.9) the direction of $\mathbf{s}^{(k+1)}$ does not depend on $\phi^{(k)}$. Also since $\mathbf{s}^{(k+1)} \propto -\mathbf{H}_{\text{BFGS}}^{(k+1)}\mathbf{g}^{(k+1)}$ it follows that the direction of $\mathbf{s}^{(k+1)}$ does not depend on $\phi^{(1)}, \phi^{(2)}, \ldots, \phi^{(k-1)}$, by (3.4.20). Thus by the exact line search, $\mathbf{x}^{(k+2)}$ (and hence $\boldsymbol{\delta}^{(k+1)}$ and $\boldsymbol{\gamma}^{(k+1)}$) are independent of $\phi^{(1)}, \phi^{(2)}, \ldots, \phi^{(k)}$. Now

$$\mathbf{H}_{\text{BFGS}}^{(k+2)} = \left(\mathbf{I} - \frac{\boldsymbol{\delta}^{(k+1)}\boldsymbol{\gamma}^{(k+1)\mathrm{T}}}{\boldsymbol{\delta}^{(k+1)\mathrm{T}}\boldsymbol{\gamma}^{(k+1)}}\right)\mathbf{H}^{(k+1)}\left(\mathbf{I} - \frac{\boldsymbol{\gamma}^{(k+1)}\boldsymbol{\delta}^{(k+1)\mathrm{T}}}{\boldsymbol{\gamma}^{(k+1)\mathrm{T}}\boldsymbol{\delta}^{(k+1)}}\right) + \frac{\boldsymbol{\delta}^{(k+1)}\boldsymbol{\delta}^{(k+1)\mathrm{T}}}{\boldsymbol{\delta}^{(k+1)\mathrm{T}}\boldsymbol{\gamma}^{(k+1)}}$$

and

$$\mathbf{H}^{(k+1)} = \mathbf{H}_{\text{BFGS}}^{(k+1)} + (\phi^{(k)} - 1)\mathbf{v}^{(k)}\mathbf{v}^{(k)\mathrm{T}}.$$

Since $(\mathbf{I} - \boldsymbol{\delta}^{(k+1)}\boldsymbol{\gamma}^{(k+1)\mathrm{T}}/\boldsymbol{\delta}^{(k+1)\mathrm{T}}\boldsymbol{\gamma}^{(k+1)})$ annihilates $\boldsymbol{\delta}^{(k+1)}$ and hence $\mathbf{v}^{(k)}$ (by (3.4.9)), it follows that $\mathbf{H}_{\text{BFGS}}^{(k+2)}$ can be defined in terms of $\mathbf{H}_{\text{BFGS}}^{(k+1)}$, $\boldsymbol{\delta}^{(k+1)}$, and $\boldsymbol{\gamma}^{(k+1)}$ alone, and so is independent of $\phi^{(1)}, \phi^{(2)}, \ldots, \phi^{(k)}$ by (3.4.20) and the definition of $\mathbf{H}_{\text{BFGS}}^{(k+1)}$. This completes the induction. □

This result is important not only in its own right, but as a means of extending convergence and order of convergence results for any one formula to cover the whole Broyden family. One such result due to Powell (1971, 1972b) is that if $f(\mathbf{x})$ is *convex*, then the DFP method with exact line searches converges globally, with superlinear convergence if $\mathbf{G}^*$ is positive definite. As a consequence, for general functions there usually exists a neighbourhood of the solution $\mathbf{x}^*$ in which $f(\mathbf{x})$ is locally convex, and in which the above result is applicable. What happens outside this neighbourhood is an open question, although for $n=2$ Powell (1972b) has proved global convergence of the DFP method. Theorem 3.4.3 extends these results to any Broyden method, in the general function/exact line search case.

All these results are idealized in that an exact line search cannot be achieved in practice for general functions, so it is important to consider what results are common to all Broyden methods when inexact line searches are used. For quadratic functions Theorem 3.2.1 shows that the rank one method will terminate. However, for all other formulae, including DFP and BFGS, termination does *not* generally occur, since the exact line search condition (2.3.3) is needed in the proof. However, Powell (1972c) has some termination results for Broyden methods when some searches are exact and some are not, the nub of which is that an inexact search cannot delay termination for more than one iteration. This is important since in practice inexact search algorithms will carry out an interpolation (and hence give an exact line minimum for a quadratic) if the reduction in f is not satisfactory. Thus although it may be possible to suggest modified algorithms which retain termination (see Section 3.6), it may be that the lack of a termination proof is not disadvantageous.

All the results so far emphasize the unity of the Broyden family, whereas what is required in practice is to know which particular $\phi^{(k)}$ to use on any iteration. Of course it is preferable to choose $\phi^{(k)} > \bar{\phi}^{(k)}$ so that positive definite $\mathbf{H}^{(k)}$ matrices, and hence the descent property, are preserved; but some more specific indications are required. For quadratic functions Fletcher (1970a) shows that the eigenvalues of $\mathbf{G}^{1/2}\mathbf{H}^{(k)}\mathbf{G}^{1/2}$ tend monotonically to 1 (but not strictly) if and only if the $\phi^{(k)}$ lie in $[0, 1]$ (the *convex class* of formulae). This suggests that stable updating formulae are confined to the convex class, the extreme elements of which are the DFP and BFGS formulae. Fletcher also suggests *switching* between the DFP or BFGS formulae according to a test which has some interesting interpretations. The Hoshino (1972) formula is also in the convex class since $\phi = \boldsymbol{\delta}^T\boldsymbol{\gamma}/(\boldsymbol{\delta}^T\boldsymbol{\gamma} + \boldsymbol{\gamma}^T\mathbf{H}\boldsymbol{\gamma})$. For general functions a proof of superlinear convergence for the DFP and BFGS method has been given by Broyden, Dennis and Moré (1973) when the choice $\alpha^{(k)} = 1$ is made at each iteration. Stoer (1975) gives similar results for the convex class when an inexact search is used which satisfies (2.5.5). Thus it seems likely that the practical choice of an updating formula should be from the convex class.

There are however some indications that the BFGS formula alone may be preferable. For general functions and exact line searches Theorem 3.4.3 shows the unique feature of the BFGS formula, that $\mathbf{H}_{\text{BFGS}}^{(k+1)}$ is independent of $\phi^{(1)}$, $\phi^{(2)}, \ldots, \phi^{(k-1)}$, so that an application of this formula cancels out the effect of any previous 'bad' choices of ϕ. For inexact searches subject to conditions (2.5.3) and (2.5.5), Powell (1976) proves global convergence for a convex objective function. He also shows that the order of convergence is superlinear if $\mathbf{G}^*$ is positive definite. As yet this global convergence result has not been demonstrated for other formulae. Another pointer to the use of the BFGS formula is given by Powell (1985d) who shows that the worst case behaviour of the DFP method on certain simple quadratic functions is significantly inferior.

All these theoretical pointers are nonetheless tentative, and the question of global convergence of any Broyden method for general functions is as yet open. The best practical choice of method therefore depends strongly on experimentation, and this is taken up in the next section.

3.5 NUMERICAL EXPERIMENTS

One of the most important aspects of algorithmic design is to try out algorithms on well chosen test problems. In this section some such experiments are described and, even though they are not very extensive, they give a good guide in answering such questions as which updating formula to use, or how accurate a line search to take, and so on. The results also show some of the difficulties of assessing numerical evidence. Three test functions have been used which are well known and are not atypical of real applications. Rosenbrock's function (1.2.2) has only two variables but is difficult to minimize on account of a steep sided curving valley. The other functions are the Chebyquad function (Fletcher, 1965), which arises from an application to Chebyshev quadrature, and the trigonometric function (Fletcher and Powell, 1963). These latter problems have the feature that the dimension n of the problem can be altered at will. In particular this feature has been used with the trigonometric function to explore the performance of methods on relatively large problems.

To avoid testing all possible combinations of updating formula and line search, a standard method has been taken and the effect of changing this in different ways is examined. In the standard method the BFGS formula is used, together with the line search of (2.6.2) and (2.6.4). Cubic (where possible) or quadratic interpolating polynomials are used as described in Section 2.6 to choose points in the allowed intervals. The parameter values are those given in Section 2.6 except that $\tau_2 = 0.05$ is used. The convergence test (2.3.7) is used with $\varepsilon = 10^{-8}$. The initial estimate α_1 is determined by (2.6.8).

First of all, some explanation of the tables is in order. Each entry consists of three integers which give the numbers of (iterations, f evaluations, $\mathbf{g}$ evaluations) required to solve the problem. Also if the accuracy $f - f^* \leqslant 10^{-8}$ is not achieved a code letter is attached: a if $f - f^* \in (10^{-8}, 10^{-6}]$, b if $f - f^* \in (10^{-6}, 10^{-4}]$, and so on. A bar (for example $\bar{\text{d}}$) indicates that the iteration has been terminated by having reached the maximum permitted number of line searches. The presence of a letter thus indicates either a low accuracy solution (a), or complete failure (e), or something in between. The three integers represent some measure of the effort required to obtain the solution. Each iteration requires two matrix multiplications and a matrix update, so the number of iterations ($\times n^2$) is a measure of the housekeeping operations required by the minimization subroutine. The numbers of f and $\mathbf{g}$ evaluations are a measure of the effort taken in the user subroutines. How these figures are assessed is of course open to debate, but it is usual to discount the housekeeping costs since these are not usually dominant, and to weight the $\mathbf{g}$ figure more heavily than that for f. The more simple measure of comparative computer times is not usually used, since test functions are often atypical in being easy to evaluate, and this distorts the comparison in favour of the housekeeping costs.

Table 3.5.1 summarizes different experiments in which a fairly accurate line search is used ($\sigma = 0.1$). Here the BFGS and DFP methods are shown to perform very similarly and the same is also true (not shown) if other formulae are used. This supports the predictions of Dixon's result (Theorem 3.4.3) that the iterative sequence $\{\mathbf{x}^{(k)}\}$ is the same in the limit. It also shows that the cost of finding the line minima does not vary significantly from method to method. In the other experiment in Table 3.5.1 the Wolfe–Powell conditions are used in which the two sided test (2.5.6) is replaced by the weaker one-sided test (2.5.4). There is

Table 3.5.1 Results with fairly accurate line search ($\sigma = 0.1$)

Problem *n*	BFGS			DFP			BFGS + Wolfe–Powell conditions		
Rosenbrock									
2	21	56	50	21	59	54	21	41	37
Chebyquad									
2	3	6	5	3	6	5	5	7	7
4	7	13	11	8	14	12	7	13	11
6	9	20	18	16	19	17	12	21	17
8	18	42	32	17	37	33	17	31	26
Trigonometric									
2	5	9	8	5	9	8	6	8	7
4	11	22	18	11	21	18	11	21	17
6	14	29	26	14	28	25	14	29	24
8	14	29	23	14	29	24	21	37	31
10	18	37	30	17	34	28	20	32	29
20	29	58	47	30	61	50	43	70	64
30	45	89	81	45	93	85	73	124	114
40	56	112	102	57	128	118	65	112	102
50	61	115	108	61	117	110	112	187	177

Table 3.5.2 Inexact line search results ($\sigma = 0.9$)

Problem *n*	BFGS			DFP			Fletcher switch			Hoshino		
Rosenbrock												
2	36	45	40	100	118	112$\bar{d}$	39	45	41	38	44	40
Chebyquad												
2	3	6	5	3	6	5	3	6	5	3	6	5
4	10	13	11	10	13	11	10	13	11	10	13	11
6	18	22	19	22	25	23	18	21	19	19	22	20
8	19	29	22	34	39	35	25	30	26	20	26	23
Trigonometric												
2	6	8	7	6	8	7	6	8	7	6	8	7
4	14	19	16	18	23	20	14	19	15	15	20	17
6	16	22	19	17	22	19	16	21	18	16	21	18
8	25	31	26	25	33	29	26	32	28	27	33	28
10	22	30	26	24	31	28	22	29	26	22	30	27
20	44	57	49	109	121	114	46	58	51	48	60	53
30	80	93	86	300	311	304$\bar{\bar{e}}$	86	99	91	85	97	90
40	91	109	99	81	104	86d	151	165	156	91	102	94
50	120	130	124	482	491	486a	205	214	209	122	133	127

then some indication that the method performs less well. The main conclusion is that for $\sigma = 0.1$ the standard method is as good as but not significantly better than most other options.

Table 3.5.2 summarizes experiments with an inexact line search ($\sigma = 0.9$), the conditions of which are almost always satisfied whenever f is reduced. Therefore the change from (2.5.6) from (2.5.4) is not examined since it would have negligible effect. The main factor here lies in the choice of the updating formula. In particular, for large or difficult problems the DFP formula is shown to require considerable extra effort, and may even fail to solve the problem. This is the feature that has caused the BFGS formula to be preferred as the standard. The Fletcher switch strategy, whilst it solves all problems to the required accuracy, also exhibits poorer performance on the $n = 40$ and $n = 50$ problems. Only the Hoshino formula is comparable to the BFGS formula; perhaps this is an indication that the former deserves more study. An explanation of these results is not easy to come by, since methods which use inexact line searches have not received much theoretical attention. However, the conclusion that the BFGS formula is preferable is currently well accepted.

Another well-accepted hypothesis is that inexact line searches are generally more efficient than near exact line searches with the BFGS method. The rationale for this is that more frequent changes are made to the $\mathbf{H}^{(k)}$ matrices, so that a better approximation to $\mathbf{G}^{-1}$ is obtained. It can indeed be observed from the tables that fewer f and $\mathbf{g}$ calls per iteration are required when $\sigma = 0.9$, although on the other hand more iterations are required. However for the BFGS method, the *total* number of f and $\mathbf{g}$ calls in the $\sigma = 0.1$ and $\sigma = 0.9$ cases are very similar. It may be that these results are atypical, or it may be that recent advances in line search design have reduced the cost of calculating near exact line searches. As far as I know however the BFGS results for $\sigma = 0.1$ given here are as good as can be obtained by any other implementation. If these results are typical, then the $\sigma = 0.1$ line search is to be preferred in that it cuts down the housekeeping costs without affecting the number of f and $\mathbf{g}$ calls. My computer times for the above experiments show a 10 per cent improvement for the $\sigma = 0.1$ run.

Finally there is some interest to know whether the performance of the DFP

Table 3.5.3 Degradation of DFP method with weaker line searches

Problem n	$\sigma = 0.1$			$\sigma = 0.5$			$\sigma = 0.9$		
Rosenbrock									
2	21	59	54	27	55	49	100	118	112d̄
Trigonometric									
20	30	61	50	43	72	66	109	121	114
30	45	93	85	80	140	134	300	311	304ē
40	57	128	118	84	144	128	81	104	86d
50	61	117	110	168	198	190a	482	491	486a

method worsens gradually or suddenly as σ is increased from 0.1 to 0.9. A sudden deterioration might suggest the breakdown of the algorithm, for example because $\mathbf{H}^{(k)}$ becomes numerically singular. Three different values are tried in Table 3.5.3, and the results rather suggest that the onset is gradual. I cannot offer an explanation for this.

3.6 OTHER FORMULAE

The literature on quasi-Newton formulae is extensive and impossible to cover thoroughly within one chapter. This final section points out some other promising lines of research. One recurrent theme has been the aim of preserving a termination property when inexact line searches are used. The rank one formula (3.2.7) has already been mentioned in this respect but causes other difficulties. Hestenes (1969) suggests using the projection formula (4.1.10), but from any positive definite $\mathbf{H}^{(1)}$. This generates conjugate directions of search. Furthermore the fact that

$$\sum_{k=1}^{n} \frac{\boldsymbol{\delta}^{(k)}\boldsymbol{\delta}^{(k)\mathrm{T}}}{\boldsymbol{\delta}^{(k)\mathrm{T}}\boldsymbol{\gamma}^{(k)}} = \mathbf{G}^{-1} \tag{3.6.1}$$

for a quadratic function shows that if $\mathbf{H}^{(n+1)}$ is re-set to the left-hand side of (3.6.1), then termination on iteration $n+1$ can be achieved. Another method is due to Dixon (1973) who defines a search direction by

$$\mathbf{s}^{(k)} = -\mathbf{H}^{(k)}\mathbf{g}^{(k)} + \mathbf{w}^{(k)}. \tag{3.6.2}$$

Initially $\mathbf{w}^{(1)} = \mathbf{0}$ and $\mathbf{w}^{(k)}$ is chosen so that $\mathbf{s}^{(k)}$ is conjugate to $\mathbf{s}^{(1)}, \mathbf{s}^{(2)}, \ldots, \mathbf{s}^{(k-1)}$. This is achieved by the update

$$\mathbf{w}^{(k+1)} = \mathbf{w}^{(k)} + \frac{\boldsymbol{\delta}^{(k)}\boldsymbol{\delta}^{(k)\mathrm{T}}\mathbf{g}^{(k+1)}}{\boldsymbol{\delta}^{(k)\mathrm{T}}\boldsymbol{\gamma}^{(k)}}. \tag{3.6.3}$$

Dixon shows for a quadratic function that $\mathbf{H}^{(k+1)} = \mathbf{G}^{-1}$ so that termination follows. The vector $\mathbf{w}^{(n+1)}$ is reset to zero. Dixon reports that practical performance is comparable to the BFGS method. Both these methods are cyclic methods in which a fresh start is made every n iterations. This is not elegant and can be disadvantageous, for instance when symmetry considerations dictate that $m \ll n$ iterations would be appropriate (see Theorem 3.4.2 in the quadratic case). Another simple method for retaining termination with inexact line searches is given by Sinclair (1979). This is not a cyclic method, and Sinclair reports good practical results, especially for large n.

An idea that commanded more attention for a while is given by Davidon (1975). He suggests a family of formulae

$$\mathbf{H}^{(k+1)} = \mathbf{H} + \frac{\mathbf{z}\mathbf{z}^{\mathrm{T}}}{\mathbf{z}^{\mathrm{T}}\mathbf{y}} - \frac{\mathbf{H}\mathbf{y}\mathbf{y}^{\mathrm{T}}\mathbf{H}}{\mathbf{y}^{\mathrm{T}}\mathbf{H}\mathbf{y}} + \phi\mathbf{y}^{\mathrm{T}}\mathbf{H}\mathbf{y}\left(\frac{\mathbf{z}}{\mathbf{z}^{\mathrm{T}}\mathbf{y}} - \frac{\mathbf{H}\mathbf{y}}{\mathbf{y}^{\mathrm{T}}\mathbf{H}\mathbf{y}}\right)\left(\frac{\mathbf{z}}{\mathbf{z}^{\mathrm{T}}\mathbf{y}} - \frac{\mathbf{H}\mathbf{y}}{\mathbf{y}^{\mathrm{T}}\mathbf{H}\mathbf{y}}\right)^{\mathrm{T}} \tag{3.6.4}$$

(cf. (3.4.3)) in which $\mathbf{z}$ and $\mathbf{y}$ are no longer $\boldsymbol{\delta}$ and $\boldsymbol{\gamma}$. The quasi-Newton condition $\mathbf{H}^{(k+1)}\boldsymbol{\gamma} = \boldsymbol{\delta}$ determines $\mathbf{y}$ once $\mathbf{z}$ has been selected. The vector $\mathbf{z}$ is chosen so that in the quadratic case the hereditary property (3.4.10) is maintained for inexact line searches. One choice for $\mathbf{z}$ leads to the rank one formula (3.2.7). However, Davidon gives another choice for $\mathbf{z}$ which does not degenerate in this way, and it is then always possible to choose ϕ so that $\mathbf{H}^{(k+1)}$ is positive definite. Also the formula is invariant in the sense of Section 3.3. A good description of the details of how $\mathbf{z}$ is chosen is given by Brodlie (1977). In some cases it may be that $\mathbf{z} = \mathbf{0}$ would be chosen, in which case an alternative selection for $\mathbf{z}$ must be made. Powell (1977a) shows that this does not delay termination for more than one iteration. Davidon's paper contains other features of interest, one of which is that ϕ is chosen so that $\mathbf{H}^{(k+1)}$ is optimally conditioned in a certain limited sense. In view of the additional complexity of Davidon's method, it will be necessary to show strong practical reasons why it should displace the BFGS method. Current experience is limited, although encouraging enough to merit further research. Little is known about convergence or order of convergence. A more recent contribution of Davidon (1980) is the use of *conic functions* and this is explored in Question 3.20.

Another interesting line of research has been the development of updating formulae $\mathbf{H}^{(k+1)} = \mathbf{H}^{(k)} + \mathbf{E}^{(k)}$ in which the correction $\mathbf{E}^{(k)}$ is minimal in some sense, subject to the quasi-Newton condition being satisfied. Greenstadt (1970) first introduced this idea and suggests a rank two formula in which the Euclidean norm of the correction $\sqrt{(\sum_{ij}E_{ij}^2)}$ is minimized. Powell (1970a) suggests the formula

$$\mathbf{B}^{(k+1)} = \mathbf{B} + \frac{\boldsymbol{\eta}\boldsymbol{\delta}^{\mathrm{T}} + \boldsymbol{\delta}\boldsymbol{\eta}^{\mathrm{T}}}{\boldsymbol{\delta}^{\mathrm{T}}\boldsymbol{\delta}} - \frac{\boldsymbol{\eta}^{\mathrm{T}}\boldsymbol{\delta}}{\boldsymbol{\delta}^{\mathrm{T}}\boldsymbol{\delta}^2}\boldsymbol{\delta}\boldsymbol{\delta}^{\mathrm{T}} \tag{3.6.5}$$

where $\boldsymbol{\eta} = \boldsymbol{\gamma} - \mathbf{B}\boldsymbol{\delta}$, and which is dual to Greenstadt's formula. It can be shown by following Greenstadt's approach that the formula (3.6.5) minimizes the Euclidean norm of the correction $\mathbf{B}^{(k+1)} - \mathbf{B}^{(k)}$: an outline of how this is done is set out in Question 3.16. The Greenstadt and Powell formulae are not invariant in the sense of Section 3.3, nor do they retain positive definite matrices, and practical performance has been disappointing although the Powell formula does possess some theoretical stability properties. Another interesting observation described in more detail in Section 3.3 is that it is possible to interpret the DFP and BFGS formulae as having minimal correction in a weighted Euclidean norm (Goldfarb, 1970). Another potentially valuable application of minimum norm corrections is to problems with *sparse Hessian matrices* (Toint, 1977). He shows that if in addition the sparsity conditions ($B_{ij} = 0$ for certain indices i, j) are imposed, then a minimum norm correction can readily be determined. One disadvantage of the algorithm is that positive definite matrices $\mathbf{B}^{(k)}$ are not maintained. Nonetheless the idea could be valuable in practice. More recently the application of sparse updates to partially separable optimization has allowed some very large structured problems to be tackled successfully (see Griewank and Toint (1984) who give further references).

Finally another interesting development has been the idea of *self-scaling* variable metric methods (see Oren, 1974, for example). These methods use a formula in the two-parameter (μ, τ) family

$$\mathbf{H}^{(k+1)} = \mu\left(\mathbf{H} - \frac{\mathbf{H}\gamma\gamma^{\mathrm{T}}\mathbf{H}}{\gamma^{\mathrm{T}}\mathbf{H}\gamma}\right) + \frac{\delta\delta^{\mathrm{T}}}{\delta^{\mathrm{T}}\gamma} + \tau\mathbf{v}\mathbf{v}^{\mathrm{T}} \tag{3.6.6}$$

where $\mathbf{v}$ is defined in (3.4.5). This can be regarded as being derived from the matrix $\mathbf{H}^{(k+1)}$ given by the symmetric Huang formula (Section 3.4) by multiplication by $\mu = 1/\rho$. Thus on each iteration the multiplier μ has the effect of *prescaling* the matrix $\mathbf{H}^{(k)}$, followed by a Broyden family update. The formulae are invariant under linear transformations provided that μ and τ are constructed from invariant scalars. Other properties, such as termination and conjugate gradients, are also preserved. Likewise positive definite matrices are preserved for $\mu > 0$ and sufficiently large τ. The main disparity seems to be that the property $\mathbf{H}^{(n+1)} = \mathbf{G}^{-1}$ for quadratics is not preserved. In general this appears to be disadvantageous, although if $\mathbf{H}^{(1)}$ is a poor approximation to $\mathbf{G}$, then pre-scaling might be useful. Numerical evidence for self-scaling is inconclusive, especially in comparison with the BFGS method in which the initial matrix $\mathbf{H}^{(1)}$ is pre-scaled (Shanno and Phua, 1978).

QUESTIONS FOR CHAPTER 3

3.1. Verify that both $\mathbf{x}' = \mathbf{0}$ and $\mathbf{x}'' = (-\sqrt{7/12}, \sqrt{7/12})^{\mathrm{T}}$ are stationary points of the function

$$f(\mathbf{x}) = (x_1 + x_2)^2 + (2(x_1^2 + x_2^2 - 1) - \tfrac{1}{3})^2$$

and identify the type of stationary point in each case. Show that Newton's method can be applied from $\mathbf{x}^{(1)} = (\sqrt{7/6}, 0)^{\mathrm{T}}$, and verify that $\mathbf{x}^{(2)}$ lies along the line which passes through $\mathbf{x}'$ and $\mathbf{x}''$. (Do not carry out a line search.)

3.2. Consider the function $f(\mathbf{x})$ in Question 2.5, and find the largest open ball about $\mathbf{x}^* = \mathbf{0}$ in which $\mathbf{G}(\mathbf{x})$ is positive definite. For what values of $\mathbf{x}^{(1)}$ in this ball does Newton's method converge, assuming that $x_1^{(1)} = x_2^{(1)}$?

3.3. Derive expressions for all first and second partial derivatives of the function

$$f(\mathbf{x}) = \tfrac{1}{2}(x_1^2 + x_2^2)\exp(x_1^2 - x_2^2)$$

and verify that $\mathbf{x}^* = \mathbf{0}$ is a local minimizer of this function. Evaluate all the derivatives at $\mathbf{x}' = (1, 1)^{\mathrm{T}}$ and show that the Hessian matrix $\mathbf{G}'$ is not positive definite. Show that $\nu = 3$ is the smallest positive integer for which $\mathbf{G}' + \nu\mathbf{I}$ is positive definite, and use the resulting matrix to carry out one iteration of Newton's method, starting from $\mathbf{x}'$. (Do not carry out a line search.)

3.4. Apply Newton's method to the function

$$f(\mathbf{x}) = \tfrac{11}{546}x^6 - \tfrac{38}{364}x^4 + \tfrac{1}{2}x^2$$

from $\mathbf{x}^{(1)} = 1.01$. Verify that $\mathbf{G}^{(k)}$ is always positive definite and that $f^{(k)}$ is monotonic decreasing. Show nonetheless that $\mathbf{x}^{\infty} = \pm 1$ are the accumulation points of the sequence $\{\mathbf{x}^{(k)}\}$, and that $\mathbf{g}^{\infty} \neq \mathbf{0}$. Verify that for any fixed positive ρ, however small, inequality (2.5.3) is violated for all k sufficiently large (except when round-off errors dominate).

3.5. Consider Newton's method for the unconstrained minimization of a $\mathbb{C}^2$ function $f(\mathbf{x})$. If the Hessian $\mathbf{G}$ is indefinite a search direction $\mathbf{p}$ can be defined alternatively by

$$\mathbf{p} = \sigma \mathbf{s}/\|\mathbf{s}\|_2 \tag{a}$$

where $\mathbf{s}$ solves

$$\mathbf{L}^{\mathrm{T}}\mathbf{s} = \mathbf{a}, \quad \text{where } a_i = 1, \quad \text{if } d_i \leqslant 0$$
$$a_i = 0, \quad \text{if } d_i > 0$$

and where $\sigma = \pm 1$ is chosen so that $\mathbf{p}^{\mathrm{T}}\mathbf{g} \leqslant 0$. You may assume that $\mathbf{G}$ always has factors $\mathbf{G} = \mathbf{LDL}^{\mathrm{T}}$, where $\mathbf{L}$ is unit lower triangular, and $\mathbf{D} = \operatorname{diag} d_i$. Show that $\mathbf{p}$ has negative curvature and is downhill.

Let $\mathbf{x}(\theta)$ be a continuous trajectory defined by a parameter θ, in the neighbourhood of $\theta = \bar{\theta}$, and assume that $\mathbf{G}(\mathbf{x}(\theta))$ is positive definite for $\theta < \bar{\theta}$ and has one negative eigenvalue for $\theta > \bar{\theta}$. If $\theta < \bar{\theta}$ let a direction $\mathbf{p}_\theta$ be defined by $\mathbf{p} = \mathbf{s}/\|\mathbf{s}\|_2$, where $\mathbf{s} = -\mathbf{G}^{-1}\mathbf{g}$, and if $\theta \geqslant \bar{\theta}$, let $\mathbf{p}_\theta$ be defined by method (a). Show that $\mathbf{p}_\theta$ is continuous at $\bar{\theta}$. You may assume that the factors are such that $d_n < d_i$ for all $i \neq n$, and that $\mathbf{Lu} = -\mathbf{g}(\mathbf{x}(\bar{\theta}))$ has a solution for which $u_n \neq 0$. Illustrate your answer in regard to the function (1.2.2) and the trajectory $\mathbf{x}(\theta) = \begin{pmatrix} \theta \\ \frac{1}{2} \end{pmatrix}$. What is the value $\bar{\theta}$?

3.6. For the function in Question 2.2, if $\mathbf{x}^{(1)} = \mathbf{0}$ why is it that Newton's method cannot be applied satisfactorily? If the search direction $\mathbf{s} = -\mathbf{G}^{-1}\mathbf{g}$ is used at $\mathbf{x}^{(1)}$, show that neither $\pm\mathbf{s}$ is downhill, and that f cannot be reduced. If a modified Newton method uses the matrix $\mathbf{G} + \nu\mathbf{I}$, what range of ν makes $\mathbf{G} + \nu\mathbf{I}$ positive definite? What step is obtained by using $\nu = 1$? (Do not carry out a line search.)

3.7. Verify that $\mathbf{H}_{\mathrm{BFGS}}^{(k+1)}\boldsymbol{\gamma} = \boldsymbol{\delta}$, $\mathbf{H}_{\mathrm{DFP}}^{(k+1)}\boldsymbol{\gamma} = \boldsymbol{\delta}$, $\mathbf{B}_{\mathrm{BFGS}}^{(k+1)}\boldsymbol{\delta} = \boldsymbol{\gamma}$, and $\mathbf{B}_{\mathrm{BFGS}}^{(k+1)}\mathbf{H}_{\mathrm{BFGS}}^{(k+1)} = \mathbf{I}$, where $\mathbf{B} = \mathbf{H}^{-1}$.

3.8. Apply the DFP method with exact line searches to the problem in Question 2.11 using $\mathbf{H}^{(1)} = \mathbf{I}$. Verify the properties of quadratic termination in n line searches, $\mathbf{H}^{(n+1)} = \mathbf{G}^{-1}$, and equivalence to the conjugate gradient method.

3.9. Apply the BFGS algorithm to the problem in Question 2.7, choosing $\mathbf{x}^{(1)} = \mathbf{0}$ and $\mathbf{H}^{(1)} = \mathbf{I}$. Show that the line $\mathbf{x} = \mathbf{x}^{(2)} + \alpha\mathbf{s}^{(2)}$ passes through $\mathbf{x}^*$, thus guaranteeing termination in two steps.

3.10. Let $\mathbf{H}$ be a positive definite symmetric matrix. Show that $\mathbf{H} - \mathbf{H}\boldsymbol{\gamma}\boldsymbol{\gamma}^{\mathrm{T}}\mathbf{H}/\boldsymbol{\gamma}^{\mathrm{T}}\mathbf{H}\boldsymbol{\gamma}$ is singular and positive semi-definite. (You may use $\mathbf{H} = \mathbf{LL}^{\mathrm{T}}$ and Cauchy's inequality $\mathbf{z}^{\mathrm{T}}\mathbf{z}\mathbf{y}^{\mathrm{T}}\mathbf{y} \geqslant (\mathbf{y}^{\mathrm{T}}\mathbf{z})^2$.) Hence show that $\mathbf{H}_{\mathrm{DFP}}^{(k+1)}$ is positive definite if and only if $\boldsymbol{\delta}^{\mathrm{T}}\boldsymbol{\gamma} > 0$. Now let $\mathbf{H}$ be symmetric positive semi-definite and singular, so that there exists $\mathbf{u} \neq \mathbf{0}$ such that

$\mathbf{Hu} = \mathbf{0}$. Show that $\mathbf{H}_{\text{DFP}}^{(k+1)}$ is singular. If $\mathbf{H}^{(1)}\mathbf{u} = \mathbf{0}$ show that the DFP method cannot locate any solution $\mathbf{x}^*$ unless $\mathbf{x}^* - \mathbf{x}^{(1)}$ is orthogonal to $\mathbf{u}$.

3.11. Consider the BFGS method. If the ith row and column of $\mathbf{H}^{(1)}$ is zeroed, show that the property is preserved in all matrices $\mathbf{H}^{(k)}$ and that $x_i^{(k)} = x_i^{(1)}$ for all $k > 1$. It then follows that the objective function is minimized subject to the constraint $x_i = x_i^{(1)}$.

3.12. Show that the rank one formula is in the Broyden family, and that its value of ϕ is not in $[0, 1]$, assuming that $\boldsymbol{\delta}^{\mathrm{T}}\boldsymbol{\gamma} > 0$ and $\mathbf{H}$ is positive definite.

3.13. The modification of an $n \times n$ matrix $\mathbf{A}$ by a change of rank m ($\leqslant n$) can be written generally as $\mathbf{A}' = \mathbf{A} + \mathbf{RST}^{\mathrm{T}}$ where $\mathbf{R}$ and $\mathbf{T}$ are $n \times m$ and $\mathbf{S}$ is $m \times m$. Verify that $\mathbf{A}^{-1}$ is modified by $(\mathbf{A}')^{-1} = \mathbf{A}^{-1} - \mathbf{A}^{-1}\mathbf{RU}^{-1}\mathbf{T}^{\mathrm{T}}\mathbf{A}^{-1}$, where $\mathbf{U} = \mathbf{S}^{-1} + \mathbf{T}^{\mathrm{T}}\mathbf{A}^{-1}\mathbf{R}$ (the *Sherman–Morrison formula*). Use this formula with $m = 2$ and $\mathbf{R} = \mathbf{T}$ to obtain an expression for $\mathbf{B}_{\text{DFP}}^{(k+1)}$ in terms of $\mathbf{B}(= \mathbf{H}^{-1})$, $\boldsymbol{\gamma}$, and $\boldsymbol{\delta}$, and verify that it is dual to the formula for $\mathbf{H}_{\text{BFGS}}^{(k+1)}$.

3.14. A general formula in the Broyden family can be written

$$\mathbf{H}_\phi^{(k+1)} = \mathbf{H}_{\text{DFP}}^{(k+1)} + \phi \mathbf{v}\mathbf{v}^{\mathrm{T}}$$

where

$$\mathbf{v} = \boldsymbol{\gamma}^{\mathrm{T}}\mathbf{H}\boldsymbol{\gamma}^{1/2}\left(\frac{\boldsymbol{\delta}}{\boldsymbol{\delta}^{\mathrm{T}}\boldsymbol{\gamma}} - \frac{\mathbf{H}\boldsymbol{\gamma}}{\boldsymbol{\gamma}^{\mathrm{T}}\mathbf{H}\boldsymbol{\gamma}}\right).$$

Verify that the corresponding expression for $\mathbf{B}_\phi^{(k+1)}$ has the form

$$\mathbf{B}_\phi^{(k+1)} = \mathbf{B}_{\text{BFGS}}^{(k+1)} + \theta \mathbf{w}\mathbf{w}^{\mathrm{T}} = \mathbf{B}_{\text{DFP}}^{(k+1)} + (\theta - 1)\mathbf{w}\mathbf{w}^{\mathrm{T}}$$

where

$$\mathbf{w} = \boldsymbol{\delta}^{\mathrm{T}}\mathbf{B}\boldsymbol{\delta}^{1/2}\left(\frac{\boldsymbol{\gamma}}{\boldsymbol{\delta}^{\mathrm{T}}\boldsymbol{\gamma}} - \frac{\mathbf{B}\boldsymbol{\delta}}{\boldsymbol{\delta}^{\mathrm{T}}\mathbf{B}\boldsymbol{\delta}}\right).$$

Show that $\theta = (\phi - 1)/(\phi - 1 - \phi\mu)$ where $\mu = \boldsymbol{\gamma}^{\mathrm{T}}\mathbf{H}\boldsymbol{\gamma}\boldsymbol{\delta}^{\mathrm{T}}\mathbf{B}\boldsymbol{\delta}/\boldsymbol{\delta}^{\mathrm{T}}\boldsymbol{\gamma}^2$. Find the value of $\bar{\phi}$ for which $\mathbf{H}_{\bar{\phi}}^{(k+1)}$ is singular. If $\mathbf{H}$ is positive definite, show using Cauchy's inequality that $\bar{\phi}$ is negative. Show also that the formulae generated by $\phi = \boldsymbol{\delta}^{\mathrm{T}}\boldsymbol{\gamma}/(\boldsymbol{\delta}^{\mathrm{T}}\boldsymbol{\gamma} \pm \boldsymbol{\gamma}^{\mathrm{T}}\mathbf{H}\boldsymbol{\gamma})$ are self-dual.

3.15. Show that the steepest descent method is invariant under an orthogonal transformation of variables.

3.16. Consider a quasi-Newton formula in which $\mathbf{B}$ approximates the Hessian matrix, and is updated by $\mathbf{B}^{(k+1)} = \mathbf{B} + \mathbf{E}$. Find the symmetric correction $\mathbf{E}$ which satisfies $\mathbf{B}^{(k+1)}\boldsymbol{\delta} = \boldsymbol{\gamma}$ and which has least Euclidean norm. (Minimize $\frac{1}{2}\Sigma_{ij}E_{ij}^2$ with respect to $\mathbf{E}$, subject to $E_{ij} = E_{ji}$ for all $i \geqslant j$, and $(\mathbf{B} + \mathbf{E})\boldsymbol{\delta} = \boldsymbol{\gamma}$. Solve by introducing Lagrange multipliers Λ_{ij} and γ, respectively. Obtain an expression for $\mathbf{E}$ in terms of $\boldsymbol{\Lambda}$ and $\boldsymbol{\lambda}$, and eliminate the constraint equations.) Show that the resulting formula is given by (3.6.5) and that this formula is not in the Broyden family. Show that when applied to a quadratic function with Hessian $\mathbf{G}$, then

$$\mathbf{B}^{(k+1)} - \mathbf{G} = \left(\mathbf{I} - \frac{\boldsymbol{\delta}\boldsymbol{\delta}^{\mathrm{T}}}{\boldsymbol{\delta}^{\mathrm{T}}\boldsymbol{\delta}}\right)(\mathbf{B} - \mathbf{G})\left(\mathbf{I} - \frac{\boldsymbol{\delta}\boldsymbol{\delta}^{\mathrm{T}}}{\boldsymbol{\delta}^{\mathrm{T}}\boldsymbol{\delta}}\right). \qquad \text{(a)}$$

Show also that if $\mathbf{B}=\begin{bmatrix}2 & 1\\ 1 & 1\end{bmatrix}$, $\boldsymbol{\delta}=\begin{pmatrix}1\\0\end{pmatrix}$ and $\boldsymbol{\gamma}=\begin{pmatrix}1\\1\end{pmatrix}$, then $\boldsymbol{\eta}=-\boldsymbol{\delta}$ and hence $\mathbf{B}^{(k+1)}=\begin{bmatrix}1 & 1\\ 1 & 1\end{bmatrix}$. What implications does (a) and the result following it have for the use of the formula in a quasi-Newton method? Verify that the dual of (3.6.5) is the formula given by Greenstadt (1970).

3.17. Continue the analysis of Theorem 3.3.2 to find the minimum value of the norm in (3.3.7). Show that it can be expressed as

$$\|\mathbf{E}\|_{\mathbf{W}}^2 = 2\frac{\boldsymbol{\gamma}^{\mathrm{T}}\mathbf{HWH}\boldsymbol{\gamma}}{\boldsymbol{\delta}^{\mathrm{T}}\boldsymbol{\gamma}} - a^2 - 2a + 1$$

where $a=\boldsymbol{\gamma}^{\mathrm{T}}\mathbf{H}\boldsymbol{\gamma}/\boldsymbol{\delta}^{\mathrm{T}}\boldsymbol{\gamma}$. Simplify this expression in the case that $\mathbf{W}=\mathbf{B}_{\mathrm{BFGS}}^{(k+1)}$.

3.18. Derive an updating formula by following Theorem 3.3.2 but without the symmetry constraint. Show that the resulting formula is

$$\mathbf{H}^{(k+1)} = \mathbf{H} + (\boldsymbol{\delta} - \mathbf{H}\boldsymbol{\gamma})\boldsymbol{\delta}^{\mathrm{T}}/\boldsymbol{\delta}^{\mathrm{T}}\boldsymbol{\gamma}.$$

If $\mathbf{s}=-\mathbf{H}^{\mathrm{T}}\mathbf{g}$ is used to generate search directions, show that the resulting quasi-Newton method with exact line searches generates hereditary properties and conjugate directions and hence terminates when applied to a quadratic function. (Use Theorem 2.5.1.)

3.19. Consider a Newton-like method for unconstrained minimization in which an exact line search is used, and the search direction is calculated from

$$\mathbf{B}\mathbf{s}^{(k)} = -\mathbf{g}^{(k)}$$

where $\mathbf{B}$ is a fixed matrix. The method is applied to the problem

$$\min \tfrac{1}{2}(x_1^2 + x_2^2) \qquad \mathbf{x} \in \mathbb{R}^2$$

from an initial point $\mathbf{x}^{(1)}=(a, ta)^{\mathrm{T}} (a>0, t>0)$ with $\mathbf{B}=\begin{bmatrix}1 & 0\\ 0 & b\end{bmatrix} (0<b<1)$. Find the value of t such that $|x_2^{(k)}| = t|x_1^{(k)}|$ for all k. Analyse the type of convergence shown by this example.

3.20. The use of the conic function

$$f(\mathbf{x}) = \tfrac{1}{2}\frac{\mathbf{x}^{\mathrm{T}}\mathbf{G}\mathbf{x}}{(1-\mathbf{c}^{\mathrm{T}}\mathbf{x})^2} + \frac{\mathbf{a}^{\mathrm{T}}\mathbf{x}}{(1-\mathbf{c}^{\mathrm{T}}\mathbf{x})} + b,$$

where $\mathbf{G}$ is symmetric positive definite, has been proposed by Davidon (1980) as a model function for unconstrained optimization. Clearly the transformation $\mathbf{y}=\mathbf{y}(\mathbf{x})=\mathbf{x}/(1-\mathbf{c}^{\mathrm{T}}\mathbf{x})$ transforms f into the quadratic function

$$q(\mathbf{y}) = \tfrac{1}{2}\mathbf{y}^{\mathrm{T}}\mathbf{G}\mathbf{y} + \mathbf{a}^{\mathrm{T}}\mathbf{y} + b.$$

Obtain the inverse transformation $\mathbf{x}=\mathbf{x}(\mathbf{y})$. Show that these transformations map any line $\mathbf{x}=\mathbf{x}'+\alpha\mathbf{s}$ in $\mathbf{x}$-space into a line $\mathbf{y}=\mathbf{y}'+\beta\mathbf{d}$ in $\mathbf{y}$-space, and vice versa, and give the relationship between $\mathbf{s}$ and $\mathbf{d}$.

Given **G**-conjugate search directions $\mathbf{d}^{(1)}, \mathbf{d}^{(2)}, \ldots, \mathbf{d}^{(n)}$ then an exact line search method in **y**-space terminates when applied to $q(\mathbf{y})$. Hence write down an equivalent line search method in **x**-space which terminates when applied to $f(\mathbf{x})$ (given $\mathbf{d}^{(1)}, \ldots, \mathbf{d}^{(n)}$ and **c**).

Obtain the relationship between ∇_x and ∇_y (derivative operators in **x**-space and **y**-space respectively) and hence say whether the steepest descent methods for $f(\mathbf{x})$ in **x**-space and for $q(\mathbf{y})$ in **y**-space are equivalent in general.

3.21. (i) This question concerns the BFGS method and explores an interesting idea of Powell (1985e) concerning how the approximate Hessian matrix **B** should be represented. Consider the representation

$$\mathbf{Z}\mathbf{Z}^{\mathrm{T}} = \mathbf{B}^{-1} \tag{a}$$

where **Z** is a non-singular $n \times n$ matrix. Show that the columns of **Z** can be interpreted as a set of **B**-conjugate directions. Show that the BFGS formula can be written in symmetric product form

$$\mathbf{H}^{(k+1)} = (\mathbf{I} - \boldsymbol{\delta}\mathbf{p}^{\mathrm{T}})\mathbf{H}(\mathbf{I} - \mathbf{p}\boldsymbol{\delta}^{\mathrm{T}})$$

(omitting superscript k) for some vector **p** (see also Brodlie, Gourlay and Greenstadt (1973) for a general treatment of product-form representations). Hence write down a corresponding formula for updating the matrix **Z** in (a).

(ii) The representation (a) is not unique and the matrix **Z** can be replaced by any matrix $\bar{\mathbf{Z}} = \mathbf{Z}\mathbf{Q}$ where **Q** is an orthogonal matrix, so that

$$\bar{\mathbf{Z}}\bar{\mathbf{Z}}^{\mathrm{T}} = \mathbf{Z}\mathbf{Z}^{\mathrm{T}} = \mathbf{B}^{-1}.$$

Deduce a more general updating formula for **Z**. In particular consider the case that the first column of **Q** satisfies

$$\mathbf{Q}\mathbf{e}_1 = \pm \mathbf{u}/\|\mathbf{u}\|_2 \tag{b}$$

where $\mathbf{u} = \mathbf{Z}^{\mathrm{T}}\mathbf{g}$. Show that the first column of the updated matrix $\mathbf{Z}^{(k+1)}$ is proportional to $\boldsymbol{\delta}$, and use the $\mathbf{B}^{(k+1)}$-conjugacy condition to determine the normalizing constant. Hence show using **B**-conjugacy properties of the columns of $\bar{\mathbf{Z}}$ that the general updating formula can be written

$$\bar{\mathbf{z}}_i^{(k+1)} = \begin{cases} \boldsymbol{\delta}/\sqrt{\boldsymbol{\delta}^{\mathrm{T}}\boldsymbol{\gamma}} & i = 1 \\ \left(\mathbf{I} - \dfrac{\boldsymbol{\delta}\boldsymbol{\gamma}^{\mathrm{T}}}{\boldsymbol{\delta}^{\mathrm{T}}\boldsymbol{\gamma}}\right)\bar{\mathbf{z}}_i & i = 2, 3, \ldots, n \end{cases} \tag{c}$$

where $\bar{\mathbf{z}}_i$, $i = 1, 2, \ldots, n$ are the columns of $\bar{\mathbf{Z}}$.

(iii) Let **Q** be orthogonal and satisfy (b), and also be a lower Hessenberg matrix with the property that

$$Q_{i,i+1} = \pm 1 \qquad i = 1, 2, \ldots, q$$

if $u_i = 0$, $i = 1, 2, \ldots, q$. (This property is readily achieved using plane rotation matrices.) Consider applying the BFGS method based on (a) and

(c) to minimize a positive definite quadratic function, using exact line searches. By making use of the conditions $\boldsymbol{\gamma}^{(k)^{\mathrm{T}}}\boldsymbol{\delta}^{(i)} = 0$ and $\mathbf{g}^{(k)^{\mathrm{T}}}\boldsymbol{\delta}^{(i)} = 0$ for all $i < k$ that apply in this case, show by induction that after iteration k, the first k columns of $\mathbf{Z}^{(k+1)}$ are multiples of the vectors $\boldsymbol{\delta}^{(k)}, \ldots, \boldsymbol{\delta}^{(2)}, \boldsymbol{\delta}^{(1)}$ respectively (note the reverse order).

Chapter 4

Conjugate Direction Methods

4.1 CONJUGATE GRADIENT METHODS

In view of Theorem 2.4.1, which equates conjugacy and exact line searches with quadratic termination, it is attractive to try to associate conjugacy properties with the steepest descent method in an attempt to achieve both efficiency and reliability. This is the aim of the *conjugate gradient method.* The method is described first of all in an idealized way insofar as it applies to a quadratic function. In this case the method is an exact line search method (2.3.2) in which

$$\mathbf{s}^{(1)} = -\mathbf{g}^{(1)} \tag{4.1.1}$$

and for $k \geqslant 1$

$$\mathbf{s}^{(k+1)} = \text{component of } -\mathbf{g}^{(k+1)} \text{ conjugate to } \mathbf{s}^{(1)}, \mathbf{s}^{(2)}, \ldots, \mathbf{s}^{(k)}$$

Since the conjugacy conditions (2.4.1) can be written as

$$\mathbf{s}^{(i)\mathrm{T}} \boldsymbol{\gamma}^{(j)} = 0 \qquad j \neq i \tag{4.1.2}$$

using (3.2.8), it follows that the Gram–Schmidt process can be used to express $\mathbf{s}^{(k+1)} = -\mathbf{g}^{(k+1)} + \sum_j \beta_j \mathbf{s}^{(j)}$. In fact for quadratic functions, $\beta_j = 0$ for all $j < k$, and the expression simplifies to

$$\mathbf{s}^{(k+1)} = -\mathbf{g}^{(k+1)} + \beta^{(k)} \mathbf{s}^{(k)} \tag{4.1.3}$$

with $\beta^{(0)} = 0$, where

$$\beta^{(k)} = \frac{\mathbf{g}^{(k+1)\mathrm{T}} \mathbf{g}^{(k+1)}}{\mathbf{g}^{(k)\mathrm{T}} \mathbf{g}^{(k)}} \tag{4.1.4}$$

This is the basis of the conjugate gradient minimization method of Fletcher and Reeves (1964), developed directly from the conjugate gradient method of Hestenes and Stiefel (1952) for solving linear systems.

An illustration of the Fletcher–Reeves method as applied to the quadratic

Table 4.1.1 The Fletcher–Reeves method

k	$\mathbf{x}^{(k)}$	$\mathbf{g}^{(k)}$	$\beta^{(k-1)}$	$\mathbf{s}^{(k)}$	$\alpha^{(k)}$
1	$\frac{1}{10}$ 1	2 2	(0)	-2 -2	$\frac{1}{11}$
2	$-\frac{9}{110}$ $\frac{9}{11}$	$-\frac{18}{11}$ $\frac{18}{11}$	$\frac{81}{121}$	$\frac{36}{121}$ $-\frac{360}{121}$	$\frac{11}{40}$
3	0 0	0 0			

function $f(\mathbf{x}) = 10x_1^2 + x_2^2$ is given in Table 4.1.1. The quantities are tabulated in the order in which they are evaluated, and exact line searches are used. This table also illustrates that the method terminates for quadratic functions. This result, and the validity of the simpler expression (4.1.3), is justified by the following theorem in which m is the largest integer for which $\mathbf{g}^{(m)} \neq \mathbf{0}$.

Theorem 4.1.1

The Fletcher–Reeves method with exact line searches terminates at a stationary point $\mathbf{x}^{(m+1)}$ after $m \leqslant n$ iterations, and the following hold for all i, $1 \leqslant i \leqslant m$,

$$\mathbf{s}^{(i)T}\mathbf{G}\mathbf{s}^{(j)} = 0 \qquad\qquad (4.1.5)$$

$$j = 1, 2, \ldots, i-1$$

$$\mathbf{g}^{(i)T}\mathbf{g}^{(j)} = 0 \qquad\qquad (4.1.6)$$

$$\mathbf{s}^{(i)T}\mathbf{g}^{(i)} = -\mathbf{g}^{(i)T}\mathbf{g}^{(i)}. \qquad\qquad (4.1.7)$$

(These are conjugate direction, orthogonal gradient, and descent conditions respectively. Note that (4.1.7) implies that $\mathbf{s}^{(i)} \neq \mathbf{0}$.)

Proof

Trivial for $m = 0$. For $m \geqslant 1$ an inductive proof is used. For $i = 1$, (4.1.5) and (4.1.6) are vacuous and (4.1.7) holds since $\mathbf{s}^{(1)} = -\mathbf{g}^{(1)}$. Let these equations hold for some $i < m$. Then by (1.2.15)

$$\mathbf{g}^{(i+1)} = \mathbf{g}^{(i)} + \alpha^{(i)}\mathbf{G}\mathbf{s}^{(i)}. \qquad\qquad (4.1.8)$$

and $\alpha^{(i)}$ is then determined by (2.3.3), giving

$$\alpha^{(i)} = \frac{-\mathbf{g}^{(i)T}\mathbf{s}^{(i)}}{\mathbf{s}^{(i)T}\mathbf{G}\mathbf{s}^{(i)}} = \frac{\mathbf{g}^{(i)T}\mathbf{g}^{(i)}}{\mathbf{s}^{(i)T}\mathbf{G}\mathbf{s}^{(i)}} \neq 0. \qquad\qquad (4.1.9)$$

Using (4.1.8) and then (4.1.3)

$$\begin{aligned}\mathbf{g}^{(i+1)^{\mathrm{T}}}\mathbf{g}^{(j)} &= \mathbf{g}^{(i)^{\mathrm{T}}}\mathbf{g}^{(j)} + \alpha^{(i)}\mathbf{s}^{(i)^{\mathrm{T}}}\mathbf{G}\mathbf{g}^{(j)} \\ &= \mathbf{g}^{(i)^{\mathrm{T}}}\mathbf{g}^{(j)} - \alpha^{(i)}\mathbf{s}^{(i)^{\mathrm{T}}}\mathbf{G}(\mathbf{s}^{(j)} - \beta^{(j-1)}\mathbf{s}^{(j-1)}).\end{aligned}$$

When $j = i$, this is zero by (4.1.5) and (4.1.9), and when $j < i$, it is zero by (4.1.6) and (4.1.5). Thus (4.1.6) is true with $i + 1$ replacing i. Using (4.1.3) and then (4.1.8)

$$\begin{aligned}\mathbf{s}^{(i+1)^{\mathrm{T}}}\mathbf{G}\mathbf{s}^{(j)} &= -\mathbf{g}^{(i+1)^{\mathrm{T}}}\mathbf{G}\mathbf{s}^{(j)} + \beta^{(i)}\mathbf{s}^{(i)^{\mathrm{T}}}\mathbf{G}\mathbf{s}^{(j)} \\ &= \mathbf{g}^{(i+1)^{\mathrm{T}}}(\mathbf{g}^{(j)} - \mathbf{g}^{(j+1)})/\alpha^{(j)} + \beta^{(i)}\mathbf{s}^{(i)^{\mathrm{T}}}\mathbf{G}\mathbf{s}^{(j)}.\end{aligned}$$

When $j = i$, this is zero by (4.1.6), (4.1.8), and (4.1.4), and when $j < i$, it is zero by (4.1.6) and (4.1.5). Thus (4.1.5) is true with $i + 1$ replacing i. Finally,

$$\begin{aligned}-\mathbf{g}^{(i+1)^{\mathrm{T}}}\mathbf{s}^{(i+1)} &= \mathbf{g}^{(i+1)^{\mathrm{T}}}\mathbf{g}^{(i+1)} - \beta^{(i)}\mathbf{g}^{(i+1)^{\mathrm{T}}}\mathbf{s}^{(i)} \\ &= \mathbf{g}^{(i+1)^{\mathrm{T}}}\mathbf{g}^{(i+1)}\end{aligned}$$

by (4.1.3) and (2.3.3), so (4.1.7) is true with $i + 1$ replacing i. This completes the induction. Termination with $m \leqslant n$ then follows because when $m \geqslant n$, (4.1.6) implies that $\mathbf{g}^{(n+1)} = \mathbf{0}$ (or equivalently from Theorem 2.5.1). □

Other properties of the method as it applies to quadratic functions are given in Theorem 3.4.2. These include a specific value for m and equivalence with quasi-Newton methods. All these properties are illustrated by Table 4.1.1 and Table 3.2.1. Note also that the gradient of the quadratic function (1.2.11) is $\mathbf{g}(\mathbf{x}) = \mathbf{G}\mathbf{x} + \mathbf{b}$, which is also the residual of the linear system $\mathbf{G}\mathbf{x} = -\mathbf{b}$ in which $\mathbf{G}$ is positive definite. It is this problem which is solved by the method of Hestenes and Stiefel (1952), the only difference being that $\alpha^{(i)}$ is calculated directly from (4.1.9) rather than being computed by a line search.

It is now appropriate to consider how the basic method (a line search method with $\mathbf{s}^{(k)}$ given by (4.1.3)) can be applied to the minimization of non-quadratic functions. A number of questions arise which are by no means easily resolved. Immediately there is the question of what sort of line search to use, and how accurate it should be. Since the method is expected to be a descent method the ideas of Sections 2.5 and 2.6 are relevant, and since a fairly accurate search turns out to work well, a search based on satisfying (2.5.1) and (2.5.6) with $\sigma = 0.1$ is recommended.

Another question arises because the iteration no longer terminates for general non-quadratic functions. There is then the possibility of either continuing to use (4.1.3) for all k or of resetting $\mathbf{s}^{(k)}$ periodically to the steepest descent direction. For instance $\mathbf{s}^{(k)}$ could be reset every n iterations by $\mathbf{s}^{(cn+1)} = -\mathbf{g}^{(cn+1)}$ where $c = 1, 2, \ldots$, in which case a cyclic type of method is obtained. An apparent advantage of this latter strategy is that if the iterates progress from a non-quadratic region into a neighbourhood of the solution in which $f(\mathbf{x})$ is closely approximated by a quadratic, then the reset method can be expected to converge rapidly, whereas the non-reset method may not. In fact for some large problems with certain types of symmetry it might be appropriate to reset more frequently than on every n iterations. This question is taken up again at a later stage.

Yet another possibility is that of using different formulae (which would be equivalent for quadratics) to replace (4.1.3) and (4.1.4) in the calculation of $\mathbf{s}^{(k)}$. For instance (4.1.1) can be implemented directly by using a symmetric projection matrix $\mathbf{P}^{(k+1)}$ which annihilates vectors $\gamma^{(1)},\ldots,\gamma^{(k)}$. Initially $\mathbf{P}^{(1)}=\mathbf{I}$ and $\mathbf{P}^{(k)}$ is updated by

$$\mathbf{P}^{(k+1)}=\mathbf{P}^{(k)}-\frac{\mathbf{P}^{(k)}\gamma^{(k)}\gamma^{(k)\mathrm{T}}\mathbf{P}^{(k)}}{\gamma^{(k)\mathrm{T}}\mathbf{P}^{(k)}\gamma^{(k)}}. \tag{4.1.10}$$

Then $\mathbf{s}^{(k)}$ is determined by $\mathbf{s}^{(k)}=-\mathbf{P}^{(k)}\mathbf{g}^{(k)}$ for all $k=1,2,\ldots,n$. Since $\mathbf{P}^{(n+1)}=\mathbf{0}$, $\mathbf{P}^{(cn+1)}$ must be reset to $\mathbf{I}$ for all $c=1,2,\ldots$. However this projection method is not a serious competitor since it loses a main advantage of the Fletcher–Reeves method which is the simplicity of (4.1.3), in which no matrix calculations are required. It does however have the descent property that $\mathbf{s}^{(k)\mathrm{T}}\mathbf{g}^{(k)}\leqslant\mathbf{0}$. It is possible to preserve the descent property and yet use the simple formula (4.1.3) if $\beta^{(k)}$ is evaluated from

$$\beta^{(k)}=-\frac{\mathbf{g}^{(k+1)\mathrm{T}}\mathbf{g}^{(k+1)}}{\mathbf{g}^{(k)\mathrm{T}}\mathbf{s}^{(k)}}. \tag{4.1.11}$$

It then follows from (2.4.8) that a stronger descent property holds, namely that $\mathbf{s}^{(k)\mathrm{T}}\mathbf{g}^{(k)}<0$ if $\mathbf{g}^{(k)}\neq\mathbf{0}$. This might be called the *conjugate descent formula* and is clearly equivalent to (4.1.4) for quadratic functions by virtue of (4.1.7). As it happens recent research has shown that there is no real need to use (4.1.11). It is easy to show that the Fletcher–Reeves formula (4.1.4) always gives descent when the line searches are exact, and this result has been extended to an inexact line search by Al-Baali (1985) under mild conditions (see Question 4.7). In fact there are many other rearrangements of (4.1.4) which are equivalent for quadratic functions (but not in general) and the difficulty is to find good reasons for preferring any one in particular. However the alternative

$$\beta^{(k)}=\frac{(\mathbf{g}^{(k+1)}-\mathbf{g}^{(k)})^{\mathrm{T}}\mathbf{g}^{(k+1)}}{\mathbf{g}^{(k)\mathrm{T}}\mathbf{g}^{(k)}} \tag{4.1.12}$$

due to Polak and Ribiere (Polak, 1971) which is equivalent to (4.1.4) for quadratics by virtue of (4.1.6), does turn out to be important.

To test some of these alternatives, various numerical experiments are given in Table 4.1.2 on the test problems of Section 3.5. The line search parameters are the same as those used in Section 3.5 except that $\tau_2=0.01$ is found to be slightly preferable. The convergence test (2.3.7) with $\varepsilon=10^{-8}$ is used, although it could be argued that an additional test like (2.3.4) on $\|\mathbf{g}\|$ might be desirable. However, a restart along the steepest descent direction is always tried before terminating from test (2.3.7), although my impression is that this restart is not effective. The results are given in Table 4.1.2 and the conventions in use are those described in Section 3.5.

Table 4.1.2 Various conjugate gradient methods

Problem n	Polak–Ribiere no reset $\sigma=0.1$			Fletcher–Reeves no reset $\sigma=0.1$			Conjugate descent no reset $\sigma=0.1$			Polak–Ribiere reset every n $\sigma=0.1$			Polak–Ribiere no reset $\sigma=0.9$		
Rosenbrock															
2	28	75	54	57	116	98a	61	125	106a	32	83	62	100	171	121$\bar{\text{b}}$
Chebyquad															
2	4	12	5	4	12	5	4	12	5	3	10	5	5	13	6
4	7	18	10	20	42	28	18	37	25	7	17	11	15	26	17
6	20	44	28	30	61	54$\bar{\text{c}}$	30	61	54$\bar{\text{c}}$	27	56	42	30	48	38$\bar{\text{a}}$
8	20	48	33	40	81	69$\bar{\text{c}}$	40	84	73$\bar{\text{c}}$	40	83	68a	40	65	43
Trigonometric															
2	5	14	8	9	20	9	9	20	9	7	18	7	11	21	10
4	23	51	34a	40	85	67$\bar{\text{b}}$	40	87	67$\bar{\text{b}}$	40	68	62$\bar{\text{d}}$	40	55	42$\bar{\text{d}}$
6	60	121	100$\bar{\text{a}}$	58	120	86a	56	114	80a	60	104	89$\bar{\text{d}}$	60	84	66$\bar{\text{d}}$
8	40	81	58a	61	125	93a	57	114	86a	80	153	139$\bar{\text{d}}$	80	117	93$\bar{\text{d}}$
10	58	113	86	54	104	75a	98	190	149a	100	175	168$\bar{\text{c}}$	100	130	104$\bar{\text{b}}$
20	156	320	267b	131	259	221b	171	320	278b		not tested			not tested	
30	239	446	410b	149	292	252b	185	359	309b						
40	97	189	158d	91	178	142d	80	160	134d						
50	187	358	326c	242	544	414b	216	427	379b						

It can be seen that the relative performance of the different methods is erratic, and also that in some cases problems are solved to low accuracy, or even not solved at all. The methods in which $\mathbf{s}^{(k)}$ is reset to the steepest descent direction every n iterations happen to give very similar performance, so only the results for the Polak–Ribiere formula are tabulated. Since the methods perform so badly on the trigonometric problems, the higher dimensional problems have not been attempted, and these results generally seem to indicate that resetting is not too favourable a strategy, contrary to what might have been expected. When no resetting takes place the numerical results are inconclusive, although there is perhaps a marginal preference for the Polak–Ribiere formula. All these results use a fairly accurate line search with $\sigma = 0.1$. The effect of a less accurate line search ($\sigma = 0.9$) is illustrated in the last column of Table 4.1.2; it is clear that performance is degraded and the method fails to solve the trigonometric problems in a reasonable time.

When compared with the results for quasi-Newton methods in Section 3.5 it is clear that conjugate gradient methods are both less efficient and less robust, and therefore would not be preferred in normal circumstances. However there is one particular advantage of conjugate gradient methods, namely the particularly simple form of (4.1.3) which requires no matrix operations to form $\mathbf{s}^{(k)}$. In fact the Fletcher–Reeves and conjugate descent methods require only three n-vectors of storage for their implementation, and the Polak–Ribiere formula requires four. Thus conjugate gradient methods may be the only methods which are applicable to *large problems*, that is problems with hundreds or thousands of variables. It is therefore of particular importance to bear this in mind when examining any evidence relating to conjugate gradient methods. For example the results of Table 4.1.2 are of diminished importance since they do not relate to large problems.

In view of Table 4.1.2 it might seem surprising that conjugate gradient methods can solve large problems effectively. Yet I have known problems in atomic structures which typically might have 3000 variables, yet which are solved in about 50 gradient evaluations. Also Reid (1971a) describes linear partial differential equation problems in 4000 variables which are solved in 40 iterations. In view of the quadratic termination requirement of up to n iterations, it is clear that some symmetry must be present which renders the effective value of m to be much smaller than n. The following analysis in the quadratic case provides some explanation. It is possible to write

$$\mathbf{g}^{(k+1)} = P_k(\mathbf{G})\mathbf{g}^{(1)}, \qquad i \leqslant m \tag{4.1.13}$$

where P_k is a polynomial of degree k for which $P_k(0) = 1$. Now if the eigenvalues and vectors of $\mathbf{G}$ are $\{\lambda_i\}$ and $\{\boldsymbol{\xi}_i\}$, respectively, then $\mathbf{g}^{(1)}$ can be expanded amongst these eigenvectors as

$$\mathbf{g}^{(1)} = \sum_{j=1}^{n} \rho_j \boldsymbol{\xi}_j$$

and hence

$$\mathbf{g}^{(k+1)} = \sum_{j=1}^{n} \rho_j P_k(\lambda_j)\xi_j. \tag{4.1.14}$$

Thus $\mathbf{g}^{(m+1)}$ can be close to zero for $m \ll n$ if either

(i) $\mathbf{x}^{(1)}$ is such that $\rho_j \approx 0$ for many values of j, or
(ii) the eigenvalues of $\mathbf{G}$ fall into *multiple* or *close groups* so that a good approximation to zero can be obtained by a low order polynomial $P_m(\lambda)$ over the points $\{\lambda_j\}$.

Since the method uses the 'best' polynomial in a certain sense, such a polynomial can be expected to be chosen if it exists. Thus the possibility arises of effectively converging in $m \ll n$ iterations. Unfortunately it is not usually possible to test these criteria a priori, and the easiest way of finding if they apply is to try out the method and see if it works well. However it does provide some explanation as to why the method gives good results on some large problems.

The above discussion does suggest that on large problems for which the conjugate gradient method is suitable, resetting $\mathbf{s}^{(k)}$ to the steepest descent direction every n iterations is an irrelevant consideration. What might be preferable is to reset every $m \ll n$ iterations where m is somehow determined by the symmetry. Unfortunately it is not obvious how to choose such an m in the algorithm. However Powell (1977b) makes the following observation. Let the algorithm be making little progress, so that $\mathbf{g}^{(k+1)} \approx \mathbf{g}^{(k)}$. Then the Polak–Ribiere formula (4.1.12) leads to $\beta^{(k)} \approx 0$ and hence $\mathbf{s}^{(k+1)} \approx -\mathbf{g}^{(k+1)}$, so that in these circumstances the algorithm tends to reset automatically to the steepest descent direction. For the Fletcher–Reeves method $\beta^{(k)} \approx 1$ so this interpretation does not hold. Powell refers to a practical large-scale example in atomic structures for which the Polak–Ribiere formula does give a staggering improvement in performance. Thus it seems that this formula should be used when solving large problems.

Another idea which is also potentially valuable for accelerating conjugate gradient methods is that of *preconditioning.* The idea originated in partial differential equation research, but is applicable more generally and concerns the possibility of an initial linear transformation of variables. In Section 3.3 it is shown that if factors $\mathbf{G} = \mathbf{L}\mathbf{L}^{\mathrm{T}}$ are available then a quadratic problem can be reduced to one with a unit Hessian matrix, and the conjugate gradient method will solve this in one step. Unfortunately calculating $\mathbf{L}$ from $\mathbf{G}$ when $\mathbf{G}$ is sparse causes substantial fill-in in the sparsity pattern, so this approach is not advantageous. However, it is possible to calculate *incomplete* (i.e. approximate) factors $\bar{\mathbf{L}}$ of $\mathbf{G}$ which have the same sparsity pattern as $\mathbf{G}$. A pre-transformation using $\mathbf{A} = \bar{\mathbf{L}}^{\mathrm{T}}$ (see Section 3.3) then yields a system for which the Hessian is much closer to a unit matrix. The reader is referred to Kershaw (1978) for an example: in particular he shows numerically that the transformation has the property of grouping most of the eigenvalues of the transformed Hessian close to unity, and hence that the conjugate gradient method solves the reduced system very rapidly.

There have been many other researches into conjugate gradient methods. The relationship of some other methods is discussed by Fletcher (1972a). There are also some other theoretical results: for instance reset methods are convergent (because the steepest descent direction is used regularly) and exhibit n-step superlinear convergence

$$\|\mathbf{x}^{(k+n)} - \mathbf{x}^*\| = o(\|\mathbf{x}^{(k)} - \mathbf{x}^*\|), \qquad k = cn + 1, \qquad c = 0, 1, \ldots$$

(McCormick and Pearson, 1969). However these results are not really relevant to the practical solution of large problems. A more interesting possibility due to Beale (1972) is that of restarting along directions other than the steepest descent direction. At the expense of increasing the storage requirement, this has been incorporated into an efficient algorithm by Powell (1977b). However the most interesting of all recent theoretical research results about conjugate gradient methods for minimization is that of Al-Baali (1985) who shows that the non-reset Fletcher–Reeves method with an inexact line search is globally convergent.

4.2 DIRECTION SET METHODS

In this section some other conjugate direction methods are described which require no derivative information to be available. The methods are examples of *direction set methods* in which a set of independent directions $\mathbf{s}^{(1)}, \mathbf{s}^{(2)}, \ldots, \mathbf{s}^{(n)}$ is maintained, and successive line searches are taken along the $\mathbf{s}^{(i)}$ (see (2.3.2)) in a cyclical manner. The alternating variable method and other early *ad hoc* methods are examples of direction set methods. In this section however the additional property holds that the directions are chosen to be conjugate when the method is applied to a quadratic function. It is necessary to have some property of quadratic functions which enables conjugate directions to be determined in the absence of any explicit derivative information. Such a property is the *parallel subspace property* which can be stated as follows.

Theorem 4.2.1

Consider two parallel subspaces S_1 and S_2, generated by independent directions $\mathbf{s}^{(1)}, \mathbf{s}^{(2)}, \ldots, \mathbf{s}^{(k)}$, where $k < n$, from the points $\mathbf{v}^{(1)}$ and $\mathbf{v}^{(2)}$ respectively. (That is $S_1 = \{\mathbf{z} \mid \mathbf{z} = \mathbf{v}^{(1)} + \sum_{i=1}^{k} \alpha_i s^{(i)}\ \forall \alpha_i\}$, S_2 similarly, and $S_1 \neq S_2$.) Denote the points which minimize the quadratic function for $\mathbf{x} \in S_1$ and $\mathbf{x} \in S_2$ by $\mathbf{z}^{(1)}$ and $\mathbf{z}^{(2)}$ respectively. Then $\mathbf{z}^{(2)} - \mathbf{z}^{(1)}$ is conjugate to $\mathbf{s}^{(1)}, \mathbf{s}^{(2)}, \ldots, \mathbf{s}^{(k)}$.

Proof

The situation is illustrated in Figure 4.2.1. It is easy to prove the result because, by

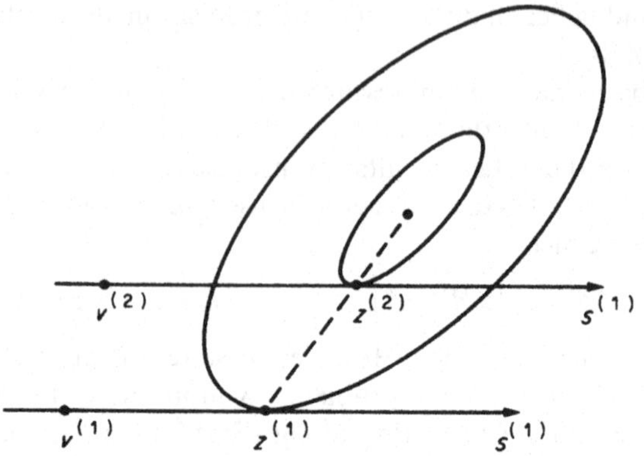

Figure 4.2.1 The parallel subspace property. (Reproduced with permission from Murray (1972). Copyright by Academic Press Inc. (London) Ltd)

definition of a minimum,

$$\mathbf{g}(\mathbf{z}^{(2)})^{\mathrm{T}}\mathbf{s}^{(i)} = \mathbf{g}(\mathbf{z}^{(1)})^{\mathrm{T}}\mathbf{s}^{(i)} = 0, \qquad i = 1, 2, \ldots, k$$

and hence

$$\gamma^{\mathrm{T}}\mathbf{s}^{(i)} = 0$$

where $\gamma = \mathbf{g}(\mathbf{z}^{(2)}) - \mathbf{g}(\mathbf{z}^{(1)})$. By virtue of equation (1.2.15), $\gamma = \mathbf{G}(\mathbf{z}^{(2)} - \mathbf{z}^{(1)})$ from which

$$(\mathbf{z}^{(2)} - \mathbf{z}^{(1)})^{\mathrm{T}}\mathbf{G}\mathbf{s} = 0 \qquad i = 1, 2, \ldots, k$$

follows, proving the theorem. □

A method based on using this property to obtain conjugacy is suggested by Smith (1962). It assumes that a line search subrountine is available which uses only function values, and that, in addition to the initial approximation $\mathbf{x}^{(1)}$, the user can supply independent directions $\mathbf{d}^{(1)}, \mathbf{d}^{(2)}, \ldots, \mathbf{d}^{(n)}$ and constants $\beta_1, \beta_2, \ldots, \beta_n$ ($\beta_i > 0$), chosen so that roughly speaking $\beta_i \mathbf{d}^{(i)}$ represents a moderate change in the variables. The algorithm as it would apply to a quadratic function can be stated as follows.

(i) Define $\mathbf{s}^{(1)} = \mathbf{d}^{(1)}$ and let $\mathbf{x}^{(2)} = \mathbf{x}^{(1)} + \alpha^{(1)}\mathbf{s}^{(1)}$ be found by a line search along $\mathbf{s}^{(1)}$.

(ii) For $i = 2, 3, \ldots, n$ note that $\mathbf{x}^{(i)}$ is the minimum point in the subspace generated by $\mathbf{x}^{(1)}$ and directions $\mathbf{s}^{(1)}, \mathbf{s}^{(2)}, \ldots, \mathbf{s}^{(i-1)}$, and carry out the following operations (a)–(c) for each i in turn.

(a) Displace $\mathbf{x}^{(i)}$ to a point $\mathbf{v} = \mathbf{x}^{(i)} + \beta_i \mathbf{d}^{(i)}$ which can be considered as an arbitrary point in a parallel subspace.

(b) Displace $\mathbf{v}$ by making successive line searches along directions $\mathbf{s}^{(1)}, \mathbf{s}^{(2)}, \ldots, \mathbf{s}^{(i-1)}$ in turn, giving a point $\mathbf{z} = \mathbf{v} + \sum_{j=1}^{i-1} \alpha_j \mathbf{s}^{(j)}$ which is the minimum point in the parallel subspace by virtue of Theorem 2.5.1.

(c) Define $\mathbf{s}^{(i)} = \mathbf{z} - \mathbf{x}^{(i)}$ and calculate $\mathbf{x}^{(i+1)} = \mathbf{x}^{(i)} + \alpha^{(i)}\mathbf{s}^{(i)}$ by a line search along $\mathbf{s}^{(i)}$. Note that $\mathbf{s}^{(i)}$ is a new conjugate direction by virtue of Theorem 4.2.1, and hence that $\mathbf{x}^{(i+1)}$ is the minimum point in the subspace generated by $\mathbf{x}^{(1)}$ and the directions $\mathbf{s}^{(1)}, \mathbf{s}^{(2)}, \ldots, \mathbf{s}^{(i)}$.

When step (ii) above has been completed, $\mathbf{x}^{(n+1)}$ has essentially been found by minimizing from $\mathbf{x}^{(1)}$ along a sequence of conjugate directions $\mathbf{s}^{(1)}, \mathbf{s}^{(2)}, \ldots, \mathbf{s}^{(n)}$, and hence is the required minimum of the quadratic function. Notice that this has been achieved by using $\frac{1}{2}n(n+1)$ line searches.

Unfortunately difficulties arise when applying the method iteratively to general functions of many variables. The obvious approach, as suggested by Smith, is merely to repeat the whole cycle described above in an iterative manner. Practical experience (Fletcher, 1965) suggests that unless n is small ($\leqslant 4$ say) then this method is inferior to other methods, and the situation gets worse as n increases. Various modifications have been tried, for instance replacing step (ii(a)) by a line search along $\mathbf{d}^{(i)}$ so as to remove the need to choose β_i, and also orthogonalizing the $\mathbf{s}^{(i)}$ after each cycle so as to provide new orthogonal directions for the next cycle (Fletcher, 1965), but neither has improved performance.

On examining the detailed progress of Smith's method, it is clear that the poor progress is caused by the fact that the directions of search receive unequal treatment in the number of times which they are used. In particular $\mathbf{s}^{(1)}$ is used n times per cycle, whereas $\mathbf{s}^{(n)}$ is used once. For general functions, because $\mathbf{x}$ is distant from the minimum in the direction $\mathbf{d}^{(n)}$, because the function varies non-quadratically along $\mathbf{d}^{(n)}$, and because a change in the direction $\mathbf{d}^{(n)}$ is not made until the final step (ii) is carried out, many of the searches along $\mathbf{s}^{(1)}$ for instance are of little value. Powell (1964) makes a significant advance in suggesting how searches can be incorporated into Smith's method, so as to treat all directions equally, whilst retaining the property of quadratic termination. In the context of Smith's method as described above, the modification consists of replacing step (ii, (a)), which is an arbitrary displacement, by a sequence of line searches along $\mathbf{d}^{(i)}, \mathbf{d}^{(i+1)}, \ldots, \mathbf{d}^{(n)}$ in turn, all of which, excepting $\mathbf{d}^{(i)}$, would not otherwise be introduced on that particular step (ii). Since the displacement of step (ii, (a)) is arbitrary, it is seen that the quadratic termination property, as proved for Smith's method, is still valid. Furthermore it is possible to rewrite the iteration in a very simple form by coalescing (ii, (a)) and (ii, (b)), with the consequence that step (ii) permits of being continued, for $i > n$, if the function is not quadratic, as follows.

(i) Given independent directions $\mathbf{s}^{(1)}, \mathbf{s}^{(2)}, \ldots, \mathbf{s}^{(n)}$, let $\mathbf{x}^{(1)}$ be the minimum point along $\mathbf{s}^{(n)}$ by using a line search if necessary.

(ii) Repeat (a)–(d) for $i = 1, 2, \ldots, \infty$.

 (a) Note that for $i \leqslant n$ the last i directions $\mathbf{s}^{(n-i+1)}, \ldots, \mathbf{s}^{(n)}$ have been replaced by conjugate directions.

 (b) Find $\mathbf{z} = \mathbf{x}^{(i)} + \sum_{j=1}^{n} \alpha_j \mathbf{s}^{(j)}$ by making optimum line searches along $\mathbf{s}^{(1)}, \mathbf{s}^{(2)}, \ldots, \mathbf{s}^{(n)}$.

(c) For $j = 1, 2, \ldots, n-1$ set $\mathbf{s}^{(j)} = \mathbf{s}^{(j+1)}$.
(d) Define $\mathbf{s}^{(n)} = \mathbf{z} - \mathbf{x}^{(i)}$ as a new conjugate direction and let $\mathbf{x}^{(i+1)} = \mathbf{z} + \alpha\mathbf{s}^{(n)}$ be found by a line search along $\mathbf{s}^{(n)}$.

The iteration terminates in $n^2 + O(n)$ line searches which is about twice as many as for Smith's method. However, the method now not only treats directions equally, but allows an interpretation in terms of 'pseudo-conjugate directions' which are retained for general functions when $i > n$ and are modified from iteration to iteration. Later directions in this sequence correspond to information which has been introduced more recently. Steps (c) and (d) above correspond to rejecting the oldest conjugate direction $\mathbf{s}^{(1)}$ and adding a new one as $\mathbf{s}^{(n)}$.

Unfortunately it is found on some problems that this basic algorithm is unsatisfactory in that the set $\{\mathbf{s}^{(i)}\}$ tends to become linearly dependent. When this occurs certain directions in n-space are no longer represented, and so the minimizer can no longer be reached by the algorithm. However, Powell also proposes an ingenious solution to this problem, which depends on the following theorem and lemma.

Theorem 4.2.2

If $\mathbf{S}$ *is any matrix with columns* $\mathbf{s}^{(1)}, \mathbf{s}^{(2)}, \ldots, \mathbf{s}^{(n)}$, *scaled with respect to a positive definite matrix* $\mathbf{G}$ *so that* $\mathbf{s}^{(i)T}\mathbf{G}\mathbf{s}^{i} = 1$ *for all i, then* $|\det(\mathbf{S})|$ *is maximized if and only if the vectors* $\{\mathbf{s}^{(i)}\}$ *are* **G**-*conjugate.*

Lemma

If $\mathbf{M}$ *is any positive definite symmetric matrix such that* $m_{ii} = 1$ *for all i, then* $\det(\mathbf{M})$ *is maximized uniquely by* $\mathbf{M} = \mathbf{I}$.

Proof

The eigenvalues $\{\lambda_i\}$ of $\mathbf{M}$ satisfy $\lambda_i > 0$, $\sum_i \lambda_i = \mathrm{tr}(\mathbf{M}) = n$, and $\prod_i \lambda_i = \det(\mathbf{M})$. The arithmetic/geometric mean inequality states that

$$\sqrt[n]{\prod_i \lambda_i} \leqslant \sum_i \lambda_i / n$$

with equality iff $\lambda_i = 1$ for all i. Since $\mathbf{M} = \mathbf{I}$ iff $\lambda_i = 1$ for all i, the lemma follows.

Proof of the theorem

Let $\mathbf{S}$ and $\bar{\mathbf{S}}$ be two sets of directions scaled as above, let $\bar{\mathbf{S}}$ have **G**-conjugate columns, and let $\mathbf{S}$ be expressed as $\mathbf{S} = \bar{\mathbf{S}}\mathbf{U}$ for some non-singular $\mathbf{U}$. Since $\bar{\mathbf{S}}$ is conjugate

$$\mathbf{S}^T\mathbf{G}\mathbf{S} = \mathbf{U}^T\bar{\mathbf{S}}^T\mathbf{G}\bar{\mathbf{S}}\mathbf{U} = \mathbf{U}^T\mathbf{U}$$

and because $\mathbf{S}$ is scaled, so $\mathbf{U}^T\mathbf{U}$ satisfies the conditions of the lemma. Hence

$\det(\mathbf{U}^T\mathbf{U}) \leqslant 1$, so that $|\det(\mathbf{U})| \leqslant 1$, with equality iff $\mathbf{U}^T\mathbf{U} = \mathbf{I}$. Equivalently $|\det(\mathbf{S})| \leqslant |\det(\bar{\mathbf{S}})|$ with equality iff the columns of $\mathbf{S}$ are conjugate. □

The usefulness of this theorem lies in the fact that it gives through $|\det(\mathbf{S})|$ some measure of the conjugacy of a set of directions, which does not require explicit evaluation of $\mathbf{G}$. This is because the terms $\mathbf{s}^{(i)T}\mathbf{G}\mathbf{s}^{(i)}$ are available from the line search along $\mathbf{s}^{(i)}$, and so enable the scaling of the directions to be performed readily. Also it is very simple to update an expression for $\det(\mathbf{S})$ every time a change is made like (c) and (d) in the Powell algorithm. Therefore Powell (1964) also proposes a modification of (c) and (d) in which any direction $\mathbf{s}^{(i)}$ can be rejected (rather than just $\mathbf{s}^{(1)}$), where i is chosen to maximize the new $|\det(\mathbf{S})|$. If no i exists for which $|\det(\mathbf{S})|$ increases, then no change is made to the direction set. This strategy also solves the problem of maintaining independence in the direction set.

This then is the method which Powell (1964) proposes, and it proved to be far superior to Smith's method, and also better than the available *ad hoc* methods, and was for some time thought to be the best method for no-derivative problems. However, the method does seem less effective if n is at all large (> 15 say), and this has been ascribed to the fact that no changes to the direction set are being made. Thus other modifications to the basic algorithm have also been tried. One method due to Brent (1973a) carries out a basic cycle of Powell's method, after which (for a quadratic) $\mathbf{S}^T\mathbf{G}\mathbf{S} = \mathbf{D}$ where $\mathbf{D}$ is diagonal. Then an expression for the Hessian matrix is $\mathbf{G} = \mathbf{S}^{-T}\mathbf{D}\mathbf{S}^{-1}$. In the non-quadratic case the method calculates the eigenvectors of $\mathbf{G}$ which are used as the initial direction set to carry out another cycle of the Powell procedure. The eigenvalues of $\mathbf{G}$ provide curvature information in the line search. Because of the orthogonality of the eigenvectors, difficulties due to dependent directions are avoided. A disadvantage however is that housekeeping costs are rather high.

Some other modifications are based on the following theorem due to Powell (1972d).

Theorem 4.2.3

Let $\mathbf{S}$ *be a matrix of scaled directions as in Theorem* 4.2.2, *and let* $\bar{\mathbf{S}}$ *be derived from* $\mathbf{S}$ *by*

$$\bar{\mathbf{S}} = \mathbf{S}\mathbf{Q}\mathbf{D}^{-1/2},$$

where $\mathbf{Q}$ *is an orthogonal matrix, and* $\mathbf{D}$ *is a positive diagonal matrix chosen so that the columns of* $\bar{\mathbf{S}}$ *are scaled (that is* $\bar{\mathbf{s}}^{(i)T}\mathbf{G}\bar{\mathbf{s}}^{(i)} = 1$ *for all* i*). Then* $|\det(\bar{\mathbf{S}})| \geqslant |\det(\mathbf{S})|$.

Proof

Because $\mathbf{Q}$ is orthogonal and $\mathbf{S}$ is scaled

$$\operatorname{tr}(\mathbf{Q}^T\mathbf{S}^T\mathbf{G}\mathbf{S}\mathbf{Q}) = \operatorname{tr}(\mathbf{S}^T\mathbf{G}\mathbf{S}) = n$$

so

$$\sum_i (\mathbf{Q}^T\mathbf{S}^T\mathbf{G}\mathbf{S}\mathbf{Q})_{ii} = n.$$

But since post-multiplication by $\mathbf{D}^{-1/2}$ scales $\mathbf{SQ}$, so $d_{ii} = (\mathbf{Q}^T\mathbf{S}^T\mathbf{G}\mathbf{S}\mathbf{Q})_{ii}$ and hence

$$\det(\mathbf{D}) = \prod_i (\mathbf{Q}^T\mathbf{S}^T\mathbf{G}\mathbf{S}\mathbf{Q})_{ii}.$$

The arithmetic/geometric mean inequality (see Theorem 4.2.2) applied to the d_{ii} gives $\det(\mathbf{D}) \leqslant 1$ and hence $\det(\mathbf{D}^{-1/2}) \geqslant 1$. Thus $|\det(\bar{\mathbf{S}})| = |\det(\mathbf{S})|\,|\det(\mathbf{D}^{-1/2})| \geqslant |\det(\mathbf{S})|$. □

This theorem suggests that algorithms which repeatedly make orthogonal transformations to the direction set (possibly using elementary orthogonal matrices such as plane rotations), followed by scaling, can make $|\det(\mathbf{S})|$ increase, so that the direction set continues to become 'more conjugate'. Such algorithms have been investigated by Rhead (1974) and Brodlie (1975). Also Coope (1976) gives algorithms of this type in which the orthogonal matrices are chosen in such a way as to maintain a termination property. Coope describes numerical experiments which show that one of these algorithms is comparable with those of Brent, Rhead, and Brodlie, and better than that of Powell (1964). However a quasi-Newton method with difference approximations to derivatives (Gill, Murray and Pitfield, 1972) is also shown to perform better still, which supports the current preference for such algorithms. However the difference is not that great (a factor of less than 2) so future developments may change matters.

Yet another similar development due to Powell (1975a) regards the scaled direction set as determining an approximation $\mathbf{B} = (\mathbf{S}\mathbf{S}^T)^{-1}$ to the Hessian matrix $\mathbf{G}$. Then line searches along the directions give new curvature information which is used to update $\mathbf{B}$, choosing any other freedom so as to minimize the norm of the correction (see also Section 3.6). Powell uses the Euclidean norm but Coope (1976) shows that it is preferable to use a weighted norm. This ensures positive definite $\mathbf{B}$ matrices and gives better numerical performance. Coope's results show that this approach is comparable with the better direction set methods referred to above.

QUESTIONS FOR CHAPTER 4

(Other questions about conjugate directions are given at the end of Chapter 2.)

4.1. Minimize the quadratic function in Question 2.7 by the Fletcher–Reeves method starting from $\mathbf{x}^{(1)} = \mathbf{0}$. Verify that conditions (4.1.5), (4.1.6), and (4.1.7) hold, and that the method is equivalent to the BFGS method (see Question 3.9).

4.2. Find an initial point $\mathbf{x}^{(1)}$ in Question 2.7 so that the Fletcher–Reeves method (and hence the steepest descent method) converges in one iteration. Verify that there is only one independent vector in the sequence (3.4.12).

4.3. Consider the quadratic function $\frac{1}{2}\mathbf{x}^T\mathbf{G}\mathbf{x} + \mathbf{b}^T\mathbf{x}$ in four variables where $\mathbf{G}$ is the tridiagonal matrix given in Question 2.18, and $\mathbf{b}$ is the vector $(-1, 0, 2, \sqrt{5})^T$. Apply the conjugate gradient method to this problem from $\mathbf{x}^{(1)} = \mathbf{0}$ and show that it converges in two iterations. Verify that there are just two independent vectors in the sequence (3.4.12).

4.4. Let the Fletcher–Reeves method be applied to a quadratic function for which the Hessian matrix $\mathbf{G}$ is positive definite. Using $\|\cdot\|$ to denote $\|\cdot\|_2$, and writing

$$\mathbf{R}_k = \left[\frac{-\mathbf{g}^{(1)}}{\|\mathbf{g}^{(1)}\|}, \frac{-\mathbf{g}^{(2)}}{\|\mathbf{g}^{(2)}\|}, \ldots, \frac{-\mathbf{g}^{(k)}}{\|\mathbf{g}^{(k)}\|}\right]$$

$$\mathbf{S}_k = \left[\frac{\mathbf{s}^{(1)}}{\|\mathbf{g}^{(1)}\|}, \frac{\mathbf{s}^{(2)}}{\|\mathbf{g}^{(2)}\|}, \ldots, \frac{\mathbf{s}^{(k)}}{\|\mathbf{g}^{(k)}\|}\right]$$

$$\mathbf{B}_k = \begin{bmatrix} 1 & & & & \\ -\beta_1^{1/2} & 1 & & & \\ & -\beta_2^{1/2} & 1 & & \\ & & \ddots & \ddots & \\ & & & -\beta_{k-1}^{1/2} & 1 \end{bmatrix} \quad \mathbf{D}_k = \begin{bmatrix} \alpha_1^{-1} & & & \\ & \alpha_2^{-1} & & \\ & & \ddots & \\ & & & \alpha_k^{-1} \end{bmatrix}$$

establish the equations

$$\mathbf{G}\mathbf{S}_k\mathbf{D}_k^{-1} = \mathbf{R}_k\mathbf{B}_k + \mathbf{g}^{(k+1)}\mathbf{e}_k^T/\|\mathbf{g}^{(k)}\|$$

and

$$\mathbf{S}_k\mathbf{B}_k^T = \mathbf{R}_k$$

where $\mathbf{e}_k$ is the kth coordinate vector. Using the fact that the vectors $\{\mathbf{g}^{(i)}\}$ are mutually orthogonal, show that

$$\mathbf{R}_k^T\mathbf{G}\mathbf{R}_k = \mathbf{T}_k$$

where $\mathbf{T}_k$ is a tridiagonal matrix. Show also that the eigenvalues of $\mathbf{T}_n$ are those of $\mathbf{G}$, assuming that n iterations of the method can be taken.

Establish the bound

$$\|\mathbf{x}^{(k)} - \mathbf{x}^{(n+1)}\| \leqslant \|\mathbf{G}^{-1}\| \, \|\mathbf{g}^{(k)}\|.$$

By also assuming that the extreme eigenvalues of $\mathbf{T}_k$ and $\mathbf{T}_n$ are in good agreement, show that a practical test for stopping the algorithm can be determined when the user requests a tolerance ε on the 2-norm of the error in $\mathbf{x}^{(k)}$. State how the eigenvalues which you require can be calculated, and how much extra computer storage is required to implement the test.

4.5. Consider finding the stationary point $\mathbf{x}^*$ of a given quadratic function $q(\mathbf{x})$, of which the Hessian matrix $\mathbf{G}$ is nonsingular and has only one negative eigenvalue. Let $\mathbf{s}$ be a given direction of negative curvature $\mathbf{s}^T\mathbf{G}\mathbf{s} < 0$. Let $\mathbf{x}^{(1)}$

be a given point, and let $\mathbf{x}^{(2)}$ maximize $q(\mathbf{x}^{(1)}+\alpha\mathbf{s})$ over α. If $\mathbf{Z}$ is a given $n\times(n-1)$ matrix with independent columns, such that $\mathbf{Z}^{\mathrm{T}}\mathbf{G}\mathbf{s}=\mathbf{0}$, write down the set of points X such that $\mathbf{x}-\mathbf{x}^{(2)}$ is $\mathbf{G}$-conjugate to $\mathbf{s}$. It can be shown that the matrix $\mathbf{S}=[\mathbf{s}:\mathbf{Z}]$ is non-singular and by Sylvester's Law that $\mathbf{S}^{\mathrm{T}}\mathbf{G}\mathbf{S}$ has just one negative eigenvalue. Use these results (without proving them) to show that $\mathbf{Z}^{\mathrm{T}}\mathbf{G}\mathbf{Z}$ is positive definite and consequently that $q(\mathbf{x})$ has a unique minimizing point

$$\mathbf{x}^*=\mathbf{x}^{(2)}-\mathbf{Z}(\mathbf{Z}^{\mathrm{T}}\mathbf{G}\mathbf{Z})^{-1}\mathbf{Z}^{\mathrm{T}}\mathbf{g}^{(2)}$$

for $\mathbf{x}\in X$, where $\mathbf{g}^{(2)}=\nabla q(\mathbf{x}^{(2)})$. By showing that $\mathbf{S}^{\mathrm{T}}\mathbf{g}^*=\mathbf{0}$, verify that $\mathbf{x}^*$ is also the saddle point of $q(\mathbf{x})$ in $\mathbb{R}^n$. Show that a suitable $\mathbf{Z}$ matrix can be obtained from an elementary Householder orthogonal matrix $\mathbf{Q}=\mathbf{I}-2\mathbf{w}\mathbf{w}^{\mathrm{T}}$, where $\mathbf{w}$ is a unit vector such that $\mathbf{Q}\gamma=\pm\|\gamma\|_2\mathbf{e}_1$, where $\gamma=\mathbf{G}\mathbf{s}$ and $\mathbf{e}_1$ is the first column of $\mathbf{I}$. Hence suggest a Newton algorithm for finding the saddle point of a general function $f(\mathbf{x})$ given that $\mathbf{G}^*$ is non-singular with only one negative eigenvalue, and that a negative curvature direction $\mathbf{s}$ is available. To what extent is the method assured of convergence?

Briefly discuss the possibilities for quasi-Newton and conjugate gradient methods for solving this problem (see Sinclair and Fletcher, 1974).

4.6. Apply Smith's method to minimize the quadratic function in Question 2.8 from $\mathbf{x}^{(1)}=\mathbf{0}$, choosing $\mathbf{d}^{(i)}=\mathbf{e}_i$ and $\beta_i=1$. Verify that three line searches are required to solve the problem.

4.7. The Fletcher–Reeves conjugate gradient method may be used with an inexact line search which satisfies the condition

$$|\mathbf{g}^{(k+1)\mathrm{T}}\mathbf{s}^{(k)}|\leqslant-\sigma\mathbf{g}^{(k)\mathrm{T}}\mathbf{s}^{(k)}\quad(\sigma\in(0,\tfrac{1}{2}]).$$

Assuming that $\mathbf{s}^{(1)}=-\mathbf{g}^{(1)}$, show by induction that

$$-\sum_{j=0}^{k-1}\sigma^j\leqslant\frac{\mathbf{g}^{(k)\mathrm{T}}\mathbf{s}^{(k)}}{\|\mathbf{g}^{(k)}\|_2^2}\leqslant-2+\sum_{j=0}^{k-1}\sigma^j,$$

and that the method has a descent property for all k. (Use the form of the method in which $\mathbf{s}^{(k)}$ is *not* reset to $-\mathbf{g}^{(k)}$ for any $k>1$.)

Chapter 5

Restricted Step Methods

5.1 A PROTOTYPE ALGORITHM

One motivation for the ideas of this chapter is to circumvent the difficulty caused by non-positive definite Hessian matrices in Newton's method. In this case the underlying quadratic function $q^{(k)}(\boldsymbol{\delta})$, obtained by truncating the Taylor series for $f(\mathbf{x}^{(k)}+\boldsymbol{\delta})$ in (3.1.1), does not have a unique minimizer and the method is not defined. Another way of regarding this fact is that the region about $\mathbf{x}^{(k)}$ in which the Taylor series is adequate, does not include a minimizing point of $q^{(k)}(\boldsymbol{\delta})$. A more realistic approach therefore is to assume that some neighbourhood $\Omega^{(k)}$ of $\mathbf{x}^{(k)}$ is defined in which $q^{(k)}(\boldsymbol{\delta})$ agrees with $f(\mathbf{x}^{(k)}+\boldsymbol{\delta})$ in some sense. Then it would be appropriate to choose $\mathbf{x}^{(k+1)}=\mathbf{x}^{(k)}+\boldsymbol{\delta}^{(k)}$, where the correction $\boldsymbol{\delta}^{(k)}$ minimizes $q^{(k)}(\boldsymbol{\delta})$ for all $\mathbf{x}^{(k)}+\boldsymbol{\delta}$ in $\Omega^{(k)}$. Such a method may be called a *restricted step method*: the step is restricted by the region of validity of the Taylor series. It will be shown that these methods retain the rapid rate of convergence of Newton's method, but are also generally applicable and globally convergent. The term *trust region* is sometimes used to refer to the neighbourhood $\Omega^{(k)}$.

Various problems of implementation arise. Clearly it is not likely that $\Omega^{(k)}$ can be defined in too general a manner. The information to do this is not readily available and it would be difficult to organize, in particular because the subproblem would be intractable. Thus it is convenient to consider the case in which

$$\Omega^{(k)}=\{\mathbf{x}:\|\mathbf{x}-\mathbf{x}^{(k)}\|\leqslant h^{(k)}\} \tag{5.1.1}$$

and to seek the solution $\boldsymbol{\delta}^{(k)}$ of the resulting subproblem

$$\underset{\boldsymbol{\delta}}{\text{minimize}}\; q^{(k)}(\boldsymbol{\delta}) \quad \text{subject to} \quad \|\boldsymbol{\delta}\|\leqslant h^{(k)} \tag{5.1.2}$$

which can readily be solved for certain types of norm. Immediately there is the problem of how the radius $h^{(k)}$ of the neighbourhood (5.1.1) shall be chosen.

To prevent undue restriction of the step, $h^{(k)}$ should be as large as possible subject to a certain measure of agreement existing between $q^{(k)}(\boldsymbol{\delta}^{(k)})$ and $f(\mathbf{x}^{(k)}+\boldsymbol{\delta}^{(k)})$. This can be quantified by defining the *actual reduction* in f on the kth step as

$$\Delta f^{(k)} = f^{(k)} - f(\mathbf{x}^{(k)} + \boldsymbol{\delta}^{(k)}) \tag{5.1.3}$$

and the corresponding *predicted reduction* as

$$\Delta q^{(k)} = q^{(k)}(\mathbf{0}) - q^{(k)}(\boldsymbol{\delta}^{(k)}) = f^{(k)} - q^{(k)}(\boldsymbol{\delta}^{(k)}). \tag{5.1.4}$$

Then the ratio

$$r^{(k)} = \Delta f^{(k)}/\Delta q^{(k)} \tag{5.1.5}$$

measures the accuracy to which $q^{(k)}(\boldsymbol{\delta}^{(k)})$ approximates $f(\mathbf{x}^{(k)}+\boldsymbol{\delta}^{(k)})$, in the sense that the closer $r^{(k)}$ is to unity, the better is the agreement. A prototype algorithm can be stated which changes $h^{(k)}$ adaptively and attempts to maintain a certain degree of agreement as measured by $r^{(k)}$, whilst keeping $h^{(k)}$ as large as possible. The kth iteration takes the following form:

(i) given $\mathbf{x}^{(k)}$ and $h^{(k)}$, calculate $\mathbf{g}^{(k)}$ and $\mathbf{G}^{(k)}$;
(ii) solve (5.1.2) for $\boldsymbol{\delta}^{(k)}$;
(iii) evaluate $f(\mathbf{x}^{(k)}+\boldsymbol{\delta}^{(k)})$ and hence $r^{(k)}$; (5.1.6)
(iv) if $r^{(k)} < 0.25$ set $h^{(k+1)} = \|\boldsymbol{\delta}^{(k)}\|/4$,
if $r^{(k)} > 0.75$ and $\|\boldsymbol{\delta}^{(k)}\| = h^{(k)}$ set $h^{(k+1)} = 2h^{(k)}$,
otherwise set $h^{(k+1)} = h^{(k)}$;
(v) if $r^{(k)} \leqslant 0$ set $\mathbf{x}^{(k+1)} = \mathbf{x}^{(k)}$ else $\mathbf{x}^{(k+1)} = \mathbf{x}^{(k)} + \boldsymbol{\delta}^{(k)}$.

The constants 0.25, 0.75, etc. are arbitrary and the algorithm is quite insensitive to their change. In practice a more sophisticated iteration has been used in which if $r^{(k)} < 0.25$, then $h^{(k+1)}$ is chosen in an interval $(0.1, 0.5)\|\boldsymbol{\delta}^{(k)}\|$ on the basis of a polynomial interpolation. Other possible changes could include a more sophisticated extrapolation strategy or the use of a line search, although I would expect any improvement to be marginal. However none of these changes affects the validity of the proofs which follow.

A particular advantage of restricted step methods is in the very strong global convergence proofs which hold, with no significant restriction on the class of problem to which they apply.

Theorem 5.1.1 (Global convergence)

For algorithm (5.1.6), *if* $\mathbf{x}^{(k)} \in B \subset \mathbb{R}^n \forall k$, *where* B *is bounded, and if* $f \in \mathbb{C}^2$ *on* B, *then* $\exists$ *an accumulation point* $\mathbf{x}^{\infty}$ *which satisfies the first and second order necessary conditions* (2.1.3) *and* (2.1.4).

Proof

By considering whether or not $\inf h^{(k)} = 0$ in the algorithm, $\exists$ a convergent subsequence $\mathbf{x}^{(k)} \to \mathbf{x}^{\infty}$ $k \in \mathscr{S}$ for which either

(i) $r^{(k)} < 0.25$, $h^{(k+1)} \to 0$ and hence $\|\boldsymbol{\delta}^{(k)}\| \to 0$, or
(ii) $r^{(k)} \geqslant 0.25$ and $\inf h^{(k)} > 0$.

In either case the necessary conditions are shown to hold. In case (i) let $\exists$ a descent direction $\mathbf{s}$ ($\|\mathbf{s}\| = 1$) at $\mathbf{x}^{\infty}$, so that

$$\mathbf{s}^{\mathrm{T}}\mathbf{g}^{\infty} = -d \qquad d > 0. \tag{5.1.7}$$

A Taylor series for $f(\mathbf{x})$ about $\mathbf{x}^{(k)}$ gives

$$f(\mathbf{x}^{(k)} + \boldsymbol{\delta}^{(k)}) = q^{(k)}(\boldsymbol{\delta}^{(k)}) + o(\|\boldsymbol{\delta}^{(k)}\|^2)$$

and hence

$$\Delta f^{(k)} = \Delta q^{(k)} + o(\|\boldsymbol{\delta}^{(k)}\|^2). \tag{5.1.8}$$

For $k \in \mathscr{S}$ consider a step of length $\varepsilon^{(k)} = \|\boldsymbol{\delta}^{(k)}\|$ along $\mathbf{s}$. Optimality of $\boldsymbol{\delta}^{(k)}$ in (5.1.2) and continuity yield

$$\begin{aligned}\Delta q^{(k)} \geqslant q^{(k)}(\mathbf{0}) - q^{(k)}(\varepsilon^{(k)}\mathbf{s}) &= -\varepsilon^{(k)}\mathbf{s}^{\mathrm{T}}\mathbf{g}^{(k)} + o(\varepsilon^{(k)}) \\ &= \varepsilon^{(k)}d + o(\varepsilon^{(k)})\end{aligned}$$

from (5.1.7). It follows from $\varepsilon^{(k)} \to 0$ and (5.1.8) that $r^{(k)} = 1 + o(1)$ which contradicts $r^{(k)} < 0.25$. Hence (5.1.7) is contradicted and so $\mathbf{g}^{\infty} = \mathbf{0}$.

Now let $\exists$ a second order descent direction $\mathbf{s}$ ($\|\mathbf{s}\| = 1$) at $\mathbf{x}^{\infty}$, so that

$$\mathbf{s}^{\mathrm{T}}\mathbf{G}^{\infty}\mathbf{s} = -d \qquad d > 0. \tag{5.1.9}$$

For $k \in \mathscr{S}$ consider a step of length $\varepsilon^{(k)}$ along $\sigma\mathbf{s}$, $\sigma = \pm 1$, choosing σ so that $\sigma\mathbf{s}^{\mathrm{T}}\mathbf{g}^{(k)} \leqslant 0$. Then by (5.1.2) and continuity

$$\Delta q^{(k)} \geqslant q^{(k)}(\mathbf{0}) - q^{(k)}(\varepsilon^{(k)}\sigma\mathbf{s}) \geqslant -\tfrac{1}{2}\varepsilon^{(k)2}\mathbf{s}^{\mathrm{T}}\mathbf{G}^{(k)}\mathbf{s} = \tfrac{1}{2}\varepsilon^{(k)2}d + o(\varepsilon^{(k)2}).$$

Again it follows from (5.1.8) that $r^{(k)} = 1 + o(1)$, which contradicts $r^{(k)} < 0.25$. Thus (5.1.9) is contradicted showing that $\mathbf{G}^{\infty}$ is positive semi-definite. Thus both first and second order necessary conditions are shown to hold when $\inf h^{(k)} = 0$.

For a subsequence $\mathscr{S}$ arising from case (ii), $f^{(1)} - f^{\infty} \geqslant \sum_{k \in \mathscr{S}} \Delta f^{(k)}$, and it therefore follows from $r^{(k)} \geqslant 0.25$ that $\Delta q^{(k)} \to 0$. Define $q^{\infty}(\boldsymbol{\delta}) = f^{\infty} + \boldsymbol{\delta}^{\mathrm{T}}\mathbf{g}^{\infty} + \frac{1}{2}\boldsymbol{\delta}^{\mathrm{T}}\mathbf{G}^{\infty}\boldsymbol{\delta}$. Let $\bar{h}$ satisfy $0 < \bar{h} < \inf h^{(k)}$ and let $\bar{\boldsymbol{\delta}}$ minimize $q^{\infty}(\boldsymbol{\delta})$ on $\|\boldsymbol{\delta}\| \leqslant \bar{h}$. Define $\bar{\mathbf{x}} = \mathbf{x}^{\infty} + \bar{\boldsymbol{\delta}}$. Observe that for sufficiently large k,

$$\|\bar{\mathbf{x}} - \mathbf{x}^{(k)}\| \leqslant \|\bar{\boldsymbol{\delta}}\| + \|\mathbf{x}^{(k)} - \mathbf{x}^{\infty}\| = \|\bar{\boldsymbol{\delta}}\| + o(1) \leqslant \bar{h} + o(1) \leqslant h^{(k)}$$

so that $\bar{\mathbf{x}} - \mathbf{x}^{(k)}$ is feasible in the subproblem. Thus

$$q^{(k)}(\bar{\mathbf{x}} - \mathbf{x}^{(k)}) \geqslant q^{(k)}(\boldsymbol{\delta}^{(k)}) = f^{(k)} - \Delta q^{(k)}.$$

In the limit $f^{(k)} \to f^{\infty}$, $\mathbf{g}^{(k)} \to \mathbf{g}^{\infty}$, $\mathbf{G}^{(k)} \to \mathbf{G}^{\infty}$, $\Delta q^{(k)} \to 0$ and $\bar{\mathbf{x}} - \mathbf{x}^{(k)} \to \bar{\boldsymbol{\delta}}$ so it follows that $q^{\infty}(\bar{\boldsymbol{\delta}}) \geqslant f^{\infty} = q^{\infty}(\mathbf{0})$. Thus $\boldsymbol{\delta} = \mathbf{0}$ also minimizes $q^{\infty}(\boldsymbol{\delta})$ on $\|\boldsymbol{\delta}\| \leqslant \bar{h}$, and since the latter constraint is not active, the first and second order necessary conditions $\mathbf{g}^{\infty} = \mathbf{0}$ and $\mathbf{G}^{\infty}$ is positive semi-definite are implied. $\square$

Notice that the existence of a bounded region B which the theorem requires is implied if any level set $\{\mathbf{x}:f(\mathbf{x})\leqslant f^{(k)}\}$ is bounded. Also the theorem assumes that the sequence $\{\mathbf{x}^{(k)}\}$ is infinite: if for some k it happens that $\Delta q^{(k)}=0$, then the iteration terminates, and first and second order conditions are satisfied at $\mathbf{x}^{(k)}$. Similar results can be obtained if the Hessian $\mathbf{G}^{(k)}$ is replaced by an approximate Hessian $\mathbf{B}^{(k)}$. If $\mathbf{B}^{(k)}$ is bounded then the same proof is valid, although weaker conditions on $\mathbf{B}^{(k)}$ can be used, along the lines of Fletcher (1972d) and Powell (1975b). With an additional stronger assumption the following result is valid.

Theorem 5.1.2

If the accumulation point $\mathbf{x}^{\infty}$ of Theorem 5.1.1 also satisfies the (second order sufficient) conditions for Theorem 3.1.1, then for the main sequence $r^{(k)}\to 1$, $\mathbf{x}^{(k)}\to\mathbf{x}^{\infty}$, $\inf h^{(k)}>0$, and the bound $\|\boldsymbol{\delta}\|_2\leqslant h^{(k)}$ is inactive for sufficiently large k. Also the convergence is second order.

Proof

Consider $k\in\mathscr{S}$ where $\mathscr{S}$ is the subsequence that exists in Theorem 5.1.1. Since $\mathbf{G}^{\infty}$ is positive definite, the vector $\mathbf{s}^{(k)}=-\mathbf{G}^{(k)-1}\mathbf{g}^{(k)}$ exists for k sufficiently large. Consider choosing α so that the step $\alpha\mathbf{s}^{(k)}$ minimizes $q^{(k)}(\alpha\mathbf{s}^{(k)})$ subject to $\|\alpha\mathbf{s}^{(k)}\|\leqslant h^{(k)}$. If $\|\mathbf{s}^{(k)}\|\leqslant h^{(k)}$ then $\alpha=1$ and $\boldsymbol{\delta}^{(k)}=\mathbf{s}^{(k)}$ is the step that solves (5.1.2). In this case

$$\Delta q^{(k)}=\tfrac{1}{2}\mathbf{s}^{(k)\mathrm{T}}\mathbf{G}^{(k)}\mathbf{s}^{(k)}\geqslant\tfrac{1}{2}\mu^{(k)}\|\boldsymbol{\delta}^{(k)}\|_2^2\geqslant\tfrac{1}{2}\mu^{(k)}c^2\|\boldsymbol{\delta}^{(k)}\|^2$$

where $\mu^{(k)}>0$ is the least eigenvalue of $\mathbf{G}^{(k)}$ and $c>0$ is a constant that depends on the norm. If $\|\mathbf{s}^{(k)}\|>h^{(k)}$ then $\alpha=h^{(k)}/\|\mathbf{s}^{(k)}\|<1$. For any quadratic it follows that

$$q(\alpha)=q(0)+\tfrac{1}{2}\alpha(q'(0)+q'(\alpha))\leqslant q(0)+\tfrac{1}{2}\alpha q'(0)$$

when $\alpha q'(\alpha)\leqslant 0$. Hence

$$\begin{aligned}\Delta q^{(k)}&\geqslant\tfrac{1}{2}\alpha^2\mathbf{s}^{(k)\mathrm{T}}\mathbf{G}^{(k)}\mathbf{s}^{(k)}\geqslant\tfrac{1}{2}\mu^{(k)}h^{(k)2}\|\mathbf{s}^{(k)}\|_2^2/\|\mathbf{s}^{(k)}\|^2\geqslant\tfrac{1}{2}\mu^{(k)}c^2h^{(k)2}\\&\geqslant\tfrac{1}{2}\mu^{(k)}c^2\|\boldsymbol{\delta}^{(k)}\|^2,\end{aligned}$$

so this result holds whatever the length of $\mathbf{s}^{(k)}$. It follows from (5.1.8) that $r^{(k)}\to 1$ $k\in\mathscr{S}$. Thus $\mathbf{x}^{\infty}$ cannot arise from case (i) but only from case (ii). Then if k is sufficiently large that $\|\mathbf{x}^{(k)}-\mathbf{x}^{\infty}\|\leqslant\tfrac{1}{2}\inf h^{(k)}$ and $\mathbf{x}^{(k)}$ is in the neighbourhood of $\mathbf{x}^{\infty}$ for which Theorem 3.1.1 holds, then $\mathbf{x}^{(k+1)}=\mathbf{x}^{(k)}-\mathbf{G}^{(k)-1}\mathbf{g}^{(k)}$ satisfies $\|\mathbf{x}^{(k+1)}-\mathbf{x}^{\infty}\|<\|\mathbf{x}^{(k)}-\mathbf{x}^{\infty}\|$ and is feasible in $\Omega^{(k)}$. Thus in the main sequence $\mathbf{x}^{(k)}\to\mathbf{x}^{\infty}$ and for this sequence both $r^{(k)}\to 1$ (as above) and hence $\inf h^{(k)}>0$. Second order convergence follows from Theorem 3.1.1. □

Excluding pathological cases when no level set is bounded, Theorem 5.1.1 shows the very strong result that there always exists a subsequence which converges to a point which satisfies both first and second order necessary conditions. The gap between these and the sufficient conditions of Theorem 3.1.1 is small, and if the latter hold then the main sequence is convergent and the restricted step method is seen to behave in a very desirable way. The heuristic of restricting the step by $\|\boldsymbol{\delta}\| \leqslant h^{(k)}$ ensures that significant reductions in $f(\mathbf{x})$ are made until $\mathbf{x}^{(k)}$ is close to a local solution. At this stage the restriction becomes inactive and the iteration then reduces to the basic Newton method with its rapid rate of convergence.

On the other hand it is wise not to become too enthusiastic about these theorems, even though restricted step methods usually do work well in practice. There are some difficulties, one of which is that slow progress can arise if the scaling of the norm is not chosen to reflect that of the problem. Ideally one would like to select a 'natural metric' norm (see Section 3.3 and Fletcher, 1978), although in practice the lesser aim of scaling the variables well is likely to be adequate. Another potential difficulty is that in certain cases it may not be convenient to calculate the global solutions of subproblem (5.1.2) which the convergence proof for algorithm (5.1.6) requires. However, I know of no practical examples in which the calculation of local solutions to (5.1.2) has led to non-convergence.

Most practical implementations of restricted step methods have used the L_2 norm in (5.1.2), and this very important case is treated in detail in the next section. In fact however it is perhaps even more simple to use the L_∞ norm (for example, Fletcher, 1972b). Methods of this kind are sometimes referred to as *hypercube* or *boxstep* methods. These methods can easily be modified to allow an implicit pre-scaling of variables. The subproblem (5.1.2) is a quadratic program with just simple bounds, and good algorithms exist for finding a local solution (see for example Gill and Murray, 1978a or Section 10.5 of this book). Unfortunately a global solution is more difficult to guarantee, since although this is a finite calculation it may be very expensive. For example if $\mathbf{G}^{(k)}$ is negative definite, and $\mathbf{g}^{(k)}$ is small, then all the vertices of the hypercube $\|\boldsymbol{\delta}\|_\infty \leqslant h^{(k)}$ are local solutions, and the problem of finding the best one is a combinatorial one. It should however be adequate in practice to calculate local solutions, if it is ensured that the local solution satisfies $\Delta q^{(k)} > 0$ (e.g. by using *pseudoconstraints*: see Section 8.4). Further discussion of the nature of the hypercube trajectory, along with some illustrative numerical examples, is given in Questions 5.2–5.4 at the end of this chapter.

So far it has been assumed that the Hessian matrix $\mathbf{G}^{(k)}$ is available to the algorithm. Some attempts have been made to derive restricted step types of method which avoid the direct evaluation of second derivatives. One obvious possibility when first derivatives are available is to set up finite difference approximations to $\mathbf{G}^{(k)}$ (see Section 3.1), and then proceed along similar lines. However, the disadvantage is that a lot of gradient calls are required, and the methods are unlikely to be as efficient as quasi-Newton methods. Although the restricted step method would have a guarantee of convergence, whereas this question is open for quasi-Newton methods, there is currently no practical evidence that quasi-Newton methods are likely to fail. The idea of updating $\mathbf{G}^{(k)}$ in conjunction with a restricted step method can be pursued, but is less attractive because the solution of (5.1.2) for $\boldsymbol{\delta}^{(k)}$ requires $\sim n^3$ housekeeping operations, whereas $\mathbf{s}^{(k)}$ in the quasi-Newton method is calculated in $\sim n^2$ operations. It is possible with hypercube methods to consider updating a reduced inverse Hessian approximation so as to avoid the n^3 calculation. This idea is investigated by Fletcher (1972b) but gains in efficiency are no better than marginal. An alternative idea, which does enable updating formula to be used, is to approximate the trajectory $\boldsymbol{\delta}(h)$ in such a way as to avoid the use of n^3 housekeeping operations. A suitable way of doing this is the Powell dog-leg trajectory described at the end of Section 5.2. This can be useful in some circumstances, but for general minimization using a quasi-Newton method there is no evidence that such a modification is needed.

Finally when no derivatives are available, finite difference approximations to both $\mathbf{G}^{(k)}$ and $\mathbf{g}^{(k)}$ can be made, in conjunction with the use of a restricted step method. This is very expensive in function values, however. An alternative possibility is to attempt to re-use function values and to estimate $\mathbf{G}^{(k)}$ say from the most recent $\frac{1}{2}n(n+1)$ values. Winfield (1973) describes an approach of this type which can be successful, but the housekeeping costs are high (to determine $\mathbf{G}^{(k)}$ requires $\sim n^6$ operations), so it is only practicable for problems with small n for which function calculations are very expensive. In general I would expect a no-derivative quasi-Newton method (see the end of Section 3.2) to be more efficient in most cases.

5.2 LEVENBERG–MARQUARDT METHODS

The most important type of restricted step method occurs when (5.1.2) is defined in terms of the L_2 norm. It will be shown that these methods are characterized by solving a system

$$(\mathbf{G}^{(k)} + \nu\mathbf{I})\boldsymbol{\delta}^{(k)} = -\mathbf{g}^{(k)}, \quad \nu \geqslant 0 \tag{5.2.1}$$

at some stage, in order to determine the correction $\boldsymbol{\delta}^{(k)}$. Such methods were first suggested by Levenberg (1944) and by Marquardt (1963) in the context of nonlinear least squares (see Chapter 6), when $\mathbf{G}^{(k)}$ is approximated using (6.1.8). The application to general minimization is given by Goldfeld, Quandt and Trotter (1966). This section will discuss the general case, which is more difficult because of the possibility that $\mathbf{G}^{(k)}$ can have negative eigenvalues. However, the same development applies equally to the least squares case.

The solution of (5.1.2) using the L_2 norm is therefore considered in more detail. This problem can be expressed as

$$\begin{aligned} &\underset{\boldsymbol{\delta}}{\text{minimize}} \;\; q^{(k)}(\boldsymbol{\delta}) \triangleq \tfrac{1}{2}\boldsymbol{\delta}^{\mathrm{T}}\mathbf{G}^{(k)}\boldsymbol{\delta} + \mathbf{g}^{(k)\mathrm{T}}\boldsymbol{\delta} \\ &\text{subject to } \boldsymbol{\delta}^{\mathrm{T}}\boldsymbol{\delta} \leqslant h^{(k)^2} \end{aligned} \tag{5.2.2}$$

and it is assumed that $h^{(k)} > 0$. A very comprehensive result can be given, following the previous edition and Moré and Sorensen (1982), regarding optimality conditions for (5.2.2).

Theorem 5.2.1

The correction $\boldsymbol{\delta}^{(k)}$ *is a global solution of* (5.2.2) *if and only if there exists* $\nu \geqslant 0$ *such that* (5.2.1) *holds,*

$$\nu(h^{(k)2} - \boldsymbol{\delta}^{(k)\mathrm{T}}\boldsymbol{\delta}^{(k)}) = 0 \tag{5.2.3}$$

and $\mathbf{G}^{(k)} + \nu\mathbf{I}$ *is positive semi-definite. In addition if* $\mathbf{G}^{(k)} + \nu\mathbf{I}$ *is positive definite then* $\boldsymbol{\delta}^{(k)}$ *is the unique solution of* (5.2.2.).

Proof

Let $\boldsymbol{\delta}^{(k)}$ and ν satisfy the conditions of the theorem. From (5.2.1) $\mathbf{g}^{(k)}$ is in the range of $\mathbf{G}^{(k)} + \nu\mathbf{I}$ and since this matrix is positive semi-definite, $\boldsymbol{\delta}^{(k)}$ therefore minimizes the quadratic function

$$\hat{q}(\boldsymbol{\delta}) = \tfrac{1}{2}\boldsymbol{\delta}^{\mathrm{T}}(\mathbf{G}^{(k)} + \nu\mathbf{I})\boldsymbol{\delta} + \mathbf{g}^{(k)\mathrm{T}}\boldsymbol{\delta}.$$

Thus $\hat{q}(\boldsymbol{\delta}) \geqslant \hat{q}(\boldsymbol{\delta}^{(k)})$ for all $\boldsymbol{\delta}$ which implies that

$$q^{(k)}(\boldsymbol{\delta}) \geqslant q^{(k)}(\boldsymbol{\delta}^{(k)}) + \tfrac{1}{2}\nu(\boldsymbol{\delta}^{(k)\mathrm{T}}\boldsymbol{\delta}^{(k)} - \boldsymbol{\delta}^{\mathrm{T}}\boldsymbol{\delta}).$$

It follows from $\nu \geqslant 0$ and (5.2.3) that if $\boldsymbol{\delta}^{\mathrm{T}}\boldsymbol{\delta} \leqslant h^{(k)2}$ then $q^{(k)}(\boldsymbol{\delta}) \geqslant q^{(k)}(\boldsymbol{\delta}^{(k)})$, so that $\boldsymbol{\delta}^{(k)}$ solves (5.2.2). Moreover if $\mathbf{G}^{(k)} + \nu\mathbf{I}$ is positive definite and $\boldsymbol{\delta} \neq \boldsymbol{\delta}^{(k)}$ then $\hat{q}(\boldsymbol{\delta}) > \hat{q}(\boldsymbol{\delta}^{(k)})$ and it follows similarly that $\boldsymbol{\delta}^{(k)}$ is a unique solution to (5.2.2). Conversely, let $\boldsymbol{\delta}^{(k)}$ solve (5.2.2). If $\boldsymbol{\delta}^{(k)\mathrm{T}}\boldsymbol{\delta}^{(k)} < h^{(k)2}$ then $\boldsymbol{\delta}^{(k)}$ is an unconstrained minimizer and the conditions of the theorem hold with $\nu = 0$. Otherwise the constraint can be assumed to be active. In this case, at $\boldsymbol{\delta}^{(k)}$ the normal vector for the constraint is the vector $2\boldsymbol{\delta}^{(k)}$ which is linearly independent since $h^{(k)} > 0$. This implies that Lemma 9.2.2 holds and hence the optimality conditions of Theorems 9.1.1 and 9.3.1 are valid. Thus there exists a Lagrange multiplier $\nu \geqslant 0$ such that $\nabla_{\delta}\mathscr{L}(\boldsymbol{\delta}^{(k)}, \nu) = \mathbf{0}$ where

$$\mathscr{L}(\boldsymbol{\delta}, \nu) = q^{(k)}(\boldsymbol{\delta}) + \tfrac{1}{2}\nu(\boldsymbol{\delta}^{\mathrm{T}}\boldsymbol{\delta} - h^{(k)2}) \tag{5.2.4}$$

is a suitable Lagrangian function, and this condition is just (5.2.1). The complementarity condition of Theorem 9.1.1 gives (5.2.3). For second-order conditions, Theorem 9.3.1 gives that

$$\mathbf{s}^{\mathrm{T}}(\mathbf{G}^{(k)} + \nu\mathbf{I})\mathbf{s} \geqslant 0 \quad \forall \mathbf{s}: \mathbf{s}^{\mathrm{T}}\boldsymbol{\delta}^{(k)} = 0. \tag{5.2.5}$$

Furthermore, consider any vector $\mathbf{v}$ such that $\mathbf{v}^{\mathrm{T}}\boldsymbol{\delta}^{(k)} \neq 0$. It is then possible to construct a feasible point $\boldsymbol{\delta}' = \boldsymbol{\delta}^{(k)} + \theta\mathbf{v}$, $\theta \neq 0$, on the constraint boundary, as in Figure 5.2.1. Using (5.2.1) to eliminate the term $\mathbf{g}^{(k)\mathrm{T}}\boldsymbol{\delta}$ from (5.2.4) gives

$$\mathscr{L}(\boldsymbol{\delta}', \nu) = q^{(k)}(\boldsymbol{\delta}^{(k)}) + \tfrac{1}{2}(\boldsymbol{\delta}' - \boldsymbol{\delta}^{(k)})^{\mathrm{T}}(\mathbf{G}^{(k)} + \nu\mathbf{I})(\boldsymbol{\delta}' - \boldsymbol{\delta}^{(k)}).$$

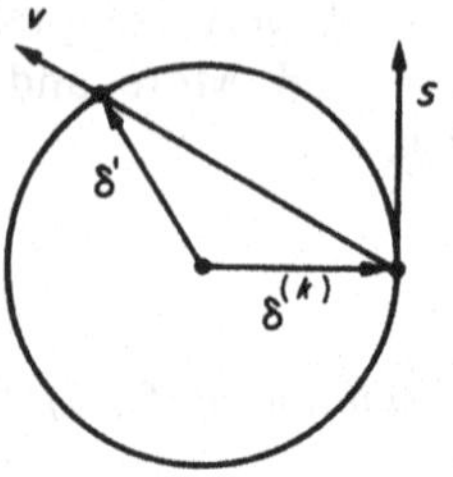

Figure 5.2.1 Construction of the vector $\boldsymbol{\delta}'$

Now $\boldsymbol{\delta}'^{\mathrm{T}}\boldsymbol{\delta}' = h^{(k)2}$ implies that $\mathscr{L}(\boldsymbol{\delta}', \nu) = q^{(k)}(\boldsymbol{\delta}') \geqslant q^{(k)}(\boldsymbol{\delta}^{(k)})$ so it follows that $\mathbf{v}^{\mathrm{T}}(\mathbf{G}^{(k)} + \nu\mathbf{I})\mathbf{v} \geqslant 0$. This together with (5.2.5) shows that $\mathbf{G}^{(k)} + \nu\mathbf{I}$ is positive semi-definite and completes the result. □

Corollary

The same results apply to the equality problem

$$\text{minimize } q^{(k)}(\boldsymbol{\delta}) \quad \text{subject to } \boldsymbol{\delta}^{\mathrm{T}}\boldsymbol{\delta} = h^{(k)2} \tag{5.2.6}$$

in place of (5.2.2) *if the conditions* $\nu \geqslant 0$ *and* (5.2.3) *are not required to hold.*

Proof

The same proof applies with obvious minor changes. □

Most of the content of this theorem can be obtained directly from the optimality conditions of Chapter 9, but not all. Important extra features are that Theorem 5.2.1 is concerned with *global* solutions, and that the conditions are both necessary *and* sufficient, so there is not the gap that exists say between Theorems 9.3.1 and 9.3.2.

All Levenberg–Marquardt algorithms therefore find a value $\nu \geqslant 0$ such that $\mathbf{G}^{(k)} + \nu\mathbf{I}$ is positive definite and solve (5.2.1) to determine $\boldsymbol{\delta}^{(k)}$. There are many variations on this theme and in some (earlier) methods the precise restriction $\|\boldsymbol{\delta}\|_2 \leqslant h^{(k)}$ on the length of $\boldsymbol{\delta}$ is not imposed. Instead ν is used as the controlling parameter in the iteration and the length of $\boldsymbol{\delta}^{(k)}$ is determined by whatever value ν happens to take. (Conversely, in the prototype algorithm, $h^{(k)}$ is given and one particular ν must be determined so that $\|\boldsymbol{\delta}\|_2 \leqslant h^{(k)}$ is satisfied.) Since it happens that increases in ν cause $\|\boldsymbol{\delta}\|_2$ to decrease, and vice versa (see below), it is possible to have a similar algorithm to (5.1.6) except that changes to $h^{(k)}$ in step (iv) are replaced by inverse changes to $\nu^{(k)}$. The kth iteration can thus be written

(i) given $\mathbf{x}^{(k)}$ and $\nu^{(k)}$, calculate $\mathbf{g}^{(k)}$ and $\mathbf{G}^{(k)}$;
(ii) factorize $\mathbf{G}^{(k)} + \nu^{(k)}\mathbf{I}$: if not positive definite, reset $\nu^{(k)} = 4\nu^{(k)}$ and repeat;
(iii) solve (5.2.1) to give $\boldsymbol{\delta}^{(k)}$;

(iv) evaluate $f(\mathbf{x}^{(k)} + \boldsymbol{\delta}^{(k)})$ and hence $r^{(k)}$;
(v) if $r^{(k)} < 0.25$ set $\nu^{(k+1)} = 4\nu^{(k)}$ (5.2.7)
if $r^{(k)} > 0.75$ set $\nu^{(k+1)} = \nu^{(k)}/2$
otherwise set $\nu^{(k+1)} = \nu^{(k)}$;
(vi) if $r^{(k)} \leqslant 0$ set $\mathbf{x}^{(k+1)} = \mathbf{x}^{(k)}$ else $\mathbf{x}^{(k+1)} = \mathbf{x}^{(k)} + \boldsymbol{\delta}^{(k)}$.

Initially $\nu^{(1)} > 0$ is chosen arbitrarily. This algorithm is in the same spirit as that which Marquardt (1963) proposed, although there are some differences. The parameters 0.25, 0.75, etc. are arbitrary and the algorithm is again not sensitive to their change. More sophisticated strategies have been proposed for increasing or reducing $\nu^{(k)}$ in step (v), and it can be advantageous to allow $\nu^{(k)} = 0$ to occur (see Fletcher, 1971a, for example). An alternative type of algorithm is one in which each iteration makes a search along the trajectory $\boldsymbol{\delta}(\nu)$ for $\nu \geqslant 0$ (the *Levenberg–Marquardt trajectory*), to optimize $f(\mathbf{x} + \boldsymbol{\delta}(\nu))$ (Levenberg, 1944).

However algorithm (5.2.7) does have disadvantages, not only when $\nu^{(k)}$ is close to zero, but also close to certain degenerate situations (see case (iii) below), in which $\boldsymbol{\delta}(\nu)$ is badly determined by ν. Therefore more recent Levenberg–Marquardt type methods (Hebden, 1973; Moré, 1978) follow the model algorithm (5.1.6) much more closely, and control the iteration using the radius $h^{(k)}$. In this case equation (5.2.1) is regarded as defining a trajectory $\boldsymbol{\delta}(\nu)$, and the precise value of ν which makes $\|\boldsymbol{\delta}(\nu)\|_2 = h^{(k)}$ is sought. In fact it is possible to state the global solution to (5.2.2) in a quite general way, although there may still be some computational difficulties. To show this, it is necessary to analyse the Levenberg–Marquardt trajectory $\boldsymbol{\delta}(\nu)$. Let the eigenvalues of $\mathbf{G}^{(k)}$ be $\lambda_1 \geqslant \lambda_2 \geqslant \cdots \geqslant \lambda_n$, with orthogonal eigenvectors $\mathbf{v}_1, \mathbf{v}_2, \ldots, \mathbf{v}_n$ ($\|\mathbf{v}_i\|_2 = 1$), and let $\mathbf{g}^{(k)} = \sum_i \alpha_i \mathbf{v}_i$, where $\alpha_i = \mathbf{v}_i^{\mathrm{T}} \mathbf{g}^{(k)}$. Then (5.2.1) gives

$$\boldsymbol{\delta}(\nu) = \textstyle\sum_i \beta_i \mathbf{v}_i \tag{5.2.8}$$

where

$$\beta_i = -\alpha_i/(\lambda_i + \nu). \tag{5.2.9}$$

Assuming that $\alpha_i \neq 0$ for some $i \in I \triangleq \{i : \lambda_i = \lambda_n\}$, then $\|\boldsymbol{\delta}(\nu)\|_2 = (\sum_i \beta_i)^{1/2}$ increases monotonically from 0 to ∞ as ν decreases from ∞ to $-\lambda_n$. Therefore it is possible to find a unique value of ν in this interval for which $\|\boldsymbol{\delta}(\nu)\|_2 = h^{(k)}$. The vector $\boldsymbol{\delta}^{(k)} = \boldsymbol{\delta}(\nu)$ satisfies (5.2.3), and also (5.2.1) by virtue of (5.2.8). Also $\nu \geqslant -\lambda_n$ shows that $\mathbf{G}^{(k)} + \nu\mathbf{I}$ is positive semi-definite so $\boldsymbol{\delta}(\nu)$ satisfies the conditions of Theorem 5.2.1 and its corollary. Thus $\boldsymbol{\delta}(\nu)$ is a solution of the equality problem (5.2.6), and also the solution of (5.2.2) if $\nu \geqslant 0$. Values of $\nu < -\lambda_n$ need not be examined because this implies that $\mathbf{G}^{(k)} + \nu\mathbf{I}$ is not positive semi-definite so that the solutions to (5.2.1) are not global solutions to (5.2.2) or (5.2.6) (but they could be local solutions: see Question 5.7). However, if $\nu < 0$ or $\alpha_i = 0\ \forall i \in I$ then other features come into play. In general there are three cases.

(i) $\lambda_n > 0$. Values of $\boldsymbol{\delta}(\nu)$ as defined by (5.2.8) for $\nu \in [-\lambda_n, \infty)$ correspond to solutions of (5.2.6), and for $\nu \geqslant 0$ these are also solutions of (5.2.2). Now $\boldsymbol{\delta}(0)$ is just the Newton step $-\mathbf{G}^{(k)^{-1}}\mathbf{g}^{(k)}$ and this exists because

$\lambda_n > 0$ implies that $\mathbf{G}^{(k)}$ is positive definite. Therefore for $\nu < 0$, $\|\boldsymbol{\delta}(\nu)\|_2 \geqslant \|\boldsymbol{\delta}(0)\|_2$, and because the Newton step is the unconstrained minimizer of $q^{(k)}(\boldsymbol{\delta})$ it follows that $\boldsymbol{\delta}(0)$ solves (5.2.2) and $h^{(k)} \geqslant \|\boldsymbol{\delta}(0)\|_2$.

(ii) $\lambda_n \leqslant 0$ and $\alpha_i \neq 0$ for some $i \in I$. For $\mathbf{G}^{(k)} + \nu\mathbf{I}$ to be positive semi-definite requires $\nu \geqslant -\lambda_n \geqslant 0$ which therefore implies that the constraint in (5.2.2) is active. Thus $\boldsymbol{\delta}(\nu)$ as defined by (5.2.8) solves both (5.2.2) and (5.2.6).

(iii) $\lambda_n \leqslant 0$ and $\alpha_i = 0$ $\forall i \in I$. As ν decreases from ∞ to $-\lambda_n$ then $\|\boldsymbol{\delta}(\nu)\|_2$ increases from 0 to $\bar{h}$, say, where $\bar{h} = \|\bar{\boldsymbol{\delta}}\|_2$, and $\bar{\boldsymbol{\delta}} = \sum_{i \leqslant m} \beta_i \mathbf{v}_i$ where $m \notin I$ is the greatest index for which $\alpha_i \neq 0$. Thus for $\nu \in (-\lambda_n, \infty)$ the vector $\boldsymbol{\delta}(\nu)$ defined by (5.2.8) solves both (5.2.2) and (5.2.6) and $h^{(k)} < \bar{h}$. For $h^{(k)} \geqslant \bar{h}$ the solution of (5.2.2) or (5.2.6) must have $\nu = -\lambda_n$ for $\mathbf{G}^{(k)} + \nu\mathbf{I}$ to be positive semi-definite. In this case $\mathbf{G}^{(k)} + \nu\mathbf{I}$ is singular and a solution is readily seen to be any vector of the form $\boldsymbol{\delta} = \bar{\boldsymbol{\delta}} + \sum_{i \in I} \mu_i \mathbf{v}_i$ which satisfies the constraint of (5.2.2) or (5.2.6), as appropriate.

The three cases are illustrated in Figure 5.2.2. Numerical examples of the three cases are discussed in Questions 5.6–5.9 at the end of the chapter. Another example of case (iii) arises when $\mathbf{x}^{(k)}$ is a saddle-point, and this corresponds to $\bar{h} = 0$ and $\bar{\boldsymbol{\delta}} = \mathbf{0}$. Then $\nu = -\lambda_n$ and $\boldsymbol{\delta}^{(k)}$ is a multiple of $\mathbf{v}_n$ of length $h^{(k)}$.

Excluding case (i) when $\nu = 0$ or case (iii) when $h^{(k)} \geqslant \bar{h}$, then to calculate the global solution of (5.2.2) it is necessary to find the value of ν for which $\mathbf{G}^{(k)} + \nu\mathbf{I}$ is positive definite and which solves the nonlinear equation $\|\boldsymbol{\delta}(\nu)\|_2 = h^{(k)}$, where $\boldsymbol{\delta} = \boldsymbol{\delta}(\nu)$ is defined by solving the system (5.2.1). Since a nonlinear equation cannot in general be solved exactly, it is convenient to accept any solution for which $|\|\boldsymbol{\delta}\|_2 - h^{(k)}| \leqslant \varepsilon h^{(k)}$, where ε is some tolerance, typically $\varepsilon = 0.1$. It is straightforward to show that Theorems 5.1.1 and 5.1.2 still apply. Case (i) with $\nu = 0$ is no problem since the Newton step $\boldsymbol{\delta}(0)$ is the solution to (5.2.2), but case (iii) with $h^{(k)} \geqslant \bar{h}$ is more difficult. Immediately there is the problem of recognizing numerically that $\alpha_i = 0$ for all $i \in I$. Solving the above nonlinear equation is not possible since solutions for all $h^{(k)} \geqslant \bar{h}$ correspond to $\nu = -\lambda_n$. Use of $\boldsymbol{\delta} = \bar{\boldsymbol{\delta}} + \sum_{i \in I} \mu_i \mathbf{v}_i$ requires an eigenvector calculation which is undesirable. One possibility is just to perturb $\mathbf{g}^{(k)}$ arbitrarily so that $\alpha_i \neq 0$ for some $i \in I$ so that $\nu > -\lambda_n$ can be used. However there are also difficulties when the α_i, $i \in I$, are small, due to round-off errors being magnified by ill-conditioning. A modification of the prototype algorithm (5.1.6) which allows approximate solutions of (5.2.2) to be calculated and avoids the difficulties with case (iii) situations for which $h^{(k)} \geqslant \bar{h}$, has been given by Moré and Sorensen (1982).

The solution of the nonlinear equation $\boldsymbol{\delta}^{\mathrm{T}}\boldsymbol{\delta} = h^{(k)^2}$ where $\boldsymbol{\delta} = \boldsymbol{\delta}(\nu)$ can be carried out with any standard method, but Hebden (1973) states a very efficient method using the explicit structure of (5.2.9) (see also Question 5.15). The method gives an iterative sequence $\nu^{(1)}, \nu^{(2)}, \ldots$ and at a general value $\nu^{(r)}$, the function $\|\boldsymbol{\delta}(\nu)\|_2$ is modelled by the function $\Delta(\nu) = \beta/(\lambda + \nu)$ by analogy with (5.2.9). The parameters β and λ are chosen so that $\Delta(\nu^{(r)}) = \|\boldsymbol{\delta}(\nu^{(r)})\|_2$ and $\Delta'(\nu^{(r)}) = \|\boldsymbol{\delta}(\nu^{(r)})\|_2'$, where a prime indicates a derivative with respect to ν. This gives $\lambda = -\nu^{(r)} - \|\boldsymbol{\delta}(\nu^{(r)})\|_2 / \|\boldsymbol{\delta}(\nu^{(r)})\|_2'$ and $\beta = (\lambda + \nu^{(r)})\|\boldsymbol{\delta}(\nu^{(r)})\|_2$. The next iterate is

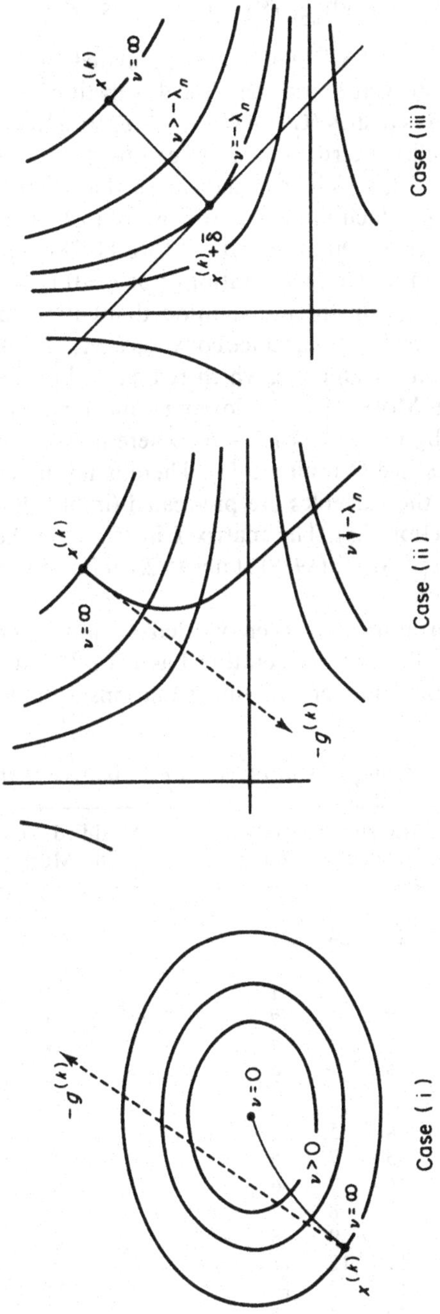

Figure 5.2.2 Levenberg–Marquardt trajectories for (5.2.2)

chosen so that $\Delta(v^{(r+1)}) = h^{(k)}$, which gives the iteration formula

$$v^{(r+1)} = v^{(r)} + (1 - \| \boldsymbol{\delta}(v^{(r)}) \|_2 / h^{(k)}) \| \boldsymbol{\delta}(v^{(r)}) \|_2 / \| \boldsymbol{\delta}(v^{(r)}) \|_2'. \tag{5.2.10}$$

For any $v^{(r)}$, $\mathbf{LL}^{\mathrm{T}}$ factors of $\mathbf{G}^{(k)} + v^{(r)}\mathbf{I}$ are calculated, so that (5.2.1) can be solved for $\boldsymbol{\delta}(v^{(r)})$ by forward and backward substitution. Also $\| \boldsymbol{\delta}(v) \|_2' = \boldsymbol{\delta}^{\mathrm{T}}\boldsymbol{\delta}' / \| \boldsymbol{\delta}(v) \|_2$, where $\boldsymbol{\delta}'$ satisfies $(\mathbf{G}^{(k)} + v\mathbf{I})\boldsymbol{\delta}' = -\boldsymbol{\delta}$, so $\boldsymbol{\delta}^{\mathrm{T}}\boldsymbol{\delta}' = -\mathbf{w}^{\mathrm{T}}\mathbf{w}$, where $\mathbf{w}$ solves $\mathbf{L}^{\mathrm{T}}\mathbf{w} = \boldsymbol{\delta}$. Thus all the terms in (5.2.10) can be determined. In some cases $\mathbf{LL}^{\mathrm{T}}$ factors may not exist, which is an indication that $v^{(r)}$ must be increased.

A numerical example which illustrates the use of (5.2.10) and the rapid rate of convergence is given in Question 5.13. Moré (1978) reports that (5.2.10) converges on average in under two iterations (for $\varepsilon = 0.1$) and this fits in with my own experience, except in the ill-conditioned situations arising when case (iii) above is almost satisfied. To guarantee convergence, (5.2.10) must be adjoined to some simple sectioning technique, which reduces a bracket on the solution uniformly to zero (see Moré, 1978). A lower bound for this is given by $v = 0$ and an upper bound by $v = \| \mathbf{g}^{(k)} \|_2 / h^{(k)} - u_n$, where $u_n \leqslant \lambda_n$ can be obtained by using Gershgorin discs (see Question 5.11). There is no difficulty in modifying the algorithm so that the variables are pre-scaled (implicitly), and this can be advantageous (see Section 3.3). The matrix $\mathbf{I}$ in (5.2.1) is then replaced by a suitable diagonal matrix. Moré (1978) states (5.2.10) in a form suitable for this modification.

A FORTRAN subroutine has been written to implement the prototype algorithm (5.1.6), and the Hebden iteration based on (5.2.10) is used to solve the subproblem (5.2.2). However case (iii) situations with $h^{(k)} > \bar{h}$ have been

Table 5.2.1 Comparison of second derivative methods

Problem n	Restricted step method of Hebden–Moré type			Modified Newton method of Gill, Murray, and Picken		
Rosenbrock						
2	21	24	21	24	24	23
Chebyqyad						
2	4	5	4	5	5	5
4	7	8	7	9	9	7
6	7	9	7	14	14	12
8	9	10	9	22	22	17
Trigonometric						
2	4	5	4	4	4	4
4	5	7	5	11	11	10
6	6	7	6	8	8	8
8	7	8	7	8	8	8
10	8	9	8	9	9	9
20	11	12	11	38	38	38
30	11	12	11	23	23	22
40	12	14	12	59	59	59
50	12	13	12	50	50	50

accounted for in a crude and rather unsatisfactory way, albeit one which would only fail outright if $\mathbf{g}^{(k)} = \mathbf{0}$. (It is interesting to note that case (iii) situations do arise in practice: for instance in the Chebyquad $n = 2$ problem – see Question 5.9.) The factorization used to solve (5.2.1) is an $\mathbf{LL}^{\mathrm{T}}$ factorization, although it can be more efficient to use a $\mathbf{QTQ}^{\mathrm{T}}$ factorization ($\mathbf{Q}$ orthogonal, $\mathbf{T}$ tridiagonal) so that (5.2.1) can be solved repeatedly for many values of v without re-computing the factorization. The results are tabulated in Table 5.2.1 for the standard set of test problems alongside those for a modified Newton method from the NAG subroutine library (Gill, Murray and Picken, 1972). The latter method computes a factorization of a matrix $\mathbf{G}^{(k)} + \mathbf{D}$, where $\mathbf{D}$ is diagonal and $d_{ii} > 0$ is chosen if the factorization would otherwise break down. An advantage of this approach is that the factorization can be computed in one pass through the matrix. Unfortunately the results for this method seem to be somewhat inferior to those for the restricted step method, especially for large problems. It seems appropriate to conclude therefore that the use of an (essentially) arbitrary modification matrix $\mathbf{D}$ slows down the practical rate of convergence, when compared with the systematic choice $v^{(k)}\mathbf{I}$ which is made in restricted step method.

Finally, another idea related indirectly to the use of (5.2.1) is worth mentioning. In case (i) situations $\boldsymbol{\delta}(0)$ is the Newton step and as $v \to \infty$ so $\boldsymbol{\delta}(v) \to -\mathbf{g}^{(k)}/v$, an incremental steepest descent step (see Figure 5.2.2(i)). In the general case $\boldsymbol{\delta}(v)$ can be interpreted as interpolating these extreme cases. It is therefore possible to approximate $\boldsymbol{\delta}(v)$ by a *dog-leg trajectory* (Powell, 1970b) of the type illustrated in Figure 5.2.3. If $\mathbf{B}^{(k)}$ approximates $\mathbf{G}^{(k)}$, then the trajectory is made up of two line segments, one joining $\boldsymbol{\delta} = \mathbf{0}$ to the estimated optimum correction $\boldsymbol{\delta}_s$ along the steepest descent direction, and the other joining $\boldsymbol{\delta}_s$ to the Newton-like correction $\boldsymbol{\delta}_{\mathrm{N}}$. ($\boldsymbol{\delta}_s$ and $\boldsymbol{\delta}_{\mathrm{N}}$ are defined by $\boldsymbol{\delta}_s = -\mathbf{g}^{(k)}\|\mathbf{g}^{(k)}\|_2^2/\mathbf{g}^{(k)\mathrm{T}}\mathbf{B}^{(k)}\mathbf{g}^{(k)}$ and $\boldsymbol{\delta}_{\mathrm{N}} = -\mathbf{B}^{(k)^{-1}}\mathbf{g}^{(k)}$, respectively.) The resulting trajectory is used in an algorithm similar to (5.1.6), but the computation of $\boldsymbol{\delta}(v)$ from (5.2.1) is replaced by using the point $\boldsymbol{\delta}$ of length $h^{(k)}$ on the dog-leg trajectory. If $h^{(k)} \geqslant \|\boldsymbol{\delta}_{\mathrm{N}}\|_2$ then $\boldsymbol{\delta}_{\mathrm{N}}$ is chosen (as in case (i)). Use of this trajectory enables convergence to a stationary point to be established. It is applicable to any Newton-like method (see Section 6.3) for which non-convergence is otherwise possible and for which a simple computational scheme, avoiding the solution of (5.2.1), is required.

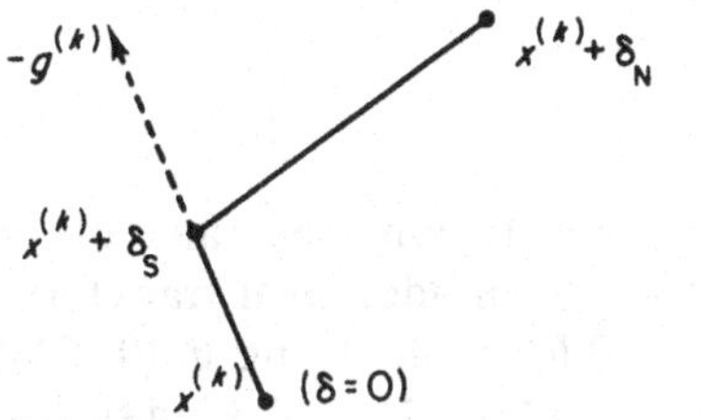

Figure 5.2.3 Powell's dog-leg trajectory

QUESTIONS FOR CHAPTER 5

5.1. Show that if algorithm (5.1.6) is applied to a positive definite quadratic function, then $r^{(k)} = 1$ and $h^{(k+1)} = 2h^{(k)}$ for all k, and the algorithm terminates at the solution in a finite number of steps.

5.2. The hypercube trajectory is the set of solutions $\boldsymbol{\delta}(h)$ to (5.1.2) for $h > 0$ when the L_∞ norm is used. By considering what happens when a fixed set of bounds $\delta_i \leqslant h$ or $-h \leqslant -\delta_i$ are active, show that the trajectory is composed of straight line segments. If $g_i^{(k)} \neq 0$ for all i, show that for small h the trajectory is along the line $\mathbf{x}^{(k)} + \alpha\mathbf{s}$, where $s_i = -\operatorname{sign} g_i^{(k)}$.

5.3. Find the hypercube trajectory for the function $f(\mathbf{x}) = \frac{1}{2}(x_2^2 - x_1^2)$ at $\mathbf{x}^{(1)} = (\frac{1}{2}, -1)^{\mathrm{T}}$. Show that the trajectory has two parts: the line segment joining $\mathbf{x}^{(1)}$ to the point $(1\frac{1}{2}, 0)^{\mathrm{T}}$ for $h \leqslant 1$, and the half line $(1\frac{1}{2}, 0)^{\mathrm{T}} + (h-1)\mathbf{e}_1$ for $h \geqslant 1$. Also show that for $h > \frac{1}{2}$ there is an additional trajectory of local minima which is piecewise linear.

5.4. Examine the hypercube trajectories for the function $f(\mathbf{x}) = x_1 x_2$ at $\mathbf{x}^{(1)} = (\frac{1}{2}, 1)^{\mathrm{T}}$. Show that the trajectory of global solutions is discontinuous at $h = \frac{1}{2}$.

5.5. For the function in Question 2.2 at $\mathbf{x}^{(1)} = \mathbf{0}$, find the range of v values for which $\mathbf{G}^{(1)} + v\mathbf{I}$ is positive definite. Evaluate the resulting correction in (5.2.1) for $v = 1$ and verify that $f(\mathbf{x})$ is reduced. Verify that the range of values of v for which $f(\mathbf{x})$ is reduced is $v \geqslant 0.9$, and that the optimum reduction occurs at $v = 1.2$ (approximately). Find the corresponding value of $r^{(1)}$ in equation (5.1.5). (See also Question 3.3.)

5.6. Find the Levenberg–Marquardt trajectory for the quadratic function defined in Question 2.8 at $\mathbf{x}^{(1)} = \mathbf{0}$. Show that a case (i) situation exists and that the trajectory can be expressed as

$$\boldsymbol{\delta}(v) = (-v, 2+v)^{\mathrm{T}}/(\tfrac{1}{2}v^2 + 3v + 2) \tag{a}$$

for v decreasing in $(\infty, 0)$. Sketch the trajectory and verify that the initial direction is the steepest descent direction and that $\|\boldsymbol{\delta}\|_2$ is a monotonically increasing function of v. For what values of h is the constraint $\|\boldsymbol{\delta}\|_2 \leqslant h$ inactive? How can solutions of (a) be interpreted for $v \in (\sqrt{5}-3, 0)$, $v \in (-3-\sqrt{5}, \sqrt{5}-3)$, and $v < -3-\sqrt{5}$, respectively?

5.7. Find the Levenberg–Marquardt trajectory for $f(\mathbf{x}) = x_1 x_2$ at $\mathbf{x}^{(1)} = (1, \frac{1}{2})^{\mathrm{T}}$. Show that a case (ii) situation exists and that the trajectory can be expressed as

$$\boldsymbol{\delta}(v) = \tfrac{1}{2}(2-v, 1-2v)^{\mathrm{T}}/(v^2 - 1) \tag{b}$$

for v decreasing in $(\infty, 1)$. Find the value of v for which $h = \|\boldsymbol{\delta}\|_2 = \frac{1}{2}$. Show that there exists an additional trajectory of local but not global minima given by (b) for v increasing in $(0.350667, 1)$ where 0.350667 is the only root of $5v^3 - 12v^2 + 15v - 4$. Find such a local solution for which $h = 1$. What is the largest value of h below which no such local solutions exist? How can solutions of (b) be interpreted for $\nu \in (0, 0.350667)$, $\nu \in (-1, 0)$, and $\nu \in (-\infty, -1)$, respectively?

5.8. Find the Levenberg–Marquardt trajectory for $f(\mathbf{x}) = x_1 x_2$ at $\mathbf{x}^{(1)} = (1, 1)^T$. Show that a case (iii) situation exists with $\|\bar{\boldsymbol{\delta}}\|_2 = \sqrt{2}/2$ and express $\boldsymbol{\delta}$ in terms of ν for $h \leqslant \sqrt{2}/2$. For $h > \sqrt{2}/2$ show that the trajectory is expressed by $\boldsymbol{\delta} = \bar{\boldsymbol{\delta}} \pm (2h^2 - 1)(\frac{1}{2}, -\frac{1}{2})^T$.

5.9. The Chebyquad $n = 2$ problem is to minimize $f(\mathbf{x}) = (x_1 + x_2 - 1)^2 + ((2x_1 - 1)^2 + (2x_2 - 1)^2 - \frac{2}{3})^2$ from $\mathbf{x}^{(1)} = (\frac{1}{3}, \frac{2}{3})^T$. Show that the Levenberg–Marquardt trajectory derived from a quadratic model exhibits a case (iii) situation with $\bar{\boldsymbol{\delta}} = (8, -8)^T/21$ and a critical value of $\nu = -\lambda_2 = 28/9$.

5.10. Investigate the conjecture that if the trajectory of global solutions $\boldsymbol{\delta}(h)$ to (5.1.2) in any fixed norm is uniquely defined for all $h > 0$, then the trajectory is a continuous function of h.

5.11. Show using the triangle inequality that if the solution of (5.2.1) has $\|\boldsymbol{\delta}\|_2 = h^{(k)}$, and if $u_n \leqslant \lambda_n \leqslant 0$ where λ_n is the smallest eigenvalue of $\mathbf{G}^{(k)}$, then $\nu \leqslant \|\mathbf{g}^{(k)}\|_2 / h^{(k)} - u_n$.

5.12. Show that if $\alpha_i \neq 0$ for one index i only, then the Hebden iteration (5.2.10) terminates.

5.13. Apply the Hebden iteration (5.2.10) to the problem described in Question 5.6 above with $h = \frac{1}{2}$ and $\nu^{(1)} = 3$. Verify that the method converges rapidly with $\nu^{(2)} = 1.8908$, $\nu^{(3)} = 1.9967$, and $\|\boldsymbol{\delta}^{(k)}\|_2 = 0.3187$, 0.5405, 0.5010 for $k = 1, 2, 3$, respectively.

5.14. By considering the angle between the two line segments of the Powell dog-leg trajectory, show using Cauchy's inequality that if $\mathbf{B}^{(k)}$ is positive definite then $\|\boldsymbol{\delta}\|_2$ is monotonically increasing along the entire trajectory, and hence that the value of $\|\boldsymbol{\delta}\|_2 = h^{(k)} \leqslant \|\boldsymbol{\delta}_N\|_2$ is well defined.

5.15. Show that the Hebden iteration (5.2.10) can be derived by writing down the Newton–Raphson method for ν to solve the equation

$$\frac{1}{\|\boldsymbol{\delta}(\nu)\|_2} = \frac{1}{h^{(k)}}.$$

This result is due to Reinsch (see Moré and Sorensen, 1982).

Chapter 6

Sums of Squares and Nonlinear Equations

6.1 OVER-DETERMINED SYSTEMS

This chapter is devoted to problems in which the objective function is a sum of m squared terms

$$f(\mathbf{x}) = \sum_{i=1}^{m} [r_i(\mathbf{x})]^2 = \mathbf{r}^T\mathbf{r} \tag{6.1.1}$$

where $\mathbf{r} = \mathbf{r}(\mathbf{x})$, because special methods exist which use this structure advantageously. This is the so-called *nonlinear least squares problem.* An alternative way of viewing such problems is that they arise from an attempt to solve the system of m equations

$$r_i(\mathbf{x}) = 0, \qquad i = 1, 2, \ldots, m. \tag{6.1.2}$$

In this case the functions $r_i(\mathbf{x})$ can be interpreted as the *residuals* of the equations. When $m > n$ the system is referred to as being *over-determined*, and it is usually not possible to obtain an exact solution. Therefore one possibility is to seek a *least squares solution* to (6.1.2), which is obtained by minimizing the function $f(\mathbf{x})$ in (6.1.1). Problems in which the functions $r_i(\mathbf{x})$ are all linear are relatively straightforward, so this section concentrates on the more difficult case of finding least squares solutions to *nonlinear* over-determined systems of equations. One area in which such problems arise is in *data fitting*, in which m is often much larger than n. A different situation arises when $m = n$, in which case the system of equations is said to be *well-determined*, and an exact solution of (6.1.2) can be expected. Special purpose methods have been devised for such problems and these are considered in the next section. However, it is also often convenient to attack these problems by attempting to minimize the sum of squares function (6.1.1). The basic methods for both types of problem require derivative information about the components $r_i(\mathbf{x})$, and in some applications this may not be convenient to obtain. Thus no-derivative methods also form an important area of study, which is taken up in Section 6.3.

One last general point concerns the choice of the least squares function (6.1.1) as the criterion for a best solution. In data fitting problems this has some statistical relevance, which is described below, but on some occasions it can be advantageous to choose different criteria. One approach is to find solutions which minimize the function

$$f(\mathbf{x}) = \|\mathbf{r}(\mathbf{x})\|, \tag{6.1.3}$$

in particular using either the L_1 or L_∞ norms. Unfortunately the resulting functions are *non-smooth* and not within the scope of methods that have been considered in Part 1. Nonetheless such functions do have valuable properties. For instance the L_1 and L_∞ norms have a statistical interpretation, although perhaps not as relevant as that arising from (6.1.1). Also the L_1 norm has an advantage in being able to disregard the effects of bias caused by *wild points* in data fitting problems. There are also certain advantages in using the L_1 norm to solve well-determined problems, and this is described in more detail in Sections 12.3 and 14.3 as a special case of nonlinear constrained optimization.

Perhaps the most frequently solved of all optimization problems are *data fitting problems*. Typically m data values $d_1, d_2, \ldots, d_m$ are given, which have been sampled for values $t_1, t_2, \ldots, t_m$ of some independent variable t. It is then desired to fit a function $\phi(t, \mathbf{x})$ which has n adjustable parameters $\mathbf{x}$, to be chosen so that the function best fits the data. Therefore the residuals are given by

$$r_i(\mathbf{x}) = \phi(t_i, \mathbf{x}) - d_i, \qquad i = 1, 2, \ldots, m, \tag{6.1.4}$$

and a least squares solution is sought by minimizing $f(\mathbf{x})$ in (6.1.1). An example in which decay rates are being calculated is to fit $\phi(t, \mathbf{x}) = x_1 e^{-t/x_2} + x_3 e^{-t/x_4}$ to measured data, so as to determine the amplitudes (x_1, x_3) and time constants (x_2, x_4) of the two separate species which are producing the decay. A more complicated situation arises when $\phi(t, \mathbf{x})$ is defined indirectly as the solution of a given ordinary differential equation, and this is explored in Question 6.11.

For a nonlinear least squares problem, derivatives of $f(\mathbf{x})$ are given (see Question 1.1) by

$$\mathbf{g}(\mathbf{x}) = 2\mathbf{A}\mathbf{r} \tag{6.1.5}$$

and

$$\mathbf{G}(\mathbf{x}) = 2\mathbf{A}\mathbf{A}^{\mathrm{T}} + 2\sum_{i=1}^{m} r_i \nabla^2 r_i \tag{6.1.6}$$

where $r_i = r_i(\mathbf{x})$, etc., and where

$$\mathbf{A}(\mathbf{x}) = [\nabla r_1, \nabla r_2, \ldots, \nabla r_m] \tag{6.1.7}$$

is the $n \times m$ *Jacobian matrix*, the columns of which are the first derivative vectors ∇r_i of the components of $\mathbf{r}$ ($A_{ij} = \partial r_j / \partial x_i$). It is possible to use these formulae in a conventional way, that is either (6.1.5) in conjunction with a quasi-Newton method, or both (6.1.5) and (6.1.6) with a modified Newton method. However, in the latter case it is necessary to have expressions for the Hessian matrices $\nabla^2 r_i$ for all the functions $r_i(\mathbf{x})$, $i = 1, 2, \ldots, m$, and this can be disadvantageous.

Since $\mathbf{r}(\mathbf{x})$ is being minimized in the least squares sense, it is often the case that the components r_i are small. This suggests that a good approximation to $\mathbf{G}(\mathbf{x})$ might be obtained by neglecting the final term in (6.1.6) to give

$$\mathbf{G}(\mathbf{x}) \approx 2\mathbf{A}\mathbf{A}^{\mathrm{T}}. \tag{6.1.8}$$

This is equivalent to making a *linear approximation* to the residuals $r_i(\mathbf{x})$. It is in this way that the structure of (6.1.1) can be taken into account to improve the performance of first derivative methods. The important feature is that using only the information ($\mathbf{r}$ and $\mathbf{A}$) required to determine the first derivative vector $\mathbf{g}(\mathbf{x})$, it is possible to approximate the second derivative matrix $\mathbf{G}(\mathbf{x})$. Whereas a quasi-Newton method might take n iterations to estimate $\mathbf{G}$ satisfactorily, here the approximation is immediately available. Thus there is the possibility of more rapid convergence.

There are also some interpretations of (6.1.8) which are useful from a *statistical* point of view. In a linear model the function $\phi(t, \mathbf{x})$ which arises in (6.1.4) can be expressed as

$$\phi(t, \mathbf{x}) = \sum_{j=1}^{n} x_j \psi_j(t).$$

Then the hypothesis is made that the data values d_i behave according to

$$\mathbf{d} = \mathbf{A}^{\mathrm{T}}\mathbf{x} + \mathbf{e}$$

where $A_{ji} = \partial r_i / \partial x_j = \psi_j(t_i)$. The vector $\mathbf{e}$ represents the errors e_i in d_i which are assumed to be independent and normally distributed with mean 0 and constant variance σ^2, and $\mathbf{x}$ is some fixed 'underlying' solution. Then the least squares solution of (6.1.2) and (6.1.4) ($\hat{\mathbf{x}}$ say) is a maximum likelihood solution and has expectation value $\mathscr{E}(\hat{\mathbf{x}}) = \mathbf{x}$. The inverse of the matrix (6.1.8) is a multiple of the *variance–covariance matrix* $\mathbf{V}$, which can be shown to satisfy

$$\mathbf{V} = \mathscr{E}((\hat{\mathbf{x}} - \mathbf{x})(\hat{\mathbf{x}} - \mathbf{x})^{\mathrm{T}}) = (\mathbf{A}\mathbf{A}^{\mathrm{T}})^{-1}\sigma^2.$$

The diagonal elements of $\mathbf{V}$ give the variances of the elements $\hat{\mathbf{x}}_i$ $i = 1, 2, \ldots, n$ in the maximum likelihood solution, and the off-diagonal elements give the covariances between $\hat{x}_i$ and $\hat{x}_j$. It can also be shown that $\mathscr{E}(\hat{f}) = (m - n)\sigma^2$ so that an estimator for σ^2 is $\hat{f}/(m - n)$, where $\hat{f}$ is the maximum likelihood sum of squares obtained by minimizing $f(\mathbf{x})$ in (6.1.1). This fact, and the use of the matrix (6.1.8), thus enables $\mathbf{V}$ to be determined and gives useful statistical information about the distribution of the least squares solution.

The main aim of this section however is to discuss numerical methods for solving (6.1.1) which make use of the estimate (6.1.8) of the Hessian matrix. These methods can be viewed as modifications of the second derivative methods of Sections 3.1 and 5.2. Thus the basic Newton's method (3.1.2) becomes the basic *Gauss–Newton method* (or *generalized least squares method*) when (6.1.8)

Table 6.1.1 Gauss–Newton method for a simple problem

k	1	2	3	4	5	6
$x^{(k)}$	1	0.131148	0.013635	0.001369	0.000137	0.000014

is used to approximate $\mathbf{G}^{(k)}$. Using the derivative expressions (6.1.5) and (6.1.6) the kth iteration of the basic Gauss–Newton method is therefore

$$\begin{aligned}&\text{(a) solve } \mathbf{A}^{(k)}\mathbf{A}^{(k)T}\boldsymbol{\delta} = -\mathbf{A}^{(k)}\mathbf{r}^{(k)} \quad \text{for } \boldsymbol{\delta} = \boldsymbol{\delta}^{(k)}\\ &\text{(b) set } \mathbf{x}^{(k+1)} = \mathbf{x}^{(k)} + \boldsymbol{\delta}^{(k)}\end{aligned} \tag{6.1.9}$$

Equation (6.1.9(a)) is analogous to the *normal equations* of linear least squares. A simple example is to apply the method to the problem (due to Powell) with two equations ($m = 2$) and one variable ($n = 1$) given by

$$\begin{aligned}r_1(x) &= x + 1\\ r_2(x) &= \lambda x^2 + x - 1\end{aligned} \tag{6.1.10}$$

and for which $x^* = 0$. If $\lambda = 0.1$ is chosen, then the progress of iteration is that given in Table 6.1.1; this illustrates that if the degree of non-linearity in $\mathbf{r}(\mathbf{x})$ is small, then the method can work well. Another feature which this example makes clear is that the order of convergence is usually linear, and in this case the errors satisfy $h^{(k+1)} \approx 0.1\, h^{(k)}$.

In fact it is possible to be much more precise about how the errors behave. If $\mathbf{B}^{(k)} = 2\mathbf{A}^{(k)}\mathbf{A}^{(k)T}$ is the approximation to $\mathbf{G}^{(k)}$, then it is possible by following a proof very similar to Theorem 3.1.1 to show that

$$\|\mathbf{h}^{(k+1)}\| \leqslant 2\|\mathbf{B}^{*-1}(\Sigma r_i^* \nabla^2 r_i^*)\|\,\|\mathbf{h}^{(k)}\| + O(\|\mathbf{h}^{(k)}\|^2). \tag{6.1.11}$$

This suggests that the order of convergence, if it occurs, is no worse than linear. However when the matrix $\Sigma r_i^* \nabla^2 r_i^* = \mathbf{0}$ (for example when $\mathbf{r}^* = \mathbf{0}$ holds as in well-determined problems) then (6.1.11) shows that the convergence is second order. As in Theorem 3.1.1 it then follows that $\mathbf{x}^{(k)} \to \mathbf{x}^*$ if some iterate is sufficiently close to $\mathbf{x}^*$. For nonlinear data fitting problems however it is more usual that the matrix $\Sigma r_i^* \nabla^2 r_i^*$ is non-zero, in which case the order of convergence is usually no better than linear, and indeed if the matrix $\Sigma r_i^* \nabla^2 r_i^*$ is sufficiently large (in that $2\mathbf{B}^{*-1}\Sigma r_i^* \nabla^2 r_i^*$ has eigenvalues greater in modulus than unity) then the iteration does not generally converge. This is true *however close* $\mathbf{x}^{(1)}$ *is to* $\mathbf{x}^*$. An example of this fact is provided by problem (6.1.10), for it can be shown that

$$x^{(k+1)} = \lambda x^{(k)} + O(x^{(k)2}), \tag{6.1.12}$$

and it follows that the basic Gauss–Newton method fails to converge to x^* when $|\lambda| > 1$. Thus (6.1.9) is valuable only when both $\mathbf{x}^{(1)}$ is close to $\mathbf{x}^*$ and the matrix $\Sigma r_i^* \nabla^2 r_i^*$ is small, either because the r_i^* are small or because the degree of non-linearity is small.

In view of these observations, modification of the Gauss–Newton method to incorporate a *line search* is therefore particularly desirable and subsequently

'Gauss–Newton method' will be taken to imply the use of a line search. In this case the search direction $\mathbf{s}^{(k)}$ is found by solving the linear system

$$\mathbf{A}^{(k)}\mathbf{A}^{(k)\mathrm{T}}\mathbf{s}^{(k)} = -\mathbf{A}^{(k)}\mathbf{r}^{(k)} \tag{6.1.13}$$

and the algorithm determines $\mathbf{x}^{(k+1)}$ by using any suitable line search strategy, for instance that described in Section 2.6. Before going on to consider the properties of the Gauss–Newton method with line search, a general observation is first made about line search methods for nonlinear least squares. In a line search it is usual to make interpolations or extrapolations based on a polynomial model. For nonlinear least squares it is advantageous to make polynomial approximations *to each individual function* $r_i(\mathbf{x})$ rather than directly to $f(\mathbf{x})$ itself. For example if $\mathbf{r}(\mathbf{x}^{(k)} + \alpha\mathbf{s}^{(k)})$ has been evaluated at $\alpha = 0$ and $\alpha = 1$, and $\mathbf{A}^{(k)}$ is known, then a quadratic model $r_i(\alpha) \approx r_i(0) + \alpha r_i'(0) + \alpha^2(r_i(1) - r_i(0) - r_i'(0))$ can be used where $r_i'(0) = \mathbf{a}_i^{(k)\mathrm{T}}\mathbf{s}^{(k)}$. The corresponding approximation to $f(\mathbf{x})$ is then found by summing the squares of these model polynomials. For example when the $r_i(\alpha)$ are modelled by quadratic functions, the corresponding approximation to $f(\mathbf{x})$ is a quartic. Although this is more difficult to handle than the cubic or quadratic approximations used in Section 2.6, a better fit usually results which can result in a worthwhile saving of function and derivative evaluations. For example, it can be seen by comparing the Rosenbrock and Chebyquad 6, 8 results for the BFGS method in Table 6.1.2 with those in Table 3.5.1 that a significant saving of Jacobian evaluations is obtained.

Returning to the properties of the Gauss–Newton method, Table 6.1.2 gives some idea of how the method compares with the BFGS quasi-Newton method of Chapter 3, on some standard zero or small residual test problems. Apart from the Chebyquad $n = 8, 9$ and 10 cases (see below) the Gauss–Newton method

Table 6.1.2 Line search methods for nonlinear least squares: zero and small residual test problems

Problem	n	m	Gauss–Newton			BFGS			FX hybrid		
Rosenbrock	2	2	16	24	16*	15	26	16	16	27	16
Chebyquad	6	6	5	16	7	8	20	10	5	16	7
	8	8		fails		10	24	13	13	36	17
	9	9		fails		12	22	15	6	11	8
	10	10		fails		14	29	17	12	26	12
Watson	6	31	5	7	7	15	26	15	5	7	6
	9	31	5	6	5	18	28	18	5	6	5
	12	31	5	8	7	21	41	22	5	8	7
	20	31	5	7	6		fails		5	7	6
Kowalik	4	11	6	20	10	10	23	13	6	16	8
Osborne 1	5	33	6	14	9	29	55	37	10	17	12
Osborne 2	11	65	9	19	10	24	43	26	9	19	10

* Figures are numbers of iterations, residual ($\mathbf{r}$) evaluations and Jacobian ($\mathbf{A}$) evaluations respectively.

is markedly better overall. This can be ascribed to the suitability of the approximate Hessian (6.1.8), for the reasons given. The satisfactory results on the Watson, Kowalik and Osborne problems are typical of how the method performs in general. A feature which appears to contribute to the reliability of the Gauss–Newton method is that the matrix $2\mathbf{A}\mathbf{A}^T$ is positive semi-definite and usually positive definite, so that there is no need to worry about indefinite $\mathbf{G}^{(k)}$ matrices such as may arise in Newton's method with line search. In fact, if the condition number $\kappa(2\mathbf{A}\mathbf{A}^T)$ is uniformly bounded away from zero then the Gauss–Newton method is globally convergent by virtue of Theorem 2.5.1 and the subsequent remarks of Section 2.5. Since $\mathbf{A}$ will usually be bounded above, this essentially requires only that $\mathbf{A}$ has full rank at a limit point. Nonetheless examples do occur which illustrate that rank deficiency can cause the method to fail. Powell (1970b) gives an example using exact arithmetic in which the iterates converge to a non-stationary limit point at which $\mathbf{A}$ is rank deficient. Also in practice if $\mathbf{x}^*$ is stationary but $\mathbf{A}^*$ is rank deficient, the search directions $\mathbf{s}^{(k)}$ can become numerically orthogonal to $\mathbf{g}^{(k)}$ at some distance from the solution and no progress can be made in the line search. Thus a very poor estimate of the solution is obtained. Test problems illustrating rank deficiency in $\mathbf{A}^*$ are Chebyquad 8 and 10 (9 is near deficient), Freudenstein and Roth, and Jennrich. Rank deficiency is more common in well-determined systems (singularity in $\mathbf{A}^*$) than in overdetermined systems arising from data fitting, which explains why the Gauss–Newton method generally has a good reputation for being reliable, despite these remarks.

Another negative aspect of the Gauss–Newton method occurs for those nonlinear problems in which $\mathbf{r}^*$ is relatively large at the solution (so-called *large residual problems*). This feature is significant insofar as it implies that the term $\Sigma r_i^* \nabla^2 r_i^*$ in $\mathbf{G}$ is substantial. Because the actual step in the method is $\boldsymbol{\delta}^{(k)} = \alpha^{(k)}\mathbf{s}^{(k)}$, the effective Hessian approximation is $\mathbf{B}^{(k)} = \alpha^{(k)-1}(2\mathbf{A}^{(k)}\mathbf{A}^{(k)T})$. Thus much as in (6.1.11), the extent to which $\mathbf{B}^{(k)}$ approximates $\mathbf{G}^*$ determines the rate of convergence (the order of convergence being linear). Because there is a limit to how well $\mathbf{G}^*$ can be satisfactorily approximated merely by scaling $\mathbf{A}^*\mathbf{A}^{*T}$, it therefore follows that the rate of convergence of the Gauss–Newton method with line search may be poor when applied to some large residual problems. To gauge the extent to which this factor might be significant, the practical performance of the method on some large residual problems is shown in Table 6.1.3. This table also shows the performance of the BFGS method (with the same line search strategy), starting from an initial matrix $\mathbf{B}^{(1)} = 2\mathbf{A}^{(1)}\mathbf{A}^{(1)T}$, which is often recommended as a more reliable method for solving large residual problems when only first derivatives are available. Compared with the BFGS method, the Gauss–Newton method is spectacularly poor for the Brown problem and marginally worse for the Trigonometric problems (in addition to failing for the rank deficient problems). However, the BFGS method is worse for the other problems in this set. Overall one might conclude for large residual problems that the Gauss–Newton method with line search is seriously inadequate only on a few occasions.

Table 6.1.3 Line search methods for nonlinear least squares: large residual test problems

Problem	n	m	Gauss–Newton			BFGS			FX Hybrid		
Freudenstein	2	2		fails		6	9	6	6	10	6
Meyer	3	16	8	28	11	64	154	67	7	28	10
Jennrich	2	10		fails		9	16	9	6	15	7
Brown	4	20	254	627	256	10	19	10	10	24	10
Signomial	6	18	30	59	31	26	56	28	16	33	17
	6	30	23	53	27	34	79	37	16	33	17
	8	24	19	44	19	45	95	47	13	22	13
	8	40	81	138	83	56	130	61	26	55	27
	10	30	17	37	17	88	193	95	13	26	13
	10	50	53	113	58	100	227	107	19	41	21
Trigonometric	6	18	37	95	38	19	52	20	12	22	12
	6	30	21	54	22	25	59	25	10	27	12
	8	24	26	73	27	27	57	27	11	28	14
	8	40	31	71	32	31	64	31	15	31	15
	10	30	33	91	37	34	85	35	14	35	16
	10	50	47	107	49	33	91	37	18	52	21

Nonetheless because the Gauss–Newton method can fail or can converge slowly, it is desirable to seek other methods. A method which does converge globally is the restricted step algorithm (5.1.6), using the L_2 in norm (5.1.2), and it is possible to use this method in conjunction with approximation (6.1.8). From (5.2.1), methods of this type determine the correction $\boldsymbol{\delta}^{(k)}$ by solving the system

$$(\mathbf{A}^{(k)}\mathbf{A}^{(k)\mathrm{T}} + \nu\mathbf{I})\boldsymbol{\delta}^{(k)} = -\mathbf{A}^{(k)}\mathbf{r}^{(k)} \qquad \nu \geqslant 0. \tag{6.1.14}$$

Such methods were suggested at an early date by Levenberg (1944) and Marquardt (1963) and have already been discussed in Section 5.2 in a wider context. Many different algorithms of this type have been suggested; some control the iteration using $\nu^{(k)}$ directly as in algorithm (5.2.7); others (Moré, 1978) control the iteration as in algorithm (5.1.6) by using the radius $h^{(k)}$ of (5.1.1), and choose $\nu^{(k)}$ so that $\|\boldsymbol{\delta}^{(k)}\|_2 = h^{(k)}$ by an inner iteration such as (5.2.10). Convergence to a stationary point is assured as in Theorem 5.1.1 because the proof that $\mathbf{g}^{\infty} = \mathbf{0}$ only requires the Taylor expansion $\Delta f^{(k)} = \Delta q^{(k)} + o(\|\boldsymbol{\delta}^{(k)}\|_1)$, which is true even if the second derivative approximation is not exact. In general (if the matrix $\Sigma_i r_i^{\infty}\nabla^2 r_i^{\infty}$ is not zero) then the results relating to second order conditions and second order convergence do not hold. Because of these considerations the Levenberg–Marquardt methods for nonlinear least squares problems are robust and often work well. However, the lack of a superlinear convergence property, and the bias of $\boldsymbol{\delta}^{(k)}$ towards the steepest descent direction which (6.1.14) provides, can have the effect of inducing a slow rate of convergence, much as for the Gauss–Newton method. Thus although a Levenberg–Marquardt method is often considered to be a good method for nonlinear least squares problems, it is by no means entirely satisfactory.

More recent research into methods for nonlinear least squares has concentrated on trying to improve on the extent to which the matrix $2\mathbf{A}\mathbf{A}^T$ approximates $\mathbf{G}$, in the hope of getting something akin to superlinear convergence. Gill and Murray (1976a) make finite difference estimates of the curvature in a subspace to improve those parts of $\mathbf{A}\mathbf{A}^T$ which most significantly affect the search direction $\mathbf{s}$. Brown and Dennis (1971) approximate the matrices $\nabla^2 r_i$ by making quasi-Newton updates, and Dennis, Gay and Welsch (1981) describe an update scheme for the composite matrix $\Sigma r_i \nabla^2 r_i$. In both cases an estimate of $\mathbf{G}$ in (6.1.6) is then constructed for use in a restricted step method. However, Dennis, Gay and Welsch also find that they need to allow for the possibility of taking Gauss–Newton steps, for the method to be fully practicable, which leads to a rather cumbersome method, albeit one which has performed well on a range of test problems. My current preference for nonlinear least squares is for a *hybrid method* between the Gauss–Newton and BFGS method (Al-Baali and Fletcher, 1985; Fletcher and Xu, 1986). This type of method is a line search descent method using positive definite approximate Hessian matrices $\mathbf{B}^{(k)}$. Depending upon the outcome of a certain test, the method chooses $\mathbf{B}^{(k)}$ to be *either* the Gauss–Newton matrix (6.1.8) *or* the result of applying the BFGS formula to $\mathbf{B}^{(k-1)}$. Initially $\mathbf{B}^{(1)}$ is the Gauss–Newton matrix. A line search suitable for nonlinear least squares is used. One small difference from the normal BFGS method is that the difference vector

$$\gamma^{(k)} = 2\mathbf{A}^{(k+1)}\mathbf{A}^{(k+1)\mathrm{T}}\delta^{(k)} + (\mathbf{A}^{(k+1)} - \mathbf{A}^{(k)})\mathbf{r}^{(k+1)} \qquad (6.1.15)$$

is usually used in place of (3.2.3) as it seems to give better results in practice. However, it is necessary to safeguard this choice if $\mathbf{B}^{(k+1)}$ would otherwise not be positive definite. In the Fletcher–Xu hybrid method the simple test

$$f^{(k)} - f^{(k+1)} \geqslant \tau f^{(k)} \qquad (6.1.16)$$

is used as the indication to take a Gauss–Newton step, where $\tau \in (0, 1)$ is a preset constant ($\tau = 0.2$ is recommended). The motivation for this test is that if $\mathbf{r}^* \neq \mathbf{0}$ so that $f^* \neq 0$, then $(f^{(k)} - f^{(k+1)})/f^{(k)} \to 0$ and therefore the test forces BFGS steps to be taken ultimately. This enables superlinear convergence for large residual problems to be obtained. Conversely if $\mathbf{r}^* = \mathbf{0}$ then both methods converge superlinearly, so that $f^{(k+1)} = o(f^{(k)})$, which implies that $(f^{(k)} - f^{(k+1)})/f^{(k)} \to 1$. Under these circumstances the Gauss–Newton step is taken ultimately, with the advantages of second-order convergence and numerical stability. Some numerical results for this hybrid method are also given in Tables 6.1.2 and 6.1.3. The method matches (almost) or improves on the better of the Gauss–Newton and BFGS methods for every test problem and therefore gives a reliable, superlinearly convergent method that contains the best features of both the Gauss–Newton and BFGS methods.

Another way in which it might be possible to improve reliability or efficiency when solving certain data fitting problems arises when some of the variables x_i occur *linearly* in the approximating function $\phi(t, \mathbf{x})$. Then for any given values of the nonlinear variables, the linear variables can be determined by the solution to a finite linear least squares problem. This solution can be used to eliminate the linear variables from the problem and gives rise to a nonlinear problem in fewer variables. References and more details of this are given by Dennis (1977). However whilst this approach is well worth trying in any particular case, there is no guarantee that it will yield superior results in general. An illustrative example is given in Question 6.3.

Some numerical aspects of these methods are also worthy of comment. In the Gauss–Newton method the search direction $\mathbf{s}^{(k)}$ is found by solving (6.1.13), and this is most conveniently done if $\mathbf{LL}^{\mathrm{T}}$ factors of $\mathbf{A}^{(k)}\mathbf{A}^{(k)\mathrm{T}}$ are available, where $\mathbf{L}$ is lower triangular. It is important to realize that substantial loss of precision can occur in the numerical computation, if this is done by first 'squaring' $\mathbf{A}$ to give $\mathbf{AA}^{\mathrm{T}}$ (omitting superscript (k)) and then factorizing $\mathbf{AA}^{\mathrm{T}}$ directly. To see this, consider the solution of the square system $\mathbf{Bs} = -\mathbf{g}$ when errors $\boldsymbol{\delta}\mathbf{B}$ in $\mathbf{B}$ induce errors $\boldsymbol{\delta}\mathbf{s}$ in $\mathbf{s}$. It is straightforward to show that

$$\frac{\|\boldsymbol{\delta}\mathbf{s}\|_2}{\|\mathbf{s}+\boldsymbol{\delta}\mathbf{s}\|_2} \leqslant \kappa(\mathbf{B})\frac{\|\boldsymbol{\delta}\mathbf{B}\|_2}{\|\mathbf{B}\|_2}$$

where $\kappa(\mathbf{B})$ is the *spectral condition number* of $\mathbf{B}$, and this bound is realistic in practice when the errors $\boldsymbol{\delta}\mathbf{B}$ can be attributed to round-off. The magnitude of the relative errors in $\mathbf{B}$ is $\sim \varepsilon$, the relative precision of the computation, and the resulting vector $\mathbf{s}$ is seen to contain relative errors which have been magnified by a factor $\kappa(\mathbf{B})$. An alternative formulation of (6.1.13) uses the QR factors

$$\mathbf{A}^{\mathrm{T}} = \mathbf{Q}\begin{bmatrix}\mathbf{L}^{\mathrm{T}}\\ \mathbf{0}\end{bmatrix} = \mathbf{Q}_1\mathbf{L}^{\mathrm{T}}$$

where $\mathbf{Q}$ is an $m \times m$ orthogonal matrix and $\mathbf{L}$ is $n \times n$ lower triangular. Then $\mathbf{AA}^{\mathrm{T}} = \mathbf{LL}^{\mathrm{T}}$ follows and the system

$$\mathbf{LL}^{\mathrm{T}}\mathbf{s} = -\tfrac{1}{2}\mathbf{g} = -\mathbf{Ar}$$

simplifies to give

$$\mathbf{L}^{\mathrm{T}}\mathbf{s} = -\mathbf{Q}_1^{\mathrm{T}}\mathbf{r}. \tag{6.1.17}$$

It can be shown that $\kappa(\mathbf{L}) = \kappa(\mathbf{B})^{1/2}$ and therefore the magnification of errors that is obtained by using (6.1.17) is much less. (For example if $\kappa(\mathbf{B}) = 10^6$, then $\kappa(\mathbf{L}) = 10^3$ and only 3 significant figures are likely to be lost by using (6.1.17), as against 6 significant figures by using $\mathbf{B} = \mathbf{LL}^{\mathrm{T}}$.) It is an advantage of the Gauss–Newton method that (6.1.13) permits the use of QR factors so as to be able to solve for $\mathbf{s}$ from (6.1.17). Conversely in the BFGS method, $\mathbf{s}$ is computed from $\mathbf{LL}^{\mathrm{T}}\mathbf{s} = -\mathbf{g}$ (the Choleski factors are updated on each iteration so $\mathbf{L}^{\mathrm{T}}$ is not the QR factor) and so this ability to improve the conditioning is not present. In fact, the Watson 20 test problem in Table 6.1.2 is an example in which $\kappa(\mathbf{B})$ is sufficiently large that all significance is lost in single precision, so the BFGS method fails, whereas the Gauss–Newton method is able to solve the problem in single precision. The ability to take Gauss–Newton steps is therefore an important attribute of the hybrid methods referred to above.

The most well known technique for calculating QR factors is the use of Householder elementary matrices and is quite practicable. It requires $n^2(m - \frac{1}{3}n) + 0(mn)$ multiplications. However, there is also an alternative orthogonal reduction technique due to Gentleman (1973), which can be thought of as a square root free Given's method (see Wilkinson, 1977). This operates on a column of $\mathbf{A}$ at a time, and can be more convenient in data fitting applications since each column corresponds to one observation which can be processed and then ignored, without storing the entire matrix $\mathbf{A}$. Asymptotically the housekeeping requirements of these two methods are the same. For the Levenberg–Marquardt type of method using (6.1.14), addition of the term $\nu\mathbf{I}$ improves the condition number of $\mathbf{A}\mathbf{A}^{\mathrm{T}}$ so the effects of rounding errors may not be as severe. However, since small or zero values of ν can be used, it is still advisable not to form $\mathbf{A}\mathbf{A}^{\mathrm{T}} + \nu\mathbf{I}$ directly. The stable alternative is to apply the Householder or Gentleman technique to the matrix $[\mathbf{A}:\nu^{1/2}\mathbf{I}]$. Another possibility for Marquardt–Levenberg methods is to factorize $\mathbf{A} = \mathbf{Q}\mathbf{B}\mathbf{P}^{\mathrm{T}}$ ($\mathbf{B}$ bi-diagonal, $\mathbf{Q}, \mathbf{P}$ orthogonal). Then repeated solution of (6.1.14) can be undertaken for different ν without re-computing the factors. A final possibility with the same feature is to compute the singular value decomposition of $\mathbf{A}$ (as in Moré, 1978) but this is considerably more expensive.

6.2 WELL-DETERMINED SYSTEMS

Although a well-determined system of nonlinear equations

$$r_i(\mathbf{x}) = 0, \qquad i = 1, 2, \ldots, n \tag{6.2.1}$$

(that is (6.1.2) with $m = n$) can be treated by the methods of the previous section, it is an important problem in its own right, and there are some simplifications which arise and some further possibilities for methods which merit attention. The obvious approach is to linearize (6.2.1) giving

$$l_i^{(k)}(\boldsymbol{\delta}) = r_i^{(k)} + \mathbf{a}_i^{(k)\mathrm{T}}\boldsymbol{\delta} = 0, \qquad i = 1, 2, \ldots, n \tag{6.2.2}$$

and to solve this linear system to obtain a correction $\boldsymbol{\delta}^{(k)}$. This is the *Newton–Raphson method* in which the kth iteration can be written

$$\begin{aligned} &\text{(a) solve } \mathbf{A}^{(k)\mathrm{T}}\boldsymbol{\delta} = -\mathbf{r}^{(k)} \quad \text{for } \boldsymbol{\delta} = \boldsymbol{\delta}^{(k)} \\ &\text{(b) set } \mathbf{x}^{(k+1)} = \mathbf{x}^{(k)} + \boldsymbol{\delta}^{(k)}. \end{aligned} \tag{6.2.3}$$

Step (a) involves the solution of a non-symmetric linear system of equations, which is best solved by making an LU decomposition of $\mathbf{A}^{(k)}$ with partial pivoting. This requires $\frac{1}{3}n^3 + O(n^2)$ multiplications. (The terminology Newton–Raphson is standard when $n = 1$, and it seems beneficial to extend this to all n so as to avoid confusion with the Newton method (3.1.2) applied to the function $\mathbf{r}(\mathbf{x})^{\mathrm{T}}\mathbf{r}(\mathbf{x})$, which gives a different method requiring higher derivatives $(\nabla^2 r_i)$.) In fact the Newton–Raphson method is equivalent to the basic Gauss–Newton method (6.1.9) in the special case that $m = n$. This follows by multiplying

Table 6.2.1 An application of the Newton–Raphson method

k	1	2	3	4	5
$x_1^{(k)}$	6	2.612245	4.826283	4.998717	5
$x_2^{(k)}$	5	4.244898	4.020244	4.000156	4
$r_1^{(k)}$	−17	−5.271248	−0.379029	−0.002841	$-0.17_{10}-6$
$r_2^{(k)}$	57	8.692322	0.681847	0.005262	$0.32_{10}-6$
$h_1^{(k)}$	1	−2.387755	−0.173717	−0.001283	$-0.768_{10}-7$
$h_2^{(k)}$	1	0.244898	0.020244	0.000156	$0.93_{10}-8$

the equation in (6.1.9(a)) by $\mathbf{A}^{(k)-1}$ which gives equation (6.2.3(a)) directly. The only difference is the possibility of using the more efficient LU factorization to solve (6.2.3(a)). If a QR factorization is used for this, the methods are identical.

An application of the Newton–Raphson method to the system

$$\begin{aligned} r_1(\mathbf{x}) &= x_1 - x_2^3 + 5x_2^2 - 2x_2 - 13 \\ r_2(\mathbf{x}) &= x_1 + x_2^3 + x_2^2 - 14x_2 - 29 \end{aligned} \tag{6.2.4}$$

(due to Freudenstein and Roth) from $\mathbf{x}^{(1)} = (6, 5)^T$ and working to about ten significant digits is given in Table 6.2.1. In this table the error $\mathbf{h}^{(k)}$ is defined as in (2.3.1). The rapid convergence of $\mathbf{h}^{(k)} \to \mathbf{0}$ when $\mathbf{x}^{(1)}$ is close to $\mathbf{x}^*$ is typical of a Newton method (see Table 3.1.1) and can be established rigorously.

Theorem 6.2.1

If $\mathbf{r} \in \mathbb{C}^1$, if $\mathbf{A}^*$ is non-singular and its elements satisfy a Lipschitz condition, and if $\mathbf{x}^{(k)}$ is sufficiently close to $\mathbf{x}^*$ for some k, then the Newton–Raphson method is well defined for all k and converges at second order.

Proof

As for Theorem 3.1.1 with $\mathbf{A}^{(k)T}$ replacing $\mathbf{G}^{(k)}$ and $\mathbf{r}^{(k)}$ replacing $\mathbf{g}^{(k)}$. □

An illustration of the second order convergence is also provided by Table 6.2.1, in that successive values of $\|\mathbf{h}^{(k+1)}\|_2/\|\mathbf{h}^{(k)}\|_2^2$ for $k = 2, 3, 4$ are 0.0304, 0.0423, 0.0463 which indicates a constant $c \approx 0.05$ in (3.1.4). As in Section 3.1, variations of the Newton–Raphson method exist in which factors of previous $\mathbf{A}$ matrices can be used to replace $\mathbf{A}^{(k)}$, and similar comments apply.

The development so far closely parallels that in Section 3.1 for Newton's method for minimization, and it is likewise true that the resulting method is not suitable for a general purpose algorithm. For example with $\mathbf{x}^{(1)} = (15, -2)^T$ in (6.2.4), the Newton–Raphson method shows no sign of converging after 50 iterations. Again some sort of line search is required, but since $\mathbf{r}(\mathbf{x})$ is not in general the gradient of a function $f(\mathbf{x})$, it is necessary to construct some scalar measure of how close $\mathbf{r}(\mathbf{x})$ is to zero. The most obvious choice therefore is to use a line search based on minimizing $\mathbf{r}(\mathbf{x})^T\mathbf{r}(\mathbf{x})$. This possibility is discussed in the previous section, and it is pointed out there that if $\mathbf{A}^{(k)}$ loses rank in the limit then convergence to a non-stationary point can occur (Powell 1970b). The situation may therefore be more severe than with Newton's method for minimization, for which no such example (with $\mathbf{x}^{(k)} \to \mathbf{x}'$, $\mathbf{g}' \neq \mathbf{0}$, $\{\mathbf{G}^{(k)}\}$ positive definite and $\mathbf{G}'$ singular) has been developed to my knowledge. Thus the Newton–Raphson method with line search on $\mathbf{r}^T\mathbf{r}$ does not provide a suitable general purpose algorithm. The currently accepted remedy is to use Levenberg–Marquardt or hybrid methods etc. as described in Section 6.1. However whilst I concur with this to some extent, I think that better methods will ultimately be developed, with more rapid convergence. In particular developments using $\|\mathbf{r}(\mathbf{x})\|_1$ as a measure offer exciting possibilities, especially for ill-conditioned problems since the squaring of $\mathbf{r}$ is avoided, and also for large problems since the sparsity of $\mathbf{A}$ can more readily be used to advantage. It is currently an open question of some interest as to whether the L_1 or least squares functions are most effective as merit functions in this context.

Another disadvantage associated with minimizing a scalar measure of $\mathbf{r}(\mathbf{x})$ is that local minima might be introduced which do *not* correspond to solutions of $\mathbf{r}(\mathbf{x}) = \mathbf{0}$. This is illustrated by (6.2.4) for which $\mathbf{r}^T\mathbf{r}$, $\|\mathbf{r}\|_1$, and $\|\mathbf{r}\|_\infty$ all have local minima at $\mathbf{x} = (53 - 4\sqrt{22},\ 2 - \sqrt{22})^T/3$. This situation must be regarded as inevitable; for example, if $r(x)$ is defined as in Figure 6.2.1, it is hard to construct any measure of how close r is to zero which avoids introducing a local solution at x'. The problem of moving from x' to x^* is then a global minimization problem which is mentioned in Section 1.1, and is very difficult. In fact if $r(x) = 0$ has no real solution, as indicated by the broken line in Figure 6.2.1, then clearly situations like x' are the best that can be achieved. A more realistic attitude therefore is to accept that local solutions to $\mathbf{r}^T\mathbf{r}$ (or $\|\mathbf{r}\|_1$, etc.) can exist, and ensure that the algorithms are able to locate them effectively. In this situation however another disadvantage of the Newton–Raphson method with line search becomes apparent. This arises because if $\mathbf{x}'$ satisfies $\mathbf{g}' = \mathbf{0}$ and $\mathbf{r}' \neq \mathbf{0}$, it follows from (6.1.5) that $\mathbf{A}'$ is singular. In fact the Newton–Raphson method is unstable near such a point and only the line search ensures that the method does not

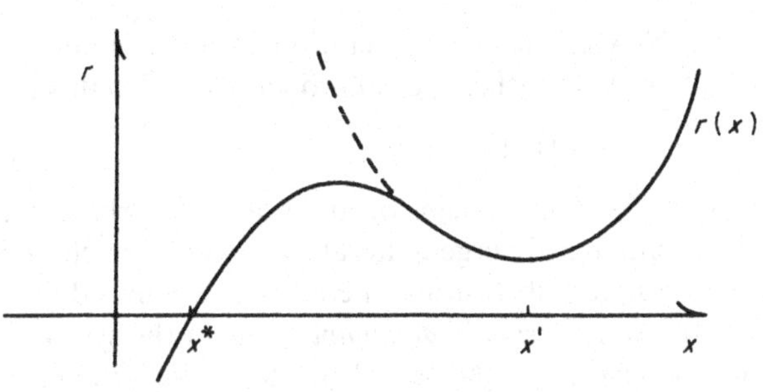

Figure 6.2.1 Local minima in $\|\mathbf{r}\|$

diverge. In practice the rate of convergence can be very slow. An example of this can be found when solving (6.2.4) from $\mathbf{x}^{(1)} = (15, -2)^T$. The Newton–Raphson method with line search fails to make any further progress at a point $\mathbf{x}^{(7)} = (13.54, -0.8968)^T$ at which $f^{(7)} - f^* \approx 9$. However the restricted step Newton minimization method obtains an estimate of $\mathbf{x}^*$ after seven iterations which gives f^* to the full computer precision (7–8 digits). Another example is provided by the Chebyquad $n = 8$ problem in which the equations $\mathbf{r}(\mathbf{x}) = \mathbf{0}$ have no real solution, and the best value of $\mathbf{r}(\mathbf{x})^T\mathbf{r}(\mathbf{x})$ is $0.3516873_{10}{-2}$. It is shown in Table 5.2.1 that the restricted step Newton method finds this solution without difficulty, whereas the Newton–Raphson method (that is the Gauss–Newton method) fails to make any further progress from a point at which $\mathbf{r}(\mathbf{x})^T\mathbf{r}(\mathbf{x}) \approx 0.03$.

Although Newton methods in some form or another give rise to the most suitable methods for general use, there are other techniques which are applicable in certain special situations. If the system $\mathbf{r}(\mathbf{x}) = \mathbf{0}$ is rearranged in the form

$$\mathbf{x} = \boldsymbol{\phi}(\mathbf{x}) \tag{6.2.5}$$

then $\mathbf{x}^*$ is a solution of $\mathbf{r}(\mathbf{x}) = \mathbf{0}$ if and only if

$$\mathbf{x}^* = \boldsymbol{\phi}(\mathbf{x}^*) \tag{6.2.6}$$

The *successive substitution method* is an iterative method based on (6.2.5) which uses

$$\mathbf{x}^{(k+1)} = \boldsymbol{\phi}(\mathbf{x}^{(k)}) \tag{6.2.7}$$

to generate the sequence $\{\mathbf{x}^{(k)}\}$ from a given initial point $\mathbf{x}^{(1)}$. Equation (6.2.6) expresses the fact that $\mathbf{x}^*$ is a so-called *fixed point* of the iteration formula (6.2.7). The rearrangement in (6.2.5) can be done in many ways and this is exemplified by the one variable equation

$$r(x) = 1 + x - e^{2x} = 0 \tag{6.2.8}$$

which might be arranged as

$$x = e^{2x} - 1 \tag{6.2.9}$$

or as

$$x = \tfrac{1}{2}\log(1 + x). \tag{6.2.10}$$

Furthermore the Newton–Raphson iteration formula is always a valid rearrangement, albeit one with rather special properties, and in this case it reduces to

$$x = (e^{2x} - 2xe^{2x} - 1)/(1 - 2e^{2x}). \tag{6.2.11}$$

Unfortunately (especially for systems of equations) any arbitrary rearrangement will usually give either a divergent iteration or one which converges rather slowly. Thus successive substitution methods are only advisable when it is possible to identify some *strongly dominant terms* in the system $\mathbf{r}(\mathbf{x}) = \mathbf{0}$ and to rearrange these to give $\mathbf{x}$ on the left of (6.2.5). A well-known example is the linear system $\mathbf{r}(\mathbf{x}) = \mathbf{A}\mathbf{x} - \mathbf{b} = \mathbf{0}$, with $\mathbf{A} = \mathbf{D} + \mathbf{L} + \mathbf{U}$, where $\mathbf{D}$ is a dominant diagonal matrix and $\mathbf{L}$ and $\mathbf{U}$ are strictly lower and upper triangular. Then

$$\mathbf{x} = \mathbf{D}^{-1}(\mathbf{b} - \mathbf{L}\mathbf{x} - \mathbf{U}\mathbf{x}) \tag{6.2.12}$$

and

$$\mathbf{x} = (\mathbf{D} + \mathbf{L})^{-1}(\mathbf{b} - \mathbf{U}\mathbf{x}) \tag{6.2.13}$$

are two suitable rearrangements which correspond to the Jacobi and the Gauss–Seidel iterative methods.

Nonetheless successive substitution methods are often used when this ideal situation does not apply, and the resulting iteration may converge quite slowly, or not at all. In my experience users do not always appreciate the fact that there is the alternative possibility of using a Newton-like method, with a potentially much more rapid rate of convergence. These possibilities are illustrated for the simple problem (6.2.8) in Table 6.2.2, from $x^{(1)} = 1$.

Table 6.2.2 Solution of (6.2.8) by method of successive substitution from $x^{(1)} = 1$

	k	2	3	4	5	6	7
error	(6.2.9)	6.389	$3.5_{10}5$	fails			
$h^{(k)}$	(6.2.10)	0.34657	0.14878	0.06935	0.03353	0.01649	0.00818
using	(6.2.11)	0.60887	0.30141	0.10331	0.01681	0.00054	$\sim 10^{-7}$

There is some theoretical analysis for successive substitution methods which nicely backs up these practical considerations. It would be expected from practical experience (e.g. (6.2.10) in Table 6.2.2) that the order of convergence is usually linear and that the rate depends upon how nearly $\boldsymbol{\phi}(\mathbf{x})$ is constant relative to $\mathbf{x}$. This latter property is quantified by considering how small is the matrix

$$\mathbf{U}(\mathbf{x}) = \nabla \boldsymbol{\phi}^{\mathrm{T}}(\mathbf{x}) \tag{6.2.14}$$

relative to the unit matrix. The smaller (in modulus) the eigenvalues of $\mathbf{U}$, the more rapid is the rate of convergence. The case in which $\mathbf{U}(\mathbf{x})$ has extreme eigenvalues around ± 1 corresponds to an iteration which neither diverges nor converges. Larger values of $\mathbf{U}$ correspond to divergent iterations. This is expressed by the following result.

Theorem 6.2.2

If $\boldsymbol{\phi} \in \mathbb{C}^1$ *and* $\|\mathbf{U}^*\| = \mu < 1$ *then there exists a neighbourhood of* $\mathbf{x}^*$ *in which iteration* (6.2.7) *converges, and the order is at worst linear.*

Proof

By (6.2.7) and a Taylor expansion for $\boldsymbol{\phi}$ about $\mathbf{x}^*$ it follows that

$$\mathbf{x}^{(k+1)} = \boldsymbol{\phi}^{(k)} = \boldsymbol{\phi}^* + \mathbf{U}^*\mathbf{e}^{(k)} + o(\|\mathbf{h}^{(k)}\|)$$

where the error $\mathbf{h}^{(k)} = \mathbf{x}^{(k)} - \mathbf{x}^*$. Thus from (6.2.6),

$$\mathbf{h}^{(k+1)} = \mathbf{U}^*\mathbf{h}^{(k)} + o(\|\mathbf{h}^{(k)}\|)$$

and hence

$$\begin{aligned}\|\mathbf{h}^{(k+1)}\| &\leqslant \|\mathbf{U}^*\|\,\|\mathbf{h}^{(k)}\| + o(\|\mathbf{h}^{(k)}\|)\\ &= \mu\|\mathbf{h}^{(k)}\| + o(\|\mathbf{h}^{(k)}\|).\end{aligned}$$

Since $\mu < 1$ it follows for sufficiently small $\mathbf{h}^{(k)}$ that both local convergence occurs and the order is linear. □

In fact when $\boldsymbol{\phi}(\mathbf{x})$ is linear in $\mathbf{x}$ then convergence only requires the weaker result that the spectral radius of $\mathbf{U}$ is less than 1 (Varga, 1962). However this does not carry over to nonlinear situations because the error in this case may not contract uniformly (as in Theorem 6.2.2). However, this analysis does show that the error $\mathbf{h}^{(k)}$ is usually dominated by the component along the eigenvector which has the largest modulus eigenvalue, and that this eigenvalue is the asymptotic constant μ in $\|\mathbf{h}^{(k+1)}\| \approx \mu\|\mathbf{h}^{(k)}\|$ which determines the rate of convergence. These features are also illustrated by Table 6.2.2. For iteration (6.2.9), $\mathbf{U}^* = [2]$, so this iteration can never converge to $x^* = 0$. For iteration (6.2.10), $\mathbf{U}^* = [\frac{1}{2}]$ and it can be verified that the order of convergence is linear with rate constant $\mu = \frac{1}{2}$. Finally for the Newton–Raphson iteration (6.2.11) $\mathbf{U}^* = [0]$, which implies superlinear, and in this case quadratic, convergence. An example of the use of a substitution method to solve large systems of nonlinear equations is given by Concus (1967), and some analysis of this type of method is provided by Sherman (1978). Arising from this theory it might be possible to accelerate the linear rate of convergence by selecting from a family of possible rearrangements that one which minimizes some measure of the matrix $\mathbf{U}$. Well-known examples of this idea for linear systems are the SOR method and the Chebyshev semi-iteration method (Varga, 1962).

Another way of presenting the theory is to replace $\|\mathbf{U}^*\| < 1$ by the assumption that there exists a constant $\mu < 1$ such that

$$\|\boldsymbol{\phi}(\mathbf{x}) - \boldsymbol{\phi}(\mathbf{y})\| \leqslant \mu\|\mathbf{x} - \mathbf{y}\| \tag{6.2.15}$$

for all $\mathbf{x}$ and $\mathbf{y}$ in some domain which includes the solution. In this case $\boldsymbol{\phi}$ is said to be a *contraction mapping*. Then the error can be shown to contract uniformly on the domain (Rall, 1969). However (6.2.15) is a stronger assumption than $\|\mathbf{U}^*\| < 1$ and is less easily verified in general.

The observation that the Newton–Raphson method can be regarded as a limiting case of a successive substitution method as $\|\mathbf{U}\| \to 0$ is reinforced by another remarkable result due to Dennis and Moré (1974). Consider *any method at all* for solving a nonlinear system in which a convergent iterative sequence $\{\mathbf{x}^{(k)}\}$ is generated, and let $\boldsymbol{\delta}^{(k)} = \mathbf{x}^{(k+1)} - \mathbf{x}^{(k)}$. Then the sequence converges superlinearly if and only if $\boldsymbol{\delta}^{(k)}$ is asymptotically equal to the Newton–Raphson step, $\boldsymbol{\delta}_{\mathrm{NR}}^{(k)}$ say, that would be calculated by equation (6.2.3(a)) at $\mathbf{x}^{(k)}$. In practice those methods which work really well are invariably those which have the property of superlinear convergence. This result tells us that we must only look at those methods which are asymptotically equivalent to the Newton–Raphson method if this desirable property is to be attained. Such methods include Newton's method, quasi-Newton methods and finite difference Newton methods but exclude arbitrary successive substitution methods with $\mathbf{U}^* \neq \mathbf{0}$. The result is applicable to nonlinear systems arising in a variety of optimization problems. These include unconstrained optimization (the stationary point condition $\mathbf{g}(\mathbf{x}) = \mathbf{0}$, Newton's method), constrained optimization (the KT conditions, Lagrange–Newton–SQP method), non-smooth optimization (structure functions, SNQP method) and many others. Before stating the result, a preliminary lemma in required.

Lemma 6.2.1

If $\mathbf{h}^{(k)} \to \mathbf{0}$ *and* $\boldsymbol{\delta}^{(k)} = \mathbf{h}^{(k+1)} - \mathbf{h}^{(k)}$ *then the following are all equivalent definitions of superlinear convergence.*

(i) $\mathbf{h}^{(k+1)} = o(\|\mathbf{h}^{(k)}\|)$ (ii) $\boldsymbol{\delta}^{(k)} = -\mathbf{h}^{(k)} + o(\|\mathbf{h}^{(k)}\|)$

(iii) $\mathbf{h}^{(k+1)} = o(\|\boldsymbol{\delta}^{(k)}\|)$ (iv) $\boldsymbol{\delta}^{(k)} = -\mathbf{h}^{(k)} + o(\|\boldsymbol{\delta}^{(k)}\|)$.

Proof

The definition $\|\mathbf{h}^{(k+1)}\| / \|\mathbf{h}^{(k)}\| \to 0$ of superlinear convergence is equivalent to (i) by definition of $o(\cdot)$ and to (ii) by definition of $\boldsymbol{\delta}^{(k)}$. It follows from (ii) that $\|\boldsymbol{\delta}^{(k)}\| / \|\mathbf{h}^{(k)}\| \to 1$ from which (iii) and (iv) then follow. Conversely (iii)$\Leftrightarrow$(iv)$\Rightarrow$ $\|\boldsymbol{\delta}^{(k)}\| / \|\mathbf{h}^{(k)}\| \to 1 \Rightarrow$(ii)$\Rightarrow$(i). □

Theorem 6.2.3 (Dennis–Moré Characterization Theorem)

Let $\mathbf{r}(\mathbf{x}) \in \mathbb{C}^1$ *and let* $\mathbf{x}^{(k)} \to \mathbf{x}^*$ *where* $\mathbf{r}^* = \mathbf{0}$ *and* $\mathbf{A}^*$ *is non-singular. Then the sequence converges superlinearly iff* $\boldsymbol{\delta}^{(k)} (= \mathbf{x}^{(k+1)} - \mathbf{x}^{(k)})$ *satisfies*

$$\boldsymbol{\delta}^{(k)} = \boldsymbol{\delta}^{(k)}_{\mathrm{NR}} + o(\|\boldsymbol{\delta}^{(k)}_{\mathrm{NR}}\|) \tag{6.2.16}$$

where $\boldsymbol{\delta}^{(k)}_{\mathrm{NR}} = -\mathbf{A}^{(k)^{-\mathrm{T}}} \mathbf{r}^{(k)}$.

Proof

As in Lemma 6.2.1, (6.2.16) is equivalent to

$$\boldsymbol{\delta}^{(k)} = \boldsymbol{\delta}^{(k)}_{\mathrm{NR}} + o(\|\boldsymbol{\delta}^{(k)}\|). \tag{6.2.17}$$

Because $\mathbf{A}^*$ is non-singular and by continuity of $\mathbf{A}, \mathbf{A}^{(k)^{-1}}$ is bounded in a neighbourhood of $\mathbf{x}^*$, so (6.2.17) is equivalent to

$$\mathbf{r}^{(k)} + \mathbf{A}^{(k)\mathrm{T}} \boldsymbol{\delta}^{(k)} = o(\|\boldsymbol{\delta}^{(k)}\|) \tag{6.2.18}$$

and hence by a Taylor series about $\mathbf{x}^{(k)}$ to

$$\mathbf{r}^{(k+1)} = o(\|\boldsymbol{\delta}^{(k)}\|). \tag{6.2.19}$$

Another Taylor series about $\mathbf{x}^{(k+1)}$ gives $\mathbf{r}^{(k+1)} = \mathbf{A}^{(k+1)\mathrm{T}}\mathbf{h}^{(k+1)} + o(\|\mathbf{h}^{(k+1)}\|)$ which can be used as above to deduce either $\mathbf{r}^{(k+1)} = O(\|\mathbf{h}^{(k+1)}\|)$ or $\mathbf{h}^{(k+1)} = O(\|\mathbf{r}^{(k+1)}\|)$. Using these, (6.2.19) is equivalent to $\mathbf{h}^{(k+1)} = o(\|\boldsymbol{\delta}^{(k)}\|)$ and hence finally to $\mathbf{h}^{(k+1)} = o(\|\mathbf{h}^{(k)}\|)$ by Lemma 6.2.1. □

Corollary

For a Newton-like method in which $\boldsymbol{\delta}^{(k)}$ satisfies an equation $\alpha^{(k)}\mathbf{B}^{(k)\mathrm{T}}\boldsymbol{\delta}^{(k)} = -\mathbf{r}^{(k)}$ for all k, then (6.2.16) *is equivalent to*

$$\lim \frac{\|(\alpha^{(k)}\mathbf{B}^{(k)} - \mathbf{A}^*)^{\mathrm{T}}\boldsymbol{\delta}^{(k)}\|}{\|\boldsymbol{\delta}^{(k)}\|} = 0. \tag{6.2.20}$$

Proof

Substitute $\mathbf{r}^{(k)} = -\alpha^{(k)}\mathbf{B}^{(k)\mathrm{T}}\boldsymbol{\delta}^{(k)}$ in (6.2.18) and use the equivalences in the main proof. □

For a unit step method ($\alpha^{(k)} = 1$), (6.2.20) becomes

$$\lim \frac{\|(\mathbf{B}^{(k)} - \mathbf{A}^*)^{\mathrm{T}}\boldsymbol{\delta}^{(k)}\|}{\|\boldsymbol{\delta}^{(k)}\|} = 0 \tag{6.2.21}$$

which is the way in which the Dennis–Moré result is often expressed. Note that it is not necessary to make an assumption that the matrices $\mathbf{B}^{(k)}$ are non-singular for these results to hold. Another interesting result about Newton-like line search methods following from (6.2.20), is that if $\mathbf{B}^{(k)} \to \mathbf{A}^*$ then superlinear convergence occurs iff $\alpha^{(k)} \to 1$.

Another interesting idea of a completely different type is the *Davidenko path method* or *continuation method.* This method aims to extend the domain of convergence of Newton-like methods for solving systems of nonlinear equations. It can be regarded most simply as a way of converting the original problem into a sequence of problems, for each of which a good initial approximation of the solution is available. Then the Newton-like method can be expected to solve each of these problems rapidly. An auxiliary system $\mathbf{h}(\mathbf{x}, \theta)$ is constructed, depending on an additional scalar parameter θ, and which satisfies

$$\mathbf{h}(\mathbf{x}, 0) = \mathbf{r}(\mathbf{x}) \tag{6.1.22}$$

and

$$\mathbf{h}(\mathbf{x}^{(1)}, 1) = \mathbf{0} \tag{6.2.23}$$

where $\mathbf{x}^{(1)}$ is given. It is not difficult to construct functions of this type, the most simple being

$$\mathbf{h}(\mathbf{x}, \theta) = \mathbf{r}(\mathbf{x}) - \theta\mathbf{r}^{(1)}. \tag{6.2.24}$$

If $\mathbf{x}(\theta)$ solves the system

$$\mathbf{h}(\mathbf{x}, \theta) = \mathbf{0} \tag{6.2.25}$$

then $\mathbf{x}(\theta)$ can be regarded as a trajectory parameterized by θ, which satisfies $\mathbf{x}(1) = \mathbf{x}^{(1)}$, $\mathbf{x}(0) = \mathbf{x}^*$, and is known as the *Davidenko path.* The method assumes that $\mathbf{x}(\theta)$ is continuous for θ in $[0, 1]$, and attempts to locate the trajectory for certain discrete values of θ in turn, say $\theta = 0.9, 0.8, \ldots, 0.1, 0$. The solution of $\mathbf{h}(\mathbf{x}^{(1)}, 1) = \mathbf{0}$ (which is $\mathbf{x}^{(1)}$ by (6.2.23)) is used as the initial estimate for the iteration to solve $\mathbf{h}(\mathbf{x}, 0.9) = \mathbf{0}$. Then the solution of this system is used as the initial estimate for the iteration to solve $\mathbf{h}(\mathbf{x}, 0.8) = \mathbf{0}$, and so on. The solution of $\mathbf{h}(\mathbf{x}, 0) = \mathbf{0}$ is the solution to the original system $\mathbf{r}(\mathbf{x}) = \mathbf{0}$ by (6.2.22). Provided the increment in θ is sufficiently small, and the Jacobian does not become singular, then the initial approximation for any auxiliary problem can be made arbitrarily good, so convergence is guaranteed by virtue of Theorem 6.2.1.

This simple-minded approach to the use of the Davidenko path can work, especially when the Jacobian determinant has the same sign at $\mathbf{x}(\theta)$ for all $\theta \in [0, 1]$. However, the standard Newton–Raphson method with line search is also adequate when the Jacobian matrix is non-singular in a neighbourhood of the solution which includes $\mathbf{x}^{(1)}$, and it is not clear that much is gained by going to the increased complexity of the Davidenko path method.

It seems clear then that if this method is to be valuable it must be able to solve problems in which the determinant of the Jacobian matrix changes sign. To obtain some intuition in such cases the Davidenko paths and contours of $\|\mathbf{r}\|_2$ have been illustrated in Figure 6.2.2 for the Freudenstein and Roth problem (6.2.4). In this case the Jacobian determinant is zero at $x_2 = (2 \pm \sqrt{22})/3 = -0.90$ or $+2.23$. Yet the Davidenko trajectories pass smoothly through these surfaces and in this example it is always possible to follow such a path to the global solution. For instance from $\mathbf{x}^{(1)} = (15, -2)^{\mathrm{T}}$, $\mathbf{r}^{(1)} = (34, 10)^{\mathrm{T}}$, it is possible to solve (6.2.25) in terms of x_2 to give

$$\theta = (8 + 6x_2 + 2x_2^2 - x_2^3)/12$$
$$x_1 = 10\theta - x_2^3 - x_2^2 + 14x_2 + 29.$$

Thus the trajectory $\mathbf{x}(\theta)$ has θ (and hence $\mathbf{r}$ from (6.2.24)) decreasing until $x_2 = (2 - \sqrt{22})/3$, when $\theta = 0.41$. Then θ and $\mathbf{r}$ increase until $x_2 = (2 + \sqrt{22})/3$ when $\theta = 1.69$, and finally θ and $\mathbf{r}$ decrease to zero at the global solution $x_2 = 5$. This example illustrates the main difficulty associated with these trajectories in that although a path to the global solution exists, it may be necessary to *increase* θ along parts of the trajectory and this corresponds to allowing an increase in $\|\mathbf{r}\|$. Thus descent methods of solution are immediately excluded. Furthermore the direction in which the trajectory should be followed is not defined. For instance a trajectory starting at $\mathbf{x}^{(1)} = \mathbf{0}$ must initially be followed for increasing θ to arrive at the solution. Thus any numerical method based on these ideas must be able to abandon a trajectory when it is judged that it is not likely to be converging to a solution. In general the situation is even more complicated than this example suggests; it is possible to have trajectories which do not pass through the point of attraction at a global solution and either go from ∞ to ∞, or form closed loops. Thus in no way does the method provide a guarantee of finding a global solution.

Nonetheless Figure 6.2.2 does show that it is possible to solve some problems in which a descent method would locate only a local solution. (Indeed, the trajectory which joins the local solution to the global solution in Figure 6.2.2 also passes through the saddle-point at $(23.92, 2.23)^T$, and hence minimizes the amount by which $\|\mathbf{r}\|$ must be increased, although this is not true in general.) It is well worth considering therefore how it might be possible to follow Davidenko paths in practice. Differentiating with respect to θ in (6.2.24) and (6.2.25) and using the chain rule, it follows that

$$\mathbf{A}^T \frac{d\mathbf{x}}{d\theta} - \mathbf{r}^{(1)} = \mathbf{0}$$

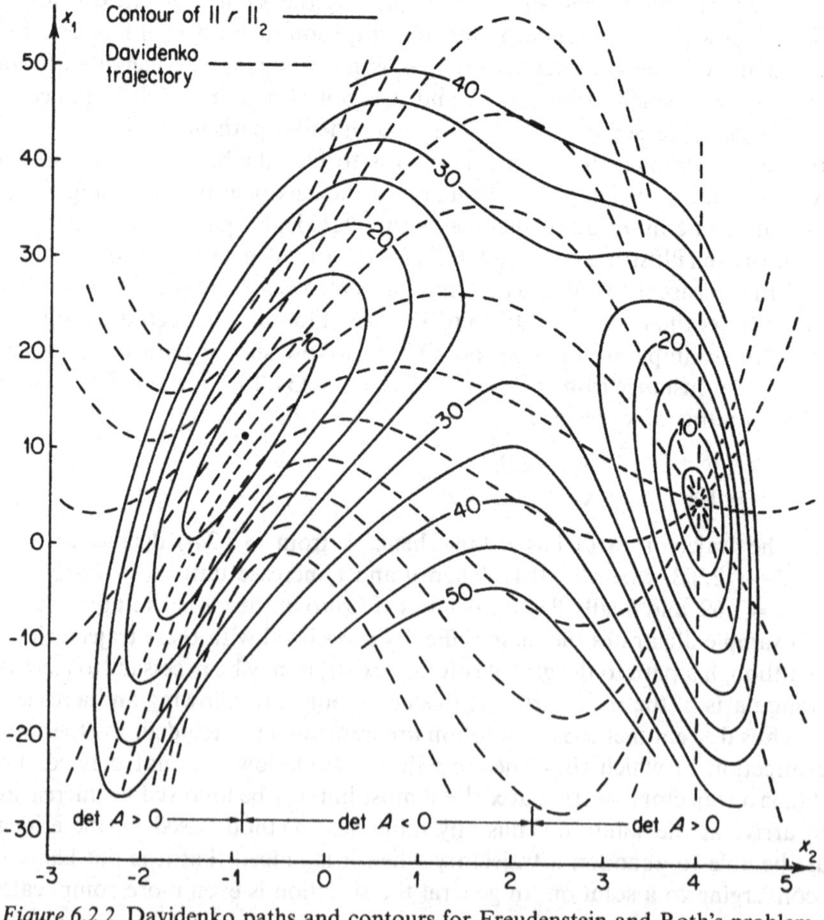

Figure 6.2.2 Davidenko paths and contours for Freudenstein and Roth's problem

and hence from (6.2.24) that

$$\frac{d\mathbf{x}}{d\theta} = \theta^{-1}\mathbf{A}^{-T}\mathbf{r}|_{\mathbf{x}(\theta)}. \tag{6.2.26}$$

Thus the Davidenko path lies along the Newton–Raphson direction (cf. (6.2.3)) and following the trajectory is the problem of solving the system of ordinary differential equations defined by (6.2.26) with the initial conditions (6.2.23). When a point on the trajectory is reached at which $\mathbf{A}$ becomes singular, then it is necessary to reverse the direction in which θ changes, as described above, in order to continue to follow the trajectory. This is pointed out by Branin and Hoo (1972) who also suggest a different parameterization in which $\theta = e^{-t}$, so that the system of differential equations (6.2.26) becomes

$$\frac{d\mathbf{x}}{dt} = -\mathbf{A}^{-T}\mathbf{r}\Big|_{\mathbf{x}(t)}. \tag{6.2.27}$$

This is an example of an *autonomous* system. The precise advantage of (6.2.27) as against (6.2.26) is not clear to me, although it may be more convenient to integrate through the singularity using (6.2.27), even though the direction in which t changes must also be reversed as above. Branin and Hoo (1972) essentially use an Euler method plus a corrector step to follow the trajectory.

It should be clear the Davidenko paths are in the nature of a global optimization technique since they do not feature the descent property and since, for sums of squares problems, a non-global minimizer does not act as a point of attraction for the trajectories. A similar technique can also be used for global minimization of general differentiable functions, although in this case both local and global minimizers act as points of attraction and the aim is to follow a Davidenko path from one such point to another. Just as for sums of squares problems, it is necessary to change the direction in which θ or t is integrated when $\mathbf{G}$ becomes singular. Branin and Hoo (1972) discuss possible strategies in more detail.

6.3 NO-DERIVATIVE METHODS

It is also important to study modifications of nonlinear least squares algorithms which avoid computing the first derivatives ∇r_i, since in practice these derivatives may not be conveniently available. The most straightforward idea is just to estimate derivatives by forward differences. Then the ith column of $\mathbf{A}^{(k)T}$ is replaced by $(\mathbf{r}(\mathbf{x}^{(k)} + h\mathbf{e}_i) - \mathbf{r}^{(k)})/h$ and is used in a Gauss–Newton or Newton–Raphson type of method (see the first paragraph of Section 3.1). Unfortunately this method can be expensive because it requires the residual vector $\mathbf{r}$ to be evaluated many times, although the option of re-using factors of previous approximate $\mathbf{A}$ matrices can be advantageous. However for *sparse systems* the number of evaluations of $\mathbf{r}$ can be reduced (see Curtis *et al.*, 1974), and the method is most useful in these circumstances.

Another similar idea is to find a linear function $\mathbf{l}(\mathbf{x})(\mathbf{l}\in\mathbb{R}^m)$ which collocates $\mathbf{r}(\mathbf{x})$ at $n+1$ points $\mathbf{x}^{(1)},\mathbf{x}^{(2)},\ldots,\mathbf{x}^{(n+1)}$ (not coplanar), and then to take a Gauss–Newton step which minimizes $\mathbf{l}^T\mathbf{l}$. For systems in which $m=n$ this is the *secant method* of Wolfe (1959). Methods of this type can be rearranged into the following form (the *generalized secant method*) which is equivalent but more convenient. Let differences

$$\mathbf{y}^{(k)}=\mathbf{r}^{(k+1)}-\mathbf{r}^{(k)} \tag{6.3.1}$$

(where $\mathbf{r}^{(k)}=\mathbf{r}(\mathbf{x}^{(k)})$) and

$$\boldsymbol{\delta}^{(k)}=\mathbf{x}^{(k+1)}-\mathbf{x}^{(k)} \tag{6.3.2}$$

be defined, for $k=1,2,\ldots,n$. Then a matrix $\mathbf{J}^{(n+1)}$ which approximates $\mathbf{A}$ can be calculated, which satisfies

$$\mathbf{J}^{(n+1)\mathrm{T}}\boldsymbol{\delta}^{(k)}=\mathbf{y}^{(k)},\qquad k=1,2,\ldots,n, \tag{6.3.3}$$

by using the updating formula due to Barnes (1965)

$$\mathbf{J}^{(k+1)}=\mathbf{J}^{(k)}+\frac{\mathbf{w}^{(k)}(\mathbf{y}^{(k)}-\mathbf{J}^{(k)\mathrm{T}}\boldsymbol{\delta}^{(k)})^{\mathrm{T}}}{\mathbf{w}^{(k)\mathrm{T}}\boldsymbol{\delta}^{(k)}} \tag{6.3.4}$$

where $\mathbf{w}^{(k)}$ is any vector which is orthogonal to $\boldsymbol{\delta}^{(1)},\boldsymbol{\delta}^{(2)},\ldots,\boldsymbol{\delta}^{(k-1)}$. This result is readily established by induction. Formula (6.3.4) is closely analogous to the updating formula in quasi-Newton methods (Section 3.2), and the equation $\mathbf{J}^{(k+1)\mathrm{T}}\boldsymbol{\delta}^{(k)}=\mathbf{y}^{(k)}$ corresponds to the quasi-Newton condition. This choice of $\mathbf{w}^{(k)}$ ensures that the updated matrix satisfies a hereditary property. When $\mathbf{r}(\mathbf{x})$ is a linear system, and the $\boldsymbol{\delta}^{(k)}$ are independent, it follows that $\mathbf{J}^{(n+1)}=\mathbf{A}$ so that the resulting Gauss–Newton step solves the problem exactly. Thus a termination result holds. All these results are closely related to those for the rank one formula (3.2.7) for minimization. An illustrative numerical example is given in Question 6.9.

An advantage of secant methods over the basic finite difference method is that they enable the iteration to proceed beyond the $n+1$th step without re-calculating all the information in $\mathbf{A}$, and this is considerably more efficient. On iteration $n+1$ the resulting difference pair $\mathbf{y}^{(n+1)}$, $\boldsymbol{\delta}^{(n+1)}$ yields too much information to determine $\mathbf{A}$. However using (6.3.4) and making $\mathbf{w}^{(n+1)}$ orthogonal to $\boldsymbol{\delta}^{(2)},\boldsymbol{\delta}^{(3)},\ldots,\boldsymbol{\delta}^{(n)}$ ensures that $\mathbf{J}^{(n+2)}\boldsymbol{\delta}^{(j)}=\mathbf{y}^{(j)}$ for $j=2,3,\ldots,n+1$. Essentially therefore the difference pair $\mathbf{y}^{(1)},\boldsymbol{\delta}^{(1)}$ is dropped out and replaced by $\mathbf{y}^{(n+1)}$, $\boldsymbol{\delta}^{(n+1)}$. Thus for any $k>n$ an approximate matrix $\mathbf{J}^{(k+1)}$ can be calculated which satisfies $\mathbf{J}^{(k+1)\mathrm{T}}\boldsymbol{\delta}^{(j)}=\mathbf{y}^{(j)}$ for the *most recent n* difference pairs. The resulting matrix can then be used to replace the true Jacobian matrix $\mathbf{A}$ in the Newton–Raphson method (6.2.3). This approach requires of order mn computer operations to update $\mathbf{J}^{(k)}$ on each iteration.

It is also worthwhile to consider alternative representations of $\mathbf{J}^{(k)}$ which enable the Gauss–Newton or Newton–Raphson steps (that is (6.1.9(a)) or (6.2.3(a)) with $\mathbf{A}^{(k)}$ approximated by $\mathbf{J}^{(k)}$) to be computed efficiently. Direct use of (6.1.9(a)) requires at least $\frac{1}{6}n^3+\frac{1}{2}mn^2+O(mn)$ multiplications, whereas it is possible to use the fact that $\mathbf{J}^{(k)}$ is modified by the rank one update (6.3.4) to reduce this figure to order mn. One possibility is to define matrices $[\mathbf{y}]^{(k)}$ and $[\boldsymbol{\delta}]^{(k)}$ whose columns contain the most recent n (or k if $k < n$) difference pairs (6.3.1) and (6.3.2). If these matrices have full rank ($\min(k, n)$), and suppressing superscript (k), then the representation

$$\mathbf{J}^{\mathrm{T}} = [\mathbf{y}][\boldsymbol{\delta}]^{+} \tag{6.3.5}$$

satisfies (6.3.3) and in fact corresponds to the use of (6.3.4) when $\mathbf{w}^{(k)}$ is the component of $\boldsymbol{\delta}^{(k)}$ orthogonal to $\boldsymbol{\delta}^{(1)}, \boldsymbol{\delta}^{(2)}, \ldots, \boldsymbol{\delta}^{(k-1)}$ (if $k < n$). The matrix $[\boldsymbol{\delta}]^{+}$ denotes the full rank generalized inverse $([\boldsymbol{\delta}]^{\mathrm{T}}[\boldsymbol{\delta}])^{-1}[\boldsymbol{\delta}]^{\mathrm{T}}$. It can then be verified from the Penrose conditions that

$$\mathbf{K}^{\mathrm{T}} = [\boldsymbol{\delta}][\mathbf{y}]^{+} \tag{6.3.6}$$

is a representation of $\mathbf{J}^{+\mathrm{T}}$, and that the Gauss–Newton search direction can be written as

$$\mathbf{s}^{(k)} = -\mathbf{K}^{(k)\mathrm{T}}\mathbf{r}^{(k)}. \tag{6.3.7}$$

In practice therefore the generalized secant method uses this formula in a line search algorithm, and some representation of $\mathbf{K}^{(k)}$ is updated to correspond with (6.3.4). An algorithm along these lines is suggested by Powell (1965) which uses the representation (6.3.6) for $\mathbf{K}$, and updates the matrix $([\mathbf{y}]^{\mathrm{T}}[\mathbf{y}])^{-1}$ which appears in the definition of $[\mathbf{y}]^{+}$. However this approach is likely to lose precision because of the 'squaring' effect referred to in Section 6.1, and this can easily be avoided by updating the matrix $[\mathbf{y}]^{+}$ directly, using the formulae given by Fletcher (1969). A more stable although more cumbersome alternative is to update factors $\mathbf{J}^{\mathrm{T}} = \mathbf{QR}$, where $\mathbf{Q}$ is orthogonal and $\mathbf{R}$ is upper triangular (for example, Gill and Murray, 1978a), and to use the representation $\mathbf{K}^{\mathrm{T}} = \mathbf{R}^{-1}\mathbf{Q}^{\mathrm{T}}$ in (6.3.7.). More recently, however, a stable method for updating $\mathbf{LU}$ factors of $\mathbf{J}$ with partial pivoting has been developed by Fletcher and Matthews (1985) and this now appears to be the most satisfactory way of maintaining an invertible representation of $\mathbf{J}$.

Some other practical points concerning generalized secant methods include the following. An initial choice $\mathbf{K}^{(1)} = \mathbf{0}$ means that (6.3.7) cannot be used for $k = 1$, so some thought must be given to initializing the algorithm. The most simple possibility is to make searches along coordinate directions for the first n steps and to use (6.3.7) thereafter, although other strategies spring to mind. Another point concerns the stability of the algorithm which can break down when the columns of $[\boldsymbol{\delta}]^{(k)}$ are close to being dependent. This can arise (Powell 1972e) and so some modification is desirable. Powell (1965) modifies the rule for rejecting the oldest column of $[\boldsymbol{\delta}]^{(k)}$ in favour of the new vector $\boldsymbol{\delta}^{(k+1)}$. and in fact allows *any* column to be rejected, using the resulting freedom to maintain independence. Finally it might be thought desirable to include a Levenberg–Marquardt modification so as to ensure a convergence proof. However this obviates the possibility of updating representations of $\mathbf{K}$ efficiently. In this case a Powell dog-leg modification (Section 5.2) can be used, although since the gradient vector $\mathbf{g}^{(k)}$ is only estimated by $2\mathbf{J}^{(k)}\mathbf{r}^{(k)}$ a convergence result does not follow directly.

An alternative to generalized secant methods is given by Broyden (1965) who uses the updating formula (6.3.4), but with $\mathbf{w}^{(k)} = \boldsymbol{\delta}^{(k)}$. For linear systems this generally destroys the hereditary property (6.3.3) and hence the result $\mathbf{J}^{(n+1)} = \mathbf{A}$ which implies termination. Also certain invariance properties possessed by (6.3.4) are lost. However, this updating formula does have the nice property that

$$\mathbf{J}^{(k+1)} - \mathbf{A} = \left(\mathbf{I} - \frac{\boldsymbol{\delta}^{(k)}\boldsymbol{\delta}^{(k)\mathrm{T}}}{\boldsymbol{\delta}^{(k)\mathrm{T}}\boldsymbol{\delta}^{(k)}} \right)(\mathbf{J}^{(k)} - \mathbf{A})$$

so that the error in the first derivative approximation is usually reduced, because it is multiplied by a symmetric projection matrix. This updating formula also has the advantage that the denominator of the correction in (6.3.4) is $\boldsymbol{\delta}^{(k)\mathrm{T}}\boldsymbol{\delta}^{(k)}$ so the correction itself cannot become arbitrarily large, which can happen with the Barnes secant method. It has also been shown (Gay, 1979; Gerber and Luk, 1980) that methods based on using Broyden's formula terminate in at most $2n$ iterations—a surprising result when it was first discovered. Broyden reports good performance with this type of update and it has since been used successfully in a number of methods. A modification of Broyden's update suitable for sparse matrix applications is given by Schubert (1970) and is described in Question 6.10.

Another interesting finite difference approach is the Brown–Brent type of method (Brent, 1973b). This introduces an inner iteration in which the elimination steps implicit in the LU factorization of $\mathbf{A}^{(k)}$ are carried out successively rather than simultaneously. Specifically at $\mathbf{x}^{(k,1)} = \mathbf{x}^{(k)}$ the first equation $r_1(\mathbf{x})$ only is considered, and the linearization

$$\mathbf{a}_1^{(k,1)\mathrm{T}}(\mathbf{x} - \mathbf{x}^{(k,1)}) + r_1^{(k,1)} = 0 \tag{6.3.8}$$

defines a hyperplane. A new point $\mathbf{x}^{(k,2)}$ which satisfies (6.3.8) is chosen, and (6.3.8) is used implicity to eliminate one variable from the problem. Then the reduced problem in $n-1$ equations and $n-1$ variables is treated in an analogous manner from $\mathbf{x}^{(k,2)}$, and so on. The method has no advantages when derivatives are available, but for a no-derivative method at $\mathbf{x}^{(k,l)}$ the resulting reduced problem has only $n-l+1$ equations and variables and therefore it is possible to estimate the reduced normal vector $\mathbf{a}^{(k,l)}$ by making only $n-l+1$ difference calculations. Thus to solve a linear system requires $\frac{1}{2}n^2 + O(n)$ evaluations of $r_i(\mathbf{x})$ rather than n^2, and the method is correspondingly more efficient than the basic finite difference method. Any linear equations should be taken first, so that the elimination which they imply need not be repeated for different k. The method is implemented by finding elementary matrices, the first of which causes $\mathbf{M}_1\mathbf{a}_1 = u_{11}\mathbf{e}_1$, and which correspond to the reduction $\mathbf{M}_{n-1}\cdots\mathbf{M}_2\mathbf{M}_1\mathbf{A} = \mathbf{U}$. Different methods are possible because a different choice of $\mathbf{M}_1$ determines a different point $\mathbf{x}^{(k,2)}$ which solves (6.3.1), and so on. The most stable choice is to use elementary Householder orthogonal matrices, as suggested by Brent (1973b), where more details can be found. In practice I would expect the method not to be as efficient as the generalized secant or Broyden methods, because the feature of preserving information in **A** is not present. However, I know of no comparative study which confirms this observation.

QUESTIONS FOR CHAPTER 6

6.1. Consider fitting the function $\phi(t, \mathbf{x}) = x_1 e^{-x_2 t}$ at points $t_i = -1, 0, 1, 2$ to data values $d_i = 2.7, 1, 0.4, 0.1$ which represent the true values rounded to one decimal place. Carry out one iteration of the Gauss–Newton method from $\mathbf{x}^{(1)} = (1, 1)^T$, and also calculate approximately the correction for the second iteration using the factors of $\mathbf{A}^{(1)}\mathbf{A}^{(1)T}$. Verify that the linear rate of convergence is rapid. Estimate the standard deviation of errors in the data values and show that it is consistent with how the errors are generated. Calculate the variance–covariance matrix and the standard deviation of the variables x_1 and x_2, and verify that the first iteration gives a solution whose error is insignificant. Calculate approximately the matrices $\nabla^2 r_i$ and the matrix $2\mathbf{B}^{-1}\Sigma r_i \nabla^2 r_i$, and hence verify that the linear rate of convergence is consistent with (6.1.11).

6.2. Find the solution to the following data fitting problem (Walsh, 1975) in which

$$\phi(t, \mathbf{x}) = (1 - x_1 t / x_2)^{(1/(x_1 c)) - 1}$$

where $c = 96.05$, and the data values are

t_i	2000	5000	10000	20000	30000	50000
d_i	0.9427	0.8616	0.7384	0.5362	0.3739	0.3096

Minimize the sum of squares of the residuals $r_i(\mathbf{x})$ defined by (6.1.4). Use a computer subroutine which implements the Gauss–Newton method with line search. Find the standard deviations of the data d_i and the variables x_1 and x_2.

6.3. Consider the data fitting problem in Question 6.1. Show that if x_2 is fixed then the linear least squares solution for x_1 is $x_1 = \Sigma d_i e^{-x_2 t_i} / \Sigma e^{-2x_2 t_i}$ with best sum of squares $f = \Sigma d_i^2 - x_1 \Sigma d_i e^{-x_2 t_i}$. Hence solve the data-fitting problem by regarding x_1 and f as functions of x_2 and by carrying out a line search on x_2. Verify that the solution is the same as that obtained in Question 6.1.

6.4. Carry out three iterations of the Newton–Raphson method applied to the Chebyquad $n = 2$ equations

$$r(\mathbf{x}) = x_1 + x_2 - 1 = 0$$
$$r_2(\mathbf{x}) = (2x_1 - 1)^2 + (2x_2 - 1)^2 - \tfrac{2}{3} = 0$$

from $\mathbf{x}^{(1)} = (\tfrac{1}{3}, \tfrac{2}{3})^T$. Show that the corrections are consistent with a quadratic order of convergence and hence estimate the error in $\mathbf{x}^{(4)}$. Show that if $\mathbf{x}^{(1)}$ is any vector which satisfies $r_1(\mathbf{x}) = 0$ and $x_1 < x_2$, then the Newton–Raphson method will converge.

6.5. Consider the system of equations

$$r_1(\mathbf{x}) = x_1^3 - x_2 - 1 = 0$$
$$r_2(\mathbf{x}) = x_1^2 - x_2 = 0.$$

Show that $f(\mathbf{x}) = \mathbf{r}(\mathbf{x})^T\mathbf{r}(\mathbf{x})$ has a global minimizer at $\mathbf{x} = (1.46557, 2.14790)^T$ which solves the above system, and a local minimizer at $\mathbf{x} = (0, -\frac{1}{2})^T$ which does not. Show that there is also a saddle point at $\mathbf{x} = (\frac{2}{3}, -\frac{7}{54})^T$. Show that the Newton–Raphson method converges rapidly from $\mathbf{x}^{(1)} = (1.5, 2.25)^T$ and verify that the order of convergence is quadratic. If $\mathbf{x}^{(1)}$ is chosen close to the local minimizer, observe that the Newton–Raphson method behaves wildly, but ultimately converges to the global solution.

6.6. Verify that the equation $r_2(\mathbf{x}) = 0$ in Question 6.4 can be rearranged as

$$x_2 - x_1 = \tfrac{1}{2}\{\tfrac{2}{3} - 2(2x_1 - 1)(2x_2 - 1)\}^{1/2}.$$

Hence, using $r_1(\mathbf{x}) = 0$, obtain a successive substitution iteration formula and show that it converges from $\mathbf{x}^{(1)} = (\frac{1}{3}, \frac{2}{3})^T$. Verify that the errors are consistent with a linear order of convergence and estimate the rate constant. Evaluate the matrix $\mathbf{U}^*$ and check that the rate of convergence is consistent with Theorem 6.2.2.

6.7. For the system in Question 6.5, and from $\mathbf{x} = (0.67, -0.13)^T$, consider computing the $\mathbf{LL}^T$ factorization of $\mathbf{AA}^T$ in three different ways.

(i) Compute $\mathbf{A}$ and find the orthogonal matrix

$$\mathbf{Q} = \begin{bmatrix} \cos\theta & \sin\theta \\ \sin\theta & -\cos\theta \end{bmatrix}$$

for which $\mathbf{AQ} = \mathbf{L}$, working to high precision.

(ii) Repeat case (i) but round-off the result of *every* arithmetic operation to four significant decimal digits.

(iii) Compute $\mathbf{A}$, $\mathbf{AA}^T$, and then factorize the resulting matrix to give $\mathbf{LL}^T$, working again to four digits.

Observe that the element l_{22} in (iii) has no significance, whereas that in case (ii) has 1–2 significant figures remaining. Hence show that the correction $\boldsymbol{\delta}$ in case (ii) has similar accuracy (computed from $\mathbf{L}^T\boldsymbol{\delta} = -\mathbf{Qr}$), whereas that in case (iii) is incorrect by a factor of 100 (computed from $\mathbf{LL}^T\boldsymbol{\delta} = -\mathbf{Ar}$). This example illustrates the loss of precision caused by the 'squaring' effect in case (iii).

6.8. Consider solving the system of equations in Question 6.5 by the Davidenko path method. Verify that the Jacobian matrix is singular for all $\mathbf{x}$ such that $x_1 = 0$ or $x_1 = \frac{2}{3}$. Take $\mathbf{x}^{(1)} = (-1, 1)^T$ and show that the trajectory $\mathbf{x}(\theta)$ which solves (6.2.25) and (6.2.24) exists for all x_1, and that x_2 and θ are given by $x_2 = x_1^2$ and $\theta = (1 + x_1^2 - x_1^3)/3$. Verify that the Davidenko path leads to the solution but that θ (and hence $\|\mathbf{r}\|$) is not a decreasing function along the whole path. Verify also that there exists a Davidenko path between the local and global minimizers of $\|\mathbf{r}\|_2^2$ which passes through the saddle-point (see Question 6.5).

6.9. Consider applying a secant method to solve the system in Question 6.4. Let $\mathbf{x}^{(1)} = (1,0)^T$, $\mathbf{x}^{(2)} = (\frac{1}{2},0)^T$, and $\mathbf{x}^{(3)} = (\frac{1}{4},\frac{3}{4})^T$, and evaluate $\mathbf{r}^{(1)}, \mathbf{r}^{(2)}$, and $\mathbf{r}^{(3)}$. Find the linear function $\mathbf{l}(\mathbf{x})$ which collocates these values and find the point $\mathbf{x}^{(4)}$ which solves $\mathbf{l}(\mathbf{x}) = \mathbf{0}$.

Also evaluate the differences in (6.3.1) and (6.3.2) and apply the Barnes formula (6.3.4) for $k = 1$ and 2. Verify that $\mathbf{J}^{(3)}$ is the Jacobian matrix of $\mathbf{l}(\mathbf{x})$. Also apply formula (6.2.3(a)) to $\boldsymbol{\delta}^{(3)}$ (using $\mathbf{J}^{(3)}$ to approximate $\mathbf{A}^{(3)}$) and verify that $\mathbf{x}^{(3)} + \boldsymbol{\delta}^{(3)}$ is the above point $\mathbf{x}^{(4)}$. Also use the Broyden formula ((6.3.4) with $\mathbf{w}^{(k)} = \boldsymbol{\delta}^{(k)}$) to calculate $\mathbf{J}^{(3)}$ and verify that a different matrix is obtained.

6.10. Consider updating the approximate Jacobian matrix $\mathbf{J}^{(k)}$ for a sparse system of equations, and let $\mathbf{j}_i^{(k)}$ denote the ith column of $\mathbf{J}^{(k)}$. Let $\mathbf{a}_i$ be the ith column of the true Jacobian matrix. Show that if $\mathbf{j}_i^{(k)}$ is updated by

$$\mathbf{j}_i^{(k+1)} = \mathbf{j}_i^{(k)} + w\mathbf{d}_i$$

then $\mathbf{d}_i$ must have a zero element whenever the corresponding element of $\mathbf{a}_i$ is any known constant (including zero), assuming that the corresponding element of $\mathbf{j}_i^{(k)}$ takes this constant value. Hence show that the quasi-Newton condition $\mathbf{J}^{(k+1)T}\boldsymbol{\delta} = \mathbf{y}$ is satisfied by the update

$$\mathbf{j}_i^{(k+1)} = \mathbf{j}_i + \frac{\mathbf{d}_i(y_i - \boldsymbol{\delta}^T\mathbf{j}_i)}{\boldsymbol{\delta}^T\mathbf{d}_i}, \qquad i = 1, 2, \ldots, m$$

(suppressing superscript (k)). In practice Schubert (1970) suggests choosing $\mathbf{d}_i$ by zeroing the appropriate components of $\boldsymbol{\delta}$, which reduces to the Broyden formula ((6.3.4) with $\mathbf{w}^{(k)} = \boldsymbol{\delta}^{(k)}$) when no constant values are known. Notice that in general the resulting correction to $\mathbf{J}^{(k)}$ is not a matrix of rank one.

6.11. It is required to fit a function $y(t, \mathbf{x})$ ($\mathbb{R}^1 \times \mathbb{R}^n \to \mathbb{R}^1$) to given data values $y_1, y_2, \ldots, y_m$ which correspond to given data points $t_1, t_2, \ldots, t_m$. The parameter vector $\mathbf{x}$ is unknown and must be determined so that $y(t, \mathbf{x})$ best fits the data in the least squares sense. In this case, however, $y(t, \mathbf{x})$ is not known directly but is defined for all t and fixed $\mathbf{x}$ by solving the ordinary differential equation

$$y' = f(y, t, \mathbf{x}) \qquad y(0) = h(\mathbf{x}).$$

The functions f and h are given, and f is such that the differential equation must be solved numerically. What further information is required to enable the Gauss–Newton method to be used to solve the data fitting problem? Show that this information can be obtained by solving a certain initial value system of first-order differential equations, defined in terms of a vector $\mathbf{z}(t, \mathbf{x})$ ($\mathbf{z} \in \mathbb{R}^{n+1}$) where $z_1 = y$ above, and $z_{1+i} = \partial y/\partial x_i$, $i = 1, 2, \ldots, n$.

As an example, consider the differential equation

$$y' = -x_1 y/(x_2 + y) \qquad y(0) = x_3$$

and the data values

y_i	24.44	19.44	15.56	10.56	9.07	6.85	4.07	1.67
t_i	0	23.6	49.1	74.5	80.0	100.0	125.5	147.3

Write a computer program to obtain a best least squares fit in the above way (subroutines for the Gauss–Newton method and the solution of a system of ordinary differential equations are required). Attempt to obtain the sum of squares correct to five decimal places. An initial estimate for the unknown parameter vector is $\mathbf{x} = (0.22, 3.27, 24.44)^T$. Comment on the accuracy obtained in this vector and the extent to which the Gauss–Newton method is successful in solving the problem. Estimate the effect of truncation error in the initial value problem by repeating the calculation with a smaller step length.

PART 2

CONSTRAINED OPTIMIZATION

Chapter 7
Introduction

7.1 PREVIEW

The motivation for studying constrained minimization has been discussed at some length in Chapter 1 of this book. The mathematical background given there, and indeed many of the concepts which arise in unconstrained optimization, are important in the study of constrained optimization. In Part 2, the selection of material from the extensive literature which exists has again been done with the main theme of practicality in mind. Thus topics such as reliability and effectiveness are uppermost; to some extent these are measured by convergence and order of convergence results. These aspects are therefore studied in some detail, and together with the subject of optimality conditions they provide good material for an academic course. However, the use of *experimentation* to validate the properties of an algorithm is still of paramount importance. In fact the study of constrained optimization is by no means as well advanced as for the unconstrained case. The writing of software is a much more complex task, and so comparative experimental results are much less widely available. Often there is even a lack of suitable test problems. Also many more special cases arise and the problem of assessing numerical evidence is more difficult. For all these reasons Part 2 departs from the feature in Part 1 of presenting detailed numerical evidence. Nonetheless important experimental results do exist in the literature and the selection of material is guided by such results. This lack of certainty also shows up in that the decision as to precisely what algorithm to recommend in any one case is often not clear. For this reason the availability of good well-documented library software is often poor. Thus I appreciate the fact that many algorithms are necessarily used which are not ideal and I have tried to make users aware of defects in these algorithms and to enable them to mitigate their worst effects.

The structure of most constrained optimization problems is essentially contained in the following:

$$\begin{array}{lll} \text{minimize} & f(\mathbf{x}) & \mathbf{x}\in\mathbb{R}^n \\ \text{subject to} & c_i(\mathbf{x})=0, & i\in E \\ & c_i(\mathbf{x})\geqslant 0, & i\in I. \end{array} \qquad (7.1.1)$$

As in Part 1, $f(\mathbf{x})$ is the *objective function*, but there are additional *constraint functions* $c_i(\mathbf{x}), i=1,2,\ldots,p$. E is the index set of equations or equality constraints in the problem, I is the set of inequality constraints, and both these sets are finite. More general constraints can usually be put into this form: for example $c_i(\mathbf{x})\leqslant b$ becomes $b-c_i(\mathbf{x})\geqslant 0$. If any point $\mathbf{x}'$ satisfies all the constraints in (7.1.1) it is said to be a *feasible point* and the set of all such points is referred to as the *feasible region* R. As in Part 1, maximization problems are easily handled by the transformation $\max f(\mathbf{x}) = -\min -f(\mathbf{x})$. Also, a local minimizer or solution (referred to by $\mathbf{x}^*$) is looked for, rather than a global minimizer, the computation of which can be difficult. The definition of a constrained local minimizer $\mathbf{x}^*$ is that $f(\mathbf{x}^*)\leqslant f(\mathbf{x})$ for all feasible $\mathbf{x}$ sufficiently close to $\mathbf{x}^*$. Definitions of strict or isolated local minimizers can be made in a similar way to Section 2.1.

It is possible to illustrate the effect of the constraints when $n=2$ by drawing the zero contour of each constraint function. For an equality constraint, the line itself is the set of feasible points; for an inequality constraint the line marks the boundary of the feasible region and the infeasible side is conventionally shaded. This is shown in Figure 7.1.1. Case (i) has constraints $x_2=x_1^2$, $\mathbf{x}\geqslant\mathbf{0}$, which can be written $c_1(\mathbf{x})=x_2-x_1^2$, $c_2(\mathbf{x})=x_1$, $c_3(\mathbf{x})=x_2$ and $E=\{1\}$, $I=\{2,3\}$. Case (ii) has constraints $x_2\geqslant x_1^2$, $x_1^2+x_2^2\leqslant 1$ which can be written as $c_1(\mathbf{x})=x_2-x_1^2$, $c_2(\mathbf{x})=1-x_1^2-x_2^2$, $I=\{1,2\}$, and E empty. Formulation (7.1.1) covers most types of problem; however the condition that some variables x_i take only discrete values is not included. This type of condition is covered in *integer programming* which is largely beyond the scope of this book. However, a useful general purpose algorithm is the branch and bound method which enables the problem to be reduced to a sequence of smooth problems and hence solved by other techniques given in this book. This is described in Section 13.1. Another type of condition which is not included in (7.1.1) is a constraint of the

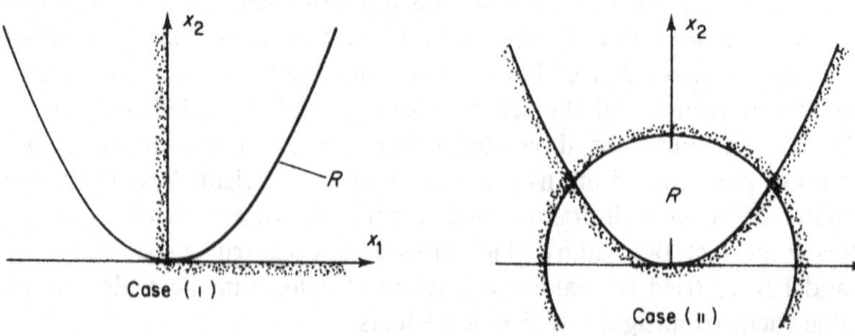

Figure 7.1.1 Examples of feasible regions

form $c_i(\mathbf{x}) > 0$; something more is said about this case in Section 7.2. In fact there is often some choice in how best to pose the problem in the first instance and a number of possibilities of this type are discussed in Section 7.2.

It is assumed in (7.1.1) that the functions $c_i(\mathbf{x})$ are continuous which implies that R is closed. It is also assumed that $f(\mathbf{x})$ is continuous for all $\mathbf{x} \in R$ and preferably for all $\mathbf{x} \in \mathbb{R}^n$. If in addition the feasible region is non-empty and bounded ($\exists a > 0$ such that $\|\mathbf{x}\| \leqslant a \ \forall \mathbf{x} \in R$), then it follows that a solution $\mathbf{x}^*$ exists. If not then the problem may be *unbounded* ($f(\mathbf{x}) \to -\infty, \mathbf{x} \in R$) or may not have a minimizing point. The problem also has no solution when R is empty, that is when the constraints are inconsistent. In fact most practical methods require the stronger assumption that the objective and constraint functions are also smooth in that their first and often second continuous derivatives exist ($f, c_i \in \mathbb{C}^1$ or $\mathbb{C}^2$). The notation $\nabla f (= \mathbf{g})$ and $\nabla^2 f (= \mathbf{G})$ for the gradient vector and Hessian matrix of f is described in Chapter 1. The notation ∇c_i and $\nabla^2 c_i$ is used to denote the corresponding first and second derivatives of any constraint function c_i. The vector ∇c_i is also denoted by $\mathbf{a}_i$ and is referred to as the *normal vector* of the constraint c_i. Note that $\mathbf{a}_i$ refers to the ith vector in a set and not to the ith component of $\mathbf{a}$. These vectors are sometimes collected into columns of the *Jacobian matrix* $\mathbf{A}$ (although this rule is contradicted in the simplex method (Chapter 8) in which the normal vectors are the rows of $\mathbf{A}$). The vector $\mathbf{a}_i'$ (that is $\mathbf{a}_i(\mathbf{x})$ evaluated at $\mathbf{x} = \mathbf{x}'$) is the direction of greatest increase of $c_i(\mathbf{x})$ at $\mathbf{x}'$, and if $c_i = 0$ and $i \in I$ then the direction is on the feasible side of the constraint (see Figure 7.1.2) and is at right angles to the zero contour. Most of what follows assumes the existence of these derivatives which can be used, for example, to characterize optimality conditions, as described in Chapter 9, which generalize the results for unconstrained minimization given in Section 2.1. This is not to say necessarily that user supplied formulae for these derivatives are required in any method. Mostly, however, formulae for first derivatives are required, and in some cases formulae for second derivatives also. Methods which require no derivative information have not been studied to any great extent and the obvious advice is to estimate these derivatives by finite differences (see (2.3.8)), although the resulting algorithm is likely to be less robust and effective when this is done. A different situation arises when the functions f and c_i do not have continuous derivatives, which is referred to as *non-smooth* or

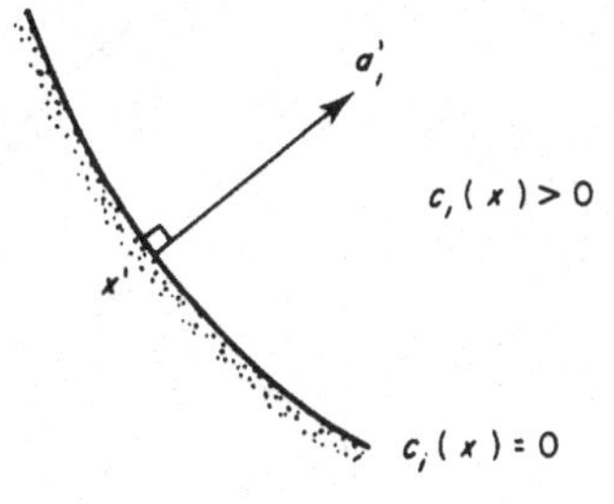

Figure 7.1.2 The normal vector

non-differentiable optimization. In this case methods for smooth problems are not appropriate and special attention must be given to the surfaces of non-differentiability. These behave somewhat like the boundary of a constraint in (7.1.1) and it is therefore appropriate to discuss the problem under the heading of constrained optimization. This is done for unconstrained non-smooth optimization in Chapter 14. These ideas are also generalized to allow non-smooth constraint functions.

Another important concept is that of an *active* or *binding constraint.* Active constraints at any point $\mathbf{x}'$ are defined by the index set

$$\mathscr{A}' = \mathscr{A}(\mathbf{x}') = \{i: c_i(\mathbf{x}') = 0\} \tag{7.1.2}$$

so that any constraint is active at $\mathbf{x}'$ if $\mathbf{x}'$ is on the boundary of its feasible region. If $\mathbf{x}'$ is feasible then $\mathscr{A}' \supset E$ clearly follows. In particular the set $\mathscr{A}^*$ of active constraints at the solution of (7.1.1) is of some importance. If this set is known then the remaining constraints can be ignored (locally) and the problem can be treated as an equality constraint problem with $E = \mathscr{A}^*$. Also constraints with $i \notin \mathscr{A}^*$ can be perturbed by small amounts without affecting the local solution whereas this is not usually true for an active constraint. An example is given by the problem: minimize $f(\mathbf{x}) = -x_1 - x_2$ subject to $x_2 \geqslant x_1^2$ and $x_1^2 + x_2^2 \leqslant 1$. Clearly from Figure 7.1.3 the solution is achieved at $\mathbf{x}^* = (1/\sqrt{2}, 1/\sqrt{2})^{\mathrm{T}}$ when the contour of $f(\mathbf{x})$ is a tangent to the unit circle. Thus the active set in the notation of Figure 7.1.1(ii) is $\mathscr{A}^* = \{2\}$, and the circle constraint $c_2(\mathbf{x})$ is active. Likewise the parabola constraint $c_1(\mathbf{x})$ is inactive and can be perturbed or removed from the problem without changing $\mathbf{x}^*$. A further refinement of this definition to include strongly active and weakly active constraints is given in Figure 9.1.2.

Methods for the solution of (7.1.1) are usually iterative so that a sequence $\mathbf{x}^{(1)}, \mathbf{x}^{(2)}, \mathbf{x}^{(3)}, \ldots$, say, is generated from a given point $\mathbf{x}^{(1)}$, hopefully converging to $\mathbf{x}^*$. If $\mathbf{x}^*$ is a member of the sequence then the method is said to *terminate.* Some early methods for constrained optimization were developed in an *ad hoc* way and are not strongly supported theoretically. Because of this these methods

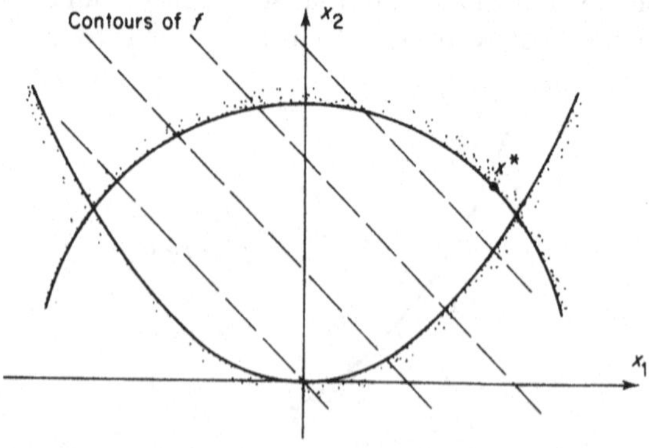

Figure 7.1.3 Active and inactive constraints

are often unreliable and expensive for problems of any size and they are not described here. However a review of what has been attempted is given by Swann (1974). The subject of constrained optimization splits into two main parts, *linear constraint programming* and *nonlinear programming* which have quite different features. In linear constraint programming each constraint is a *linear function* $c_i(\mathbf{x}) = \mathbf{a}_i^T\mathbf{x} - b_i$. (More precisely an *affine function*: see the remarks in Section 1.2.) The boundary of the feasible region for any one such constraint is a hyperplane, and the normal vector ∇c_i is constant and is again the vector $\mathbf{a}_i$. Linear constraint problems can be handled by a combination of an elimination method and an active set method (see Section 7.2) and the iterates $\mathbf{x}^{(k)}$ are always feasible points. The simplest cases are when the objective function is either linear or quadratic (*linear programming* or *quadratic programming*—Chapters 8 and 10 respectively) in both of which cases algorithms which terminate can be determined. The application to a general objective function is given in Chapter 11, and in this case many of the possibilities for unconstrained optimization carry over directly. For example there are analogues of Newton's method, quasi-Newton methods, the Gauss–Newton method, restricted step methods, and no-derivative methods which use finite difference approximations. Similar considerations hold in regard to using line searches and in regard to deciding what type of convergence test to use to terminate the iteration. Special cases of a linear constraint are the bounds $x_i \geqslant l_i$ or $x_i \leqslant u_i$ in which $\mathbf{a}_i$ is $\pm\mathbf{e}_i$, the ith coordinate vector, and it is particularly simple to handle such constraints. It is important that algorithms should take this into account. A special type of linearly constrained problem is a *network program* in which the constraints represent conservation of flow in a network. Very efficient methods exist which can take advantage of the network structure. The subject also has an elegant theoretical structure associated with graph theory and is described in Section 13.3.

The most difficult type of smooth constrained minimization problem is nonlinear programming (Chapter 12) in which there exist some nonlinear constraint functions in the problem. In this case a completely satisfactory general purpose method has yet to be agreed upon and the subject is one of intense research activity. Of course if the nonlinear constraints can be rearraned so as to be eliminated directly then this should be done, but is not usually possible. Indirect elimination by solving a system of equations numerically is possible (Sections 7.2 and 12.5) but is not usually efficient and other difficulties exist; this idea is closely related to another approach known as a *feasible direction method*. A different approach is to attempt to transform the problem to one of unconstrained minimization by using *penalty functions* (Sections 12.1, 12.2, 12.3 and 14.3). Efficiency depends on exactly how this is done, but it seems inevitable that some sort of penalty function must be used to get good global convergence properties. In many algorithms the iteration is determined by modelling the original problem in a suitable way. In particular a *linearization* of the constraint functions is often used. This is a first order Taylor series approximation about the current iterate $\mathbf{x}^{(k)}$:

$$c_i(\mathbf{x}^{(k)} + \boldsymbol{\delta}) \approx l_i^{(k)}(\boldsymbol{\delta}) = c_i^{(k)} + \mathbf{a}_i^{(k)T}\boldsymbol{\delta}, \quad i = 1, 2, \ldots, p. \tag{7.1.3}$$

The linearized function $l_i^{(k)}$ is defined in terms of the correction $\boldsymbol{\delta}$ to $\mathbf{x}^{(k)}$, the superscript k indicating that it is made on iteration k. This approximation enables linear constraint subproblems to be solved on each iteration. As in Part 1, it is also possible to make a quadratic model of the objective function. However to take constraint curvature correctly into account it is appropriate to modify the quadratic term in a suitable way. Methods of this type are very important, although to obtain good global properties they must be incorporated with some type of penalty function (Section 12.4). A special case of nonlinear programming is *geometric programming* in which the functions f and c_i have a polynomial type structure. It is possible to reduce this problem to a linear constraint problem which is more readily solved (Section 13.2). Often linear constraints and bounds arise in nonlinear programming problems: it is usually possible to take advantage of this fact to make the algorithm more efficient.

These algorithms, in particular those for nonlinear programming, depend on a study of optimality conditions for problem (7.1.1), and this theory is set out in Chapter 9. In Part 1, I tried to write parts of the book in simple terms, avoiding the use of too much theory. To some extent I have done this here, for example in the presentation of linear and quadratic programming. However, constrained minimization problems are much more complex than unconstrained problems and it is important for the user to have some grasp of this theory. This is especially true in regard to *Lagrange multipliers* and *first order conditions* and I have tried to give a simple semi-rigorous introduction in Section 9.1 showing how these multipliers arise and can be interpreted. A more rigorous presentation then follows. The same is true in regard to second order conditions in Section 9.3. Some simple notions of convexity and duality for smooth problems appear in Section 9.4 and 9.5. The subject of *non-smooth optimization* is arguably the most difficult that I have tried to cover in this book. Some presentations of this subject are extremely theoretical although I have tried to avoid this as much as possible. However, at the expense of introducing a little more theory concerning optimality conditions for non-smooth convex functions (Section 14.2), a reasonably elegant and not too difficult treatment can be given. A discussion of algorithms (Sections 14.4 and 14.5) is also given. The main thrust of Chapter 14 is to show that a wider class of non-smooth problem, that is *composite non-smooth optimization*, which includes many practical applications, can be handled without significant extra complication. The generalization to constrained composite non-smooth optimization is also given.

7.2 ELIMINATION AND OTHER TRANSFORMATIONS

There is considerable scope for making transformations to a constrained minimization problem which reduce it to a form that is more readily solved. This can be advantageous and a number of such possibilities are discussed. However, it is important to be aware from the outset that this procedure is not entirely without risk and that solutions of the original and transformed problems may not correspond on a one-to-one basis or that methods may not perform adequately on the transformed problem. A number of examples of this are given in this section and elsewhere in the book, and the user should be on his guard. The most simple possibility for equality constraints is to use the equations to eliminate some of the variables in the problem (*elimination*). If there are just m equations $\mathbf{c}(\mathbf{x})=\mathbf{0}$ which can be rearranged directly to give

$$\mathbf{x}_1 = \boldsymbol{\phi}(\mathbf{x}_2) \tag{7.2.1}$$

where $\mathbf{x}_1$ and $\mathbf{x}_2$ are partitions of $\mathbf{x}$ in $\mathbb{R}^m$ and $\mathbb{R}^{n-m}$, then the original objective function $f(\mathbf{x}_1, \mathbf{x}_2)$ is replaced by

$$\psi(\mathbf{x}_2) = f(\boldsymbol{\phi}(\mathbf{x}_2), \mathbf{x}_2) \tag{7.2.2}$$

and $\psi(\mathbf{x}_2)$ is minimized over $\mathbf{x}_2$ without any constraints. A simple example is given in Question 7.3. Derivatives of ψ are readily obtained from those of f and $\mathbf{c}$ (see Question 7.5). Some care has to be taken to avoid an ill-conditioned rearrangement when forming (7.2.1), for instance with linear constraints it is advisable to use some sort of pivoting on the variables. In some cases the method may fail completely, as shown in Question 7.4. In fact it is possible to discuss elimination in more general terms, implicitly by first making a linear transformation of variables; this is described in Section 10.1. In cases where no direct rearrangement like (7.2.1) is available, it is possible to regard $\mathbf{c}(\mathbf{x}_1, \mathbf{x}_2) = \mathbf{0}$ as a system of nonlinear equations which can be solved by the Newton–Raphson method (Section 6.2). In doing this $\mathbf{x}_2$ remains fixed and a vector $\mathbf{x}_1$ is determined which solves the equations. Thus $\mathbf{x}_1$ depends on $\mathbf{x}_2$ and so the process implicitly defines a function $\mathbf{x}_1 = \boldsymbol{\phi}(\mathbf{x}_2)$. This method is outlined in a more general form in Section 12.5; however the process is not always the most efficient and there can be difficulties in getting the Newton–Raphson method to converge. An alternative transformation for the equality constraint problem is the *method of Lagrange multipliers* (Section 9.1) in which the system of nonlinear equations (9.1.5) is solved which arises from the first order necessary conditions. Except in special cases this system must be solved numerically, which may cause difficulties. The method can also fail, not only when the solution of (9.1.5) corresponds to a constrained maximum point or saddle point, but also when the regularity condition (9.2.4) does not hold (see Question 9.14).

Elimination methods are not directly applicable to inequality constraint problems unless the set of active constraints $\mathscr{A}^*$ is known. However, it is possible to use a trial and error sort of method in which a guess $\mathscr{A}$ is made at the set of active constraints, and constraints in $\mathscr{A}$ are then treated as equalities, neglecting the remaining inequality constraints. The resulting equality constraint problem is then solved by elimination or by the method of Lagrange multipliers, giving a solution $\hat{\mathbf{x}}$. It is necessary to check that $\hat{\mathbf{x}}$ is feasible with respect to the constraints which have been ignored. If not, one of these is added to the active set and the above process is repeated. If $\hat{\mathbf{x}}$ is feasible then it is also necessary to check that the first order conditions are satisfied. To do this requires the calculation of the corresponding Lagrange multiplier vector $\hat{\boldsymbol{\lambda}}$. Since

$\lambda_i = \partial f/\partial c_i$ to first order measures the effect of perturbations in the c_i on f, it is necessary for an inequality constraint $c_i(\mathbf{x}) \geqslant 0$ that $\lambda_i \geqslant 0$ at the solution, for otherwise a feasible perturbation would reduce f. Thus if there are any $\hat{\lambda}_i < 0$, one such constraint must be removed from the active set and the process repeated again. On the other hand, if $\hat{\boldsymbol{\lambda}} \geqslant \mathbf{0}$ then the required solution is located. Methods of this type can be used in an informal way on small problems. However, they are most useful in solving all types of linear constraint problem when systematic procedures can be devised. Such methods include the *simplex method* for linear programming and the *active set method* for all types of linear constraint programming. So-called *exchange algorithms* for best linear L_1 and L_∞ data fitting are also examples of this type of procedure. Systematic procedures using active set methods for nonlinear constraints based on solving the equality constraint problems by implicit elimination can also be devised (Section 12.5) but there are some difficulties which are not readily overcome. It is difficult to handle constraints of the form $c_i(\mathbf{x}) > 0$ in an active set method because the feasible region is not closed and the constraint cannot be active at a solution. However, it can be useful to include them in the problem via the transformation $c_i(\mathbf{x}) \geqslant \varepsilon > 0$, possibly solving a sequence of problems in which $\varepsilon \downarrow 0$ if the constraints happen to be active. The reason for doing this might be to prevent or dissuade $f(\mathbf{x})$ being evaluated at an infeasible point at which it is not defined (for example the problem: $\min x \log_e x$ subject to $x > 0$). It may not be satisfactory just to ignore the constraints because the problem may then become unbounded or have a global solution with $c_i(\mathbf{x}) \leqslant 0$, which is of no interest.

Some other transformations are worthy of note which relate equality and inequality constraint problems. For example a constraint $c_i(\mathbf{x}) = 0$ can be equivalently replaced by two opposite inequality constraints $c_i(\mathbf{x}) \geqslant 0$ and $-c_i(\mathbf{x}) \geqslant 0$. This enables (7.1.1) to be reduced to an inequality constraint problem. However, there are some practical disadvantages due to degeneracy and other reasons and the idea is best avoided, although it can occasionally be useful for theoretical purposes. The alternative possibility is to write $c_i(\mathbf{x}) \geqslant 0$ as the equality constraint $\min(c_i(\mathbf{x}), 0) = 0$. Unfortunately this function is not a $\mathbb{C}^1$ function and so is usually excluded on this count. Another possibility is to replace a constraint $c_i(\mathbf{x}) \geqslant 0$ by adding an extra variable, z say, giving an equality constraint $c_i(\mathbf{x}) = z$ and a bound $z \geqslant 0$. The variable z is referred to as a *slack variable* since it measures the slack in the inequality constraint. This transformation is most useful in the simplex method for linear programming which requires all general inequality constraints to be handled in this way, but is not necessary in active set methods which treat inequalities of any type directly. Furthermore, following an idea introduced later in this section, it is possible to do away with the need for the bound $z \geqslant 0$. This is done by adding a *quadratic slack variable* y and replacing $c_i(\mathbf{x}) \geqslant 0$ by the (nonlinear) equality constraint $c_i(\mathbf{x}) = y^2$. This removes the need to treat inequality constraints directly. However this transformation does cause some distortion as explained below and in this case it may be somewhat dangerous, in particular because of the following feature. Let for example $c(\mathbf{x}) \geqslant 0$ be the only constraint and let $\mathbf{x}'$

be such that $c' = 0$ and $\mathbf{g}' = \mathbf{a}'\lambda'$ where $\lambda' < 0$. Then $\mathbf{x}'$ and λ' do not satisfy first order conditions for a solution. Yet the vector $\mathbf{x}'$ augmented by $y' = 0$ does satisfy first order conditions in the transformed equality constraint problem with the same λ'. Thus the transformation does not seem to be able to distinguish whether or not constraints are active on the basis of first order information. I have also heard bad reports of quadratic slacks in practice which might well be accountable for in this way.

A similar transformation is useful, however, when $f(\mathbf{x})$ is not defined for all $\mathbf{x} \in \mathbb{R}^n$. In this case it may be possible to add extra variables and equations to give a problem which is more tractable. For example if $f(\mathbf{x}) = (x_1^3 + x_2^3)^{1/2}$ then by introducing a new variable x_3 and a new equation $x_3 = x_1^3 + x_2^3$, $f(\mathbf{x})$ can be replaced by $f(\mathbf{x}) \triangleq x_3^{1/2}$. Thus it is only necessary for f to be defined on $x_3 \geqslant 0$ and it is usually straightforward to modify a method so that iterates are feasible with respect to the simple bounds in the problem.

Many other useful transformations arise in constrained optimization and are used in subsequent chapters. Perhaps the most well-known idea is the use of *penalty functions* for nonlinear programming. The idea is to transform the problem to one of unconstrained optimization by adding to the objective function a penalty term which weights constraint violations. In *sequential penalty functions* $\mathbf{x}^*$ is found as the limit of the minimizing points of a sequence of penalty functions, as some controlling parameter is changed. More recently the value has been realized of an *exact penalty function* which has $\mathbf{x}^*$ as its local minimizer. These transformations are described in some detail in Sections 12.1, 12.2, 12.3, 12.6 and 14.3. Other transformations of some importance are those arising in duality (Section 9.5), integer programming (Section 13.1), and geometric programming (Section 13.2), amongst others.

It is also possible to make transformations of variables in an attempt to simplify the problem. For example the bound $x_i \geqslant 0$ can be removed by defining a new variable y_i which replaces x_i, such that $x_i = y_i^2$. Then for any y_i in $(-\infty, \infty)$ it follows that $x_i \geqslant 0$ so the bound does not need to be explicitly enforeced. Another similar transformation for $l_i \leqslant x_i \leqslant u_i$ is to let y_i satisfy $x_i = l_i + (u_i - l_i)\sin^2 y_i$. For strict constraints $x_i > 0$ it is possible to use $x_i = e^{y_i}$. The advantage of these transformations is that they do extend the range of problems which can be handled by an unconstrained minimization routine. This is not to say that minimization with simple bounds $l_i \leqslant x_i \leqslant u_i$ is at all difficult; in fact the opposite is true and it is probably more efficient to treat the problem directly. It is simply that subroutines which minimize functions subject only to bounds are much less readily available to the user at present. These ideas can also be used to transform inequality constraints to equalities (see above in regard to quadratic slacks), although this possibility should be viewed with some suspicion for the reason given above. These transformations do cause some distortion which often may not be favourable. For example the problem $\min x^2$ subject to $x \geqslant 0$, after transforming $x = y^2$, becomes $\min y^4$. This has a singular Hessian at the solution which causes any standard minimization method based on a quadratic model to converge slowly. Another example is the convex programming problem $\min(x-1)^2$ subject to $x \geqslant 0$. Although the transformation is well behaved at the solution $x^* = 1$, it induces a stationary point with a non-positive-definite Hessian matrix at $x = 0$ and both these features could possibly cause difficulties (see also Question 7.6). Thus although such transformations can be useful, the user should be aware that they are not entirely risk free.

Another transformation which enables $|x_i|$ functions to be handled is to replace the variable x_i by two non-negative variables, x_i^+ and x_i^- representing the positive and negative parts of x_i (that is $\max(x_i, 0)$ and $\max(-x_i, 0)$). The conditions $x_i^+ \geqslant 0$ and $x_i^- \geqslant 0$ are explicitly included in the problem; also whenever x_i appears in the problem it is replaced by $x_i^+ - x_i^-$ and similarly $|x_i|$ is replaced by $x_i^+ + x_i^-$ (see Question 8.12). This latter replacement is only valid if one of x_i^+ or x_i^- is zero, which can sometimes be guaranteed, for example in otherwise linear problems when both x_i^+ and x_i^- together cannot be basic (Chapter 8). This transformation can also be used to handle unbounded variables in a linear programming problem. An alternative technique for handling $|x_i|$ terms is described in Section 8.4. These ideas can be extended to functions $|c_i(x)|$ by adding extra variables y_i and the equality constraint $c_i(\mathbf{x}) = y_i$ (see Question 8.11), thus enabling L_1 approximation problems to be handled by smooth techniques. Similar ideas for minimizing max functions or L_∞ functions can be tackled by introducing an extra variable v as described in Section 14.1. However, all these techniques are really attempting to solve non-differentiable optimization problems as smooth problems. In the current state of the art this can be useful, but when software becomes readily available for some of the better more direct methods described in Sections 14.4 and 14.5, these should be preferred.

Finally, the very important transformation of *scaling* either the constraints or the variables in the problem is discussed. Scaling of a constraint set is achieved by multiplying each constraint function by a constant chosen so that the value of each constraint function, evaluated for typical values of $\mathbf{x}$, is of the same order of magnitude. This can be important in that this scales the Lagrange multipliers (inversely) and so can make more reliable the test on the magnitude of a multiplier which is used in some algorithms. A well-scaled matrix is also important in some linear algebra routines when pivoting tests are made. Moreover when using penalty functions which involve quantities like $\mathbf{c}^T\mathbf{c}$ or $\|\mathbf{c}\|_1$, it is important that constraints are scaled. In a similar way scaling of the variables can sometimes be important. This again arises when pivoting tests on the variables are made, or when implicitly using some norm of the variables, for example in restricted step methods or methods with a bias towards steepest descent (see Part 1). In practice variables are usually scaled by multiplying each one by a suitable constant. However, a nonlinear scaling which can be useful for variables $x_i > 0$ is to use the transformation $x_i = e^{y_i}$. Then variables of magnitudes 10^{-6}, 10^{-3}, 10^0, $10^3, \ldots$, say, which typically can occur in kinetics problems, are transformed into logarithmic variables with magnitudes which are well scaled.

QUESTIONS FOR CHAPTER 7

7.1. Calculate the Jacobian matrix $\nabla \mathbf{l}^T$ of the linear system $\mathbf{l}(\mathbf{x}) = \mathbf{A}^T\mathbf{x} - \mathbf{b}$. If $\mathbf{l}$ is obtained by linearizing a nonlinear system as in (7.1.3) show that both systems have the same Jacobian matrix.

7.2. Obtain the gradient vector and the Hessian matrix of the functions $f(\mathbf{x}) + h(\mathbf{c}(\mathbf{x}))$ and $f(\mathbf{x}) + \boldsymbol{\lambda}^T\mathbf{c}(\mathbf{x})$. In the latter case treat the cases both where $\boldsymbol{\lambda}$ is a constant vector and where it is a function $\boldsymbol{\lambda}(\mathbf{x})$.

7.3. Find the solution of the problem minimize $-x-y$ subject to $x^2 + y^2 = 1$ by graphical means and also by eliminating x, and show that the same solution is obtained. Discuss what happens, however, if the square root which is required is chosen to have a negative sign.

7.4. Solve the problem minimize $x^2 + y^2$ subject to $(x-1)^3 = y^2$ both graphically and also by eliminating y. In the latter case show that the resulting function of x has no minimizer and explain this apparent contradiction. What happens if the problem is solved by eliminating x?

7.5. Consider finding derivatives of the functions $\boldsymbol{\phi}(\mathbf{x}_2)$ and $\psi(\mathbf{x}_2)$ defined in (7.2.1) and (7.2.2). Define partitions

$$\nabla = \begin{pmatrix} \nabla_1 \\ \nabla_2 \end{pmatrix}, \qquad \mathbf{g} = \begin{pmatrix} \mathbf{g}_1 \\ \mathbf{g}_2 \end{pmatrix}, \qquad \mathbf{A} = \begin{bmatrix} \mathbf{A}_1 \\ \mathbf{A}_2 \end{bmatrix}$$

and show by using the chain rule that $\nabla_2\boldsymbol{\phi}^T = -\mathbf{A}_2\mathbf{A}_1^{-1}$ and hence that $\nabla_2\psi = \mathbf{g}_2 - \mathbf{A}_2\mathbf{A}_1^{-1}\mathbf{g}_1$. Second derivatives of ψ are most conveniently obtained as in (12.5.6) and (12.5.7) by setting $\mathbf{V}^T = [\mathbf{0}:\mathbf{I}]$ and hence $\mathbf{Z}^T = [-\mathbf{A}_2\mathbf{A}_1^{-1}:\mathbf{I}]$ and $\mathbf{Y}^T = [\mathbf{A}_1^{-1}:\mathbf{0}]$.

7.6. Consider the problem minimize $f(x_1, x_2)$ subject to $x_1 \geq 0, x_2 \geq 0$ when the transformation $x = y^2$ is used. Show that $\mathbf{x}' = \mathbf{0}$ is a stationary point of the transformed function, but is not minimal if any $g_i' \leq 0$. If $\mathbf{g}' = \mathbf{0}$ then second order information in the original problem would usually enable the question of whether $\mathbf{x}'$ is a minimizer to be determined. Show that in the transformed problem this cannot be done on the basis of second order information.

Chapter 8
Linear Programming

8.1 STRUCTURE

The most simple type of constrained optimization problem is obtained when the functions $f(\mathbf{x})$ and $c_i(\mathbf{x})$ in (7.1.1) are all linear functions of $\mathbf{x}$. The resulting problem is known as a *linear programming* (LP) problem. Such problems have been studied since the earliest days of electronic computers, and the subject is often expressed in a quasi-economic terminology which to some extent obscures the basic numerical processes which are involved. This presentation aims to make these processes clear, whilst retaining some of the traditional nomenclature which is widely used. One main feature of the traditional approach is that linear programming is expressed in the *standard form*

$$\begin{aligned} &\underset{\mathbf{x}}{\text{minimize}} \quad f(\mathbf{x}) \triangleq \mathbf{c}^{\mathrm{T}}\mathbf{x} \\ &\text{subject to } \mathbf{A}\mathbf{x} = \mathbf{b}, \qquad \mathbf{x} \geqslant \mathbf{0}, \end{aligned} \tag{8.1.1}$$

where $\mathbf{A}$ is an $m \times n$ matrix, and $m \leqslant n$ (usually $<$). Thus the allowable constraints on the variables are either linear equations or non-negativity bounds. The coefficients $\mathbf{c}$ in the linear objective function are often referred to as *costs*. An example with four variables ($n = 4$) and two equations ($m = 2$) is

$$\begin{aligned} \text{minimize} \quad & x_1 + 2x_2 + 3x_3 + 4x_4 \\ \text{subject to} \quad & x_1 + x_2 + x_3 + x_4 = 1 \\ & x_1 \qquad\quad + x_3 - 3x_4 = \tfrac{1}{2}, \\ & x_1 \geqslant 0,\ x_2 \geqslant 0,\ x_3 \geqslant 0,\ x_4 \geqslant 0. \end{aligned} \tag{8.1.2}$$

More general LP problems can be reduced to standard form without undue difficulty, albeit with some possible loss of efficiency. For instance a general linear inequality $\mathbf{a}^{\mathrm{T}}\mathbf{x} \leqslant b$ can be transformed using a slack variable z (see Section 7.2) to the equation $\mathbf{a}^{\mathrm{T}}\mathbf{x} + z = b$ and the bound $z \geqslant 0$. Alternatively the dual transformation can sometimes be used advantageously to obtain a standard form and this is described in more detail in Section 9.5. More general bounds $x_i \geqslant l_i$ can be dealt with by a shift of origin, and if no bound exists at all on x_i in the original problem, then the standard form can be reached by introducing non-negative variables x_i^+ and x_i^-, as described in Section 7.2. In fact very little is lost in complexity if the bounds in (8.1.1) are expressed as

$$\mathbf{l} \leqslant \mathbf{x} \leqslant \mathbf{u}, \tag{8.1.3}$$

as Question 8.8 illustrates. The merit of using other possible standard forms is discussed in more detail in Section 8.3. However, for the most part this text will concentrate on the solution of problems which are already in the standard form (8.1.1).

It is important to realize that a problem in standard form may have no solution, either because there is no feasible point (the problem is *infeasible*), or because $f(\mathbf{x}) \to -\infty$ for $\mathbf{x}$ in the feasible region (the problem is *unbounded*). However, it is shown that there is no difficulty in detecting these situations, and so the text concentrates on the usual case in which a solution exists (possibly not unique). It is also convenient to assume that the equations are independent, so that they have no trivial linear combination. In theory this situation can always be achieved, either by removing dependent equations or by adding artificial variables (see Section 8.4 and Question 8.21), although in practice there may be numerical difficulties if this dependence is not recognized.

If (8.1.1) is considered in more detail, it can be seen that if $m = n$, then the equations $\mathbf{Ax} = \mathbf{b}$ determine a unique solution, and the objective function $\mathbf{c}^T\mathbf{x}$ and the bounds $\mathbf{x} \geqslant \mathbf{0}$ play no part. In most cases however $m < n$, so that the system $\mathbf{Ax} = \mathbf{b}$ is underdetermined and $n - m$ degrees of freedom remain. In particular the system can determine only m variables, given values for the remaining $n - m$ variables. For example the equations $\mathbf{Ax} = \mathbf{b}$ in (8.1.2) can be rearranged as

$$\begin{aligned} x_1 &= \tfrac{1}{2} - x_3 + 3x_4 \\ x_2 &= \tfrac{1}{2} \qquad\quad - 4x_4 \end{aligned} \tag{8.1.4}$$

which determines x_1 and x_2 given values for x_3 and x_4, or alternatively as

$$\begin{aligned} x_1 &= \tfrac{7}{8} - \tfrac{3}{4}x_2 - x_3 \\ x_4 &= \tfrac{1}{8} - \tfrac{1}{4}x_2 \end{aligned} \tag{8.1.5}$$

which determines x_1 and x_4 from x_2 and x_3, and in other ways as well. It is important to consider what values these remaining $n - m$ variables can take in the standard form problem. The objective function $\mathbf{c}^T\mathbf{x}$ is linear and so contains no curvature which can give rise to a minimizing point. Hence such a point must be created by the conditions $x_i \geqslant 0$ becoming active on the boundary of the feasible region. For example if (8.1.5) is used to eliminate the variables x_1 and x_4 from the problem (8.1.2), then the objective function can be expressed as

$$f = x_1 + 2x_2 + 3x_3 + 4x_4 = \tfrac{11}{8} + \tfrac{1}{4}x_2 + 2x_3. \tag{8.1.6}$$

Clearly this function has no minimum value unless the conditions $x_2 \geqslant 0$, $x_3 \geqslant 0$

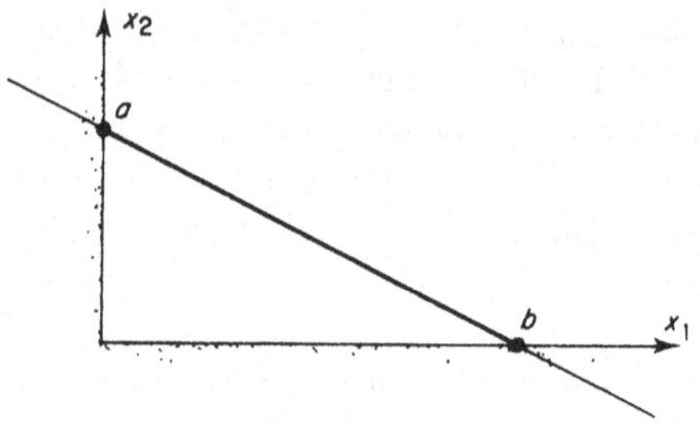

Figure 8.1.1 Constraints for a simple LP problem

are imposed, in which case the minimum occurs when $x_2 = x_3 = 0$. An illustration is given in Figure 8.1.1 for the more simple conditions $x_1 + 2x_2 = 1$ and $x_1 \geqslant 0$, $x_2 \geqslant 0$. The feasible region is the line joining the points $\mathbf{a} = (0, \frac{1}{2})^T$ and $\mathbf{b} = (1, 0)^T$. When the objective function $f(\mathbf{x})$ is linear the solution must occur at either **a** or **b** with either $x_1 = 0$ or $x_2 = 0$ (try different linear functions, for example $f = x_1 + x_2$ or $f = x_1 + 3x_2$). If however $f(\mathbf{x}) = x_1 + 2x_2$ then any point on the line segment is a solution and this includes both **a** and **b**. This corresponds to the existence of a non-unique solution.

To summarize therefore, a solution of an LP problem in standard form always exists at one particular *extreme point* or *vertex* of the feasible region, with at least $n - m$ variables having *zero value*, and the remaining m variables being uniquely determined by the equations $\mathbf{Ax} = \mathbf{b}$ and taking non-negative values. This result is fundamental to the development of LP methods, and can be established rigorously using the notions of convexity (Section 9.4). The proof is sketched out in some detail in Questions 9.20 to 9.22.

The main difficulty in linear programming is to find which $n - m$ variables take zero value at the solution. The earliest method for solving this problem is the *simplex method*, which tries different sets of possibilities in a systematic way. This method is described in Section 8.2 and is still predominant today, albeit often in more sophisticated forms. Different variations of the method exist, depending upon exactly which intermediate quantities are computed. The earliest *tableau form* became superseded by the more efficient *revised simplex method*, both of which are described in Section 8.2. More recently methods based on using matrix factorizations have been suggested in order to control round-off errors more effectively. For large sparse LP problems, *product form methods* have enabled problems of up to 10^5 variables to be solved in practice. Both these developments are described in Section 8.5. In fact even larger LP problems can be solved arising from *linear network flow* which is described in Section 13.3. An apparently different approach to LP is the active set method described in Section 8.3 which, however, turns out to be equivalent to the simplex method with slack variables, although different intermediate matrices are stored. The problem of calculating initial feasible points for LP and other linear constraint problems is described in Section 8.4. All these methods have one possible situation in which they can fail to solve a problem which has a well-defined solution. This is referred to as *degeneracy* and is described in Section 8.6. In Section 8.7 the recent research topic of *polynomial time algorithms* for LP is discussed. These methods differ from the simplex method and are able to improve on its worst case behaviour. There is some controversy as to whether the methods are suitable for general purpose use.

8.2 THE SIMPLEX METHOD

The simplex method for solving an LP problem in standard form generates a sequence of feasible points $\mathbf{x}^{(1)}, \mathbf{x}^{(2)}, \ldots$ which terminates at a solution. Since there exists an extreme point at which the solution occurs, each iterate $\mathbf{x}^{(k)}$ is an extreme point. Thus $n-m$ of the variables have zero value at $\mathbf{x}^{(k)}$ and are referred to as *nonbasic variables* (the index set $N^{(k)}$). The remaining m variables have a non-negative value (usually positive) and are referred to as *basic variables* (the index set $B^{(k)}$). The simplex method makes systematic changes to these sets after each iteration, in order to find the choice which gives the optimal solution. The superscript (k) is often omitted for clarity. At each iteration it is convenient to assume that the variables are permuted so that the basic variables are the first m elements of $\mathbf{x}$. Then $\mathbf{x}^T = (\mathbf{x}_B^T, \mathbf{x}_N^T)$ can be written, where $\mathbf{x}_B$ and $\mathbf{x}_N$ refer collectively to the basic and nonbasic variables respectively. The matrix $\mathbf{A}$ in (8.1.1) can also be partitioned similarly into $\mathbf{A} = [\mathbf{A}_B : \mathbf{A}_N]$ where $\mathbf{A}_B$ (the *basis matrix*) is $m \times m$ and $\mathbf{A}_N$ is $m \times (n-m)$. The equations $\mathbf{Ax} = \mathbf{b}$ can thus be written

$$[\mathbf{A}_B : \mathbf{A}_N]\begin{pmatrix} \mathbf{x}_B \\ \mathbf{x}_N \end{pmatrix} = \mathbf{A}_B \mathbf{x}_B + \mathbf{A}_N \mathbf{x}_N = \mathbf{b}. \tag{8.2.1}$$

At an extreme point it is always possible to find a partitioning into B and N such that $\mathbf{A}_B$ is non-singular (assuming that $\mathbf{A}$ has full rank; see Question 9.20). Also since $\mathbf{x}_N^{(k)} = \mathbf{0}$ it is possible to write

$$\mathbf{x}^{(k)} = \begin{pmatrix} \mathbf{x}_B \\ \mathbf{x}_N \end{pmatrix}^{(k)} = \begin{pmatrix} \hat{\mathbf{b}} \\ \mathbf{0} \end{pmatrix} \tag{8.2.2}$$

where $\hat{\mathbf{b}} = \mathbf{A}_B^{-1}\mathbf{b}$. Since the basic variables must take non-negative values it is required that $\hat{\mathbf{b}} \geqslant \mathbf{0}$. The partitioning $B^{(k)}$ and $N^{(k)}$ and the extreme point $\mathbf{x}^{(k)}$ with the above properties ($\mathbf{x}_B^{(k)} = \hat{\mathbf{b}} \geqslant \mathbf{0}$, $\mathbf{x}_N^{(k)} = \mathbf{0}$, $\mathbf{A}_B$ non-singular) is referred to as a *basic feasible solution* (*b.f.s.*). An example is provided by the LP problem (8.1.2) with the choice $B = \{1, 2\}$ and $N = \{3, 4\}$ (that is the variables x_1 and x_2 are basic, and x_3 and x_4 are nonbasic). Then

$$\mathbf{A}_B = \begin{bmatrix} 1 & 1 \\ 1 & 0 \end{bmatrix}, \qquad \mathbf{A}_N = \begin{bmatrix} 1 & 1 \\ 1 & -3 \end{bmatrix},$$

and

$$\hat{\mathbf{b}} = \mathbf{A}_B^{-1}\mathbf{b} = \begin{bmatrix} 0 & 1 \\ 1 & -1 \end{bmatrix}\begin{pmatrix} 1 \\ \frac{1}{2} \end{pmatrix} = \begin{pmatrix} \frac{1}{2} \\ \frac{1}{2} \end{pmatrix} \geqslant \mathbf{0}.$$

Since $\mathbf{A}_B$ is non-singular and $\hat{\mathbf{b}} \geqslant \mathbf{0}$, this choice of B and N determines a basic feasible solution. It is also of interest to know the value ($\hat{f}$ say) of the objective function at a basic feasible solution. Partitioning $\mathbf{c}^T = (\mathbf{c}_B^T, \mathbf{c}_N^T)$ and using (8.2.2), it follows that

$$\hat{f} = \mathbf{c}^T\mathbf{x}^{(k)} = \mathbf{c}_B^T\hat{\mathbf{b}}.$$

In the example above, $\mathbf{c}_B = (1,2)^T$ so that the objective function has the value $\hat{f} = 1\frac{1}{2}$. The nomenclature 'basic feasible solution' can be somewhat confusing (basic feasible point would be better) since 'solution' refers to a solution of the constraint equations $\mathbf{Ax} = \mathbf{b}$ and $\mathbf{x} \geqslant \mathbf{0}$, and not to the overall solution of (8.1.1), which is therefore referred to as an *optimal* b.f.s. It should be noticed that not every possible choice of B and N gives rise to a b.f.s. For instance in (8.1.2), $B = \{1,3\}$ gives a singular matrix $\mathbf{A}_B$, and $B = \{2,4\}$ gives a vector $\hat{\mathbf{b}}$ for which $\hat{b}_4 < 0$. (Note that indices such as $\hat{b}_4$ refer to the variable to which the element corresponds, and not to the position in the vector $\hat{\mathbf{b}}$.) Consequently the determination of an initial b.f.s. $B^{(1)}$, $N^{(1)}$ and $\mathbf{x}^{(1)}$ is a non-trivial task. Suitable methods exist which incorporate the logic of the simplex method, and these are discussed in Section 8.4.

It is a simple matter to discover whether a basic feasible solution is optimal. Equation (8.2.1) is used to eliminate the basic variables from the objective function, yielding a reduced objective function $f(\mathbf{x}_N)$ of the nonbasic variables only, whose coefficients give this information directly. For example in the LP problem (8.1.2) with the choice $B = \{1,2\}$, $N = \{3,4\}$, (8.2.1) can be rearranged as shown in (8.1.4) and substituted into the objective function in (8.1.1) to give

$$f(x_3, x_4) = 1\tfrac{1}{2} + 2x_3 - x_4. \tag{8.2.3}$$

Now at the b.f.s. x_3 and x_4 take the values $x_3 = x_4 = 0$, but in general may satisfy $x_3 \geqslant 0$ and $x_4 \geqslant 0$. Thus only an *increase* in the value of any nonbasic variable is allowed. Furthermore f can be decreased from its value of $\hat{f} = 1\frac{1}{2}$ by increasing x_4, since x_4 has a negative coefficient in (8.2.3), and so it follows that this b.f.s. is not optimal. On the other hand, if the coefficients of the nonbasic variables in the reduced objective function are *all non-negative* as in (8.1.6), then the corresponding b.f.s. is *optimal* since there is no feasible change to the nonbasic variables which will reduce $f(\mathbf{x}_N)$.

In general terms the elimination of the basic variables uses the rearrangement of (8.2.1) given by

$$\mathbf{x}_B = \mathbf{A}_B^{-1}(\mathbf{b} - \mathbf{A}_N\mathbf{x}_N) = \hat{\mathbf{b}} - \mathbf{A}_B^{-1}\mathbf{A}_N\mathbf{x}_N. \tag{8.2.4}$$

The reduced objective function can then be written as

$$\begin{aligned} f(\mathbf{x}_N) &= \mathbf{c}_B^T\mathbf{x}_B + \mathbf{c}_N^T\mathbf{x}_N \\ &= \mathbf{c}_B^T(\hat{\mathbf{b}} - \mathbf{A}_B^{-1}\mathbf{A}_N\mathbf{x}_N) + \mathbf{c}_N^T\mathbf{x}_N \\ &= \hat{f} + \hat{\mathbf{c}}_N^T\mathbf{x}_N \end{aligned} \tag{8.2.5}$$

say, where the coefficients $\hat{\mathbf{c}}_N$ are defined by

$$\hat{\mathbf{c}}_N = \mathbf{c}_N - \mathbf{A}_N^T\boldsymbol{\pi}, \tag{8.2.6}$$

and where

$$\boldsymbol{\pi} = \mathbf{A}_B^{-T}\mathbf{c}_B. \tag{8.2.7}$$

In the example above $\boldsymbol{\pi} = (2, -1)^T$ and $\hat{\mathbf{c}}_N = (2, -1)^T$. The coefficients $\hat{\mathbf{c}}_N$ are known as the *reduced costs* at the basic feasible solution. By virtue of the discussion in the previous paragraph, the basic feasible solution is optimal if the reduced costs satisfy the *optimality test*

$$\hat{\mathbf{c}}_N \geqslant \mathbf{0}. \tag{8.2.8}$$

The first part of a simplex method iteration therefore is to determine the reduced costs. If (8.2.8) holds then the basic feasible solution is optimal and the method terminates. If (8.2.8) also contains some element $\hat{c}_q = 0$, then a similar argument shows that x_q can be increased with f staying constant, so that the solution is *non-unique* (assuming that a non-zero step is permitted—for example if $\hat{\mathbf{b}} > \mathbf{0}$—see next paragraph and Question 8.2).

Usually (8.2.8) is not satisfied, in which case the simplex method proceeds to find a new basic feasible solution which has a lower value of the objective function. Firstly a variable x_q, $q \in N$, is chosen for which $\hat{c}_q < 0$, and $f(\mathbf{x}_N)$ is decreased by increasing x_q whilst the other nonbasic variables retain their zero value. Usually the most negative $\hat{c}_q$ is chosen, that is

$$\hat{c}_q = \min_{i \in N} \hat{c}_i$$

although other selections have been investigated (Goldfarb and Reid, 1977). As x_q increases, the values of the basic variables $\mathbf{x}_B$ change as indicated by (8.2.4), in order to keep the system $\mathbf{Ax} = \mathbf{b}$ satisfied. Usually the need to keep $\mathbf{x}_B \geqslant \mathbf{0}$ to maintain feasibility limits the amount by which x_q can be increased. In the above example $\hat{c}_4$ is the only negative reduced cost, so x_4 is chosen to be increased, whilst x_3 retains its zero value. The effect on x_1 and x_2 is indicated by (8.1.4), and x_1 increases like $\frac{1}{2} + 3x_4$ whilst x_2 decreases like $\frac{1}{2} - 4x_4$. Thus x_2 becomes zero when x_4 reaches the value $\frac{1}{8}$, and no further increase in x_4 is permitted.

In general, because x_q is the only nonbasic variable which changes, it follows from (8.2.4) that the effect on $\mathbf{x}_B$ is given by

$$\begin{aligned} \mathbf{x}_B &= \hat{\mathbf{b}} - \mathbf{A}_B^{-1}\mathbf{a}_q x_q \\ &= \hat{\mathbf{b}} + \mathbf{d}x_q \end{aligned} \tag{8.2.9}$$

say, where $\mathbf{a}_q$ is the column of $\mathbf{A}$ corresponding to variable q, and where

$$\mathbf{d} = -\mathbf{A}_B^{-1}\mathbf{a}_q \tag{8.2.10}$$

can be thought of as the derivative of the basic variables with respect to changes in x_q. In particular if any d_i has a *negative* value then an increase in x_q causes a reduction in the value of the basic variable x_i. From (8.2.9), x_i becomes zero when $x_q = \hat{b}_i / -d_i$. The amount by which x_q can be increased is limited by the *first* basic variable, x_p say, to become zero, and the index p is therefore that for which

$$\frac{\hat{b}_p}{-d_p} = \min_{\substack{i \in B \\ d_i < 0}} \frac{\hat{b}_i}{-d_i}. \tag{8.2.11}$$

It may be, however, that there are no indices $i \in B$ such that $d_i < 0$. In this case $f(\mathbf{x})$ can be decreased without limit, and this is the manner in which an unbounded solution is indicated. In practical terms this usually implies a mistake in setting up the problem in which some restriction on the variables has been omitted. Equation (8.2.11) is referred to as the *ratio test.*

In geometric terms, the increase of x_q and the corresponding change to $\mathbf{x}_B$ in (8.2.9) causes a move from the extreme point $\mathbf{x}^{(k)}$ along an *edge* of the feasible region. The termination of this move because x_p, $p \in B$, becomes zero indicates that a new extreme point of the feasible region is reached. Correspondingly there are again $n - m$ variables with zero value (x_p and the x_i, $i \in N^{(k)}$, $i \neq q$), and these are the nonbasic variables at the new b.f.s. Thus the new sets $N^{(k+1)}$ and $B^{(k+1)}$ are obtained by replacing q by p in $N^{(k)}$, and p by q in $B^{(k)}$, respectively. The form of the iteration ensures that $\hat{\mathbf{b}} \geqslant \mathbf{0}$ at the new b.f.s. and it is possible to show also that $\mathbf{A}_B$ is non-singular (see (8.2.15)), so that the conditions for a b.f.s. remain satisfied. In the example above the nonbasic variable x_4 is increased and the basic variable x_2 becomes zero. Thus 4 and 2 are interchanged between the sets $B = \{1, 2\}$ and $N = \{3, 4\}$ giving rise to a new basic feasible solution determined by $B = \{1, 4\}$ and $N = \{2, 3\}$. It is shown by virtue of (8.1.5) and (8.1.6) that the resulting b.f.s. is in fact optimal.

With the determination of a new b.f.s., the description of an iteration of the simplex method is complete. The method repeats the sequence of calculations until an optimal b.f.s. is recognized. Usually each iteration reduces $f(\mathbf{x})$, and since the number of vertices is finite, it follows that the iteration must terminate. There is one case, however, in which this may not be true, and this is when $\hat{b}_i = 0$ and $d_i < 0$ occurs. Then no increase in x_q is permitted, and although a new partitioning B and N can be made as before, no decrease in $f(\mathbf{x})$ is made. In fact it is possible for the algorithm to *cycle* by returning to a previous set B and N and so fail to terminate even though a well-determined solution does exist. The possibility that $\hat{b}_i = 0$ is known as *degeneracy* and the whole subject is discussed in more detail in Section 8.6.

For small illustrative problems ($n = 2$ or 3) it is straightforward to follow the above scheme of computation. That is given B and N, then

$$\mathbf{A}_B^{-1} = \frac{\text{adjoint}(\mathbf{A})}{\det(\mathbf{A})} \tag{8.2.12}$$

is calculated, and hence $\hat{\mathbf{b}} = \mathbf{A}_B^{-1}\mathbf{b}$. Then (8.2.7) and (8.2.6) enable the reduced costs $\hat{\mathbf{c}}_N$ to be obtained. Either the algorithm terminates by (8.2.8) or the least $\hat{c}_q$ determines the index q. Computation of $\mathbf{d}$ by (8.2.10) is followed by the test (8.2.11) to determine p. Finally p and q are interchanged giving a new B and N with which to repeat the process.

For larger problems some more efficient scheme is desirable which avoids the computation of $\mathbf{A}_B^{-1}$ from (8.2.12) on each iteration. The earliest method was the *tableau form* of the simplex method. In this method the data is arranged in a tableau:

$$
\begin{array}{r|c|c|}
\hline
f\{ & \mathbf{c}^{\mathrm{T}} & 0 \\
\hline
B\{ & \mathbf{A} & \mathbf{b} \\
\hline
\end{array}
\qquad \text{(columns: } \underbrace{\;B\;}\;\underbrace{\;N\;}\text{)} .
$$

Then by making *row operations* on the tableau (adding or subtracting multiples of one row (not the f row) from another, or by scaling any row (not the f row)), it is possible to reduce the tableau to the form:

$$
\begin{array}{r|c:c|c|}
\hline
f\{ & \mathbf{0}^{\mathrm{T}} & \hat{\mathbf{c}}_N^{\mathrm{T}} & -\hat{f} \\
\hline
B\{ & \mathbf{I} & \hat{\mathbf{A}}_N & \hat{\mathbf{b}} \\
\hline
 & \underbrace{\;B\;} & \underbrace{\;N\;} &
\end{array}
\tag{8.2.13}
$$

where $\hat{\mathbf{A}}_N = \mathbf{A}_B^{-1}\mathbf{A}_N$, which represents the *canonical* or *reduced form* of the LP problem. Essentially the process is that used in obtaining (8.1.4) and (8.2.3) from (8.1.2): the operations on the B rows are equivalent to premultiplying by $\mathbf{A}_B^{-1}$, and those on the f row are equivalent to premultiplying by $\mathbf{c}_B^{\mathrm{T}}$ in the new B rows and subtracting this row from the f row. In fact an initial canonical form (8.2.13) is usually available directly as a result of the operations to find the initial b.f.s. (see Table 8.4.1). Tests (8.2.8) and (8.2.11) can be carried out directly on information in the tableau (note that $\mathbf{d} = -\hat{\mathbf{a}}_q$) so that the indices p and q can be determined. A new tableau is then obtained by first scaling row p so that the element in column q is unity. Multiples of row p are then subtracted from other rows so that the remaining elements in column q become zero, so that column q becomes a column of the unit matrix. The previous column (p) with this property becomes full during these operations. Then columns p and q of (8.2.13) are interchanged, giving a new tableau in canonical form. In fact this interchange need not take place, in which case the columns of the unit matrix in (8.2.13) no longer occur in the first n columns of the tableau. The iteration is then repeated on the new tableau. An example of the tableau form applied to (8.1.2) is given in Table 8.2.1. In general the quantity $\hat{a}_{pq}(= -d_p)$ must be non-zero (see (8.2.11)) and plays the role of *pivot* in these row operations, in a similar way to the pivot in Gaussian elimination. In Table 8.2.1 the pivot element is circled.

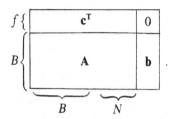

Table 8.2.1 The tableau form applied to problem (8.1.2)

		x_1	x_2	x_3	x_4		
Data:		1	2	3	4	0	
		1	1	1	1	1	
		1	0	1	-3	$\frac{1}{2}$	
$k=1$:	f			2	-1	$-1\frac{1}{2}$	(see (8.2.3))
	x_1	1		1	-3	$\frac{1}{2}$	(see (8.1.4))
	x_2		1	0	④	$\frac{1}{2}$	
$k=2$:	f		$\frac{1}{4}$	2		$-1\frac{3}{8}$	(see (8.1.6))
	x_1	1	$\frac{3}{4}$	1		$\frac{7}{8}$	(see (8.1.5))
	x_4		$\frac{1}{4}$	0	1	$\frac{1}{8}$	

(blanks correspond to zeros in columns marked B in (8.2.13))

It became apparent that the tableau form is inefficient in that it updates the whole tableau $\hat{\mathbf{A}}_N$ at each iteration. The earlier part of this section shows that the simplex method can be carried out with an explicit knowledge of $\mathbf{A}_B^{-1}$, and it is possible to carry out updating operations on this matrix. Since $\mathbf{A}_B^{-1}$ is often smaller than $\hat{\mathbf{A}}_N$, the resulting method, known as the *revised simplex method*, is usually more efficient. The effect of a basis change is to replace the column $\mathbf{a}_p$ by $\mathbf{a}_q$ in $\mathbf{A}_B^{(k)}$, which can be written as the rank one change

$$\mathbf{A}_B^{(k+1)} = \mathbf{A}_B + (\mathbf{a}_q - \mathbf{a}_p)\mathbf{e}_p^{\mathrm{T}} \tag{8.2.14}$$

(suppressing superscript (k)). By using the Sherman–Morrison formula (Question 3.13) it follows that

$$[\mathbf{A}_B^{(k+1)}]^{-1} = [\mathbf{A}_B]^{-1} - \frac{(\mathbf{d} + \mathbf{e}_p)\mathbf{v}_p^{\mathrm{T}}}{d_p} \tag{8.2.15}$$

where $\mathbf{v}_p^{\mathrm{T}}$ is the row of $\mathbf{A}_B^{-1}$ corresponding to x_p. The fact that $d_p \neq 0$ is implied by (8.2.11) ensures that $\mathbf{A}_B^{(k+1)}$ is non-singular. It is also possible to update the vectors $\boldsymbol{\pi}$ and $\hat{\mathbf{b}}$ by using (8.2.15).

A disadvantage of representing $\mathbf{A}_B^{-1}$ in the revised simplex method is that it is not adequate for solving large problems. Here the matrix $\mathbf{A}$ is sparse, and the problem is only manageable if this sparsity can be taken into account. Storing $\mathbf{A}_B^{-1}$ which is usually a full matrix is no longer possible. Also is has more recently been realized that representing $\mathbf{A}_B^{-1}$ directly can sometimes cause numerical problems due to magnification of round-off errors. Thus there is currently much research into other representations of $\mathbf{A}_B$ in the form of matrix factorizations which are easily invertible, numerically stable, and which enable sparsity to be exploited. These developments are described further in Section 8.5.

8.3 OTHER LP TECHNIQUES

It has already been remarked that general LP problems can be reduced to the standard form (8.1.1) by adding extra variables. This section reviews other ways in which a general LP problem can be tackled. One possibility is to use *duality* results to transform the problem. These are dealt with in more detail in Section 9.5 where three different LP duals of differing complexity are given. The resulting dual problem is then solved by the simplex method, possibly after adding extra variables to give a standard form. The aim in using a dual is to generate a transformed problem which has a more favourable structure, in particular one with fewer rows in the constraint matrix. Another type of transformation which can be used is known as *decomposition* in which a large structured problem is decomposed into a more simple master problem which is defined in terms of several smaller subproblems. The aim is to solve a sequence of smaller problems in such a way as to determine the solution of the original large problem. An example of this type of transformation is given in more detail in Section 8.5.

A different technique, known as *parametric programming*, concerns finding the solution of an LP problem which is a perturbation of another problem for which the solution is available. This may occur, for example, when the effect of changes in the data ($\mathbf{A}$, $\mathbf{b}$, or $\mathbf{c}$) is considered, or when an extra constraint or variable is added to the problem. The effect of changes in the data gives useful information regarding the *sensitivity* of various features in the model, and may often be more important than the solution itself. The rate of change of the values of the function and the basic variables with respect to $\mathbf{b}$ or $\mathbf{c}$ can be determined directly from $\hat{\mathbf{b}} = \mathbf{A}_B^{-1}\mathbf{b}$ and $\hat{f} = \mathbf{c}_B^T\hat{\mathbf{b}}$ (see Question 8.14). To find the effect of small finite non-zero changes, the aim is to retain as much information as possible from the unperturbed solution when re-solving the problem. If only $\mathbf{b}$ or $\mathbf{c}$ is changed then a representation of $\mathbf{A}_B^{-1}$ is directly available. If $\mathbf{A}_B$ is changed then the partition into B and N can still be valuable. The only difficulty is that changes to $\mathbf{b}$ or $\mathbf{A}_B$ might result in the condition $\hat{\mathbf{b}} \geqslant \mathbf{0}$ being violated. It is important to be able to restore feasibility without restarting *ab initio*, and this is discussed further in Section 8.4. Once feasibility is obtained then simplex steps are taken to achieve optimality. Another type of parametric programming arises when extra constraints are added (see Section 13.1 for example) which cut out the solution. The same approach of taking simplex steps to restore first feasibility and then optimality can be used. An alternative possibility is to use the dual formulation, when extra constraints in the primal just correspond to extra variables in the dual. The previous solution is then still dual feasible, and all that is required is to restore optimality.

Another way of treating the general LP problem

$$\begin{aligned} &\underset{\mathbf{x}}{\text{minimize}} \quad f(\mathbf{x}) \triangleq \mathbf{c}^{\mathrm{T}}\mathbf{x} \\ &\text{subject to} \quad \mathbf{a}_i^{\mathrm{T}}\mathbf{x} = b_i, \quad i \in E \\ &\qquad\qquad\quad \mathbf{a}_i^{\mathrm{T}}\mathbf{x} \geqslant b_i, \quad i \in I \end{aligned} \tag{8.3.1}$$

for finite index sets E and I, is the *active set method*, which devolves from ideas used in more general linear constraint problems (Chapters 10 and 11), and which is perhaps more natural than the traditional approach via (8.1.1). Similar introductory considerations apply as in Section 8.1, and in particular the solution usually occurs at an extreme point or vertex of the feasible region. (When this situation does not hold, for example in $\min x_1 + x_2$ subject to $x_1 + x_2 = 1$, then it is possible to create a vertex by removing undetermined variables, or equivalently by having *pseudoconstraints* $x_i = 0$ in the active set—see Section 8.4.) Each iterate $\mathbf{x}^{(k)}$ in the method is therefore a feasible vertex defined by the equations

$$\mathbf{a}_i^{\mathrm{T}}\mathbf{x} = b_i, \quad i \in \mathscr{A} \tag{8.3.2}$$

where $\mathscr{A}$ is the *active set* of n constraint indices, and where the vectors $\mathbf{a}_i$ $i \in \mathscr{A}$ are independent. Except in degenerate cases (see Section 8.6), $\mathscr{A}$ is the set of active constraints $\mathscr{A}(\mathbf{x}^{(k)})$ in (7.1.2).

Each iteration in the active set method consists of a move from one vertex to another along a common edge. At $\mathbf{x}^{(k)}$ the Lagrange multiplier vector

$$\boldsymbol{\lambda} = \mathbf{A}^{-1}\nabla f = \mathbf{A}^{-1}\mathbf{c} \tag{8.3.3}$$

is evaluated, where $\mathbf{A}$ now denotes the $n \times n$ matrix with columns $\mathbf{a}_i$, $i \in \mathscr{A}$. Let the columns of $\mathbf{A}^{-\mathrm{T}}$ be written $\mathbf{s}_1, \mathbf{s}_2, \ldots, \mathbf{s}_n$. Then the construction (9.1.14) shows that the vectors $\mathbf{s}_i$, $i \in \mathscr{A} \cap I$ are the directions of all feasible edges at $\mathbf{x}^{(k)}$, and the multipliers λ_i are the slopes of $f(\mathbf{x})$ along these edges. Thus if

$$\lambda_i \geqslant 0, \qquad i \in \mathscr{A} \cap I \tag{8.3.4}$$

then no feasible descent directions exist, so $\mathbf{x}^{(k)}$ is optimal and the iteration terminates. Otherwise the most negative λ_i, $i \in \mathscr{A} \cap I$ (λ_q say) is chosen, and a search is made along the downhill edge

$$\mathbf{x} = \mathbf{x}^{(k)} + \alpha\mathbf{s}_q, \qquad \alpha \geqslant 0. \tag{8.3.5}$$

The ith constraint has the value $c_i^{(k)} = \mathbf{a}_i^{\mathrm{T}}\mathbf{x}^{(k)} - b_i$ at $\mathbf{x}^{(k)}$ and the value $c_i(\mathbf{x}) = c_i^{(k)} + \alpha\mathbf{a}_i^{\mathrm{T}}\mathbf{s}_q$ at points along the edge. The search is terminated by the first inactive constraint (p say) to become active. Candidates for p are therefore indices $i \notin \mathscr{A}$ with $\mathbf{a}_i^{\mathrm{T}}\mathbf{s}_q < 0$ and any such constraint function becomes zero when

$$\alpha = \frac{b_i - \mathbf{a}_i^{\mathrm{T}}\mathbf{x}^{(k)}}{\mathbf{a}_i^{\mathrm{T}}\mathbf{s}_q}.$$

Thus the index p and the corresponding value of α are defined by

$$\alpha = \min_{\substack{i:i\notin\mathcal{A},\\ \mathbf{a}_i^T\mathbf{s}_q<0}} \frac{b_i - \mathbf{a}_i^T\mathbf{x}^{(k)}}{\mathbf{a}_i^T\mathbf{s}_q}. \tag{8.3.6}$$

The corresponding point defined by (8.3.5) is the new vertex $\mathbf{x}^{(k+1)}$, and the new active set is obtained by replacing q by p in $\mathcal{A}$. The iteration is then repeated from this new vertex until termination occurs. Only one column of $\mathbf{A}$ is changed at each iteration, so $\mathbf{A}^{-1}$ can be readily updated as in (8.2.15).

In fact the active set method is closely related to the simplex method, when the slack variables are added to (8.3.1). It is not difficult (see Question 8.17) to show that the Lagrange multiplier vector in (8.3.3) is the reduced costs vector in (8.2.6), so that the optimality tests (8.3.4) and (8.2.8) are identical. Likewise the index q which is determined at a non-optimal solution is the same in both cases. Moreover the tests (8.3.6) and (8.2.11) which determine p are also equivalent, so the same interchanges in active constraints (nonbasic slack variables) are made. In fact the only difference between the two methods lies in how the inverse matrix information is represented, because the active set method updates the $n \times n$ matrix $\mathbf{A}^{-1}$ whereas the revised simplex method updates the $m \times m$ matrix $\mathbf{A}_B^{-1}$. Thus for problems in standard form (8.1.1) which only have a few equations ($m \ll n$), a smaller matrix is updated by the revised simplex method, which is therefore preferable. On the other hand, if the problem is like (8.3.1) and there are $m \gg n$ constraints (mostly inequalities), then the simplex method requires many slack variables to be added, and the active set method updates the smaller inverse matrix and is therefore preferable.

In both cases the inefficiency of the worse method is caused by the fact that columns of the unit matrix occur in the matrix ($\mathbf{A}$ or $\mathbf{A}_B$) whose inverse is being updated. For example if bounds $\mathbf{x} \geqslant \mathbf{0}$ (or more generally $\mathbf{l} \leqslant \mathbf{x} \leqslant \mathbf{u}$) are recognized in the active set method, then the matrix $\mathbf{A}$ can be partitioned (after a suitable permutation of variables) into the form

$$\begin{array}{cc} \scriptstyle n-p & \scriptstyle p \end{array}$$
$$\mathbf{A} = \begin{bmatrix} \mathbf{A}_1 & \mathbf{0} \\ \mathbf{A}_2 & \mathbf{I} \end{bmatrix} \begin{matrix} \scriptstyle n-p \\ \scriptstyle p \end{matrix}$$

where p is the number of active bounds. Then only $\mathbf{A}_1^{-1}$ need be recurred, and $\mathbf{A}^{-1}$ can be recovered from the expression

$$\mathbf{A}^{-1} = \begin{bmatrix} \mathbf{A}_1^{-1} & \mathbf{0} \\ -\mathbf{A}_2\mathbf{A}_1^{-1} & \mathbf{I} \end{bmatrix}.$$

In this way the active set method becomes as efficient as possible, if sparsity in the remaining constraint normals can be ignored. However, a similar modification to the revised simplex method can be made when basic slack variables are present. Then $\mathbf{A}_B$ also contains columns of the unit matrix, which enables a smaller submatrix to be recurred. If this is done then the methods are entirely equivalent. However, as a personal choice I prefer the description of linear programming afforded by the active set method, since it is more natural and direct, and does not rely on introducing extra variables to solve general LP problems.

8.4 FEASIBLE POINTS FOR LINEAR CONSTRAINTS

In this section methods are discussed for finding feasible points for a variety of problems having linear constraints. First of all the *method of artificial variables* is described for finding an initial basic feasible solution for use in the simplex method. The underlying idea is of wide generality and it is subsequently shown that it is possible to determine a similar but more flexible and efficient method for use in other situations. These include the active set methods for minimization subject to linear constraints described in Sections 8.3 (linear objective function), 10.3 (quadratic objective) and 11.2 (general objective). The resulting methods are also very suitable for restoring feasibility when doing parametric programming. The idea of an L_1 exact penalty function for nonlinear constraints (Section 14.3) is also seen to be a generalization of these ideas.

In the first instance the problem is considered of finding a basic feasible solution to the constraints

$$\mathbf{A}\mathbf{x} = \mathbf{b}, \qquad \mathbf{x} \geqslant \mathbf{0} \tag{8.4.1}$$

which arise in the standard form of linear programming (8.1.1). Extra variables $\mathbf{r}$ (often referred to as *artificial variables*) are introduced into the problem, such that

$$\mathbf{r} = \mathbf{b} - \mathbf{A}\mathbf{x} \tag{8.4.2}$$

These variables can be interpreted as *residuals* of the equations $\mathbf{A}\mathbf{x} = \mathbf{b}$. In the method of artificial variables it is first ensured that $\mathbf{b} \geqslant \mathbf{0}$ by first reversing the sign of any row of $\mathbf{A}$ and $\mathbf{b}$ for which $b_i < 0$. Then the auxiliary problem

$$\begin{aligned} &\underset{\mathbf{x},\mathbf{r}}{\text{minimize}} \quad \sum_i r_i \\ &\text{subject to } \mathbf{A}\mathbf{x} + \mathbf{r} = \mathbf{b}, \qquad \mathbf{x} \geqslant \mathbf{0}, \quad \mathbf{r} \geqslant \mathbf{0} \end{aligned} \tag{8.4.3}$$

is solved in an attempt to reduce the residuals to zero. Let $\mathbf{x}'$ and $\mathbf{r}'$ solve (8.4.3); clearly if $\mathbf{r}' = \mathbf{0}$ then $\mathbf{x}'$ is feasible in (8.4.1) as required, whereas if $\mathbf{r}' \neq \mathbf{0}$ then there is *no feasible point* for (8.4.1). (This latter observation is true since if $\mathbf{x}'$ is feasible in (8.4.1) then $\mathbf{x}', \mathbf{0}$ is a feasible point of (8.4.3) with $\sum r_i = 0$ which contradicts the optimality of $\mathbf{r}' \neq \mathbf{0}$.) Also (8.4.3) is a linear program in standard form with variables $\mathbf{x}, \mathbf{r}$ for which the coefficient matrix is $[\mathbf{A}:\mathbf{I}]$ and the costs are $\mathbf{0}, \mathbf{e}$ where $\mathbf{e}$ is a vector of ones. An initial basic feasible solution is given by having the $\mathbf{r}$ variables basic and the $\mathbf{x}$ variables nonbasic so that $\mathbf{r}^{(1)} = \mathbf{b}$ and $\mathbf{x}^{(1)} = \mathbf{0}$. The initial basis matrix is simply $\mathbf{A}_B = \mathbf{I}$. Thus the simplex method itself can be used to solve (8.4.3) directly from this b.f.s.

A feature of the calculation is that if the artificial variable r_i becomes nonbasic then the constraint $(\mathbf{Ax}-\mathbf{b})_i = 0$ becomes satisfied and need never be relaxed, that is r_i need never become basic again. Thus once an artificial variable becomes nonbasic, it may be removed from the computation entirely, and some effort may be saved. If a feasible point exists ($\mathbf{r}' = \mathbf{0}$) the optimum basic feasible solution to (8.4.3) has no basic r_i variables (assuming no degeneracy) so that all the artificial variables must have been removed. Therefore the remaining variables $\mathbf{x}^T = (\mathbf{x}_B^T, \mathbf{x}_N^T)$ are partitioned to give a basic feasible solution to (8.4.1). The same tableau $\hat{\mathbf{A}}_N$ or inverse basis matrix $\mathbf{A}_B^{-1}$ is directly available to be used in the solution of the main LP (8.1.1). Because the solution of (8.4.3) is a preliminary to the solution of (8.1.1), the two are sometimes referred to as *phase I* and *phase II* of the simplex method respectively. In some degenerate cases (for example when the constraint matrix $\mathbf{A}$ is rank deficient) it is possible that some artificial variables may remain in the basis after (8.4.3) is solved. Since $\mathbf{r} = \mathbf{0}$ nothing is lost by going on to solve the main problem with the artificial variables remaining in the basis, essentially just as a means of enabling the basic variables to be eliminated (see Question 8.21).

An example of the method to find a basic feasible solution to the constraints in (8.1.2) is described. Since $\mathbf{b} \geqslant \mathbf{0}$ no sign change is required. Artificial variables r_1 and r_2 (x_5 and x_6 say) are added to give the phase I problem

$$\begin{aligned}
&\underset{\mathbf{x}}{\text{minimize}} \quad x_5 + x_6 \\
&\text{subject to} \quad x_1 + x_2 + x_3 + x_4 + x_5 \qquad = 1 \\
&\qquad\qquad\quad x_1 \qquad + x_3 - 3x_4 \quad + x_6 = \tfrac{1}{2} \\
&\qquad\qquad\quad x_i \geqslant 0, \qquad i = 1, 2, \ldots, 6.
\end{aligned}$$

A basic feasible solution for this problem is $B = \{5, 6\}$, $N = \{1, 2, 3, 4\}$, with $\mathbf{A}_B = \mathbf{I}$ by construction. A calculation shows that $\hat{\mathbf{c}}_N = (-2, -1, -2, 2)^T$ so that x_1 ($q = 1$) is chosen to be increased. Then $\mathbf{d} = (-1, -1)^T$ so both x_5 and x_6 are reduced, but x_6 reaches zero first. Thus x_1 is made basic and x_6 nonbasic. Since x_6 is artificial it is removed from the computation, and a smaller problem is solved with $B = \{1, 5\}$, $N = \{2, 3, 4\}$. In this case the calculation shows that x_4 is increased and x_5 goes to zero. When x_4 is made basic and x_5 removed, there remains the partition $B = \{1, 4\}$, $N = \{2, 3\}$. Both the artificial variables have been removed and so phase I is complete. The basic feasible solution is carried forward into phase II and by chance is found to be optimal (see Section 8.2), so no further basis changes are made. The tableau method applied to this problem is shown in Table 8.4.1.

The way in which an artificial variable is introduced is very similar to that in which a slack variable is introduced (see Sections 8.1 and 7.2). In fact if a slack variable has already been added into an inequality $\mathbf{a}_i^T\mathbf{x} \leqslant b_i$, and if $b_i \geqslant 0$, then it is possible to use a slack variable in the initial basis and not add an artificial variable as well. For example consider the system

$$\begin{aligned}
&x_1 + x_2 + x_3 + \ x_4 \leqslant 1 \\
&x_1 \qquad + x_3 - 3x_4 = \tfrac{1}{2}, \qquad \mathbf{x} \geqslant \mathbf{0}.
\end{aligned}$$

Table 8.4.1 The tableau method in phase I

Phase I

		x_1	x_2	x_3	x_4	x_5	x_6	
Data:		0	0	0	0	1	1	0
	x_5	1	1	1	1	1		1
	x_6	1	0	1	-3		1	$\frac{1}{2}$

$k=1$:	f	-2	-1	-2	2			$-1\frac{1}{2}$
	x_5	1	1	1	1	1		1
	x_6	(1)	0	1	-3		1	$\frac{1}{2}$

$k=2$:	f		-1	0	-4		2	$-\frac{1}{2}$
	x_5		1	0	(4)	1	-1	$\frac{1}{2}$
	x_1	1	0	1	-3		1	$\frac{1}{2}$

$k=3$:	f		0	0		1	1	0
	x_4		$\frac{1}{4}$	0	1	$\frac{1}{4}$	$-\frac{1}{4}$	$\frac{1}{8}$
	x_1	1	$\frac{3}{4}$	1		$\frac{3}{4}$	$\frac{1}{4}$	$\frac{7}{8}$

Phase II

Data:		1	2	3	4	0
	x_4		$\frac{1}{4}$	0	1	$\frac{1}{8}$
	x_4	1	$\frac{3}{4}$	1		$\frac{7}{8}$

$k=1$:	f		$\frac{1}{4}$	2		$-1\frac{3}{8}$
	x_4		$\frac{1}{4}$	0	1	$\frac{1}{8}$
	x_1	1	$\frac{3}{4}$	1		$\frac{7}{8}$

Adding slack variable x_5 gives

$$\begin{aligned} x_1 + x_2 + x_3 + \ x_4 + x_5 &= 1 \\ x_1 \quad\quad + x_3 - 3x_4 \quad\quad &= \tfrac{1}{2}, \qquad \mathbf{x} \geqslant \mathbf{0} \end{aligned} \tag{8.4.4}$$

which is in standard form. Since $b_1 = 1 \geqslant 0$ the slack variable x_5 can be used in the initial basis, and so an artificial variable (x_6 say) need only be added into the second equation. The resulting cost function is the sum of the *artificial variables only*, so in this case the phase I problem is

$$\begin{aligned} &\text{minimize} \quad x_6 \\ &\text{subject to} \quad x_1 + x_2 + x_3 + \ x_4 + x_5 \quad\quad = 1 \\ &\qquad\qquad\quad x_1 \quad\quad + x_3 - 3x_4 \quad\quad + x_6 = \tfrac{1}{2}, \qquad \mathbf{x} \geqslant \mathbf{0} \end{aligned}$$

and the initial partition is $B = \{5, 6\}$, $N = \{1, 2, 3, 4\}$. As before artificial variables are removed from the problem when they become nonbasic, but slack variables remain in the problem throughout. In this example it happens that x_6 is removed on the first iteration and the resulting partition $B = \{1, 5\}$, $N = \{2, 3, 4\}$ gives a basic feasible solution for (8.4.4), to be used in phase II. In general if $b_i < 0$ after adding a slack variable to an inequality $\mathbf{a}_i^T\mathbf{x} \leqslant b_i$, then it is also necessary to add an artificial variable to that equation. This is cumbersome and can be avoided by adopting the more general framework which follows, and which does not require any sign changes to ensure $\mathbf{b} \geqslant \mathbf{0}$.

At the expense of a small addition to the logic in phase I, a much more flexible technique for obtaining feasibility can be developed (Wolfe, 1965). This is to allow negative variables in the basis and to extend the cost function, giving the auxiliary problem on the kth step as

$$\begin{aligned} &\text{minimize} \quad \sum_{i:x_i^{(k)}<0} -x_i \ + \sum_{i:r_i^{(k)}<0} -r_i \ + \sum_{i:r_i^{(k)}>0} r_i \\ &\text{subject to} \quad \mathbf{Ax} + \mathbf{r} = \mathbf{b}, \quad x_i \geqslant 0 \qquad \forall i : x_i^{(k)} \geqslant 0. \end{aligned} \tag{8.4.5}$$

In this problem it is understood that an element of the cost vector is taken as -1 if $x_i^{(k)} < 0$ and 0 if $x_i^{(k)} \geqslant 0$, and likewise as -1 if $r_i^{(k)} < 0$ and $+1$ if $r_i^{(k)} > 0$ so that the costs may change from one iteration to another. *Any* set of basic variables can be chosen initially (assuming $\mathbf{A}_B$ is non-singular) without the need to have $\hat{\mathbf{b}} \geqslant \mathbf{0}$, and simplex steps are made as before, except that (8.2.11) must be modified in an obvious way to allow for a negative variable being increased to zero. Once x_i becomes non-negative then the condition $x_i \geqslant 0$ is subsequently enforced and artificial variables are removed once they become zero. There is no need with (8.4.5) to have both artificial and slack variables in any one equation. An example is given in Question 8.22. In fact, because any set of basic variables can be chosen, the artificial variable is added only as a means to enable an easily invertible matrix $\mathbf{A}_B$ to be chosen. Thus if such a matrix is already available then artificial variables are not required at all.

Another consequent advantage of (8.4.5) occurs in *parametric programming* (Section 8.3) when the solution from a previous problem (B and $\mathbf{A}_B^{-1}$) is used to start phase I for the new problem. It is likely that the old basis is close to a feasible basis if the changes are small. However it is quite possible that the perturbation causes feasibility ($\hat{\mathbf{b}} \geqslant \mathbf{0}$) to be lost, and it is much more efficient to use (8.4.5) in conjunction with the old basis, rather than starting the phase I procedure with a basis of artificial variables. Also a better feasible point is likely to be obtained. For example in (8.4.4) if b_2 is perturbed to $-\frac{1}{2}$ and the same partition $B = \{1, 5\}$, $N = \{2, 3, 4\}$ is used, then $\hat{\mathbf{b}} = (-\frac{1}{2}, \frac{3}{2})^{\mathrm{T}}$ can be calculated. Thus the cost function in (8.4.5) is $-x_1$ and it follows that $\hat{\mathbf{c}}_N = (0, 1, -3)^{\mathrm{T}}$ so that x_4 is chosen to be increased. Then $\mathbf{d} = (3, -4)^{\mathrm{T}}$ so the increase in x_4 causes x_1 to increase towards zero and x_5 to decrease towards zero. x_1 reaches zero first when $x_4 = \frac{1}{6}$, so the new partition is $B = \{4, 5\}$, $N = \{1, 2, 3\}$ with $\hat{\mathbf{b}} = (\frac{1}{6}, \frac{5}{6})^{\mathrm{T}}$. Since there are now no negative variables, feasibility is restored and phase II can commence.

Use of the auxiliary problem (8.4.5) is closely related to a technique for solving linear L_1 approximation problems in which $\|\mathbf{r}\|_1$ $(= \sum_i |r_i|)$ is minimized, where $\mathbf{r}$ is defined by (8.4.2). The $|r_i|$ terms are handled by introducing positive and negative variables r_i^+ and r_i^-, where $r_i^+ \geqslant 0$, $r_i^- \geqslant 0$, and by replacing $r_i = r_i^+ - r_i^-$ and $|r_i| = r_i^+ + r_i^-$ as described in Section 7.2. Barrodale and Roberts (1973) suggest a modification for improving the efficiency of the simplex method so that fewer iterations are required to solve this particular type of problem. The idea is to allow the simplex iteration to pass through a number of vertices at one time so as to minimize the total cost function. The same type of modification can be used advantageously with (8.4.5). Let the nonbasic variable x_q (either an element of $\mathbf{x}$ or $\mathbf{r}$) for which $\hat{c}_q < 0$ be chosen to be increased as in Section 8.2. If a variable $x_i < 0$ is increased to zero, then the corresponding element of $\mathbf{c}_B$ is changed from -1 to 0 and $\hat{c}_q$ is easily re-computed. If $\hat{c}_q < 0$ still holds then x_q can continue to be increased without making a pivot step to change $\mathbf{A}_B^{-1}$ or $\hat{\mathbf{A}}_N$. Only if $\hat{c}_q \geqslant 0$ is a pivot step made. Likewise if a variable $r_i < 0$ (or > 0) becomes zero then the corresponding element of $\mathbf{c}_B$ is changed from -1 to $+1$ (or $+1$ to -1) and $\hat{c}_q$ is recomputed and tested in the same way. In this modification an artificial variable need not be removed immediately it becomes zero and $\mathbf{x}$ variables may be allowed to go negative if the cost function is reduced. For best efficiency it is valuable to scale the rows of $\mathbf{Ax} = \mathbf{b}$, or equivalently to scale the elements of the cost function in (8.4.5) to balance the contribution from each term (to the same order of magnitude).

It is possible to suggest an entirely equivalent technique for calculating a feasible point for use with the active set method for linear programming (Section 8.3), or with other active set methods for linear constraints (Sections 10.3 and 11.2). The technique is described here in the case that E is empty, although it is readily modified to the more general case. The phase I auxiliary problem (Fletcher, 1970b) which is solved at $\mathbf{x}^{(k)}$ is

$$\begin{array}{ll} \text{minimize} & \sum_{i \in V^{(k)}} (b_i - \mathbf{a}_i^{\mathrm{T}}\mathbf{x}) \\ \text{subject to} & \mathbf{a}_i^{\mathrm{T}}\mathbf{x} \geqslant b_i, \quad i \notin V^{(k)} \end{array} \tag{8.4.6}$$

where $V^{(k)} = V(\mathbf{x}^{(k)})$ is the set of violated (that is infeasible) constraints at $\mathbf{x}^{(k)}$.

Thus the cost function in (8.4.6) is the sum of moduli of violated constraints. The iteration commences by making any convenient arbitrary choice of an active set $\mathscr{A}$ of n constraints, and this determines an initial vertex $\mathbf{x}^{(1)}$. At $\mathbf{x}^{(k)}$ the gradient of the cost function is $-\sum_{i\in V^{(k)}}\mathbf{a}_i$, and the method proceeds as in (8.3.3) to use this vector to calculate multipliers and hence to determine an edge $\mathbf{s}_q$ along which the cost function has negative slope. A search along $\mathbf{x}^{(k)}+\alpha\mathbf{s}_q$ is made to minimize the sum of constraint violations $\sum_{i\in V(\mathbf{x})}(b_i-\mathbf{a}_i^{\mathrm{T}}\mathbf{x})$. This search is always terminated by an inactive constraint (p say) becoming active, and p then replaces q in the active set. The method terminates when $\mathbf{x}^{(k)}$ is found to be a feasible point, and the resulting active set and the matrix $\mathbf{A}^{-1}$ in (8.3.3) are passed forward to the phase II problem (for example (8.3.1)).

To create the initial vertex $\mathbf{x}^{(1)}$ easily in the absence of any parametric programming information, it is also possible to have something similar to the artificial variable method. In fact artificial constraints or *pseudoconstraints* $x_i=0$, $i=1,2,\ldots,n$, are added to the problem, if they are not already present, and form the initial active set (Fletcher and Jackson, 1974). Pseudoconstraints with a non-zero Lagrange multiplier are always eligible for removal from the active set, and they do not contribute to the cost function in the line search. The value of this device is that the initial matrix used in (8.3.3), etc., is $\mathbf{A}=\mathbf{I}$, which is immediately available. Once a pseudoconstraint becomes inactive it is removed from the problem; this usually ensures that the pseudoconstraints are rapidly removed from the problem in favour of actual constraints. However pseudoconstraints with a zero multiplier are allowed to remain in the active set at a solution. This can be advantageous in phase I when the feasible region has no true vertex, other than at large values of $\mathbf{x}$ which might cause round-off difficulties. In phase II it also solves the problem of eliminating redundant variables (for example as in $\min x_1+x_2$ subject to $x_1+x_2=1$) in a systematic way.

One final idea related to feasible points is worthy of mention. It is applicable to both a standard form problem (8.1.1) or to the more general form (8.3.1) used in the active set method. The feasible point which is found by the methods described in this section is essentially arbitrary, and may be far from the solution. It can be advantageous to bias the phase I cost function in (8.4.5) or (8.4.6) by adding to it a multiple $\nu f(\mathbf{x})$ of the phase II cost function. For all ν sufficiently small the solution of phase I is also that of phase II so the possibility exists of combining both these stages into one. The limit $\nu\to 0$ corresponds to the phase I–phase II iteration above, and is sometimes referred to (after multiplying through by $M=1/\nu\gg 0$) as the *big M method*. However, the method is usually more efficient if larger values of ν are taken, although if ν is too large then a feasible point is not obtained. In fact it is not difficult to see that the method has become one for minimizing an L_1 *exact penalty function* for the LP problem (8.3.1), as described in Section 12.3. Thus the result of Theorem 14.3.1 indicates that the bound $\nu\leqslant 1/\|\boldsymbol{\lambda}^*\|_\infty$ limits the choice of ν where $\boldsymbol{\lambda}^*$ is the optimum Lagrange multiplier vector for (8.3.1).

This type of idea is of particular advantage for parametric programming in QP problems (Chapter 10), because the active set $\mathscr{A}$ from a previously solved

QP problem may have less than n elements and so cannot be used as the initial vertex in the phase I problem (8.4.6). However, if $\nu f(\mathbf{x})$ (now a quadratic function) is added to the phase I cost function in (8.4.6), then a QP-like problem results and the previous active set and the associated inverse information can be carried forward and used directly as initial data for this problem. No doubt a similar idea could also be used in more general linear constraint problems (Chapter 11), although I do not think that this possibility has been investigated.

8.5 STABLE AND LARGE-SCALE LINEAR PROGRAMMING

A motivation for considering alternatives to the tableau method or the revised simplex method is that the direct representation of $\mathbf{A}_B^{-1}$ or its implicit occurrence in the matrix $\hat{\mathbf{A}} = \mathbf{A}_B^{-1}\mathbf{A}_N$ can lead to difficulties over round-off error. If an intermediate b.f.s. is such that $\mathbf{A}_B$ is nearly singular then $\mathbf{A}_B^{-1}$ becomes large and representing this matrix directly introduces large round-off errors. If on a later iteration $\mathbf{A}_B$ is well conditioned then $\mathbf{A}_B^{-1}$ is not large but the previous large round-off errors remain and their relative effect is large. This problem is avoided if a stable factorization of $\mathbf{A}_B$ is used which does not represent the inverse of potentially nearly singular matrices. Typical examples of this for solving equations $\mathbf{Ax} = \mathbf{b}$ are the **LU** factors (with pivoting) or the **QR** factors (**L** is lower triangular, **U**, **R** are upper triangular, and **Q** is orthogonal). Some generalizations of these ideas in linear programming are described in this section.

Also in this section, methods are discussed which are suitable for solving large sparse LP problems because they have some features in common. In terms of (8.1.1) this usually refers to problems in which the number of equations m is greater than a few hundred, and in which the coefficient matrix **A** is sparse, that is it contains a high proportion of zero elements. The standard techniques of Section 8.2 (or 8.3) fail because the matrices such as $\mathbf{A}_B^{-1}$ or $\hat{\mathbf{A}}_N$ do not retain the sparsity of **A** and must be stored in full, which is usually beyond the capacity of the rapid access computer store. By taking the sparsity structure of **A** into account it is possible to solve large sparse problems. There are two main types of method, one of which is a *general sparse matrix method* which allows for any sparsity structure and attempts to update an invertible representation of $\mathbf{A}_B$ which maintains a large proportion of the sparsity. The other type is the *decomposition method* which attempts to reduce the problem to a number of much smaller LP problems which can be solved conventionally. Decomposition methods usually require a particular structure in the matrix **A** so are not of general applicability. One practical example of this type of method is given, and other possibilities are referenced. It is not possible to say that any one method for large-scale LP is uniformly best because comparisons are strongly dependent on problem size and structure, but my interpretation of current thinking is that even those problems which are suitable for solving by decomposition methods can also be handled efficiently by general sparse matrix methods, so a good implementation of the latter type of method is of most importance.

The earliest attempt to take sparsity into account is the *product form method.* Equation (8.2.15) for updating the matrix $\mathbf{A}_B^{-1}$ can be written

$$[\mathbf{A}_B^{(k+1)}]^{-1} = \mathbf{M}_k[\mathbf{A}_B^{(k)}]^{-1} \tag{8.5.1}$$

where $\mathbf{M}_k$ is the matrix

$$\mathbf{M}_k = \begin{bmatrix} 1 & & & -d_1/d_p & & & \\ & \ddots & & -d_2/d_p & & & \\ & & 1 & \vdots & & & \\ & & & -d_p^{-1} & & & \\ & & & \vdots & 1 & & \\ & & & \vdots & & \ddots & \\ & & & -d_n/d_p & & & 1 \end{bmatrix} \tag{8.5.2}$$

$\quad$ ↑ column i

where $\mathbf{d}$ is defined in (8.2.10). Thus if $\mathbf{A}_B^{(1)} = \mathbf{I}$, then $[\mathbf{A}_B^{(k)}]^{-1}$ can be represented by the product

$$[\mathbf{A}_B^{(k)}]^{-1} = \mathbf{M}_{k-1} \cdots \mathbf{M}_2\mathbf{M}_1. \tag{8.5.3}$$

In the product form method the non-trivial column of each $\mathbf{M}_j$, $j = 1, 2, \ldots, k-1$, is stored in a computer file in packed form, that is just the non-zero elements and their row index. These vectors were known as *eta vectors* in early implementations. Then the representation (8.5.3) is used whenever operations with $\mathbf{A}_B^{-1}$ are required. As k increases the file of eta vectors grows longer and operating with it becomes more expensive. Thus ultimately it becomes cheaper to *reinvert* the basis matrix by restarting with a unit matrix and introducing the basic columns to give new eta vectors. This process of reinversion is common to most product form methods, and the choice of iteration on which it occurs is usually made on grounds of efficiency, perhaps using readings from a computer clock.

Product form methods are successful because the number of non-zeros required to represent $\mathbf{A}_B^{-1}$ as a product can be smaller than the requirement for storing the usually full matrix $\mathbf{A}_B^{-1}$ itself. However, there are reasons why (8.5.3), known as the *Gauss–Jordan product form*, is not the most efficient. Consider the Gauss–Jordan method for eliminating variables in a set of linear equations with coefficient matrix $\mathbf{A}$ (ignoring pivoting). This can be written as the reduction of $\mathbf{A}(=\mathbf{A}^{(1)})$ to a unit matrix $\mathbf{I}(=\mathbf{A}^{(n+1)})$ by premultiplication by a sequence of Gauss–Jordan elementary matrices

$$\mathbf{I} = \mathbf{M}_n \cdots \mathbf{M}_2\mathbf{M}_1\mathbf{A} \tag{8.5.4}$$

where $\mathbf{M}_p$ has the form (8.5.2) with $d_j = -a_{jp}^{(p)}$ for all j. Another method for solving linear equations is Gaussian elimination, which is equivalent to the factorization $\mathbf{A} = \mathbf{LU}$ (again ignoring pivoting), where $\mathbf{L}$ is a unit lower triangular matrix and $\mathbf{U}$ is upper triangular. This can also be written in product form with

$$\begin{aligned} \mathbf{L} &= \mathbf{L}_1 \mathbf{L}_2 \cdots \mathbf{L}_{n-1} \\ \mathbf{U} &= \mathbf{U}_n \mathbf{U}_{n-1} \cdots \mathbf{U}_1 \end{aligned} \tag{8.5.5}$$

where $\mathbf{L}_i$ and $\mathbf{U}_i$ are the elementary matrices

$$\mathbf{L}_i = \begin{bmatrix} 1 & & & & & \\ & \ddots & & & & \\ & & 1 & & & \\ & & & 1 & & \\ & & & l_{i+1,i} & 1 & \\ & & & \vdots & & \ddots & \\ & & & l_{n,i} & & & 1 \end{bmatrix} \tag{8.5.6}$$

$\uparrow$column i

$$\mathbf{U}_i = \begin{bmatrix} 1 & & & & & \\ & \ddots & & & & \\ & & 1 & & & \\ & & & u_{ii} & \cdots & u_{in} \\ & & & & 1 & \\ & & & & & \ddots \\ & & & & & & 1 \end{bmatrix} \leftarrow \text{row } i \tag{8.5.7}$$

and where l_{ij} and u_{ij} are the elements of the factors $\mathbf{L}$ and $\mathbf{U}$ of $\mathbf{A}$. It has been observed (see various chapters in Reid (1971a,b), for example, and Question 8.23) that an **LU** factorization of a sparse matrix $\mathbf{A}$ retains much of the sparsity of $\mathbf{A}$ itself; typically the number of non-zero elements to be stored might be no more than double the number of non-zeros in $\mathbf{A}$, especially when there is some freedom to do pivoting (as in reinversion).

It is now possible to explain why the Gauss–Jordan product form is not very efficient. It can be shown (see Question 8.24) that the non-trivial elements below the diagonal in the Gauss–Jordan elementary matrices $\mathbf{M}_k$ are the corresponding elements of $\mathbf{L}$ (with opposite sign), whereas the elements on and above the diagonal are the elements of $\mathbf{U}^{-1}$. Since $\mathbf{U}^{-1}$ is generally full it follows that only a limited amount of sparsity is retained in the factors $\mathbf{M}_k$. A simple example of this result is given in Question 8.23. This analysis also shows that use of a Gauss–Jordan product form does not solve the problem of numerical stability. If $\mathbf{A}_B$ is nearly singular then $\mathbf{U}$ can also be nearly singular and $\mathbf{U}^{-1}$ is then large. Thus the potential difficulty due to the presence of large round-off errors is not solved.

In view of this situation various efforts have been made to try to achieve the advantages of both stability and sparsity possessed by an **LU** decomposition. A product form method of Forrest and Tomlin (1972) has proved very successful from a sparsity viewpoint and has been used in some commercial codes.

Unfortunately it is potentially unstable numerically and must therefore be used with care. Another product form method due to Bartels and Golub (Bartels, 1971) addresses both the sparsity and stability aspects, and uses a combination of elementary elimination and interchange operations. Assume without loss of generality that the initial basis matrix has triangular factors $\mathbf{A}_B^{(1)} = \mathbf{LU}$. Then on the kth iteration $\mathbf{A}_B^{(k)}$ has the invertible representation defined by the expression

$$(\mathbf{L}_r\mathbf{P}_r)(\mathbf{L}_{r-1}\mathbf{P}_{r-1})\cdots(\mathbf{L}_1\mathbf{P}_1)\mathbf{L}^{-1}\mathbf{A}_B^{(k)}\mathbf{Q}^{(k)} = \mathbf{U}^{(k)} \tag{8.5.8}$$

where $\mathbf{U}^{(k)}$ is upper triangular, $\mathbf{Q}^{(k)}$ is a permutation matrix, and $\mathbf{L}_i$ and $\mathbf{P}_i$ are certain elementary lower triangular and permutation matrices for $i = 1, 2, \ldots, r$. (Note that r refers to $r^{(k)}$, $\mathbf{U}^{(1)} = \mathbf{U}$, and $\mathbf{Q}^{(1)} = \mathbf{I}$.) An operation with $\mathbf{A}_B^{-1}$ is readily accomplished since from (8.5.8) it only requires permutations and operations with triangular matrices.

After a simplex iteration the new matrix $\mathbf{A}_B^{(k+1)}$ is defined by (8.2.14), and since only one column is changed, it is possible to write

$$(\mathbf{L}_r\mathbf{P}_r)(\mathbf{L}_{r-1}\mathbf{P}_{r-1})\cdots(\mathbf{L}_1\mathbf{P}_1)\mathbf{L}^{-1}\mathbf{A}_B^{(k+1)}\mathbf{Q}^{(k)} = \mathbf{S} \tag{8.5.9}$$

where $\mathbf{S}$ differs in only one column (t say) from $\mathbf{U}^{(k)}$ and is therefore the *spike matrix* illustrated in Figure 8.5.1. This is no longer upper triangular and is not readily returned to triangular form. However, if the columns t to n of $\mathbf{S}$ are cyclically permuted in the order $t \leftarrow t+1 \leftarrow \cdots \leftarrow n \leftarrow t$ then an upper Hessenberg matrix $\mathbf{H}$ is obtained (see Figure 8.5.1). If this permutation is incorporated into $\mathbf{Q}^{(k)}$ giving $\mathbf{Q}^{(k+1)}$ then (8.5.9) can be rewritten as

$$(\mathbf{L}_r\mathbf{P}_r)(\mathbf{L}_{r-1}\mathbf{P}_{r-1})\cdots(\mathbf{L}_1\mathbf{P}_1)\mathbf{L}^{-1}\mathbf{A}_B^{(k+1)}\mathbf{Q}^{(k+1)} = \mathbf{H}. \tag{8.5.10}$$

It is now possible to reduce $\mathbf{H}$ to upper triangular form $\mathbf{U}^{(k+1)}$ in a stable manner by a sequence of elementary operations. Firstly a row interchange may be required to ensure that $|H_{tt}| > |H_{t+1,t}|$ and this operation is represented by $\mathbf{P}_{r+1}$. Then the off-diagonal element in column t is eliminated by premultiplying by

$$\mathbf{L}_{r+1} = \begin{bmatrix} 1 & & & & & \\ & \ddots & & & & \\ & & 1 & & & \\ & & & 1 & & \\ & & & -l_{t+1,t} & 1 & \\ & & & & & \ddots & \\ & & & & & & 1 \end{bmatrix}$$

where $l_{t+1,t} = H_{t+1,t}/H_{tt}$ (or its inverse) and is bounded by 1 in modulus. By repeating these operations for off-diagonals $t+2, \ldots, n$, a reduction of the form

$$(\mathbf{L}_{r^{(k+1)}}\mathbf{P}_{r^{(k+1)}})\cdots(\mathbf{L}_{r+1}\mathbf{P}_{r+1})\mathbf{H} = \mathbf{U}^{(k+1)}$$

is obtained. Incorporating this with (8.5.10) shows that $\mathbf{A}_B^{(k+1)}$ has become expressed in the form (8.5.8).

The Bartels–Golub algorithm was originally proposed as a stable form of the simplex method, but because of its close relation to an **LU** factorization, it is well suited to an algorithm for sparse LP which attempts to maintain factors in compact form. Reid (1975) describes such an algorithm which builds up a file of the row operations which premultiply $\mathbf{A}_B^{(k)}$ in (8.5.8). The matrix $\mathbf{U}^{(k)}$ is also stored in compact form in random access storage. Reid also describes a further modification in which additional row and column permutations are made to shorten the length of the spike before reducing to Hessenberg form.

Another idea for stable LP is to update **QR** factors of $\mathbf{A}_B$ (or $\mathbf{A}_B^{\mathrm{T}}$), where **Q** is orthogonal and **R** is upper triangular (Gill and Murray, 1973). Since $\mathbf{A}_B$ is part of the problem data and is always available, it is also attractive to exploit the possibility of not storing **Q**, by using the representation

$$\mathbf{A}_B^{-1} = \mathbf{R}^{-1}\mathbf{Q}^{\mathrm{T}} = \mathbf{R}^{-1}\mathbf{R}^{-\mathrm{T}}\mathbf{A}_B^{\mathrm{T}}.$$

Operations with $\mathbf{R}^{-1}$ or $\mathbf{R}^{-\mathrm{T}}$ are carried out by backward or forward substitution. Also, since $\mathbf{A}_B^{\mathrm{T}}\mathbf{A}_B = \mathbf{R}^{\mathrm{T}}\mathbf{R}$, when $\mathbf{A}_B$ is updated by the rank one change (8.2.14), a rank two change in $\mathbf{A}_B^{\mathrm{T}}\mathbf{A}_B$ is made and hence **R** can be updated efficiently by using the methods described by Fletcher and Powell (1975) or Gill and Murray (1978a). A form of this algorithm suitable for large-scale LP in which the factor is updated in product form is given by Saunders (1972). Adverse features of this method are the increased ill-conditioning caused by 'squaring' $\mathbf{A}_B$, and the greater fill-in usually incurred with **QR** factors.

Many other different representations for $\mathbf{A}_B$ have been suggested but the one most commonly preferred in other circumstances, namely the factors $\mathbf{PA}_B = \mathbf{LU}$ computed by Gaussian elimination with partial pivoting, have not been used because it was widely regarded that these factors could not be updated in an adequately stable way. However, a recent algorithm proposed by Fletcher and Matthews (1984) has caused this attitude to be revised, and this method is now the best one to use with dense LP codes. It is able to update explicit LU factors without the need for a product form and with reasonable stability properties.

To explain the idea, let factors $\mathbf{PA}_B = \mathbf{LU}$ of $\mathbf{A}_B$ be available, and remove a column from $\mathbf{A}_B$ giving $\mathbf{PA}' = \mathbf{LU}'$, where $\mathbf{A}'$ and $\mathbf{U}'$ are $n \times (n-1)$ matrices and $\mathbf{U}'$ has a step on the diagonal. The first stage of the method is to restore $\mathbf{U}'$ to upper triangular form, giving factors $\mathbf{P}'\mathbf{A}' = \mathbf{L}^*\mathbf{U}^*$, and this part of the method

$$\begin{bmatrix} \times & \times & \times & \times & \times & \times & \times & \times \\ & \times & \times & \times & \times & \times & \times & \times \\ & & \times & \times & \times & \times & \times & \times \\ & & \times & \times & \times & \times & \times & \times \\ & & \times & & \times & \times & \times & \times \\ & & \times & & & \times & \times & \times \\ & & \times & & & & \times & \times \\ & & \times & & & & & \times \end{bmatrix} \qquad \begin{bmatrix} \times & \times & \times & \times & \times & \times & \times & \times \\ & \times & \times & \times & \times & \times & \times & \times \\ & & \times & \times & \times & \times & \times & \times \\ & & \times & \times & \times & \times & \times & \times \\ & & & \times & \times & \times & \times & \times \\ & & & & \times & \times & \times & \times \\ & & & & & \times & \times & \times \\ & & & & & & \times & \times \end{bmatrix}$$

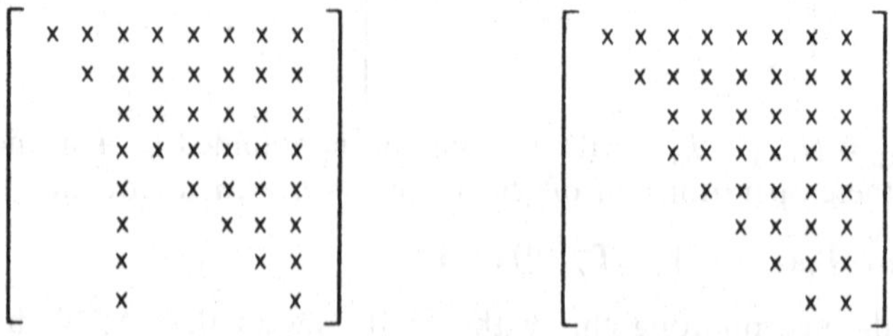

Figure 8.5.1 Spike and Hessenberg matrices in the Bartels–Golub method

is new. The second stage is to add a new column to $\mathbf{A}'$ giving

$$\mathbf{P}'\mathbf{A}_B^* = \mathbf{P}'[\mathbf{A}':\mathbf{a}] = \mathbf{L}^*[\mathbf{U}^*:\mathbf{u}],$$

and $\mathbf{u}$ is readily computed by solving $\mathbf{L}^*\mathbf{u} = \mathbf{P}'\mathbf{a}$ by forward substitution, which gives the final updated factors.

The main interest then is in how to restore $\mathbf{U}'$ to upper triangular form. Consider the (worst) case when column 1 of $\mathbf{A}_B$ is removed so that $\mathbf{U}'$ is fully upper Hessenberg. Let $\mathbf{L}$ have columns $\mathbf{l}_1, \mathbf{l}_2, \ldots, \mathbf{l}_n$ and $\mathbf{U}'$ have rows $\mathbf{u}_1, \mathbf{u}_2, \ldots, \mathbf{u}_n$ so that

$$\mathbf{P}\mathbf{A}' = \mathbf{L}\mathbf{U}' = \sum \mathbf{l}_i \mathbf{u}_i.$$

The first step of the method is to make $u'_{21} = 0$. One possibility is to write the first two terms in the sum as

$$\mathbf{l}_1\mathbf{u}_1 + \mathbf{l}_2\mathbf{u}_2 = [\mathbf{l}_1 \quad \mathbf{l}_2]\begin{bmatrix}\mathbf{u}_1\\ \mathbf{u}_2\end{bmatrix} = [\mathbf{l}_1 \quad \mathbf{l}_2]\mathbf{B}\mathbf{B}^{-1}\begin{bmatrix}\mathbf{u}_1\\ \mathbf{u}_2\end{bmatrix} \tag{8.5.11}$$

where $\mathbf{B}$ is a 2×2 matrix, and define

$$[\mathbf{l}_1^+ \quad \mathbf{l}_2^+] = [\mathbf{l}_1 \quad \mathbf{l}_2]\mathbf{B} \qquad \begin{bmatrix}\mathbf{u}_1^+\\ \mathbf{u}_2^+\end{bmatrix} = \mathbf{B}^{-1}\begin{bmatrix}\mathbf{u}_1\\ \mathbf{u}_2\end{bmatrix}.$$

To eliminate u'_{21} define $r = u'_{21}/u'_{11}$ and

$$\mathbf{B}^{-1} = \begin{bmatrix}1 & 0\\ -r & 1\end{bmatrix} \qquad \mathbf{B} = \begin{bmatrix}1 & 0\\ r & 1\end{bmatrix}. \tag{8.5.12}$$

Then $\mathbf{l}_1, \mathbf{l}_2, \mathbf{u}_1, \mathbf{u}_2$ are replaced by $\mathbf{l}_1^+, \mathbf{l}_2^+, \mathbf{u}_1^+, \mathbf{u}_2^+$ respectively. The process is repeated to eliminate $u'_{32}, u'_{43}, \ldots$ and so on, until an upper triangular matrix is obtained. This method is essentially that of Bennett (1965). Unfortunately it fails when $u'_{11} = 0$ and is numerically unstable because r can be arbitrarily large. To circumvent the difficulty, when $|u'_{21}| > |u'_{11}|$ one might consider first interchanging $\mathbf{u}_1$ and $\mathbf{u}_2$, so that

$$\begin{bmatrix}\mathbf{u}_1^+\\ \mathbf{u}_2^+\end{bmatrix} = \mathbf{B}^{-1}\mathbf{P}\begin{bmatrix}\mathbf{u}_1\\ \mathbf{u}_2\end{bmatrix} = \begin{bmatrix}1 & 0\\ -r^{-1} & 1\end{bmatrix}\begin{bmatrix}0 & 1\\ 1 & 0\end{bmatrix}\begin{bmatrix}\mathbf{u}_1\\ \mathbf{u}_2\end{bmatrix}.$$

Unfortunately the corresponding operations on $\mathbf{L}$, namely $[\mathbf{l}_1^+, \mathbf{l}_2^+] = [\mathbf{l}_1, \mathbf{l}_2]\mathbf{P}\mathbf{B}$, destroy the lower triangular structure.

The new method of Fletcher and Matthews uses the update based on (8.5.12) when r is small in a certain sense. However, in the difficult case that r is large in modulus, a different method of elimination is used. Here the idea is to interchange *rows* 1 and 2 of $\mathbf{L}$, making a corresponding change to $\mathbf{P}$ to preserve equality in $\mathbf{P}\mathbf{A}' = \mathbf{L}\mathbf{U}'$. An operation of the form $\mathbf{B}\mathbf{B}^{-1}$ is introduced as in (8.5.11), but this time $\mathbf{B}$ is defined by

$$\mathbf{B} = \begin{bmatrix}0 & 1\\ 1 & -l_{21}\end{bmatrix}\begin{bmatrix}1 & 0\\ s & 1\end{bmatrix} \qquad \mathbf{B}^{-1} = \begin{bmatrix}1 & 0\\ -s & 1\end{bmatrix}\begin{bmatrix}l_{21} & 1\\ 1 & 0\end{bmatrix} \tag{8.5.13}$$

where $s = u'_{11}/\Delta$ and $\Delta = l_{21}u'_{11} + u'_{21}$. The purpose of the left-hand matrix in $\mathbf{B}$ is to return $\mathbf{L}$ to lower triangular form, and the right-hand matrix in $\mathbf{B}^{-1}$ gives the corresponding change to $\mathbf{U}'$. This operation replaces the elements u'_{11} and u'_{21} by Δ and u'_{11} respectively. The left-hand matrix in $\mathbf{B}^{-1}$ then eliminates the new element in the u_{21} position as above.

In exact arithmetic it is easy to see that one of the above types of operation (based on either (8.5.12) or (8.5.13)) must be able to eliminate u'_{21} (if $u'_{21} \neq 0$ then r and Δ cannot both be zero). When round-off errors are present Fletcher and Matthews choose to use (8.5.13) on the basis of the test

$$|\Delta| > |u_{11}|\max(1, |l_{21}|) \tag{8.5.14}$$

which minimizes a bound on growth in the LU factors. Subsequent steps to eliminate $u'_{32}, u'_{43}, \ldots$ are made in a similar way. The cost of the double operation (8.5.13) is twice that of (8.5.12), and if p is the column of $\mathbf{A}_B$ that is removed then a total of $c(n-p)^2 + O(n)$ multiplications are required to reduce $\mathbf{U}'$ to upper triangular form, where $1 \leqslant c \leqslant 2$. This is the same order of magnitude as the cost of the forward substitution ($\frac{1}{2}n^2$ multiplications) required for the second stage. Essentially the same algorithm can also be used to update the factors when rows as well as columns of $\mathbf{A}_B$ may be replaced. This problem arises for example in some QP algorithms.

As regards numerical stability, experiments in which the updates are repeated many times show excellent round-off error control, even for ill-conditioned matrices and numerically singular matrices. In pathological cases we are familiar with growth of 2^n in the LU factors when using Gaussian elimination with partial pivoting. The same is true of the Bartels–Golub method. A similar type of result holds for the Fletcher–Matthews algorithm but the rate of growth in the worst case is more severe. Powell (1985b) has produced worst case examples in which the growth is compounded on successive updates with the method. However, the situation can be monitored, and factors can be recomputed if large growth is detected, because explicit factors are available. This is not the case for product form algorithms. On the other hand, as with Gaussian elimination, it may be that these pathological situations are so unlikely that it is not worth including special code to detect them in practice.

For dense LP the Fletcher–Matthews algorithm is currently the most attractive one for maintaining an invertible representation of the basis matrix. For sparse LP, the same may not be true because considerable effort has gone into increasing the efficiency of product form methods, and these are currently regarded as the most suitable in this case. If the Fletcher–Matthews algorithm is to be used it will be particularly important to detect zeros caused by cancellation in the updating process and so avoid unnecessary fill in. A promising approach might be

(i) Start with Markowitz factors to try to minimize fill in.
(ii) Use Fletcher–Matthews updates in sparse form.
(iii) Reinvert if the file length has increased by some factor.

The hope would be that the time between reinversions would be longer with the Fletcher–Matthews update as against using a product form method. The Fletcher–Matthews update is very convenient for coding in BLAS (Basic Linear Algebra Subroutines) and can readily take advantage of any available optimized code for these routines.

A different approach to sparsity is to attempt to find a partial *decomposition* of the problem into smaller subproblems, each of which can be solved repeatedly under the control of some master problem. The decomposition method of Dantzig and Wolfe (1960) illustrates the principles well, and is described here although some other possibilities are described by Gill and Murray (1974b). The idea is of limited applicability since it depends on the main problem having a favourable structure, and even then it is not clear that the resulting method is better than using a general sparse matrix LP package. A well-known situation which might be expected to favour decomposition is when the LP problem models a system composed of a number of quasi-independent subsystems. Each subsystem has its own equations and variables, but there are a few additional equations which interrelate all the variables and may represent, for example, the distribution of shared resources throughout the system. Let each subsystem (for $i = 1, 2, \ldots, r$) be expressed in terms of n_i variables $\mathbf{x}_i$ and be subject to the conditions

$$\mathbf{C}_i \mathbf{x}_i = \mathbf{b}_i, \quad \mathbf{x}_i \geqslant \mathbf{0}, \tag{8.5.15}$$

where $\mathbf{C}_i$ and $\mathbf{b}_i$ have m_i rows. Also let there be m_0 additional conditions

$$\sum_{i=1}^{r} \mathbf{B}_i \mathbf{x}_i = \mathbf{b}_0 \tag{8.5.16}$$

so that $\mathbf{b}_0$ and the matrices $\mathbf{B}_i$ have m_0 rows. Then the overall problem can be expressed in terms of an LP problem of the form (8.1.1) with

$$n = \sum_{i=1}^{r} n_i, \qquad m = \sum_{i=0}^{r} m_i,$$
$$\mathbf{x}^{\mathrm{T}} = (\mathbf{x}_1^{\mathrm{T}}, \mathbf{x}_2^{\mathrm{T}}, \ldots, \mathbf{x}_r^{\mathrm{T}}), \quad \mathbf{b}^{\mathrm{T}} = (\mathbf{b}_0^{\mathrm{T}}, \mathbf{b}_1^{\mathrm{T}}, \ldots, \mathbf{b}_r^{\mathrm{T}}),$$

and the coefficient matrix has the structure

$$\mathbf{A} = \begin{bmatrix} \mathbf{B}_1 & \mathbf{B}_2 & \mathbf{B}_3 & \cdots & \mathbf{B}_r \\ \mathbf{C}_1 & & & & \\ & \mathbf{C}_2 & & & \\ & & \mathbf{C}_3 & & \\ & & & \ddots & \\ & & & & \mathbf{C}_r \end{bmatrix} \tag{8.5.17}$$

The Dantzig–Wolfe decomposition method is applicable to this problem and uses the following idea. Assume for each i that (8.5.15) has p_i extreme points which are columns of the matrix $\mathbf{E}_i$, and that any feasible $\mathbf{x}_i$ can be expanded as a convex combination

$$\mathbf{x}_i = \mathbf{E}_i \boldsymbol{\theta}_i, \qquad \mathbf{e}^{\mathrm{T}} \boldsymbol{\theta}_i = 1, \qquad \boldsymbol{\theta}_i \geqslant \mathbf{0} \tag{8.5.18}$$

of these extreme points (see Section 9.4), where $\mathbf{e}$ denotes a vector of ones. This assumption is valid if the feasible region of (8.5.15) is bounded for each i, although in fact the assumption of boundedness is unnecessary. Then the main problem can be transformed to the equivalent LP problem

$$\begin{aligned} &\underset{\boldsymbol{\theta}}{\text{minimize}}\ \mathbf{f}^{\mathrm{T}}\boldsymbol{\theta} \\ &\text{subject to } \mathbf{G}\boldsymbol{\theta}=\mathbf{h}, \qquad \boldsymbol{\theta}\geqslant \mathbf{0} \end{aligned} \tag{8.5.19}$$

where

$$\boldsymbol{\theta}=\begin{pmatrix}\boldsymbol{\theta}_1\\ \boldsymbol{\theta}_2\\ \vdots\\ \boldsymbol{\theta}_r\end{pmatrix}, \qquad \mathbf{f}=\begin{pmatrix}\mathbf{f}_1\\ \mathbf{f}_2\\ \vdots\\ \mathbf{f}_r\end{pmatrix}, \qquad \mathbf{h}=\begin{pmatrix}\mathbf{b}_0\\ 1\\ 1\\ \vdots\\ 1\end{pmatrix},$$

$$\mathbf{G}=\begin{pmatrix}\mathbf{G}_1 & \mathbf{G}_2 & \mathbf{G}_3 & \cdots & \mathbf{G}_r\\ \mathbf{e}^{\mathrm{T}} & & & & \\ & \mathbf{e}^{\mathrm{T}} & & & \\ & & \mathbf{e}^{\mathrm{T}} & & \\ & & & \ddots & \\ & & & & \mathbf{e}^{\mathrm{T}}\end{pmatrix}$$

and where $\mathbf{G}_i=\mathbf{B}_i\mathbf{E}_i$ and $\mathbf{f}_i^{\mathrm{T}}=\mathbf{c}_i^{\mathrm{T}}\mathbf{E}_i$. This problem has many fewer equations (m_0+r) but very many more variables $(\sum p_i)$. However, it turns out that it is possible to solve (8.5.19) by the *revised* simplex method without explicitly evaluating all the matrices $\mathbf{E}_i$.

Consider therefore an iteration of the revised simplex method for (8.5.19). There is no difficulty in having available the basis matrix $\mathbf{G}_B$, which has m_0+r columns (one from each subsystem). Each column is calculated from one extreme point. As usual this matrix determines the values $\hat{\mathbf{h}}=\mathbf{G}_B^{-1}\mathbf{h}$ of the basic variables $\boldsymbol{\theta}_B$, and $\hat{\mathbf{h}}\geqslant\mathbf{0}$. However, the reduced costs of the nonbasic variables defined by

$$\hat{\mathbf{f}}_N=\mathbf{f}_N-\mathbf{G}_N^{\mathrm{T}}\boldsymbol{\pi}$$

where $\boldsymbol{\pi}=\mathbf{G}_B^{-\mathrm{T}}\mathbf{f}_B$ is not available since this would require all the remaining extreme points to be calculated. Nonetheless the smallest reduced cost can be calculated indirectly. Let $j\in N$ index a variable θ_j in the subvector $\boldsymbol{\theta}_i$, and let $\boldsymbol{\pi}$ be partitioned into $\boldsymbol{\pi}^{\mathrm{T}}=(\mathbf{u}^{\mathrm{T}},\mathbf{v}^{\mathrm{T}})$ where $\mathbf{v}$ has r elements. Then by definition of $\mathbf{f}_i$ and $\mathbf{G}_i$, the corresponding reduced cost $\hat{f}_j$ can be written

$$\hat{f}_j=f_j-\mathbf{g}_j^{\mathrm{T}}\mathbf{u}-v_i=(\mathbf{c}_i^{\mathrm{T}}-\mathbf{u}^{\mathrm{T}}\mathbf{B}_i)\boldsymbol{\xi}_j-v_i \tag{8.5.20}$$

where $\mathbf{g}_j$ is the corresponding column of $\mathbf{G}_i$ and $\boldsymbol{\xi}_j$ is the corresponding extreme point of (8.5.15). Consider finding the smallest $\hat{f}_j$ for all indices j in the subset i; this is equivalent to finding the extreme point of (8.5.15) which minimizes (8.5.20). Since the solution of an LP problem occurs at an extreme point, this is equivalent to solving the LP

$$\begin{aligned} &\underset{\mathbf{x}_i}{\text{minimize}} \quad (\mathbf{c}_i - \mathbf{B}_i^{\mathrm{T}}\mathbf{u})^{\mathrm{T}}\mathbf{x}_i - v_i \\ &\text{subject to} \quad \mathbf{C}_i\mathbf{x}_i = \mathbf{b}_i, \qquad \mathbf{x}_i \geqslant \mathbf{0}. \end{aligned} \tag{8.5.21}$$

Thus the smallest reduced cost for *all* j is found by solving (8.5.21) for all $i = 1, 2, \ldots, r$ so as to give the extreme point (ξ_q say) which for all i gives the smallest cost function value in (8.5.21). Thus the smallest reduced cost $\hat{f}_q$ is determined and the simplex method can proceed by increasing the variable θ_q.

If the method does not terminate, the next step is to find the basic variable θ_p which first becomes zero. Column q of $\mathbf{G}$ is readily computed using $\mathbf{g}_q = \mathbf{B}_i^{\mathrm{T}} \boldsymbol{\pi}_q$, where i now refers to the subsystem which includes the variable q. Thus $\mathbf{d}$ in (8.2.10) can be calculated so that the solution of (8.2.15) determines θ_p in the usual way. The indices p and q are interchanged, and the same column of $\mathbf{G}$ enables the matrix $\mathbf{G}_B$ to be updated using (8.2.19), which completes the description of the algorithm.

One possible situation which has not been considered is that the solution to (8.5.21) may be unbounded. This can be handled with a small addition to the algorithm as described by Hadley (1962). Another feature of note is that the sub-problem constraints (8.5.15) can be any convex polyhedron and the same development applies. A numerical example of the method in which this feature occurs is given by Hadley (1962).

As a final remark on large-scale LP, it is observed that the idea which occurs in the Dantzig–Wolfe method of solving an LP without explicitly generating all the columns in the coefficient matrix, is in fact of much wider applicability. It is possible to model some applications by LP problems in which the number of variables is extremely large, and to utilize some other feature to find which reduced cost is least. An interesting example of this is the ship scheduling problem; see Appelgren (1971), for example.

8.6 DEGENERACY

It is observed in Section 8.2 that the simplex method converges when each iteration is known to reduce the objective function. Failure to terminate can arise when an iteration occurs on which $\hat{b}_p = 0$ and $d_p < 0$. Then no change to the variable x_q is permitted, and hence no decrease in f. Although a new partitioning B and N can be made, it is possible for the algorithm to return to a previous set B and N and hence to *cycle* infinitely without locating the solution. The situation in which $\hat{b}_i = 0$ occurs is known as *degeneracy*. The same can happen in the active set method when $\mathbf{a}_i^{\mathrm{T}}\mathbf{x}^{(k)} = b_i$ for some indices $i \notin \mathscr{A}$. If $\mathbf{a}_i^{\mathrm{T}}\mathbf{s}_q < 0$ then these constraints allow no progress along the search direction $\mathbf{s}_q$. Although changes in the active set $\mathscr{A}$ can be made, the possibility of cycling again arises. This is to be expected since the algorithms are equivalent as described in Question 8.17, and so the examples of degeneracy, etc., described below apply equally to both the simplex and active set methods. Degeneracy can arise in other active set methods (Sections 10.3 and 11.2) and the remarks in this section also apply there.

An example in which cycling occurs has been given by Beale (see Hadley, 1962) and is described in Question 8.25. For cycling to occur, a number of elements $\hat{b}_i = 0$ must occur, and the way in which ties are broken in (8.2.11) is crucial. Beale's example uses the plausible rule that p is chosen as the first such p which occurs in the tableau. (This is not the same as the first in numerical order but is a well-defined rule.) In practice, however, degeneracy and cycling are rarely observed and it has become fashionable to assert that the problem can be neglected, or can be solved by making small perturbations to the data. On the other hand, I doubt that many LP codes monitor whether degeneracy or near degeneracy arises, so it may be that the problem arises more frequently and goes undetected. Graves and Brown (1979) indicate that on some large sparse LP systems, between 50 and 90 per cent of iterations are degenerate. Linear network flow problems are also well known for being highly degenerate (see Section 13.3). I have also come across degeneracy on two or three occasions in quadratic programming problems.

It is possible to state a rule (Charnes, 1952) for resolving ties in (8.2.11) which does not cause cycling, and hence which in theory resolves the degeneracy problem. (In practice this may not be the case because of round-off error difficulties, a subject which is considered further at the end of this section.) Consider perturbing the right-hand side vector $\mathbf{b}$ to give $\mathbf{b}(\varepsilon)$ where

$$\mathbf{b}(\varepsilon) = \mathbf{b} + \sum_{j=1}^{n} \varepsilon^j \mathbf{a}_j. \tag{8.6.1}$$

Multiplying through by $\mathbf{A}_B^{-1}$ and denoting $\hat{\mathbf{A}} = \mathbf{A}_B^{-1}\mathbf{A}$ gives

$$\hat{\mathbf{b}}(\varepsilon) = \hat{\mathbf{b}} + \sum_{j=1}^{n} \varepsilon^j \hat{\mathbf{a}}_j \tag{8.6.2}$$

where $\hat{\mathbf{a}}_j$ is the jth column of $\hat{\mathbf{A}}$. This perturbation breaks up the degenerate situation, giving instead a closely grouped set of non-degenerate extreme points. The tie break rule (which does not require ε to be known) comes about by finding how the simplex method solves this perturbed problem for sufficiently small ε. Each element in (8.6.1) or (8.6.2) is a polynomial of degree n in ε. The following notation is useful. If

$$u(\varepsilon) = u_0 + u_1\varepsilon + \cdots + u_n\varepsilon^n$$

is any such polynomial, $v(\varepsilon)$ similarly, then $u(\varepsilon) > v(\varepsilon)$ for all sufficiently small $\varepsilon > 0$ if and only if

$$\begin{aligned}
&u_0 > v_0 \\
\text{or}\quad &u_0 = v_0 \quad \text{and} \quad u_1 > v_1 \\
\text{or}\quad &u_0 = v_0, \quad u_1 = v_1 \quad \text{and} \quad u_2 > v_2, \\
&\vdots \\
\text{or}\quad &u_0 = v_0, \quad u_1 = v_1 \cdots u_{n-1} = v_{n-1} \quad \text{and} \quad u_n > v_n,
\end{aligned} \tag{8.6.3}$$

and $u(\varepsilon)=v(\varepsilon)$ iff $u_i=v_i$ $\forall i=0,1,\ldots,n$. Dantzig, Orden and Wolfe (1955) let (8.6.3) define a *lexicographic ordering* $\mathbf{u}\succ\mathbf{v}$ of the vectors $\mathbf{u}$ and $\mathbf{v}$ whose elements are the coefficients of the polynomial and show that the perturbation method can be interpreted in this way. I shall use the notation $u(\varepsilon)\succ v(\varepsilon)$ to mean that $u(\varepsilon)>v(\varepsilon)$ for all sufficiently small $\varepsilon>0$.

Assume for the moment that the initial vertex satisfies $\hat{\mathbf{b}}^{(1)}(\varepsilon)\succ\mathbf{0}$ (that is $\hat{\mathbf{b}}_i^{(1)}(\varepsilon)\succ 0$ $\forall i\in B^{(1)}$). The rule for updating the tableau is

$$\begin{aligned}\hat{b}_i^{(k+1)}&=\hat{b}_i^{(k)}-\frac{\hat{a}_{iq}}{\hat{a}_{pq}}\hat{b}_p^{(k)}, \qquad i\neq p\\ \hat{b}_p^{(k+1)}&=\frac{1}{\hat{a}_{pq}}\hat{b}_p^{(k)}.\end{aligned} \tag{8.6.4}$$

Also $\hat{\mathbf{a}}_q=-\mathbf{d}$ by definition, so $\hat{a}_{pq}=-d_p>0$ from (8.2.11). Thus if $\hat{\mathbf{b}}^{(k)}(\varepsilon)\succ\mathbf{0}$, then both $\hat{b}_p^{(k+1)}(\varepsilon)\succ 0$ and $\hat{b}_i^{(k+1)}(\varepsilon)\succ 0$ for all $i{:}d_i\geqslant 0$, $i\neq p$. Now let (8.2.11) be solved in the lexicographic sense

$$\frac{\hat{b}_p(\varepsilon)}{-d_p}=\min_{\substack{i\in B\\ d_i<0}}\frac{\hat{b}_i(\varepsilon)}{-d_i}. \tag{8.6.5}$$

It follows using $\hat{a}_{iq}=-d_i$ that

$$\hat{b}_i(\varepsilon)\succ\frac{\hat{a}_{iq}}{\hat{a}_{pq}}\hat{b}_p^{(k)}(\varepsilon).$$

(Equality is excluded because this would imply two equal rows in the tableau, which contradicts the independence assumption on $\mathbf{A}$.) Hence from (8.6.4), $\hat{b}_i^{(k+1)}(\varepsilon)\succ 0$ for $i{:}d_i<0$. Thus $\hat{\mathbf{b}}^{(k)}(\varepsilon)\succ\mathbf{0}$ implies $\hat{\mathbf{b}}^{(k+1)}(\varepsilon)\succ\mathbf{0}$, and so inductively the result holds for all k. Thus the points obtained when the simplex method is applied to the perturbed problem for sufficiently small ε are both feasible and non-degenerate. It therefore follows that $f^{(k)}(\varepsilon)\succ f^{(k+1)}(\varepsilon)$, and so the algorithm must terminate at the solution. There is no difficulty in ensuring that $\hat{\mathbf{b}}^{(1)}(\varepsilon)\succ\mathbf{0}$. If the initial b.f.s. is non-degenerate then this condition is implied by $\hat{\mathbf{b}}^{(1)}>\mathbf{0}$. If not, then there always exists a unit matrix in the tableau, and the columns of the tableau should be rearranged so that these columns occur first, which again implies that $\hat{\mathbf{b}}^{(1)}(\varepsilon)\succ\mathbf{0}$.

The required tie break rule in (8.2.11) is therefore the one obtained by solving (8.6.5) in the lexicographic sense of (8.6.3). Thus in finding $\min\hat{b}_i(\varepsilon)/-d_i$ $\forall i\in B$, $d_i<0$, the zero order terms $\hat{b}_i/-d_i$ are compared first. If there are any ties then the first order terms $\hat{a}_{i1}/-d_i$ are used to break the ties. If any ties still remain, then the second order terms $\hat{a}_{i2}/-d_i$ are used, and so on. The rule clearly depends on the column ordering and any initial ordering for which $\hat{\mathbf{b}}^{(1)}(\varepsilon)\succ\mathbf{0}$ may be used (see Question 8.26). However, once chosen this ordering must remain fixed, and the rearrangement of the tableau into B and N columns in (8.2.13) must be avoided (or accounted for). The method assumes that the columns stay in a fixed order and that the columns of $\mathbf{I}$ may occur anywhere in the tableau. A similar method (Wolfe, 1963b) is to introduce the virtual perturbations only as they are required, and is particularly suitable for practical use.

An equivalent rule for breaking ties in the active set method can be determined by virtue of Question 8.17. The equivalent tableau is given by

$$[\hat{\mathbf{b}}:\hat{\mathbf{A}}] = \left[\begin{array}{c|ccc} \mathbf{x}^{(k)} & \underbrace{\mathbf{I}}_{x} & & -\mathbf{A}_1^{-1} \\ \mathbf{z}_2^{(k)} & & \underbrace{\mathbf{I}}_{z_2} & \underbrace{-\mathbf{A}_2^{\mathrm{T}}\mathbf{A}_1^{-1}}_{z_1} \end{array}\right] \tag{8.6.6}$$

and if degeneracy occurs in some inactive slacks z_i then ties are resolved by examining the corresponding ith rows of $[\mathbf{I}: -\mathbf{A}_2^{\mathrm{T}}\mathbf{A}_1^{-1}]$ in accordance with some fixed ordering of the slack variables.

In practice, however, cycling is not the only difficulty caused by degeneracy, and may not even be the most important. I have also observed difficulties due to round-off errors which can cause an algorithm to break down. The situation is most easily explained in terms of the active set method and is illustrated in Figure 8.6.1. There are three constraints in $\mathbb{R}^3$ which have a common line and are therefore degenerate. Initially constraint 1 is in $\mathscr{A}$, and the search along $\mathbf{s}^{(1)}$ approaches the degenerate vertex $\mathbf{x}^{(2)}$. The tie concerning which constraint to include in $\mathscr{A}$ is resolved in favour of constraint 2. The next search is along $\mathbf{s}^{(2)}$ with constraints 1 and 2 active; all appears normal and cycling does not occur at $\mathbf{x}^{(2)}$. What can happen, however, is that due to found-off error, the slack in constraint 3 is calculated as $\sim\varepsilon$ (in place of zero) and the component of $-\mathbf{a}_3$ along $\mathbf{s}^{(2)}$ is also $\sim\varepsilon$ in place of zero. Thus constraint 3 can become active at a spurious vertex $\mathbf{x}^{(3)}$ with a value of $\alpha \sim 1$ in (8.3.6). The algorithm thus breaks down because $\mathbf{x}^{(3)}$ is not a true vertex, and the matrix of active constraints $\mathbf{A}$ in (8.3.3) is singular. An equivalent situation arises in the simplex method when true values of $\hat{b}_i$ and $-d_i$ are zero, but are perturbed to $\sim\varepsilon$ by round-off error. Then the solution of (8.2.11) can give a spurious b.f.s. in which the matrix $\mathbf{A}_B$ is singular. I have observed this type of behaviour on two or

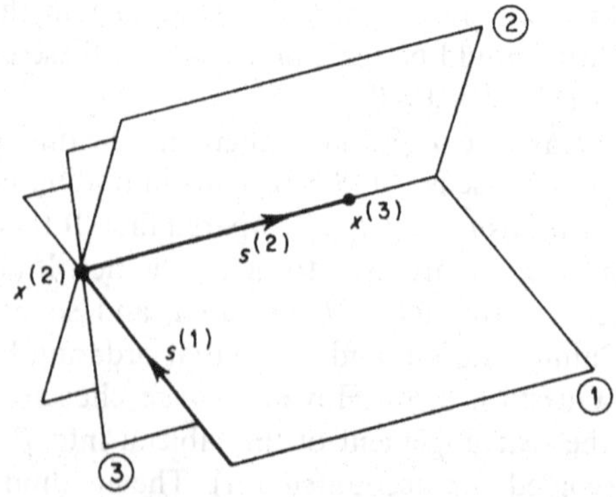

Figure 8.6.1 A practical difficulty caused by degeneracy

three occasions in practice when applying a quadratic programming package, and the only remedy has been to remove degenerate constraints from the problem. It may not be easy for the user to do this (or even to recognize that it is necessary) although it is certainly important for him or her to be aware of the possibility of failure on this account. Thus it is important that algorithms for handling degeneracy also give some attention to the possible effects of round-off error in a near degenerate situation.

It is possible to treat degeneracy by the use of a dual transformation and some tentative ideas were mentioned in the previous edition of this book. A variant of these ideas (Fletcher, 1985a) now strikes me as being promising for treating degeneracy in the presence of round-off error. It is based on a recursive method for handling degeneracy which is similar to a method of Balinski and Gomory (1963). The particular feature that makes this method suitable for round-off error control is that there is a *cost function* directly available at any level of recursion. Small changes to any cost function which are negligible at any given level of precision are taken as an indication of degeneracy and handled accordingly. When this degeneracy is resolved it is guaranteed that the cost function can be improved or optimality recognized. Hence every iteration at any level of recursion always guarantees to make progress.

The method is a phase I/phase II method and it suffices to describe phase I which aims to find a feasible vertex of a system of inequalities $\mathbf{r} \equiv \mathbf{A}^{\mathrm{T}}\mathbf{x} - \mathbf{b} \geqslant \mathbf{0}$ where $\mathbf{A}$ is an $n \times m$ matrix of rank n. This can be partitioned as

$$\begin{bmatrix} \mathbf{r}_B \\ \mathbf{r}_N \end{bmatrix} = \begin{bmatrix} \mathbf{A}_B^{\mathrm{T}} \\ \mathbf{A}_N^{\mathrm{T}} \end{bmatrix} \mathbf{x} - \begin{bmatrix} \mathbf{b}_B \\ \mathbf{b}_N \end{bmatrix} \geqslant \mathbf{0}$$

where B (basic residuals) represent *active* constraints and N (nonbasic residuals) are *inactive* constraints, the active constraints being chosen so that $\mathbf{A}_B$ is square and non-singular. The current vertex $\hat{\mathbf{x}}$ is therefore defined by $\mathbf{A}_B^{\mathrm{T}}\hat{\mathbf{x}} = \mathbf{b}_B$ (that is by $\hat{\mathbf{r}}_B = \mathbf{0}$). If $\mathbf{x}$ is eliminated by virtue of $\mathbf{x} = \hat{\mathbf{x}} + \mathbf{A}_B^{-\mathrm{T}}\mathbf{r}_B$, the problem can be expressed locally in terms of $\mathbf{r}_B$ as

$$\begin{aligned} \mathbf{r}_N \equiv \hat{\mathbf{r}}_N + \hat{\mathbf{A}}^{\mathrm{T}}\mathbf{r}_B &\geqslant \mathbf{0} \\ \mathbf{r}_B &\geqslant \mathbf{0} \end{aligned} \tag{8.6.7}$$

where $\hat{\mathbf{A}} = \mathbf{A}_B^{-1}\mathbf{A}_N$. The vector $\hat{\mathbf{r}}_N = \mathbf{A}_N^{\mathrm{T}}\hat{\mathbf{x}} - \mathbf{b}$ represents the current values of the inactive constraint residuals.

The method starts by finding the most violated inactive constraint residual, $\hat{r}_q$ say, and considers the LP

$$\begin{aligned} &\text{maximize} \quad \hat{r}_q + \hat{\mathbf{a}}_q^{\mathrm{T}}\mathbf{r}_B \\ &\text{subject to} \quad \hat{r}_j + \hat{\mathbf{a}}_j^{\mathrm{T}}\mathbf{r}_B \geqslant 0 \qquad j: \hat{r}_j \geqslant 0 \\ &\qquad\qquad\qquad \mathbf{r}_B \geqslant \mathbf{0} \end{aligned} \tag{8.6.8}$$

where $\hat{\mathbf{a}}_j$ refers to column j of $\hat{\mathbf{A}}$, in which the constraints are the currently non-violated constraints in (8.6.7). The aim is not so much to solve the LP, although this possibility does arise, but is mainly to make $\hat{r}_q$ non-negative whilst keeping feasibility for the non-violated constraints. As usual in LP there is an *optimality test* to identify some $p \in B$ for which $\hat{a}_{pq} > 0$. If so $\hat{r}_q$ can be increased by increasing r_p in a *line search*, which is terminated by a previously inactive constraint, q' say, becoming active. The iteration is completed by interchanging $p \leftrightarrow q'$ and updating the tableau. Iterations of this type are continued until

(i) The infeasibility is removed, that is $\hat{r}_q = 0$, in which case the method returns to search for further violated residuals and terminates if there are none.
(ii) Optimality is detected, that is no suitable $p \in B$ exists, which implies that there is no feasible point.
(iii) Degeneracy is detected, that is a step of zero length is taken in the line search because a degenerate inactive constraint blocks any progress.

In case (iii) an outline of the measures that are taken is the following. The problem is first *localized* by ignoring non-degenerate inactive constraints (the set $N - Z_1$ below), since the aim is to find whether or not any ascent is possible from the current vertex. After introducing multipliers $\boldsymbol{\lambda}_N$ and $\boldsymbol{\lambda}_B$ for the constraints in (8.6.7), the *dual* of the localized system is taken, giving rise to the system of inequalities

$$\boldsymbol{\lambda}_B \equiv -\hat{\mathbf{a}}_q - \hat{\mathbf{A}}\boldsymbol{\lambda}_N \geqslant \mathbf{0}$$

$$\lambda_j \begin{cases} \geqslant 0 & j \in Z_1 \\ = 0 & j \in N - Z_1 \end{cases}$$

where

$$Z_1 = \{j : j \in N, \quad \hat{r}_j = 0\}$$

is the current set of degenerate constraints (e.g. Question 8.27). This is closely related to (8.6.7), the main difference being that $\hat{\mathbf{A}}$ is no longer transposed, thus switching the status of B and N. Also there are changes in sign and some of the variables are fixed at zero. Notice that the *residuals* in the dual are $(-)$ the *costs* $\hat{\mathbf{a}}_q$ in the primal, and inequality p is violated in the dual because $\hat{a}_{pq} > 0$ in the primal optimality test. The next step is to set up a dual LP, analogous to (8.6.8), to reduce $\hat{a}_{pq}$ to zero. Iterations in the dual LP change B and N but these changes only involve indices corresponding to zero valued constraints in the primal, so do not change any values of $\hat{r}_j$ for non-degenerate constraints. These iterations are carried out until one of the following situations arises (cf. (i)–(iii) above).

(i) $\hat{a}_{pq}$ becomes zero: this implies that a negative cost has been removed in the primal, so the method returns to the primal to reassess the optimality test.
(ii) $-\hat{a}_{pq}$ is maximized in the dual but cannot be increased to zero: this implies that the degeneracy block in the primal has been removed, so the method returns to the primal to carry out the line search.
(iii) Degeneracy in the dual is detected: in this case the dual problem is

localized as described above, and a further dual is taken of the localized problem. This results in a system of inequalities to be solved which is handled in the same way as above.

It can be seen that a recursive method is set up with a stack of problems at different levels. Level 1 = Primal, Level 2 = Dual, Level 3 = Dualdual, etc.…. Each level has its own *cost function* corresponding to some infeasibility that is being reduced (in phase II, the level 1 cost function is $\mathbf{c}^T\mathbf{x}$, the usual LP cost function). In exact arithmetic a termination proof for the algorithm can be deduced, consequent on the following facts. Because each dualizing step removes at least one variable from the problem, there is a bound $L_{max} \leqslant \min(2m+1, 2n)$ on the highest level number (See Question 8.28). Also, each iteration always makes progress in that *either* a reduction in some cost function is guaranteed *or* the number of infeasibilities in some problem is reduced. Finally the number of vertices in any one problem is finite.

For inexact arithmetic, a similar termination proof follows if it is ensured that all iterations always make progress. Thus any iteration is regarded as being degenerate if a small step is taken in the line search which does not reduce the current cost function due to round-of error. The fact about the number of vertices being finite is replaced in the proof by the fact that the set of floating point numbers on a computer is finite. Some other precautions also have to be taken, for example non-violated constraints must not be allowed to become violated due to round-off error.

Despite being recursive, the method is easy to code, mainly because the *same tableau* is used at each level. Thus only a single index needs to be stacked when recursion takes place, and the overheads are minimal. Standard matrix handling techniques can be used such as those described in Section 8.5. The method extends to solving L_1LP problems (Fletcher, 1985a) and this circumvents the criticism of the above method that minimizing the maximum violation is less efficient than minimizing an L_1 sum of violations.

8.7 POLYNOMIAL TIME ALGORITHMS

In many ways one of the most interesting developments of the last decade has been that of polynomial time algorithms, which have given a new impetus and excitement to research into linear programming. The *ellipsoid algorithms*, attributed to Khachiyan (1979) but based on the work of a number of Russian authors, first attracted widespread interest. A later more efficient polynomial algorithm of Karmarkar (1984), claimed controversially to be many more times faster than the simplex method, has also encouraged research into radically different practical alternatives to the simplex method. Both these developments attracted wide publicity and were even reported in the national press, an event that is possibly without precedence in the history of numerical analysis! Yet the cautious research worker will in my opinion need more convincing evidence, and access to definitive methods or developed software, before coming to a favourable appraisal.

The reason for seeking an alternative to the simplex method is that in the worst case it can perform remarkably badly, and the number of iterations can increase *exponentially* in the number of variables n. This is despite the fact that in practice the algorithm is generally rated as being reasonably effective and reliable. This 'exponential time' behaviour was pointed out by Klee and Minty (1971) and later made explicit in an example due to Chvatal. Thus the KMC problem of order n is

$$\begin{aligned} &\text{maximize} \quad \sum_j 10^{j-1}x_j \\ &\text{subject to} \quad x_i + 2\sum_{j>i} 10^{j-i}x_j \leqslant 10^{2n-2i} \qquad i = 1, 2, \ldots n \\ &\qquad\qquad\quad \mathbf{x} \geqslant \mathbf{0}. \end{aligned}$$

It is possibly more illuminating to write down this problem for a particular value of n, for example the case $n = 5$ is

$$\begin{aligned} \text{maximize} \quad & x_1 + 10x_2 + 100x_3 + 1000x_4 + 10000x_5 \\ \text{subject to} \quad & x_1 + 20x_2 + 200x_3 + 2000x_4 + 20000x_5 \leqslant 10^8 \\ & \qquad\;\; x_2 + \;\; 20x_3 + \;\; 200x_4 + \;\; 2000x_5 \leqslant 10^6 \\ & \qquad\qquad\qquad x_3 + \quad 20x_4 + \quad 200x_5 \leqslant 10^4 \\ & \qquad\qquad\qquad\qquad\qquad x_4 + \quad\; 20x_5 \leqslant 10^2 \\ & \qquad\qquad\qquad\qquad\qquad\qquad\qquad x_5 \leqslant 10^0 \\ & \mathbf{x} \geqslant \mathbf{0}. \end{aligned}$$

For the general problem the feasible region has 2^n vertices, which is not in itself unusual. However, starting from $\mathbf{x} = \mathbf{0}$ (all slacks basic) the simplex method in exact arithmetic visits *all 2^n vertices*, hence the exponential time solution.

Khachiyan (1979) first showed that it is possible to solve LP problems in polynomial time. However, his algorithm permitted some obvious improvements and the version described by Wolfe (1980) is presented here. The basic method relates to finding a feasible point of the system of inequalities $\mathbf{A}^T\mathbf{x} \leqslant \mathbf{b}$ where $\mathbf{A}$ is an $n \times m$ matrix. The algorithm is iterative and holds a current point $\mathbf{x}_c$ which is the centre of an ellipsoid

$$E(\mathbf{x}_c, \mathbf{J}_c) = \{\mathbf{x}: \|\mathbf{J}_c^{-T}(\mathbf{x} - \mathbf{x}_c)\|_2 \leqslant 1\}$$

that contains a feasible point, if one exists. Such an ellipsoid is readily constructed initially (Gacs and Lovasz, 1979). A constraint $\mathbf{a}^T\mathbf{x} \leqslant \beta$ is found which is violated by $\mathbf{x}_c$. For the part of the ellipsoid which is cut off by this constraint, a new ellipsoid $E(\mathbf{x}_+, \mathbf{J}_+)$ is constructed which contains it, having smallest volume. This process is repeated until convergence. It can be shown that each iteration of the method reduces the volume of the ellipsoid by the factor

$$\left(\frac{1-d^2}{1-n^{-2}}\right)^{n/2}\left(\frac{(n-1)(1-d)}{(n+1)(1+d)}\right)^{1/2} < e^{-1/2n} \qquad (8.7.1)$$

so the volume converges linearly to zero with rate at least $e^{-1/2n}$. If the part of the feasible region contained in the original ellipsoid has positive volume, V_f say, and the volume of the initial ellipsoid is V_0, then the algorithm must terminate in at most $p = [p'] + 1$ iterations, where

$$V_f = (e^{-1/2n})^{p'} V_0$$

which implies that $p = O(n)$.

If the feasible region has zero volume (for example if the problem essentially contains equations) and the constraints are relaxed by an amount 2^{-L}, then it can be shown (Gacs and Lovasz, 1979) that the volume of the feasible set is at least 2^{-nL}. Thus to this accuracy, and following the same argument, the algorithm will find a solution in $p = O(n^2 L)$ iterations. By relating L to the number of bits required to determine the problem, the solvability of the unperturbed system can also be determined, but this is more of interest to workers in computational complexity. Thus the number of iterations required to solve the problem is *polynomial* in n, and this is the major advance of Khachiyan's work. However, not even Khachiyan himself has claimed that this algorithm provides a practical alternative to the simplex method. For example to improve the accuracy of each component in the solution by a factor of 10, it can be expected that the volume of the current ellipsoid must be reduced by a factor 10^{-n}. Using the above estimate, in k steps the volume will be reduced by a factor $e^{-k/2n}$, which gives $k = 4.6n^2$ as the expected number of iterations. The simplex method would normally perform much faster than this. Since both methods require $O(n^2)$ operations per iteration, the ellipsoid algorithm is not competitive in practice.

As presented the ellipsoid algorithm is a method for finding feasible points. If it is required to solve an LP problem, minimizing $\mathbf{c}^T\mathbf{x}$ subject to the same constraints, then there are at least two ways in which the ellipsoid algorithm can be used. One is to find a feasible point, $\mathbf{x}_c$ say, as above, and then cut out this point with the constraint $\mathbf{c}^T(\mathbf{x} - \mathbf{x}_c) < 0$, re-solving to find another feasible point. This is repeated until the algorithm converges. An alternative approach is to convert the LP into the primal–dual system

$$\mathbf{c}^T\mathbf{x} = \mathbf{b}^T\boldsymbol{\lambda}, \quad \mathbf{A}^T\mathbf{x} \leqslant \mathbf{b}, \quad \mathbf{x} \geqslant \mathbf{0}, \quad \mathbf{A}\boldsymbol{\lambda} \geqslant \mathbf{c}, \quad \boldsymbol{\lambda} \geqslant \mathbf{0}$$

and solve this using the ellipsoid algorithm. Some other references for the ellipsoid algorithm are given by Wolfe (1980).

A more recent polynomial time algorithm with a faster rate of convergence is given by Karmarkar (1984). This algorithm is claimed to be substantially faster then the simplex method, a claim received with enthusiasm by some members of the mathematical programming community, and scepticism by others. Unfortunately the details on which the claim is made have not been publicised, and there is the lack of a definitive method and software. Already many researchers are deriving similar algorithms: however, to revise our ideas of what is the standard algorithm for LP requires a thorough comparison that shows a distinct overall improvement on a wide range of large-scale LP test problems. In my opinion this has not yet been done, despite some interesting attempts. Whilst it is not unlikely in the future that a polynomial time algorithm will be adopted as the standard, that time is not yet here, and it is by no means certain even that such a method will be a development of the Karmarkar algorithm.

Nonetheless Karmarkar's method is of substantial interest. In its basic form it solves a very particular form of an LP problem, namely

$$\begin{aligned} &\text{minimize} \quad f \equiv \mathbf{c}^{\mathrm{T}}\mathbf{x} \qquad \mathbf{x} \in \mathbb{R}^n \\ &\text{subject to} \quad \mathbf{A}\mathbf{x} = \mathbf{0}, \quad \mathbf{e}^{\mathrm{T}}\mathbf{x} = 1, \quad \mathbf{x} \geqslant 0, \end{aligned} \tag{8.7.2}$$

where $\mathbf{e} = (1, 1, \ldots, 1)^{\mathrm{T}}$. The constraints $\mathbf{e}^{\mathrm{T}}\mathbf{x} = 1$ and $\mathbf{x} \geqslant \mathbf{0}$ define a regular simplex of dimension $n - 1$, and the feasible region is the intersection of this set with the homogeneous constraint $\mathbf{A}\mathbf{x} = \mathbf{0}$. It is assumed in the first instance that $\mathbf{A}\mathbf{e} = \mathbf{0}$, which implies that the initial point $\mathbf{x}_c = \mathbf{e}/n$ (the centre of the simplex) is feasible. It is also assumed that the solution value of the LP is zero, so that $f^* \equiv \mathbf{c}^{\mathrm{T}}\mathbf{x}^* = 0$. Conceptually the first iteration of the method is as follows. Let S_r be the sphere $\{\mathbf{x}: \|\mathbf{x} - \mathbf{x}_c\|_2 \leqslant r, \mathbf{e}^{\mathrm{T}}\mathbf{x} = 1\}$, where r is the largest value that keeps S_r inscribed within the simplex. A new point $\mathbf{x}_+$ is determined by minimizing f on the set $S_r \cap \{\mathbf{x}: \mathbf{A}\mathbf{x} = \mathbf{0}\}$. This is a relatively straightforward calculation which is considered in more detail below. Let S_R be the concentric sphere that circumscribes the simplex, so that $R/r = n - 1$, and let $\mathbf{x}_R$ minimize f on the set $S_R \cap \{\mathbf{x}: \mathbf{A}\mathbf{x} = \mathbf{0}\}$. Because f is linear, it follows that $(f_c - f_R) = (n-1)(f_c - f_+)$. But the set $S_R \cap \{\mathbf{x}: \mathbf{A}\mathbf{x} = \mathbf{0}\}$ contains the feasible region of the LP. Hence $f_R \leqslant f^* = 0$ and so

$$f_+ \leqslant (n-2) f_c/(n-1) < \mathrm{e}^{-1/(n-1)} f_c. \tag{8.7.3}$$

Comparing this with expression (8.7.1) for the ellipsoid algorithm, it can be seen that if this reduction were obtained on every iteration, then $O(nL)$ iterations would be required to get f within a tolerance 2^{-L} of f^*, and the algorithm would be polynomial.

After the first iteration, the current point $\mathbf{x}_c$ is no longer the centre of the simplex, but can be made so by transforming the problem. Karmarkar uses a *projective transformation*

$$\mathbf{x}' = \mathbf{D}^{-1}\mathbf{x}/\mathbf{e}^{\mathrm{T}}\mathbf{D}^{-1}\mathbf{x}$$

where $\mathbf{D} = \mathrm{diag}((\mathbf{x}_c)_i)$, which transform the simplex into itself and makes $\mathbf{x}_c'$ the centre of the simplex. Therefore subsequent iterations can proceed as above, except that linearity in f is not preserved by the transformation, so the linear approximation $\mathbf{c}^{\mathrm{T}}\mathbf{D}\mathbf{x}'$ is used. Karmarkar shows by means of a potential function that a similar result to (8.7.3) holds good and so the polynomial behaviour follows.

The subproblem which must be solved in the transformed variables $\mathbf{x}'$ is

$$\begin{aligned} &\text{minimize} \quad \mathbf{c}^{\mathrm{T}}\mathbf{D}\mathbf{x}' \\ &\text{subject to} \quad \mathbf{x}' \in S', \qquad \mathbf{A}\mathbf{D}\mathbf{x}' = \mathbf{0}. \end{aligned}$$

where S' is the sphere $\{\mathbf{x}: \|\mathbf{x} - \mathbf{x}_c\|_2 \leqslant r, \mathbf{e}^{\mathrm{T}}\mathbf{x} = 1\}$. It is not difficult to show that the solution to this problem is on the line through $\mathbf{x}'_c$ with direction $\mathbf{d}'$, which is defined by

$$\begin{array}{ll} \text{minimize} & \tfrac{1}{2}\mathbf{d}'^{\mathrm{T}}\mathbf{d}' \\ \text{subject to} & -\mathbf{c}^{\mathrm{T}}\mathbf{D}\mathbf{d}' = 1, \qquad \mathbf{A}\mathbf{D}\mathbf{d}' = \mathbf{0}, \qquad \mathbf{e}^{\mathrm{T}}\mathbf{d}' = 0. \end{array}$$

Introducing multipliers λ, $\lambda\boldsymbol{\pi}$ and $\lambda\phi$ respectively, the first order conditions give

$$\mathbf{d}' + \lambda(\mathbf{D}\mathbf{c} - \mathbf{D}\mathbf{A}^{\mathrm{T}}\boldsymbol{\pi} - \phi\mathbf{e}) = \mathbf{0}$$

and by normalizing $\mathbf{d}'$ suitably, it follows that

$$\mathbf{d}' = -\mathbf{D}\mathbf{c} + \mathbf{D}\mathbf{A}^{\mathrm{T}}\boldsymbol{\pi} + \phi\mathbf{e}. \tag{8.7.4}$$

Using $\mathbf{A}\mathbf{D}\mathbf{e} = \mathbf{A}\mathbf{x}_c = \mathbf{0}$, the solution separates into $\phi = \mathbf{c}^{\mathrm{T}}\mathbf{x}_c/n$ and

$$\mathbf{A}\mathbf{D}^2\mathbf{A}^{\mathrm{T}}\boldsymbol{\pi} = \mathbf{A}\mathbf{D}^2\mathbf{c}. \tag{8.7.5}$$

Thus $\boldsymbol{\pi}$ is defined by the *normal equations of a weighted least squares problem.* Well understood methods exist for its solution, and it is the dominant cost in an iteration of Karmarkar's method.

The next iterate in the method is defined by

$$\mathbf{x}'_+ = \mathbf{x}'_c + \alpha'\mathbf{d}'$$

where α' is a suitably chosen step length: its choice is not resolved with any certainty by Karmarkar. When transformed back into the original coordinate system, the new vector of variables is

$$\mathbf{x}_+ = \mathbf{D}\mathbf{x}'_+/\mathbf{e}^{\mathrm{T}}\mathbf{D}\mathbf{x}'_+.$$

Using $\mathbf{D}\mathbf{e} = \mathbf{x}_c$, $\mathbf{e}^{\mathrm{T}}\mathbf{x}_c = 1$, $\mathbf{x}'_c = \mathbf{e}/n$ and (8.7.4), it follows that

$$\mathbf{e}^{\mathrm{T}}\mathbf{D}(\mathbf{x}'_c + \alpha'\mathbf{d}') = (1 + n\alpha'(\phi - \mu_k))/n = (\gamma^{-1})/n$$

say, where $\mu_k = \mathbf{x}_c^{\mathrm{T}}\mathbf{D}(\mathbf{c} - \mathbf{A}^T\boldsymbol{\pi})$, and hence $\mathbf{x}_+$ can be rearranged as

$$\mathbf{x}_+ = \mathbf{x}_c + \gamma n\alpha'(\mu_k\mathbf{x}_c - \mathbf{D}^2(\mathbf{c} - \mathbf{A}^{\mathrm{T}}\boldsymbol{\pi})). \tag{8.7.6}$$

Thus Karmarkar's algorithm can also be regarded as a line search method in the original variables.

Karmarkar does suggest some modifications to his algorithm in order to solve a more general form of LP problem, and in particular to avoid the assumption that $f^* = 0$. Modifications with a similar aim are given by Gay (1985) and Lustig (1985). Karmarkar's idea has also engendered a number of papers proposing other interior point methods. However, a particularly perceptive paper is that of Gill *et al.* (1985) who derive a relationship between Karmarkar's method and the logarithmic barrier function (12.1.25), normally thought of as the basis of a somewhat out-of-date interior point method for nonlinear programming. For an LP problem in standard form (minimize $\mathbf{c}^{\mathrm{T}}\mathbf{x}$ subject to $\mathbf{A}\mathbf{x} = \mathbf{b}$, $\mathbf{x} \geqslant \mathbf{0}$) the bounds can be handled in a log barrier function, giving the problem

$$\begin{array}{ll} \text{minimize} & \phi(\mathbf{x}) \equiv \mathbf{c}^{\mathrm{T}}\mathbf{x} - \mu\sum \ln x_i \\ \text{subject to} & \mathbf{A}\mathbf{x} = \mathbf{b}. \end{array} \tag{8.7.7}$$

If this transformation is applied to the LP format (8.7.2) used by Karmarkar, and the iteration formula for Newton's method from an initial point $\mathbf{x}_c$ is written down, then

$$\nabla\phi = \mathbf{c} - \mu\mathbf{D}^{-1}\mathbf{e}, \qquad \nabla^2\phi = \mu\mathbf{D}^{-2}$$

and the search direction $\mathbf{d}$ is defined by solving the QP problem

$$\begin{aligned} &\text{minimize} \quad \tfrac{1}{2}\mu\mathbf{d}^T\mathbf{D}^{-2}\mathbf{d} + (\mathbf{c} - \mu\mathbf{D}^{-1}\mathbf{e})^T\mathbf{d} \\ &\text{subject to} \quad \mathbf{A}\mathbf{d} = \mathbf{0}, \quad \mathbf{e}^T\mathbf{d} = 0. \end{aligned}$$

Ignoring the constraint $\mathbf{e}^T\mathbf{d} = 0$, an explicit expression for $\mathbf{d}$ (see Section 11.1) is

$$\mathbf{d} = \mathbf{x}_c - \mu^{-1}\mathbf{D}^2(\mathbf{c} - \mathbf{A}^T\boldsymbol{\pi})$$

(using $\mathbf{De} = \mathbf{x}_c$, $\mathbf{Ax}_c = \mathbf{0}$ and (8.7.5)). The penalty parameter μ in (8.7.7) is chosen to have the value $\mu_k = \mathbf{x}_c^T\mathbf{D}(\mathbf{c} - \mathbf{A}^T\boldsymbol{\pi})$ that occurs in (8.7.6). It follows from $\mathbf{e}^T\mathbf{x}_c = 1$ and $\mathbf{De} = \mathbf{x}_c$ that $\mathbf{e}^T\mathbf{d} = 0$ is nonetheless satisfied. Hence incorporating a step length β, the next point is defined by

$$\mathbf{x}_+ = \mathbf{x}_c + \beta\mu^{-1}(\mu\mathbf{x}_c - \mathbf{D}^2(\mathbf{c} - \mathbf{A}^T\boldsymbol{\pi})).$$

Comparison with (8.7.6) shows that the search directions in both methods are parallel and the iterates are identical if the step lengths satisfy $\beta\mu^{-1} = \gamma n\alpha'$. Thus Karmarkar's method is seen to be just a special case of the log barrier function applied to the LP problem.

It would seem to follow *a-forteriori* therefore that if good results can be obtained by Karmarkar's method then it should be possible to do at least as well by using the log barrier function. Gill *et al.* (1985) investigate this possibility, solving (2.5) by means of a preconditioned conjugate gradient method. (Another possibility would be to use sparse QR factors of $\mathbf{DA}^T$, modified to treat dense columns specially.) On the basis of a number of large-scale tests, they report that the method can be competitive with the simplex method on some problems, particularly those with a favourable sparsity structure for QR. However, the overall performance of the simplex method is better, and the interior point algorithms have difficulties with degenerate problems, and in making use of a good initial estimate of the solution. Gill *et al.* are not able to reproduce the significant improvements reported by Karmarkar. Clearly more research is needed before a final evaluation of the merit of interior point methods in LP can be made.

QUESTIONS FOR CHAPTER 8

8.1. Consider the linear program

$$\begin{aligned} &\text{minimize} \quad -4x_1 - 2x_2 - 2x_3 \\ &\text{subject to} \quad 3x_1 + x_2 + x_3 = 12 \\ &\qquad\qquad\quad x_1 - x_2 + x_3 = -8, \qquad \mathbf{x} \geqslant \mathbf{0}. \end{aligned}$$

Verify that the choice $B \equiv \{1, 2\}$, $N \equiv \{3\}$ gives a basic feasible solution, and hence solve the problem.

8.2. Consider the LP problem of Question 8.1 and replace the objective function by the function $-4x_1 - 2x_3$. Show that the basic feasible solutions derived from $B = \{1, 2\}$, $N = \{3\}$ and $B = \{2, 3\}$, $N = \{1\}$ are both optimal, and that any convex combination of these solutions is also a solution to the problem.

In general describe how the fact that an LP problem has a non-unique solution can be recognized, and explain why.

8.3. Sketch the set of feasible solutions of the following inequalities:

$$\begin{aligned} x_1 \geqslant 0, \qquad x_2 &\geqslant 0 \\ x_1 + 2x_2 &\leqslant 4 \\ -x_1 + x_2 &\leqslant 1 \\ x_1 + x_2 &\leqslant 3. \end{aligned}$$

At which points of this set does the function $x_1 - 2x_2$ take (a) its maximum and (b) its minimum value?

8.4. Consider the problem

$$\begin{aligned} &\text{maximize} \quad x_2 \\ &\text{subject to} \quad 2x_1 + 3x_2 \leqslant 9 \\ &\qquad\qquad |x_1 - 2| \leqslant 1, \quad \mathbf{x} \geqslant \mathbf{0}. \end{aligned}$$

(i) Solve the problem graphically.
(ii) Formulate the problem as a standard LP problem.

8.5. Use the tableau form of the simplex method to solve the LP problem

$$\begin{aligned} &\text{minimize} \quad 5x_1 - 8x_2 - 3x_3 \\ &\text{subject to} \quad 2x_1 + 5x_2 - x_3 \leqslant 1 \\ &\qquad -3x_1 - 8x_2 + 2x_3 \leqslant 4 \\ &\qquad -2x_1 - 12x_2 + 3x_3 \leqslant 9, \qquad \mathbf{x} \geqslant \mathbf{0} \end{aligned}$$

after reducing the problem to standard form.

8.6. A manufacturer uses resources of material (m) and labour (l) to make up to four possible items (a to d). The requirements for these, and the resulting profits are given by

item a requires $4m + 2l$ resources and yields £5/item profit,
item b requires $m + 5l$ resources and yields £8/item profit,
item c requires $2m + l$ resources and yields £3/item profit,
item d requires $2m + 3l$ resources and yields £4/item profit.

There are available up to 30 units of material and 50 units of labour per day. Assuming that these resources are fully used and neglecting integrality constraints, show that the manufacturing schedule for maximum profit is an LP problem in standard form. Show that the policy of manufacturing only the two highest profit items yields a b.f.s. which is not optimal. Find the optimal schedule. Evaluate also the schedule in which equal amounts of each item are manufactured and show there is an under-use of one resource. Compare the profit in each of these cases.

8.7. Use the simplex method to show that for all μ in the range $-3<\mu<-\frac{1}{2}$, the solution of the problem

$$\begin{aligned} &\text{minimize} \quad \mu x_1 - x_2 \\ &\text{subject to} \quad x_1 + 2x_2 \leqslant 4 \\ &\qquad 6x_1 + 2x_2 \leqslant 9, \qquad \mathbf{x} \geqslant \mathbf{0} \end{aligned}$$

occurs at the same point $\mathbf{x}^*$ and find x_1^* and x_2^*.

8.8. Consider modifying the revised simplex method to solve a more general standard form of LP,

$$\begin{aligned} &\text{minimize} \quad \mathbf{c}^T\mathbf{x} \\ &\text{subject to} \quad \mathbf{A}\mathbf{x} = \mathbf{b}, \qquad \mathbf{l} \leqslant \mathbf{x} \leqslant \mathbf{u}. \end{aligned}$$

Show that the following changes to the simplex method are required. Nonbasic variables take the value $x_i^{(k)} = l_i$ or u_i, and $\hat{\mathbf{b}} = \mathbf{A}_B^{-1}(\mathbf{b} - \mathbf{A}_N\mathbf{x}_N^{(k)})$ is used to calculate $\hat{\mathbf{b}}$. The optimality test (8.2.8) is changed to require that the set

$$\{i \colon i \in N, \quad \hat{c}_i < 0 \text{ if } x_i = l_i, \quad \hat{c}_i > 0 \text{ if } x_i = u\}$$

is empty; if not $\hat{c}_q$ is chosen to $\min|\hat{c}_i|$ for i in this set. The choice of p in (8.2.11) is determined *either* when a variable in B reaches its upper or lower bound *or* when the variable x_q reaches its other bound. In the latter case no changes to B or N need to be made. Write down the resulting formula for the choice of p.

8.9. Using the method of Question 8.8., verify that $B \equiv \{1,2\}$, $N \equiv \{3,4\}$, $\mathbf{x}_N = \mathbf{0}$ gives a b.f.s. for the LP problem

$$\begin{aligned} &\text{minimize} \quad x_1 - 3x_3 \\ &\text{subject to} \quad x_1 + x_2 - x_3 + 2x_4 = 6 \\ &\qquad 2x_1 - x_3 - 2x_4 = 2 \\ &0 \leqslant x_1 \leqslant 4, \quad 0 \leqslant x_2 \leqslant 10, \quad 0 \leqslant x_3 \leqslant 4, \quad 0 \leqslant x_4 \leqslant 2 \end{aligned}$$

and hence solve the problem.

8.10. By drawing a diagram of the feasible region, solve the problems

$$\begin{aligned} &\text{minimize} \quad x_2 \\ &\text{subject to} \quad x_1 + x_2 = 1 \\ &\qquad x_1 - x_2 \leqslant 2 \\ &\qquad x_1 \geqslant 0 \end{aligned}$$

(a) when $x_2 \geqslant 0$ also applies and (b) when x_2 is unrestricted. Verify that the solutions are correct by posing each problem in standard form, determining the basic variables from the diagram, and then checking feasibility and optimality of the b.f.s. In case (b) introduce variables x_2^+ and x_2^- as described in Section 7.2.

8.11. A linear data fitting problem can be stated as finding the best solution for $\mathbf{x}$ of the system of equations $\mathbf{Ax} = \mathbf{b}$ in which $\mathbf{A}$ has more rows than columns (over-determined). In some circumstances it can be valuable to find that solution which minimizes $\sum_i |r_i|$ where $\mathbf{r} = \mathbf{Ax} - \mathbf{b}$, rather than the more usual $\sum r_i^2$ (least squares). Using the idea of replacing r_i by two variables r_i^+ and r_i^-, one of which will always be zero, show that the problem can be stated as an LP (not in standard form).

8.12. Consider the problem

$$\begin{aligned} \text{minimize} \quad & |x| + |y| + |z| \\ \text{subject to} \quad & x - 2y = 3 \\ & -y + z \leqslant 1 \\ & x, y, z \text{ unrestricted in sign.} \end{aligned}$$

Define suitable non-negative variables x^+, x^-, y^+, y^-, etc., and write down an LP problem in standard form. Verify that the variables x^+ and z^+ provide an initial basic feasible solution, and hence solve the LP problem. What are the optimum function value and variables in the original problem? Is the solution unique?

8.13. Consider the problem in Question 8.12. Without adding extra variables other than a slack variable, show that the problem can be solved (as in (8.4.5)) by allowing the coefficients in the cost function to change between ± 1 as the iteration proceeds.

8.14. Given an optimal b.f.s. to a problem in standard form, consider perturbing the problem so that

(i) $\mathbf{b}(\lambda) = \mathbf{b} + \lambda \mathbf{r}$
(ii) $\mathbf{c}_N(\lambda) = \mathbf{c}_N + \lambda \mathbf{s}$
(iii) $\mathbf{c}_B(\lambda) = \mathbf{c}_B + \lambda \mathbf{t}$

where in each case λ is increased from zero. At what value of λ, if any (in each case), will the present solution no longer be an optimal b.f.s.? Consider the solution of Question 8.6 and the effect of (i) a reduction in the availability of labour, (ii) an increase in the profit of item a, and (iii) a decrease in profit of item c. At what stage (in each case) do these changes cause a change in the type of item to be manufactured in the optimal schedule?

8.15. Use the active set method to solve the LP problem

$$\begin{aligned} \text{maximize} \quad & x_1 + x_2 \\ \text{subject to} \quad & 3x_1 + x_2 \geqslant 3 \\ & x_1 + 4x_2 \geqslant 4, \quad \mathbf{x} \geqslant \mathbf{0}. \end{aligned}$$

Illustrate the result by sketching the set of feasible solutions.

8.16. Solve the LP problem

$$\begin{array}{ll} \text{minimize} & -x_1 + x_2 \\ \text{subject to} & -x_1 + 2x_2 \leqslant 2 \\ & x_1 + x_2 \leqslant 4 \\ & x_1 \leqslant 3, \qquad \mathbf{x} \geqslant \mathbf{0} \end{array}$$

both graphically and by the active set method. Correlate the stages of the active set method with the extreme points of the feasible region.

8.17. Consider the LP problem (8.3.1) with E empty, and compare its solution by the active set method as against the simplex method. By introducing slack variables $\mathbf{z}$, write the constraints in the simplex method as $[-\mathbf{A}^{\mathrm{T}}:\mathbf{I}]\begin{pmatrix}\mathbf{x}\\ \mathbf{z}\end{pmatrix} = -\mathbf{b}$ and $\mathbf{z} \geqslant \mathbf{0}$. Use a modification of the simplex method in which the variables $\mathbf{x}$ are allowed to be unrestricted (for example as in Question 8.8 with $u_i = -l_i = K$ for sufficiently large K). Choose the nonbasic variables as the slack variables z_i, $i \in \mathscr{A}$, where $\mathscr{A}$ is the active set in the active set method. Partition $\mathbf{A} = [\mathbf{A}_1:\mathbf{A}_2]$, $\mathbf{b}^{\mathrm{T}} = (\mathbf{b}_1^{\mathrm{T}}, \mathbf{b}_2^{\mathrm{T}})$, and $\mathbf{z}^{\mathrm{T}} = (\mathbf{z}_1^{\mathrm{T}}, \mathbf{z}_2^{\mathrm{T}})$, corresponding to active and inactive constraints at $\mathbf{x}^{(k)}$ in the latter method. The quantities in the active set method are described in Section 8.3. In the simplex method, show that $\hat{\mathbf{b}}^{\mathrm{T}} = (\mathbf{x}^{(k)\mathrm{T}}, \mathbf{z}_2^{(k)\mathrm{T}})$, $\boldsymbol{\pi}^{\mathrm{T}} = (-\mathbf{c}^{\mathrm{T}}\mathbf{A}_1^{-\mathrm{T}}, \mathbf{0}^{\mathrm{T}})$, and $\hat{\mathbf{c}}_N = \mathbf{A}_1^{-1}\mathbf{c} = \boldsymbol{\lambda}$ in (8.3.3) and hence conclude that the same index q is chosen. Hence show that the vector $\mathbf{d}$ is given by $\mathbf{d}^{\mathrm{T}} = (\mathbf{s}_q^{\mathrm{T}}, \mathbf{s}_q^{\mathrm{T}}\mathbf{A}_2)$ (see (8.6.6)). Since there are no restrictions on the $\mathbf{x}$ variables, show that test (8.2.11) becomes

$$\min_{\substack{i:i\notin\mathscr{A} \\ \mathbf{a}_i^{\mathrm{T}}\mathbf{s}_q < 0}} \frac{z_i^{(k)}}{-\mathbf{a}_i^{\mathrm{T}}\mathbf{s}_q}$$

which is (8.3.6). Hence conclude that the same index p is chosen, and therefore that the sequence of values $\mathbf{x}^{(k)}$ and $\mathbf{z}^{(k)}$ in both methods is identical.

8.18. Use the method of artificial variables to find a b.f.s. for the LP problem in Question 8.1.

8.19. Find a basic feasible solution for the LP problem

$$\begin{array}{ll} \text{minimize} & -x_1 + x_2 + x_3 - 2x_4 \\ \text{subject to} & 4x_1 + x_2 - 2x_3 + 3x_4 \leqslant 8 \\ & x_1 - x_3 + x_4 = -2, \qquad \mathbf{x} \geqslant \mathbf{0} \end{array}$$

by the method of artificial variables, and hence solve the problem. Is the inequality constraint active at the solution?

8.20. Convert the LP problem

$$\begin{array}{ll} \text{minimize} & x_1 + x_2 + 2x_3 \\ \text{subject to} & x_1 + x_2 + x_3 \leqslant 9 \\ & 2x_1 - 3x_2 + 3x_3 = 1 \\ & -3x_1 + 6x_2 - 4x_3 = 3, \qquad \mathbf{x} \geqslant \mathbf{0} \end{array}$$

to standard form, and solve by a combined phase I–phase II simplex method, using the tableau form.

8.21. Convert the LP problem

$$\begin{array}{ll} \text{minimize} & -x_1 - x_2 \\ \text{subject to} & x_1 \qquad + 2x_3 \leqslant 1 \\ & \qquad x_2 - \ x_3 \leqslant 1 \\ & x_1 + x_2 + \ x_3 = 2, \qquad \mathbf{x} \geqslant \mathbf{0} \end{array}$$

to standard form and find an initial b.f.s. by the method of artificial variables. Show that an artificial variable remains in the resulting basis at zero value. Why is this so? Show that the solution to the LP problem can nonetheless be obtained.

8.22. Convert the LP problem

$$\begin{array}{ll} \text{minimize} & 2x_1 + 4x_2 + 3x_3 \\ \text{subject to} & x_1 - \ x_2 - \ x_3 \leqslant -2 \\ & 2x_1 + \ x_2 \qquad \geqslant \ 1, \qquad \mathbf{x} \geqslant \mathbf{0} \end{array}$$

to standard form. Solve using the simplex method

(i) starting with a basic feasible solution for which $x_1 = \frac{1}{2}$, $x_2 = 0$, $x_3 = 2\frac{1}{2}$,

(ii) using (8.4.5) with a basic of slack variables.

8.23. Compute $\mathbf{LL}^T$ factors of the tridiagonal matrix $\mathbf{A}$ in which $A_{11} = 1$ and $A_{ii} = 2$ and $A_{i,i-1} = A_{i-1,i} = -1 \; \forall i > 1$. Observe that no sparsity is lost in the factors, but that the inverse matrix $\mathbf{L}^{-1}$ is always a full lower triangular matrix with a simple general form.

8.24. Consider the Gauss–Jordon factors (8.5.4) of a matrix $\mathbf{A}$. Express each $\mathbf{M}_i$ as $\mathbf{M}_i = \bar{\mathbf{U}}_i\bar{\mathbf{L}}_i$ where

$$\bar{\mathbf{L}}_i = \begin{bmatrix} 1 & & & & \\ & \ddots & & & \\ & & 1 & & \\ & & -d_{i+1}/d_i & & \\ & & \vdots & \ddots & \\ & & -d_n/d_i & & 1 \end{bmatrix}, \qquad \bar{\mathbf{U}}_i = \begin{bmatrix} 1 & & -d_1/d_i & & \\ & \ddots & \vdots & & \\ & & -d_{i-1}/d_i & & \\ & & 1 & & \\ & & & \ddots & \\ & & & & 1 \end{bmatrix}$$

$\quad\llcorner$column i $\qquad\qquad\qquad\qquad\qquad\qquad \llcorner$column i

and show that $\bar{\mathbf{L}}_i\bar{\mathbf{U}}_j = \bar{\mathbf{U}}_j\bar{\mathbf{L}}_i$ for $j < i$. Hence rearrange (8.5.4) to give

$$\bar{\mathbf{U}}_n\bar{\mathbf{U}}_{n-1}\cdots\bar{\mathbf{U}}_1\bar{\mathbf{L}}_{n-1}\cdots\bar{\mathbf{L}}_2\bar{\mathbf{L}}_1\mathbf{A} = \mathbf{I}$$

or

$$\mathbf{A} = (\bar{\mathbf{L}}_1^{-1}\cdots\bar{\mathbf{L}}_{n-1}^{-1})(\bar{\mathbf{U}}_1^{-1}\cdots\bar{\mathbf{U}}_n^{-1}).$$

By comparing this with $\mathbf{A} = \mathbf{LU}$ where $\mathbf{L}$ and $\mathbf{U}$ are given in (8.5.5), show that the spike in $\bar{\mathbf{L}}_i$ is the ith column of $\mathbf{L}$, but that the same is not true for $\mathbf{U}$, since the spikes in $\mathbf{U}_i$ in (8.5.7) occur horizontally. However, observe from the last equation that

$$\mathbf{U}^{-1} = \bar{\mathbf{U}}_n\bar{\mathbf{U}}_{n-1}\cdots\bar{\mathbf{U}}_1$$

in which the spikes in the $\bar{U}_i$ are the columns of $\mathbf{U}^{-1}$. This justifies the assertion about the inefficiency of the Gauss–Jordon product form in Section 8.5.

8.25. Consider the LP problem due to Beale:

$$\begin{array}{ll} \text{minimize} & -\tfrac{3}{4}x_1 + 20x_2 - \tfrac{1}{2}x_3 + 6x_4 \\ \text{subject to} & \tfrac{1}{4}x_1 - 8x_2 - x_3 + 9x_4 \leqslant 0 \\ & \tfrac{1}{2}x_1 - 12x_2 - \tfrac{1}{2}x_3 + 3x_4 \leqslant 0 \\ & x_3 \leqslant 1, \quad \mathbf{x} \geqslant \mathbf{0}. \end{array}$$

Add slack variables x_5, x_6, x_7 and show that the initial choice $B = \{5, 6, 7\}$ is feasible but non-optimal and degenerate. Solve the problem by the tableau method, resolving ties in (8.2.11) by choosing the first permitted basic variable in the tableau. Show that after six iterations the original tableau is restored so that cycling is established. (Hadley, 1962, gives the detailed tableaux for this example.)

8.26. For the example of Question 8.25, show that initial column orderings in the tableau $\{5, 6, 7, 1, 2, 3, 4\}$ and $\{1, 2, 3, 4, 5, 6, 7\}$ both have $\hat{\mathbf{b}}^{(1)}(\varepsilon) \succ \mathbf{0}$. In the first case show that the tie break on iteration 1 is resolved by $p = 6$, which is different from the choice $p = 5$ made in the cycling iteration. Hence show that degeneracy is removed on the next iteration which solves the problem. In the second case show that the cycling iteration is followed for the first two iterations, but that the tie on iteration 3 is broken by $p = 2$ rather than by $p = 1$. Again show that degeneracy is removed on the next iteration which solves the problem. This example illustrates that different iteration sequences which resolve degeneracy can be obtained by different column orderings in the perturbation method.

8.27. Consider solving the LP in Question 8.25 by the method of (8.6.7) and (8.6.8), starting with the active constraint $\mathbf{x} \geqslant \mathbf{0}$ ($\mathbf{A}_B = \mathbf{I}$). The initial point is feasible so only phase II is required. Regard the cost function as column 0 of $\mathbf{A}$ and set up to maximize this function, with $q = 0$ in (8.6.7). Show that only one iteration in the dual problem is required to resolve degeneracy.

8.28. Consider trying to find a feasible point of the system $\mathbf{A}^T\mathbf{x} \geqslant \mathbf{b}$, $\mathbf{x} \geqslant \mathbf{0}$ by the method of (8.6.7) and (8.6.8), where

$$\mathbf{A} = \begin{bmatrix} 1 & -1 & & & \\ & 1 & -1 & & \\ & & 1 & -1 & \end{bmatrix} \qquad \mathbf{b} = \begin{bmatrix} 1 \\ 0 \\ 0 \\ 0 \end{bmatrix}$$

starting with the active constraints $\mathbf{x} \geqslant \mathbf{0}$. Show that the maximum level of recursion required is 6. This example typifies the case in which the maximum number of levels is required. The problem is in fact infeasible.

Chapter 9

The Theory of Constrained Optimization

9.1 LAGRANGE MULTIPLIERS

There have been many contributions to the theory of constrained optimization. In this chapter a number of the most important results are developed; the presentation aims towards practicality and avoids undue generality. Perhaps the most important concept which needs to be understood is the way in which so-called *Lagrange multipliers* are introduced and this is the aim of this section. In order to make the underlying structure clear, this is done in a semi-rigorous way, with a fully rigorous treatment following in the next section. In Part 1 it can be appreciated that the concept of a stationary point (for which $\mathbf{g}(\mathbf{x}) = \nabla f(\mathbf{x}) = \mathbf{0}$) is fundamental to the subject of unconstrained optimization, and is a necessary condition for a local minimizer. Lagrange multipliers arise when similar necessary conditions are sought for the solution $\mathbf{x}^*$ of the constrained minimization problem (7.1.1).

For unconstrained minimization the necessary conditions (2.1.3) and (2.1.4) illustrate the requirement for zero slope and non-negative curvature in any direction at $\mathbf{x}^*$, that is to say there is no descent direction at $\mathbf{x}^*$. In constrained optimization there is the additional complication of a feasible region. Hence a local minimizer must be a feasible point, and other necessary conditions illustrate the need for there to be *no feasible descent directions* at $\mathbf{x}^*$. However, there are some difficulties on account of the fact that the boundary of the feasible region may be curved. In the first instance the simplest case of only equality constraints is presented (that is $I = \varnothing$ (the empty set)).

Suppose that a feasible incremental step $\boldsymbol{\delta}$ is taken from the minimizing point $\mathbf{x}^*$. By a Taylor series

$$c_i(\mathbf{x}^* + \boldsymbol{\delta}) = c_i^* + \boldsymbol{\delta}^{\mathrm{T}}\mathbf{a}_i^* + o(\|\boldsymbol{\delta}\|)$$

where $\mathbf{a}_i = \nabla c_i$ and $o(\cdot)$ indicates terms which can be ignored relative to $\boldsymbol{\delta}$ in the limit. By feasibility $c_i(\mathbf{x}^* + \boldsymbol{\delta}) = c_i^* = 0$, so that any feasible incremental step lies along a *feasible direction* $\mathbf{s}$ which satisfies

$$\mathbf{s}^T\mathbf{a}_i^* = 0, \qquad \forall i \in E. \tag{9.1.1}$$

In a regular situation (for example if the vectors $\mathbf{a}_i^*$, $i \in E$, are independent), it is also possible to construct a feasible incremental step $\boldsymbol{\delta}$, given any such $\mathbf{s}$ (see Section 9.2). If in addition $f(\mathbf{x})$ has negative slope along $\mathbf{s}$, that is

$$\mathbf{s}^T\mathbf{g}^* < 0, \tag{9.1.2}$$

then the direction $\mathbf{s}$ is a descent direction and the feasible incremental step along $\mathbf{s}$ reduces $f(\mathbf{x})$. This cannot happen, however, because $\mathbf{x}^*$ is a local minimizer. Therefore there can be no direction $\mathbf{s}$ which satisfies both (9.1.1) and (9.1.2). Now this statement is clearly true *if* $\mathbf{g}^*$ is a linear combination of the vectors $\mathbf{a}_i^*$, $i \in E$, that is if

$$\mathbf{g}^* = \sum_{i \in E} \mathbf{a}_i^* \lambda_i^* = \mathbf{A}^*\boldsymbol{\lambda}^*, \tag{9.1.3}$$

and in fact *only if* this condition holds, as is shown below. Therefore (9.1.3) is a *necessary condition for a local minimizer.* The multipliers λ_i^* in this linear combination are referred to as *Lagrange multipliers*, and the superscript * indicates that they are associated with the point $\mathbf{x}^*$. $\mathbf{A}^*$ denotes the matrix with columns $\mathbf{a}_i^*$, $i \in E$. Notice that there is a multiplier associated with each constraint function. In fact if $\mathbf{A}^*$ has full rank, then from (9.1.3) $\boldsymbol{\lambda}^*$ is defined *uniquely* by the expression

$$\boldsymbol{\lambda}^* = \mathbf{A}^{*+}\mathbf{g}^*$$

where $\mathbf{A}^+ = (\mathbf{A}^T\mathbf{A})^{-1}\mathbf{A}^T$ denotes the generalized inverse of $\mathbf{A}$ (see Question 9.15). Of course when there are no constraints present, then (9.1.3) reduces to the usual stationary point condition that $\mathbf{g}^* = \mathbf{0}$.

The formal proof that (9.1.3) is necessary is by contradiction. If (9.1.3) does not hold, then $\mathbf{g}^*$ can be expressed as

$$\mathbf{g}^* = \mathbf{A}^*\boldsymbol{\lambda} + \boldsymbol{\mu} \tag{9.1.4}$$

where $\boldsymbol{\mu} \neq \mathbf{0}$ is the component of $\mathbf{g}^*$ orthogonal to the vectors $\mathbf{a}_i^*$, so that $\mathbf{A}^{*T}\boldsymbol{\mu} = \mathbf{0}$. Then $\mathbf{s} = -\boldsymbol{\mu}$ satisfies both (9.1.1) and (9.1.2). Hence by the regularity assumption there exists a feasible incremental step $\boldsymbol{\delta}$ along $\mathbf{s}$ which reduces $f(\mathbf{x})$, and this contradicts the fact that $\mathbf{x}^*$ is a local minimizer. Thus (9.1.3) is necessary. The conditions are illustrated in Figure 9.1.1. At $\mathbf{x}'$ which is not a local minimizer, $\mathbf{g}' \neq \mathbf{a}'\lambda$ and so there exists a non-zero vector $\boldsymbol{\mu}$ which is orthogonal to $\mathbf{a}'$, and an incremental step $\boldsymbol{\delta}$ along the feasible descent direction $-\boldsymbol{\mu}$ reduces $f(\mathbf{x})$. At $\mathbf{x}^*$, $\mathbf{g}^* = \mathbf{a}^*\lambda^*$ and no feasible descent direction exists. A numerical example is provided by the problem: minimize $f(\mathbf{x}) \triangleq x_1 + x_2$ subject to $c(\mathbf{x}) \triangleq x_1^2 - x_2 = 0$. In this case $\mathbf{x}^* = (-\frac{1}{2}, \frac{1}{4})^T$ so that $\mathbf{g}^* = (1, 1)^T$ and $\mathbf{a}^* = (-1, -1)^T$, and (9.1.3) is thus satisfied with $\lambda^* = -1$. The regularity assumption clearly holds because $\mathbf{a}^*$ is independent; more discussion about the regularity condition is given in the next section. The use of incremental steps

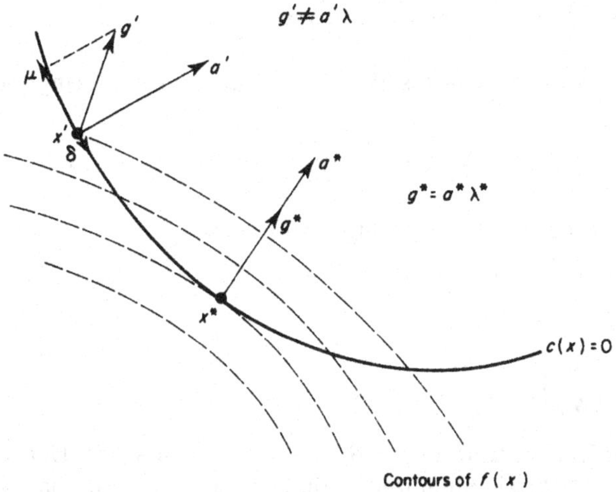

Figure 9.1.1 Existence of Lagrange multipliers

in the above can be expressed more rigorously by introducing directional sequences, as in the next section, or by means of differentiable arcs (Kuhn and Tucker, 1951). However, the aim of this section is to avoid these technicalities as much as possible.

These conditions give rise to the classical *method of Lagrange multipliers* for solving equality constraint problems. The method is to find vectors $\mathbf{x}^*$ and $\boldsymbol{\lambda}^*$ which solve the equations

$$\begin{aligned} \mathbf{g}(\mathbf{x}) &= \sum_{i\in E} \mathbf{a}_i(\mathbf{x})\lambda_i \\ c_i(\mathbf{x}) &= 0, \qquad i\in E \end{aligned} \tag{9.1.5}$$

which arise from (9.1.3) and feasibility. If there are m equality constraints then there are $n+m$ equations and $n+m$ unknowns $\mathbf{x}$ and $\boldsymbol{\lambda}$, so the system is well-determined. However the system is nonlinear (in $\mathbf{x}$) so in general may not be easy to solve (see Chapter 6), although this may be possible in simple cases. An additional objection is that no second order information is taken into account, so (9.1.5) is also satisfied at a constrained saddle point or maximizer. The example of the previous paragraph can be used to illustrate the method. There are then three equations in (9.1.5), that is

$$\begin{pmatrix}1\\1\end{pmatrix} = \begin{pmatrix}2x_1\\-1\end{pmatrix}\lambda$$

$$x_1^2 - x_2 = 0.$$

These can be solved in turn for the three variables $\lambda^* = -1$, $x_1^* = -\frac{1}{2}$, and $x_2^* = \frac{1}{4}$. It is instructive to see how the method differs from that of direct elimination. In this case $x_2 = x_1^2$ is used to eliminate x_2 from $f(\mathbf{x})$ leaving

$f(x_1)=x_1+x_1^2$ which is minimized by $x_1^*=-\frac{1}{2}$. Then back substitution gives $x_2^*=\frac{1}{4}$.

It is often convenient to restate these results by introducing the *Lagrangian function*

$$\mathscr{L}(\mathbf{x},\boldsymbol{\lambda})=f(\mathbf{x})-\sum_i \lambda_i c_i(\mathbf{x}). \tag{9.1.6}$$

Then (9.1.5) becomes the very simple expression

$$\blacktriangledown \mathscr{L}(\mathbf{x}^*,\boldsymbol{\lambda}^*)=\mathbf{0} \tag{9.1.7}$$

where

$$\blacktriangledown = \begin{pmatrix} \nabla_{\mathbf{x}} \\ \nabla_{\lambda} \end{pmatrix} \tag{9.1.8}$$

is a first derivative operator for the $n+m$ variable space. Hence a necessary condition for a local minimizer is that $\mathbf{x}^*$, $\boldsymbol{\lambda}^*$ is a stationary point of the Lagrangian function.

An alternative way of deriving these results starts by trying to find a stationary point of $f(\mathbf{x})$ subject to $\mathbf{c}(\mathbf{x})=\mathbf{0}$, that is a feasible point $\mathbf{x}^*$ at which $f(\mathbf{x}^*+\boldsymbol{\delta})=f(\mathbf{x}^*)+o(\|\boldsymbol{\delta}\|)$ for all feasible changes $\boldsymbol{\delta}$. The method of Lagrange multipliers finds a stationary point of $\mathscr{L}(\mathbf{x},\boldsymbol{\lambda})$ by solving (9.1.7) or equivalently (9.1.5). Since $\nabla_{\lambda}\mathscr{L}=\mathbf{0}$ it follows that $\mathbf{x}^*$ is feasible. Since $\nabla_x\mathscr{L}(\mathbf{x},\boldsymbol{\lambda}^*)=\mathbf{0}$ it follows that $\mathscr{L}(\mathbf{x},\boldsymbol{\lambda}^*)$ is stationary at $\mathbf{x}^*$ for all changes $\boldsymbol{\delta}$, and hence for all feasible changes. But if $\boldsymbol{\delta}$ is a feasible change, $\mathscr{L}(\mathbf{x}^*+\boldsymbol{\delta},\boldsymbol{\lambda}^*)=f(\mathbf{x}^*+\boldsymbol{\delta})$ and so $f(\mathbf{x})$ is stationary at $\mathbf{x}^*$ for all feasible changes $\boldsymbol{\delta}$. Thus if (9.1.7) can be solved, this solution $\mathbf{x}^*$ is a constrained stationary point of $f(\mathbf{x})$. The opposite may not always be true; it is possible (for example Questions 7.4 and 9.14) that $\mathbf{x}^*$ can be a minimizer but not satisfy (9.1.7). Also the examples with $f(\mathbf{x})\triangleq\pm(x-1)^3+y$ and the same constraint show that $\mathbf{x}^*=(0,1)^{\mathrm{T}}$ can be a constrained stationary point but not satisfy (9.1.7).

To get another insight into the meaning of Lagrange multipliers, consider what happens if the right-hand sides of the constraints are perturbed, so that

$$c_i(\mathbf{x})=\varepsilon_i, \qquad i\in E. \tag{9.1.9}$$

Let $\mathbf{x}(\boldsymbol{\varepsilon})$, $\boldsymbol{\lambda}(\boldsymbol{\varepsilon})$ denote how the solution and multipliers change as $\boldsymbol{\varepsilon}$ changes. The Lagrangian for this problem is

$$\mathscr{L}(\mathbf{x},\boldsymbol{\lambda},\boldsymbol{\varepsilon})=f(\mathbf{x})-\sum_{i\in E}\lambda_i(c_i(\mathbf{x})-\varepsilon_i).$$

From (9.1.9), $f(\mathbf{x}(\boldsymbol{\varepsilon}))=\mathscr{L}(\mathbf{x}(\boldsymbol{\varepsilon}),\boldsymbol{\lambda}(\boldsymbol{\varepsilon}),\boldsymbol{\varepsilon})$, so using the chain rule and then (9.1.7) it follows that

$$\frac{\mathrm{d}f}{\mathrm{d}\varepsilon_i}=\frac{\mathrm{d}\mathscr{L}}{\mathrm{d}\varepsilon_i}=\frac{\partial\mathbf{x}^{\mathrm{T}}}{\partial\varepsilon_i}\nabla_x\mathscr{L}+\frac{\partial\boldsymbol{\lambda}^{\mathrm{T}}}{\partial\varepsilon_i}\nabla_{\lambda}\mathscr{L}+\frac{\partial\mathscr{L}}{\partial\varepsilon_i}=\lambda_i. \tag{9.1.10}$$

Thus the Lagrange multiplier of any constraint measures the rate of change in the objective function, consequent upon changes in that constraint function.

This information can be valuable in that it indicates how sensitive the objective functions is to changes in the different constraints: see Question 9.4 for example.

The additional complication of having inequality constraints present is now discussed. It is important to realize that only the *active constraints* $\mathscr{A}^*$ (see (7.1.2)) at $\mathbf{x}^*$ can influence matters. Denote the active inequality constraints at $\mathbf{x}^*$ by I^* ($=\mathscr{A}^* \cap I$). Since $c_i^* = 0$ and $c_i(\mathbf{x}^* + \boldsymbol{\delta}) \geqslant 0$ for $i \in I^*$, then any feasible incremental step $\boldsymbol{\delta}$ lies along a feasible direction $\mathbf{s}$ which satisfies

$$\mathbf{s}^{\mathrm{T}}\mathbf{a}_i^* \geqslant 0, \qquad \forall i \in I^* \tag{9.1.11}$$

in addition to (9.1.1). In this case it is clear that there is no direction $\mathbf{s}$ which satisfies (9.1.1), (9.1.11), and (9.1.2) together, *if* both

$$\mathbf{g}^* = \sum_{i \in \mathscr{A}^*} \mathbf{a}_i^* \lambda_i^* \tag{9.1.12}$$

and

$$\lambda_i^* \geqslant 0, \qquad i \in I^* \tag{9.1.13}$$

hold, and in fact *only* if these conditions hold, as is shown below. Hence these are therefore necessary conditions for a local minimizer. The only extra condition beyond (9.1.3) in this case, therefore, is that the multipliers of active inequality constraints must be non-negative. These conditions can be established by contradiction when the regularity assumption is made that the set of normal vectors $\mathbf{a}_i^*$, $i \in \mathscr{A}^*$, is linearly independent. Equation (9.1.12) holds as for (9.1.3). Let (9.1.13) not hold so that there is some multiplier $\lambda_p^* < 0$. It is always possible to find a direction $\mathbf{s}$ for which $\mathbf{s}^{\mathrm{T}}\mathbf{a}_i^* = 0$, $i \in \mathscr{A}^*$, $i \neq p$, and $\mathbf{s}^{\mathrm{T}}\mathbf{a}_p^* = 1$ (for instance $\mathbf{s} = \mathbf{A}^{*+\mathrm{T}}\mathbf{e}_p$ where $\mathbf{A}^+ = (\mathbf{A}^{\mathrm{T}}\mathbf{A})^{-1}\mathbf{A}^{\mathrm{T}}$ denotes the generalized inverse of $\mathbf{A}$). Then $\mathbf{s}$ is feasible in (9.1.1) and (9.1.11), and from (9.1.12) it follows that

$$\mathbf{s}^{\mathrm{T}}\mathbf{g}^* = \mathbf{s}^{\mathrm{T}}\mathbf{a}_p^*\lambda_p^* = \lambda_p^* < 0. \tag{9.1.14}$$

Thus $\mathbf{s}$ is also downhill, and by the regularity assumption there exists a feasible incremental step $\boldsymbol{\delta}$ along $\mathbf{s}$ which reduces $f(\mathbf{x})$; this contradicts the fact that $\mathbf{x}^*$ is a local minimizer. Thus (9.1.13) is necessary. Note that this proof uses the linear independence of the vectors $\mathbf{a}_i^*$, $i \in \mathscr{A}^*$, in constructing the generalized inverse. A more general proof using Lemma 9.2.4 (Farkas' lemma) and its corollary does not make this assumption.

The need for condition (9.1.13) can also be deduced simply from (9.1.9) and (9.1.10). Let an active inequality constraint be perturbed from $c_i(\mathbf{x}) = 0$ to $c_i(\mathbf{x}) = \varepsilon_i > 0$, $i \in I^*$. This induces a feasible change in $\mathbf{x}(\boldsymbol{\varepsilon})$ so it necessary that $f(\mathbf{x}(\boldsymbol{\varepsilon}))$ does not decrease. This implies that $\mathrm{d}f^*/\mathrm{d}\varepsilon_i \geqslant 0$ at the solution and hence $\lambda_i^* \geqslant 0$. Thus the necessity of (9.1.13) has an obvious interpretation in these terms.

As an example of these conditions, consider the problem

$$\begin{aligned} &\text{minimize} \quad f(\mathbf{x}) \triangleq -x_1 - x_2 \\ &\text{subject to} \quad c_1(\mathbf{x}) \triangleq x_2 - x_1^2 \geqslant 0 \\ &\qquad\qquad\quad c_2(\mathbf{x}) \triangleq 1 - x_1^2 - x_2^2 \geqslant 0 \end{aligned} \tag{9.1.15}$$

which is illustrated in Figure 7.1.3. The solution is $\mathbf{x}^* = (1/\sqrt{2}, 1/\sqrt{2})^T$ so c_1 is not active and hence $\mathscr{A}^* = \{2\}$. Then $\mathbf{g}^* = (-1, -1)^T$ and $\mathbf{a}_2^* = (-\sqrt{2}, -\sqrt{2})^T$ so (9.1.12) and (9.1.13) are satisfied with $\lambda_2^* = 1/\sqrt{2} \geqslant 0$. It is important to remember that a general inequality constraint must be correctly rearranged into the form $c_i(\mathbf{x}) \geqslant 0$ before the condition $\lambda_i \geqslant 0$ applies.

In fact the construction of a descent direction when $\lambda_p^* < 0$ above indicates another important property of Lagrange multipliers for inequality constraints. The conditions $\mathbf{s}^T\mathbf{a}_i^* = 0$, $i \neq p$, and $\mathbf{s}^T\mathbf{a}_p^* = 1$ indicate that the resulting feasible incremental step satisfies $c_i(\mathbf{x}^* + \boldsymbol{\delta}) = 0$ for $i \neq p$, and $\mathbf{c}_p(\mathbf{x}^* + \boldsymbol{\delta}) > 0$. Thus it indicates that $f(\mathbf{x})$ can be reduced *by moving away from the boundary of constraint* p. This result also follows from (9.1.10) and is of great importance in the various active set methods (see Section 7.2) for handling inequality constraints. If conditions (9.1.13) are not satisfied then a constraint index p with $\lambda_p^* < 0$ can be removed from the active set. This result is also illustrated by the problem in the previous paragraph. Consider the feasible point $\mathbf{x}' \simeq (0.786, 0.618)$ to three decimal places, at which both constraints are active. Since $\mathbf{g}' = (-1, -1)^T$, $\mathbf{a}_1' = (-1.572, 1)^T$ and $\mathbf{a}_2' = (-1.572, -1.236)^T$, it follows that (9.1.12) is satisfied with $\boldsymbol{\lambda}' = (-0.096, 0.732)^T$. However (9.1.13) is not satisfied, so $\mathbf{x}'$ is not a local minimizer. Since $\lambda_1' < 0$ the objective function can be reduced by moving away from the boundary of constraint 1, along a direction for which $\mathbf{s}^T\mathbf{a}_1^* = 1$ and $\mathbf{s}^T\mathbf{a}_2^* = 0$. This is the direction $\mathbf{s} = (-0.352, 0.447)^T$ and is in fact the tangent to the circle at $\mathbf{x}'$, Moving round the arc of the circle in this direction leads to the solution point $\mathbf{x}^*$ at which only constraint 2 is active.

A further restatement of (9.1.12) and (9.1.13) is possible in terms of all the constraints rather than just the active ones. It is consistent to regard any inactive constraint as having a zero Lagrange multiplier, in which case (9.1.12), (9.1.13), and the feasibility conditions can be combined in the following theorem.

Theorem 9.1.1 (First order necessary conditions)

If $\mathbf{x}^*$ *is a local minimizer of problem* (7.1.1) *and if a regularity assumption* (9.2.4) *holds at* $\mathbf{x}^*$, *then there exist Lagrange multipliers* $\boldsymbol{\lambda}^*$ *such that* $\mathbf{x}^*$, $\boldsymbol{\lambda}^*$ *satisfy the following system:*

$$
\begin{aligned}
\nabla_x \mathscr{L}(\mathbf{x}, \boldsymbol{\lambda}) &= \mathbf{0} \\
c_i(\mathbf{x}) &= 0, \qquad i \in E \\
c_i(\mathbf{x}) &\geqslant 0, \qquad i \in I \\
\lambda_i &\geqslant 0, \qquad i \in I \\
\lambda_i c_i(\mathbf{x}) &= 0 \qquad \forall i.
\end{aligned}
\tag{9.1.16}
$$

These are often described as Kuhn–Tucker (KT) conditions (Kuhn and Tucker, 1951) and a point $\mathbf{x}^*$ which satisfies the conditions is sometimes referred to as *KT point*. The regularity assumption (9.2.4) is implied by the vectors $\mathbf{a}_i^*$, $i \in \mathscr{A}^*$, being independent and is discussed in detail in the next section where a more rigorous proof is given. The final condition $\lambda_i^* c_i^* = 0$ is referred to as the

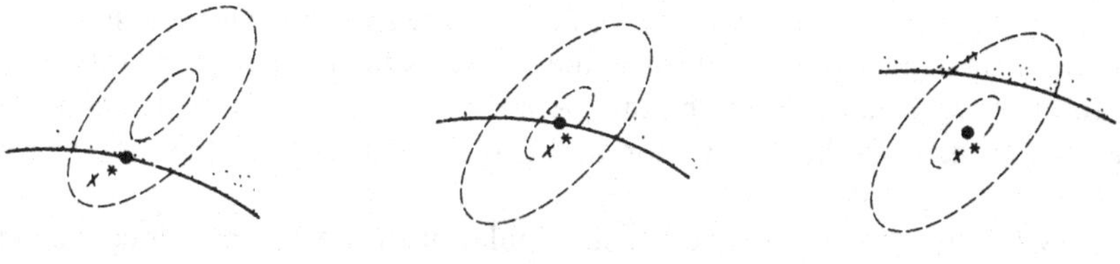

Figure 9.1.2 Complementarity

complementarity condition and states that both λ_i^* and c_i^* cannot be non-zero, or equivalently that inactive constraints have a zero multiplier. If there is no i such that $\lambda_i^* = c_i^* = 0$ then *strict complementarity* is said to hold. The case $\lambda_i^* = c_i^* = 0$ is an intermediate state between a constraint being strongly active and being inactive, as indicated in Figure 9.1.2.

So far only first order (that is first derivative) conditions have been considered. It is also possible to state second order conditions which give information about the curvature of the objective and constraint functions at a local minimizer. This subject is discussed in Section 9.3. It is also possible to make even stronger statements when the original problem is a *convex programming problem*, and the more simple results of convexity and its application to optimization theory are developed in Section 9.4. For certain convex programming problems it is possible to state useful alternative (dual) problems from which the solution to the original (primal) problem can be obtained. These problems involve the Lagrange multipliers as dual variables, much in the way that they arise in the method of Lagrange multipliers. The subject of duality is discussed further in Section 9.5. In fact the literature on convexity and duality is very extensive and often very theoretical, so as to become a branch of pure mathematics. In this volume I have attempted to describe those aspects of these subjects which are of most relevance to practical algorithms.

9.2 FIRST ORDER CONDITIONS

In this section the results of the previous section are considered in more technical detail. First of all it is important to have a more rigorous notion of what is meant by a feasible incremental step. Consider any feasible point $\mathbf{x}'$ and any infinite sequence of feasible points $\{\mathbf{x}^{(k)}\} \rightarrow \mathbf{x}'$ where $\mathbf{x}^{(k)} \neq \mathbf{x}'$ for all k. Then it is possible to write

$$\mathbf{x}^{(k)} - \mathbf{x}' = \delta^{(k)}\mathbf{s}^{(k)} \qquad \forall k \tag{9.2.1}$$

where $\delta^{(k)} > 0$ is a scalar and $\mathbf{s}^{(k)}$ is a vector of any fixed length $\sigma > 0$ ($\|\mathbf{s}^{(k)}\|_2 = \sigma$). It follows that $\delta^{(k)} \rightarrow 0$. A *directional sequence* can be defined as any such sequence for which $\mathbf{s}^{(k)} \rightarrow \mathbf{s}$. The limiting vector $\mathbf{s}$ is referred to as a *feasible direction*, and $\mathscr{F}(\mathbf{x}')$ or $\mathscr{F}'$ is used to denote the *set of feasible directions* at $\mathbf{x}'$. Taking the limit in (9.2.1) corresponds to the notion of making a feasible incremental step along $\mathbf{s}$. Clearly the length of $\mathbf{s}$ is arbitrary and it is possible to restrict the discussion to any fixed normalization, for example $\sigma = 1$. In some texts $\mathscr{F}'$ is defined so that it also includes the zero vector, and is then referred to as the *tangent cone* at $\mathbf{x}'$.

The set $\mathscr{F}'$ is not very amenable to manipulation, however, and it is convenient to consider a related set of feasible directions which are obtained if the constraints are linearized. The linearized constraint function is given by (7.1.3) so clearly the set of feasible directions for the linearized constraint set can be written as

$$\begin{aligned} F(\mathbf{x}') = F' = \{\mathbf{s} \mid \mathbf{s} \neq \mathbf{0}, \quad & \mathbf{s}^{\mathrm{T}}\mathbf{a}_i' = 0, \quad i \in E, \\ & \mathbf{s}^{\mathrm{T}}\mathbf{a}_i' \geqslant 0, \quad i \in I'\}. \end{aligned} \tag{9.2.2}$$

where $I' = \mathscr{A}' \cap I$ denotes the set of active inequality constraints at $\mathbf{x}'$. (A vector $\mathbf{s} \in F'$ corresponds to a directional sequence along a half-line through $\mathbf{x}'$ in the direction $\mathbf{s}$, which is clearly feasible, and it is straightforward to contradict feasibility in the linearized set for any directional sequence for which $\mathbf{s} \notin F'$.) It is very convenient if the sets F' and $\mathscr{F}'$ are the same, so it is important to consider the extent to which this is true.

Lemma 9.2.1

$F' \supseteq \mathscr{F}'$

Proof

Let $\mathbf{s} \in \mathscr{F}'$: then $\exists$ a directional sequence $\mathbf{x}^{(k)} \to \mathbf{x}'$ such that $\mathbf{s}^{(k)} \to \mathbf{s}$. A Taylor series about $\mathbf{x}'$ using (9.2.1) gives

$$c_i(\mathbf{x}^{(k)}) = c_i' + \delta^{(k)}\mathbf{s}^{(k)\mathrm{T}}\mathbf{a}_i' + o(\delta^{(k)}).$$

Now $c_i(\mathbf{x}^{(k)}) = c_i' = 0$ for $i \in E$ and $c_i(\mathbf{x}^{(k)}) \geqslant c_i' = 0$ for $i \in I'$, so dividing by $\delta^{(k)} > 0$ it follows that

$$\begin{aligned} \mathbf{s}^{(k)\mathrm{T}}\mathbf{a}_i' + o(1) &= 0, \qquad i \in E \\ \mathbf{s}^{(k)\mathrm{T}}\mathbf{a}_i' + o(1) &\geqslant 0, \qquad i \in I'. \end{aligned}$$

Taking limits as $k \to \infty, \mathbf{s}^{(k)} \to \mathbf{s}$, $o(1) \to 0$, then $\mathbf{s} \in F'$ from (9.2.2). $\square$

Unfortunately a result going the other way ($\mathscr{F}' \supseteq F'$) is not in general true and it is this that is required in the proof of Theorem 9.1.1. To get round this difficulty, Kuhn and Tucker (1951) make an assumption that $F' = \mathscr{F}'$, which they refer to as a *constraint qualification* at $\mathbf{x}'$. That is to say, for any $\mathbf{s} \in F'$, the existence of a feasible directional sequence with feasible direction $\mathbf{s}$ is assumed. They also give an example in which this property does not hold which is essentially that illustrated in Figure 9.2.1 Clearly at $\mathbf{x}' = \mathbf{0}$, the direction $\mathbf{s} = (-1, 0)^{\mathrm{T}}$ is in F' but there is no corresponding feasible directional sequence,

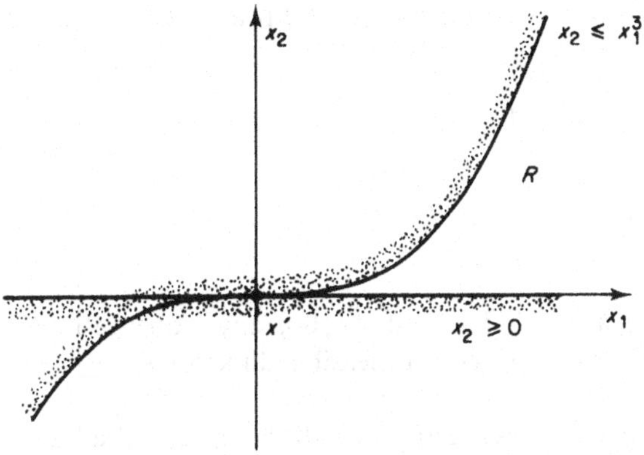

Figure 9.2.1 Failure of constraint qualification

and so $\mathbf{s} \notin \mathscr{F}'$. However, it is imporant to realize that this failing case is an unlikely situation and that it is usually valid to assume that $F' = \mathscr{F}'$. Indeed this result can be guaranteed if certain linearity or independence conditions hold, as the following lemma shows.

Lemma 9.2.2

Sufficient conditions for $F' = \mathscr{F}'$ at a feasible point $\mathbf{x}'$ are either

(i) *the constraints $i \in \mathscr{A}'$ are all linear constraints or*
(ii) *the vectors $\mathbf{a}'_i$, $i \in \mathscr{A}'$, are linearly independent.*

Proof

Case (i) is clear by definition of F'. For case (ii) a feasible directional sequence with feasible direction $\mathbf{s}$ is constructed for any $\mathbf{s} \in F'$. Let $\mathbf{s} \in F'$, and consider the nonlinear system

$$\mathbf{r}(\mathbf{x}, \theta) = \mathbf{0} \tag{9.2.3}$$

defined by

$$\begin{aligned} r_i(\mathbf{x}, \theta) &= c_i(\mathbf{x}) - \theta \mathbf{s}^{\mathrm{T}} \mathbf{a}'_i, & i &= 1, 2, \ldots, m \\ r_i(\mathbf{x}, \theta) &= (\mathbf{x} - \mathbf{x}')^{\mathrm{T}} \mathbf{b}_i - \theta \mathbf{s}^{\mathrm{T}} \mathbf{b}_i, & i &= m+1, \ldots, n \end{aligned}$$

where it is assumed that $\mathscr{A}' = \{1, 2, \ldots, m\}$. The system (9.2.3) is solved by $\mathbf{x}'$ when $\theta = 0$, and any solution $\mathbf{x}$ is also a feasible point in (7.1.1) when $\theta \geqslant 0$ is sufficiently small. Writing $\mathbf{A} = [\mathbf{a}_1, \ldots, \mathbf{a}_m]$ and $\mathbf{B} = [\mathbf{b}_{m+1}, \ldots, \mathbf{b}_n]$ then the Jacobian matrix $\mathbf{J}(\mathbf{x}, \theta) = \nabla_x \mathbf{r}^{\mathrm{T}}(\mathbf{x}, \theta) = [\mathbf{A} : \mathbf{B}]$. Since $\mathbf{A}'$ has full rank by case (ii), it is possible to choose $\mathbf{B}$ so that $\mathbf{J}'$ is non-singular. Hence by the implicit function theorem (Apostol (1957) for example) there exist open neighbourhoods Ω_x about $\mathbf{x}'$ and Ω_θ about $\theta = 0$ such that for any $\theta \in \Omega_\theta$, a unique solution $\mathbf{x}(\theta) \in \Omega_x$ exists to (9.2.3), and $\mathbf{x}(\theta)$ is a $\mathbb{C}^1$ function of θ. From (9.2.3) and using the chain rule

$$0=\frac{\mathrm{d}r_i}{\mathrm{d}\theta}=\sum_j \frac{\partial r_i}{\partial x_j}\frac{\mathrm{d}x_j}{\mathrm{d}\theta}+\frac{\partial r_i}{\partial \theta}$$

so that

$$\mathbf{0}=\mathbf{J}^{\mathrm{T}}\frac{\mathrm{d}\mathbf{x}}{\mathrm{d}\theta}-\mathbf{J}'^{\mathrm{T}}\mathbf{s}.$$

Thus $\mathrm{d}\mathbf{x}/\mathrm{d}\theta=\mathbf{s}$ at $\theta=0$. Hence if $\theta^{(k)}\downarrow 0$ is any sequence then $\mathbf{x}^{(k)}=\mathbf{x}(\theta^{(k)})$ is a feasible directional sequence with feasible direction **s**. □

In moving on to discuss necessary conditions at a local solution it is convenient to define the *set of descent directions*

$$\mathscr{D}(\mathbf{x}')=\mathscr{D}'=\{\mathbf{s}\mid \mathbf{s}^{\mathrm{T}}\mathbf{g}'<\mathbf{0}\}.$$

Then the most basic necessary condition is the following.

Lemma 9.2.3

If $\mathbf{x}^*$ *is a local minimizer, then* $\mathscr{F}^*\cap\mathscr{D}^*=\varnothing$ *(no feasible descent directions).*

Proof

Let $\mathbf{s}\in\mathscr{F}^*$ so there exists a feasible sequence $\mathbf{x}^{(k)}\to\mathbf{x}^*$ such that $\mathbf{s}^{(k)}\to\mathbf{s}$. By a Taylor series about $\mathbf{x}^*$,

$$f(\mathbf{x}^{(k)})=f^*+\delta^{(k)}\mathbf{s}^{(k)\mathrm{T}}\mathbf{g}^*+o(\delta^{(k)}).$$

Because $\mathbf{x}^*$ is a local minimizer, $f(\mathbf{x}^{(k)})\geqslant f^*$ for all k sufficiently large, so dividing by $\delta^{(k)}>0$,

$$\mathbf{s}^{(k)\mathrm{T}}\mathbf{g}^*+o(1)\geqslant 0.$$

In the limit, $\mathbf{s}^{(k)}\to\mathbf{s}$, $o(1)\to 0$ so $\mathbf{s}\notin\mathscr{D}^*$ and hence $\mathscr{F}^*\cap\mathscr{D}^*=\varnothing$. □

Unfortunately it is not possible to proceed further without making a *regularity assumption*

$$F^*\cap\mathscr{D}^*=\mathscr{F}^*\cap\mathscr{D}^*. \tag{9.2.4}$$

This assumption is clearly implied by the Kuhn–Tucker constraint qualification ($F^*=\mathscr{F}^*$) at $\mathbf{x}^*$, but (9.2.4) may hold when $F^*=\mathscr{F}^*$ does not, for example at $\mathbf{x}^*=\mathbf{0}$ in the problem: mimimize x_2 subject to the constraints of Figure 9.2.1. Also the problem: minimize x_1 subject to the same constraints, illustrates the need for a regularity assumption. Here $\mathbf{s}=(-1,0)^{\mathrm{T}}\in F^*\cap\mathscr{D}^*$ at $\mathbf{x}^*=\mathbf{0}$ so this set is not empty and in fact $\mathbf{x}^*$ is not a KT point, although it is a minimizer and $\mathscr{F}^*\cap\mathscr{D}^*$ is empty.

With assumption (9.2.4) the necessary condition from Lemma 9.2.3 becomes

$F^* \cap \mathscr{D}^* = \varnothing$ (no linearized feasible descent directions). It is now possible to relate this condition to the existence of multipliers in (9.1.12) and (9.1.13). In fact the construction given after (9.1.13) can be used to do this when the set $\mathbf{a}_i^*$, $i \in \mathscr{A}^*$, is independent. However, the following lemma shows that the result is more general than this.

Lemma 9.2.4 (Farkas' lemma)

Given any vectors $\mathbf{a}_1, \mathbf{a}_2, \ldots, \mathbf{a}_m$ *and* $\mathbf{g}$ *then the set*

$$S = \{\mathbf{s} \mid \mathbf{s}^T\mathbf{g} < 0, \tag{9.2.5}$$

$$\mathbf{s}^T\mathbf{a}_i \geqslant 0, \qquad i = 1, 2, \ldots, m\} \tag{9.2.6}$$

is empty if and only if there exist multipliers $\lambda_i \geqslant 0$ *such that*

$$\mathbf{g} = \sum_{i=1}^{m} \mathbf{a}_i \lambda_i. \tag{9.2.7}$$

Remark

In the context of this section, S is the set of linearized feasible descent directions for a problem with m active inequality constraints and no equality constraints. The extension to include equality constraints is made in the corollary.

Proof

The 'if' part is straightforward since (9.2.7) implies that $\mathbf{s}^T\mathbf{g} = \sum \mathbf{s}^T\mathbf{a}_i\lambda_i \geqslant 0$ by (9.2.6) and $\lambda_i \geqslant 0$. Thus (9.2.5) is not true and so S is empty. The converse result is established by showing that if (9.2.7) with $\lambda_i \geqslant 0$ does not hold, then there is a vector $\mathbf{s} \in S$. Geometrically, the result is easy to visualize. The set of vectors

$$C = \left\{ \mathbf{v} \mid \mathbf{v} = \sum_{i=1}^{m} \mathbf{a}_i \lambda_i, \quad \lambda_i \geqslant 0 \right\}$$

is known as a *polyhedral cone* and is closed and convex (see Section 9.4). From Figure 9.2.2 it is clear that if $\mathbf{g} \notin C$ then there exists a hyperplane with normal vector $\mathbf{s}$ which *separates* C and $\mathbf{g}$, and for which $\mathbf{s}^T\mathbf{a}_i \geqslant 0$, $i = 1, 2, \ldots, m$, and $\mathbf{s}^T\mathbf{g} < 0$. Thus $\mathbf{s} \in S$ exists and the lemma is proved. For completeness, the general proof of the existence of a separating hyperplane is given below in Lemma 9.2.5. □

Corollary

The set

$$S = \{\mathbf{s} \mid \mathbf{s}^T\mathbf{g}^* < 0$$
$$\mathbf{s}^T\mathbf{a}_i^* = 0, \quad i \in E$$
$$\mathbf{s}^T\mathbf{a}_i^* \geqslant 0, \quad i \in I^*\}$$

is empty if and only if there exist multipliers λ_i^* *such that* (9.1.12) *and* (9.1.13) *hold.*

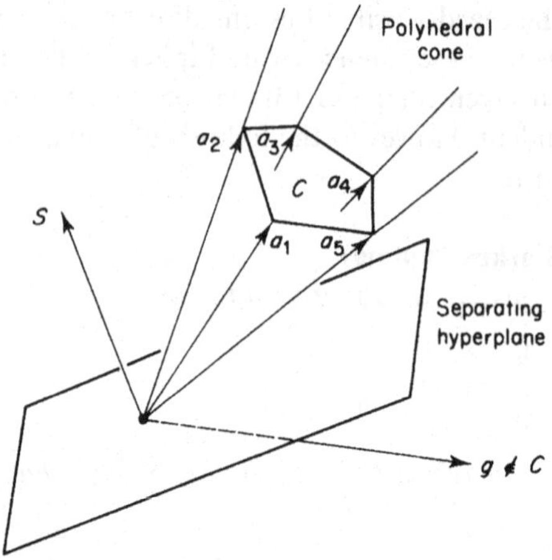

Figure 9.2.2 Existence of a separating hyperplane

Proof

Ignoring superscript *, then $\mathbf{s}^T\mathbf{a}_i = 0$, $i \in E$, can be written as $\mathbf{s}^T\mathbf{a}_i \geqslant 0$ and $-\mathbf{s}^T\mathbf{a}_i \geqslant 0$, $i \in E$. Then by Farkas' lemma, this is equivalent to the fact that there exist non-negative multipliers λ_i. $i \in I^*$, λ_i^+, λ_i^-, $i \in E$, such that

$$\mathbf{g} = \sum_{i \in I^*} \mathbf{a}_i \lambda_i + \sum_{i \in E} \mathbf{a}_i \lambda_i^+ + \sum_{i \in E} -\mathbf{a}_i \lambda_i^-.$$

But defining $\lambda_i = \lambda_i^+ - \lambda_i^-$, $i \in E$, gives (9.1.12) and (9.1.13) so the corollary is proved. □

Lemma 9.2.5 (Existence of a separating hyperplane)

There exists a hyperplane $\mathbf{s}^T\mathbf{x} = 0$ *which separates a closed convex cone C and a non-zero vector* $\mathbf{g} \notin C$.

Proof

By construction. Consider minimizing $\|\mathbf{g} - \mathbf{x}\|_2^2$ for all $\mathbf{x} \in C$, and let $\mathbf{x}_1 \in C$. Since the solution satisfies $\|\mathbf{g} - \mathbf{x}\|_2 \leqslant \|\mathbf{g} - \mathbf{x}_1\|_2$ it is bounded and so by continuity of $\|\cdot\|_2$ a minimizing point, $\hat{\mathbf{g}}$ say, exists. Because $\lambda\mathbf{g} \in C$ all $\lambda \geqslant 0$, and because $\|\lambda\hat{\mathbf{g}} - \mathbf{g}\|_2^2$ has a minimum at $\lambda = 1$, it follows by setting $\mathrm{d}/\mathrm{d}\lambda = 0$ that

$$\hat{\mathbf{g}}^T(\hat{\mathbf{g}} - \mathbf{g}) = 0. \tag{9.2.8}$$

Let $\mathbf{x} \in C$; then $\hat{\mathbf{g}} + \theta(\mathbf{x} - \hat{\mathbf{g}}) \in C$ for $\theta \in (0, 1)$ by convexity, and hence

$$\|\theta(\mathbf{x} - \hat{\mathbf{g}}) + \hat{\mathbf{g}} - \mathbf{g}\|_2^2 \geqslant \|\hat{\mathbf{g}} - \mathbf{g}\|_2^2.$$

Simplifying and taking the limit $\theta \downarrow 0$ it follows that

$$0 \leqslant (\mathbf{x} - \hat{\mathbf{g}})^T(\hat{\mathbf{g}} - \mathbf{g}) = \mathbf{x}^T(\hat{\mathbf{g}} - \mathbf{g})$$

from (9.2.8). Thus the vector $\mathbf{s}=\hat{\mathbf{g}}-\mathbf{g}$ is such that $\mathbf{s}^T\mathbf{x}\geqslant 0$ for all $\mathbf{x}\in C$. But $\mathbf{g}^T\mathbf{s}=-\mathbf{s}^T\mathbf{s}$ from (9.2.8), and $\mathbf{s}\neq\mathbf{0}$ since $\mathbf{g}\notin C$. Thus $\mathbf{g}^T\mathbf{s}<0$ and hence the hyperplane $\mathbf{s}^T\mathbf{x}=0$ separates C and $\mathbf{g}$. □

Geometrically the vector $\mathbf{s}$ is along the perpendicular from $\mathbf{g}$ to C. In Figure 9.2.2 $\hat{\mathbf{g}}$ would be a multiple of the vector $\mathbf{a}_5$ and the resulting hyperplane (not the one illustrated) would touch the cone along $\mathbf{a}_5$.

It is now possible to bring together the various aspects of this section in proving that the first order conditions (9.1.12) and (9.1.13) (or equivalently the conditions (9.1.16) of Theorem 9.1.1) are necessary at a local minimizing point $\mathbf{x}^*$. At $\mathbf{x}^*$ there are no feasible descent directions ($\mathscr{F}^*\cap\mathscr{D}^*=\emptyset$) by Lemma 9.2.3. A regularity assumption (9.2.4) is made that there are no linearized feasible descent directions. Then by the corollary to Farkas' lemma it follows that (9.1.12) and (9.1.13) hold. Thus these results have been established in a quite rigorous way.

9.3 SECOND ORDER CONDITIONS

A natural progression from the previous section is to examine the effect of second order (curvature) terms in the neighbourhood of a local solution. It can readily be seen for unconstrained optimization that the resulting sufficient condition that $\mathbf{G}^*$ is positive definite has significant implications for the design of satisfactory algorithms, and the same is true for constrained optimization. It is important to realize first of all that constraint curvature plays an important role, and that it is not possible to examine the curvature of $f(\mathbf{x})$ in isolation. For example realistic problems exist for which $\mathbf{G}^*$ is positive definite at a Kuhn–Tucker point $\mathbf{x}^*$, which is not, however, a local minimizer (see (9.3.6) below). As in Section 9.1, it is possible to present the essence of the situation in a fairly straightforward way, to be followed by a more general and more rigorous treatment later in the section. It is assumed that $f(\mathbf{x})$ and $c_i(\mathbf{x})$ for all i are $\mathbb{C}^2$ functions.

Suppose that there are only equality constraints present, and that a local solution $\mathbf{x}^*$ exists at which the vectors $\mathbf{a}_i^*$, $i\in E$, are independent, so that a unique vector of multipliers $\boldsymbol{\lambda}^*$ exists in (9.1.3). Under these conditions a feasible incremental step $\boldsymbol{\delta}$ can be taken along any feasible direction $\mathbf{s}$ and $\mathbf{x}^*$. By feasibility and (9.1.6) it follows that $f(\mathbf{x}^*+\boldsymbol{\delta})=\mathscr{L}(\mathbf{x}^*+\boldsymbol{\delta},\boldsymbol{\lambda})$. Also since $\mathscr{L}$ is stationary at $\mathbf{x}^*,\boldsymbol{\lambda}^*$ (equation (9.1.7)), a Taylor expansion of $\mathscr{L}(\mathbf{x},\boldsymbol{\lambda}^*)$ about $\mathbf{x}^*$ enables the second order terms to be isolated. Hence

$$\begin{aligned} f(\mathbf{x}^*+\boldsymbol{\delta}) &= \mathscr{L}(\mathbf{x}^*+\boldsymbol{\delta},\boldsymbol{\lambda}^*) \\ &= \mathscr{L}(\mathbf{x}^*,\boldsymbol{\lambda}^*)+\boldsymbol{\delta}^T\nabla_x\mathscr{L}(\mathbf{x}^*,\boldsymbol{\lambda}^*)+\tfrac{1}{2}\boldsymbol{\delta}^T\mathbf{W}^*\boldsymbol{\delta}+o(\boldsymbol{\delta}^T\boldsymbol{\delta}) \\ &= f^*+\tfrac{1}{2}\boldsymbol{\delta}^T\mathbf{W}^*\boldsymbol{\delta}+o(\boldsymbol{\delta}^T\boldsymbol{\delta}) \end{aligned} \tag{9.3.1}$$

where $\mathbf{W}^*=\nabla_x^2\mathscr{L}(\mathbf{x}^*,\boldsymbol{\lambda}^*)=\nabla^2 f(\mathbf{x}^*)-\sum_i\lambda_i^*\nabla^2 c_i(\mathbf{x}^*)$ denotes the Hessian matrix with respect to **x** of the Lagrangian function. It follows by the minimality of f^*, and taking the limit in (9.3.1), that

$$\mathbf{s}^T\mathbf{W}^*\mathbf{s} \geqslant 0. \tag{9.3.2}$$

As in Section 9.1 a feasible direction satisfies

$$\mathbf{a}_i^{*T}\mathbf{s} = 0, \qquad i \in E \tag{9.3.3}$$

which can be written in matrix notation as

$$\mathbf{A}^{*T}\mathbf{s} = \mathbf{0}. \tag{9.3.4}$$

Thus a *second order necessary condition for a local minimizer* is that (9.3.2) must hold for any **s** which satisfies (9.3.4). That is to say, the Lagrangian function must have non-negative curvature for all feasible directions at $\mathbf{x}^*$. Of course, when no constraints are present then (9.3.2) reduces to the usual condition that $\mathbf{G}^*$ is positive semi-definite.

As for unconstrained optimization in Section 2.1, it is also possible to state very similar conditions which are sufficient. A *sufficient condition for a strict local minimizer* is that if (9.1.3) holds at any feasible point $\mathbf{x}^*$ and if

$$\mathbf{s}^T\mathbf{W}^*\mathbf{s} > 0 \tag{9.3.5}$$

for all **s** ($\neq \mathbf{0}$) in (9.3.4), then $\mathbf{x}^*$ is a strict local minimizer. The proof of this makes use of the fact that (9.3.5) implies the existence of a constant $a > 0$ for which $\mathbf{s}^T\mathbf{W}^*\mathbf{s} \geqslant a\mathbf{s}^T\mathbf{s}$ for all **s** ($\neq \mathbf{0}$) in (9.3.4). Then for any feasible step $\boldsymbol{\delta}$, (9.3.1) holds, and if $\boldsymbol{\delta}$ is sufficiently small it follows that $f(\mathbf{x}^* + \boldsymbol{\delta}) > f^*$. No regularity assumptions are made in this proof.

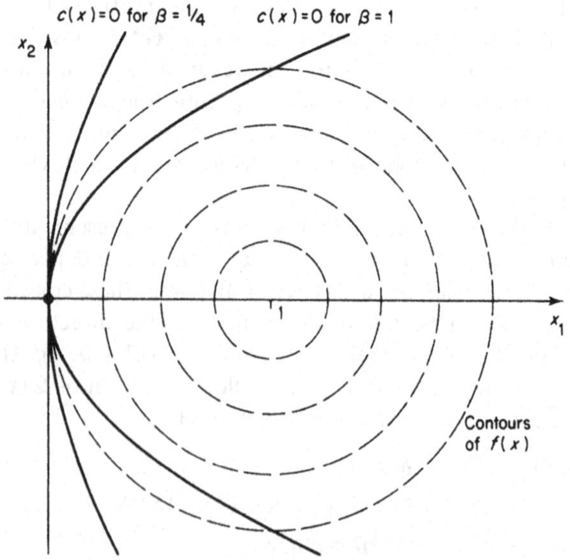

Figure 9.3.1 Second order conditions

A simple but effective illustration of these conditions is given by Fiacco and McCormick (1968). Consider the problem

$$\begin{aligned} &\text{minimize} \quad f(\mathbf{x}) \triangleq \tfrac{1}{2}((x_1 - 1)^2 + x_2^2) \\ &\text{subject to} \quad c(\mathbf{x}) \triangleq -x_1 + \beta x_2^2 = 0 \end{aligned} \tag{9.3.6}$$

where β is fixed, and examine for what values of β is $\mathbf{x}^* = \mathbf{0}$ a local minimizer. The cases $\beta = \frac{1}{4}$ ($\mathbf{x}^*$ is a local minimizer) and $\beta = 1$ (not a minimizer) are illustrated in Figure 9.3.1. At $\mathbf{x}^*$, $\mathbf{g}^* = \mathbf{a}^* = (-1, 0)^T$ so the first order conditions (9.1.3) are satisfied with $\lambda^* = 1$, and $\mathbf{x}^*$ is also feasible. The set of feasible directions in (9.3.4) is $\mathbf{s} = (0, s_2)^T$ for all $s_2 \neq 0$. Now $\mathbf{W}^* = \begin{bmatrix} 1 & 0 \\ 0 & 1 - 2\beta \end{bmatrix}$ so $\mathbf{s}^T\mathbf{W}^*\mathbf{s} = s_2^2(1 - 2\beta)$. Thus the second order necessary conditions are violated when $\beta > \frac{1}{2}$, in which case it can be concluded that $\mathbf{x}^*$ is not a local minimizer. When $\beta < \frac{1}{2}$ then the sufficient conditions (9.3.5) and (9.3.4) are satisfied and it follows that $\mathbf{x}^*$ is a local minimizer. Only when $\beta = \frac{1}{2}$ is the result not determined by the second order conditions. This corresponds to zero curvature of $\mathscr{L}$ existing along a feasible direction, so that higher order terms become significant.

An important generalization of these conditions is to allow inequality constraints to be present. Now second order conditions are only operative along feasible stationary directions ($\mathbf{s}^T\mathbf{g}^* = 0$) and not along ascent directions. If an inequality constraint $c_i(\mathbf{x}) \geqslant 0$ is present, and if its multiplier is $\lambda_i^* > 0$, then feasible directions for which $\mathbf{s}^T\mathbf{a}_i^* > 0$ are ascent directions (see the discussion regarding equation (9.1.14)). Thus usually the stationary directions will satisfy

$$\mathbf{s}^T\mathbf{a}_i^* = 0, \qquad i \in \mathscr{A}^* \tag{9.3.7}$$

and second order necessary conditions are that (9.3.2) holds for all $\mathbf{s}$ in (9.3.7). Another way of looking at this is that if $\mathbf{x}^*$ solves (7.1.1) locally, then it must solve

$$\begin{aligned} &\text{minimize} \quad f(\mathbf{x}) \\ &\text{subject to} \quad c_i(\mathbf{x}) = 0, \qquad i \in \mathscr{A}^* \end{aligned} \tag{9.3.8}$$

locally, and these conditions follow from (9.3.2) and (9.3.3). For sufficient conditions, if $\mathbf{x}^*$ is feasible, if (9.1.12) and (9.1.13) hold, and if $\lambda_i^* > 0 \, \forall i \in I^*$ (KT conditions with strict complementarity), then a sufficient condition for a strict local minimizer is

$$\mathbf{s}^T\mathbf{W}^*\mathbf{s} > 0 \qquad \forall \mathbf{s}: \mathbf{s}^T\mathbf{a}_i^* = 0 \quad i \in \mathscr{A}^*.$$

Alternatively positive curvature can be assumed on a larger subspace in which the conditions corresponding to $\lambda_i^* = 0$ are excluded. These results are justified below. An illustration of these conditions is also given by problem (9.3.6) if the constraint is changed to read $c(\mathbf{x}) \geqslant 0$. A feasible direction $\mathbf{s}$ can then have any $s_1 \leqslant 0$. However, because $\lambda^* = 1 > 0$, these directions are uphill unless $s_1 = 0$ and hence stationary feasible directions are given by $\mathbf{s} = (0, s_2)^T$ as in the equality constraint case. Thus the same conclusions about β can be deduced.

It is not difficult to make the further generalization which includes the possibility that there exists a $\lambda_i^* = 0$, $i \in I^*$, in which case a stationary direction exists for which $\mathbf{s}^T\mathbf{a}_i^* > 0$. It is also possible to allow for non-unique λ_i^* in the necessary conditions. A rigorous derivation of the second order conditions is now set out which includes these features. Given any fixed vector $\boldsymbol{\lambda}^*$, it is possible to define the set of *strictly* (or *strongly*) *active constraints*

$$\mathscr{A}_+^* = \{i \mid i \in E \quad \text{or} \quad \lambda_i^* > 0\} \tag{9.3.9}$$

which is obtained by deleting indices for which $\lambda_i^* = 0$, $i \in I^*$ from $\mathscr{A}^*$. Consider all feasible directional sequences $\mathbf{x}^{(k)} \to \mathbf{x}^*$ for which

$$c_i(\mathbf{x}^{(k)}) = 0 \qquad \forall i \in \mathscr{A}_+^* \tag{9.3.10}$$

also holds. Define $\mathscr{G}^*$ as the resulting set of feasible directions. As in Section 9.2, consider also the set of feasible directions

$$\begin{aligned} G^* = \{\mathbf{s} \mid \mathbf{s} \neq \mathbf{0}, \quad & \mathbf{a}_i^{*T}\mathbf{s} = 0, \quad i \in \mathscr{A}_+^* \\ & \mathbf{a}_i^{*T}\mathbf{s} \geqslant 0, \quad i \in \mathscr{A}^* \backslash \mathscr{A}_+^*\} \end{aligned} \tag{9.3.11}$$

in which the constraints which determine $\mathscr{G}^*$ (including (9.3.10)) are linearized. By an identical argument to Lemma 9.2.1 it follows that $G^* \supseteq \mathscr{G}^*$. However, to state the second order necessary conditions a result going the other way is required, so another regularity assumption is made, namely that

$$G^* = \mathscr{G}^*. \tag{9.3.12}$$

Again this is a reasonable assumption to make as it is also implied by the mild conditions of Lemma 9.2.2 in a similar way.

It is now possible to state the main results of this section in their full generality.

Theorem 9.3.1 (Second order necessary conditions)

If $\mathbf{x}^*$ *is a local solution* (7.1.1), *and if* (9.2.4) *holds, then there exist multipliers* $\boldsymbol{\lambda}^*$ *such that Theorem* 9.1.1 *is valid. For any such* $\boldsymbol{\lambda}^*$, *if* (9.3.12) *holds, it follows that*

$$\mathbf{s}^T\mathbf{W}^*\mathbf{s} \geqslant 0 \qquad \forall \mathbf{s} \in G^*. \tag{9.3.13}$$

Proof

Let $\mathbf{s} \in G^*$. Then by (9.3.12), $\mathbf{s} \in \mathscr{G}^*$, and $\exists$ a feasible directional sequence with $\mathbf{s}^{(k)} \to \mathbf{s}$, for which (9.3.10) holds. Since either $c_i^{(k)} = 0$ for $i \in \mathscr{A}_+^*$, or $\lambda_i^* = 0$ otherwise, it follows that $f^{(k)} = \mathscr{L}(\mathbf{x}^{(k)}, \boldsymbol{\lambda}^*)$. Using (9.2.1) and (9.1.16), a Taylor series for $\mathscr{L}(\mathbf{x}, \boldsymbol{\lambda}^*)$ about $\mathbf{x}^*$ gives

$$\begin{aligned} \mathscr{L}(\mathbf{x}^{(k)}, \boldsymbol{\lambda}^*) &= \mathscr{L}(\mathbf{x}^*, \boldsymbol{\lambda}^*) + \delta^{(k)}\mathbf{s}^{(k)T}\nabla_x\mathscr{L}(\mathbf{x}^*, \boldsymbol{\lambda}^*) \\ &\quad + \tfrac{1}{2}\delta^{(k)2}\mathbf{s}^{(k)T}\mathbf{W}^*\mathbf{s}^{(k)} + o(\delta^{(k)2}) \\ &= f^* + \tfrac{1}{2}\delta^{(k)2}\mathbf{s}^{(k)T}\mathbf{W}^*\mathbf{s}^{(k)} + o(\delta^{(k)2}). \end{aligned} \tag{9.3.14}$$

Since $\mathbf{x}^*$ is a local minimizer, it follows for all k sufficiently large that $f^{(k)} \geqslant f^*$, and hence that

$$\mathbf{s}^{(k)\mathrm{T}}\mathbf{W}^*\mathbf{s}^{(k)} + o(1) \geqslant 0.$$

Then (9.3.13) follows in the limit. □

Theorem 9.3.2 (Second order sufficient conditions)

If at $\mathbf{x}^*$ *there exists multipliers* $\boldsymbol{\lambda}^*$ *such that conditions* (9.1.16) *hold, and if*

$$\mathbf{s}^{\mathrm{T}}\mathbf{W}^*\mathbf{s} > 0 \qquad \forall \mathbf{s} \in G^*, \tag{9.3.15}$$

then $\mathbf{x}^*$ *is a strict local solution to* (7.1.1).

Proof

Assume $\mathbf{x}^*$ is not a strict local minimizer, so that $\exists$ a feasible sequence $\mathbf{x}^{(k)} \to \mathbf{x}^*$ such that $f^{(k)} \leqslant f^*$. Fixing $\|\mathbf{s}^{(k)}\| = 1$, say, in (9.2.1) then this bound implies that $\exists$ a subsequence such that $\mathbf{s}^{(k)}$ converges to $\mathbf{s}$, say. By Lemma 9.2.1, $\mathbf{s} \in F^*$, and by a similar argument to that in Lemma 9.2.3, $\mathbf{s}^{\mathrm{T}}\mathbf{g}^* \leqslant 0$. Two cases occur, both of which imply a contradiction:

(i) $\mathbf{s} \notin G^*$; then $\exists i : \lambda_i^* > 0$ and $\mathbf{a}_i^{*\mathrm{T}}\mathbf{s} > 0$ whence $0 \geqslant \mathbf{s}^{\mathrm{T}}\mathbf{g}^* = \sum \mathbf{s}^{\mathrm{T}}\mathbf{a}_i^*\lambda_i^* > 0$.

(ii) $\mathbf{s} \in G^*$; by feasibility of $\mathbf{x}^{(k)}$, $\mathscr{L}(\mathbf{x}^{(k)}, \boldsymbol{\lambda}^*) \leqslant f^{(k)}$, so from (9.3.14) it follows that $0 \geqslant f^{(k)} - f^* = \frac{1}{2}\delta^{(k)2}\mathbf{s}^{(k)\mathrm{T}}\mathbf{W}^*\mathbf{s}^{(k)} + o(\delta^{(k)2})$, and dividing by $\delta^{(k)2}$ and taking the limit contradicts (9.3.15). □

Notice that a sufficient condition for (9.3.15) is that $\mathbf{s}^{\mathrm{T}}\mathbf{W}^*\mathbf{s} > 0$ $\forall \mathbf{s} \neq \mathbf{0}$ such that $\mathbf{s}^{\mathrm{T}}\mathbf{a}_i^* = 0$, $i \in \mathscr{A}_+^*$, which is more convenient to verify in practice. Also in the statement of Theorem 9.3.2 it is instructive to observe that *strict* (i.e. $f(\mathbf{x}) > f^*$ for all feasible $\mathbf{x}$ in the neighbourhood of $\mathbf{x}^*$) is a weaker qualification than *isolated* ($\mathbf{x}^*$ is the only local solution in a neighbourhood of $\mathbf{x}^*$), in contrast to Theorem 2.1.1 for unconstrained optimization. Robinson (1982) gives an example which differentiates between the two cases, that is

$$\text{minimize } x^2 \quad \text{subject to} \quad x^6 \sin(1/x) = 0 \qquad (\sin 1/0 \triangleq 0).$$

Here the feasible region is a set of distinct points 0 and $\pm(i\pi)^{-1}$, $i = 1, 2, \ldots$ which cluster at the origin, each of which is a local solution. Clearly the example is pathological: however, if the constraint qualification $F^* = \mathscr{F}^*$ is assumed to hold then strict and isolated local minimizers are equivalent.

This treatment of second order conditions owes a lot to the presentation given by Fiacco and McCormick (1968). However, it is worth pointing out that the statement of the second order necessary conditions given here is an improvement. Fiacco and McCormick define feasible directions from arcs which satisfy the conditions $c_i(\mathbf{x}) = 0$, $i \in \mathscr{A}^*$, rather than using $\mathscr{A}_+^*$ as in (9.3.10). Although this has the advantage of not involving $\boldsymbol{\lambda}^*$ and hence f, it neglects the stronger implications which can be made when $\lambda_i^* = 0$ in regular situations.

Furthermore in a degenerate case it is extremely unlikely that any feasible arc will exist at all, so that the regularity assumption which they make (the second order constraint qualification) cannot usually hold. Both these objections are overcome when $\mathscr{A}_+^*$ is used. The situation is shown by the example.

$$\begin{aligned} \text{minimize} \quad & f(\mathbf{x}) \triangleq x_3 - \tfrac{1}{2}x_1^2 \\ \text{subject to} \quad & c_1(\mathbf{x}) \triangleq x_3 + x_2 + x_1^2 \geqslant 0 \\ & c_2(\mathbf{x}) \triangleq x_3 - x_2 + x_1^2 \geqslant 0 \\ & c_3(\mathbf{x}) \triangleq x_3 \geqslant 0 \end{aligned} \qquad (9.3.16)$$

illustrated in Figure 9.3.2. Clearly $\mathbf{x}^* = \mathbf{0}$ is not a minimizing point. Yet $F^* = \mathscr{F}^*$ and $\mathbf{g}^* = (0,0,1)^T, \mathbf{a}_1^* = (0,1,1)^T, \mathbf{a}_2^* = (0,-1,1)^T, \mathbf{a}_3^* = (0,0,1)^T$. All constraints are active and the first order conditions at $\mathbf{x}^*$ are satisfied non-uniquely by any vector $\boldsymbol{\lambda}^*$ which is a convex combination of $(\frac{1}{2},\frac{1}{2},0)^T$ and $(0,0,1)^T$. Since, however, there is no point other than $\mathbf{x}^*$ which satisfies $c_i(\mathbf{x}) = 0$ all $i \in \mathscr{A}^*$, Fiacco and McCormick's second order constraint qualification does not hold, and no implication can be made. However consider Theorem 9.3.1 using the extreme multiplier vectors. In both cases the set G^* comprises any vector $\mathbf{s}$ with $s_1 \neq 0$ and $s_2 = s_3 = 0$. For $\boldsymbol{\lambda}^* = (\frac{1}{2},\frac{1}{2},0)^T$ there is an arc (OC in Figure 9.3.2) which satisfies $c_i(\mathbf{x}) = 0$ for all $i \in \mathscr{A}_+^*$, but which is not feasible in $c_i(\mathbf{x}) \geqslant 0$, $i \in \mathscr{A}^* \backslash \mathscr{A}_+^*$. Hence (9.3.12) does not hold for this arc and so no implication can be made. However, for $\boldsymbol{\lambda}^* = (0,0,1)^T$, either of the arcs OA or OB in Figure 9.3.2, or any intermediate arc with $x_3 = 0$, provides a suitable arc when $s_1 > 0$ and an opposite arc can be used when $s_1 < 0$. Thén since $W_{11}^* = -1$, (9.3.13) does not hold, so it can be concluded correctly that $\mathbf{x}^*$ is not a local solution to problem (9.3.16).

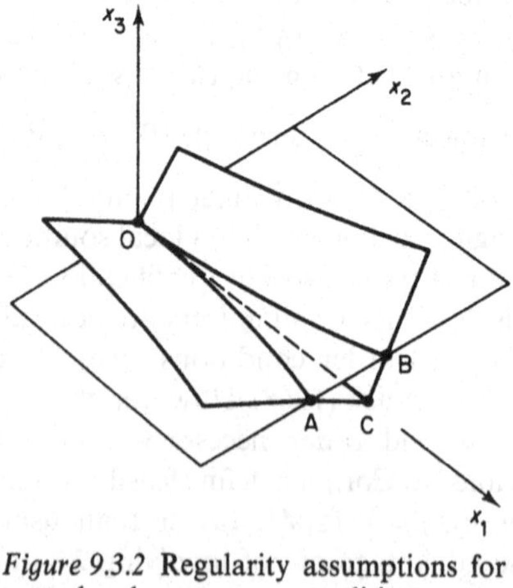

Figure 9.3.2 Regularity assumptions for second order necessary conditions

9.4 CONVEXITY

The subject of convexity is often treated quite extensively in texts on optimization. My experience, however, is that much of this theory contributes little to the development and use of optimization algorithms. Applications of convexity are expressed in terms of a so-called convex programming problem. Into this category come linear programming, certain quadratic programming problems, and some more general problems, more often with linear constraints. Unfortunately many real life problems do not fit into this category, and this is especially so when the constraint functions are nonlinear. On the other hand, it is possible to give quite strong (and simple) results for a convex programming problem about the global nature of solutions and the sufficiency of first order conditions. Therefore a fairly simple treatment of convexity is given in this section, aimed mainly at establishing these results for smooth problems. Some extensions of convexity theory which are helpful for handling non-smooth problems are given in Section 14.2.

First of all, a *convex set* K in $\mathbb{R}^n$ is defined by the property that for all $\mathbf{x}_0, \mathbf{x}_1 \in K$, it follows that $\mathbf{x}_\theta \in K$ where

$$\mathbf{x}_\theta = (1-\theta)\mathbf{x}_0 + \theta\mathbf{x}_1 \qquad \forall\theta\in[0,1]. \tag{9.4.1}$$

It follows from this that K can have no re-entrant corners (see Figure 9.4.1). A more general definition of a convex set which readily follows is that for all $\mathbf{x}_0, \mathbf{x}_1, \ldots, \mathbf{x}_m \in K$ it follows that $\mathbf{x}_\theta \in K$ where

$$\mathbf{x}_\theta = \sum_{i=0}^{m} \theta_i \mathbf{x}_i, \qquad \sum_{i=0}^{m} \theta_i = 1, \qquad \theta_i \geqslant 0. \tag{9.4.2}$$

The vector $\mathbf{x}_\theta$ in (9.4.1) or (9.4.2) is referred to as a *convex combination* of the points $\mathbf{x}_0, \mathbf{x}_1$, etc. If $\mathbf{x}_0, \mathbf{x}_1, \ldots, \mathbf{x}_m$ is a given set of points, then the set of all vectors $\mathbf{x}_\theta$ defined by (9.4.2) is a convex set referred to as the *convex hull* of the set of points. Examples of convex sets are many and include the empty set, a point, the whole of $\mathbb{R}^n$, a line or line segment, a hyperplane (or linear equation) $\mathbf{a}^T\mathbf{x} = b$, the half-space (or linear inequality) $\mathbf{a}^T\mathbf{x} \geqslant b$, the ball $\|\mathbf{x} - \mathbf{x}'\|_2 \leqslant h$, a convex cone, and many others. A simple result is the following.

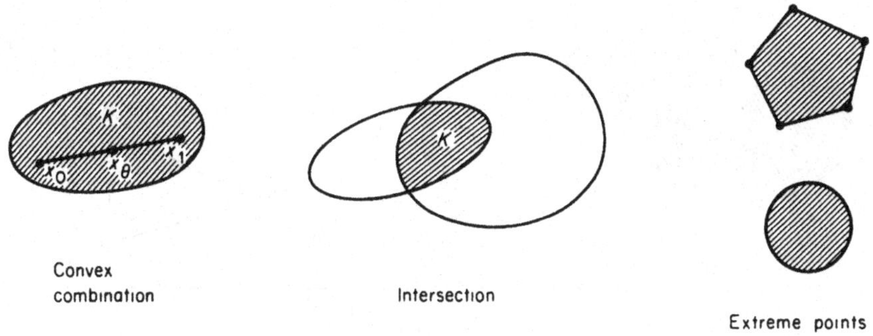

Figure 9.4.1 Convex sets

Lemma 9.4.1

If $K_i, i = 1, 2, \ldots, m$, are convex sets then the intersection $K = \bigcap_i^m K_i$ is also a convex set.

Proof

Because $\mathbf{x}_0, \mathbf{x}_1 \in K$ implies $\mathbf{x}_0, \mathbf{x}_1 \in K_i$, $i = 1, 2, \ldots, m$. □

This result is also illustrated in Figure 9.4.1. It shows that the feasible region in a linear or quadratic programming problem is a convex set because it is the intersection of hyperplanes and half-spaces.

Another useful concept is that of an *extreme point* of a convex set K. An extreme point $\mathbf{x}$ is one which may not lie interior to any line segment contained in K, that is $\mathbf{x} = (1-\theta)\mathbf{x}_0 + \theta\mathbf{x}_1$ for $\mathbf{x}_0, \mathbf{x}_1 \in K$, $\theta \in (0, 1)$ implies that $\mathbf{x} = \mathbf{x}_0 = \mathbf{x}_1$. The vertices of a regular polygon or any point on the circumference of a circle are examples of extreme points (Figure 9.4.1). Another example is the basic feasible solution $\mathbf{x}$ of the convex set K defined by the feasible region R of a linear programming problem in standard form (8.1.1). Details of the relationship between an extreme point and a b.f.s. are sketched out in Questions 9.20, 9.21, and 9.22.

The other fundamental idea is that of a *convex function*. The discussion is limited to continuous functions defined on a convex set K, to eliminate trivial cases. Essentially a convex function $f(\mathbf{x})$ is one for which the epigraph is a convex set. (The *epigraph* is the set of points in $\mathbb{R} \times \mathbb{R}^n$ that lies on or above the graph of $f(\mathbf{x})$: see Figure 9.4.2.) Thus a convex function $f(\mathbf{x})$ is defined by the condition that for any $\mathbf{x}_0, \mathbf{x}_1 \in K$ it follows that

$$f_\theta \leqslant (1-\theta)f_0 + \theta f_1 \qquad \forall \theta \in [0, 1] \tag{9.4.3}$$

where f_θ refers to $f(\mathbf{x}_\theta)$, etc., and where $\mathbf{x}_\theta$ is given by (9.4.1). The right-hand side of (9.4.3) is the chord joining $(\mathbf{x}_0, f_0)$ to $(\mathbf{x}_1, f_1)$ on the graph of $f(\mathbf{x})$ (see Figure 9.4.2), and the inequality expresses the fact that the graph of a convex function always lies below (or along) the chord. If K is an open set and $f(\mathbf{x})$ is differentiable ($\mathbb{C}^1$) on K, an equivalent definition of convexity is that for all

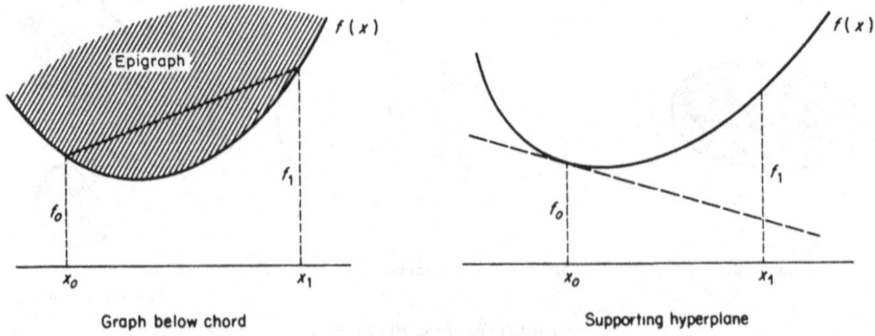

Figure 9.4.2 Convex functions

$\mathbf{x}_0, \mathbf{x}_1 \in K$ it follows that

$$f_1 \geqslant f_0 + (\mathbf{x}_1 - \mathbf{x}_0)^T \nabla f_0. \tag{9.4.4}$$

This definition shows that the graph of $f(\mathbf{x})$ must lie above (or along) the linearization of $f(\mathbf{x})$ about $\mathbf{x}_0$, and hence that this linearization acts as a *supporting hyperplane* for the convex function. The equivalence of (9.4.3) and (9.4.4) is readily demonstrated. It follows from (9.4.3) that

$$\frac{f_\theta - f_0}{\theta} \leqslant f_1 - f_0.$$

But regarding $\mathbf{x}_\theta$ as a point on the line $\mathbf{x}_\theta = \mathbf{x}_0 + \theta(\mathbf{x}_1 - \mathbf{x}_0)$, and taking the limit $\theta \downarrow 0$, then (9.4.4) follows. Conversely if (9.4.4) holds then expanding about $\mathbf{x}_\theta$,

$$f_1 \geqslant f_\theta + (\mathbf{x}_1 - \mathbf{x}_\theta)^T \nabla f_\theta$$
$$f_0 \geqslant f_\theta + (\mathbf{x}_0 - \mathbf{x}_\theta)^T \nabla f_\theta$$

so that

$$(1-\theta) f_0 + \theta f_1 \geqslant f_\theta + ((1-\theta)(\mathbf{x}_0 - \mathbf{x}_\theta) + \theta(\mathbf{x}_1 - \mathbf{x}_\theta))^T \nabla f_\theta = f_\theta$$

which is (9.4.3).

Another result which follows from (9.4.4) is that

$$(\mathbf{x}_1 - \mathbf{x}_0)^T \nabla f_1 \geqslant f_1 - f_0 \geqslant (\mathbf{x}_1 - \mathbf{x}_0)^T \nabla f_0. \tag{9.4.5}$$

This illustrates the fact that the slope of a convex function is non-decreasing along any line. In fact this result (for the directional derivative) can also be proved to hold for a non-differentiable convex function. Finally for twice differentiable ($\mathbb{C}^2$) convex functions and K open, another equivalent definition of a convex function is that

$$\nabla^2 f_0 \text{ is positive semi-definite} \qquad \forall \mathbf{x}_0 \in K. \tag{9.4.6}$$

Thus convex functions are typified by having non-negative curvature. To establish this result, let $\mathbf{s} \neq \mathbf{0}$ and let $\mathbf{x}_1 = \mathbf{x}_0 + \alpha\mathbf{s}$. Then a Taylor series for ∇f_1 gives

$$\nabla f_1 = \nabla f_0 + \alpha \nabla^2 f_0 \mathbf{s} + o(\alpha). \tag{9.4.7}$$

Substituting in (9.4.5) and taking the limit $\alpha \to 0$ gives $\mathbf{s}^T \nabla^2 f_0 \mathbf{s} \geqslant 0$, which is (9.4.6). Conversely, a Taylor series for f_1 (with $\theta \in [0, 1]$) and (9.4.6) yields

$$f_1 = f_0 + (\mathbf{x}_1 - \mathbf{x}_0)^T \nabla f_0 + \tfrac{1}{2}\alpha^2 \mathbf{s}^T \nabla^2 f_\theta \mathbf{s}$$
$$\geqslant f_0 + (\mathbf{x}_1 - \mathbf{x}_0)^T \nabla f_0,$$

which is (9.4.4).

Other definitions which are closely related to that of a convex function are the following. A *strictly convex function* is defined whenever the inequality in (9.4.3) is strict for all distinct $\mathbf{x}_0, \mathbf{x}_1$ and $\theta \in (0, 1)$. For $\mathbb{C}^1$ functions (9.4.4) again provides an equivalent definition when the inequality is strict and $\mathbf{x}_0 \neq \mathbf{x}_1$.

However, for $\mathbb{C}^2$ functions, (9.4.6) is not equivalent, since although $\nabla^2 f_0$ positive definite $\forall \mathbf{x}_0 \in K$ implies that $f(\mathbf{x})$ is strictly convex, the converse result does not hold (for example x^4 is strictly convex but has zero Hessian at $x = 0$). A *concave function* $f(\mathbf{x})$ is defined as one for which $-f(\mathbf{x})$ is convex, and so is associated with non-increasing slope or non-positive curvature. Likewise a *strictly concave function* $f(\mathbf{x})$ has $-f(\mathbf{x})$ strictly convex.

Examples of convex functions include the linear function, which is both convex and concave. A quadratic function is convex when the Hessian is positive semi-definite and strictly convex when the Hessian is positive definite. Another convex function is $\|\mathbf{x}\|$ (for any norm). However $\|\mathbf{c}(\mathbf{x})\|$, where $\mathbf{c}(\mathbf{x})$ maps $\mathbb{R}^n$ into $\mathbb{R}^m$, is not generally convex, except when $\mathbf{c}(\mathbf{x})$ is a linear function. A transformation which preserves convexity is expressed in the next result.

Lemma 9.4.2

If $f_i(\mathbf{x})$, $i = 1, 2, \ldots, m$, are convex functions on a convex set K, and if $\lambda_i \geqslant 0$, then $\sum_i \lambda_i f_i(\mathbf{x})$ is a convex function on K.

Proof

Take $\mathbf{x}_\theta$ as in (9.4.1) and use the definition of a convex function. □

The problem of minimizing a convex function on a convex set K is said to be a *convex programming problem.* Such a problem arises when (7.1.1) can be expressed as

$$\begin{aligned} &\text{minimize} \quad f(\mathbf{x}) \\ &\text{subject to} \quad \mathbf{x} \in K \triangleq \{\mathbf{x} \mid c_i(\mathbf{x}) \geqslant 0, \qquad i = 1, 2, \ldots, m\} \end{aligned} \tag{9.4.8}$$

where $f(\mathbf{x})$ is a convex function on K, and the functions $c_i(\mathbf{x})$, $i = 1, 2, \ldots, m$, are concave on $\mathbb{R}^n$. That the feasible region in (9.4.8) is a convex set is a consequence of the following lemma and of Lemma 9.4.1.

Lemma 9.4.3

If $c(\mathbf{x})$ is a concave function then the set

$$S(k) = \{\mathbf{x} \mid c(\mathbf{x}) \geqslant k\}$$

is convex.

Proof

For $\mathbf{x}_0, \mathbf{x}_1 \in S(k)$, and if $\mathbf{x}_\theta$ is given by (9.4.1), it follows by concavity that

$$c_\theta \geqslant (1 - \theta)c_0 + \theta c_1 \geqslant (1 - \theta)k + \theta k$$

by definition of $S(k)$. Thus $\mathbf{x}_\theta \in S(k)$ which is therefore convex. □

Notice that the system (9.4.8) does not allow general equality constraints, although it is possible to include any linear equality $c(\mathbf{x}) = 0$ as the intersection of $c(\mathbf{x}) \geqslant 0$ and $-c(\mathbf{x}) \geqslant 0$. An example of a convex programming problem is therefore the linear programming problem (linear objective function, linear equality and inequality constraints). Another example is the quadratic programming problem (quadratic objective function, linear constraints) when the Hessian of the quadratic function is positive semi-definite. However it should be noticed that quadratic programming problems can (and do) exist which have well-behaved local (or even unique global) solutions, yet for which the Hessian is indefinite. It is then erroneous to assume that some of the consequences of convexity (see next section) will apply in this case.

One main attraction of convexity is that it provides an overall assumption whereby the existence of local but not global solutions can be excluded, as the next theorem shows. An additional assumption, given in the corollary, enables uniqueness of global solutions to be established.

Theorem 9.4.1

Every local solution $\mathbf{x}^*$ to a convex programming problem (9.4.8) *is a global solution, and the set of global solutions S is convex.*

Proof

Let $\mathbf{x}^*$ be a local but not global solution. Then $\exists \mathbf{x}_1 \in K$ such that $f_1 < f^*$. For $\theta \in [0, 1]$, consider $\mathbf{x}_\theta = (1-\theta)\mathbf{x}^* + \theta \mathbf{x}_1 \in K$ by convexity of K. By convexity of f, $f_\theta \leqslant (1-\theta) f^* + \theta f_1 = f^* + \theta(f_1 - f^*) < f^*$. In the limit $\theta \to 0$ the local solution property is contradicted. Thus local solutions are global. Now let $\mathbf{x}_0, \mathbf{x}_1 \in S$ and define $\mathbf{x}_\theta$ by (9.4.1). By the global solution property, $f_\theta \geqslant f_0 = f_1$. By convexity $f_\theta \leqslant (1-\theta) f_0 + \theta f_1 = f_0 = f_1$. Thus $f_\theta = f_0 = f_1$ and so $\mathbf{x}_\theta \in S$, so S is convex. □

Corollary

If also $f(\mathbf{x})$ is strictly convex on K then any global solution is unique.

Proof

Let $\mathbf{x}_0 \neq \mathbf{x}_1 \in S$ and $\theta \in (0, 1)$. As above, but using strict convexity, both $f_\theta \geqslant f_0 = f_1$ and $f_\theta < f_0 = f_1$, which is a contradiction. □

A second attraction of a convexity assumption is that it provides a framework within which the first order (Kuhn–Tucker) conditions are sufficient for a global solution, as the next theorem shows. In common with other sufficient conditions (Theorem 9.3.2), no regularity assumption is required.

Theorem 9.4.2

In the convex programming problem (9.4.8), *if* $f(\mathbf{x})$ *and* $c_i(\mathbf{x})$, $i=1,2,\dots,m$, *are* $\mathbb{C}^1$ *functions on* K *and if conditions* (9.1.16) *hold at* $\mathbf{x}^*$, *then* $\mathbf{x}^*$ *is a global solution to* (9.4.8).

Proof

Let $\mathbf{x}' \in K$, $\mathbf{x}' \neq \mathbf{x}^*$. Then since $\lambda_i^* \geqslant 0$ and $c_i' \geqslant 0$,

$$f' \geqslant f' - \sum_{i=1}^{m} \lambda_i^* c_i'$$

$$\geqslant f^* + (\mathbf{x}' - \mathbf{x}^*)^{\mathrm{T}}\mathbf{g}^* - \sum \lambda_i^*(c_i^* + (\mathbf{x}' - \mathbf{x}^*)^{\mathrm{T}}\mathbf{a}_i^*),$$

using (9.4.4), since f is convex and the c_i are concave. Then from (9.1.16), $\lambda_i^* c_i^* = 0$ and $\mathbf{g}^* = \sum \mathbf{a}_i^* \lambda_i^*$ show that $f' \geqslant f^*$ and hence $\mathbf{x}^*$ is a global solution. □

To summarize, for $\mathbf{x}^*$ to solve the convex programming problem (9.4.8), conditions (9.1.16) are sufficient, and if the regularity assumption (9.2.4) holds, then conditions (9.1.16) are also necessary. Of course (9.2.4) is implied by the constraint qualification $F^* = \mathscr{F}^*$ which in turn is implied by the assumptions of Lemma 9.2.2, as before. It should be emphasized that it is not possible to dispense with the regularity assumption (9.2.4). An example of a convex feasible region in which $F^* \neq \mathscr{F}^*$ (in both $\mathbb{R}^2$ and $\mathbb{R}^3$) is given by the inequalities $x_2 \geqslant x_1^2$ and $x_2 \leqslant 0$ at $\mathbf{x}^* = \mathbf{0}$. An illustration of a (regular) convex programming problem is provided by problem (9.1.15). The objective function is linear and hence convex. The Hessian matrices $\nabla^2 c_1 = \begin{bmatrix} -2 & 0 \\ 0 & 0 \end{bmatrix}$ and $\nabla^2 c_2 = \begin{bmatrix} -2 & 0 \\ 0 & -2 \end{bmatrix}$ are negative semi-definite so the constraint functions are concave. It can be seen from Figure 7.1.3 that the feasible region is a convex set. Since first order conditions hold at $\mathbf{x}^* = (1/\sqrt{2}, 1/\sqrt{2})^{\mathrm{T}}$ it follows from Theorem 9.4.2 that $\mathbf{x}^*$ is a global solution to the problem.

It can be seen from the above theorems that a convexity assumption is in the nature of a curvature or second order assumption, in that the directional derivative is non-decreasing along any line. Therefore, although convexity gives useful results for certain special types of problem, it is an assumption which does not often hold in the general case. A weakening of the assumption is to require convexity of $f(\mathbf{x})$ and $-c_i(\mathbf{x})$, $i=1,2,\dots,m$, on a ball about $\mathbf{x}^*$. In this case Theorem 9.4.2 can be interpreted as stating that local convexity and (9.1.16) is sufficient for a local solution. Even in this form, however, the assumption is not valid for many problems. The requirement is essentially that the matrices $\nabla^2 f^*$ and $-\nabla^2 c_i^*$, $i=1,2,\dots,m$, are positive semi-definite. The second order conditions of Theorem 9.3.2 involve the much weaker assumption (9.3.14) that $\mathbf{W}^* = \nabla^2 f^* - \sum \lambda_i^* \nabla^2 c_i^*$ is positive definite only on a restricted subspace, and it is only rarely that this condition does not hold at a local solution. Thus convexity does not provide a valid model situation for the general nonlinear programming problem from which to derive an algorithm, whereas the second order conditions can do this (see Section 12.4).

9.5 DUALITY

The concept of duality occurs widely in the mathematical programming literature. The aim is to provide an alternative formulation of a mathematical programming problem which is more convenient computationally or has some theoretical significance. The original problem is referred to as the *primal* and the transformed problem as the *dual*. Usually the variables in the dual (or some of them) can be interpreted as Lagrange multipliers and take the value $\boldsymbol{\lambda}^*$ at the dual solution, where $\boldsymbol{\lambda}^*$ is a multipler vector associated with a primal solution $\mathbf{x}^*$. In this sense the method of Lagrange multipliers (Section 9.1) might be thought of as a dual method. Usually, however, there is also present an objective function (often related to the Lagrangian function (9.1.6)) which has to be optimized. Duality theory of this kind is associated with a convex programming problem as the primal, and it is important to realize that if the primal is not convex then the dual problem may well not have a solution from which the primal solution can be derived (see Question 9.23). Thus it is not valid to apply the duality transformation as a general purpose solution technique.

In this book the emphasis is on duality transformations which are convenient computationally, and it seems that these can largely be deduced as a consequence of one particular form known as the *Wolfe dual* (Wolfe, 1961). This is a very simple result, closely related to the first order conditions (9.1.16), but which replaces the constraint conditions by an optimality requirement on the Lagrangian function.

Theorem 9.5.1

If $\mathbf{x}^*$ *solves the convex programming primal problem* (9.4.8), *if* f *and* c_i, $i = 1, 2, \ldots, m$, *are* $\mathbb{C}^1$ *functions, and if the regularity assumption* (9.2.4) *holds, then* $\mathbf{x}^*$, $\boldsymbol{\lambda}^*$ *solves the dual problem*

$$\begin{aligned} &\underset{\mathbf{x},\boldsymbol{\lambda}}{\text{maximize}} \quad \mathscr{L}(\mathbf{x}, \boldsymbol{\lambda}) \\ &\text{subject to} \quad \nabla_x \mathscr{L}(\mathbf{x}, \boldsymbol{\lambda}) = \mathbf{0}, \qquad \boldsymbol{\lambda} \geqslant \mathbf{0}. \end{aligned} \tag{9.5.1}$$

Furthermore the minimum primal and maximum dual function values are equal, that is $f^* = \mathscr{L}(\mathbf{x}^*, \boldsymbol{\lambda}^*)$.

Proof

The conditions of the theorem are those of Theorem 9.1.1 so it follows that multipliers $\boldsymbol{\lambda}^* \geqslant \mathbf{0}$ exist such that $\nabla_x \mathscr{L}(\mathbf{x}^*, \boldsymbol{\lambda}^*) = \mathbf{0}$ (dual feasibility), and $\lambda_i^* c_i^* = 0$, $i = 1, 2, \ldots, m$, from which $f^* = \mathscr{L}(\mathbf{x}^*, \boldsymbol{\lambda}^*)$ follows. Now let $\mathbf{x}$, $\boldsymbol{\lambda}$ be dual feasible. Then using $\boldsymbol{\lambda} \geqslant \mathbf{0}$, convexity of $\mathscr{L}$(Lemma 9.4.2), and $\nabla_x \mathscr{L} = \mathbf{0}$ in turn,

$$\begin{aligned}\mathscr{L}(\mathbf{x}^*, \boldsymbol{\lambda}^*) &= f^* \geqslant f^* - \sum_i \lambda_i c_i^* = \mathscr{L}(\mathbf{x}^*, \boldsymbol{\lambda}) \\ &\geqslant \mathscr{L}(\mathbf{x}, \boldsymbol{\lambda}) + (\mathbf{x}^* - \mathbf{x})^{\mathrm{T}} \nabla_x \mathscr{L}(\mathbf{x}, \boldsymbol{\lambda}) \\ &= \mathscr{L}(\mathbf{x}, \boldsymbol{\lambda}).\end{aligned}$$

Hence $\mathscr{L}(\mathbf{x}^*, \boldsymbol{\lambda}^*) \geqslant \mathscr{L}(\mathbf{x}, \boldsymbol{\lambda})$ and so $\mathbf{x}^*$, $\boldsymbol{\lambda}^*$ solves the dual. □

An apparent disadvantage of the Wolfe dual is that the symmetry in which the primal is the dual of the dual is not generally present, since the dual may not even be a convex programming problem. However this does not affect the computational value of the Wolfe dual, and in fact for some of the transformed problems which can be deduced directly from the Wolfe dual, this symmetry does hold. It is also important to consider what happens in the dual if the primal has no solution; this point is taken up at the end of this section.

The first example of the dual transformation is in linear programming (LP). The primal problem

$$\begin{aligned}&\underset{\mathbf{x}}{\text{minimize}} \quad f_0 + \mathbf{c}^{\mathrm{T}}\mathbf{x} \\ &\text{subject to} \quad \mathbf{A}^{\mathrm{T}}\mathbf{x} \geqslant \mathbf{b}\end{aligned} \tag{9.5.2}$$

is not in standard form. Section 7.2 describes how to obtain a standard form by including both slack variables and also non-negative variables $\mathbf{x}^+$ and $\mathbf{x}^-$ (since the bounds $\mathbf{x} \geqslant \mathbf{0}$ are not present). However if there are many more inequalities than variables ($m \gg n$) it is much more attractive to disregard this transformation and instead to use the Wolfe dual. This is valid because the linear functions in (9.5.2) imply both that (9.5.2) is a convex programming problem, and that the regularity assumption (9.2.4) is true. The dual problem (9.5.1) becomes

$$\begin{aligned}&\underset{\mathbf{x}, \boldsymbol{\lambda}}{\text{maximize}} \quad f_0 + \mathbf{c}^{\mathrm{T}}\mathbf{x} - \boldsymbol{\lambda}^{\mathrm{T}}(\mathbf{A}^{\mathrm{T}}\mathbf{x} - \mathbf{b}) \\ &\text{subject to} \quad \mathbf{c} - \mathbf{A}\boldsymbol{\lambda} = \mathbf{0}, \qquad \boldsymbol{\lambda} \geqslant \mathbf{0}.\end{aligned}$$

On substituting for $\mathbf{c}$ in the objective function, the problem becomes independent of $\mathbf{x}$, giving

$$\begin{aligned}&\underset{\boldsymbol{\lambda}}{\text{maximize}} \quad f_0 + \mathbf{b}^{\mathrm{T}}\boldsymbol{\lambda} \\ &\text{subject to} \quad \mathbf{A}\boldsymbol{\lambda} = \mathbf{c}, \qquad \boldsymbol{\lambda} \geqslant \mathbf{0}\end{aligned} \tag{9.5.3}$$

which is an LP problem in standard form. Once this problem has been solved for $\boldsymbol{\lambda}^*$, the variables λ_i^*, $i \in B^*$, will have $\lambda_i^* > 0$ (ignoring the possibility of degeneracy) which implies that $c_i^* = 0$ from (9.1.16). Thus the solution of the square system of equations $\mathbf{a}_i^{\mathrm{T}}\mathbf{x} = b_i$, $i \in B^*$, determines the vector $\mathbf{x}^*$ which minimizes the primal.

Another example in linear programming arises by considering the primal problem

$$\begin{aligned} &\underset{\mathbf{x}}{\text{minimize}} \quad f_0 + \mathbf{c}^{\mathrm{T}}\mathbf{x} \\ &\text{subject to} \quad \mathbf{A}^{\mathrm{T}}\mathbf{x} \geqslant \mathbf{b}, \qquad \mathbf{x} \geqslant \mathbf{0}. \end{aligned} \tag{9.5.4}$$

Introducing multipliers $\boldsymbol{\lambda}$ and $\boldsymbol{\pi}$ respectively, then the Wolfe dual (9.5.1) becomes

$$\begin{aligned} &\underset{\mathbf{x},\boldsymbol{\lambda},\boldsymbol{\pi}}{\text{maximize}} \quad f_0 + \mathbf{c}^{\mathrm{T}}\mathbf{x} - \boldsymbol{\lambda}^{\mathrm{T}}(\mathbf{A}^{\mathrm{T}}\mathbf{x} - \mathbf{b}) - \boldsymbol{\pi}^{\mathrm{T}}\mathbf{x} \\ &\text{subject to} \quad \mathbf{c} - \mathbf{A}\boldsymbol{\lambda} - \boldsymbol{\pi} = \mathbf{0}, \qquad \boldsymbol{\lambda} \geqslant \mathbf{0}, \boldsymbol{\pi} \geqslant \mathbf{0}. \end{aligned}$$

Substituting for $\mathbf{c}$ in the objective eliminates both the $\mathbf{x}$ and $\boldsymbol{\pi}$ variables to give the problem

$$\begin{aligned} &\underset{\boldsymbol{\lambda}}{\text{maximize}} \quad f_0 + \mathbf{b}^{\mathrm{T}}\boldsymbol{\lambda} \\ &\text{subject to} \quad \mathbf{A}\boldsymbol{\lambda} \leqslant \mathbf{c}, \qquad \boldsymbol{\lambda} \geqslant 0. \end{aligned} \tag{9.5.5}$$

Since this problem is like (9.5.4) in having both inequality constraints and bounds it is often referred to as the *symmetric dual*. Its use is advantageous when $\mathbf{A}$ has many fewer rows than columns. Then (after adding slack variables) the standard form arising from (9.5.5) is much smaller than that arising from (9.5.4).

An extension of problem (9.5.4) is to include equality constraints as

$$\begin{aligned} &\underset{\mathbf{x}}{\text{minimize}} \quad f_0 + \mathbf{c}^{\mathrm{T}}\mathbf{x} \\ &\text{subject to} \quad \mathbf{A}_1^{\mathrm{T}}\mathbf{x} \geqslant \mathbf{b}_1 \\ &\qquad\qquad\quad \mathbf{A}_2^{\mathrm{T}}\mathbf{x} = \mathbf{b}_2, \qquad \mathbf{x} \geqslant \mathbf{0}. \end{aligned} \tag{9.5.6}$$

This problem can be reduced to (9.5.4) by writing the equality constraints as $\mathbf{A}_2^{\mathrm{T}}\mathbf{x} \geqslant \mathbf{b}_2$ and $-\mathbf{A}_2^{\mathrm{T}}\mathbf{x} \geqslant -\mathbf{b}_2$. Introducing non-negative multipliers $\boldsymbol{\lambda}_1, \boldsymbol{\lambda}_2^+$, $\boldsymbol{\lambda}_2^-$, and $\boldsymbol{\pi}$, and defining $\boldsymbol{\lambda}_2 = \boldsymbol{\lambda}_2^+ - \boldsymbol{\lambda}_2^-, \boldsymbol{\lambda} = \begin{pmatrix} \boldsymbol{\lambda}_1 \\ \boldsymbol{\lambda}_2 \end{pmatrix}, \mathbf{b} = \begin{pmatrix} \mathbf{b}_1 \\ \mathbf{b}_2 \end{pmatrix}$, and $\mathbf{A} = [\mathbf{A}_1 \vdots \mathbf{A}_2]$, then from (9.5.5) the dual can be written

$$\begin{aligned} &\underset{\boldsymbol{\lambda}}{\text{maximize}} \quad f_0 + \mathbf{b}^{\mathrm{T}}\boldsymbol{\lambda} \\ &\text{subject to} \quad \mathbf{A}\boldsymbol{\lambda} \leqslant \mathbf{c}, \qquad \boldsymbol{\lambda}_1 \geqslant \mathbf{0}. \end{aligned} \tag{9.5.7}$$

In general the strengthening of a primal linear inequality constraint to become an equality just causes the bound on the multiplier in the dual to be relaxed. It is readily observed that all these duals arising from linear programming have the symmetry property that the dual of the dual simplifies to give the primal problem.

Another useful application of the dual is to solve the primal quadratic programming (QP) problem

$$\begin{aligned} &\underset{\mathbf{x}}{\text{minimize}} \quad \tfrac{1}{2}\mathbf{x}^{\mathrm{T}}\mathbf{G}\mathbf{x} + \mathbf{g}^{\mathrm{T}}\mathbf{x} \\ &\text{subject to} \quad \mathbf{A}^{\mathrm{T}}\mathbf{x} \geqslant \mathbf{b} \end{aligned} \tag{9.5.8}$$

in which $\mathbf{G}$ is positive definite. The assumptions of Theorem 9.5.1 are again clearly satisfied, so from (9.5.1) the Wolfe dual is

$$\begin{aligned}&\underset{\mathbf{x},\boldsymbol{\lambda}}{\text{maximize}} \quad \tfrac{1}{2}\mathbf{x}^T\mathbf{G}\mathbf{x} + \mathbf{g}^T\mathbf{x} - \boldsymbol{\lambda}^T(\mathbf{A}^T\mathbf{x} - \mathbf{b})\\ &\text{subject to} \quad \mathbf{G}\mathbf{x} + \mathbf{g} - \mathbf{A}\boldsymbol{\lambda} = \mathbf{0}, \qquad \boldsymbol{\lambda} \geqslant \mathbf{0}.\end{aligned}$$

Using the constraints of this dual to eliminate $\mathbf{x}$ from the objective function gives the problem

$$\begin{aligned}&\underset{\boldsymbol{\lambda}}{\text{maximize}} \quad -\tfrac{1}{2}\boldsymbol{\lambda}^T(\mathbf{A}^T\mathbf{G}^{-1}\mathbf{A})\boldsymbol{\lambda} + \boldsymbol{\lambda}^T(\mathbf{b} + \mathbf{A}^T\mathbf{G}^{-1}\mathbf{g}) - \tfrac{1}{2}\mathbf{g}^T\mathbf{G}^{-1}\mathbf{g}\\ &\text{subject to} \quad \boldsymbol{\lambda} \geqslant \mathbf{0}.\end{aligned} \tag{9.5.9}$$

This is again a quadratic programming problem in the multipliers $\boldsymbol{\lambda}$, but subject only to the bounds $\boldsymbol{\lambda} \geqslant \mathbf{0}$. As such this can make the problem easier to solve. Once the solution $\boldsymbol{\lambda}^*$ to (9.5.9) has been found then $\mathbf{x}^*$ is obtained by solving the equations $\mathbf{G}\mathbf{x} = \mathbf{A}\boldsymbol{\lambda}^* - \mathbf{g}$ used to eliminate $\mathbf{x}$. In a similar way to (9.5.6), the addition of equality constraints into (9.5.8) causes no significant difficulty.

An example of duality for a non-quadratic objective function is in the solution of maximum entropy problems in information theory (Eriksson, 1980). The primal problem is

$$\begin{aligned}&\underset{\mathbf{x}}{\text{minimize}} \quad \sum_{i=1}^{n} x_i \log(x_i/c_i)\\ &\text{subject to} \quad \mathbf{A}^T\mathbf{x} = \mathbf{b}, \qquad \mathbf{x} \geqslant \mathbf{0}\end{aligned} \tag{9.5.10}$$

where the constants c_i are positive and $\mathbf{A}$ is $n \times m$ with $m \ll n$. One possible method of solution is to eliminate variables as described in Section 11.1. However, since $m \ll n$ this does not reduce the size of the problem very much and it is more advantageous to consider solving the dual. For $c > 0$, the function $f(x) = x\log(x/c)$ is convex on $x > 0$ since $f''(x) > 0$ hence by continuity is convex on $x \geqslant 0$ $(0\log 0 = 0)$. Thus (9.5.10) is a convex programming problem. Introducing multipliers $\boldsymbol{\lambda}$ and $\boldsymbol{\pi}$ respectively, the condition $\nabla_x \mathscr{L} = \mathbf{0}$ in (9.5.1) becomes

$$\log(x_i/c_i) + 1 - \mathbf{e}_i^T\mathbf{A}\boldsymbol{\lambda} - \pi_i = 0 \tag{9.5.11}$$

for $i = 1, 2, \ldots, n$. It follows that

$$x_i = c_i \exp(\mathbf{e}_i^T\mathbf{A}\boldsymbol{\lambda} + \pi_i - 1) \tag{9.5.12}$$

and hence that $x_i^* > 0 \ \forall i$. Thus the bounds $\mathbf{x} \geqslant \mathbf{0}$ are inactive with $\boldsymbol{\pi}^* = \mathbf{0}$ and can be ignored. After relaxing the bounds $\boldsymbol{\lambda} \geqslant \mathbf{0}$ to allow for equality constraints the Wolfe dual (9.5.1) thus becomes

$$\underset{\mathbf{x},\boldsymbol{\lambda}}{\text{maximize}} \quad \sum x_i \log(x_i/c_i) - \boldsymbol{\lambda}^T(\mathbf{A}^T\mathbf{x} - \mathbf{b})$$

subject to (9.5.11). These constraints can be written as in (9.5.12) and used to eliminate x_i from the dual objective function. Thus the optimum multipliers $\boldsymbol{\lambda}^*$

can be found by solving

$$\underset{\boldsymbol{\lambda}}{\text{minimize}} \quad h(\boldsymbol{\lambda}) \triangleq \sum_i c_i \exp(\mathbf{e}_i^T \mathbf{A}\boldsymbol{\lambda} - 1) - \boldsymbol{\lambda}^T \mathbf{b}$$

without any constraints. This is a problem in only m variables. It is easy to show that

$$\nabla_{\lambda} h = \mathbf{A}^T \mathbf{x} - \mathbf{b}$$
$$\nabla_{\lambda}^2 h = \mathbf{A}^T [\operatorname{diag} x_i] \mathbf{A}$$

where x_i is dependent on $\boldsymbol{\lambda}$ through (9.5.12). Since these derivatives are available, and the Hessian is positive definite. Newton's method with line search, (Section 3.1) can be used to solve the problem. For large sparse primal problems Eriksson (1980) has also investigated the use of conjugate gradient methods.

An example of the use of duality to solve convex programming problems in which both the objective and constraint functions are nonlinear arises in the study of geometric programming. A description of this technique, and of how the Wolfe dual enables the problem to be solved efficiently, is given in detail in Section 13.2. An interesting duality result for a non-smooth optimization problem is given in Question 12.22.

It is important to consider what happens if the primal has no solution. In particular it is desirable that the dual also does not have a solution, and it is shown that this is often but not always true. The primal problem can fail to have a solution in a number of ways and firstly the case is considered in which the primal is *unbounded* ($f(\mathbf{x}) \to -\infty, \mathbf{x} \in R$). A useful result is then as follows.

Theorem 9.5.2

If V is the infimum of $f(\mathbf{x})$ for feasible $\mathbf{x}$ in the primal problem (9.4.8), and v is the supremum of $\mathscr{L}(\mathbf{x}, \boldsymbol{\lambda})$ for feasible $\mathbf{x}$, $\boldsymbol{\lambda}$ in the dual, then $V \geqslant v$.

Proof

Let $\mathbf{x}'$ be primal feasible and $\mathbf{x}, \boldsymbol{\lambda}$ dual feasible. Then by convexity of f, dual feasibility, concavity of c_i, and non-negativity of c_i' and λ_i in turn, it follows that

$$f' - f \geqslant \mathbf{g}^T(\mathbf{x}' - \mathbf{x}) = \sum_i \lambda_i \mathbf{a}_i^T (\mathbf{x}' - \mathbf{x}).$$
$$\geqslant \sum_i \lambda_i (c_i' - c_i) \geqslant -\sum_i \lambda_i c_i.$$

Hence $f' \geqslant f - \sum_i \lambda_i c_i = \mathscr{L}$, so taking the infimum over all $\mathbf{x}'$ and the supremum over all $\mathbf{x}, \boldsymbol{\lambda}$ it follows that $V \geqslant v$. □

If the primal problem is unbounded it follows that $V = v = -\infty$ and this is not possible if any feasible $\mathbf{x}, \boldsymbol{\lambda}$ exists. Thus an *unbounded primal implies an inconsistent dual.*

Next the case in which the primal constraints are inconsistent is considered.

This result is implied by Theorem 9.5.2 if the dual is unbounded; however the converse is not always true. An example of this is given in Question 9.24 in which although the primal is inconsistent yet the dual has a solution. However, for linear constraints this possibility is excluded. In this case the constraints can be written $\mathbf{c}(\mathbf{x}) \triangleq \mathbf{A}^T\mathbf{x} - \mathbf{b} \geqslant \mathbf{0}$. Now the set $\{\mathbf{x}:\mathbf{A}^T\mathbf{x} \geqslant \mathbf{b}\}$ is empty if and only if there exists a vector $\boldsymbol{\lambda} \geqslant \mathbf{0}$ such that $\mathbf{A}\boldsymbol{\lambda} = \mathbf{0}$ and $\mathbf{b}^T\boldsymbol{\lambda} > 0$. (This result is similar to Farkas' lemma (Lemma 9.2.4) and is proved in a similar way.) Now let $\mathbf{x}, \boldsymbol{\lambda}$ be dual feasible and $\boldsymbol{\lambda}'$ be the vector which exists above. Then $(\mathbf{x}, \boldsymbol{\lambda} + \alpha\boldsymbol{\lambda}')$ is dual feasible for all $\alpha \geqslant 0$ and

$$\mathscr{L}(\mathbf{x}, \boldsymbol{\lambda} + \alpha\boldsymbol{\lambda}') = \mathscr{L}(\mathbf{x}, \boldsymbol{\lambda}) + \alpha\boldsymbol{\lambda}'^T\mathbf{b}$$

which $\to \infty$ as $\alpha \to \infty$. Thus for linearly constrained problems, *if the primal is infeasible and the dual is feasible, then the dual is unbounded.* It is also possible that both the primal and the dual may be infeasible. Thus using the Wolfe dual for linear constraint problems is always satisfactory in that a failure to solve the dual always implies that the primal has no solution.

A final possibility which should also be considered (although it cannot occur for linear or quadratic programming problems) is that the primal (or dual) problem is bounded but has no solution. In this case there are open questions about the nature of solutions to the dual (or primal) problems and the situation is described in more detail by Wolfe (1961).

QUESTIONS FOR CHAPTER 9

9.1. Verify that the points $\mathbf{x}' = \begin{pmatrix} 1 \\ 0 \\ 0 \end{pmatrix}$ and $\mathbf{x}'' = \begin{pmatrix} -\frac{1}{3} \\ \frac{2}{3} \\ \frac{2}{3} \end{pmatrix}$ satisfy Kuhn–Tucker (first order necessary) conditions for the problem

$$\begin{aligned} &\text{minimize} && x_2 + x_3 \\ &\text{subject to} && x_1 + x_2 + x_3 = 1 \\ & && x_1^2 + x_2^2 + x_3^2 = 1 \end{aligned}$$

and evaluate the corresponding Lagrange multipliers.

9.2. By drawing a diagram of the feasible region and the contours of $f(\mathbf{x})$ determine the solution of the problem

$$\begin{aligned} &\text{minimize} && f(\mathbf{x}) \triangleq -x_1 + x_2 \\ &\text{subject to} && 0 \leqslant x_1 \leqslant a \\ & && 0 \leqslant x_2 \leqslant 1 \\ & && x_2 \geqslant x_1^2 \end{aligned}$$

where a is a fixed positive constant. Show that the set of active constraints at the solution differs according to whether or not a is greater than a certain fixed value $\bar{a}$, and determine $\bar{a}$. Obtain the Lagrange multipliers of the active constraints in both cases and verify that the KT conditions are satisfied.

9.3. A parcel has its longest side of length x_1 and its two other sides are of length x_2 and x_3. Postage regulations are that each dimension should be no greater than 42 in, and that the total girth (that is $2(x_2 + x_3)$) plus length should be no greater than 72 in. State the constrained minimization problem which determines the parcel of maximum volume which is permissible. Use the symmetry between x_2 and x_3 to eliminate x_3, draw a diagram of the feasible region, and show (approximately) how the objective function behaves. Identify two possibilities for the set of active constraints at the solution. Solve these (as if equality constraints) for $\mathbf{x}$ and $\boldsymbol{\lambda}$, and determine at which point the multipliers satisfy KT conditions for inequality constraints.

9.4. It is desired to build a warehouse of width x_1, height x_2, and length x_3 (in metres), with capacity 1500 m^3. Building costs per square metre are: walls £4, roof £6, floor plus land £12. For aesthetic reasons, the width should be twice the height. State the problem which determines the dimensions of the warehouse of minimum cost and write down the KT conditions. By eliminating x_1 and x_3, show that to the nearest metre, $x_2 = 10$ minimizes the cost, and hence find x_1 and x_3. Determine the optimum multipliers in the KT conditions.

It can be shown that changing $c_i(\mathbf{x}) = 0$ to $c_i(\mathbf{x}) = \varepsilon_i$ in the problem induces a change $\lambda_i\varepsilon_i$ (to first order) in $f(\mathbf{x})$ at the resulting solution. Estimate the change in cost on reducing the required capacity by 10 per cent.

9.5. List all the stationary points of the function

$$f(\mathbf{x}) = -x_1^2 - 4x_2^2 - 16x_3^2,$$

subject to the constraint $c(\mathbf{x}) = 0$, where $c(\mathbf{x})$ is given in turn by

(i) $c(\mathbf{x}) = x_1 - 1$,
(ii) $c(\mathbf{x}) = x_1x_2 - 1$,
(iii) $c(\mathbf{x}) = x_1x_2x_3 - 1$.

9.6. Solve the problem

$$\begin{aligned} &\text{minimize} \quad f(x, y) \triangleq x^2 + y^2 + 3xy + 6x + 19y \\ &\text{subject to} \quad 3y + x = 5. \end{aligned}$$

9.7. a, b, and c are positive constants. Find the least value of a sum of three positive numbers x, y, and z subject to the constraint

$$\frac{a}{x} + \frac{b}{y} + \frac{c}{z} = 1$$

by the method of Lagrange multipliers, assuming that the positivity conditions are not active.

9.8. Consider the problem

$$\underset{\mathbf{x}}{\text{minimize}} \quad \tfrac{1}{2}\alpha(x_1 - 1)^2 - x_1 - x_2$$

$$\text{subject to} \quad x_1 \geqslant x_2^2, \quad x_1^2 + x_2^2 \leqslant 1.$$

Find all the points at which both the constraints are active. One of these points is $\mathbf{x}' = (0.618, 0.786)^T$ to three decimal places. Working to this accuracy, find the range of values of α for which $\mathbf{x}'$ satisfies the KT conditions.

9.9. Consider the problem

$$\begin{aligned} \text{maximize} \quad & x_2 \\ \text{subject to} \quad & (3-x_1)^3 - (x_2-2) \geqslant 0 \\ & 3x_1 + x_2 \geqslant 9. \end{aligned}$$

(i) Derive the KT conditions for this problem, and find all solutions of these.

(ii) Solve the problem graphically.

(iii) Repeat the analysis of (i) and (ii) for the same problem with the additional constraint

$$2x_1 - 3x_2 \geqslant 0.$$

9.10. Under what conditions on the problem are the KT conditions (a) necessary (b) sufficient, (c) necessary *and* sufficient for the solution of an inequality constrained optimization problem?

Form the KT conditions for the problem

$$\begin{aligned} \text{maximize} \quad & (x+1)^2 + (y+1)^2 \\ \text{subject to} \quad & x^2 + y^2 \leqslant 2 \\ & y \leqslant 1 \end{aligned}$$

and hence determine the solution.

9.11. Find the point on the ellipse defined by the intersection of the surfaces $x + y = 1$ and $x^2 + 2y^2 + z^2 = 1$ which is nearest to the origin. Use (i) the method of Lagrange multipliers, (ii) direct elimination.

9.12. A bookmaker offers odds of r_i:1 against each of n runners in a race. A punter bets a proportion x_i of his total stake t on each runner. Assume that only one runner can win and that $r_1 > r_2 > \cdots > r_n > 0$. Clearly the punter can guarantee not to lose money if and only if

$$r_i x_i \geqslant \sum_{j \neq i} x_j, \qquad i = 1, 2, \ldots, n.$$

Show that this situation can arise if $\sum_i 1/(r_i + 1) \leqslant 1$. (Consider $x_i = c/(r_i + 1)$ where $1/c = \sum_i 1/(r_i + 1)$.)

If this condition holds, and if it is equally likely that any runner can win, the expected profit to the punter is $(t\sum_i (r_i + 1)x_i) - t$. Show that the choice of the x_i which gives maximum expected profit, yet which guarantees no loss of money, can be posed as a constrained minimization problem involving only bounds on the variables and an equality constraint. By showing that the KT conditions are satisfied, verify that the solution is

$$x_1 = 1 - \sum_{i>1} \frac{1}{r_i + 1} \qquad x_i = \frac{1}{r_i + 1}, \qquad i > 1.$$

9.13. Consider the relationship between the method of Lagrange multipliers in Section 9.1 and the direct elimination method in Section 7.2 and Question 7.5. In the latter notation, show that (9.1.3) can be rearranged to give $\boldsymbol{\lambda}^* = [\mathbf{A}_1^*]^{-1}\mathbf{g}_1^*$. Hence use the result of Question 7.5 to show that if $\mathbf{x}^*$ satisfies (9.1.3) then it is a stationary point of the function $\psi(\mathbf{x}_2)$ in (7.2.2).

9.14. Attempt to solve the problem in Question 7.4 by the method of Lagrange multipliers. Show that either $y=0$ or $\lambda=-1$ and both of these imply a contradiction, so no solution of (9.1.5) exists. Explain this fact in terms of the regularity assumption (9.2.4) and the independence of the vectors $\mathbf{a}_i^*$, $i \in E$.

9.15. Show that if the matrix $\mathbf{A}^*$ in (9.1.3) has full rank then the Lagrange multipliers $\boldsymbol{\lambda}^*$ are uniquely defined by $\boldsymbol{\lambda}^* = \mathbf{A}^{*+}\mathbf{g}^*$ or by solving any $m \times m$ subsystem $\mathbf{A}_1^*\boldsymbol{\lambda}^* = \mathbf{g}_1^*$ where $\mathbf{A}_1^*$ is non-singular (see Question 9.13). Computationally the former is most stable if the matrix factors $\mathbf{A}^* = \mathbf{QR}$ ($\mathbf{Q}$ orthogonal, $\mathbf{R}$ upper triangular) are calculated and $\boldsymbol{\lambda}^*$ is obtained by back substitution in $\mathbf{R}\boldsymbol{\lambda}^* = \mathbf{Q}^T\mathbf{g}^*$.

9.16. By examining second order conditions, determine whether or not each of the points $\mathbf{x}'$ and $\mathbf{x}''$ are local solutions to the problem in Question 9.1.

9.17. By examining second order conditions, determine the nature (maximizer, minimizer, or saddle-point) of each of the stationary points obtained in Question 9.5.

9.18. Given an optimal b.f.s. to an LP, show that the reduced costs $\hat{\mathbf{c}}_N$ are the Lagrange multipliers to the LP after having used the equations $\mathbf{Ax} = \mathbf{b}$ to eliminate the basic variables.

9.19. For the LP in Question 8.1, illustrate (for non-negative $\mathbf{x}$) the plane $3x_1 + x_2 + x_3 = 12$ and the line along which it intersects the plane $x_1 - x_2 + x_3 = -8$. Hence show that the feasible region is a convex set and give its extreme points. Do the same for the LP which results from deleting the condition $x_1 - x_2 + x_3 = -8$.

9.20. For the LP (8.1.1) in standard form, prove that

(i)	$\exists$ a solution	$\Rightarrow \exists$ an extreme point which is a solution,
(ii)	$\mathbf{x}$ is an extreme point	$\Rightarrow \mathbf{x}$ has $p \leqslant m$ positive components,
(iii)	$\mathbf{x}$ is an extreme point and $\mathbf{A}$ has full rank	
		$\Rightarrow \exists$ a b.f.s. at $\mathbf{x}$ (degenerate iff $p < m$),
(iv)	$\exists$ a b.f.s. at $\mathbf{x}$	$\Rightarrow \mathbf{x}$ is an extreme point.

(By a b.f.s. at $\mathbf{x}$ is meant that there exists a partition into B and N variables such that $\mathbf{x}_B \geqslant \mathbf{0}$, $\mathbf{x}_N = \mathbf{0}$, and $\mathbf{A}_B$ is non-singular (see Section 8.2).) Part (i) follows from the convexity of S in Theorem 9.4.1, the result of Question 9.21, and the fact that an extreme point of S must be an extreme point of (8.1.1) by the linearity of f. The remaining results follow from Question 9.22. Part (ii) is true since if $\mathbf{A}_P$ has rank p then $p \leqslant m$. Part (iii) is derived by including all positive components of $\mathbf{x}$ in B. If $p < m$, other independent columns of $\mathbf{A}$ exists which can augment $\mathbf{A}_P$ to become a square non-singular matrix, and the corresponding variables (with zero value, hence degeneracy) are added to B. In part (iv) the non-singularity of $\mathbf{A}_B$ implies that $\mathbf{A}_P$ has full rank and hence $\mathbf{x}$ is extreme.

9.21. Consider a closed convex set $K \subset \mathbb{R}^{\oplus} = \{\mathbf{x}:\mathbf{x} \geqslant \mathbf{0}\}$. Show that K has an extreme point in any supporting hyperplane $S = \{\mathbf{x}:\mathbf{c}^T\mathbf{x} = z\}$. Show that $T \triangleq K \cap S$ is not empty. Since $T \subset \mathbb{R}^{\oplus}$, construct an extreme point in the following way. Choose the set $T_1 \subset T$ such that the first component t_1 of all vectors $\mathbf{t} \in T_1$ is smallest. Then choose a set $T_2 \subset T_1$ with the smallest second component, and so on. Show that a unique point $\mathbf{t}^*$ is ultimately determined which is extreme in T. Hence show that $\mathbf{t}^*$ is extreme in K (Hadley, 1962, chapter 2, Theorem III).

9.22. Let $\mathbf{x}$ be any feasible point in (8.1.1), with p positive components, let $\mathbf{A}_P$ be the matrix with columns $\mathbf{a}_i \; \forall i : x_i > 0$ and $\mathbf{x}_P$ likewise. Show that $\mathbf{x}$ is an extreme point of (8.1.1) iff $\mathbf{A}_P$ has full rank, in the following way. If $\mathbf{A}_P$ has not full rank there exists $\mathbf{u} \neq \mathbf{0}$ such that $\mathbf{A}_P\mathbf{u} = \mathbf{0}$. By examining $\mathbf{x}_P \pm \varepsilon\mathbf{u}$ show that $\mathbf{x}$ is not extreme. If $\mathbf{A}_P$ has full rank, show that $\mathbf{x}_P = \mathbf{A}_P^+\mathbf{b}$ is uniquely defined. Let $\mathbf{x} = (1-\theta)\mathbf{x}_0 + \theta\mathbf{x}_1$, $\theta \in (0, 1)$, $\mathbf{x}_0, \mathbf{x}_1$ feasible, and show that the zero components of $\mathbf{x}$ must be zero in $\mathbf{x}_0$ and $\mathbf{x}_1$. Hence show that the remaining components are defined by $\mathbf{A}_P^+\mathbf{b}$ and hence $\mathbf{x}_0 = \mathbf{x}_1 = \mathbf{x}$, so that $\mathbf{x}$ is extreme.

9.23. Show that the dual of the problem

$$\begin{array}{ll} \text{minimize} & \tfrac{1}{2}\sigma x_1^2 + \tfrac{1}{2}x_2^2 + x_1 \\ \text{subject to} & x_1 \geqslant 0 \end{array}$$

is a maximization problem in terms of a Lagrange multiplier λ. For the cases $\sigma = +1$ and $\sigma = -1$, investigate whether the local solution of the dual gives the multiplier λ^* which exists at the local solution to the primal, and explain the difference between the two cases.

9.24. Consider the problem

$$\begin{array}{ll} \text{minimize} & f(x) \triangleq 0 \\ \text{subject to} & c(x) \triangleq -e^x \geqslant 0. \end{array}$$

Verify that the constraint is concave but inconsistent, so that the feasible region is empty. Set up the dual problem and show that it is solved by $\lambda = 0$ and any x.

9.25. Consider finding KT points of the problem

$$\text{maximize } \tfrac{1}{3}\sum_{i=1}^{n} x_i^3 \quad \text{subject to} \quad \sum_{i=1}^{n} x_i = 0, \quad \sum_{i=1}^{n} x_i^2 = n$$

for any $n > 2$. Use the method of Lagrange multipliers (with multipliers λ and μ respectively) to determine the general form of a KT point for the problem (for general n). Find the largest value of the objective function that occurs at a KT point, and give all the corresponding KT points. By examining second order sufficient conditions, show that these KT points are all local maximizers.

Chapter 10

Quadratic Programming

10.1 EQUALITY CONSTRAINTS

Like linear programming problems, another optimization problem which can be solved in a finite number of steps is a *quadratic programming* (*QP*) *problem*. In terms of (7.1.1) this is a problem in which the objective function $f(\mathbf{x})$ is quadratic and the constraint functions $\mathbf{c}_i(\mathbf{x})$ are linear. Thus the problem is to find a solution $\mathbf{x}^*$ to

$$\begin{aligned} &\underset{\mathbf{x}}{\text{minimize}} && q(\mathbf{x}) \triangleq \tfrac{1}{2}\mathbf{x}^{\mathrm{T}}\mathbf{G}\mathbf{x} + \mathbf{g}^{\mathrm{T}}\mathbf{x} \\ &\text{subject to} && \mathbf{a}_i^{\mathrm{T}}\mathbf{x} = b_i, \quad i \in E \\ & && \mathbf{a}_i^{\mathrm{T}}\mathbf{x} \geqslant b_i, \quad i \in I, \end{aligned} \tag{10.1.1}$$

where it is always possible to arrange that the matrix $\mathbf{G}$ is symmetric. As in linear programming, the problem may be infeasible or the solution may be unbounded; however these possibilities are readily detected in the algorithms, so for the most part it is assumed that a solution $\mathbf{x}^*$ exists. If the Hessian matrix $\mathbf{G}$ is positive semi-definite, $\mathbf{x}^*$ is a global solution, and if $\mathbf{G}$ is positive definite, $\mathbf{x}^*$ is also unique. These results follow from the (strict) convexity of $q(\mathbf{x})$, so that (10.1.1) is a convex programming problem and Theorem 9.4.1 and its corollary apply. When the Hessian $\mathbf{G}$ is indefinite then local solutions which are not global can occur, and the computation of any such local solution is of interest (see Section 10.4). A modern computer code for QP needs to be quite sophisticated, so a simplified account of the basic structure is given first, and is amplified or qualified later. In the first case it is shown in Sections 10.1 and 10.2 how equality constraint problems can be treated. The generalization of these ideas to handle inequality constraints is by means of an active set strategy and is described in Section 10.3. More advanced features of a QP algorithm which handle an indefinite Hessian and allow a sequence of problems to be solved efficiently are given in Section 10.4. QP problems with a special structure such as having only bounded constraints, or such as arise from least squares problems, are considered in Section 10.5. Early work on QP was often presented as a modification of the tableau form of the simplex method for linear programming. This work is described in Section 10.6 and has come to be expressed in terms of a *linear complementarity problem.* The equivalence between these methods and the active set methods is described and reasons for preferring the latter derivation are put forward.

Quadratic programming differs from linear programming in that it is possible to have meaningful problems (other than an $n \times n$ system of equations) in which there are no inequality constraints. This section therefore studies how to find a solution $\mathbf{x}^*$ to the equality constraint problem

$$\begin{aligned} \underset{\mathbf{x}}{\text{minimize}} \quad & q(\mathbf{x}) \triangleq \tfrac{1}{2}\mathbf{x}^{\mathrm{T}}\mathbf{G}\mathbf{x} + \mathbf{g}^{\mathrm{T}}\mathbf{x} \\ \text{subject to} \quad & \mathbf{A}^{\mathrm{T}}\mathbf{x} = \mathbf{b}. \end{aligned} \tag{10.1.2}$$

It is assumed that there are $m \leqslant n$ constraints so that $\mathbf{b} \in \mathbb{R}^m$, and $\mathbf{A}$ is $n \times m$ and collects the column vectors $\mathbf{a}_i$, $i \in E$, in (10.1.1). It is assumed that $\mathbf{A}$ has rank m; if the constraints are consistent this can always be achieved by removing dependent constraints, although there may be numerical difficulties in recognizing this situation. This assumption also ensures that unique Lagrange multipliers $\boldsymbol{\lambda}^*$ exist, and calculation of these quantities, which may be required for sensitivity analysis or for use in an active set method, is also considered in this section.

A straightforward way of solving (10.1.2) is to use the constraints to eliminate variables. If the partitions

$$\mathbf{x} = \begin{pmatrix} \mathbf{x}_1 \\ \mathbf{x}_2 \end{pmatrix}, \quad \mathbf{A} = \begin{pmatrix} \mathbf{A}_1 \\ \mathbf{A}_2 \end{pmatrix}, \quad \mathbf{g} = \begin{pmatrix} \mathbf{g}_1 \\ \mathbf{g}_2 \end{pmatrix}, \quad \mathbf{G} = \begin{bmatrix} \mathbf{G}_{11} & \mathbf{G}_{12} \\ \mathbf{G}_{21} & \mathbf{G}_{22} \end{bmatrix}$$

are defined, where $\mathbf{x}_1 \in \mathbb{R}^m$ and $\mathbf{x}_2 \in \mathbb{R}^{n-m}$, etc., then the equations in (10.1.2) become $\mathbf{A}_1^{\mathrm{T}}\mathbf{x}_1 + \mathbf{A}_2^{\mathrm{T}}\mathbf{x}_2 = \mathbf{b}$ and are readily solved (by Gaussian elimination, say) to give $\mathbf{x}_1$ in terms of $\mathbf{x}_2$; this is conveniently written

$$\mathbf{x}_1 = \mathbf{A}_1^{-\mathrm{T}}(\mathbf{b} - \mathbf{A}_2^{\mathrm{T}}\mathbf{x}_2). \tag{10.1.3}$$

Substituting into $q(\mathbf{x})$ gives the problem: minimize $\psi(\mathbf{x}_2)$, $\mathbf{x}_2 \in \mathbb{R}^{n-m}$ where $\psi(\mathbf{x}_2)$ is the quadratic function

$$\begin{aligned} \psi(\mathbf{x}_2) = {} & \tfrac{1}{2}\mathbf{x}_2^{\mathrm{T}}(\mathbf{G}_{22} - \mathbf{G}_{21}\mathbf{A}_1^{-\mathrm{T}}\mathbf{A}_2^{\mathrm{T}} - \mathbf{A}_2\mathbf{A}_1^{-1}\mathbf{G}_{12} + \mathbf{A}_2\mathbf{A}_1^{-1}\mathbf{G}_{11}\mathbf{A}_1^{-\mathrm{T}}\mathbf{A}_2^{\mathrm{T}})\mathbf{x}_2 \\ & + \mathbf{x}_2^{\mathrm{T}}(\mathbf{G}_{21} - \mathbf{A}_2\mathbf{A}_1^{-1}\mathbf{G}_{11})\mathbf{A}_1^{-\mathrm{T}}\mathbf{b} + \tfrac{1}{2}\mathbf{b}^{\mathrm{T}}\mathbf{A}_1^{-1}\mathbf{G}_{11}\mathbf{A}_1^{-\mathrm{T}}\mathbf{b} \\ & + \mathbf{x}_2^{\mathrm{T}}(\mathbf{g}_2 - \mathbf{A}_2\mathbf{A}_1^{-1}\mathbf{g}_1) + \mathbf{g}_1^{\mathrm{T}}\mathbf{A}_1^{-\mathrm{T}}\mathbf{b}. \end{aligned} \tag{10.1.4}$$

A unique minimizer $\mathbf{x}_2^*$ exists if the Hessian $\nabla^2\psi$ in the quadratic term is positive definite, in which case $\mathbf{x}_2^*$ is obtained by solving the linear system $\nabla\psi(\mathbf{x}_2) = \mathbf{0}$. Then $\mathbf{x}_1^*$ is found by substitution in (10.1.3). The Lagrange multiplier vector $\boldsymbol{\lambda}^*$ is defined by $\mathbf{g}^* = \mathbf{A}\boldsymbol{\lambda}^*$ (Section 9.1) where $\mathbf{g}^* = \nabla q(\mathbf{x}^*)$, and can be calculated by solving the first partition $\mathbf{g}_1^* = \mathbf{A}_1\boldsymbol{\lambda}^*$. By definition of $q(\mathbf{x})$ in (10.1.2), $\mathbf{g}^* = \mathbf{g} + \mathbf{G}\mathbf{x}^*$ so an explicit expression for $\boldsymbol{\lambda}^*$ is

$$\boldsymbol{\lambda}^* = \mathbf{A}_1^{-1}(\mathbf{g}_1 + \mathbf{G}_{11}\mathbf{x}_1^* + \mathbf{G}_{12}\mathbf{x}_2^*) \tag{10.1.5}$$

An example of the method is given to solve the problem

$$\begin{aligned}\underset{\mathbf{x}}{\text{minimize}} \quad & q(\mathbf{x}) \triangleq x_1^2 + x_2^2 + x_3^2 \\ \text{subject to} \quad & x_1 + 2x_2 - x_3 = 4 \\ & x_1 - x_2 + x_3 = -2.\end{aligned} \tag{10.1.6}$$

To eliminate x_3, the equations are written

$$\begin{aligned} x_1 + 2x_2 &= 4 + x_3 \\ x_1 - x_2 &= -2 - x_3 \end{aligned}$$

and are readily solved by Gaussian elimination giving

$$x_1 = 0 - \tfrac{1}{3}x_3, \qquad x_2 = 2 + \tfrac{2}{3}x_3. \tag{10.1.7}$$

It is easily verified that this solution corresponds to (10.1.3) with $\mathbf{x}_1 = \begin{pmatrix} x_1 \\ x_2 \end{pmatrix}$, $\mathbf{x}_2 = (x_3)$, $\mathbf{A}_1 = \begin{bmatrix} 1 & 1 \\ 2 & -1 \end{bmatrix}$, and $\mathbf{A}_2 = [-1 \quad 1]$. Substituting (10.1.7) into $q(\mathbf{x})$ in (10.1.6) gives

$$\psi(x_3) = \tfrac{14}{9}x_3^2 + \tfrac{8}{3}x_3 + 4 \tag{10.1.8}$$

which corresponds to (10.1.4) with $\mathbf{G}_{11} = \begin{bmatrix} 2 & 0 \\ 0 & 2 \end{bmatrix}$, $\mathbf{G}_{12} = \mathbf{G}_{21}^{\mathrm{T}} = \mathbf{0}, \mathbf{G}_{22} = [2]$, and $\mathbf{g} = \mathbf{0}$. The Hessian matrix in (10.1.8) is $[\frac{28}{9}]$ which is positive definite, so the minimizer is obtained by setting $\nabla\psi = \mathbf{0}$ and is $x_3^* = -\frac{6}{7}$. Back substitution in (10.1.6) gives $x_1^* = \frac{2}{7}$ and $x_2^* = \frac{10}{7}$. The system $\mathbf{g}^* = \mathbf{A}\boldsymbol{\lambda}^*$ becomes

$$\tfrac{2}{7}\begin{pmatrix} 2 \\ 10 \\ -6 \end{pmatrix} = \begin{bmatrix} 1 & 1 \\ 2 & -1 \\ -1 & 1 \end{bmatrix} \begin{pmatrix} \lambda_1^* \\ \lambda_2^* \end{pmatrix}$$

and solving the first two rows gives $\lambda_1^* = \frac{8}{7}$ and $\lambda_2^* = -\frac{4}{7}$, and this is consistent with the third row.

Direct elimination of variables is not the only way of solving (10.1.2) nor may it be best. A *generalized elimination method* is possible in which essentially a linear transformation of variables is made initially. Let $\mathbf{Y}$ and $\mathbf{Z}$ be $n \times m$ and $n \times (n - m)$ matrices respectively such that $[\mathbf{Y}:\mathbf{Z}]$ is non-singular, and in addition let $\mathbf{A}^{\mathrm{T}}\mathbf{Y} = \mathbf{I}$ and $\mathbf{A}^{\mathrm{T}}\mathbf{Z} = \mathbf{0}$, $\mathbf{Y}^{\mathrm{T}}$ can be regarded as a left generalized inverse for $\mathbf{A}$ so that a solution of $\mathbf{A}^{\mathrm{T}}\mathbf{x} = \mathbf{b}$ is given by $\mathbf{x} = \mathbf{Y}\mathbf{b}$. However this solution is non-unique in general and other feasible points are given by $\mathbf{x} = \mathbf{Y}\mathbf{b} + \boldsymbol{\delta}$ where $\boldsymbol{\delta}$ is in the null column space of $\mathbf{A}$. This is the linear space

$$\{\boldsymbol{\delta}\colon \mathbf{A}^{\mathrm{T}}\boldsymbol{\delta} = \mathbf{0}\} \tag{10.1.9}$$

which has dimension $n - m$. The purpose of the matrix $\mathbf{Z}$ is that it has linearly independent columns $\mathbf{z}_1, \mathbf{z}_2, \ldots, \mathbf{z}_{n-m}$ which are in this null space and therefore act as basis vectors (or reduced coordinate directions) for the null space. That

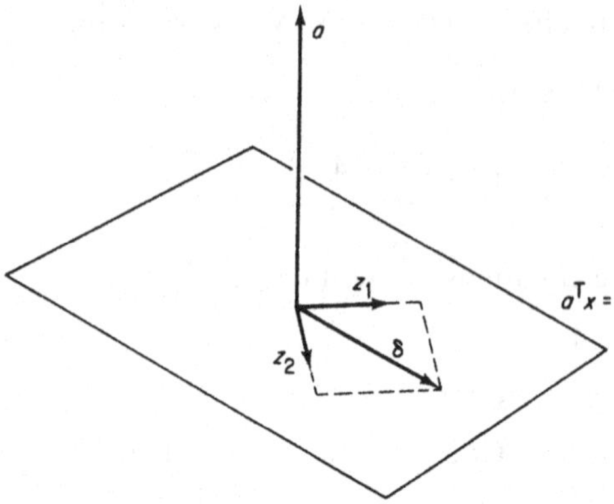

Figure 10.1.1 Reduced coordinates for the feasible region

is to say, at any feasible point $\mathbf{x}$ any feasible correction $\boldsymbol{\delta}$ can be written as

$$\boldsymbol{\delta} = \mathbf{Z}\mathbf{y} = \sum_{i=1}^{n-m} \mathbf{z}_i y_i \tag{10.1.10}$$

where $y_1, y_2, \ldots, y_{n-m}$ are the components (or reduced variables) in each reduced coordinate direction (see Figure 10.1.1). Thus any feasible point $\mathbf{x}$ can be written

$$\mathbf{x} = \mathbf{Y}\mathbf{b} + \mathbf{Z}\mathbf{y}. \tag{10.1.11}$$

This can be interpreted (Figure 10.1.2) as a step from the origin to the feasible point $\mathbf{Yb}$, followed by a feasible correction $\mathbf{Zy}$ to reach the point $\mathbf{x}$. Thus (10.1.11) provides way of eliminating the constraints $\mathbf{A}^T\mathbf{x} = \mathbf{b}$ in terms of the vector of reduced variables $\mathbf{y}$ which has $n - m$ elements, and is therefore a generalization of (10.1.3). Substituting into $q(\mathbf{x})$ gives the reduced quadratic function

$$\psi(\mathbf{y}) = \tfrac{1}{2}\mathbf{y}^T\mathbf{Z}^T\mathbf{G}\mathbf{Z}\mathbf{y} + (\mathbf{g} + \mathbf{G}\mathbf{Y}\mathbf{b})^T\mathbf{Z}\mathbf{y} + \tfrac{1}{2}(\mathbf{g} + \mathbf{G}\mathbf{Y}\mathbf{b})^T\mathbf{Y}\mathbf{b}. \tag{10.1.12}$$

If $\mathbf{Z}^T\mathbf{G}\mathbf{Z}$ is positive definite then a unique minimizer $\mathbf{y}^*$ exists which (from $\nabla\psi(\mathbf{y}) = \mathbf{0}$) solves the linear system

$$(\mathbf{Z}^T\mathbf{G}\mathbf{Z})\mathbf{y} = -\mathbf{Z}^T(\mathbf{g} + \mathbf{G}\mathbf{Y}\mathbf{b}). \tag{10.1.13}$$

The solution is best achieved by computing $\mathbf{LL}^T$ or $\mathbf{LDL}^T$ factors of $\mathbf{Z}^T\mathbf{G}\mathbf{Z}$ which also enables the positive definite condition to be checked. Then $\mathbf{x}^*$ is obtained by substituting ino (10.1.11). The matrix $\mathbf{Z}^T\mathbf{G}\mathbf{Z}$ in (10.1.12) is often referred to as the *reduced Hessian matrix* and the vector $\mathbf{Z}^T(\mathbf{g} + \mathbf{G}\mathbf{Y}\mathbf{b})$ as the *reduced gradient vector*. Notice that $\mathbf{g} + \mathbf{G}\mathbf{Y}\mathbf{b} = \nabla q(\mathbf{Y}\mathbf{b})$ is the gradient vector of $q(\mathbf{x})$ at $\mathbf{x} = \mathbf{Y}\mathbf{b}$ (just as $\mathbf{g}$ is $\nabla q(\mathbf{0})$), so reduced derivatives are obtained by a matrix operation with $\mathbf{Z}^T$. In addition, premultiplying by $\mathbf{Y}^T$ in $\mathbf{g}^* = \mathbf{A}\boldsymbol{\lambda}^*$ gives the equation

$$\boldsymbol{\lambda}^* = \mathbf{Y}^T\mathbf{g}^* = \mathbf{Y}^T(\mathbf{G}\mathbf{x}^* + \mathbf{g}) \tag{10.1.14}$$

which can be used to calculate Lagrange multipliers. Explicit expressions for

$\mathbf{x}^*$ and $\boldsymbol{\lambda}^*$ in terms of the original data are

$$\mathbf{x}^* = \mathbf{Y}\mathbf{b} - \mathbf{Z}(\mathbf{Z}^T\mathbf{G}\mathbf{Z})^{-1}\mathbf{Z}^T(\mathbf{g} + \mathbf{G}\mathbf{Y}\mathbf{b}) \tag{10.1.15}$$

and

$$\begin{aligned}\boldsymbol{\lambda}^* &= \mathbf{Y}^T(\mathbf{g} + \mathbf{G}\mathbf{x}^*)\\ &= (\mathbf{Y}^T - \mathbf{Y}^T\mathbf{G}\mathbf{Z}(\mathbf{Z}^T\mathbf{G}\mathbf{Z})^{-1}\mathbf{Z}^T)\mathbf{g}\\ &\quad + \mathbf{Y}^T(\mathbf{G} - \mathbf{G}\mathbf{Z}(\mathbf{Z}^T\mathbf{G}\mathbf{Z})^{-1}\mathbf{Z}^T\mathbf{G})\mathbf{Y}\mathbf{b}.\end{aligned} \tag{10.1.16}$$

These formulae would not be used directly in computation but are useful in showing the relationship with the alternative method of Lagrange multipliers for solving equality constraint QP problems, considered in Section 10.2.

Depending on the choice of **Y** and **Z**, a number of methods exist which can be interpreted in this way, and a general procedure is described below for constructing matrices **Y** and **Z** with the correct properties. However, one choice of particular importance is obtained by way of any QR factorization (Householder's, for example) of the matrix **A**. This can be written

$$\mathbf{A} = \mathbf{Q}\begin{bmatrix}\mathbf{R}\\ \mathbf{0}\end{bmatrix} = [\mathbf{Q}_1\mathbf{Q}_2]\begin{bmatrix}\mathbf{R}\\ \mathbf{0}\end{bmatrix} = \mathbf{Q}_1\mathbf{R} \tag{10.1.17}$$

where $\mathbf{Q}$ is $n \times n$ and orthogonal, $\mathbf{R}$ is $m \times m$ and upper triangular, and $\mathbf{Q}_1$ and $\mathbf{Q}_2$ are $n \times m$ and $n \times (n-m)$ respectively. The choices

$$\mathbf{Y} = \mathbf{A}^{+T} = \mathbf{Q}_1\mathbf{R}^{-T}, \qquad \mathbf{Z} = \mathbf{Q}_2 \tag{10.1.18}$$

are readily observed to have the correct properties. Moreover the vector $\mathbf{Yb}$ in (10.1.11) and Figure 10.1.2 is observed to be orthogonal to the constraint manifold, and the reduced coordinate directions $\mathbf{z}_i$ are also mutually orthogonal. The solution $\mathbf{x}^*$ is obtained as above by setting up and solving (10.1.13) for $\mathbf{y}^*$ and then substituting into (10.1.11). $\mathbf{Yb}$ is calculated by forward substitution in $\mathbf{R}^T\mathbf{u} = \mathbf{b}$ followed by forming $\mathbf{Yb} = \mathbf{Q}_1\mathbf{u}$. The multipliers $\boldsymbol{\lambda}^*$ in (10.1.14) are

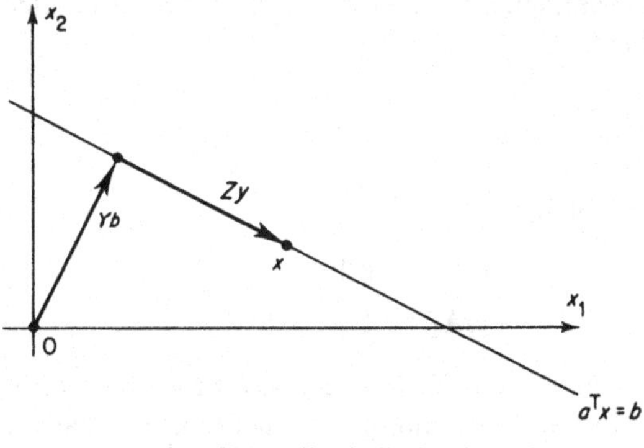

Figure 10.1.2 Generalized elimination in the special case (10.1.18)

calculated by backward substitution in

$$\mathbf{R}\boldsymbol{\lambda}^* = \mathbf{Q}_1^{\mathrm{T}}\mathbf{g}^*. \tag{10.1.19}$$

This scheme is due to Gill and Murray (1974a) and I shall refer to it as the *orthogonal factorization method.* It is recommended for non-sparse problems because it is stable in regard to propagation of round-off errors (see below).

An example is given of the method applied to problem (10.1.6). The matrices involved are

$$\mathbf{Y} = \tfrac{1}{14}\begin{bmatrix} 5 & 8 \\ 4 & -2 \\ -1 & 4 \end{bmatrix}, \quad \mathbf{Z} = \begin{bmatrix} 1 \\ -2 \\ -3 \end{bmatrix}$$

(unnormalized) and these can be represented by the QR decomposition

$$\mathbf{A} = \begin{bmatrix} 1/\sqrt{6} & 4/\sqrt{21} & 1/\sqrt{14} \\ 2/\sqrt{6} & -1/\sqrt{21} & -2/\sqrt{14} \\ 1/\sqrt{6} & 2/\sqrt{21} & -3/\sqrt{14} \end{bmatrix}\begin{bmatrix} \sqrt{6} & -\sqrt{6}/3 \\ 0 & \sqrt{21}/3 \\ 0 & 0 \end{bmatrix}.$$

The vector $\mathbf{Yb} = \frac{1}{7}(2, 10, -6)^{\mathrm{T}}$, and since $\mathbf{g} = \mathbf{0}$ and $\mathbf{GYb} = 2\mathbf{Yb}$ it follows that $\mathbf{Z}^{\mathrm{T}}(\mathbf{g} + \mathbf{GYb}) = \mathbf{0}$. Hence $\mathbf{y}^* = \mathbf{0}$ and $\mathbf{x}^* = \mathbf{Yb} = \frac{1}{7}(2, 10, -6)^{\mathrm{T}}$. Also $\mathbf{g}^* = \mathbf{g} + \mathbf{GYb} = \frac{2}{7}(2, 10, -6)^{\mathrm{T}}$ so $\boldsymbol{\lambda}^* = (\frac{8}{7}, -\frac{4}{7})^{\mathrm{T}}$ and these results agree with those obtained by direct elimination.

A general scheme for computing suitable $\mathbf{Y}$ and $\mathbf{Z}$ matrices is the following. Choose any $n \times (n-m)$ matrix $\mathbf{V}$ such that $[\mathbf{A}:\mathbf{V}]$ is non-singular. Let the inverse be expressed in partitioned form

$$[\mathbf{A}:\mathbf{V}]^{-1} = \begin{bmatrix} \mathbf{Y}^{\mathrm{T}} \\ \mathbf{Z}^{\mathrm{T}} \end{bmatrix} \tag{10.1.20}$$

where $\mathbf{Y}$ and $\mathbf{Z}$ are $n \times m$ and $n \times (n-m)$ respectively. Then it follows that $\mathbf{Y}^{\mathrm{T}}\mathbf{A} = \mathbf{I}$ and $\mathbf{Z}^{\mathrm{T}}\mathbf{A} = \mathbf{0}$ so these matrices are suitable for use in the generalized elimination method. The resulting method can also be interpreted as one which makes a linear transformation with the matrix $[\mathbf{A}:\mathbf{V}]$, as described at the start of Section 12.5. The methods described earlier can all be identified as special cases of this scheme. If

$$\mathbf{V} = \begin{bmatrix} \mathbf{0} \\ \mathbf{I} \end{bmatrix} \tag{10.1.21}$$

is chosen, then the identity

$$\begin{bmatrix} \mathbf{A}_1 & \mathbf{0} \\ \mathbf{A}_2 & \mathbf{I} \end{bmatrix}^{-1} = \begin{bmatrix} \mathbf{A}_1^{-1} & \mathbf{0} \\ -\mathbf{A}_2\mathbf{A}_1^{-1} & \mathbf{I} \end{bmatrix} = \begin{bmatrix} \mathbf{Y}^{\mathrm{T}} \\ \mathbf{Z}^{\mathrm{T}} \end{bmatrix} \tag{10.1.22}$$

gives expressions for $\mathbf{Y}$ and $\mathbf{Z}$. It can easily be verified that in this case the method reduces to the direct elimination method. Alternatively if the choice

$$\mathbf{V} = \mathbf{Q}_2 \tag{10.1.23}$$

is made, where $\mathbf{Q}_2$ is defined by (10.1.17), then the above orthogonal factorization method is obtained. This follows by virtue of the identity

$$[\mathbf{A}:\mathbf{V}]^{-1} = [\mathbf{Q}_1\mathbf{R}:\mathbf{Q}_2]^{-1} = \begin{bmatrix} \mathbf{R}^{-1}\mathbf{Q}_1^{\mathrm{T}} \\ \mathbf{Q}_2^{\mathrm{T}} \end{bmatrix}, \tag{10.1.24}$$

which can also be expressed as

$$[\mathbf{A}:\mathbf{V}]^{-1} = \begin{bmatrix} \mathbf{A}^{+} \\ \mathbf{V}^{+} \end{bmatrix}, \tag{10.1.25}$$

where $\mathbf{A}^{+} = (\mathbf{A}^{\mathrm{T}}\mathbf{A})^{-1}\mathbf{A}^{\mathrm{T}}$ is the full rank Penrose generalized inverse matrix, and (10.1.14) can be written

$$\boldsymbol{\lambda}^* = \mathbf{A}^{+}\mathbf{g}^* \tag{10.1.26}$$

in this case. In fact for any given $\mathbf{Y}$ and $\mathbf{Z}$, a unique $\mathbf{V}$ must exist by virtue of (10.1.20). However, if $\mathbf{Z}$ is given but $\mathbf{Y}$ is arbitrary then some freedom is possible in the choice of $\mathbf{V}$; this point is discussed further in Question 10.6. The generalization expressed by (10.1.20) is not entirely an academic one; it may be preferable not to form $\mathbf{Y}$ and $\mathbf{Z}$ but to perform any convenient stable factorization $[\mathbf{A}:\mathbf{V}]$ and use this to generate $\mathbf{Y}$ and $\mathbf{Z}$ indirectly: in particular the simple form of (10.1.21) and (10.1.22) can be advantageous for larger problems or for large sparse problems.

Another method which can be described within this framework is the *reduced gradient method* of Wolfe (1963a). In terms of the active set method (Section 10.3) the matrix $\mathbf{V}$ is formed from normal vectors $\mathbf{a}_i$ of *inactive* constraints which have previously been active. Thus when a constraint becomes inactive the column $\mathbf{a}_i$ in $\mathbf{A}$ is transferred to $\mathbf{V}$ so that $[\mathbf{A}:\mathbf{V}]^{-1}$ need not be recomputed and only the partition line is repositioned. When an inactive constraint becomes active, the incoming vector $\mathbf{a}_i$ replaces one column of $\mathbf{V}$. The choice of column is arbitrary and can be made so as to keep $[\mathbf{A}:\mathbf{V}]$ well conditioned. This exchange of columns is analogous to that in linear programming (see equation (8.2.14)) and developments such as those described in Section 8.5 for taking account of sparsity or maintaining stability can be taken over into this method.

Other methods for QP also can be interpreted in these terms. Murray (1971) gives a method in which a matrix $\mathbf{Z}$ is constructed for which

$$\mathbf{Z}^{\mathrm{T}}\mathbf{G}\mathbf{Z} = \mathbf{D} \tag{10.1.27}$$

where $\mathbf{D}$ is diagonal. In this method columns of $\mathbf{Z}$ can be interpreted as conjugate directions, and are related to the matrix ($\hat{\mathbf{Z}}$ say) in the orthogonal factorization method by $\mathbf{Z} = \hat{\mathbf{Z}}\mathbf{L}^{-1}$, where $\hat{\mathbf{Z}}^{\mathrm{T}}\mathbf{G}\hat{\mathbf{Z}}$ has factors $\mathbf{L}\mathbf{D}\mathbf{L}^{\mathrm{T}}$. Because of this the method essentially represents inverse matrix information and may be doubtful on grounds of stability. However, it does have the advantage that the solution of (10.1.13) becomes trivial and no storage of factors of $\mathbf{Z}^{\mathrm{T}}\mathbf{G}\mathbf{Z}$ is required. It is also possible to consider Beale's (1959) method (see Section 10.6) partly within this framework since it can be interpreted as selecting conjugate search directions from the rows of $[\mathbf{A}:\mathbf{V}]^{-1}$. In this case the matrix $\mathbf{V}$ is formed from gradient differences $\gamma^{(j)}$ on previous iterations. However, the method also has some disadvantageous features.

It is not easy to give a unique best choice amongst these many methods, because this depends on the type of problem and also on whether the method is to be considered as part of an active set method for inequality constraints (Section 10.3). Moreover there are other methods described in Section 10.2, some of which are of interest. However, the orthogonal factorization method is advantageous in that calculating $\mathbf{Z}$ involves operations with elementary orthogonal matrices which are very stable numerically. Also this choice ($\mathbf{Z} = \mathbf{Q}_2$) gives the best possible bound

$$\kappa(\mathbf{Z}^T\mathbf{G}\mathbf{Z}) \leqslant \kappa(\mathbf{G}) \tag{10.1.28}$$

on the condition number $\kappa(\mathbf{Z}^T\mathbf{G}\mathbf{Z})$. This is not to say that it gives the best $\kappa(\mathbf{Z}^T\mathbf{G}\mathbf{Z})$ itself, as Gill and Murray (1974a) erroneously claim. Indeed a trivial modification to Murray's (1971) method above with $\mathbf{D} = \mathbf{I}$ enables a $\mathbf{Z}$ matrix to be calculated for which $\kappa(\mathbf{Z}^T\mathbf{G}\mathbf{Z}) = 1$. In fact the relevance of the condition number is doubtful in that it can be changed by symmetric row and column scaling in $\mathbf{Z}^T\mathbf{G}\mathbf{Z}$ without changing the propagation of errors. Insofar as (10.1.28) does have some meaning it indicates that $\kappa(\mathbf{Z}^T\mathbf{G}\mathbf{Z})$ cannot become arbitrarily large. However this conclusion can also be obtained in regard to other methods if careful attention is given to pivoting, and I see no reason why these methods could not be implemented in a reasonably stable way. On the other hand, an arbitrary choice of $\mathbf{V}$ which makes the columns of $\mathbf{Z}$ very close to being dependent might be expected to induce unpleasantly large growth of round-off errors.

10.2 LAGRANGIAN METHODS

An alternative way of deriving the solution $\mathbf{x}^*$ to (10.1.2) and the associated multipiers $\boldsymbol{\lambda}^*$ is by the *method of Lagrange multipliers* (9.1.5). The Lagrangian function (9.1.6) becomes

$$\mathscr{L}(\mathbf{x}, \boldsymbol{\lambda}) = \tfrac{1}{2}\mathbf{x}^T\mathbf{G}\mathbf{x} + \mathbf{g}^T\mathbf{x} - \boldsymbol{\lambda}^T(\mathbf{A}^T\mathbf{x} - \mathbf{b}) \tag{10.2.1}$$

and the stationary point condition (9.1.7) yields the equations

$$\begin{aligned} \nabla_x\mathscr{L} = \mathbf{0}: &\quad \mathbf{G}\mathbf{x} + \mathbf{g} - \mathbf{A}\boldsymbol{\lambda} = \mathbf{0} \\ \nabla_\lambda\mathscr{L} = \mathbf{0}: &\quad \mathbf{A}^T\mathbf{x} - \mathbf{b} = \mathbf{0} \end{aligned}$$

which can be rearranged to give the linear system

$$\begin{bmatrix} \mathbf{G} & -\mathbf{A} \\ -\mathbf{A}^T & \mathbf{0} \end{bmatrix} \begin{pmatrix} \mathbf{x} \\ \boldsymbol{\lambda} \end{pmatrix} = -\begin{pmatrix} \mathbf{g} \\ \mathbf{b} \end{pmatrix}. \tag{10.2.2}$$

The coefficient matrix is referred to as the *Lagrangian matrix* and is symmetric but not positive definite. If the inverse exists and is expressed as

$$\begin{bmatrix} \mathbf{G} & -\mathbf{A} \\ -\mathbf{A}^T & \mathbf{0} \end{bmatrix}^{-1} = \begin{bmatrix} \mathbf{H} & -\mathbf{T} \\ -\mathbf{T}^T & \mathbf{U} \end{bmatrix} \tag{10.2.3}$$

then the solution to (10.2.2) can be written

$$\mathbf{x}^* = -\mathbf{H}\mathbf{g} + \mathbf{T}\mathbf{b} \tag{10.2.4}$$

$$\boldsymbol{\lambda}^* = \mathbf{T}^T\mathbf{g} - \mathbf{U}\mathbf{b}. \tag{10.2.5}$$

These relationships are used by Fletcher (1971b) to solve the equality constraint problem (10.3.1) which arises in the active set method. Since $\mathbf{b} = \mathbf{0}$ in that case only the matrices $\mathbf{H}$ and $\mathbf{T}$ need to be stored. Explicit expressions for $\mathbf{H}, \mathbf{T}$, and $\mathbf{U}$ when $\mathbf{G}^{-1}$ exists are

$$\begin{aligned} \mathbf{H} &= \mathbf{G}^{-1} - \mathbf{G}^{-1}\mathbf{A}(\mathbf{A}^T\mathbf{G}^{-1}\mathbf{A})^{-1}\mathbf{A}^T\mathbf{G}^{-1} \\ \mathbf{T} &= \mathbf{G}^{-1}\mathbf{A}(\mathbf{A}^T\mathbf{G}^{-1}\mathbf{A})^{-1} \\ \mathbf{U} &= -(\mathbf{A}^T\mathbf{G}^{-1}\mathbf{A})^{-1} \end{aligned} \tag{10.2.6}$$

and are readily verified by multiplying out the Lagrangian matrix and its inverse. Murtagh and Sargent (1969) suggest methods for linear constraints which use this representation by storing $(\mathbf{A}^T\mathbf{G}^{-1}\mathbf{A})^{-1}$ and $\mathbf{G}^{-1}$. However it is not necessary for $\mathbf{G}^{-1}$ to exist for the Lagrangian matrix to be non-singular. Neither of the above methods are recommended in practice since they represent inverses directly and so have potential stability problems. In fact if $\mathbf{Y}$ and $\mathbf{Z}$ are defined by (10.1.20) then an alternative representation of the inverse Lagrangian matrix is

$$\begin{aligned} \mathbf{H} &= \mathbf{Z}(\mathbf{Z}^T\mathbf{G}\mathbf{Z})^{-1}\mathbf{Z}^T \\ \mathbf{T} &= \mathbf{Y} - \mathbf{Z}(\mathbf{Z}^T\mathbf{G}\mathbf{Z})^{-1}\mathbf{Z}^T\mathbf{G}\mathbf{Y} \\ \mathbf{U} &= \mathbf{Y}^T\mathbf{G}\mathbf{Z}(\mathbf{Z}^T\mathbf{G}\mathbf{Z})^{-1}\mathbf{Z}^T\mathbf{G}\mathbf{Y} - \mathbf{Y}^T\mathbf{G}\mathbf{Y}. \end{aligned} \tag{10.2.7}$$

This can be verified by using relationships derived from (10.1.20) (see Question 10.7). Thus it can be regarded that the computation of $\mathbf{Y}, \mathbf{Z}$, and the $\mathbf{L}\mathbf{L}^T$ factors of $\mathbf{Z}^T\mathbf{G}\mathbf{Z}$ in any of the elimination methods is essentially a subtle way of factorizing the Lagrangian matrix in a Lagrangian method. Equations (10.2.7) prove that the Lagrangian matrix is non-singular if and only if there exists $\mathbf{Z}$ such that the matrix $\mathbf{Z}^T\mathbf{G}\mathbf{Z}$ is non-singular (see Question 10.11). Furthermore $\mathbf{x}^*$ is a unique local minimizer if and only if $\mathbf{Z}^T\mathbf{G}\mathbf{Z}$ is positive definite by virtue of the second order conditions (Section 2.1) applied to the quadratic function (10.1.12).

These different representations of the inverse of the Lagrangian matrix have their parallels in two different types of method that have been proposed for solving (10.2.4) and (10.2.5). Most general is the *null space method* in which $\mathbf{Y}$ and $\mathbf{Z}$ and Choleski factors $\mathbf{Z}^T\mathbf{G}\mathbf{Z} = \mathbf{L}\mathbf{L}^T$ are assumed to exist. The null space method is associated with representation (10.2.7) which is used to indicate how (10.2.4) and (10.2.5) are solved. Of course intermediate expressions are simplified where possible and inverse operations with triangular matrices are performed by forward or backward substitution. In fact this just gives what has been called the generalized elimination method in Section 10.1. Various different null space methods exist according to different ways of choosing $\mathbf{Y}$ and $\mathbf{Z}$ and their relative merits are discussed in Section 10.1. The alternative approach is a *range space method* which requires the Hessian matrix to be positive definite, and makes use of its Choleski factors, $\mathbf{G} = \mathbf{L}\mathbf{L}^T$ say. The range space method is associated with the representation (10.2.6). The use of Choleski factors enables (10.2.6) to be rearranged, and it is readily observed that the product $\mathbf{L}^{-1}\mathbf{A}$ occurs frequently. Recent range space methods are based on using QR factors of $\mathbf{L}^{-1}\mathbf{A}$, that is

$$\mathbf{L}^{-1}\mathbf{A} = [\mathbf{Q}_1 \vdots \mathbf{Q}_2]\begin{bmatrix}\mathbf{R}\\ \mathbf{0}\end{bmatrix} = \mathbf{Q}_1\mathbf{R},$$

and the matrices in (10.2.6) are determined from

$$\begin{aligned}\mathbf{H} &= \mathbf{L}^{-T}(\mathbf{I} - \mathbf{Q}_1\mathbf{Q}_1{}^T)\mathbf{L}^{-1}\\ \mathbf{T} &= \mathbf{L}^{-T}\mathbf{Q}_1\mathbf{R}^{-T}\\ \mathbf{U} &= -\mathbf{R}^{-1}\mathbf{R}^{-T}.\end{aligned}$$

Gill *et al.* (1984a) give one such method in which $\mathbf{Q}_1$ (but not $\mathbf{Q}_2$) and $\mathbf{R}$ are updated explicitly. Goldfarb and Idnani (1983) give a related method in which they introduce the matrix $\mathbf{S} = \mathbf{L}^{-T}\mathbf{Q}$, and they store and update $\mathbf{S}$ and $\mathbf{R}$. Clearly the formulae above allow $\mathbf{L}^{-T}\mathbf{Q}_1$ to be replaced by $\mathbf{S}_1$: it may be that the method could be improved by only storing and updating $\mathbf{S}_1$ and $\mathbf{R}$. Powell (1985a) compares the Goldfarb and Idnani representation with a direct QR representation, and concludes that, whilst the former may lack numerical stability in certain pathological cases, it is entirely adequate for all practical purposes. A further matrix scheme might be called the *low storage method* because it just updates $\mathbf{R}$, replacing $\mathbf{Q}_1$ by the product $\mathbf{L}^{-1}\mathbf{A}\mathbf{R}^{-1}$ in the above formulae. This transformation squares the condition number of the problem (e.g. Björck, 1985) and so is undesirable as it stands. However, it may be that iterative refinement could be used to advantage in this context, along the lines given by Björck. Range space methods are of most use when the Choleski factor of $\mathbf{G}$ is given a priori and does not have to be calculated. The operation counts for these methods are most favourable when there are few constraints, and the matrix operations do not allow much advantage to be taken of simple bounds.

Some other observations on the structure of the inverse in (10.2.3) are the following. Firstly $\mathbf{T}^T\mathbf{A} = \mathbf{I}$ so $\mathbf{T}^T$ is a left generalized inverse for $\mathbf{A}$. If $\mathbf{Z}^T\mathbf{G}\mathbf{Z}$ is positive definite, $\mathbf{H}$ is positive semi-definite with rank $n - m$. It also satisfies $\mathbf{H}\mathbf{A} = \mathbf{0}$ so projects any vector $\mathbf{v}$ into the constraint manifold (since $\mathbf{A}^T\mathbf{H}\mathbf{v} = \mathbf{0}$). If $\mathbf{x}^{(k)}$ is any feasible point then it follows that $\mathbf{A}^T\mathbf{x}^{(k)} = \mathbf{b}$, and $\mathbf{g}^{(k)} = \mathbf{g} + \mathbf{G}\mathbf{x}^{(k)}$ is the gradient vector of $q(\mathbf{x})$ at $\mathbf{x}^{(k)}$. Then using (10.2.6) it follows that (10.2.4) and (10.2.5) can be rearranged as

$$\mathbf{x}^* = \mathbf{x}^{(k)} - \mathbf{H}\mathbf{g}^{(k)} \tag{10.2.8}$$

$$\boldsymbol{\lambda}^* = \mathbf{T}^T\mathbf{g}^{(k)}. \tag{10.2.9}$$

Equation (10.2.8) has a close relationship with Newton's method and shows that $\mathbf{H}$ contains the correct curvature information for the feasible region and so can be regarded as a reduced inverse Hessian matrix. Therefore these formulae can be considered to define a projection method with respect to a metric $\mathbf{G}$ (if positive definite); that is if $\mathbf{G}$ is replaced by $\mathbf{I}$ then $\mathbf{H} = \mathbf{A}\mathbf{A}^+ = \mathbf{P}$ which is an orthogonal projection matrix, and the gradient projection method of Section 11.1 is obtained. In fact it can be established that $\mathbf{H} = (\mathbf{PGP})^+$, the generalized inverse of the projected Hessian matrix.

There are methods for factorizing a general non-singular symmetric matrix which is not positive definite, initially due to Bunch and Parlett (1971), which can also be used. This involves calculating $\mathbf{LDL}^{\mathrm{T}}$ factors of the Lagrangian matrix (with some symmetric pivoting) in which $\mathbf{D}$ is block diagonal with a mixture of 1×1 and 2×2 blocks. This therefore provides a stable means of solving (10.2.2), especially for problems in which $\mathbf{G}$ is indefinite. However, the method does ignore the zero matrix which exists in (10.2.2) and so does not take full advantage of the structure. Also since pivoting is involved the factors are not very convenient for being updated in an activé set method. On the other hand, the Lagrangian matrix is non-singular if and only if $\mathbf{A}$ has full rank, so generalized elimination methods which form $\mathbf{Y}$ and $\mathbf{Z}$ matrices in a reasonably stable way when $\mathbf{A}$ has full rank cannot be unstable as a means of factorizing the Lagrangian matrix as long as $\mathbf{Z}^{\mathrm{T}}\mathbf{GZ}$ is positive definite. Thus the stable generalized elimination methods seem to be preferable.

Yet another way of solving the Lagrangian system (10.2.2) is to factorize the Lagrangian matrix forward. Firstly $\mathbf{LDL}^{\mathrm{T}}$ factors of $\mathbf{G}$ are calculated which is stable so long as $\mathbf{G}$ is positive definite. Then these factors are used to eliminate the off-diagonal partitions ($-\mathbf{A}$ and $-\mathbf{A}^{\mathrm{T}}$) in the Lagrangian matrix. The $\mathbf{0}$ partition then becomes changed to $-\mathbf{A}^{\mathrm{T}}\mathbf{G}^{-1}\mathbf{A}$ which is negative definite and so can be factorized as $-\hat{\mathbf{L}}\hat{\mathbf{D}}\hat{\mathbf{L}}^{\mathrm{T}}$ with $\hat{\mathbf{D}} > \mathbf{0}$. The resulting factors $\bar{\mathbf{L}}\bar{\mathbf{D}}\bar{\mathbf{L}}^{\mathrm{T}}$ of the Lagrangian matrix are therefore given by

$$\bar{\mathbf{L}} = \begin{bmatrix} \mathbf{L} & \\ \mathbf{B}^{\mathrm{T}} & \hat{\mathbf{L}} \end{bmatrix}, \quad \bar{\mathbf{D}} = \begin{bmatrix} \mathbf{D} & \\ & -\hat{\mathbf{D}} \end{bmatrix} \tag{10.2.10}$$

where $\mathbf{B}$ is defined by $\mathbf{LDB} = -\mathbf{A}$ and is readily computed by forward substitution. To avoid loss of precision in forming $\mathbf{A}^{\mathrm{T}}\mathbf{G}^{-1}\mathbf{A}$ (the 'squaring' effect), $\hat{\mathbf{L}}$ and $\hat{\mathbf{D}}$ are best calculated by forming $\mathbf{D}^{-1/2}\mathbf{L}^{-1}\mathbf{A}$ and using the QR method to factorize this matrix into $\mathbf{Q}\hat{\mathbf{D}}^{1/2}\hat{\mathbf{L}}$. This method is most advantageous when $\mathbf{G}$ has some structure (for example a band matrix) which can be exploited and when the number of constraints m is small (see Question 10.3). It is not entirely obvious that the method is stable with regard to round-off-error, especially when $\mathbf{G}$ is nearly singular, but the method has been used successfully. The requirement that $\mathbf{G}$ is positive definite is important, however, in that otherwise computation of the $\mathbf{LDL}^{\mathrm{T}}$ factors may not be possible or may be unstable. The method is equivalent to the range space method using $\mathbf{QR}$ factors as described above. It is also closely related to the dual transformation described in (9.5.8) and (9.5.9) in the case that the constraints are equalities.

10.3 ACTIVE SET METHODS

Most QP problems involve inequality constraints and so can be expressed in the form given in (10.1.1) This section describes how methods for solving equality constraints can be generalized to handle the inequality problem by means of an *active set method* similar to that described for LP problems in Section 8.3. Most common is the *primal active set method* and most of the section is devoted to this method. It is described in the case that the Hessian matrix $\mathbf{G}$ is positive definite which ensures that any solution is a unique global minimizer, and that some potential difficulties are avoided. However, it is possible with care to handle the more general case which is described in Section 10.4. Later in the section the possibility of a *dual active set method* is considered, although this is only applicable to the case that $\mathbf{G}$ is positive definite.

In the primal active set method certain constraints, indexed by the *active set* $\mathscr{A}$, are regarded as equalities whilst the rest are temporarily disregarded, and the method adjusts this set in order to identify the correct active constraints at the solution to (10.1.1). Because the objective function is quadratic there may be m $(0 \leqslant m \leqslant n)$ active constraints, in contrast to linear programming when $m = n$. On iteration k a feasible point $\mathbf{x}^{(k)}$ is known which satisfies the active constraints as equalities, that is $\mathbf{a}_i^{\mathrm{T}}\mathbf{x}^{(k)} = b_i$, $i \in \mathscr{A}$. Also except in degenerate cases $\mathbf{a}_i^{\mathrm{T}}\mathbf{x}^{(k)} > b_i$, $i \notin \mathscr{A}$, so that the current active set $\mathscr{A}$ is equivalent to the set of active constraints $\mathscr{A}^{(k)}$ defined in (7.1.2). Each iteration attempts to locate the solution to an *equality problem* (EP) in which only the active constraints occur. This is most conveniently done by shifting the origin to $\mathbf{x}^{(k)}$ and looking for a correction $\boldsymbol{\delta}^{(k)}$ which solves

$$\begin{aligned} &\underset{\boldsymbol{\delta}}{\text{minimize}} \quad \tfrac{1}{2}\boldsymbol{\delta}^{\mathrm{T}}\mathbf{G}\boldsymbol{\delta} + \boldsymbol{\delta}^{\mathrm{T}}\mathbf{g}^{(k)} \\ &\text{subject to} \quad \mathbf{a}_i^{\mathrm{T}}\boldsymbol{\delta} = 0, \qquad i \in \mathscr{A} \end{aligned} \tag{10.3.1}$$

where $\mathbf{g}^{(k)}$ is defined by $\mathbf{g}^{(k)} = \mathbf{g} + \mathbf{G}\mathbf{x}^{(k)}$ and is $\nabla q(\mathbf{x}^{(k)})$ for the function $q(\mathbf{x})$ defined in (10.1.1). This of course is basically in the form of (10.1.2) so can be solved by any of the methods of Sections 10.1 and 10.2. If $\boldsymbol{\delta}^{(k)}$ is feasible with regard to the constraints not in $\mathscr{A}$, then the next iterate is taken as $\mathbf{x}^{(k+1)} = \mathbf{x}^{(k)} + \boldsymbol{\delta}^{(k)}$. If not then a line search is made in the direction of $\boldsymbol{\delta}^{(k)}$ to find the best feasible point as described in (8.3.6). This can be expressed by defining the solution of (10.3.1) as a search direction $\mathbf{s}^{(k)}$, and choosing the step $\alpha^{(k)}$ to solve

$$\alpha^{(k)} = \min\left(1, \min_{\substack{i: i \notin \mathscr{A}, \\ \mathbf{a}_i^{\mathrm{T}}\mathbf{s}^{(k)} < 0}} \frac{b_i - \mathbf{a}_i^{\mathrm{T}}\mathbf{x}^{(k)}}{\mathbf{a}_i^{\mathrm{T}}\mathbf{s}^{(k)}}\right) \tag{10.3.2}$$

so that $\mathbf{x}^{(k+1)} = \mathbf{x}^{(k)} + \alpha^{(k)}\mathbf{s}^{(k)}$. If $\alpha^{(k)} < 1$ in (10.3.2) then a new constraint becomes active, defined by the index (p say) which achieves the min in (10.3.2), and this index is added to the active set $\mathscr{A}$.

If $\mathbf{x}^{(k)}$ (that is $\boldsymbol{\delta} = \mathbf{0}$) solves the current EP (10.3.1) then it is possible to compute multipliers ($\boldsymbol{\lambda}^{(k)}$ say) for the active constraints as described in Sections 10.1 and

10.2. The vectors $\mathbf{x}^{(k)}$ and $\boldsymbol{\lambda}^{(k)}$ then satisfy all the first order conditions (9.1.16) for the inequality constraint problem except possibly the dual feasibility conditions that $\lambda_i \geqslant 0$, $i \in I$. Thus a test is made to determine whether $\lambda_i^{(k)} \geqslant 0$ $\forall i \in \mathscr{A}^{(k)} \cap I$; If so then the first order conditions are satisfied and these are sufficient to ensure a global solution since $q(\mathbf{x})$ is convex. Otherwise there exists an index, q say, $q \in \mathscr{A}^{(k)} \cap I$, such that $\lambda_q^{(k)} < 0$. In this case, following the discussion in Section 9.1, it is possible to reduce $q(\mathbf{x})$ by allowing the constraint q to become inactive. Thus q is removed from $\mathscr{A}$ and the algorithm continues as before by solving the resulting EP (10.3.1). If there is more than one index for which $\lambda_q^{(k)} < 0$ then it is usual to select q to solve

$$\min_{i \in \mathscr{A} \cap I} \lambda_i^{(k)}. \tag{10.3.3}$$

This selection works quite well and is convenient, so is usually used. One slight disadvantage is that it is not invariant to scaling of the constraints so some attention to scaling may be required. An invariant but more complex test can be devised based on the expected reduction in $q(\mathbf{x})$ (Fletcher and Jackson, 1974). To summarize the algorithm, therefore, if $\mathbf{x}^{(1)}$ is a given feasible point and $\mathscr{A} = \mathscr{A}^{(1)}$ is the corresponding active set then the primal active set method is defined as follows.

(a) Given $\mathbf{x}^{(1)}$ and $\mathscr{A}$, set $k = 1$.
(b) If $\boldsymbol{\delta} = \mathbf{0}$ does not solve (10.3.1) go to (d).
(c) Compute Lagrange multipliers $\boldsymbol{\lambda}^{(k)}$ and solve (10.3.3); if $\lambda_q^{(k)} \geqslant 0$ terminate with $\mathbf{x}^* = \mathbf{x}^{(k)}$, otherwise remove q from $\mathscr{A}$.
(d) Solve (10.3.1) for $\mathbf{s}^{(k)}$ (10.3.4)
(e) Find $\alpha^{(k)}$ to solve (10.3.2) and set $\mathbf{x}^{(k+1)} = \mathbf{x}^{(k)} + \alpha^{(k)}\mathbf{s}^{(k)}$.
(f) If $\alpha^{(k)} < 1$, add p to $\mathscr{A}$.
(g) Set $k = k + 1$ and go to (b).

An illustration of the method for a simple QP problem is shown in Figure 10.3.1. In this QP the constraints are the bounds $\mathbf{x} \geqslant \mathbf{0}$ and a general constraint $\mathbf{a}^T\mathbf{x} \geqslant b$.

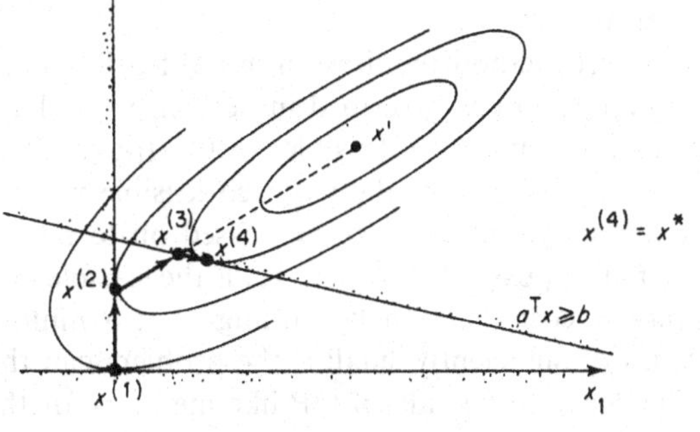

Figure 10.3.1 The primal active set method for a simple QP problem

At $\mathbf{x}^{(1)}$ the bounds $x_1 \geqslant 0$ and $x_2 \geqslant 0$ are both active and so $\mathbf{x}^{(1)}$ (that is $\boldsymbol{\delta} = \mathbf{0}$) solves the current EP (10.3.2). Calculating $\boldsymbol{\lambda}^{(1)}$ shows that the constraint $x_2 \geqslant 0$ has negative multiplier and so becomes inactive, so that only $x_1 \geqslant 0$ is active. The corresponding EP is now solved by $\mathbf{x}^{(2)}$ (or $\boldsymbol{\delta} = \mathbf{x}^{(2)} - \mathbf{x}^{(1)}$) which is feasible and so becomes the next iterate. Calculating $\boldsymbol{\lambda}^{(2)}$ indicates that the constraint $x_1 \geqslant 0$ has a negative multiplier so this too becomes inactive leaving $\mathscr{A}$ empty. The corresponding EP is now solved by $\mathbf{x}'$ which is infeasible, so the search direction $\mathbf{s}^{(2)} = \mathbf{x}' - \mathbf{x}^{(2)}$ is calculated from (10.3.1) and a line search along $\mathbf{s}^{(2)}$ by solving (10.3.2) gives $\mathbf{x}^{(3)}$ as the best feasible point. The general constraint $\mathbf{a}^{\mathrm{T}}\mathbf{x} \geqslant b$ becomes active at $\mathbf{x}^{(3)}$ so is added to $\mathscr{A}$. Since $\mathbf{x}^{(3)}$ does not solve the current EP, multipliers are not calculated, but instead the EP is solved, yielding $\mathbf{x}^{(4)}$ as the next iterate. Calculation of multipliers at $\mathbf{x}^{(4)}$ indicates that $\mathbf{x}^{(4)}$ is the solution of the inequality QP problem, and the algorithm terminates. A numerical example with similar properties is given in Question 10.4.

Termination of the algorithm in general can be proved if each step $\alpha^{(k)} \neq 0$, in which case $q(\mathbf{x})$ is reduced on each iteration. It is a consequence of (10.3.2) (see Question 10.10) that the vectors $\mathbf{a}_i$, $i \in \mathscr{A}$, are independent so the EP in (10.3.1) is always well defined. The termination proof relies on the fact that there is a subsequence $\{\mathbf{x}^{(k)}\}$ of iterates which solve the current EP. (Only when $\alpha^{(k)} < 1$ in (10.3.2) is $\mathbf{x}^{(k+1)}$ not a solution to the EP. In this case an index p is added to $\mathscr{A}$. This can happen at most n times in which case $\mathbf{x}^{(k)}$ is then a vertex, so solves the corresponding EP.) Since the number of possible EPs is finite, since each $\mathbf{x}^{(k)}$ in the subsequence is the unique global solution of an EP, and since $q(\mathbf{x}^{(k)})$ is monotonically decreasing, it follows that termination must occur. This proof fails when steps of zero length are taken and $q(\mathbf{x}^{(k)})$ does not decrease, in which case the algorithm can *cycle* by returning to a previous active set in the sequence. This is caused by *degeneracy* in the constraint set and the situation is similar to that described in Section 8.6. There is the additional possibility of a tie occurring when the value $\alpha^{(k)} = 1$ (for which $\mathbf{s}^{(k)}$ solves (10.3.1)) is also the value for which inactive constraints become active in (8.3.6). It is possible to give perturbation results which enable these ties to be broken in such a way as to theoretically avoid cycling, as in Section 8.6, but in practice there are some difficulties (again see Section 8.6) and most current algorithms ignore the fact that degeneracy can occur.

The primal active set method requires an initial feasible point $\mathbf{x}^{(1)}$; this can be calculated by the techniques described in Section 8.4, solving the auxiliary problem (8.4.6). This calculates a feasible vertex (possibly including some pseudoconstraints), which becomes the required feasible point $\mathbf{x}^{(1)}$. Factors of the $\mathbf{A}$ matrix for the active set can also be passed on to be used in the main algorithm. An alternative possibility which avoids the feasible point requirement is to bias the phase I cost function by adding to it a multiple $vq(\mathbf{x})$ of the quadratic function. For sufficiently small v, the problem can then be solved in a single phase. This leads to a study of QP-like methods for the problem

$$\text{minimize } \psi(\mathbf{x}) \triangleq q(\mathbf{x}) + \|\mathbf{l}(\mathbf{x})^-\|_1 \tag{10.3.5}$$

(for inequality constraints $\mathbf{l}(\mathbf{x}) \geqslant \mathbf{0}$), where $\mathbf{l}(\mathbf{x}) = \mathbf{A}^T\mathbf{x} - \mathbf{b}$ and $a^- = \max(-a, 0)$. This is the L_1QP *problem* which is an example of the use of an L_1 exact penalty function, described in more detail in Section 12.3. An active set method is also possible here which differs from (10.3.4) in a number of ways. One is that the line search (10.3.2) is changed to search for a minimizer of (10.3.5): if a constraint $l_p(\mathbf{x})$ becomes active (zero) then it is added to $\mathscr{A}$ as before. Another difference is that the first order conditions for (10.3.5) are that multipliers $\boldsymbol{\lambda}$ exist which satisfy $0 \leqslant \lambda_i \leqslant 1$ (see (12.3.9) and later in Section 12.3). Thus test (10.3.3) is changed in an obvious way to choose λ_q as that multiplier which violates these conditions the most. Likewise inactive constraints have their corresponding multiplier either 0 or 1 according to whether or not $l_i(\mathbf{x}) > 0$ or < 0. The EP which is solved is

$$\begin{aligned} &\underset{\mathbf{x}}{\text{minimize}} \quad q(\mathbf{x}) - \sum_{i \notin \mathscr{A}} \lambda_i l_i(\mathbf{x}) \\ &\text{subject to} \quad l_i(\mathbf{x}) = 0, \qquad i \in \mathscr{A}, \end{aligned} \tag{10.3.6}$$

and the techniques of Sections 10.1 and 10.2 are again relevant, together with suitable updating methods.

The relationship between the active set method (10.3.4) and some methods derived as extensions of linear programming is described in Section 10.6. Possible variations on (10.3.4) which have been suggested include the removal of more than one constraint (with $\lambda_i^{(k)} < 0$) from $\mathscr{A}$ in step (c) (Goldfarb, 1972), or the modification of step (b) so as to accept an approximate solution to the EP. Both these possibilities tend to induce smaller active sets. Goldfarb (1972) argues that this is advantageous, although no extensive experience is available and I would expect the amount of improvement to be small. The modification in which more than one constraint is removed also creates a potential difficulty for indefinite QP associated with the need to maintain stable factors of the curvature matrix $\mathbf{Z}^T\mathbf{G}\mathbf{Z}$.

A more recent method for convex QP is suggested by Goldfarb and Idnani (1983) and can be interpreted as a *dual active set method.* In it a sequence of vectors $\mathbf{x}$, $\boldsymbol{\lambda}$ is calculated which satisfy the KT conditions except for *primal* feasibility. Initially $\mathbf{x} = -\mathbf{G}^{-1}\mathbf{g}$ is the unconstrained minimizer of the quadratic function, $\mathscr{A} = \varnothing$, and $\boldsymbol{\lambda} = \mathbf{0}$ is a vertex in the dual space. A major iteration of the method consists of the following steps.

(i) Pick some q such that constraint q is infeasible in (10.1.1) and add q to $\mathscr{A}$.
(ii) If $\mathbf{a}_q \in \operatorname{span}\{\mathbf{a}_i, i \in \mathscr{A}\}$ then drop from $\mathscr{A}$ the index of an inequality constraint that will then become positive in step (iii).
(iii) Move towards the solution of the EP.
(iv) If any $\lambda_i \downarrow 0$ $i \in \mathscr{A} \cap I$ then remove i from $\mathscr{A}$ and go to (iii).

The major iteration is completed when the solution of the EP is reached in step (iii). Major iterations are continued until primal feasibility is recognized in step (i) ($\mathbf{x}$ is an optimal solution) or until no suitable index in $\mathscr{A}$ to drop can be found in step (ii) (the QP is infeasible). An unbounded QP cannot occur because **G** is positive definite. If the QP problem that is dual to (10.1.1) is set up (see (9.5.8) and (9.5.9)) then the Goldfarb and Idnani method is equivalent to the primal active set method being applied to this dual problem, and it is instructive to compare the properties of these primal and dual active set methods with this in mind. Although the Goldfarb and Idnani method looks for and removes primal infeasibilities, it is not necessarily the case that a primal constraint stays feasible once it is made feasible. The termination property of the method is assured by the fact that each iteration increases both the primal and dual objective functions. Dual feasibility is maintained throughout.

Good features of the method are that it can readily take advantage of the availability of Choleski factors of the matrix **G** such as can occur in some SQP methods for nonlinear programming (see Section 12.4). The method is most effective when there are only a few active constraints at the solution, and is also able to take advantage of a good estimate of $\mathscr{A}$. Degeneracy of the dual constraints $\lambda_i \geqslant 0$ $i \in I$ is not possible. The only difficulty caused by degeneracy in the primal constraints arises near the solution and it is claimed that this can readily be handled by the use of tolerances. Adverse features of the method include its lack of generality, in that the matrix **G** must be positive definite. In this respect, it is also the case that difficulties will arise when **G** is ill-conditioned, for example large elements and hence large round-off errors can arise in the vector $\mathbf{x} = -\mathbf{G}^{-1}\mathbf{g}$. However, Powell (1985a) claims that with care these difficulties can be circumvented. The method is most readily associated with the use of range space methods for solving the EP. It is also possible to develop other what might be called *primal-dual* methods for convex *QP* which allow both primal and dual infeasibilities, and take steps similar to those in the methods above (e.g. Goldfarb and Idnani, 1981). The transformation of a QP problem to a linear complementarity problem as described in Section 10.6 can also be regarded as giving rise to a number of primal-dual methods.

An important feature of any active set method concerns efficient solution of the EP (e.g. (10.3.1)) when changes are made to $\mathscr{A}$. As in linear programming it is possible to avoid refactorizing the Lagrangian matrix on each iteration (requiring $O(n^3)$ housekeeping operations) and instead to *update* the factors whenever a change is made to $\mathscr{A}$ ($O(n^2)$ operations). Some care has to be taken in the different methods in choosing how the various factors are represented to ensure that this updating is possible and efficient. These computations are quite intricate to write down and explain, even for just one method, so I shall not give details. Different ideas are used in different methods; amongst these are simplex-type updates and the idea for representing **QR** factors of **A** without requiring **Q**, described in Sections 8.5 and 10.2. Use of rank one updates and use of elementary Givens rotations often occur (Fletcher and Powell, 1975; Gill and Murray, 1978a). Permutations (interchange or cyclic) in the constraint or variable orderings are often required to achieve a favourable structure (for example Fletcher and Jackson, 1974). A review of a number of these techniques is given by Gill and Murray (1978a) for dense QP and by Gill *et al.* (1984b) for sparse *QP*. A description of the details in the case that the orthogonal factorization method in Section 10.1 is used to solve the EP is given by Gill and Murray (1978b). Another important feature of a similar nature is that the algorithms should treat bounds (when $\mathbf{a}_i = \pm \mathbf{e}_i$) in an efficient way. This is most readily achieved in null space methods. For instance if the first $p(\leqslant m)$ active constraints are lower bounds on the variables 1 to p, then the matrices $\mathbf{A}$, $\mathbf{Y}$, and $\mathbf{Z}$ have the structure

$$\mathbf{A} = \underset{\begin{matrix} p & m-p \end{matrix}}{\begin{bmatrix} \mathbf{I} & \mathbf{A}_1 \\ \mathbf{0} & \mathbf{A}_2 \end{bmatrix}} \begin{matrix} p \\ n-p \end{matrix}, \quad \mathbf{Y} = \underset{\begin{matrix} p & m-p \end{matrix}}{\begin{bmatrix} \mathbf{I} & \mathbf{0} \\ -\mathbf{Y}_2^{\mathrm{T}}\mathbf{A}_1^{\mathrm{T}} & \mathbf{Y}_2^{\mathrm{T}} \end{bmatrix}}, \quad \mathbf{Z} = \underset{n-m}{\begin{bmatrix} \mathbf{0} \\ \mathbf{Z}_2 \end{bmatrix}}$$

where $\mathbf{Y}_2$ is any left generalized for $\mathbf{A}_2$ ($\mathbf{Y}_2^{\mathrm{T}}\mathbf{A}_2 = \mathbf{I}$). This structure cuts down storage and operations requirements but makes the updating more complicated as there are four cases to consider (adding or removing either a bound or a general constraint from $\mathscr{A}$) instead of two.

10.4 ADVANCED FEATURES

A modern code for QP should be of wider application than has so far been considered; it should be able to handle an indefinite QP problem, and it should enable the user to cary forward efficiently information from a previous QP calculation when a sequence of problems is being solved. These extensions are considered in this section. Indefinite QP refers to the case in which the Hessian $\mathbf{G}$ is not required to be positive definite, so will often be indefinite, although the negative and positive semi-definite cases are also included. For equality constrained QP no difficulties arise and a unique global solution exists if and only if the reduced Hessian matrix $\mathbf{Z}^{\mathrm{T}}\mathbf{G}\mathbf{Z}$ is positive definite. However when inequalities are present, and excluding the positive semi-definite case, it is possible for local solutions to exist which are not global. The problem of global optimization, here as elsewhere in the book, is usually very difficult so the algorithms only aim to find a local solution. In fact a modification of the active set method is used which only guarantees to find a KT point for which the reduced Hessian $\mathbf{Z}^{\mathrm{T}}\mathbf{G}\mathbf{Z}$ is positive definite. If $\lambda_i^* > 0$, $i \in I^*$, then these conditions are sufficient for a local solution. However, if there exists $\lambda_i^* = 0$, $i \in I^*$, then it is possible that the method will find a KT point which is not a local solution. In this case there usually exist arbitrarily small perturbations of the problem, (a) which make the KT point a local solution or (b) which make the point no longer a KT point so that the active set method continues to find a better local solution. Both these features are illustrated in Question 10.9. However, although it is theoretically possible for these situations to arise, in practice (especially with round-off errors) it is unlikely and the algorithm is usually successful in finding a local (and often global) solution.

The main difficulty in solving indefinite QP problems arises in a different way. The possibility exists for an arbitrary choice of $\mathscr{A}$ that the resulting reduced Hessian matrix $\mathbf{Z}^{\mathrm{T}}\mathbf{G}\mathbf{Z}$ may be indefinite so that its $\mathbf{LDL}^{\mathrm{T}}$ factors may not exist or may be unstable in regard to round-off propagation. (Of course if $\mathbf{G}$ is positive definite then so is $\mathbf{Z}^T\mathbf{GZ}$ and no problems arise.) To some extent a null space method avoids the difficulty. Initially $\mathbf{x}^{(1)}$ is a vertex so the null space (10.1.9) has dimension zero and the difficulty does not arise. Also if $\mathbf{Z}^T\mathbf{GZ}$ is positive definite for some given $\mathscr{A}$ and a constraint is added to $\mathscr{A}$, then the new null space is a subset of the previous one, so the new matrix $\mathbf{Z}^T\mathbf{GZ}$ is of one less dimension and is also positive definite.

The difficulty arises therefore in step (c) of (10.3.4). In this case $\mathbf{x}^{(k)}$ solves the current EP and $\mathbf{Z}^T\mathbf{GZ}$ is positive definite. However, when a constraint index q is removed from $\mathscr{A}$, the dimension of the null space is increased and an extra column is adjoined to $\mathbf{Z}$. This causes an extra row and column to be adjoined to $\mathbf{Z}^T\mathbf{GZ}$, and it is possible that the resulting matrix is not positive definite. Computing the new $\mathbf{LDL}^T$ factors is just the usual step for extending a Choleski factorization, but when $\mathbf{Z}^T\mathbf{GZ}$ is not positive definite the new element of $\mathbf{D}$ is either negative or zero. (The latter case corresponds to a singular Lagrangian matrix.) This implies that the correction $\boldsymbol{\delta}^{(k)}$ which is a stationary point of (10.3.1) is no longer a minimizer, so some thought must be given to the choice of the search direction $\mathbf{s}^{(k)}$ in step (d) of (10.3.4). Since $\mathbf{x}^{(k)}$ is not a KT point, feasible descent directions exist and any such direction can be chosen for $\mathbf{s}^{(k)}$. One possibility is to consider the line caused by increasing the slack on constraint q at $\mathbf{x}^{(k)}$. Thus b_q is changed to $b_q + \alpha$ and from (10.2.4) the resulting search direction is $\mathbf{s}^{(k)} = \mathbf{T}\mathbf{e}_q$, the column of $\mathbf{T}$ corresponding to constraint q. In fact when the updated $\mathbf{D}$ has a negative element this is equivalent to choosing $\mathbf{s}^{(k)} = -\boldsymbol{\delta}^{(k)}$ where $\boldsymbol{\delta}^{(k)}$ is the stationary point of (10.3.1) for the updated problem. In this case the unit step $\alpha^{(k)} = 1$ in (10.3.2) no longer has significance. Thus the required modification to (10.3.4) when $\mathbf{D}$ has a negative element is to choose $\mathbf{s}^{(k)} = -\boldsymbol{\delta}^{(k)}$ in step (d), and to solve (8.3.6) rather than (10.3.2) to obtain $\alpha^{(k)}$ in step (e). If $\mathbf{D}$ has a zero element it can be replaced by $-\varepsilon$ where $\varepsilon > 0$ is negligible. This causes $\mathbf{s}^{(k)}$ to be very large but the length of $\mathbf{s}^{(k)}$ is unimportant in solving (8.3.6). It may be that (8.3.6) has no solution; this is the way in which an unbounded solution of an indefinite QP is indicated.

It can be seen therefore that in exact arithmetic, with the above changes, there is no real difficulty at all. The situation in which $\mathbf{D}$ has one negative element is only temporary since the algorithm proceeds to add constraints, reducing the dimension of the null space, until the solution of an EP is found, in which case the matrix $\mathbf{Z}^T\mathbf{GZ}$ and hence $\mathbf{D}$ is positive definite. No further forward extension of the $\mathbf{LDL}^T$ factorization is made which might break down. The only adverse possibility is that there is no bound on the possible size of the negative element of $\mathbf{D}$. Thus when this element is computed the resulting round-off error can be significant. Therefore Gill and Murray (1978b) suggest that this possibility is monitored, and if a large negative element is observed, then the factors are recomputed at a later stage.

An alternative possibility which avoids this round-off problem is the following. When a constraint q is determined as in step (c), then it can be left in $\mathscr{A}$ as a *pseudoconstraint*. The right-hand side b_q is allowed to increase (implicitly) to

$b_q + \alpha$ and the search direction $\mathbf{s}^{(k)} = \mathbf{Te}_q$ described above is followed. If a new constraint p becomes active in the line search, then p is added to $\mathscr{A}$ and the process is repeated. Only when the solution of an EP is located in a subsequent line search is the index q removed from $\mathscr{A}$. This approach has the advantage that all the reduced Hessian matrices $\mathbf{Z}^T\mathbf{GZ}$ which are computed are positive definite. The disadvantage is that additional degenerate situations arise when the incoming vector $\mathbf{a}_p$ in step (f) of (10.3.4) is dependent on the vectors $\mathbf{a}_i$, $i \in \mathscr{A} \cup q$. This arises most obviously when moving from one vertex to another, as in linear programming. A possible remedy is to have formulae for updating the factors of the Lagrangian matrix when an *exchange* of constraint indices in $\mathscr{A}$ is made (as in the simplex method for LP). An idea of this type is used by Fletcher (1971b) and should be practicable with more stable methods but has not, I believe, been investigated.

Another important feature of a QP code is that it should allow information to be passed forward from a previous QP calculation when the problem has only been perturbed by a small amount. This is an example of *parametric programming* (see Section 8.3 for the LP case). In this case it can be advantageous to solve an initial EP with the active set $\mathscr{A}$ from the previous QP calculation. The resulting point $\mathbf{x}^{(1)}$ will often solve the new QP problem. If $\mathbf{A}$ is unchanged then $\mathbf{Z}$ need not be recomputed in most methods, and if $\mathbf{G}$ is also unchanged then $\mathbf{Z}^T\mathbf{GZ}$ does not change so its factors need not be recomputed. There are, however, some potential difficulties. If $\mathbf{A}$ changes it may be that it comes close to losing rank, so that round-off difficulties occur. In indefinite QP problems, if $\mathbf{A}$ or $\mathbf{G}$ change it may be that $\mathbf{Z}^T\mathbf{GZ}$ is no longer positive definite. However, there are possible solutions to these difficulties without having to abandon the attempt to use the previous active set. Another possibility is that the initial solution $\mathbf{x}^{(1)}$ of the EP is infeasible with respect to certain constraints $i \in \mathscr{A}$. In this case it is preferable not to have to resort to trying to find a feasible vertex as in Section 8.4. A better idea is to replace the right-hand sides b_i of the violated constraints by $b_i + \theta$ and to trace the solution of the QP problem as $\theta \downarrow 0$. An alternative approach is based on using an exact penalty function as described at the end of Section 8.4 and in (10.3.5) and (12.3.19).

Another aspect of parametric programming is in finding the sensitivity of the solution with respect to changes in $\mathbf{g}$ and $\mathbf{b}$. The resulting rate of change of $\mathbf{x}^*$ and $\boldsymbol{\lambda}^*$ can be obtained by substituting in (10.2.4) and (10.2.5), and the effect on $q(\mathbf{x}^*)$ by using the definition of $q(\mathbf{x})$ in (10.1.1). Notice however that $\mathrm{d}q/\mathrm{d}b_i$ is approximated by λ_i^* by virtue of (9.1.10). These estimates may be invalid if there exist multipliers for which $\lambda_i^* = 0$, $i \in I^*$.

10.5 SPECIAL QP PROBLEMS

In some cases a QP problem has a special structure which can be utilized in order to gain improved efficiency or stability. A number of these possibilities are described in this section. An important case arises when the only constraints in the problem are the bounds

$$l_i \leqslant x_i \leqslant u_i, \qquad i = 1, 2, \ldots, n \tag{10.5.1}$$

where any u_i or $-l_i$ may be ∞ and $l_i \leqslant u_i$. Thus the vectors $\mathbf{a}_i$ are $\pm\mathbf{e}_i$ and it is possible to express the EP (after permuting the variables so that the active bounds are those on the first m variables) with

$$\mathbf{A} = \mathbf{Y} = \begin{bmatrix} \mathbf{I} \\ \mathbf{0} \end{bmatrix}\begin{matrix} m \\ n-m \end{matrix}, \qquad \mathbf{Z} = \mathbf{V} = \begin{bmatrix} \mathbf{0} \\ \mathbf{I} \end{bmatrix}\begin{matrix} m \\ n-m \end{matrix}, \tag{10.5.2}$$

and $\mathbf{Z}^{\mathrm{T}}\mathbf{G}\mathbf{Z} = \mathbf{G}_{22}$ (see the definition before (10.1.3)). Thus no calculations on $\mathbf{Z}$ are required and this can be exploited in a special purpose method. The only updating which is required is to the $\mathbf{LDL}^{\mathrm{T}}$ factors of $\mathbf{Z}^{\mathrm{T}}\mathbf{G}\mathbf{Z}$. Also from (10.1.14) the multipliers $\boldsymbol{\lambda}^{(k)}$ at the solution to an EP are just the corresponding elements of $\mathbf{g}^{(k)}$ by virtue of the above $\mathbf{Y}$. Another feature of this problem is that degeneracy in the constraints cannot arise. Also there are no difficulties in dealing with indefinite QP using the second method (based on pseudoconstraints) described in Section 10.4. The only exchange formula which is needed is that for replacing a lower bound on x_i by its upper bound, or vice versa, which is trivial. Pseudoconstraints (Section 8.4) are also important for calculating an initial vertex for the problem, especially if u_i and $-l_i$ are large. Methods for this problem are given by Gill and Murray (1976b) and by Fletcher and Jackson (1974). The latter method updates a partial factorization of the matrix $\begin{bmatrix} \mathbf{G}_{22} & \mathbf{G}_{12} \\ \mathbf{G}_{21} & \mathbf{G}_{11} \end{bmatrix}$; it now seems preferable to me to update merely the $\mathbf{LDL}^{\mathrm{T}}$ factors of $\mathbf{G}_{22}$ as mentioned above.

Another special case of QP arises when $\mathbf{G}$ is a positive definite diagonal matrix. By scaling the variables the objective function can be written with $\mathbf{G} = \mathbf{I}$ as

$$q(\mathbf{x}) = \tfrac{1}{2}\mathbf{x}^{\mathrm{T}}\mathbf{x} + \mathbf{g}^{\mathrm{T}}\mathbf{x}. \tag{10.5.3}$$

The resulting problem is known as a *least distance problem* since the solution $\mathbf{x}^*$ in (10.1.1) can be interpreted as the feasible point which is the least distance (in L_2) from the point $\mathbf{x} = -\mathbf{g}$ (after writing (10.5.3) as $\tfrac{1}{2}\|\mathbf{x} + \mathbf{g}\|_2^2$ and ignoring a constant term). Again a special purpose method is appropriate since the choice $\mathbf{Z} = \mathbf{Q}_2$ is advantageous in that $\mathbf{Z}^{\mathrm{T}}\mathbf{G}\mathbf{Z} = \mathbf{Z}^{\mathrm{T}}\mathbf{Z} = \mathbf{I}$. Thus no special provision need be made for updating factors of $\mathbf{Z}^{\mathrm{T}}\mathbf{G}\mathbf{Z}$ and only the updating operations on $\mathbf{Z}$ are required, for which the formulation of Gill and Murray (1978b) can be used. Since $\mathbf{G}$ is positive definite, no special difficulties arise.

Yet another special case of QP occurs when the problem is to find a *least squares solution* of the linear equations

$$\mathbf{B}\mathbf{x} = \mathbf{c} \tag{10.5.4}$$

where $\mathbf{B}$ is $p \times n$ and $\mathbf{c} \in \mathbb{R}^p$, subject to the constraints in (10.1.1). In this case the objective function is the sum of squares of residuals defined by

$$q(\mathbf{x}) = (\mathbf{B}\mathbf{x} - \mathbf{c})^{\mathrm{T}}(\mathbf{B}\mathbf{x} - \mathbf{c}). \tag{10.5.5}$$

It is possible to proceed as in the standard method with $\mathbf{G} = \mathbf{B}^T\mathbf{B}$ and $\mathbf{g} = -\mathbf{B}^T\mathbf{c}$, after dividing by 2 and omitting the constant term. However, it is well known that to form $\mathbf{B}^T\mathbf{B}$ explicitly in a linear least squares calculation causes the 'squaring effect' to occur which can cause the loss of twice as many significant figures as are necessary. It is therefore again advantageous to have a special purpose routine which avoids the difficulty. This can be done by updating QR factors of the matrix $\mathbf{BZ}$. That is to say,

$$\mathbf{BZ} = \mathbf{P}\begin{bmatrix} \mathbf{U} \\ \mathbf{0} \end{bmatrix} = \mathbf{P}_1\mathbf{U} \tag{10.5.6}$$

is defined where $\mathbf{P}$ is $p \times p$ orthogonal, $\mathbf{U}$ is $(n-m) \times (n-m)$ upper triangular, and $\mathbf{P}_1$ is the first $n-m$ columns of $\mathbf{P}$. It follows that $\mathbf{U}^T\mathbf{U}$ is the Choleski factorization of $\mathbf{Z}^T\mathbf{GZ}(=\mathbf{Z}^T\mathbf{B}^T\mathbf{BZ})$ and can thus be used to solve (10.1.3) or (10.3.1). A null space algorithm of this type described by Schittkowski and Stoer (1979) in which $\mathbf{Z}$ is defined from the QR factors of $\mathbf{A}$. Alternatively if $\mathbf{B} = \mathbf{Q}_1\mathbf{L}^T$ is the QR factorization of $\mathbf{B}$ itself, then $\mathbf{L}$ is the Choleski factor of $\mathbf{G}$ and could therefore be used in any of the range space methods described above.

Finally the special case of least squares QP subject only to bounds on the variables is considered. It is possible to introduce a further special purpose method, similar to those described above in which, using (10.5.2) and (10.5.6), $\mathbf{B}$ is written $[\mathbf{B}_1 : \mathbf{B}_2]$ and $\mathbf{P}_1\mathbf{U}$ factors of $\mathbf{B}_2$ are used to solve the EP. An alternative possibility is to transform the problem into a least distance problem using the Wolfe dual (Theorem 9.5.1). For example the QP

$$\begin{aligned} &\underset{\mathbf{x}}{\text{minimize}} \quad \tfrac{1}{2}\mathbf{x}^T\mathbf{B}^T\mathbf{B}\mathbf{x} + \mathbf{g}^T\mathbf{x} \\ &\text{subject to} \quad \mathbf{x} \geqslant \mathbf{0} \end{aligned} \tag{10.5.7}$$

becomes

$$\underset{\mathbf{x},\boldsymbol{\lambda}}{\text{maximize}} \quad \tfrac{1}{2}\mathbf{x}^T\mathbf{B}^T\mathbf{B}\mathbf{x} + \mathbf{g}^T\mathbf{x} - \boldsymbol{\lambda}^T\mathbf{x} \tag{10.5.8}$$

$$\text{subject to} \quad \mathbf{B}^T\mathbf{B}\mathbf{x} + \mathbf{g} - \boldsymbol{\lambda} = \mathbf{0} \tag{10.5.9}$$

$$\boldsymbol{\lambda} \geqslant \mathbf{0} \tag{10.5.10}$$

using (9.5.1). Eliminating $\boldsymbol{\lambda}$ from (10.5.8) using (10.5.9) and writing $\mathbf{Bx} = \mathbf{u}$ yields the least distance problem

$$\begin{aligned} &\underset{\mathbf{u}}{\text{minimize}} \quad \tfrac{1}{2}\mathbf{u}^T\mathbf{u} \\ &\text{subject to} \quad \mathbf{B}^T\mathbf{u} \geqslant -\mathbf{g}. \end{aligned} \tag{10.5.11}$$

The system can be solved by the method described above giving a solution $\mathbf{u}^*$. The solution $\mathbf{x}^*$ of (10.5.7) can then be recovered in the following way. By applying the dual transformation (Theorem 9.5.1) to (10.5.11) it is easily seen that the dual of (10.5.11) is (10.5.7) so that the vector $\mathbf{x}$ is the multiplier vector for the constraints in (10.5.11). Thus $x_i^* = 0$ if i corresponds to an inactive constraint at the solution of (10.5.11). Furthermore a QR factorization of the columns of $\mathbf{B}$ corresponding to the active constraints is available (see (10.1.17)) so the remaining elements of $\mathbf{x}^*$ can be determined from $\mathbf{B}\mathbf{x}^* = \mathbf{u}^*$. Thus bounded least squares problems can be reduced to least distance problems, and *vice versa.* This result is a special case of the duality between (9.5.8) and (9.5.9) given in Section 9.5. Another example of this correspondence is explored in Question 10.15

10.6 COMPLEMENTARY PIVOTING AND OTHER METHODS

A number of methods for QP have been suggested as extensions of the simplex method for LP. The earliest of these is probably the Dantzig–Wolfe algorithm (Dantzig, 1963) which solves the KT conditions for the QP

$$\begin{aligned} &\underset{\mathbf{x}}{\text{minimize}} \quad \tfrac{1}{2}\mathbf{x}^T\mathbf{G}\mathbf{x} + \mathbf{g}^T\mathbf{x} \\ &\text{subject to} \quad \mathbf{A}^T\mathbf{x} \geqslant \mathbf{b}, \qquad \mathbf{x} \geqslant \mathbf{0}. \end{aligned} \tag{10.6.1}$$

Introducing multipliers $\boldsymbol{\lambda}$ for the constraints $\mathbf{A}^T\mathbf{x} \geqslant \mathbf{b}$ and $\boldsymbol{\pi}$ for the bounds $\mathbf{x} \geqslant \mathbf{0}$ gives the Lagrangian function

$$\mathscr{L}(\mathbf{x}, \boldsymbol{\lambda}, \boldsymbol{\pi}) = \tfrac{1}{2}\mathbf{x}^T\mathbf{G}\mathbf{x} + \mathbf{g}^T\mathbf{x} - \boldsymbol{\lambda}^T(\mathbf{A}^T\mathbf{x} - \mathbf{b}) - \boldsymbol{\pi}^T\mathbf{x}. \tag{10.6.2}$$

Defining slack variables $\mathbf{r} = \mathbf{A}^T\mathbf{x} - \mathbf{b}$, the KT conditions (9.1.16) become

$$\begin{aligned} &\boldsymbol{\pi} - \mathbf{G}\mathbf{x} + \mathbf{A}\boldsymbol{\lambda} = \mathbf{g} \\ &\mathbf{r} - \mathbf{A}^T\mathbf{x} = -\mathbf{b} \\ &\boldsymbol{\pi}, \mathbf{r}, \mathbf{x}, \boldsymbol{\lambda} \geqslant \mathbf{0}, \qquad \boldsymbol{\pi}^T\mathbf{x} = 0, \qquad \mathbf{r}^T\boldsymbol{\lambda} = 0. \end{aligned} \tag{10.6.3}$$

The method assumes that $\mathbf{G}$ is positive definite which is a sufficient condition that any solution to (10.6.3) solves (10.6.1). The system (10.6.3) can be expressed in the form

$$\begin{aligned} &\mathbf{w} - \mathbf{M}\mathbf{z} = \mathbf{q} \\ &\mathbf{w} \geqslant \mathbf{0}, \qquad \mathbf{z} \geqslant \mathbf{0}, \qquad \mathbf{w}^T\mathbf{z} = 0 \end{aligned} \tag{10.6.4}$$

where

$$\mathbf{w} = \begin{pmatrix} \boldsymbol{\pi} \\ \mathbf{r} \end{pmatrix}, \quad \mathbf{z} = \begin{pmatrix} \mathbf{x} \\ \boldsymbol{\lambda} \end{pmatrix}, \quad \mathbf{M} = \begin{bmatrix} \mathbf{G} & -\mathbf{A} \\ \mathbf{A}^T & \mathbf{0} \end{bmatrix}, \quad \mathbf{q} = \begin{pmatrix} \mathbf{g} \\ -\mathbf{b} \end{pmatrix}. \tag{10.6.5}$$

This has come to be referred to as a *linear complementarity problem* (*LCP*) and there are other problems in game theory and boundary value calculations which can also be expressed in this way. The Dantzig–Wolfe method for solving (10.6.3) has evolved to give the *principal pivoting method* for solving (10.6.5) which is outlined below. However all methods for solving (10.6.4) have some features in common. They carry out row operations on the equations in (10.6.4) in a closely related way to LP (Section 8.2). Thus the variables $\mathbf{w}$ and $\mathbf{z}$ together are rearranged into basic (B) and nonbasic (N) variables $\mathbf{x}_B$ and $\mathbf{x}_N$. At a general stage, a tableau representation

$$[\mathbf{I}:\hat{\mathbf{A}}]\begin{pmatrix}\mathbf{x}_B\\ \mathbf{x}_N\end{pmatrix}=\hat{\mathbf{b}} \tag{10.6.6}$$

of the equations in (10.6.4) is available and the variables have the values $\mathbf{x}_B=\hat{\mathbf{b}}$ and $\mathbf{x}_N=\mathbf{0}$. A variable x_q, $q\in N$, is chosen to be increased and the effect on $\mathbf{x}_B$ from (10.6.6) is given by

$$\mathbf{x}_B=\hat{\mathbf{b}}-\hat{\mathbf{a}}_q x_q \tag{10.6.7}$$

where $\hat{\mathbf{a}}_q$ is the column of $\hat{\mathbf{A}}$ corresponding to variable x_q. An element x_i, $i\in B$, becomes zero therefore when $x_q=\hat{b}_i/\hat{a}_{iq}$, and this is used in a test similar to (8.2.11). This determines some variable x_p which has become zero, and so p and q are interchanged between B and N as in LP. The tableau is then rearranged by making row operations which reduce the old x_q column of the tableau to a unit vector (the *pivot step* of LP – see Section 8.2 and Table 8.2.1). The algorithms terminate when $\hat{\mathbf{b}}\geqslant\mathbf{0}$ and when the solution is *complementary*, that is $z_i\in B\Leftrightarrow w_i\in N$ and $z_i\in N\Leftrightarrow w_i\in B$, for all i.

The principal pivoting method is initialized by $\mathbf{x}_B=\mathbf{w}$, $\mathbf{x}_N=\mathbf{z}$, $\hat{\mathbf{A}}=-\mathbf{M}$, and $\hat{\mathbf{b}}=\mathbf{q}$. This is complementary so the algorithm terminates if $\hat{\mathbf{b}}\geqslant\mathbf{0}$. If not, there

Table 10.6.1 Tableaux for the principal pivoting method

B	π_1	π_2	r_1	x_1	x_2	λ_1	$\hat{\mathbf{b}}$	
π_1	1			-2	1	-1	-3	
π_2		1		$\textcircled{1}$	-2	-1	0	$\uparrow x_1 \quad \pi_2\downarrow 0$
r_1			1	1	1	0	2	
π_1	1	2			-3	-3	-3	
x_1		1		1	-2	-1	0	$\uparrow x_2 \quad r_1\downarrow 0$
r_1		-1	1		$\textcircled{3}$	1	2	
π_1	1	1	1			$\textcircled{-2}$	-1	
x_1		$\frac{1}{3}$	$\frac{2}{3}$	1		$-\frac{1}{3}$	$\frac{4}{3}$	$\uparrow \lambda_1 \quad \pi_1\downarrow 0$
x_2		$-\frac{1}{3}$	$\frac{1}{3}$		1	$\frac{1}{3}$	$\frac{2}{3}$	
λ_1	$-\frac{1}{2}$	$-\frac{1}{2}$	$-\frac{1}{2}$			1	$\frac{1}{2}$	
x_1	$-\frac{1}{6}$	$\frac{1}{6}$	$\frac{1}{2}$	1			$\frac{3}{2}$	
x_2	$\frac{1}{6}$	$-\frac{1}{6}$	$\frac{1}{2}$		1		$\frac{1}{2}$	

is an element $x_t \in B$ for which $\hat{b}_t < 0$ (the most negative is chosen if more than one exists). The complementary variable $x_q \in N$ is then chosen to be increased. (x_t and x_q are complementary iff (x_t, x_q) can be identified as (w_i, z_i) or (z_i, w_i) for some i.) The effect on the basic variables is observed by virtue of (10.6.7). As long as all non-negative variables in B stay non-negative, then x_q is increased until $x_t \uparrow 0$. Then x_q and x_t are interchanged in B and N and the tableau is updated; the resulting tableau is again complementary so the iteration can be repeated. It may be, however, that when x_q is increased then a basic variable $x_p \downarrow 0$. In this case a pivot interchange is made to give a non-complementary tableau. On the next iteration the complement of x_p is chosen to be increased. If this causes $x_t \uparrow 0$ then complementarity is restored and the algorithm can proceed as described above. However, a different x_p may become zero in which case the same operation of increasing the complement is repeated. Excluding degeneracy, each iteration increases x_t so eventually complementarity is always restored. An example of the method applied to the problem of Question 10.4 is given in Table 10.6.1. The Dantzig–Wolfe method differs in only one respect from the principal pivoting method, in that it allows basic variables that correspond to $\boldsymbol{\lambda}$ or $\boldsymbol{\pi}$ variables in (10.6.3) to go negative during the search. The Dantzig–Wolfe method is illustrated in Table 10.6.2 and the differences

Table 10.6.2 Tableaux for the Dantzig–Wolfe method

B	π_1	π_2	r_1	x_1	x_2	λ_1	$\hat{b}$	
π_1	1			(−2)	1	−1	−3	
π_2		1		1	−2	−1	0	$\uparrow x_1 \quad \pi_1 \downarrow 0$
r_1			1	1	1	0	2	
x_1	$-\frac{1}{2}$			1	$-\frac{1}{2}$	$\frac{1}{2}$	$\frac{3}{2}$	
π_2	$\frac{1}{2}$	1			$-\frac{3}{2}$	$-\frac{3}{2}$	$-\frac{3}{2}$	$\uparrow x_2 \quad r_1 \downarrow 0$
r_1	$\frac{1}{2}$		1		($\frac{3}{2}$)	$-\frac{1}{2}$	$\frac{1}{2}$	
x_1	$-\frac{1}{3}$		$\frac{1}{3}$	1		$\frac{1}{3}$	$\frac{5}{3}$	
π_2	1	1	1			(−2)	−1	$\uparrow \lambda_1 \quad \pi_2 \downarrow 0$
x_2	$\frac{1}{3}$		$\frac{2}{3}$		1	$-\frac{1}{3}$	$\frac{1}{3}$	
x_1	$-\frac{1}{6}$	$\frac{1}{6}$	$\frac{1}{2}$	(1)			$\frac{3}{2}$	
λ_1	$-\frac{1}{2}$	$-\frac{1}{2}$	$-\frac{1}{2}$			1	$\frac{1}{2}$	
x_2	$\frac{1}{6}$	$-\frac{1}{6}$	$\frac{1}{2}$		1		$\frac{1}{2}$	

with Table 10.6.1 can be seen. It is not difficult along the lines of Question 8.17 to show that the Dantzig–Wolfe method is equivalent to the primal active set method (10.3.4). A complementary solution is equivalent to the solution of an EP, and choosing the most negative x_t is analogous to (10.3.3). The case when $x_t \uparrow 0$ is the same as choosing $\alpha^{(k)} = 1$ in (10.3.2), and the case $x_p \downarrow 0$ is equivalent to $\alpha^{(k)} < 1$ when an inactive constraint becomes active in the line search. These features can all be observed in Table 10.6.2. Solving the problem in the tableau form has the advantage of simplicity for a small problem and can give a compact code for microcomputer applications. Otherwise, however, the tableau form is disadvantageous; it does not exploit efficient factorizations of the Lagrangian matrix and instead updates essentially an inverse matrix which can be numerically unstable. Slack variables for all the inequalities are required so that the tableau can be large. Only a restricted form (10.6.1) of the QP problem is solved and the method may fail for indefinite QP. Thus in general a modern code for the active set method is preferable.

A different algorithm for solving the linear complementarity problem (10.6.4) is due to Lemke (1965) (see the review of Cottle and Dantzig, 1968). In this method the tableau is extended by adding an extra variable z_0 and a corresponding extra column $-\mathbf{e} = -(1, 1, \ldots, 1)^{\mathrm{T}}$ to $\hat{\mathbf{A}}$. Problem (10.6.4) is initialized by $\mathbf{x}_B = \mathbf{w}$, $\mathbf{x}_N = (z_0, \mathbf{z})$, $\hat{\mathbf{A}} = (-\mathbf{e}, \ -\mathbf{M})$, $\hat{\mathbf{b}} = \mathbf{q}$, and the values of the variables are $x_N = \mathbf{0}$ and $\mathbf{x}_B = \hat{\mathbf{b}}$. If $\hat{\mathbf{b}} \geqslant \mathbf{0}$ then this is the solution; otherwise z_0 is increased in (10.6.6) until all the variables $\mathbf{w}$ are non-negative.The algorithm now attempts to reduce z_0 down to zero whilst retaining feasibility in (10.6.4). The effect of the initial step is to drive a variable, x_p say, to zero (corresponding to the most negative q_i). Then x_p and z_0 are interchanged in the pivot step and the complement of x_p is increased at the start of the second iteration. In general the variable being increased and the variable which is driven to zero are interchanged at the pivot step. The next variable to be increased is then the complement of the one which is driven to zero. The algorithm terminates when z_0 becomes zero, which gives a solution to (10.6.4). An example of the method applied to the problem in Question 10.4 is given in Table 10.6.3. An observation about the method is that including the z_0 term makes each tableau in the method a solution of a modified form of problem (10.6.1) in which $\mathbf{g}$ is replaced by $\mathbf{g} + z_0\mathbf{e}$ and $\mathbf{b}$ by $\mathbf{b} - z_0\mathbf{e}$. Thus Lemke's method can be regarded as a parametric programming solution to (10.6.1) in which the solution is traced out as the parameter $z_0 \downarrow 0$. A step in which $\pi_i \downarrow 0$ or $\lambda_i \downarrow 0$ corresponds to an active bound or constraint becoming inactive as z_0 is decreased. A step in which $x_i \downarrow 0$ and $r_i \downarrow 0$ corresponds to an inactive bound or constraint becoming active. An advantage that is claimed for Lemke's algorithm is that it does not require special techniques to resolve degeneracy. A form of Lemke's algorithm without the z_0 variable arises if there is some column $\mathbf{m}_q \geqslant \mathbf{0}$ in $\mathbf{M}$ such that $m_{jq} > 0$ when $q_j < 0$. Then an initial step to increase z_q is possible which from (10.6.6) makes $\mathbf{w} \geqslant \mathbf{0}$. Then the algorithm proceeds as above and terminates when complementarity is achieved (see Question 10.13).

Table 10.6.3 Tableaux for Lemke's method

B	π_1	π_2	r_1	z_0	x_1	x_2	λ_1	$\mathbf{b}$	
π_1	1			$\textcircled{-1}$	-2	1	-1	-3	
π_2		1		-1	1	-2	-1	0	$\uparrow z_0 \quad \pi_1 \uparrow 0$
r_1			1	-1	1	1	0	2	
z_0	-1			1	2	-1	1	3	
π_2	-1	1			$\textcircled{3}$	-3	0	3	$\uparrow x_1 \quad \pi_2 \downarrow 0$
r_1	-1		1		3	0	1	5	
z_0	$-\frac{1}{3}$	$-\frac{2}{3}$		1		1	1	1	
x_1	$\frac{1}{3}$	$\frac{1}{3}$			1	-1	0	1	$\uparrow x_2 \quad r_1 \downarrow 0$
r_1	0	-1	1			$\textcircled{3}$	1	2	
z_0	$-\frac{1}{3}$	$-\frac{1}{3}$	$-\frac{1}{3}$	1			$\textcircled{\frac{2}{3}}$	$\frac{1}{3}$	
x_1	$-\frac{1}{3}$	0	$\frac{1}{3}$		1		$\frac{1}{3}$	$\frac{5}{3}$	$\uparrow \lambda_1 \quad z_0 \downarrow 0$
x_2	0	$-\frac{1}{3}$	$\frac{1}{3}$			1	$\frac{1}{3}$	$\frac{2}{3}$	
λ_1	$-\frac{1}{2}$	$-\frac{1}{2}$	$-\frac{1}{3}$	$\frac{3}{2}$			1	$\frac{1}{2}$	
x_1	$-\frac{1}{6}$	$\frac{1}{6}$	$\frac{1}{2}$	$-\frac{1}{2}$	1			$\frac{3}{2}$	
x_2	$\frac{1}{6}$	$-\frac{1}{6}$	$\frac{1}{2}$	$-\frac{1}{2}$		1		$\frac{1}{2}$	

There is an interesting more general form of an LCP problem. The L_1QP problem referred to in (10.3.5) can also be expressed using an LCP tableau, but with lower and upper bounds on both primal and dual variables. This form is discussed in more detail in Section 12.3 and it is shown that it has an elegant symmetric dual property, in addition to its potential for solving the L_1QP problem.

A popular early technique for QP was *Beale's method* (Beale, 1959) which was also developed as an extension of linear programming. However the method can also be described briefly as an active set method (Goldfarb, private communication) in the following way. Consider the active set method for LP in Section 8.3 applied to minimize a quadratic function. If the kth line search is terminated by reaching an unconstrained minimum, then the difference in gradients $\gamma^{(k)} = \mathbf{g}^{(k+1)} - \mathbf{g}^{(k)}$ is exchanged with the outgoing constraint normal in the matrix $\mathbf{A}$ in (8.3.3). After a sequence of steps in which constraints are removed from $\mathscr{A}$ the effect is that the search direction $\mathbf{s}^{(k+1)}$ is orthogonal to the vectors $\gamma^{(1)},\ldots,\gamma^{(k)}$ and hence conjugate to $\mathbf{s}^{(1)},\ldots,\mathbf{s}^{(k)}$. Thus the possibility of quadratic termination is introduced (see Section 2.5). Unfortunately if a previously inactive constraint then becomes active, the sequence of conjugate directions is broken. To recover from this Beale's method requires some unproductive iterations in which the now irrelevant $\gamma^{(j)}$ vectors are removed from the **A** matrix. The possibility of rearranging the tableau in order to avoid these iterations has been studied by Benveniste (1979). The method is then likely to be closely related to Murray's (1971) method which also recurs conjugate directions. Since essentially an inverse matrix is updated, the more stable factorizations described in Secton 10.1 are preferred. The relationship between the active set method, the Dantzig–Wolfe method, Beale's method, and others is discussed more extensively by Goldfarb (1972).

QUESTIONS FOR CHAPTER 10

10.1. Consider the equality quadratic programming problem

$$\text{minimize} \quad q(\mathbf{x}) \triangleq \tfrac{1}{2}\mathbf{x}^{\mathrm{T}}\begin{bmatrix} 3 & -1 & 0 \\ -1 & 2 & -1 \\ 0 & -1 & 1 \end{bmatrix}\mathbf{x} + \begin{pmatrix} 1 \\ 1 \\ 1 \end{pmatrix}^{\mathrm{T}}\mathbf{x}$$

$$\text{subject to} \quad x_1 + 2x_2 + x_3 = 4.$$

Eliminate x_1 and express the resulting function in the form $\frac{1}{2}\mathbf{y}^{\mathrm{T}}\mathbf{U}\mathbf{y} + \mathbf{v}^{\mathrm{T}}\mathbf{y}$, where $\mathbf{y} = \begin{pmatrix} x_2 \\ x_3 \end{pmatrix}$, **U** is a constant symmetric matrix, and **v** is a constant vector. Hence find the solution $\mathbf{x}^*$ to the QP. Find the Lagrange multiplier λ^* of the equality constraint. Does $\mathbf{x}^*$ solve the QP

$$\text{minimize} \quad q(\mathbf{x})$$

$$\text{subject to} \quad x_1 + 2x_2 + x_3 \geqslant 4, \qquad \mathbf{x} \geqslant \mathbf{0}?$$

10.2. Solve the equality constraint QP problem

$$\underset{\delta}{\text{minimize}} \quad \delta_1^2 - \delta_1\delta_2 + \delta_2^2 - \delta_2\delta_3 + \delta_3^2 + 2\delta_1 - \delta_2$$

$$\text{subject to} \quad 3\delta_1 - \delta_2 + \delta_3 = 0$$
$$2\delta_1 - \delta_2 - \delta_3 = 0$$

by eliminating the variables δ_1 and δ_2. Exhibit the matrices $\mathbf{A}, \mathbf{A}_1, \mathbf{A}_2$ and $\mathbf{Z}$, and a matrix $\mathbf{Y}$ such that $\mathbf{Y}^{\mathrm{T}}\mathbf{A} = \mathbf{I}$. Hence find the multipliers of the two constraints at the solution.

10.3. Solve the equality constraint QP

$$\text{minimize} \quad \sum_{i=1}^{n} i x_i^2$$

$$\text{subject to} \quad \sum_{i=1}^{n} x_i = K \qquad (K > 0)$$

by the method of Lagrange multipliers. Does this solution minimize the function subject to the constraints

$$x_i \geqslant 0 \quad i = 1, 2, \ldots, n \quad \text{and} \quad \sum_{i=1}^{n} x_i \geqslant K?$$

Give the explicit solution of the equality constraint problem when $n = 3$, $K = 10$.

10.4. Starting from $\mathbf{x}^{(1)} = \mathbf{0}$, illustrate the steps taken by the primal active set method to solve the QP problem

$$\underset{\mathbf{x}}{\text{minimize}} \quad x_1^2 - x_1x_2 + x_2^2 - 3x_1$$

$$\begin{aligned} \text{subject to} \qquad x_1 &\geqslant 0 \\ x_2 &\geqslant 0 \\ -x_1 - x_2 &\geqslant -2. \end{aligned}$$

For simplicity, solve each EP without making a shift of variables. Verify that the method is equivalent to the Dantzig–Wolfe method (Table 10.6.2).

10.5. Illustrate the matrices $\mathbf{V}$, $\mathbf{Y}$, and $\mathbf{Z}$ when solving the QP problem in Question 9.6 by (a) direct elimination, (b) orthogonal factorization.

10.6. Given matrices $\mathbf{A}$ and $\mathbf{Z}$ which are $n \times m$ and $n \times (n-m)$ respectively $(1 < m < n)$ consider how to choose an $n \times (n-m)$ matrix $\mathbf{V}$ such that (10.1.20) holds. Let the matrices be partitioned after the mth row as

$$\mathbf{A} = \begin{bmatrix} \mathbf{A}_1 \\ \mathbf{A}_2 \end{bmatrix}, \quad \mathbf{V} = \begin{bmatrix} \mathbf{V}_1 \\ \mathbf{V}_2 \end{bmatrix}, \quad \mathbf{Y} = \begin{bmatrix} \mathbf{Y}_1 \\ \mathbf{Y}_2 \end{bmatrix}, \quad \mathbf{Z} = \begin{bmatrix} \mathbf{Z}_1 \\ \mathbf{Z}_2 \end{bmatrix}.$$

Show that $\mathbf{Y}_1 = \mathbf{A}_1^{-\mathrm{T}}(\mathbf{I} - \mathbf{A}_2^{\mathrm{T}}\mathbf{Y}_2)$, $\mathbf{V}_1 = -\mathbf{A}_1\mathbf{Y}_2^{\mathrm{T}}\mathbf{Z}_2^{-\mathrm{T}}\mathbf{Z}_2^{-\mathrm{T}}$ and $\mathbf{V}_2 = \mathbf{Z}_2^{-\mathrm{T}}(\mathbf{I} - \mathbf{Z}_1^{\mathrm{T}}\mathbf{V}_1)$ define $\mathbf{Y}_1$, $\mathbf{V}_1$, and hence $\mathbf{V}_2$ in terms of $\mathbf{Y}_2$. It follows that $\mathbf{V}$ is under-determined to the extent of an arbitrary choice of the $(n-m) \times m$ matrix $\mathbf{Y}_2$. Investigate the choice $\mathbf{Y}_2 = \mathbf{0}$.

10.7. Let matrices $\mathbf{G}$, $\mathbf{A}$, $\mathbf{Y}$, and $\mathbf{Z}$ be $n \times n$, $n \times m$, $n \times m$, and $n \times (n-m)$ respectively $(m \leqslant n)$, let $\mathbf{Y}^{\mathrm{T}}\mathbf{A} = \mathbf{I}$, $\mathbf{Z}^{\mathrm{T}}\mathbf{A} = \mathbf{0}$, and let the matrix $[\mathbf{Y}:\mathbf{Z}]$ be nonsingular. Show that there exists a unique $n \times (n-m)$ matrix $\mathbf{V}$ such that $[\mathbf{A}:\mathbf{V}]^{-1} = [\mathbf{Y}:\mathbf{Z}]^{\mathrm{T}}$, and show that $\mathbf{Y}^{\mathrm{T}}\mathbf{V} = \mathbf{0}$, $\mathbf{Z}^{\mathrm{T}}\mathbf{V} = \mathbf{I}$, and $\mathbf{A}\mathbf{Y}^{\mathrm{T}} + \mathbf{V}\mathbf{Z}^{\mathrm{T}} = \mathbf{I}$. Hence show that (10.2.3) holds where $\mathbf{H}, \mathbf{T}$, and $\mathbf{U}$ are defined by (10.2.7). Show how the problem

$$\underset{\mathbf{x}}{\text{minimize}} \quad \tfrac{1}{2}\mathbf{x}^{\mathrm{T}}\mathbf{G}\mathbf{x} + \mathbf{g}^{\mathrm{T}}\mathbf{x}$$

$$\text{subject to} \quad \mathbf{A}^{\mathrm{T}}\mathbf{x} = \mathbf{b}$$

can be solved by the method of Lagrange multipliers, and use (10.2.3) and (10.2.7) to obtain explicit expressions for the solution $\mathbf{x}^*$ and the associated Lagrange multiplier $\boldsymbol{\lambda}^*$. Verify that these reduce to (10.1.15) and (10.1.16) respectively.

10.8. Establish equations (10.2.8) and (10.2.9) and show that the matrix $\mathbf{H}$ in (10.2.3) satisfies $\mathbf{H} = (\mathbf{PGP})^+$ where $\mathbf{P} = \mathbf{I} - \mathbf{AA}^+$.

10.9. Consider the indefinite QP problem

$$\underset{\mathbf{x}}{\text{minimize}} \quad -x_1 x_2$$

$$\text{subject to} \quad 2 \geqslant x_1 + x_2 \geqslant 0$$
$$2 \geqslant x_1 - x_2 \geqslant -2$$

and show that if the active set method starts at the vertex $\mathbf{x}^{(1)} = (1, -1)^T$ then the method terminates at a point $\mathbf{x}'$ which is not a local solution to the problem. Show that there exist arbitrary small perturbations to the constraints such that (a) $\mathbf{x}'$ is a local solution, (b) the algorithm does not terminate at $\mathbf{x}'$ but finds the global solution.

10.10. In the active set method assume that the initial active set $\mathscr{A}^{(1)}$ has the property that the set of vectors $\mathbf{a}_i$, $i \in \mathscr{A}^{(1)}$, are independent. Show using equation (10.3.2) that any vector $\mathbf{a}_p$ which is added to the set is not dependent on the other vectors in the set, and hence prove inductively that the independence condition is retained.

10.11. If the Lagrangian matrix in (10.2.2) in non-singular, show $\mathbf{A}$ has full rank and hence that a matrix $\mathbf{Z}$ can be chosen as in (10.1.18) with orthogonal columns. If $\mathbf{Z}^T\mathbf{GZ}$ is singular, consider $\mathbf{G} + \nu\mathbf{I}$ as $\nu \to 0$ and show that (10.2.7) becomes unbounded whereas (10.2.3) does not. This is a contradiction and proves that $\mathbf{Z}^T\mathbf{GZ}$ is non-singular.

10.12. Show that the least distance to a fixed point $\mathbf{p}$ of any point $\mathbf{x}$ which satisfies the constaints $\mathbf{A}^T\mathbf{x} \geqslant \mathbf{b}$ can be found by solving a certain QP problem. If all the constraints are active, show by the method of Lagrange multipliers that the problem reduces to the solution of a system of linear equations. In general, when there may be inactive constraints, use the Wolfe dual (Theorem 9.5.1) to show that the optimum multipliers satisfy a QP problem with simple bounds only. How does the solution of this problem determine the least distance solution?

10.13. State the linear complementarity problem arising from Question 10.4 and show that it satisfies the conditions for the modified form of Lemke's method (Section 10.6) to be applicable, without introducing the variable z_0. Hence solve the problem in this way.

10.14. Show that the problem of finding the point of least Euclidean distance from the origin (in $\mathbb{R}^3$) subject to

$$x_1 + 2x_2 - x_3 \geqslant 4$$
$$-x_1 + x_2 - x_3 \leqslant 2$$

can be formulated as a quadratic programming problem. By eliminating x_1 and x_2, calculate a solution to the equality problem in which both constraints are active. Is this the solution to the least distance quadratic programming problem?

10.15. The nearest point to the origin on a polytope

$$P = \{\mathbf{x}:\mathbf{x} = \mathbf{A}\mathbf{y},\quad \mathbf{e}^T\mathbf{y} = 1,\qquad \mathbf{y} \geqslant \mathbf{0}\}$$

(the convex hull of vectors $\mathbf{a}_1, \ldots, \mathbf{a}_m$ which are columns of an $n \times m$ matrix $\mathbf{A}$) is given by the QP

$$\begin{array}{ll} \text{minimize} & \|\mathbf{A}\mathbf{y}\|_2^2 \\ \text{subject to} & \mathbf{y} \geqslant \mathbf{0} \quad \text{and} \quad \mathbf{e}^T\mathbf{y} = 1. \end{array}$$

The solution $\mathbf{y}^*$ of this problem can be obtained by solving the problem

$$\begin{array}{ll} \text{minimize} & \|\mathbf{A}\mathbf{y}\|_2^2 + (\mathbf{e}^T\mathbf{y} - 1)^2 \\ \text{subject to} & \mathbf{y} \geqslant \mathbf{0} \end{array}$$

giving a vector $\mathbf{y}'$ say, and normalizing to get $\mathbf{y}^* = \mathbf{y}'/\mathbf{e}^T\mathbf{y}'$. Derive the least distance problem obtained by taking the dual of this problem.

Chapter 11
General Linearly Constrained Optimization

11.1 EQUALITY CONSTRAINTS

The next class of problem which it is convenient to consider is that in which the objective function is general but the constraints remain linear, that is

$$\begin{aligned}
&\underset{\mathbf{x}}{\text{minimize}} \quad f(\mathbf{x}) \\
&\text{subject to} \quad \mathbf{a}_i^{\mathrm{T}}\mathbf{x} = b_i, \qquad i \in E \\
&\qquad\qquad\quad\; \mathbf{a}_i^{\mathrm{T}}\mathbf{x} \geqslant b_i, \qquad i \in I.
\end{aligned} \tag{11.1.1}$$

The methods for handling the linear constraints are largely those used in quadratic programming, but the non-quadratic objective function introduces an extra level of difficulty. One aspect of this is that in general the problem can no longer be solved finitely so the solution $\mathbf{x}^*$ is obtained in the limit of some iterative sequence $\{\mathbf{x}^{(k)}\}$. The equality constraint problem ($I = \varnothing$) is considered in this section and is readily handled by generalized elimination, so that the main features are essentially those relating unconstrained optimization as described in Part 1. Methods for calculating Lagrange multipliers at the solution of an equality constraint problem are also considered at the end of this section. Inequality constraints can be handled by an active set method (Section 11.2) which is generalization of that used in quadratic programming. A new decision which occurs is how accurately to solve each equality problem which arises: a poor decision in this respect can lead to the phenomenon of *zigzagging* which can slow down the rate of convergence appreciably. Methods for overcoming this problem are described in Section 11.3. In addition, some alternative possibilities for handling inequality constraints by way of a reduction to a sequence of quadratic programming problems are described at the end of Section 11.2.

The rest of this section is devoted to finding the solution $\mathbf{x}^*$ of the equality constraint problem

$$\begin{aligned} &\underset{\mathbf{x}}{\text{minimize}} \quad f(\mathbf{x}) \\ &\text{subject to} \quad \mathbf{A}^{\mathrm{T}}\mathbf{x} = \mathbf{b} \end{aligned} \tag{11.1.2}$$

in which $\mathbf{A}$ is $n \times m$ and $\mathbf{b} \in \mathbb{R}^m$. As in Section 10.1 it is assumed that rank $(\mathbf{A}) = m$. The *generalized elimination method* of Section 10.1 is used to provide a reduction to an unconstrained problem and includes both the elimination of variables method and the orthogonal factorization method, amongst others. Thus matrices $\mathbf{Y}$ and $\mathbf{Z}$ are introduced such that $\mathbf{Y}^{\mathrm{T}}\mathbf{A} = \mathbf{I}$, $\mathbf{Z}^{\mathrm{T}}\mathbf{A} = \mathbf{0}$, and $[\mathbf{Y}:\mathbf{Z}]$ is non-singular. If the current iterate is the feasible point $\mathbf{x}^{(k)}$ then a general feasible point can be expressed as

$$\mathbf{x} = \mathbf{x}^{(k)} + \boldsymbol{\delta} = \mathbf{x}^{(k)} + \mathbf{Z}\mathbf{y} \tag{11.1.3}$$

where $\boldsymbol{\delta} = \mathbf{Z}\mathbf{y}$ is a feasible correction in the null space of $\mathbf{A}$ (see (10.1.9) and (10.1.10)). Thus an equivalent form of (11.1.2) is to solve the reduced unconstrained problem

$$\underset{\mathbf{y}}{\text{minimize}} \quad \psi(\mathbf{y}) \triangleq f(\mathbf{x}^{(k)} + \mathbf{Z}\mathbf{y}). \tag{11.1.4}$$

For computational convenience and stability it is usually best to define a new reduced function on each iteration as (11.1.4) indicates. By using the chain rule in (11.1.3) it follows that

$$\nabla_y = \mathbf{Z}^{\mathrm{T}}\nabla_x \tag{11.1.5}$$

and so derivatives of (11.1.4) are given by

$$\nabla_y \psi(\mathbf{y}) = \mathbf{Z}^{\mathrm{T}}\mathbf{g}(\mathbf{x}) \tag{11.1.6}$$

which is the *reduced gradient vector*, and by

$$\nabla_y^2 \psi(\mathbf{y}) = \mathbf{Z}^{\mathrm{T}}[\mathbf{G}(\mathbf{x})]\mathbf{Z} \tag{11.1.7}$$

which is the *reduced Hessian matrix*.

It is possible to apply any appropriate technique for unconstrained optimization described in Part 1 to solve the reduced problem (11.1.4) and most of the remarks in Chapter 2 about the structure of methods apply here also. From (11.1.6) and (11.1.7), sufficient conditions for $\mathbf{x}^*$ to be optimal are that $\mathbf{Z}^{\mathrm{T}}\mathbf{g}^* = \mathbf{0}$ and $\mathbf{Z}^{\mathrm{T}}\mathbf{G}^*\mathbf{Z}$ is positive definite, and necessary conditions are that $\mathbf{Z}^{\mathrm{T}}\mathbf{g}^* = \mathbf{0}$ or $\mathbf{Z}^{\mathrm{T}}\mathbf{G}^*\mathbf{Z}$ is positive semi-definite. These conditions are readily shown to be equivalent to those described in Sections 9.1 and 9.3 (see Question 11.1). The initial feasible point $\mathbf{x}^{(1)}$ can be $\mathbf{x}^{(1)} = \mathbf{Y}\mathbf{b}$, which is the closest feasible point to the origin in the orthogonal factorization method, or more generally $\mathbf{x}^{(1)} = \mathbf{x}' + \mathbf{Y}(\mathbf{b} - \mathbf{A}^{\mathrm{T}}\mathbf{x}')$ which is the closest feasible point to any given point $\mathbf{x}'$. As in Chapter 3 it is important to start by considering *Newton's method* which is based on the quadratic model obtained by truncating the Taylor series for $\psi(\mathbf{y})$ about $\mathbf{y} = \mathbf{0}$, that is

$$\psi(\mathbf{y}) \approx q^{(k)}(\mathbf{y}) = f^{(k)} + \mathbf{y}^{\mathrm{T}}\mathbf{Z}^{\mathrm{T}}\mathbf{g}^{(k)} + \tfrac{1}{2}\mathbf{y}^{\mathrm{T}}\mathbf{Z}^{\mathrm{T}}\mathbf{G}^{(k)}\mathbf{Z}\mathbf{y}. \tag{11.1.8}$$

The origin in **y**-space corresponds to the point $\mathbf{x}^{(k)}$ in **x**-space (see (11.1.3)) so $f^{(k)}, \mathbf{g}^{(k)}$, and $\mathbf{G}^{(k)}$ refer to quantities evaluated at $\mathbf{x}^{(k)}$. The model quadratic $q^{(k)}(\mathbf{y})$ refers to the function obtained on iteration k. Thus the basic Newton's method chooses $\mathbf{y}^{(k)}$ to minimize $q^{(k)}(\mathbf{y})$. A unique minimizer exists if and only if $\mathbf{Z}^T\mathbf{G}^{(k)}\mathbf{Z}$ is positive definite and is obtained by making $\nabla q^{(k)} = \mathbf{0}$, that is by solving

$$(\mathbf{Z}^T\mathbf{G}^{(k)}\mathbf{Z})\mathbf{y} = -\mathbf{Z}^T\mathbf{g}^{(k)} \tag{11.1.9}$$

for $\mathbf{y} = \mathbf{y}^{(k)}$, in which case the next iterate is $\mathbf{x}^{(k+1)} = \mathbf{x}^{(k)} + \mathbf{Z}\mathbf{y}^{(k)}$ from (11.1.3). If any $\mathbf{x}^{(k)}$ is sufficiently close to $\mathbf{x}^*$ then the method converges and the order is second order. However the method can fail to converge, and moreover is undefined when $\mathbf{Z}^T\mathbf{G}^{(k)}\mathbf{Z}$ is not positive definite. Changes to correct these disadvantages include the use of a line search and the possibility of modifying the reduced Hessian matrix to become positive definite. The latter includes the ideas of restricted step and Levenberg–Marquardt methods. All of this development is given in Section 3.1 and Chapter 5, and the reader is referred there for further details. One point of interest is that a step restriction $\|\mathbf{y}\|_2 \leqslant h^{(k)}$ can be handled by a modification $\mathbf{Z}^T\mathbf{G}^{(k)}\mathbf{Z} + \nu\mathbf{I}$ for some $\nu \geqslant 0$ as in Section 5.2 and corresponds to a restriction $\|\mathbf{x} - \mathbf{x}^{(k)}\|_2 \leqslant h^{(k)}$ if the orthogonal factorization method is used to reduce the problem. However for other generalized elimination methods this correspondence does not hold.

It is often the case that the user is unable or does not wish to supply formulae for second derivatives so it is important to consider methods which require only first derivative information or even no derivatives at all. When first derivatives are available, one possibility is a finite difference Newton method (see Section 3.2). In this case the differences are best taken in **y**-space so that the ith column of the reduced Hessian matrix is defined by

$$\mathbf{Z}^T(\mathbf{g}(\mathbf{x}^{(k)} + \mathbf{z}_i h) - \mathbf{g}^{(k)})/h \tag{11.1.10}$$

for some small h, and after symmetrization this matrix is used to replace $\mathbf{Z}^T\mathbf{G}^{(k)}\mathbf{Z}$ in (11.1.9). Only $n - m$ additional gradient evaluations are required per iteration so the method might be useful when $n - m$ is small, especially as a way to estimate an initial reduced Hessian approximation.

However for the reasons described in Section 3.2 it is usually preferable to select a quasi-Newton method to solve the reduced problem (11.1.4). Various suggestions (see below) have been made but the most satisfactory in view of the development given here is due to Gill and Murray (1974c). The reduced Hessian matrix is approximated by a positive definite matrix $\mathbf{M}^{(k)}$ given in factored form

$$\mathbf{Z}^T\mathbf{G}^{(k)}\mathbf{Z} \simeq \mathbf{M}^{(k)} = \mathbf{L}^{(k)}\mathbf{D}^{(k)}\mathbf{L}^{(k)T}. \tag{11.1.11}$$

The search direction in **y**-space is then defined by solving

$$\mathbf{M}^{(k)}\mathbf{p} = -\mathbf{Z}^T\mathbf{g}^{(k)} \tag{11.1.12}$$

for $\mathbf{p} = \mathbf{p}^{(k)}$, by analogy with (11.1.9), and $f(\mathbf{x})$ is minimized approximately by a search along the line $\mathbf{x}^{(k)} + \alpha\mathbf{s}^{(k)}$, where $\mathbf{s}^{(k)} = \mathbf{Z}\mathbf{p}^{(k)}$. The conditions for terminating the line search are those described in Sections 2.5 and 2.6. More in keeping with the 'old-fashioned' development of Section 3.2 is to represent $\mathbf{M}^{(k)}$ by the inverse reduced Hessian approximation

$$\mathbf{H}^{(k)} = [\mathbf{M}^{(k)}]^{-1} \approx [\mathbf{Z}^{\mathrm{T}}\mathbf{G}^{(k)}\mathbf{Z}]^{-1} \tag{11.1.13}$$

and to compute $\mathbf{s}^{(k)}$ from

$$\mathbf{s}^{(k)} = -\mathbf{Z}\mathbf{H}^{(k)}\mathbf{Z}^{\mathrm{T}}\mathbf{g}^{(k)}. \tag{11.1.14}$$

In view of the remarks at the end of Section 3.2 it may be that (11.1.13) is no less stable than (11.1.11) whilst being somewhat more convenient to use. After each iteration $\mathbf{H}^{(k)}$ (or the factorization of $\mathbf{M}^{(k)}$) is updated to include the additional curvature information obtained in the line search. For the reasons described in Chapter 3 the BFGS formula is currently preferred. The required qunatities $\boldsymbol{\gamma}^{(k)}$ and $\boldsymbol{\delta}^{(k)}$ (in (3.2.12) say) become differences in reduced gradients and variables, that is $\boldsymbol{\gamma}^{(k)} = \mathbf{Z}^{\mathrm{T}}(\mathbf{g}^{(k+1)} - \mathbf{g}^{(k)})$ and $\boldsymbol{\delta}^{(k)} = \mathbf{y}^{(k)} - \mathbf{0} = \mathbf{y}^{(k)}$. The initial choice $\mathbf{H}^{(1)}$ can be any positive definite matrix; $\mathbf{H}^{(1)} = \mathbf{I}$ is usually chosen in the absence of any other information. The properties described in Section 3.2 relating amongst others to the maintaining of positive definite matrices $\mathbf{H}^{(k)}$ and to quadratic termination (here in $n-m$ iterations) also hold good.

Another possibility for a suitable algorithm is to apply a conjugate gradient method (Section 4.1) to the reduced problem. This is likely to be preferable only for large problems in which the matrix $\mathbf{H}^{(k)}$ cannot conveniently be stored. In this case it is also important to give some thought to storage for the matrices $\mathbf{Y}$ and $\mathbf{Z}$. The best possibility is to define these implicitly by factorizing the matrix $[\mathbf{A}:\mathbf{V}]$ in such a way as to retain sparsity as much as possible. Likewise a choice of $\mathbf{V}$ which retains sparsity should be made, for example that given in (10.1.21) corresponding to elimination of variables. For no-derivative problems the experience of Section 3.2 again suggests using a quasi-Newton method with difference approximations to derivatives. These are best calculated in the space of the reduced variables so that the ith element of $\mathbf{Z}^{\mathrm{T}}\mathbf{g}^{(k)}$ is estimated by

$$(f(\mathbf{x}^{(k)} + \mathbf{z}_i h) - f^{(k)})/h \tag{11.1.15}$$

for some small interval h, which requires only $n-m$ additional function evaluations. Near the solution it is preferable to use the corresponding central difference formula

$$\tfrac{1}{2}(f(\mathbf{x}^{(k)} + \mathbf{z}_i h) - f(\mathbf{x}^{(k)} - \mathbf{z}_i h))/h. \tag{11.1.16}$$

Conjugate direction set methods as in Section 4.2 are also possible (Buckley, 1975) but there is no evidence to suggest that they are preferable. Least squares problems ($f(\mathbf{x}) \triangleq \mathbf{r}(\mathbf{x})^{\mathrm{T}}\mathbf{r}(\mathbf{x})$) with linear constraints can be handled readily: the Jacobian matrix in the reduced coordinates is $\nabla_y(\mathbf{r}^{(k)\mathrm{T}}) = \mathbf{Z}^{\mathrm{T}}\mathbf{A}^{(k)}$ where $\mathbf{A} = \nabla_x \mathbf{r}^{\mathrm{T}}$, and analogues of the Gauss–Newton method, etc., are obtained by using the estimate $\mathbf{Z}^{\mathrm{T}}\mathbf{G}^{(k)}\mathbf{Z} = 2\mathbf{Z}^{\mathrm{T}}\mathbf{A}^{(k)}\mathbf{A}^{(k)\mathrm{T}}\mathbf{Z}$ in (11.1.9); the resulting properties are the same as those described in Section 6.1.

Historically the earliest methods for solving (11.1.2) used steepest descent searches as typified by Rosen's (1960) *gradient projection method*. This is equivalent to choosing a search direction $\mathbf{p}^{(k)} = -\nabla_y \psi^{(k)} = -\mathbf{Z}^T\mathbf{g}^{(k)}$ as the steepest descent vector in the reduced coordinates and hence $\mathbf{s}^{(k)} = -\mathbf{Z}\mathbf{Z}^T\mathbf{g}^{(k)}$ as the search direction in **x**-space. When **Z** is defined by the orthogonal factorization method it follows that $\mathbf{s}^{(k)} = -\mathbf{P}\mathbf{g}^{(k)}$ where $\mathbf{P} = \mathbf{Z}\mathbf{Z}^T = \mathbf{I} - \mathbf{A}\mathbf{A}^+$ is a projection matrix which projects into the null space of **A**. Rosen suggests calculating this matrix although it is preferable to use the implicit definition $\mathbf{P} = \mathbf{Z}\mathbf{Z}^T$. The idea for using an active set method also derives from Rosen (1960). Quasi-Newton methods for linearly constrained problems were first considered by Goldfarb (1969) although he updates the $n \times n$ matrix

$$\bar{\mathbf{H}}^{(k)} = \mathbf{Z}\mathbf{H}^{(k)}\mathbf{Z}^T \tag{11.1.17}$$

which is positive semi-definite with rank $n - m$ and is related to the **H** matrix in (10.2.3). It is easy to show that updating $\mathbf{H}^{(k)}$ by any formula in the Broyden family, using reduced differences as above, is equivalent to updating $\bar{\mathbf{H}}^{(k)}$ by the same formula with unreduced differences (Question 11.2) The choice $\mathbf{H}^{(1)} = \mathbf{I}$ is equivalent to choosing $\bar{\mathbf{H}}^{(1)} = \mathbf{Z}\mathbf{Z}^T$ which is the projection matrix **P** when the orthogonal factorization method defines **Z**. However computing the search direction using $\mathbf{s}^{(k)} = -\bar{\mathbf{H}}^{(k)}\mathbf{g}^{(k)}$ causes problems due to round-off errors in that $\mathbf{s}^{(k)}$ no longer remains in the null space of **A** when $\mathbf{x}^{(k)}$ is close to $\mathbf{x}^*$, and so the implicit representation through (11.1.14) is preferable. Another early quasi-Newton method is due to Murtagh and Sargent (1969), and updates estimates of $\mathbf{G}^{-1}$ and $(\mathbf{A}^T\mathbf{G}^{-1}\mathbf{A})^{-1}$ using the rank one formula, and uses the representation (10.2.6) to solve the quadratic/linear approximating subproblem. However, neither matrix needs to be positive definite and the rank one formula is potentially unstable, so again this method would not be recommended. The observation that a variety of methods can be classified as generalized elimination methods is due to Fletcher (1972c) and the implementation via **Y** and **Z** matrices to Gill and Murray (1974a).

Finally it is of interest to consider the computation of the Lagrange multiplier vector $\boldsymbol{\lambda}^*$ at the solution to (11.1.2). This information is required when the equality problem is a subproblem in the active set method, but can also be useful in a sensitivity analysis as described in Section 9.1. Essentially $\boldsymbol{\lambda}^*$ is defind by $\mathbf{g}^* = \mathbf{A}^*\boldsymbol{\lambda}^*$ and is readily computed by (10.1.14) or (10.1.19). There is the additional complication however that $\mathbf{x}^*$ and hence $\mathbf{g}^*$ cannot be computed exactly due to the non-finite nature of methods for solving (11.1.2). Thus the effect of calculating approximate multipliers $\boldsymbol{\lambda}^{(k)}$ at a point $\mathbf{x}^{(k)}$ which approximates $\mathbf{x}^*$ is considered. In particular in the active set method $\mathbf{x}^{(k)}$ might be quite a poor approximation to $\mathbf{x}^*$ so the errors involved may not be negligible. The obvious possibility from (10.1.14) is to compute $\boldsymbol{\lambda}^{(k)}$ from

$$\boldsymbol{\lambda}^{(k)} = \mathbf{Y}^T\mathbf{g}^{(k)} \tag{11.1.18}$$

which can be regarded as a first order estimate of $\boldsymbol{\lambda}^*$ since $\mathbf{g}^{(k)} - \mathbf{g}^* = O(\|\mathbf{x}^{(k)} - \mathbf{x}^*\|)$. Unlike (10.1.14) the resulting $\boldsymbol{\lambda}^{(k)}$ depends on how **Y** is chosen because the equations $\mathbf{g}^{(k)} = \mathbf{A}\boldsymbol{\lambda}^{(k)}$ are no longer consistent. The orthogonal factorization method in which $\mathbf{Y}^T = \mathbf{A}^+$ has a nice interpretation in that it gives the least squares solution of these equations. Another left inverse for $\mathbf{A}$ with special properties is the matrix $\mathbf{T}^{\mathrm{T}}$ defined in (10.2.3). Thus another possibility is to compute $\boldsymbol{\lambda}^{(k)}$ from

$$\boldsymbol{\lambda}^{(k)} = \mathbf{T}^{\mathrm{T}}\mathbf{g}^{(k)}. \tag{11.1.19}$$

$\mathbf{T}$ includes curvature information about the problem and it can be shown that the resulting $\boldsymbol{\lambda}^{(k)}$ is a second order estimate of $\boldsymbol{\lambda}^*$ (see (10.2.9) and Question 11.3). Given that $\mathbf{T}$ is available and that $\mathbf{x}^{(k)}$ is sufficienty close to $\mathbf{x}^*$, then (11.1.19) gives a more accurate estimate of $\boldsymbol{\lambda}^*$. However for reasons of convenience or numerical stability it may be preferable to use (11.1.18).

For no-derivative problems, if differencing is carried out as in (11.1.15) or (11.1.16) to estimate reduced derivatives, then no information is available to compute Lagrange multipliers. In this case an additional m function evaluations are required to estimate the Lagrange multipliers $\lambda_i^{(k)}$ by using

$$\lambda_i^{(k)} \approx (f(\mathbf{x})^{(k)} + \mathbf{y}_i h) - f^{(k)}/h \tag{11.1.20}$$

where $\mathbf{y}_i$ is the corresponding column of the matrix $\mathbf{Y}$.

11.2 INEQUALITY CONSTRAINTS

Most problems of interest involve inequality constraints and so can be expressed as in (11.1.1). This section describes how the methods of the previous section can be generalized to handle this problem by means of a primal active set method similar to that described in Section 10.3. Some alternative possibilities are also discussed at the end of the section. In the primal active set method certain constraints indexed by the active set $\mathscr{A}$ are regarded as equalities and the rest are temporarily disregarded as described in Section 10.3. Each iterate $\mathbf{x}^{(k)}$ is a feasible point and each iteration attempts to locate the solution of the EP

$$\begin{aligned} &\underset{\boldsymbol{\delta}}{\text{minimize}} \quad f(\mathbf{x}^{(k)} + \boldsymbol{\delta}) \\ &\text{subject to} \quad \mathbf{a}_i^{\mathrm{T}}\boldsymbol{\delta} = 0, \qquad i \in \mathscr{A} \end{aligned} \tag{11.2.1}$$

obtained by shifting the origin to the point $\mathbf{x}^{(k)}$. The initial point $\mathbf{x}^{(1)}$ can be found by the methods of Sections 8.4 or 10.3. The solution of (11.2.1) yields a search direction $\mathbf{s}^{(k)}$ according to the method used, and $\mathbf{x}^{(k+1)}$ is taken (ideally) as the best feasible point on the line $\mathbf{x}^{(k)} + \alpha\mathbf{s}^{(k)}$. In quadratic programming there is a clear distinction either that $\mathbf{x}^{(k+1)}$ solves the EP or that a previously inactive constraint becomes active in the line search. This is not so in the more general case (11.1.1) and the minimizer of the EP is only located in the limit of a sequence of iterations with the same $\mathscr{A}$. Thus on each iteration a decision must be taken as to whether $\mathbf{x}^{(k)}$ (that is $\boldsymbol{\delta} = \mathbf{0}$) is an acceptable solution of the EP; if not, then one or more iterations are carried out with the same active set. The definition of an acceptable solution must be made with some care and is considered in more detail in Section 11.3. If $\mathbf{x}^{(k)}$ is accepted as the solution to the EP, then Lagrange multipliers $\boldsymbol{\lambda}^{(k)}$ are examined to determine whether or not $\mathbf{x}^{(k)}$ is a KT point. If so then the iteration terminates; otherwise the index q of an inequality constraint for which $\lambda_q^{(k)} < 0$ is least is removed from $\mathscr{A}$ as in (10.3.3).

The line search for finding the best feasible point in this more general case is also not a finite process and is therefore more complicated than (10.3.2). As in Sections 2.5 and 2.6 it is necessary to choose conditions which define a range of acceptable α-values, and to locate such a value by a combination of interpolation and sectioning in such a way that the iteration is guaranteed to terminate. Exactly the same holds good here, but there is the additional complication that α must not exceed the value $\bar{\alpha}^{(k)}$, say, defined as in (8.3.6) by

$$\bar{\alpha}^{(k)} = \min_{\substack{i: i \notin \mathscr{A} \\ \mathbf{a}_i^{\mathrm{T}}\mathbf{s}^{(k)} < 0}} \frac{b_i - \mathbf{a}_i^{\mathrm{T}}\mathbf{x}^{(k)}}{\mathbf{a}_i^{\mathrm{T}}\mathbf{s}^{(k)}}.$$

Thus the initial choice for α in the line search, or any α-values determined by extrapolation, must be reduced to $\bar{\alpha}^{(k)}$ if they would otherwise exceed this value. It may be that $\bar{\alpha}^{(k)}$ is the end point of a bracket which cannot be guaranteed to contain acceptable α-values in the sense of Section 2.6 (i.e. conditions (2.6.3) are not satisfied). In this case $\alpha^{(k)} = \bar{\alpha}^{(k)}$ is chosen. It is in this way that the value $\alpha^{(k)}$ in the line search in step (e) of (11.2.2) below is determined. Thus the primal active set method for solving (11.1.1) can be summarized as follows.

(a) Given $\mathbf{x}^{(1)}$ and $\mathscr{A} = \mathscr{A}(\mathbf{x}^{(1)})$, set $k = 1$.
(b) If $\boldsymbol{\delta} = \mathbf{0}$ is not an acceptable solution of (10.3.1), go to (d).
(c) Let $\lambda_q^{(k)}$ solve $\min \lambda_i^{(k)}, i \in I \cap \mathscr{A}$; if $\lambda_q^{(k)} \geqslant 0$, terminate with $\mathbf{x}^* = \mathbf{x}^{(k)}$, otherwise remove q from $\mathscr{A}$.
(d) Solve (11.1.1) for $\mathbf{s}^{(k)}$. (11.2.2)
(e) Choose $\mathbf{x}^{(k+1)} = \mathbf{x}^{(k)} + \alpha^{(k)}\mathbf{s}^{(k)}$ as a near best feasible point along the line $\mathbf{x}^{(k)} + \alpha\mathbf{s}^{(k)}$.
(f) If $\alpha^{(k)} = \bar{\alpha}^{(k)}$, add p to $\mathscr{A}$.
(g) Set $k = k + 1$ and go to (b).

Note that after an index q is removed from $\mathscr{A}$ in step (c) it is assumed that the search direction $\mathbf{s}^{(k)}$ which is calculated in step (d) is downhill ($\mathbf{s}^{(k)\mathrm{T}}\mathbf{g}^{(k)} < 0$) and strictly feasible with respect to the deleted constraint ($\mathbf{s}^{(k)\mathrm{T}}\mathbf{a}_q > 0$). This can be done in most common methods (see Question 11.7). The choice of method for solving (11.1.1) in step (d) is otherwise any of the methods considered in Section 11.1, and depends on what derivative information is available, amongst other things. Possible difficulties caused by degeneracy are no less severe than those described in Section 10.3 and elsewhere and it is usual, albeit somewhat unsatisfactory, to assume that it does not occur. In this case convergence of the algorithm depends on how the choice of an acceptable solution in step (b) is made, and this is considered in more detail in Section 11.3.

An important feature of an active set method concerns the efficient solution of the EP (11.1.2) when changes are made to $\mathscr{A}$. It is possible to avoid recalculating $\mathbf{Y}$ and $\mathbf{Z}$ in $O(n^3)$ operations and instead to update these matrices in $O(n^2)$ operations, taking advantage of the fact that a column is either added to or removed from $\mathbf{A}$. Updating of other matrices may also be possible. Again details of these computations are not given but the references towards the end of Section 10.3 are relevant. It is interesting however to review the changes to the reduced Hessian matrix $\mathbf{Z}^T\mathbf{G}^{(k)}\mathbf{Z}$ when a change is made to the active set $\mathscr{A}$, and hence to $\mathbf{Z}$. For the basic Newton method $\mathbf{G}^{(k)}$ also changes in a general way so no advantage can be taken of updating techniques. However the situation is more favourable when using a quasi-Newton method. First of all consider the case when $\alpha^{(k)} = \bar{\alpha}^{(k)}$ in step (e) of (11.2.2) so that a constraint becomes active. It is possible that the curvature estimate $\gamma^T\delta \leqslant 0$ may occur (see Section 3.2) in which case the matrix $\mathbf{H}^{(k)}$ must not be updated. The new $\mathbf{Z}$ matrix has one fewer column than the old matrix; by making a linear transformation of the columns of $\mathbf{Z}$ it is possible to arrange matters so that column $\mathbf{z}_{n-m}$ is removed from $\mathbf{Z}$. In this case the new matrix $\mathbf{M}^{(k)}$ in (11.1.11) is obtained by removing row and column $n-m$ from the old matrix $\mathbf{M}^{(k)}$ (after transformation). This is the same operation that is carried out in quadratic programming and details are given for example by Gill and Murray (1978b). A corresponding formula for updating $\mathbf{H}^{(k)}$ can also be obtained (Question 11.4); in terms of the matrix $\bar{\mathbf{H}}^{(k)}$ used in Goldfarb's method (see (11.1.17)) the recurrence relation is

$$\bar{\mathbf{H}} := \bar{\mathbf{H}} - \frac{\bar{\mathbf{H}}\mathbf{a}\mathbf{a}^T\bar{\mathbf{H}}}{\mathbf{a}^T\bar{\mathbf{H}}\mathbf{a}} \tag{11.2.3}$$

suppressing superscript k, where $\mathbf{a}$ is the column that is added to $\mathbf{A}$. When a constraint index is removed from $\mathscr{A}$ as in step (c) of (11.2.2), an extra column $\mathbf{z}_{n-m+1}$ is adjoined to $\mathbf{Z}$. This extends the space of the free variables to one higher dimension, and no curvature information is available in the new direction. Hence $\mathbf{M}^{(k)}$ must be extended in an arbitrary way: the most convenient way is to assign

$$\mathbf{M}^{(k)} := \begin{bmatrix} \mathbf{M}^{(k)} & \mathbf{0} \\ \mathbf{0}^T & 1 \end{bmatrix} \tag{11.2.4}$$

which is in the spirit of an initial choice of $\mathbf{M}^{(1)} = \mathbf{I}$. $\mathbf{H}^{(k)}$ is updated in an analogous way and corresponds to an update

$$\bar{\mathbf{H}}^{(k)} := \bar{\mathbf{H}}^{(k)} + \mathbf{z}_{n-m+1}\mathbf{z}_{n-m+1}^T \tag{11.2.5}$$

in $\bar{\mathbf{H}}^{(k)}$.

It is interesting to consider quadratic termination results for the active set/quasi-Newton methods when the above updating formulae are used. If the active set is changed c times in all, and m_i, $i = 0, 1, \ldots, c$, is the number of active constraints after the ith change, then an obvious bound for the total number t of exact line searches is

$$t \leqslant \sum_{i=0}^{c} (n - m_i).$$

However, Powell (1972c) shows that this is unduly pessimistic and that a better bound is

$$t \leqslant c + n,$$

which shows that introducing arbitrary information when using (11.2.4) does not increase the bound by more than one. This result also allows for additional inexact line searches to be mixed with exact line searches. It seems that the tighter bound

$$t \leqslant c + n - \max_{0 \leqslant i \leqslant c} m_i$$

also follows from Powell's paper, which can be important for large almost linear problems. These bounds all beg the question as to whether the algorithm does terminate, that is whether there exists a finite c. This is by no means obvious since the algorithm can return to a previous active set, so the finiteness of all possible active sets does not immediately imply termination.

Another observation relating to the active set method when using a Levenberg–Marquardt parameter is also of interest. It is described in Section 5.2 how it is possible to control the iteration in two ways, either using the parameter $h^{(k)}$ (algorithm (5.1.6)) or the parameter $\nu^{(k)}$ (algorithm (5.2.7)). In an active set method it is far preferable to control the algorithm using $h^{(k)}$ (Holt and Fletcher, 1979), since a suitable value of $h^{(k)}$ is likely to be little affected by changes in active set, whereas this is not true for the $\nu^{(k)}$ parameter. An interesting example of the potential difficulties is described in Question 11.5.

Finally a method for solving (11.1.1) is described which avoids using the active set method. The method is a version of Newton's method in which at an iterate $\mathbf{x}^{(k)}$, a quadratic Taylor series approximation

$$f(\mathbf{x}^{(k)} + \boldsymbol{\delta}) \approx q^{(k)}(\boldsymbol{\delta}) = f^{(k)} + \boldsymbol{\delta}^{\mathrm{T}}\mathbf{g}^{(k)} + \tfrac{1}{2}\boldsymbol{\delta}^{\mathrm{T}}\mathbf{G}^{(k)}\boldsymbol{\delta} \tag{11.2.6}$$

is made, and the subproblem

$$\begin{aligned} &\underset{\boldsymbol{\delta}}{\text{minimize}} \quad q^{(k)}(\boldsymbol{\delta}) \\ &\text{subject to} \quad \mathbf{a}_i^{\mathrm{T}}\boldsymbol{\delta} = b_i - \mathbf{a}_i^{\mathrm{T}}\mathbf{x}^{(k)}, \qquad i \in E \\ &\phantom{\text{subject to}} \quad \mathbf{a}_i^{\mathrm{T}}\boldsymbol{\delta} \geqslant b_i - \mathbf{a}_i^{\mathrm{T}}\mathbf{x}^{(k)}, \qquad i \in I \end{aligned} \tag{11.2.7}$$

derived directly from (11.1.1) is solved. To ensure global convergence a step restriction

$$\|\boldsymbol{\delta}\| \leqslant h^{(k)} \tag{11.2.8}$$

is used in an algorithm like (5.1.6). Usually the L_∞ norm is chosen in (11.2.8) so that the method solves a sequence of quadratic programming subproblems with inequality constraints. The method is a special case of the SQP method described in Section 12.4. A disadvantage is that each iteration is relatively expensive and advantage cannot usually be taken of the updating techniques of quasi-Newton methods to reduce the overall operation count to $O(n^2)$

operations per iteration. However the solution of this subproblem does enable the correct active set to be determined quickly and also avoids the problem of zigzagging (see the theorems at the end of Section 11.3) so may converge more rapidly when remote from the solution. Certainly if the function and derivatives are expensive to compute, this approach can be preferable. A quasi-Newton version of the algorithm is given by Fletcher (1972b), although it might be better to use as an updating formula the version of the BFGS formula due to Powell (1978a) based on (12.4.18).

11.3 ZIGZAGGING

A feature which can adversely affect the rate of convergence of any type of method for handling inequality constraints is known as *zigzagging*. Although the set of active constraints $\mathscr{A}^*$ at a local solution is well defined, it may be that the sets $\mathscr{A}^{(k)}$ (as defined in (7.1.2)) obtained at the iterates $\mathbf{x}^{(k)}$ in some method do not settle down (so that $\mathscr{A}^{(k)} = \mathscr{A}^*$ for all $k \geqslant K$ where K is sufficiently large) but oscillate between different subsets of the constraints in the problem. For linear constraints, this corresponds to zigzagging between different linear manifolds corresponding to the feasible region in (11.2.1) for different $\mathscr{A}$ (see Figure 11.3.1). If the active set does settle down (with $\mathscr{A}^{(k)} = \mathscr{A}^* \; \forall k \geqslant K$) then the order of convergence becomes that for an equality constraint method (Section 11.1) which is usually superlinear for most good methods. However if zigzagging occurs then the order can degenerate to being linear with a rate constant which is not small, and in some cases the method can fail to converge to a solution.

For the active set method of algorithm (11.2.2), the likelihood of zigzagging is correlated with the test for an acceptable solution implied in step (b). If it is possible to solve any EP exactly in a finite number of steps, and if step (b) tests for an exact solution to the EP (as in quadratic programming, (10.3.4), step (b)), then it is not possible for zigzagging to occur (excluding degenerate cases). This

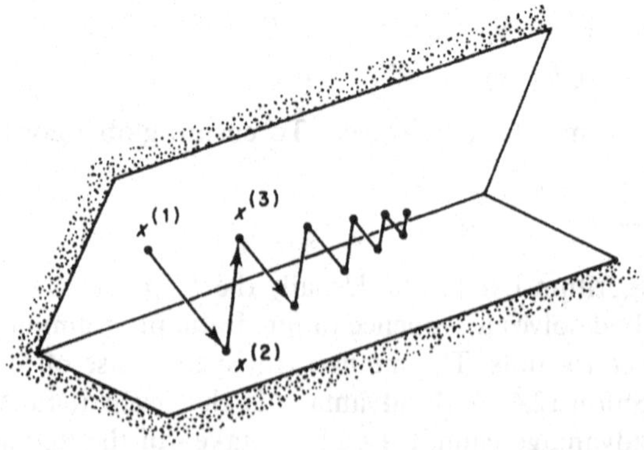

Figure 11.3.1 Zigzagging

is because the algorithm terminates for the reasons described in Section 10.3; once a constraint index is removed from the active set $\mathscr{A}^{(k)}$ in step (c), then the algorithm can no longer return to that active set by virtue of the optimality of $\mathbf{x}^{(k)}$ and the fact that $f^{(k)}$ is monotonically decreasing. However, it is not possible to solve any EP exactly when $f(\mathbf{x})$ is a general function, and it is inefficient to find the solution to high accuracy because the solution to the EP may not correspond to the solution of (11.1.1). The opposite possibility therefore is to allow step (c) to be taken on every iteration. It is this strategy which is most likely to cause zigzagging, especially when the multiplier estimates are poor. An example is given by Wolfe (1972) in which zigzagging causes non-convergence to a solution (see Question 11.6). The example is somewhat pathological because the objective function is not $\mathbb{C}^2$ and the steepest descent method is used to determine $\mathbf{s}^{(k)}$ in (11.2.2) step (d). Nonetheless it has been observed in practice with quasi-Newton methods and smooth functions that this strategy can induce a slow rate of linear convergence. Therefore what is required on step (b) is to have some compromise between these two extremes which eliminates the possibility of zigzagging and yet which enables an active set $\mathscr{A}^{(k)}$ to be changed when it has become likely that it is not the same as $\mathscr{A}^*$.

Various *ad hoc* rules to prevent zigzagging are reviewed by Fletcher (1972c) and can work well enough in practice. However it seems preferable in view of the above discussion to try to eliminate zigzagging by looking at the accuracy to which the current EP has been solved. This is done by Rosen (1960). Goldfarb (1969), and Murtagh and Sargent (1969). They all suggest similar strategies which are typified by, although different to, the following (Fletcher, 1972c). At $\mathbf{x}^{(k)}$, a second order estimate of the reduction in $f(\mathbf{x})$ by keeping the same active set is given by

$$\Delta^{(k)} = \tfrac{1}{2}\mathbf{g}^{(k)\mathrm{T}}\mathbf{Z}\mathbf{H}^{(k)}\mathbf{Z}^{\mathrm{T}}\mathbf{g}^{(k)} \tag{11.3.1}$$

by virtue of (11.1.8) and (11.1.9). $\mathbf{H}^{(k)}$ is positive definite and is either equal to or an approximation to the inverse reduced Hessian matrix, as in (11.1.13). If $\boldsymbol{\lambda}^{(k)}$ is a second order multiplier estimate, and $\lambda_i^{(k)} < 0$, then the additional function reduction obtained by removing the ith constraint index from $\mathscr{A}^{(k)}$ (assuming $i \in I$) is

$$\Delta_i^{(k)} = \frac{\tfrac{1}{2}\lambda_i^{(k)2}}{-u_{ii}^{(k)}} \tag{11.3.2}$$

where $u_{ii}^{(k)}$ is the diagonal element of the matrix $\mathbf{U}^{(k)}$ defined by (10.2.3) at $\mathbf{x}^{(k)}$. This result is valid when $u_{ii}^{(k)} < 0$ which is implied by (10.2.6) when $\mathbf{G}^{(k)}$ is positive definite. Hence a possible strategy is to find the integer p which maximizes $\Delta_i^{(k)}$ over the set $\{i: \lambda_i^{(k)} < 0\}$ and then to define $\mathbf{x}^{(k)}$ as an acceptable solution (and hence remove the pth constraint) if $\Delta_p^{(k)} \geqslant \Delta^{(k)}$. The motivation for such a test is that a greater reduction in $f(\mathbf{x})$ is likely to occur by removing the constraint than by not doing so. For a quasi-Newton method a recurrence relation for the quantities $u_{ii}^{(k)}$ can be determined. An advantage of this test is that it is invariant under both linear transformations of the variables and scaling of the constraints. If $\mathbf{G}^{(k)}$ (or its approximation) is not positive definite, something has to be done to generalize the definitions of $\Delta^{(k)}$ and $\Delta_i^{(k)}$. One possibility is to define $\Delta^{(k)} = \infty$ if $\mathbf{H}^{(k)}$ is not positive definite and to have $\mathbf{x}^{(k)}$ as not being an acceptable solution. Otherwise, if $u_{ii}^{(k)} \geqslant 0$, then $\Delta_i^{(k)} = \infty$ is defined and p is chosen to minimize $\lambda_i^{(k)}$ over the set $\{i: \lambda_i^{(k)} < 0, \Delta_i^{(k)} = \infty\}$. Strategies based on comparing quantities like (11.3.1) and (11.3.2) have been used successfully in practice.

The dependence of the above test on the quantities $u_{ii}^{(k)}$ makes it somewhat unwieldy and it is preferable to have a more simple test for an acceptable solution with good invariance properties. It is also important to consider whether it can be proved that the resulting method coverges to the solution of (11.1.1) and avoids zigzagging. A test which meets these criteria to a certain extent can be derived in the following way. Let $\{l(k)\}, k = 1, 2, \ldots,$ be a sequence of integers such that $l(k)$ $(<k)$ is the greatest previous iteration index on which a constraint is removed from $\mathscr{A}$, or $l(k) = 1$ if no such index exists. Then a point $\mathbf{x}^{(k)}$ is acceptable in step (b) if

$$\Delta^{(k)} \leqslant f^{(l(k))} - f^{(k)}, \tag{11.3.3}$$

that is if the predicted function reduction for the current $\mathscr{A}$ is less than the total reduction since a constraint was last removed from $\mathscr{A}$. An alternative possibility is to define the right-hand side of (11.3.3) as the total function reduction since $\mathscr{A}$ was last changed. The motivation for this test is that when zigzagging occurs the right-hand side of (11.3.3) goes to zero, which ensures that $\Delta^{(k)}$ goes to zero, which usually ensures that convergence occurs for the subsequence of points with the same $\mathscr{A}$. If $\mathbf{x}^{(k)}$ is accepted as a solution of the EP then the decision as to which constraint to remove is that given in (11.2.2), step (c). If the constraints are prescaled then these tests are invariant under linear transformations of variables and constraint scalings. The test must also be incorporated with the termination test which is conveniently chosen to be $\Delta^{(k)} \leqslant \varepsilon$ where $\varepsilon > 0$ is some preset tolerance on the accuracy required in f^*. The detailed form of steps (b) and (c) is therefore

(b) if $\Delta^{(k)} > f^{(l(k))} - f^{(k)}$ then go to (d);
(c) (i) calculate $\boldsymbol{\lambda}^{(k)}$ from (11.1.18) or (11.1.19) and
  let $\lambda_q^{(k)}$ minimize $\lambda_i^{(k)}$, $i \in I \cap \mathscr{A}$; (11.3.4)
  (ii) if $\lambda_q^{(k)} \geqslant 0$ then (if $\Delta^{(k)} \leqslant \varepsilon$ then terminate else go to (d)) else remove q from $\mathscr{A}$.

Note that no special attention is required to detect the case that $\mathbf{x}^{(k)}$ is a vertex. Here $\Delta^{(k)} = 0$, (11.3.3) is true, and hence step (c) in (11.3.4) is always carried out.

In some cases it is possible to prove that convergence without zigzagging takes place when (11.3.4) is used in (11.2.2). Such a result is given below. To some extent this depends on the properties of the method used for finding the solution and multipliers of the EP. To make the result easy to state, it is assumed that the feasible region R is not empty, that $f(\mathbf{x})$ is $\mathbb{C}^2$ on R, and that the smallest eigenvalue μ_n of the Hessian matrix $\mathbf{G}(\mathbf{x})$ satisfies $\mu_n(\mathbf{x}) \geqslant a > 0$ on R, an assumption which implies strict convexity of f on R. Newton's method with line search is used and the search can be terminated by any of the conditions described in Section 2.5, so that convergence results like Theorem 2.5.1 can be applied. It is also assumed for any $\mathbf{x}\in R$ that the vectors $\mathbf{a}_i$, $i\in\mathscr{A}(\mathbf{x})$, are independent. In this case either (11.1.18) or (11.1.19) can be used to calculate $\boldsymbol{\lambda}^{(k)}$; all that is required is that if $\mathbf{x}^{(k)}\to\mathbf{x}^*$ for some fixed $\mathscr{A}$ then $\boldsymbol{\lambda}^{(k)}\to\boldsymbol{\lambda}^*$.

Theorem 11.3.1

Under the above assumptions then the above method converges to the solution of (11.1.1) *from any initial feasible point* $\mathbf{x}^{(1)}$. *In addition if strict complementarity holds at* $\mathbf{x}^*$ *then no zigzagging can occur and* $\mathscr{A}^{(k)}=\mathscr{A}^*$ *for all* k *sufficiently large.*

Proof

If $\mathscr{A}^{(k)}$ is constant for all k sufficiently large then $\mathbf{x}^{(k)}\to\mathbf{x}^*$ by Theorem 2.5.1 applied to the reduced problem. Otherwise the integers $l(k)\to\infty$; also the assumptions imply that $f(\mathbf{x})$ is bounded below so $f^{(k)}\downarrow f^{\infty}$ and $f^{(l(k))}-f^{(k)}\to 0$ which implies that $\Delta^{(k)}\to 0$ on the subsequence of iterations for which (11.3.3) holds. Let K be sufficiently large such that any set $\mathscr{A}^{(k)}$ $k\geqslant K$ occurs infinitely often and restrict attention to $k\geqslant K$. Let $\mathscr{S}$ be the corresponding subsequence of iteration indices. Using positive definiteness and independence assumptions it follows from $\Delta^{(k)}\to 0$ that $\mathbf{Z}^{\mathrm{T}}\mathbf{g}^{(k)}\to 0$ for $k\in\mathscr{S}$ and hence that there exists a unique limit point $\mathbf{x}^{\mathscr{S}}$ say, which minimizes $f(\mathbf{x})$ on $\mathbf{a}_i^{\mathrm{T}}\mathbf{x}=b_i$, $i\in\mathscr{A}^{(k)}$, $k\in\mathscr{S}$, and that $\mathbf{g}^{\mathscr{S}}=\sum_{i\in\mathscr{A}^{(k)}}\mathbf{a}_i\lambda_i^{\mathscr{S}}$. Also $\lambda_i^{(k)}\to\lambda_i^{\mathscr{S}}$ for all $i\in\mathscr{A}^{(k)}$, $k\in\mathscr{S}$.

Since the number of possible different active sets is finite, it follows for $k\geqslant K$ that on the basis of $\mathscr{A}^{(k)}$ the main sequence can be divided into a finite number of disjoint subsequences $\mathscr{S}_1,\mathscr{S}_2,\ldots,\mathscr{S}_t$ which have different limit points $\mathbf{x}^{\mathscr{S}_i}$ $i=1,2,\ldots,t$. It is proved by contradiction that $t=1$. Let $t>1$, and let $\mathscr{S}_i$ and $\mathscr{S}_j$ be any two subsequences ($i\neq j$) such that $k\in\mathscr{S}_i$ and $k+1\in\mathscr{S}_j$ an infinite number of times. Since $\mathbf{x}^{(k+1)}$ is acceptable in the line search, (2.5.3) holds, and from a Taylor series and $\mu_n\geqslant a$

$$f^{(k+1)}\geqslant f^{(k)}+\mathbf{g}^{(k)\mathrm{T}}(\mathbf{x}^{(k+1)}-\mathbf{x}^{(k)})+\tfrac{1}{2}a\,\|\mathbf{x}^{(k+1)}-\mathbf{x}^{(k)}\|_2^2.$$

Substituting this into (2.5.3) and rearranging gives

$$-(1-\rho)\mathbf{g}^{(k)\mathrm{T}}(\mathbf{x}^{(k+1)}-\mathbf{x}^{(k)})\geqslant\tfrac{1}{2}a\,\|\mathbf{x}^{(k+1)}-\mathbf{x}^{(k)}\|_2^2$$

and substituting into (2.5.3) again gives

$$f^{(k)}\geqslant f^{(k+1)}+\tfrac{1}{2}a\frac{\rho}{1-\rho}\|\mathbf{x}^{(k+1)}-\mathbf{x}^{(k)}\|_2^2.$$

In the limit $f^{(k)}\to f^{\infty}$, $f^{(k+1)}\to f^{\infty}$ and $\mathbf{x}^{(k+1)}-\mathbf{x}^{(k)}\to\mathbf{x}^{\mathscr{S}_j}-\mathbf{x}^{\mathscr{S}_i}\neq\mathbf{0}$ which contradicts the inequality. Thus $t=1$ and the main sequence converges.

Let the limit point be $\mathbf{x}^{\infty}$. By independence and the equation for $\mathbf{g}^{\mathscr{S}}$ above, it follows that $\mathbf{g}^{\infty}=\sum_{i\in\bigcap_k\mathscr{A}^{(k)}}\mathbf{a}_i\lambda_i^{\infty}$ and so $\lambda_i^{\infty}=0$ for all $i\in\bigcup_k\mathscr{A}^{(k)}\backslash\bigcap_k\mathscr{A}^{(k)}$. Also by continuity and since $\lambda_q^{(k)}$ is also chosen as in (11.3.4), step (c) (i), it follows that $\lambda_i^\infty \geqslant 0\ i\in I\bigcap_k \mathscr{A}^{(k)}$. Hence $\mathbf{x}^\infty, \boldsymbol{\lambda}^\infty$ is a KT point and by convexity solves (11.1.1). If strict complementarity $\lambda_i^* > 0\ i\in \mathscr{A}^* \cap I$ holds then the set $\bigcup_k \mathscr{A}^{(k)} \backslash \bigcap_k \mathscr{A}^{(k)}$ is empty and hence $\mathscr{A}^{(k)}$ is constant for sufficiently large k. Also $\mathscr{A}^{(k)} = \mathscr{A}^*$ follows since otherwise $\lambda_i^* = 0\ i\in \mathscr{A}^* \backslash \mathscr{A}^{(k)}$ is implied which is a contradiction. □

This proof corrects a non-trivial error in the version given in the previous edition. Note that the proof avoids difficulties over degeneracy, cycling, linear dependence, etc. by making a blanket assumption about independence. This is an unsatisfactory feature in a theorem because it tends to hide potential situations in which the algorithm might fail. A somewhat more general result is given by Byrd and Shultz (1982) but in general degenerate situations can be expected to cause difficulties for this algorithm.

The rest of this section concerns the algorithm defined in (11.2.7) and (11.2.8) (based on extending the trust region algorithm (5.1.6)). An advantage of this algorithm is that a much stronger result about global convergence can be proved, requiring no assumptions about the linear independence of vectors $\mathbf{a}_i\ i\in \mathscr{A}^\infty$ at a limit point $\mathbf{x}^\infty$. This shows that the algorithm cannot converge to a non-stationary limit point of this nature, so that the algorithm is less likely to fail.

Theorem 11.3.2

For the algorithm of (11.2.7), *and* (11.2.8), *if* $\mathbf{x}^{(k)}\in B \subset \mathbb{R}^n\ \forall k$ *where B is bounded, and if* $f\in \mathbb{C}^2$ *on B, then there exists an accumulation point* $\mathbf{x}^\infty$ *which satsifies first order (KT) necessary conditions and a weak form of the second order necessary conditions, that is*

$$\mathbf{s}^{\mathrm{T}}\mathbf{G}^\infty \mathbf{s} \geqslant 0 \qquad \forall \mathbf{s}: \mathbf{a}_i^{\mathrm{T}}\mathbf{s} = 0 \quad i\in \mathscr{A}^\infty. \tag{11.3.5}$$

Proof

By considering whether or not $\inf h^{(k)} = 0$, $\exists$ a convergent subsequence $\mathbf{x}^{(k)} \to \mathbf{x}^\infty$ $k\in \mathscr{S}$ for which either

(i) $r^{(k)} < 0.25$, $h^{(k+1)} \to 0$ and hence $\|\boldsymbol{\delta}^{(k)}\| \to 0$, or
(ii) $r^{(k)} \geqslant 0.25$ and $\inf h^{(k)} > 0$.

In either case the necessary conditions are shown to hold. In case (i) let $\exists$ a feasible direction $\mathbf{s}$ ($\|\mathbf{s}\| = 1$) at $\mathbf{x}^\infty$, so that $\mathbf{s}\in F^\infty$ (see (9.2.2)) and

$$\mathbf{s}^{\mathrm{T}}\mathbf{g}^\infty = -d \qquad d > 0. \tag{11.3.6}$$

A Taylor series for $f(\mathbf{x})$ about $\mathbf{x}^{(k)}$ gives

$$f(\mathbf{x}^{(k)} + \boldsymbol{\delta}^{(k)}) = q^{(k)}(\boldsymbol{\delta}^{(k)}) + o(\|\boldsymbol{\delta}^{(k)}\|^2)$$

and hence

$$\Delta f^{(k)} = \Delta q^{(k)} + o(\|\boldsymbol{\delta}^{(k)}\|^2). \tag{11.3.7}$$

For $k\in \mathscr{S}$ consider a step of length $\varepsilon^{(k)} = \|\boldsymbol{\delta}^{(k)}\|$ along $\mathbf{s}^{(k)}$. For k sufficiently large, $\mathscr{A}^{(k)} \subset \mathscr{A}^{\infty}$ and hence $F^{(k)} \supset F^{\infty}$. Thus $\mathbf{s} \in F^{(k)}$ and $\varepsilon^{(k)}\mathbf{s}^{(k)}$ is feasible in (11.2.8). Optimality of $\boldsymbol{\delta}^{(k)}$ in (11.2.7) and continuity yield

$$\Delta q^{(k)} \geqslant q^{(k)}(\mathbf{0}) - q^{(k)}(\varepsilon^{(k)}\mathbf{s}) = -\varepsilon^{(k)}\mathbf{s}^{\mathrm{T}}\mathbf{g}^{(k)} + o(\varepsilon^{(k)}) = \varepsilon^{(k)}d + o(\varepsilon^{(k)})$$

from (11.3.6). It follows from (5.1.8) that $r^{(k)} = 1 + o(1)$ which contradicts $r^{(k)} < 0.25$. Thus (11.3.6) is contradicted and so $\mathbf{s}^{\mathrm{T}}\mathbf{g}^{\infty} \geqslant 0$ for all $\mathbf{s} \in F^{\infty}$. It follows from the corollary to Farkas' Lemma (Lemma 9.2.4) that first order (KT) conditions are satisfied at $\mathbf{x}^{\infty}$.

Now let $\exists$ a second order feasible descent direction $\mathbf{s}$ ($\|\mathbf{s}\| = 1$) at $\mathbf{x}^{\infty}$, so that

$$\mathbf{s}^{\mathrm{T}}\mathbf{G}^{\infty}\mathbf{s} = -d, \quad d > 0, \qquad \mathbf{a}_i^{\mathrm{T}}\mathbf{s} = 0 \quad \forall i \in \mathscr{A}^{\infty}. \tag{11.3.8}$$

For $k \in \mathscr{S}$ consider a step of length $\varepsilon^{(k)}$ along $\sigma\mathbf{s}$, choosing σ so that $\sigma\mathbf{s}^{\mathrm{T}}\mathbf{g}^{(k)} \leqslant 0$. For sufficiently large k, $\mathscr{A}^{(k)} \subset \mathscr{A}^{\infty}$ so $\sigma\mathbf{s} \in F^{(k)}$, and the step $\sigma\varepsilon^{(k)}\mathbf{s}$ is feasible in the subproblem. Then by optimality and continuity, it is possible as in Theorem 5.1.2 to deduce that $r^{(k)} = 1 + o(1)$ and contradict $r^{(k)} < 0.25$. Thus there is no $\mathbf{s}$ satisfying (11.3.8) and so (11.3.5) follows. Thus both first and second order necessary conditions are shown to hold when $\inf h^{(k)} = 0$.

In case (ii), $f^{(1)} - f^{\infty} \geqslant \sum_{k \in \mathscr{S}} \Delta f^{(k)}$ and so $r^{(k)} \geqslant 0.25$ implies that $\Delta q^{(k)} \to 0$ $k \in \mathscr{S}$. Define $q^{\infty}(\boldsymbol{\delta}) = f^{\infty} + \boldsymbol{\delta}^{\mathrm{T}}\mathbf{g}^{\infty} + \frac{1}{2}\boldsymbol{\delta}^{\mathrm{T}}\mathbf{G}^{\infty}\boldsymbol{\delta}$. Let $\bar{h}$ satisfy $0 < \bar{h} < \inf h^{(k)}$ and let $\bar{\boldsymbol{\delta}}$ minimize $q^{\infty}(\boldsymbol{\delta})$ subject to $\|\boldsymbol{\delta}\| \leqslant \bar{h}$ and $\mathbf{x}^{\infty} + \boldsymbol{\delta} \in R$ (R denotes the feasible region of the constraints in (11.1.1)). Define $\bar{\mathbf{x}} = \mathbf{x}^{\infty} + \bar{\boldsymbol{\delta}}$. Observe that for sufficiently large k, $\bar{\mathbf{x}} - \mathbf{x}^{(k)}$ is feasible in the subproblem (11.2.7) and (11.2.8) (since $\bar{\mathbf{x}} \in R$, $\mathbf{x}^{(k)} \in R$ and $\|\bar{\mathbf{x}} - \mathbf{x}^{(k)}\| \leqslant \|\bar{\boldsymbol{\delta}}\| + o(1) \leqslant \bar{h} + o(1) \leqslant h^{(k)}$) so

$$q^{(k)}(\bar{\mathbf{x}} - \mathbf{x}^{(k)}) \geqslant q^{(k)}(\boldsymbol{\delta}^{(k)}) = f^{(k)} - \Delta q^{(k)}.$$

As in Theorem 5.1.1, taking the limit shows that $\boldsymbol{\delta} = \mathbf{0}$ also minimizes $q^{\infty}(\boldsymbol{\delta})$ on $\mathbf{x}^{\infty} + \boldsymbol{\delta} \in R$ and $\|\boldsymbol{\delta}\| \leqslant \bar{h}$. As the latter condition is inactive, and the constraints that define R are linear (needed for regularity), it follows that the first and second order necessary conditions also hold in this case. □

The remarks following Theorem 5.1.1 are equally valid here. With additional mild assumptions at $\mathbf{x}^{\infty}$ and for sufficiently large k, the possibility of zigzagging can be excluded, the trust region bound is shown to be inactive, and the algorithm becomes equivalent to Newton's method applied to a reduced optimization problem.

Theorem 11.3.3

Let the accumulation point $\mathbf{x}^{\infty}$ of Theorem 11.3.2 have linearly independent vectors $\mathbf{a}_i$ $i \in \mathscr{A}^{\infty}$. This implies the existence of unique multipliers $\boldsymbol{\lambda}^{\infty}$: it is assumed that these satisfy strict complementarity $\lambda_i^{\infty} > 0$ $i \in I \cap \mathscr{A}^{\infty}$. Assume also that $\mathbf{s}^{\mathrm{T}}\mathbf{G}^{\infty}\mathbf{s} > 0$ for all $\mathbf{s} \neq \mathbf{0}$ such that $\mathbf{s}^{\mathrm{T}}\mathbf{a}_i = 0$ $i \in \mathscr{A}^{\infty}$ (a strong form of the second order sufficient conditions). Then for k in the main sequence $\mathbf{x}^{(k)} \to \mathbf{x}^{\infty}$, $r^{(k)} \to 1$, $\inf h^{(k)} > 0$, and for k sufficient large $\mathscr{A}^{(k)} = \mathscr{A}^{\infty}$ and the algorithm is equivalent to Newton's method applied to the reduced optimization problem defined by eliminating the constraints indexed by $i \in \mathscr{A}^{\infty}$.

Proof

Let $\mathbf{Y}$ and $\mathbf{Z}$ be the generalized elimination matrices of Section 10.1 for the matrix $\mathbf{A}$ whose columns are the vectors $\mathbf{a}_i$ $i \in \mathscr{A}^\infty$. Denote $\mathbf{c}^{(k)}$ as the vector of elements $\mathbf{a}_i^T\mathbf{x}^{(k)} - b_i$ $i \in \mathscr{A}^\infty$, and $I^\infty = I \cap \mathscr{A}^\infty$. Feasibility of $\mathbf{x}^{(k)}$ is maintained by the algorithm which implies that $c_i^{(k)} = 0$ $i \in E$ and $c_i^{(k)} \geqslant 0$ $i \in I^\infty$. Consider a sufficiently large index $k \in \mathscr{S}$. If $\mathbf{c}^{(k)} = \mathbf{0}$ then it is possible to proceed as in Theorem 5.1.2 applied to the reduced problem: minimize $q^{(k)}(\mathbf{Zy})$. Additional observations are required that constraints $i \notin \mathscr{A}^\infty$ do not affect feasibility of the step $\alpha\mathbf{s}^{(k)}$ in the subproblem, and that the Newton step for the reduced problem is also that which solves the subproblem, by continuity of the multipliers (see (11.3.9) below). The results of Theorem 5.1.2 are therefore deduced, and also the fact that $\mathbf{c}^{(k)} = \mathbf{0}$ for all k in the main sequence, which implies that $\mathscr{A}^{(k)} = \mathscr{A}^*$ for such k.

Consider therefore the case that $\mathbf{c}^{(k)} \neq \mathbf{0}$ for sufficient large $k \in \mathscr{S}$ and consider a step—$\alpha^{(k)}\mathbf{Yc}^{(k)}$ in the subproblem, where $\alpha^{(k)} = \min(1, h^{(k)}/\|\mathbf{Yc}^{(k)}\|)$. Note that $\mathbf{c}(\mathbf{x}^{(k)} - \alpha^{(k)}\mathbf{Yc}^{(k)}) = \mathbf{c}^{(k)} - \alpha^{(k)}\mathbf{A}^T\mathbf{Yc}^{(k)} = (1 - \alpha^{(k)})\mathbf{c}^{(k)}$ showing that the step remains feasible for constraints $i \in \mathscr{A}^\infty$, and $\|-\alpha^{(k)}\mathbf{Yc}^{(k)}\| \leqslant h^{(k)}$ showing that the step remains feasible in the trust region bound. Since $\mathbf{c}^{(k)} \to \mathbf{0}$ the step is also feasible for constraints $i \in \mathscr{A}^\infty$ for k sufficiently large. Thus the step is feasible in the subproblem, so by optimality of $\boldsymbol{\delta}^{(k)}$ and continuity

$$\Delta q^{(k)} \geqslant q^{(k)}(0) - q^{(k)}(-\alpha^{(k)}\mathbf{Yc}^{(k)}) = \alpha^{(k)}\mathbf{g}^{(k)T}\mathbf{Yc}^{(k)} + o(h^{(k)}).$$

The definition of $\alpha^{(k)}$ ensures that

$$\alpha^{(k)}\mathbf{g}^{(k)T}\mathbf{Yc}^{(k)} \geqslant h^{(k)}\mathbf{g}^{(k)T}\mathbf{Yc}^{(k)}/\|\mathbf{Yc}^{(k)}\|.$$

Since $\mathbf{Y}^T\mathbf{g}^{(k)} \to \boldsymbol{\lambda}^\infty$ it follows that

$$\Delta q^{(k)} \geqslant h^{(k)}\boldsymbol{\lambda}^{\infty T}\mathbf{c}^{(k)}/\|\mathbf{Yc}^{(k)}\| + o(h^{(k)}).$$

Now $\mathbf{c}^{(k)} \neq \mathbf{0}$ implies $c_i^{(k)} \neq 0$ for at least one $i \in I^\infty$, and since $\lambda_i^\infty > 0$ for $i \in I^\infty$ it is possible to bound $\boldsymbol{\lambda}^{\infty T}\mathbf{c}^{(k)}/\|\mathbf{Yc}^{(k)}\| \geqslant d > 0$ where d is independent of k. Then $\Delta q^{(k)} \geqslant h^{(k)}d + o(h^{(k)})$ and it follows as in Theorem 11.2.2 that $r^{(k)} = 1 + o(1)$ and therefore that $\mathscr{S}$ arises from case (ii) and the trust region bound is inactive. Thus second order conditions and strict complemetarity at $\mathbf{x}^\infty$, $\boldsymbol{\lambda}^\infty$ ensure that $\boldsymbol{\delta}^{(k)}$, $\boldsymbol{\lambda}^{(k)}$ obtained from

$$\begin{bmatrix} \mathbf{G}^{(k)} & -\mathbf{A}^{(k)} \\ -\mathbf{A}^{(k)T} & \mathbf{0} \end{bmatrix} \begin{bmatrix} \boldsymbol{\delta}^{(k)} \\ \boldsymbol{\lambda}^{(k)} \end{bmatrix} = \begin{bmatrix} -\mathbf{g}^{(k)} \\ \mathbf{c}^{(k)} \end{bmatrix} \qquad (11.3.9)$$

give the minimizer and multipliers for the subproblem for k sufficiently large, and it follows that $\boldsymbol{\delta}^{(k)} \to \mathbf{0}$ (i.e. $\mathbf{x}^{(k+1)} \to \mathbf{x}^\infty$) for $k \in \mathscr{S}$. Also $\mathbf{c}^{(k+1)} = \mathbf{0}$ by linearity. Hence it is again possible to consider the reduced problem on iteration $k+1$ as described at the start of the this proof, and deduce the conclusions of theorem.

Corollary

If in addition $\mathbf{x}^\infty$ *is a vertex* ($|\mathscr{A}^\infty| = n$) *then the algorithm terminates finitely.*

Proof

This follows because there is a neighbourhood of $\mathbf{x}^\infty$ in which $\mathbf{x}^\infty$ is the only point for which $\mathscr{A}(\mathbf{x}) = \mathscr{A}^\infty$. Thus $\mathscr{A}^{(k)} = \mathscr{A}^\infty$ cannot occur infinitely often. □

Thus the algorithm is shown to behave in a very satisfactory way and the remarks after Theorem 5.1.2 are again valid. Finally the possibility of approximating $\mathbf{G}^{(k)}$ in (11.2.6) by a matrix $\mathbf{B}^{(k)}$ which is updated after each iteration is discussed. Under mild conditions [$\mathbf{B}^{(k)}$ bounded is sufficient, but weaker conditions are possible, e.g. (Fletcher, 1972b)] global convergence to a KT point (the first part of Theorem 11.3.2) and vertex termination (corollary of Theorem 11.3.3) are proved in a similar way. Necessary and sufficient conditions on $\mathbf{B}^{(k)}$ for superlinear convergence are $\|\mathbf{Z}^T\mathbf{B}^{(k)}\mathbf{Z}\mathbf{y}^{(k)}\|/\|\mathbf{y}^{(k)}\| \to 0$ where $\boldsymbol{\delta}^{(k)} = \mathbf{Z}\mathbf{y}^{(k)}$, following the Dennis and Moré theorem (Theorem 6.2.3) for the reduced problem.

QUESTIONS FOR CHAPTER 11

11.1. Relate the conditions $\mathbf{Z}^T\mathbf{g}^* = \mathbf{0}$ and $\mathbf{Z}^T\mathbf{G}^*\mathbf{Z}$ positive definite arising from (11.1.6) and (11.1.7) to the first and second order conditions of Sections 9.1 and 9.3 applied to problem (11.1.2) when rank $(\mathbf{A}) = m$. Show that $\mathbf{Z}^T\mathbf{g}^* = \mathbf{0}$ is equivalent to $\mathbf{g}^* = \mathbf{A}\boldsymbol{\lambda}^*$. Also show that the columns of $\mathbf{Z}$ are a basis for the linear space (9.3.4) and hence that (9.3.5) is equivalent to the condition that $\mathbf{Z}^T\mathbf{G}^*\mathbf{Z}$ is positive definite.

11.2. Consider any Broyden method with parameters $\phi^{(k)}$, $k = 1, 2, \ldots$, applied to solve the reduced problem (11.1.4) using the positive definite inverse reduced Hessian matrix $\mathbf{H}^{(k)}$ in (11.1.13). Show for the original problem (11.1.2) that an equivalent sequence $\{\mathbf{x}^{(k)}\}$ is obtained if the same $\phi^{(k)}$ are used in a Broyden family update of the positive semi-definite matrix $\bar{\mathbf{H}}^{(k)}$defined in (1.1.17), when $\mathbf{s}^{(k)}$ is calculated from $\mathbf{s}^{(k)} = -\bar{\mathbf{H}}^{(k)}\mathbf{g}^{(k)}$. Assume that the initial matrices are related by $\bar{\mathbf{H}}^{(1)} = \mathbf{Z}\mathbf{H}^{(1)}\mathbf{Z}^T$.

11.3. If $\mathbf{h}^{(k)} = \mathbf{x}^{(k)} - \mathbf{x}^*$ where $\mathbf{x}^*$ solves (11.1.2) show that the vector $\boldsymbol{\lambda}^{(k)}$ in (11.1.18) satisfies $\boldsymbol{\lambda}^{(k)} = \boldsymbol{\lambda}^* + O(\|\mathbf{h}^{(k)}\|)$ and is therefore a first order estimate of $\boldsymbol{\lambda}^*$. If $\boldsymbol{\lambda}^{(k)}$ is computed from (11.1.19) show that $\boldsymbol{\lambda}^{(k)} = \boldsymbol{\lambda}^* + O(\|\mathbf{h}^{(k)}\|^2)$. Use a result like (10.2.9) and assume that rank$(\mathbf{A}) = m$, that $\mathbf{Z}^T\mathbf{G}^*\mathbf{Z}$ is positive definite, and that the expansion $\mathbf{g}^{(k)} = \mathbf{g}^* + \mathbf{G}^*\mathbf{h}^{(k)} + O(\|\mathbf{h}^{(k)}\|^2)$ is valid.

11.4. Assume that $n \times n$ matrices are related by

$$\hat{\mathbf{M}} = \begin{bmatrix} \mathbf{M} & \mathbf{b} \\ \mathbf{b}^T & \beta \end{bmatrix}$$

and let $\hat{\mathbf{H}} = \hat{\mathbf{M}}^{-1}$and $\mathbf{H} = \mathbf{M}^{-1}$. Show that $\hat{\mathbf{H}}$ and $\mathbf{H}$ are related by

$$\hat{\mathbf{H}} - \mu\mathbf{u}\mathbf{u}^T = \begin{bmatrix} \mathbf{H} & \mathbf{0} \\ \mathbf{0}^T & 0 \end{bmatrix}$$

where either $\mathbf{u}^T = (\mathbf{b}^T\mathbf{H}: -1)$ and $\mu = 1/(\beta - \mathbf{b}^T\mathbf{H}\mathbf{b})$ or $\mathbf{u} = \hat{\mathbf{H}}\mathbf{e}_n$ and $\mu = u_n^{-1}$, depending whether it is $\hat{\mathbf{H}}$ or $\mathbf{H}$ that is known. Hence deduce (11.2.3).

11.5. For fixed $\mathbf{x}'$, show that adding a linear equality constraint to problem (11.1.2) cannot increase the condition number of the reduced Hessian matrix at $\mathbf{x}'$, assuming that $\mathbf{x}'$ is feasible in both problems. It might be thought in general therefore that adding linear constraints cannot degrade the conditioning of an optimization problem. This is not so as the following example shows. Consider best least squares data fitting with a sum of exponentials $ae^{\alpha t} + be^{\beta t}$. The unknown parameters a, α, b, β must be chosen to best fit some given data. The problem usually has a well-defined (albeit ill-conditioned) solution. Consider adding the linear constraint $\alpha = \beta$. Then the parameters a and b become underdetermined and the reduced Hessian matrix is singular at any feasible point. Explain this apparent contradiction.

11.6. Consider the linear constraint problem

$$\underset{\mathbf{x}}{\text{minimize}} \quad \tfrac{4}{3}(x_1^2 - x_1x_2 + x_1^2)^{3/4} - x_3$$

$$\text{subject to} \quad x_3 \leqslant 2, \qquad \mathbf{x} \geqslant \mathbf{0}$$

due to Wolfe (1972). Show that the objective function is convex but not $\mathbb{C}^2$ and that the solution is $\mathbf{x}^* = (0, 0, 2)^T$. Solve the problem from $\mathbf{x}^{(1)} = (0, a, 0)^T$ where $0 < a \leqslant \sqrt{2}/4$ using algorithm (11.2.2). Allow any point to be an acceptable solution in step (b) and use the steepest descent method in step (d). Show that $\mathbf{x}^{(2)} = \frac{1}{2}(a, 0, \sqrt{a})^T$ and hence that, for $k \geqslant 2$,

$$\mathbf{x}^{(k)} = \begin{cases} (0, \alpha, \beta)^T & \text{if } k \text{ is odd} \\ (\alpha, 0, \beta)^T & \text{if } k \text{ is even} \end{cases}$$

where $\alpha = (\frac{1}{2})^{k-1}a$ and $\beta = \frac{1}{2}\sum_{j=0}^{k-2}(a/2^j)^{1/2}$. Hence show that $\mathbf{x}^{(k)} \to (0, 0, (1 + \frac{1}{2}\sqrt{2})\sqrt{2a})^T$ which is neither optimal nor a KT point.

11.7. After a constraint index q is removed from $\mathscr{A}$ in step (c) of the active set method, consider proving that the subsequent search direction $\mathbf{s}^{(k)}$ is down hill ($\mathbf{s}^{(k)T}\mathbf{g}^{(k)} < 0$) and strictly feasible ($\mathbf{s}^{(k)T}\mathbf{a}_q > 0$). The $\mathbf{Z}$ matrix is augmented by a column vector $\mathbf{z}$ as $\mathbf{Z} := [\mathbf{Z}:\mathbf{z}]$. Show that the required conditions are obtained if $\mathbf{Z}^T\mathbf{G}^{(k)}\mathbf{Z}$ is positive definite for the new $\mathbf{Z}$ matrix. If $\mathbf{Z}^T\mathbf{G}^{(k)}\mathbf{Z}$ is positive definite for the old $\mathbf{Z}$ matrix only, show that the choice $\mathbf{s}^{(k)} = \mathbf{T}\mathbf{e}_q$ described in Section 10.4 has the required properties. In both cases assume that a second order multiplier estimate (11.1.19) is used.

Consider proving the same result for a quasi-Newton method when update (11.2.3) is used, when a first order estimate (11.1.18) for $\lambda^{(k)}$ is used, and when the orthogonal factorization method defines $\mathbf{Y}$ and $\mathbf{Z}$. Show that the column $\mathbf{z}$ which augments $\mathbf{Z}$ is $\mathbf{z} = \mathbf{a}_q^+$ which is the column of $\mathbf{A}^{+T}(= \mathbf{Y})$ corresponding to constraint q. Hence show that $\mathbf{s}^{(k)T}\mathbf{g}^{(k)} < 0$ and $\mathbf{s}^{(k)T}\mathbf{a}_q = 1$, which satisfies the required conditions.

Chapter 12

Nonlinear Programming

12.1 PENALTY AND BARRIER FUNCTIONS

Nonlinear programming is the general case of (7.1.1) in which both the objective and constraint functions may be nonlinear, and is the most difficult of the smooth optimization problems. Indeed there is no general agreement on the best approach and much research is still to be done. Historically the earliest developments were *sequential minimization methods* based on the use of *penalty* and *barrier functions* as described in Sections 12.1 and 12.2. These methods suffer from some computational disadvantages and are not entirely efficient. Nonetheless, especially for no-derivative problems in the absence of alternative software, the methods of Section 12.2 can still be recommended. Also the simplicity of the methods in this section (especially the *shortcut method*—see later) will continue to attract the unsophisticated user. Thus sequential penalty function techniques still have a useful part to play and are described in some detail. Another apparently attractive idea is to define an *exact penalty function* in which the minimizer of the penalty function and the solution of the nonlinear programming problem coincide. This avoids the inefficiency inherent in sequential techniques. Most popular of these is the *L_1 exact penalty function* described in Section 12.3. However, this is a non-smooth function which cannot adequately be minimized by techniques for smooth functions (Part 1), and the best way of using this penalty function is currently being researched. Smooth exact penalty functions are also possible (Section 12.6) but there are some attendant disadvantages.

Penalty and barrier functions constitute a global approach to nonlinear programming and an alternative way to proceed is to consider local methods which perform well in a neighbourhood of the solution as described in Section 12.4. Applying Newton's method to the first order conditions that arise in the method of Lagrange multipliers. (Section 9.1) is a key idea. (Indeed it is a consequence of the Dennis–Moré theorem (Theorem 6.2.3) that any nonlinear programming method converges superlinearly if and only if it is asymptotically equivalent to this method.) It is shown in Section 12.4 that this method generalizes to give the *Sequential Quadratic Programming (SQP) method.* This method converges locally at second order and has the same standing for nonlinear programming as Newton's method does for unconstrained minimization. Thus quasi-Newton variants of the SQP method have been suggested when second derivatives are not available. The global properties of the SQP method are improved by associating it with an exact penalty function, and it is shown that this can be done in a number of ways. When a non-smooth penalty function is used then the presence of derivative discontinuities can cause slow convergence, for example in the *Maratos effect*, and ways of avoiding these difficulties are discussed. Current research interest in all these aspects is strong and further developments can be expected, although some software for this type of method is already available. No-derivative methods, possibly using finite differences to obtain derivative estimates, can be expected to follow once the best approach has been determined. However, it is likely that the effort involved in deriving expressions for first derivatives will usually pay off in terms of the efficiency and reliability of the resulting software.

Another idea which has attracted a lot of attention is that of a *feasible direction method* which generalizes the active set type of method for linear constraints (Chapters 10, 11), and aims to avoid the use of a penalty function. This is described in Section 12.5 and it is shown that the idea is essentially equivalent to a nonlinear generalized elimination of variables. Although software is available for this type of method, there are nonetheless difficulties in determining a fully reliable method. Some other interesting, but not currently favoured approaches to the solution of nonlinear programming problems are reviewed in Section 12.6.

To simplify the presentation, methods are discussed either in terms of the equality constraint problem

$$\begin{aligned} &\underset{\mathbf{x}}{\text{minimize}} \quad f(\mathbf{x}) \\ &\text{subject to} \quad \mathbf{c}(\mathbf{x}) = \mathbf{0} \end{aligned} \tag{12.1.1}$$

where $\mathbf{c}(\mathbf{x})$ is $\mathbb{R}^n \to \mathbb{R}^m$, or the inequality constraint problem

$$\begin{aligned} &\underset{\mathbf{x}}{\text{minimize}} \quad f(\mathbf{x}) \\ &\text{subject to} \quad c_i(\mathbf{x}) \geqslant 0, \quad i = 1, 2, \ldots, m. \end{aligned} \tag{12.1.2}$$

Usually the generalization to solve the mixed problem (7.1.1) is straightforward.

When solving a general nonlinear programming problem in which the constraints cannot easily be eliminated, it is necessary to balance the aims of reducing the objective function and staying inside or close to the feasible region, in order to induce global convergence (that is convergence to a local solution from any initial approximation). This inevitably leads to the idea of a *penalty function* which is some combination of f and $\mathbf{c}$ which enables f to be minimized whilst controlling constraint violations (or near constraint violations) by penalizing them. Early penalty functions were smooth so as to enable efficient techniques

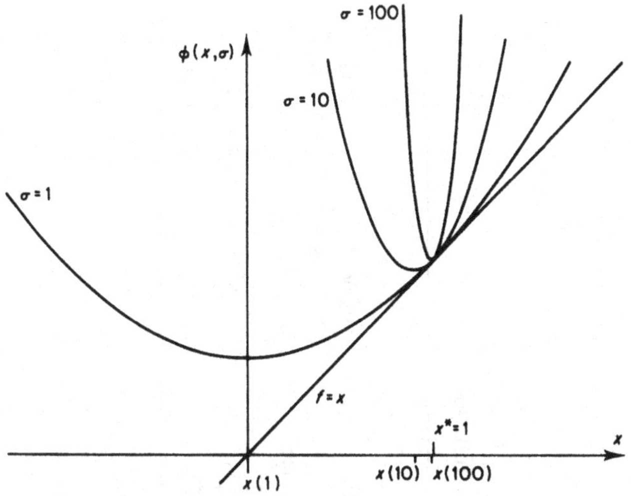

Figure 12.1.1 Convergence of the Courant penalty function

for smooth unconstrained optimization to be used. For the equality problem the earliest penalty function (Courant, 1943) is

$$\begin{aligned}\phi(\mathbf{x},\sigma) &= f(\mathbf{x}) + \tfrac{1}{2}\sigma\sum_i (c_i(\mathbf{x}))^2 \\ &= f(\mathbf{x}) + \tfrac{1}{2}\sigma\mathbf{c}(\mathbf{x})^{\mathrm{T}}\mathbf{c}(\mathbf{x}).\end{aligned} \tag{12.1.3}$$

The penalty is formed from a sum of squares of constraint violations and the parameter σ determines the amount of the penalty. Some graphs of $\phi(\mathbf{x},\sigma)$ are given in Figure 12.1.1 for the trivial problem: minimize x subject to $c(x) \triangleq x - 1 = 0$ for which $\phi = x + \frac{1}{2}\sigma(x-1)^2$. If the solution $x^* = 1$ is compared with the points which minimize $\phi(x,\sigma)$, it is clear that x^* is a limit point of the latter as $\sigma \to \infty$. Thus the technique of solving a *sequence* of minimization problems is suggested. This is traditionally implemented as follows.

(i) Choose a fixed sequence $\{\sigma^{(k)}\} \to \infty$, typically $\{1, 10, 10^2, 10^3, \ldots\}$.
(ii) For each $\sigma^{(k)}$ find a local minimizer, $\mathbf{x}(\sigma^{(k)})$ say, to $\min_{\mathbf{x}}\phi(\mathbf{x},\sigma^{(k)})$.
(iii) Terminate when $\mathbf{c}(\mathbf{x}(\sigma^{(k)}))$ is sufficiently small. (12.1.4)

The effect of this iteration on the problem

$$\begin{aligned}&\text{minimize} && -x_1 - x_2 \\ &\text{subject to} && 1 - x_1^2 - x_2^2 = 0\end{aligned} \tag{12.1.5}$$

for which the solution and optimum Lagrange multiplier is $x_1^* = x_2^* = \lambda^* = 1/\sqrt{2}$, is shown in Table 12.1.1 It can be seen that $\mathbf{x}(\sigma^{(k)}) \to \mathbf{x}^*$ and that linear convergence is obtained with one extra decimal place being obtained at each iteration. This behaviour can in fact be justified for all problems, and this is done in Theorem 12.1.2, equation (12.1.13), below. It must be emphasized that in practice step (ii) in (12.1.4) is done *numerically*, that is by the application

Table 12.1.1 Application of the Courant penalty function

k	1	2	3	4	5	6
$\sigma^{(k)}$	1	10	100	1000	10000	100000
$x_1^{(k)} = x_2^{(k)}$	0.8846462	0.7308931	0.7095936	0.7073566	0.7071318	0.7071093
$c^{(k)}$	−0.5651978	−0.0684094	−0.0070462	−0.0007067	−0.0000708	−0.0000071
$\lambda^{(k)}$	0.5651978	0.684094	0.70462	0.7067	0.708	0.71
$\phi^{(k)}$	−1.609568	−1.438387	−1.416705	−1.414463	−1.414239	−1.414216
$\phi^{(k)} - \frac{1}{2}\boldsymbol{\lambda}^{(k)\mathrm{T}}\mathbf{c}^{(k)}$	−1.449844	−1.414988	−1.414222	−1.414213	−1.414214	−1.414213

of an unconstrained minimization method. The choice of this method will depend on whether or not derivatives are available and on the size of the problem (see Part 1). Often $\mathbf{x}(\sigma^{(k)})$ is used as an initial approximation when minimizing $\phi(\mathbf{x}, \sigma^{(k+1)})$, and other information such as inverse Hessian approximations can also be passed forward from one iteration to the next. In fact algorithm (12.1.4) is idealized in that step (ii) cannot be solved exactly in a finite number of operations. It is assumed that $\mathbf{x}(\sigma^{(k)})$ is obtained as accurately as possible, although bounds on the accuracy can be given (equation (12.1.21) below) which guarantee convergence. It has been assumed that the local minimizer $\mathbf{x}(\sigma^{(k)})$ exists. This may not be so, not only when the nonlinear programming problem is unbounded, but also in cases when local solutions exist. In the latter case a remedy (not guaranteed to work) is to increase the initial value $\sigma^{(1)}$ and repeat.

A variety of results relating to the convergence of this sequential penalty function can be given. In doing this this, quantities derived from $\sigma^{(k)}$ like $\mathbf{x}(\sigma^{(k)})$, $f(\mathbf{x}(\sigma^{(k)}))$, etc., are denoted by $\mathbf{x}^{(k)}, f^{(k)}$, etc. It is assumed for the first theorem that $f(\mathbf{x})$ is bounded below on the (non-emtpy) feasible region so that

$$f^* = \inf f(\mathbf{x}) \qquad \forall\{\mathbf{x}\colon \mathbf{c}(\mathbf{x}) = \mathbf{0}\} \tag{12.1.6}$$

exists. If global solutions can be computed in (12.1.4), step (ii), then the most simple result is the following.

Theorem 12.1.1 (Penalty function convergence)

If $\{\sigma^{(k)}\}\uparrow\infty$ *then*

(i) $\{\phi(\mathbf{x}^{(k)}, \sigma^{(k)})\}$ *is non-decreasing,*
(ii) $\{\mathbf{c}^{(k)\mathrm{T}}\mathbf{c}^{(k)}\}$ *is non-increasing,*
(iii) $\{f^{(k)}\}$ *is non-decreasing.*

Also $\mathbf{c}^{(k)} \to \mathbf{0}$ *and any accumulation point* $\mathbf{x}^*$ *of* $\{\mathbf{x}^{(k)}\}$ *solves* (12.1.1).

Proof

Let $\sigma^{(k)} < \sigma^{(l)}$; then from the definition of $\mathbf{x}^{(k)}$ and (12.1.3),

$$\phi(\mathbf{x}^{(k)}, \sigma^{(k)}) \leqslant \phi(\mathbf{x}^{(l)}, \sigma^{(k)}) \leqslant \phi(\mathbf{x}^{(l)}, \sigma^{(l)}) \leqslant \phi(\mathbf{x}^{(k)}, \sigma^{(l)}).$$

The first two inequalities give case (i). Subtracting the inner and outer inequalities gives

$$(\sigma^{(l)} - \sigma^{(k)})(\mathbf{c}^{(k)\mathrm{T}}\mathbf{c}^{(k)} - \mathbf{c}^{(l)\mathrm{T}}\mathbf{c}^{(l)}) \geqslant 0$$

and hence case (ii). Substituting in the first inequality then gives case (iii). By definition of $\mathbf{x}^{(k)}$,

$$\phi(\mathbf{x}^{(k)}, \sigma^{(k)}) \leqslant \inf_{\mathbf{x}:\mathbf{c}(\mathbf{x})=\mathbf{0}} \phi(\mathbf{x}, \sigma^{(k)}) = f^* \tag{12.1.7}$$

by (12.1.6). Thus using case (iii) above in (12.1.3), $\sigma^{(k)}\uparrow\infty$ implies $\mathbf{c}^{(k)} \to \mathbf{0}$. If

$\mathbf{x}^{(k)} \rightarrow \mathbf{x}^*$ it follows that $\mathbf{c}(\mathbf{x}^*) = \mathbf{0}$, so $f(\mathbf{x}^*) \geqslant f^*$ as defined in (12.1.6). But from (12.1.3) and (12.1.7), $f^{(k)} \leqslant f^*$ so it follows that $f(\mathbf{x}^*) = f^*$, that is $\mathbf{x}^*$ solves (12.1.1). □

It is interesting to observe that this result is obtained in absence of differentiability or Kuhn–Tucker regularity assumptions.

It is also possible to prove similar results when local minima are computed (with different assumptions on the problem), and also to get asymptotic estimates of the rate of convergence. In doing this the vector

$$\boldsymbol{\lambda}^{(k)} = -\sigma^{(k)}\mathbf{c}^{(k)} \tag{12.1.8}$$

is defined, and can be regarded as a *Lagrange multiplier estimate* by virtue of (12.1.9). The notation $\mathbf{h}^{(k)} = \mathbf{x}^{(k)} - \mathbf{x}^*$, $\mathbf{a}_i = \nabla c_i$, and $\mathbf{g} = \nabla f$, etc., as in Chapter 9 is used.

Theorem 12.1.2

If $\sigma^{(k)} \rightarrow \infty$ and $\mathbf{x}^{(k)} \rightarrow \mathbf{x}^*$ is any accumulation point, and if rank $\mathbf{A}^* = m$, then $\mathbf{x}^*$ is a KT point and it follows that

$$\boldsymbol{\lambda}^{(k)} = \boldsymbol{\lambda}^* + o(1) \tag{12.1.9}$$

$$\mathbf{c}^{(k)} = -\boldsymbol{\lambda}^*/\sigma^{(k)} + o(1/\sigma) \tag{12.1.10}$$

$$\sigma^{(k)}\mathbf{c}^{(k)\mathrm{T}}\mathbf{c}^{(k)} = \boldsymbol{\lambda}^{*\mathrm{T}}\boldsymbol{\lambda}^*/\sigma^{(k)} + o(1/\sigma). \tag{12.1.11}$$

Furthermore if second order sufficient conditions (9.3.5) *hold at $\mathbf{x}^*$, $\boldsymbol{\lambda}^*$ then*

$$f^* = \phi^* = \phi^{(k)} + \tfrac{1}{2}\sigma^{(k)}\mathbf{c}^{(k)\mathrm{T}}\mathbf{c}^{(k)} + o(1/\sigma) \tag{12.1.12}$$

$$\mathbf{h}^{(k)} = -\mathbf{T}^*\boldsymbol{\lambda}^*/\sigma^{(k)} + o(1/\sigma) \tag{12.1.13}$$

where $\mathbf{T}^*$ is defined by

$$\begin{bmatrix} \mathbf{W}^* & -\mathbf{A}^* \\ -\mathbf{A}^{*\mathrm{T}} & \mathbf{0} \end{bmatrix}^{-1} = \begin{bmatrix} \mathbf{H}^* & -\mathbf{T}^* \\ -\mathbf{T}^{*\mathrm{T}} & \mathbf{U}^* \end{bmatrix}. \tag{12.1.14}$$

Proof

The fact that $\mathbf{x}^{(k)}$ solves (12.1.4), step (ii), implies that

$$\nabla\phi(\mathbf{x}^{(k)}, \sigma^{(k)}) = \mathbf{g}^{(k)} + \sigma^{(k)}\mathbf{A}^{(k)}\mathbf{c}^{(k)} = \mathbf{0} \tag{12.1.15}$$

and hence from (12.1.8) that

$$\mathbf{g}^{(k)} = \mathbf{A}^{(k)}\boldsymbol{\lambda}^{(k)}. \tag{12.1.16}$$

Since rank $\mathbf{A}^* = m$ it follows that $\mathbf{A}^{(k)+}$ exists and is bounded for all k sufficiently large and hence that

$$\boldsymbol{\lambda}^{(k)} = \mathbf{A}^{(k)+}\mathbf{g}^{(k)} = \boldsymbol{\lambda}^* + o(1) \tag{12.1.17}$$

where λ^* is defined by $\lambda^* = \mathbf{A}^{*+}\mathbf{g}^*$. It also follows from (12.1.16) by continuity that $\mathbf{g}^* = \mathbf{A}^*\lambda^*$, and from (12.1.17) and (12.1.8) that $\mathbf{c}^{(k)} = -\lambda^*/\sigma^{(k)} + o(1/\sigma)$ so that $\mathbf{c}^* = \mathbf{0}$ in the limit $\sigma^{(k)} \to \infty$. Thus $\mathbf{x}^*$ satisfies KT conditions (Section 9.1) and (12.1.9) and (12.1.10) are established. Equation (12.1.11) follows directly and shows that $\lim \phi^{(k)} \triangleq \phi^* = f^*$. Without assuming second order conditions it is also possible to show from a Taylor series for $f(\mathbf{x})$ about $\mathbf{x}^{(k)}$ and (12.1.16) that

$$\begin{aligned} f^* &= f^{(k)} - \mathbf{h}^{(k)T}\mathbf{g}^{(k)} + o(\|\mathbf{h}^{(k)}\|) \\ &= f^{(k)} - \mathbf{h}^{(k)T}\mathbf{A}^{(k)}\lambda^{(k)} + o(\|\mathbf{h}^{(k)}\|) \\ &= f^{(k)} - \mathbf{c}^{(k)T}\lambda^{(k)} + o(\|\mathbf{h}^{(k)}\|) \end{aligned}$$

using a similar Taylor series for $\mathbf{c}(\mathbf{x})$. Thus

$$\phi^* = \phi^{(k)} + \tfrac{1}{2}\sigma^{(k)}\mathbf{c}^{(k)T}\mathbf{c}^{(k)} + o(\|\mathbf{h}^{(k)}\|) \tag{12.1.18}$$

from (12.1.4) and (12.1.8). Second order sufficient conditions for the equality constraint problem (9.3.5) and rank $\mathbf{A}^* = m$ imply that the Lagrangian matrix is nonsingular at $\mathbf{x}^*, \lambda^*$ so that the inverse in (12.1.14) exists (see Question 12.4). An expansion of the Lagrangian function about $\mathbf{x}^*, \lambda^*$, and using (9.1.7) and (12.1.16) gives

$$\begin{pmatrix} \mathbf{0} \\ \mathbf{c}^{(k)} \end{pmatrix} = \begin{bmatrix} \mathbf{W}^* & -\mathbf{A}^* \\ -\mathbf{A}^{*T} & \mathbf{0} \end{bmatrix} \begin{pmatrix} \mathbf{h}^{(k)} \\ \lambda^{(k)} - \lambda^* \end{pmatrix} + o(\max(\|\mathbf{h}^{(k)}\|, \|\lambda^{(k)} - \lambda^*\|)).$$

It follows from (12.1.10) and the existence of (12.1.14) that $\mathbf{h}^{(k)} = O(1/\sigma)$ which extends (12.1.18) to give (12.1.12). Multiplying on the left by (12.1.14) and using (12.1.8) then gives (12.1.13). □

The convergence of (12.1.9) and (12.1.10) can be observed easily in Table 12.1.1 and also the convergence of $\phi^{(k)} \to f^* = \sqrt{2}$ which (12.1.11) implies. The other results of Theorem 12.1.2 can be used in a more sophisticated algorithm. Equation (12.1.12) gives a $o(1/\sigma)$ estimate of f^* which is better than the $O(1/\sigma)$ estimate given by $\phi^{(k)}$ itself. The asymptotic form of $\mathbf{h}^{(k)}$ in (12.1.13) can be used in an extrapolation scheme to estimate $\mathbf{x}^*$ (as given by Fiacco and McCormick (1968) in the context of barrier functions.) These estimates can be used to terminate the penalty function iteration and also to provide better initial approximations when minimizing $\phi(\mathbf{x}, \sigma^{(k)})$. It is also interesting to observe that the rank assumption on $\mathbf{A}^*$ cannot easily be relaxed. For example if (12.1.3) has no feasible point then $\mathbf{c}^* \neq \mathbf{0}$ must occur and so as $\sigma \to \infty$ it is necessary from (12.1.8) and (12.1.15) that $\mathbf{A}^{(k)}\mathbf{c}^{(k)} \to \mathbf{0}$, that is $\mathbf{A}^*$ has dependent columns.

This well-developed theoretical background may make it appear that, apart from the inefficiency of sequential minimization, the method is a robust one which can be used with confidence. In fact this is not true at all and there are severe numerical difficulties which arise when the method is used in practice. These are caused by the fact that as $\sigma^{(k)} \to \infty$, it is increasingly difficult to solve the minimization problem in (12.1.4), step (ii). To illustrate this behaviour, the contours of $\phi(\mathbf{x}, \sigma)$ in (12.1.3) for the problem (12.1.5) are shown in Figure 12.1.2

for increasing values of σ. For $\sigma = 100$, it can be seen that whilst the solution $\mathbf{x}(100)$ is well determined in a radial direction, this is not so tangential to the constraint boundary, so that the exact location of $\mathbf{x}(100)$ is very difficult to determine numerically. This is expressed mathematically by the fact that (for $0 < m < n$) the Hessian matrix $\nabla^2\phi(\mathbf{x}^{(k)}, \sigma^{(k)})$ becomes increasing ill conditioned as $\sigma^{(k)} \to \infty$. This result follows by virtue of

$$\nabla^2\phi(\mathbf{x}^{(k)}, \sigma^{(k)}) = \mathbf{W}^{(k)} + \sigma^{(k)}\mathbf{A}^{(k)}\mathbf{A}^{(k)\mathrm{T}} \tag{12.1.19}$$

where (12.1.8) is used in the definition

$$\mathbf{W}^{(k)} = \nabla_x^2\mathscr{L}(\mathbf{x}^{(k)}, \boldsymbol{\lambda}^{(k)}). \tag{12.1.20}$$

The $\sigma\mathbf{A}\mathbf{A}^{\mathrm{T}}$ term in (12.1.19) has rank m and so there are m eigenvalues of $\nabla^2\phi$

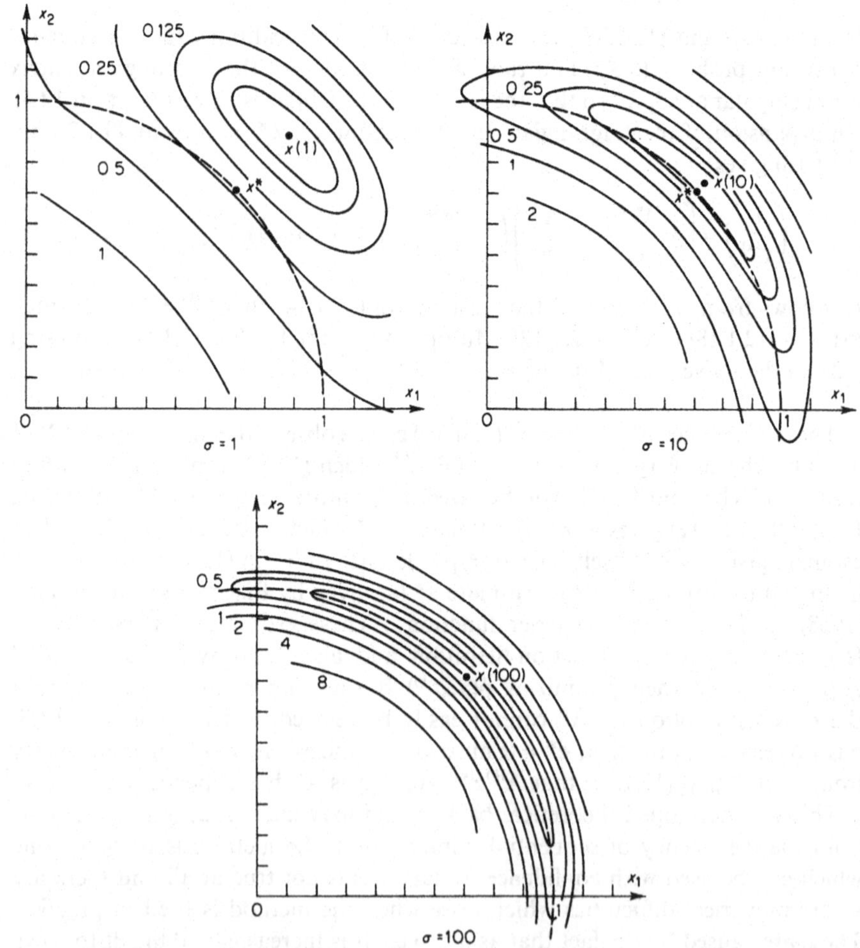

Figure 12.1.2 Increasing ill-conditioning in penalty functions (contours of $\phi - \phi_{\min}$ in powers of 2)

which approach ∞ as $\sigma^{(k)} \to \infty$. That the remaining eigenvalues are bounded is a consequence (see Lancaster, 1969) of the Courant–Fisher theorem. Thus the condition number of $\nabla^2\phi$ approaches ∞. In practice this shows up in that large values of $\nabla\phi$ are obtained whilst the minimization routine is unable to make progress in reducing ϕ.

These remarks have implications for the choice of the sequence $\{\sigma^{(k)}\}$. Choosing a very large $\sigma^{(1)}$, or increasing σ very rapidly in the sequence, gives minimization problems which are very difficult to solve accurately. The alternative of choosing $\sigma^{(1)}$ small and increasing σ slowly keeps $\mathbf{x}^{(k)}$ close to the minimizer of $\phi(\mathbf{x}, \sigma^{(k+1)})$ and makes it easier to get accurate solutions, but is very inefficient. The typical sequence in (12.1.4), step (i), is a trade-off between these two effects. Probably $\sigma^{(1)}$ should be chosen to balance the f and $\frac{1}{2}\sigma\mathbf{c}^{\mathrm{T}}\mathbf{c}$ terms in (12.1.3) or to minimize the magnitude of $\nabla\phi$. This discussion also highlights the fact that it may not be possible to make good use of an accurate estimate of $\mathbf{x}^*$ in algorithm (12.1.4) since $\mathbf{x}^{(1)}$ will be remote from $\mathbf{x}^*$. In fact many users do not carry out the sequential technique at all, but carry out a *shortcut method* in which they minimize (12.1.3) for a single largish value of σ. In doing this they must be prepared to accept errors in the extent to which first order conditions are satisfied. Such users would be well advised to observe the constraint errors, and to estimate both the error in objective function using (12.1.12) and the error in the KT condition $\mathbf{g}^{(k)} = \mathbf{A}^{(k)}\boldsymbol{\lambda}^{(k)}$, in order to decide whether these errors are acceptable. If not, then it is quite easy to go on to use a multiplier penalty function as in Section 12.2. However, if derivatives and software permit, a Lagrangian method (Section 12.4) is likely to prove to be more efficient in the long run.

As mentioned earlier, algorithm (12.1.4) and Theorem 12.1.1 are idealized in that they assume exact minimization of the penalty function. In fact it is straightforward to give the same results as in Theorem 12.1.2 when approximate minimization is allowed. To do this let the test for terminating the minimization of $\phi(\mathbf{x}, \sigma^{(k)})$ with an approximate minimizer $\mathbf{x}^{(k)}$ be

$$\|\nabla\phi(\mathbf{x}^{(k)}, \sigma^{(k)})\| \leqslant \nu\|\mathbf{c}^{(k)}\|, \tag{12.1.21}$$

where $\nu > 0$ is pre-set and where $\mathbf{c}^{(k)} = \mathbf{c}(\mathbf{x}^{(k)})$, etc. With the same assumptions as in Theorem 12.1.2, and using the fact that there exist a constant $\alpha > 0$ such that

$$\|\mathbf{A}^{(k)}\mathbf{c}^{(k)}\| \geqslant \alpha\|\mathbf{c}^{(k)}\| \tag{12.1.22}$$

for k sufficiently large (see Question 12.5), it follows from (12.1.21) and (12.1.15) that

$$\nu\|\mathbf{c}^{(k)}\| \geqslant \|\mathbf{g}^{(k)} + \sigma^{(k)}\mathbf{A}^{(k)}\mathbf{c}^{(k)}\| \geqslant \sigma^{(k)}\alpha\|\mathbf{c}^{(k)}\| - \|\mathbf{g}^{(k)}\|.$$

This shows that $\mathbf{c}^{(k)} \to \mathbf{0}$. Then (12.1.21) yields

$$\mathbf{g}^{(k)} = \mathbf{A}^{(k)}\boldsymbol{\lambda}^{(k)} + o(1)$$

and the rest of the results in Theorem 12.1.2 follow as before. It is interesting to relate (12.1.21) to the practical difficulties caused by ill-conditioning, in that

(12.1.21) definitely limits the extent to which large gradients can be accepted at an approximate minimizer of $\phi(\mathbf{x}, \sigma^{(k)})$.

A penalty function for the inequality constraint problem (12.1.2) can be given in an analogous way to (12.1.3) by

$$\phi(\mathbf{x}, \sigma) = f(\mathbf{x}) + \tfrac{1}{2}\sigma \sum_i [\min(c_i(\mathbf{x}), 0)]^2. \tag{12.1.23}$$

An illustration is provided by the trivial problem: min x subject to $x \geqslant 1$. Then Figure 12.1.1 also illustrates the graph of ϕ, except that ϕ and f are equal for $x \geqslant 1$. Likewise if the constraint in (12.1.5) is replaced by $1 - x_1^2 - x_2^2 \geqslant 0$ then Figure 12.1.2 illustrates contours of ϕ, except that inside the unit circle the linear contours of $-x_1 - x_2$ must be drawn. These figures illustrate the jump discontinuity in the second derivative of (12.1.23) when $\mathbf{c} = \mathbf{0}$ (for example at $\mathbf{x}^*$). They also illustrate that $\mathbf{x}^{(k)}$ approaches $\mathbf{x}^*$ from the infeasible side of the inequality constraints, which leads to the term *exterior penalty function* for (12.1.23). Exactly similar results to those in Theorems 12.1.1 and 12.1.2 follow on replacing $c_i^{(k)}$ by $\min(c_i^{(k)}, 0)$.

Another class of sequential minimization techniques is available to solve the inequality constraint problem (12.1.2), known as *barrier function methods.* These are characterized by their property of preserving strict constraint feasibility at all times, by using a barrier term which is infinite on the constraint boundaries. This can be advantageous if the objective function is not defined when the constraints are violated. The sequence of minimizers is also feasible, therefore, and hence the techniques are sometimes referred to as *interior point methods.* The two most important cases are the inverse barrier function

$$\phi(\mathbf{x}, r) = f(\mathbf{x}) + r \sum_i [c_i(\mathbf{x})]^{-1} \tag{12.1.24}$$

(Carroll, 1961) and the logarithmic barrier function

$$\phi(\mathbf{x}, r) = f(\mathbf{x}) - r \sum_i \log(c_i(\mathbf{x})) \tag{12.1.25}$$

(Frisch, 1955). As with σ in (12.1.3) the coefficient r is used to control the barrier function iteration. In this case a sequence $\{r^{(k)}\} \rightarrow 0$ is chosen, which ensures that the barrier term becomes more and more negligible except close to the boundary. Also $\mathbf{x}(r^{(k)})$ is defined as the minimizer of $\phi(\mathbf{x}, r^{(k)})$. Otherwise the procedure is the same as that in (12.1.4). Typical graphs of (12.1.24) for a sequence of values of $r^{(k)}$ are illustrated in Figure 12.1.3, and it can be seen that $\mathbf{x}(r^{(k)}) \rightarrow \mathbf{x}^*$ as $r^{(k)} \rightarrow 0$. This behaviour can be established rigorously in a similar way to Theorem 12.1.1 (Osborne, 1972). Other features such as estimates of Lagrange multipliers and the asymptotic behaviour of $\mathbf{h}^{(k)}$ also follow, which can be used to determine a suitable sequence $\{r^{(k)}\}$ and to allow the use of extrapolation techniques (see Questions 12.6 and 12.7).

Unfortunately barrier function algorithms suffer from the same numerical difficulties in the limit as the penalty function algorithm does, in particular the badly determined nature of $\mathbf{x}(r^{(k)})$ tangential to the constraint surface, and the

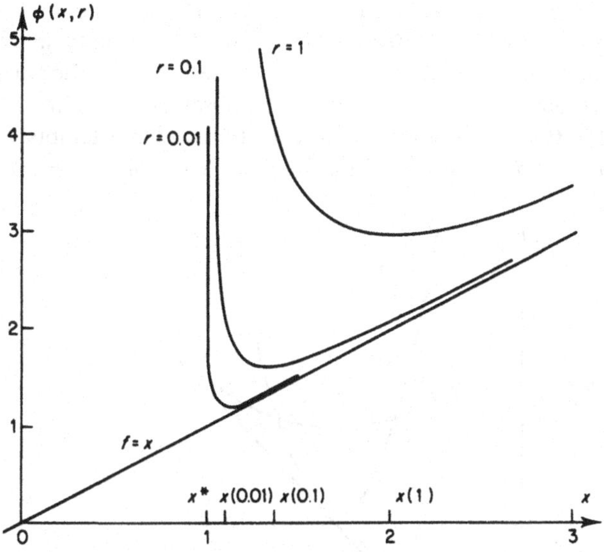

Figure 12.1.3 Increasing ill-conditioning in the barrier function (12.1.24)

difficulty of locating the minimizer due to ill-conditioning and large gradients. Moreover there are additional problems which arise. The barrier function is undefined for infeasible points, and the simple expedient of setting it to infinity can make the line search inefficient. Also the singularity makes conventional quadratic or cubic interpolations in the line search work less efficiently. Thus special purpose line searches are required (Fletcher and McCann, 1969), and the aim of a simple-to-use algorithm is lost. Another difficulty is that an initial interior feasible point is required, and this in itself is a non-trivial problem involving the strict solution of a set of inequalities. In view of these difficulties and the general inefficiency of sequential techniques, barrier functions currently attract little interest for nonlinear programming. However, recent developments in polynomial time algorithms for linear programming (Section 8.7) do use techniques closely related to barrier functions.

12.2 MULTIPLIER PENALTY FUNCTIONS

The way in which the penalty function (12.1.3) is used to solve (12.1.1) can be envisaged as an attempt to create a local minimizer at $\mathbf{x}^*$ in the limit $\sigma^{(k)} \to \infty$ (see Figures 12.1.1 and 12.1.2). However $\mathbf{x}^*$ can be made to minimize ϕ for finite σ by changing the origin of the penalty term. This suggests using the function

$$\begin{aligned}\phi(\mathbf{x}, \boldsymbol{\theta}, \boldsymbol{\sigma}) &= f(\mathbf{x}) + \tfrac{1}{2}\textstyle\sum \sigma_i(c_i(\mathbf{x}) - \theta_i)^2 \\ &= f(\mathbf{x}) + \tfrac{1}{2}(\mathbf{c}(\mathbf{x}) - \boldsymbol{\theta})^{\mathrm{T}}\mathbf{S}(\mathbf{c}(\mathbf{x}) - \boldsymbol{\theta}) \qquad (12.2.1)\end{aligned}$$

(Powell, 1969), where $\boldsymbol{\theta}, \boldsymbol{\sigma} \in \mathbb{R}^m$ and $\mathbf{S} = \operatorname{diag} \sigma_i$. The parameters θ_i correspond to shifts of origin and the $\sigma_i \geqslant 0$ control the size of the penalty, like σ in (12.1.3). For the trivial problem: minimize x subject to $x = 1$ again, if the correct shift θ is chosen (which depends on σ), then it can be observed in Figure 12.2.1 that $\mathbf{x}^*$ minimizes $\phi(\mathbf{x}, \boldsymbol{\theta}, \boldsymbol{\sigma})$. This suggests an algorithm which attempts to locate the optimum value of $\boldsymbol{\theta}$ whilst keeping $\boldsymbol{\sigma}$ finite and so *avoids the ill-conditioning in*

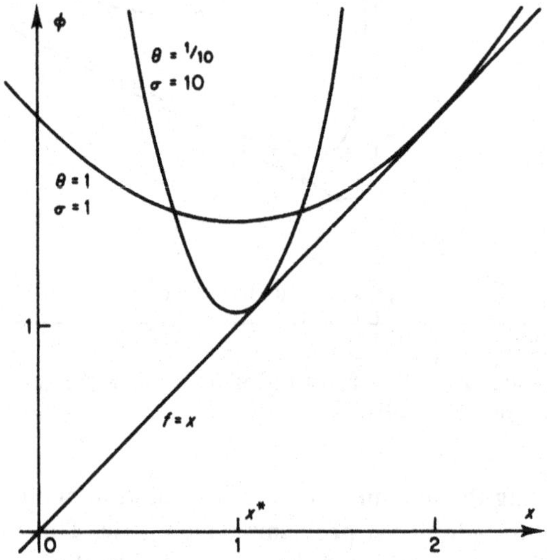

Figure 12.2.1 Multiplier penalty functions

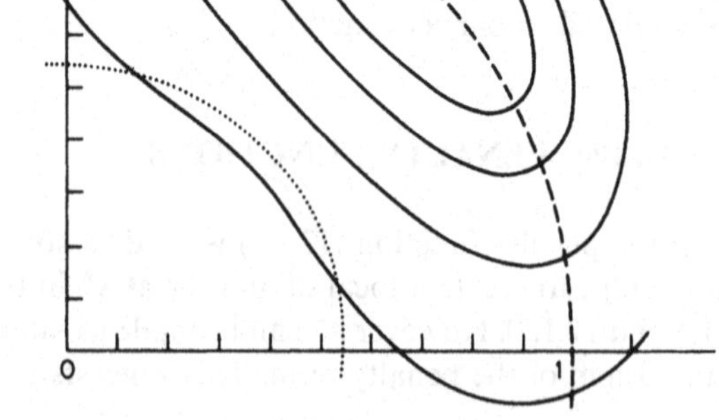

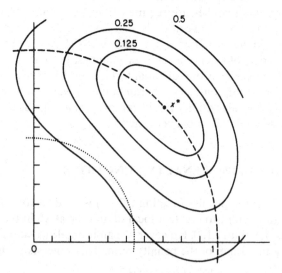

Figure 12.2.2 A multiplier penalty function ($\sigma = 1, \theta = 1/\sqrt{2}$) (contours of $\phi - \phi_{\min}$ in powers of 2)

the limit $\sigma \to \infty$. This is also illustrated for problem (12.1.5) with $\sigma = 1$, in which the optimum shift is $\theta = 1/\sqrt{2}$. The resulting contours are shown in Figure 12.2.2 and the contrast with Figure 12.1.2 as $\sigma^{(k)} \to \infty$ can be seen.

In fact it is more convenient to introduce different parameters

$$\lambda_i = \theta_i \sigma_i, \qquad i = 1, 2, \ldots, m \tag{12.2.2}$$

and ignore the term $\frac{1}{2}\sum \sigma_i \theta_i^2$ (independent of $\mathbf{x}$) giving

$$\phi(\mathbf{x}, \boldsymbol{\lambda}, \boldsymbol{\sigma}) = f(\mathbf{x}) - \boldsymbol{\lambda}^{\mathrm{T}}\mathbf{c}(\mathbf{x}) + \tfrac{1}{2}\mathbf{c}(\mathbf{x})^{\mathrm{T}}\mathbf{S}\mathbf{c}(\mathbf{x}) \tag{12.2.3}$$

(Hestenes, 1969). There exists a corresponding optimum value of $\boldsymbol{\lambda}$, for which $\mathbf{x}^*$ minimizes $\phi(\mathbf{x}, \boldsymbol{\lambda}, \boldsymbol{\sigma})$, which in fact is the Lagrange multiplier $\boldsymbol{\lambda}^*$ at the solution to (12.1.1). This result is now true independent of $\boldsymbol{\sigma}$, so it is usually convenient to ignore the dependence on $\boldsymbol{\sigma}$ and write $\phi(\mathbf{x}, \boldsymbol{\lambda})$ in (12.2.3), and to use $\boldsymbol{\lambda}$ as the control parameter in a sequential minimization algorithm as follows.

$$\begin{array}{ll} \text{(i)} & \text{Determine a sequence } \{\boldsymbol{\lambda}^{(k)}\} \to \boldsymbol{\lambda}^*. \\ \text{(ii)} & \text{For each } \boldsymbol{\lambda}^{(k)} \text{ find a local minimizer, } \mathbf{x}(\boldsymbol{\lambda}^{(k)}) \text{ say,} \\ & \text{to } \min_{\mathbf{x}} \phi(\mathbf{x}, \boldsymbol{\lambda}). \\ \text{(iii)} & \text{Terminate when } \mathbf{c}(\mathbf{x}(\boldsymbol{\lambda}^{(k)})) \text{ is sufficiently small.} \end{array} \tag{12.2.4}$$

The main difference between this algorithm and (12.1.4) is that $\boldsymbol{\lambda}^*$ is not known in advance so the sequence in step (i) cannot be predetermined. However, it is shown below how such a sequence can be constructed. Because (12.2.3) is obtained from (12.1.3) by adding a multiplier term $-\boldsymbol{\lambda}^{\mathrm{T}}\mathbf{c}$, (12.2.3) is often referred to as a *multiplier penalty function*. Alternatively (12.2.3) is the Lagrangian function (9.1.6) in which the objective f is augmented by the term $\frac{1}{2}\mathbf{c}^{\mathrm{T}}\mathbf{S}\mathbf{c}$. Hence the term *augmented Lagrangian function* is also used to describe (12.2.3).

The result that $\boldsymbol{\lambda}^*$ is the optimum choice of the control parameter vector in (12.2.3) is expressed in the following.

Theorem 12.2.1

If second order sufficient conditions hold at $\mathbf{x}^*$, $\boldsymbol{\lambda}^*$ *then there exists* $\boldsymbol{\sigma}' \geqslant \mathbf{0}$ *such that for any* $\boldsymbol{\sigma} > \boldsymbol{\sigma}'$, $\mathbf{x}^*$ *is an isolated local minimizer of* $\phi(\mathbf{x}, \boldsymbol{\lambda}^*, \boldsymbol{\sigma})$, *that is* $\mathbf{x}^* = \mathbf{x}(\boldsymbol{\lambda}^*)$.

Proof

Differentiating (12.2.3) gives

$$\nabla\phi(\mathbf{x}, \boldsymbol{\lambda}^*, \boldsymbol{\sigma}) = \mathbf{g} - \mathbf{A}\boldsymbol{\lambda}^* + \mathbf{A}\mathbf{S}\mathbf{c}. \tag{12.2.5}$$

The second order conditions require $\mathbf{x}^*$, $\boldsymbol{\lambda}^*$ to be a KT point, that is $\mathbf{g}^* = \mathbf{A}^*\boldsymbol{\lambda}^*$ and $\mathbf{c}^* = \mathbf{0}$, so it follows that $\nabla\phi(\mathbf{x}^*, \boldsymbol{\lambda}^*, \boldsymbol{\sigma}) = \mathbf{0}$. Differentiating (12.2.5) gives

$$\nabla^2\phi(\mathbf{x}, \boldsymbol{\lambda}, \boldsymbol{\sigma}) = \mathbf{W} + \mathbf{A}\mathbf{S}\mathbf{A}^{\mathrm{T}} \triangleq \mathbf{W}_\sigma \tag{12.2.6}$$

say, where $\mathbf{W} = \nabla^2 f - \sum_i(\lambda_i - \sigma_i c_i)\nabla^2 c_i$, and hence

$$\mathbf{W}_\sigma^* \triangleq \nabla^2\phi(\mathbf{x}^*, \boldsymbol{\lambda}^*, \boldsymbol{\sigma}) = \mathbf{W}^* + \mathbf{A}^*\mathbf{S}\mathbf{A}^{*\mathrm{T}}. \tag{12.2.7}$$

Let rank $\mathbf{A}^* = r \leqslant m$ and let $\mathbf{B}$ $(n \times r)$ be an orthonormal basis matrix ($\mathbf{B}^\mathrm{T}\mathbf{B} = \mathbf{I}$) for $\mathbf{A}^*$, so that $\mathbf{A}^* = \mathbf{B}\mathbf{C}$ where $\mathbf{C}(= \mathbf{B}^\mathrm{T}\mathbf{A}^*)$ has rank r. Consider any vector $\mathbf{u} \neq \mathbf{0}$ and let $\mathbf{u} = \mathbf{v} + \mathbf{B}\mathbf{w}$ where $\mathbf{B}^\mathrm{T}\mathbf{v} = \mathbf{0} = \mathbf{A}^{*\mathrm{T}}\mathbf{v}$. Then

$$\mathbf{u}^\mathrm{T}\mathbf{W}_\sigma^*\mathbf{u} = \mathbf{v}^\mathrm{T}\mathbf{W}^*\mathbf{v} + 2\mathbf{v}^\mathrm{T}\mathbf{W}^*\mathbf{B}\mathbf{w} + \mathbf{w}^\mathrm{T}\mathbf{B}^\mathrm{T}\mathbf{W}^*\mathbf{B}\mathbf{w} + \mathbf{w}^\mathrm{T}\mathbf{C}\mathbf{S}\mathbf{C}^\mathrm{T}\mathbf{w}.$$

From (9.3.5) and (9.3.4) there exists a constant $a > 0$ such that $\mathbf{v}^\mathrm{T}\mathbf{W}^*\mathbf{v} \geqslant a\|\mathbf{v}\|_2^2$. Let b be the greatest singular value of $\mathbf{W}^*\mathbf{B}$ and let $d = \|\mathbf{B}^\mathrm{T}\mathbf{W}^*\mathbf{B}\|_2$. Let $\sigma = \min_i \sigma_i \geqslant 0$ and let $\mu > 0$ be the smallest eigenvalue of $\mathbf{C}\mathbf{C}^\mathrm{T}$. Then

$$\mathbf{u}^\mathrm{T}\mathbf{W}_\sigma^*\mathbf{u} \geqslant a\|\mathbf{v}\|_2^2 - 2b\|\mathbf{v}\|_2\|\mathbf{w}\|_2 + (\sigma\mu - d)\|\mathbf{w}\|_2^2.$$

Since $\|\mathbf{v}\| = \|\mathbf{w}\| = 0$ cannot hold, if $\sigma > \sigma'$ where $\sigma' = (d + b^2/a)/\mu$ then it follows that $\mathbf{u}^\mathrm{T}\mathbf{W}_\sigma^*\mathbf{u} > 0$. Thus both $\nabla\phi(\mathbf{x}^*, \boldsymbol{\lambda}^*, \boldsymbol{\sigma}) = \mathbf{0}$ and $\nabla^2\phi(\mathbf{x}^*, \boldsymbol{\lambda}^*, \boldsymbol{\sigma})$ is positive definite so $\mathbf{x}^*$ is an isolated local minimizer of $\phi(\mathbf{x}, \boldsymbol{\lambda}^*, \boldsymbol{\sigma})$ for $\boldsymbol{\sigma}$ sufficiently large. □

The assumption of second order conditions is important and not easily relaxed. For example if $f = x_1^4 + x_1x_2$ and $c = x_2$ then $\mathbf{x}^* = \mathbf{0}$ solves (12.1.1) with the unique multiplier $\lambda^* = 0$. However second order conditions are not satisfied, and in fact $\mathbf{x}^* = \mathbf{0}$ does not minimize $\phi(\mathbf{x}, 0, \sigma) = x_1^4 + x_1x_2 + \frac{1}{2}\sigma x_2^2$ for any value of σ. Henceworth it is assumed that second order sufficient conditions hold and $\boldsymbol{\sigma}$ is sufficiently large.

The minimizer $\mathbf{x}(\boldsymbol{\lambda})$ of $\phi(\mathbf{x}, \boldsymbol{\lambda})$ can also be regarded as having been determined by solving the nonlinear equations

$$\nabla\phi(\mathbf{x}, \boldsymbol{\lambda}) = \mathbf{0}. \tag{12.2.8}$$

Because $\nabla^2\phi(\mathbf{x}^*, \boldsymbol{\lambda}^*)$ is positive definite and is the Jacobian matrix of this system, it follows from the implicit function theorem (Apostol, 1957) that there exist open neighbourhoods $\Omega_\lambda \subset \mathbb{R}^m$ about $\boldsymbol{\lambda}^*$ and $\Omega_x \subset \mathbb{R}^n$ about $\mathbf{x}^*$ and a $\mathbb{C}^1$ function $\mathbf{x}(\boldsymbol{\lambda})(\Omega_\lambda \to \Omega_x)$ such that $\nabla\phi(\mathbf{x}(\boldsymbol{\lambda}), \boldsymbol{\lambda}) = \mathbf{0}$. Also $\nabla^2\phi(\mathbf{x}, \boldsymbol{\lambda})$ is positive definite for all $\mathbf{x} \in \Omega_x$ and $\boldsymbol{\lambda} \in \Omega_\lambda$, so $\mathbf{x}(\boldsymbol{\lambda})$ is the minimizer of $\phi(\mathbf{x}, \boldsymbol{\lambda})$. It may be that $\phi(\mathbf{x}, \boldsymbol{\lambda})$ has local minima so that various solutions to (12.2.8) exist: it is assumed that consistent choice of $\mathbf{x}(\boldsymbol{\lambda})$ is made by the minimization routine as that solution which exists in Ω_x. The vector $\mathbf{x}(\boldsymbol{\lambda})$ can also be interpreted in yet another way as the solution to a neighbouring problem to (12.1.1) (see Question 12.14) and this can sometimes be convenient.

It is important to examine the function

$$\psi(\boldsymbol{\lambda}) \triangleq \phi(\mathbf{x}(\boldsymbol{\lambda}), \boldsymbol{\lambda}) \tag{12.2.9}$$

which is derived from $\boldsymbol{\lambda}$ by first finding $\mathbf{x}(\boldsymbol{\lambda})$. By the local optimality of $\mathbf{x}(\boldsymbol{\lambda})$ on Ω_x, for any $\boldsymbol{\lambda}$ in Ω_λ it follows that

$$\psi(\boldsymbol{\lambda}) = \phi(\mathbf{x}(\boldsymbol{\lambda}), \boldsymbol{\lambda}) \leqslant \phi(\mathbf{x}^*, \boldsymbol{\lambda}) = \phi(\mathbf{x}^*, \boldsymbol{\lambda}^*) = \psi(\boldsymbol{\lambda}^*) \tag{12.2.10}$$

(using $\mathbf{c}^* = \mathbf{0}$). Thus $\boldsymbol{\lambda}^*$ *is a local unconstrained maximizer of* $\psi(\boldsymbol{\lambda})$. This result is also true globally if $\mathbf{x}(\boldsymbol{\lambda})$ is a global maximizer of $\phi(\mathbf{x}, \boldsymbol{\lambda})$. Thus methods for generating a sequence $\boldsymbol{\lambda}^{(k)} \to \boldsymbol{\lambda}^*$ can be derived by applying unconstrained minimization methods to $-\psi(\boldsymbol{\lambda})$. To do this requires formulae for the derivatives $\nabla\psi$ and $\nabla^2\psi$ of ψ with respect to $\boldsymbol{\lambda}$. Using the matrix notation $[\partial\mathbf{x}/\partial\boldsymbol{\lambda}]_{ij}$ to denote $\partial x_i/\partial\lambda_j$, it follows from the chain rule that

$$[\mathrm{d}\psi/\mathrm{d}\boldsymbol{\lambda}] = [\partial\phi/\partial\mathbf{x}][\partial\mathbf{x}/\partial\boldsymbol{\lambda}] + [\partial\phi/\partial\boldsymbol{\lambda}],$$

using the total derivative to denote variations through both $\mathbf{x}(\boldsymbol{\lambda})$ and $\boldsymbol{\lambda}$. Since $[\partial\phi/\partial\mathbf{x}] = \mathbf{0}$ from (12.2.8) and $\partial\phi/\partial\lambda_i = -c_i$ from (12.2.3) it follows that

$$\nabla\psi(\boldsymbol{\lambda}) = -\mathbf{c}(\mathbf{x}(\boldsymbol{\lambda})). \qquad (12.2.11)$$

Also by the chain rule

$$[\mathrm{d}\mathbf{c}/\mathrm{d}\boldsymbol{\lambda}] = [\partial\mathbf{c}/\partial\mathbf{x}][\partial\mathbf{x}/\partial\boldsymbol{\lambda}] = \mathbf{A}^{\mathrm{T}}[\partial\mathbf{x}/\partial\boldsymbol{\lambda}].$$

Operating on (12.2.8) by $[\mathrm{d}/\mathrm{d}\boldsymbol{\lambda}]$ gives

$$[\mathrm{d}\nabla\phi(\mathbf{x}(\boldsymbol{\lambda}), \boldsymbol{\lambda})/\mathrm{d}\boldsymbol{\lambda}] = [\partial\nabla\phi/\partial\mathbf{x}][\partial\mathbf{x}/\partial\boldsymbol{\lambda}] + [\partial\nabla\phi/\partial\boldsymbol{\lambda}] = \mathbf{0}.$$

But $[\partial\nabla\phi/\partial\mathbf{x}] = \nabla^2\phi(\mathbf{x}(\boldsymbol{\lambda}), \boldsymbol{\lambda}) = \mathbf{W}_\sigma$ and $[\partial\nabla\psi/\partial\boldsymbol{\lambda}] = -\mathbf{A}$ so it follows that

$$\nabla^2\psi(\boldsymbol{\lambda}) = -[\mathrm{d}\mathbf{c}/\mathrm{d}\boldsymbol{\lambda}] = -\mathbf{A}^{\mathrm{T}}\mathbf{W}_\sigma^{-1}\mathbf{A}|_{\mathbf{x}(\boldsymbol{\lambda})}. \qquad (12.2.12)$$

Since $\mathbf{c}(\mathbf{x}(\boldsymbol{\lambda}^*)) = \mathbf{c}(\mathbf{x}^*) = \mathbf{0}$ and $\mathbf{W}_\sigma^*$ is positive definite it follows that $\nabla\psi(\boldsymbol{\lambda}^*) = \mathbf{0}$ and (when rank $\mathbf{A}^* = m$) that $\nabla^2\psi(\boldsymbol{\lambda}^*)$ is negative definite, which reinforces the maximization result in (12.1.10).

The most obvious sequence $\{\boldsymbol{\lambda}^{(k)}\}$ to be used in step (i) of algorithm (12.2.4) is obtained by using Newton's method from an initial estimate $\boldsymbol{\lambda}^{(1)}$, giving the iteration

$$\boldsymbol{\lambda}^{(k+1)} = \boldsymbol{\lambda}^{(k)} - (\mathbf{A}^{\mathrm{T}}\mathbf{W}_\sigma^{-1}\mathbf{A})^{-1}\mathbf{c}|_{\mathbf{x}(\boldsymbol{\lambda}^{(k)})} \qquad (12.2.13)$$

(Buys, 1972). This method requires $\mathbf{W}_\sigma$, and hence explicit formulae for second derivatives, which is disadvantageous. However when only first derivatives are available and a quasi-Newton method is used to find $\mathbf{x}(\boldsymbol{\lambda}^{(k)})$, then the resulting $\mathbf{H}$ matrix (see Section 3.2) is a very good approximation to $\mathbf{W}_\sigma^{-1}$. Using this matrix in (12.2.13) (Fletcher, 1975) enables the advantages of Newton's method to be obtained whilst only requiring first derivatives. A different method is suggested by Powell (1969) and by Hestenes (1969), and is best motivated by the fact that for large $\boldsymbol{\sigma}$,

$$(\mathbf{A}^{\mathrm{T}}\mathbf{W}_\sigma^{-1}\mathbf{A})^{-1} \approx \mathbf{S} \qquad (12.2.14)$$

(see Question 12.12). When this approximation is used in (12.2.13) the iteration

$$\boldsymbol{\lambda}^{(k+1)} = \boldsymbol{\lambda}^{(k)} - \mathbf{S}\mathbf{c}^{(k)} \qquad (12.2.15)$$

is obtained. No derivatives are required by this formula so it is particularly convenient when the routine which minimizes $\phi(\mathbf{x}, \boldsymbol{\lambda})$ does not calculate or estimate derivatives. Furthermore by making $\mathbf{S}$ sufficiently large, an arbitrarily fast rate of linear convergence of $\boldsymbol{\lambda}^{(k)}$ to $\boldsymbol{\lambda}^*$ can be obtained (see Question 12.13).

An illustration of the methods based on using (12.2.13) or (12.2.15), applied to solve the problem (9.1.15), is given in Table 12.2.1, starting from $\lambda^{(1)} = 0$. It can be observed that $\lambda^{(k)} \to \lambda^*$ and $c^{(k)} \to 0$ in all cases. The Powell–Hestenes method exhibits linear convergence at a rate 0.26 when $\sigma = 1$ and 0.034 when $\sigma = 10$. This illustrates the fact that increasing σ_i by 10 will asymptotically reduce the rate of convergence of c_i by one-tenth. Newton's method is seen to converge at an order which is approximately quadratic and is a little better than the Powell–Hestenes method with $\sigma = 10$.

Although these methods have good local convergence properties, they must be supplemented in a general algorithm to ensure global convergence. This can be done by an algorithm due to Powell (1969).

(i) Initially set $\boldsymbol{\lambda} = \boldsymbol{\lambda}^{(1)}$, $\boldsymbol{\sigma} = \boldsymbol{\sigma}^{(1)}$, $k = 0$, $\|\mathbf{c}^{(0)}\|_\infty = \infty$.
(ii) Find the minimizer $\mathbf{x}(\boldsymbol{\lambda}, \boldsymbol{\sigma})$ of $\phi(\mathbf{x}, \boldsymbol{\lambda}, \boldsymbol{\sigma})$ and denote $\mathbf{c} = \mathbf{c}(\mathbf{x}(\boldsymbol{\lambda}, \boldsymbol{\sigma}))$.
(iii) If $\|\mathbf{c}\|_\infty > \frac{1}{4}\|\mathbf{c}^{(k)}\|_\infty$ set $\sigma_i = 10\sigma_i$ $\forall i$: $|c_i| > \frac{1}{4}\|\mathbf{c}^{(k)}\|_\infty$ and go to (ii). (12.2.16)
(iv) Set $k = k + 1$, $\boldsymbol{\lambda}^{(k)} = \boldsymbol{\lambda}, \boldsymbol{\sigma}^{(k)} = \boldsymbol{\sigma}, \mathbf{c}^{(k)} = \mathbf{c}$.
(v) Set $\boldsymbol{\lambda} = \boldsymbol{\lambda}^{(k)} - \mathbf{S}^{(k)}\mathbf{c}^{(k)}$ and go to (ii).

The aim of the algorithm is to achieve linear convergence at a rate of 0.25 or better. If any component c_i is not reduced at this rate, the corresponding penalty σ_i is increased tenfold which induces more rapid convergence (see Question 12.12). A simple proof that $\mathbf{c}^{(k)} \to \mathbf{0}$ can be given along the following lines. Clearly this happens unless the inner iteration ((iii) → (ii)) fails to terminate. In this iteration $\boldsymbol{\lambda}$ is fixed, and if for any $i, |c_i| > \frac{1}{4}\|\mathbf{c}^{(k)}\|_\infty$ occurs infinitely often, then $\sigma_i \to \infty$. As in Theorem 12.1.1 this implies $c_i \to 0$, which contradicts the infinitely often assumption and proves termination. This convergence result is true whatever formula is used in step (v) of (12.2.16). It follows as in Theorem 12.1.2 that any limit point is a KT point of (12.1.1) with $\mathbf{x}^{(k)} \to \mathbf{x}^*$ and $\boldsymbol{\lambda}^{(k)} - \mathbf{S}^{(k)}\mathbf{c}^{(k)} \to \boldsymbol{\lambda}^*$. For formulae (12.2.13) and (12.2.15), the required rate of convergence is obtained when $\boldsymbol{\sigma}$ is sufficiently large, and then the basic iteration takes over in which $\boldsymbol{\sigma}$ stays constant and only the $\boldsymbol{\lambda}$ parameters are changed.

In practice this proof is not as powerful as it might seem. Unfortunately increasing $\boldsymbol{\sigma}$ can lead to difficulties caused by ill-conditioning as described in Section 12.1, in which case accuracy in the solution is lost. Furthermore when no feasible point exists, this situation is not detected, and $\boldsymbol{\sigma}$ is increased without bound which is an unsatisfactory situation. I think there is scope for other algorithms to be determined which keep $\boldsymbol{\sigma}$ fixed at all times and induce convergence by ensuring that $\psi(\boldsymbol{\lambda})$ is increased sufficiently at each iteration, for example by using a restricted step modification of Newton's method (12.2.13), as described in Chapter 5.

There is no difficulty in modifying these penalty functions to handle inequality constraints. For the inequality constraint problem (12.1.2) a suitable modifica-

Table 12.2.1 Different formula for changing $\lambda^{(k)}$

	Newton's method (12.2.13), $\sigma = 1$		Powell–Hestenes method (12.2.15) $\sigma = 1$		$\sigma = 10$	
Iteration	$\lambda^{(k)}$	$c^{(k)}$	$\lambda^{(k)}$	$c^{(k)}$	$\lambda^{(k)}$	$c^{(k)}$
1	0	−0.5651977	0	−0.5651977	0	−0.0684095
2	0.6672450	−0.0296174	0.5651977	−0.1068981	0.6840946	−0.0022228
3	0.7068853	−0.0001637	0.6720958	−0.0259956	0.7063222	−0.0000758
4	0.7071068	$-0.149_{10}-7$	0.6980914	−0.0066692	0.7070801	−0.0000026
5			0.7047606	−0.0017339	0.7071058	$-0.894_{10}-7$
6			0.7064945	−0.0004524		
7			0.7069469	−0.0001181		
8			0.7070650	−0.0000308		

tion of (12.2.3) (Fletcher, 1975) is

$$\phi(\mathbf{x}, \boldsymbol{\theta}, \boldsymbol{\sigma}) = f(\mathbf{x}) + \tfrac{1}{2}\sum_i \sigma_i (c_i(\mathbf{x}) - \theta_i)_-^2 \tag{12.2.17}$$

where $a_- = \min(a, 0)$. The effect of this on the problem: minimize x subject to $x \geqslant 1$ can also be seen in Figure 12.2.1, the only difference being that for $c \geqslant \theta$ (that is $x \geqslant 1 + \theta$) the graph of ϕ is identical with that of f. Thus although $\phi(\mathbf{x}, \boldsymbol{\theta}, \boldsymbol{\sigma})$ has second derivative jump discontinuities at $c_i(\mathbf{x}) = \theta_i$, these are usually remote from the solution and in practice do not appear to affect the performance of the unconstrained minimization routine adversely. Another example of this is the inequality constraint version of problem (12.1.5). The surface $c(\mathbf{x}) = \theta = 1/\sqrt{2}$ on which this discontinuity occurs is illustrated by the dotted circle in Figure 12.2.2, which is remote from $\mathbf{x}^*$. The contours of (12.2.17) differ from those given only within this dotted circle, where they become the unpenalized linear contours of $f(\mathbf{x})$. As with (12.2.1) it is possible to rearrange the function, using (12.2.2) and omitting terms independent of $\mathbf{x}$, to give the multiplier penalty function

$$\phi(\mathbf{x}, \boldsymbol{\lambda}, \boldsymbol{\sigma}) = f(\mathbf{x}) + \sum_i \begin{cases} -\lambda_i c_i + \tfrac{1}{2}\sigma_i c_i^2 & \text{if } c_i \leqslant \lambda_i/\sigma_i \\ -\tfrac{1}{2}\lambda_i^2/\sigma_i & \text{if } c_i \geqslant \lambda_i/\sigma_i \end{cases} \tag{12.2.18}$$

where $c_i = c_i(\mathbf{x})$. Rockafeller (1974) first suggested this type of function, and as with (12.2.3) it is the most convenient form for developing the theory.

Most of the theoretical results can be extended to the inequality case. If strict complementarity holds then the extension of Theorem 12.2.1 is immediate, although the result can also be proved in the absence of this condition (Fletcher, 1975). The dual function $\psi(\boldsymbol{\lambda})$ is again defined by (12.2.9) and an analogous global result to (12.2.10) is

$$\begin{aligned}
\psi(\boldsymbol{\lambda}) &= \phi(\mathbf{x}(\boldsymbol{\lambda}), \boldsymbol{\lambda}) \leqslant \phi(\mathbf{x}^*, \boldsymbol{\lambda}) \\
&= f^* + \sum_i \begin{cases} -c_i^*\lambda_i + \tfrac{1}{2}\sigma_i c_i^{*2} & c_i^* \leqslant \lambda_i/\sigma_i \\ -\tfrac{1}{2}\lambda_i^2/\sigma_i & c_i^* \geqslant \lambda_i/\sigma_i \end{cases} \\
&\leqslant f^* + \sum_i \begin{cases} -\tfrac{1}{2}\sigma_i c_i^{*2} \\ -\tfrac{1}{2}\lambda_i^2/\sigma_i \end{cases} \\
&\leqslant f^* = \phi(\mathbf{x}^*, \boldsymbol{\lambda}^*) = \psi(\boldsymbol{\lambda}^*).
\end{aligned} \tag{12.2.19}$$

This result is also true locally if strict complementarity holds and can probably be extended in the absence of this condition, again by following Fletcher (1975). Derivative expressions analogous to (12.2.11) and (12.2.12) are easily obtained as

$$\mathrm{d}\psi(\boldsymbol{\lambda})/\mathrm{d}\lambda_i = -\min(c_i, \theta_i), \qquad i = 1, 2, \ldots, m \tag{12.2.20}$$

where $c_i = c_i(\mathbf{x}(\boldsymbol{\lambda}))$ and $\theta_i = \lambda_i/\sigma_i$, and

$$\nabla^2\psi(\boldsymbol{\lambda}) = \begin{bmatrix} -\mathbf{A}^{\mathrm{T}}\mathbf{W}_\sigma^{-1}\mathbf{A} & \mathbf{0} \\ \mathbf{0} & -\mathbf{S}^{-1} \end{bmatrix}_{\mathbf{x}(\boldsymbol{\lambda})} \tag{12.2.21}$$

where the columns of $\mathbf{A}$ correspond to indices $i{:}c_i < \theta_i$ and those of $\mathbf{S}^{-1}$ to indices $i{:}c_i \geqslant \theta_i$. Algorithms for determining the sequence $\{\boldsymbol{\lambda}^{(k)}\}$ in step (i) of algorithm (12.2.4) can be determined using these derivative expressions. An equivalent form of Newton's method (12.2.13) is possible, although in view of the implicit inequalities $\boldsymbol{\lambda} \geqslant \mathbf{0}$ it is probably preferable to choose $\boldsymbol{\lambda}^{(k+1)}$ to solve a subproblem

$$\begin{aligned}&\underset{\boldsymbol{\lambda}}{\text{maximize}} \quad q^{(k)}(\boldsymbol{\lambda})\\ &\text{subject to} \quad \boldsymbol{\lambda} \geqslant \mathbf{0}\end{aligned} \tag{12.2.22}$$

where $q^{(k)}(\boldsymbol{\lambda})$ is obtained by truncating a Taylor series for $\psi(\boldsymbol{\lambda})$ about $\boldsymbol{\lambda}^{(k)}$ after the quadratic term. In fact the simple structure of (12.2.21) enables a more simple problem to be solved in terms of just the indices $i{:}c_i < \theta_i$. The result in (12.2.14) can also be used to determine an extension of (12.2.15), that is

$$\lambda_i^{(k+1)} = \lambda_i^{(k)} - \min(\sigma_i c_i^{(k)}, \lambda_i^{(k)}), \qquad i = 1, 2, \ldots, m. \tag{12.2.23}$$

These formulae can be incorporated in a globally convergent algorithm like (12.2.16) by using $\|\nabla\psi\|_\infty$ in place of $\|\mathbf{c}\|_\infty$ to monitor the rate of convergence.

A selection of numerical experiments is given by Fletcher (1975), which seems to indicate that whilst both the Newton-like formulae or the Powell–Hestenes formulae for updating $\boldsymbol{\lambda}^{(k)}$ are effective, the Newton-like method is somewhat more efficient. Local convergence is rapid and high accuracy can usually be achieved in about four to six minimizations. When this occurs with modest values of $\boldsymbol{\sigma}$ then no difficulties due to ill-conditioning and loss of accuracy are observed. Furthermore the Hessian matrix can be carried forward, and updated if σ_i is increased, so the computational effort for the successive minimizations goes down rapidly. Since $\phi(\mathbf{x}, \boldsymbol{\lambda}, \boldsymbol{\sigma})$ is always well defined, there is no difficulty in coping with infeasible points, and it is easy to program the method using an existing quasi-Newton subroutine. The main disadvantage is that the sequential nature of the method is less efficient than the more direct approach of Section 12.4. Also the global convergence result based on increasing $\boldsymbol{\sigma}$, whilst powerful in theory, does not always work well in practice, and there are practical applications in which it has caused ill-conditioning and low accuracy.

Two final points are worthy of note. Firstly if the problem is a mixture of linear and nonlinear constraints, then it may be worth incorporating only the nonlinear constraints into the penalty function, and the linear constraints can be included when $\phi(\mathbf{x}, \boldsymbol{\lambda}, \boldsymbol{\sigma})$ is minimized, for example at step (ii) in algorithm (12.2.4). This is especially true for bounds on the variables, since minimization with bounds is not a significant complication on an unconstrained minimization routine. Another point is that approximate minimization of the penalty function (see (12.1.21) for example) can also be considered, and a review of recent work in this area is given by Coope and Fletcher (1980). In this case the algorithms start to become more like the direct methods of Section 12.4 and Coope and Fletcher give an algorithm which incorporates the Lagrangian correction defined in (12.4.5).

12.3 THE L_1 EXACT PENALTY FUNCTION

An attractive approach to nonlinear programming is to attempt to determine an *exact penalty function* $\phi(\mathbf{x})$ by which is meant a function, defined in terms of $f(\mathbf{x})$, $\mathbf{c}(\mathbf{x})$ and possibly derivatives thereof, which is minimized locally by the solution $\mathbf{x}^*$ to (7.1.1). This holds out the possibility that the solution can be found by a single application of an unconstrained minimization technique to $\phi(\mathbf{x})$, as against the sequential processes described in Sections 12.1 and 12.2. The simplest exact penalty function is that described in this section in which constraint violations are penalized by weighted L_1 terms. Unfortunately this function is *non-smooth* or *non-differentiable* so that the many effective techniques for smooth minimization described in Chapters 3, 4 and 5 cannot adequately be used. A more realistic approach is to use the exact penalty function as a criterion function to be used in conjunction with other iterative methods for nonlinear programming such as those described in the next section. The most satisfactory approach of all is to apply methods of non-smooth optimization which take account of the special structure of $\phi(\mathbf{x})$ and such methods are described in Chapter 14.

This section considers the most frequently used type of exact penalty function which is the L_1 *exact penalty function.* This function has been researched widely, for example by Pietrzykowski (1969), Conn (1973), Han (1977), Coleman and Conn (1982a, b), Fletcher (1981) and Mayne (1980) in nonlinear programming applications and by Barrodale (1970) in L_1 data-fitting problems, amongst others. For the nonlinear programming problem (7.1.1) with both equations and inequalities the associated L_1 penalty function is

$$\phi(\mathbf{x}) = \nu f(\mathbf{x}) + \sum_{i\in E} |c_i(\mathbf{x})| + \sum_{i\in I} c_i(\mathbf{x})^- \tag{12.3.1}$$

where $a^- \triangleq \max(-a, 0)$. The parameter ν ($\nu > 0$) provides a means of weighting the relative contribution of $f(\mathbf{x})$ and the penalty terms: equivalently a parameter $\sigma = \nu^{-1}$ can be used which multiplies the penalty terms. As an illustration, the nonlinear programming problem: minimize $-x_1 - x_2$ subject to $1 - x_1^2 - x_2^2 \geqslant 0$ with solution $x_1^* = x_2^* = \lambda^* = 1/\sqrt{2}$ is considered. The corresponding L_1 exact penalty function is

$$\phi(\mathbf{x}) = \nu(-x_1 - x_2) + \max(x_1^2 + x_2^2 - 1, 0) \tag{12.3.2}$$

where ν takes any value in the range $0 < \nu < 1/\lambda^* = \sqrt{2}$. The contours of $\phi(\mathbf{x})$ for $\nu = 1$ are shown in Figure 12.3.1 and it is clear that $\mathbf{x}^*$ minimizes $\phi(\mathbf{x})$. The non-smooth nature of $\phi(\mathbf{x})$ having a curved 'groove' on the unit circle (broken line) can also be seen.

The penalty function (12.3.1) is exact in the sense that for sufficiently small ν, local solutions of the nonlinear programming problem are equivalent to local minimizers of (12.3.1) to a large extent. Practical considerations in choosing ν are discussed at the end of this section. The main situation in which there is a discrepancy between solutions of the nonlinear programming problem and the

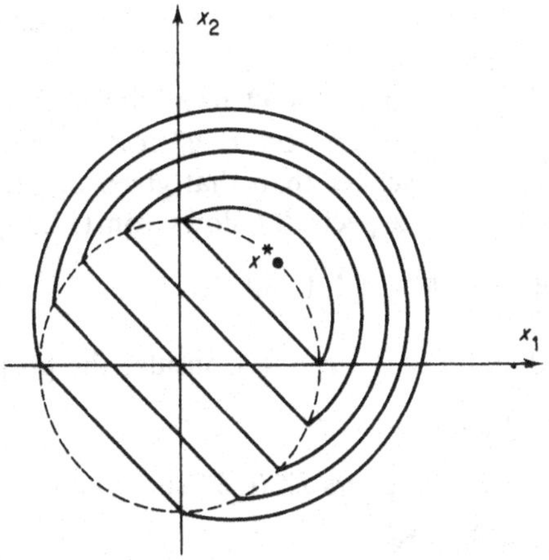

Figure 12.3.1 Contours for the exact penalty function (12.3.2)

exact penalty function problem arises when the minimizer $\mathbf{x}^*$ of the latter is not feasible in the nonlinear programming problem, even though feasible points do exist. This is the same situation as that illustrated in Figure 6.2.1: it is an inevitable consequence of using the penalty approach as a way of inducing global convergence. To circumvent the difficulty is a global minimization problem and hence largely impracticable. Corresponding advantages are that *best* solutions can be detemined when no feasible point exists in the nonlinear programming problem, and the difficulty of finding an initial feasible point is avoided. In practice the most likely unfavourable situation which arises when actually applying an exact penalty function is that a sequence is calculated for which $\phi^{(k)} \to -\infty$, which is an indication that the calculation should be repeated with a smaller value of v.

The main aim of this section is to provide a simple derivation of optimality conditions, both first and second order, suitable for an introductory study of the L_1 penalty function. A more general treatment is given in Section 14.3. Also in this section the equivalence of local solutions of the nonlinear programming problem and local minimizers of the exact penalty function is considered. Finally the use of (12.3.1) in linear and quadratic programming and the equivalence to linear complementarity problems is discussed. To simplify the presentation it is assumed that the set E in (12.3.1) is empty, and that the function has arisen as an exact penalty function for the inequality problem (12.1.2). Results applicable to the general case are also stated. The notation for the directional derivative of ϕ at $\mathbf{x}'$

$$D_s\phi(\mathbf{x}') \triangleq \lim_{\theta \downarrow 0} \frac{\phi(\mathbf{x}(\theta)) - \phi(\mathbf{x}')}{\theta}$$

is used, where $\mathbf{x}(\theta) = \mathbf{x}' + \theta\mathbf{s}$ is a ray from $\mathbf{x}'$ or more generally may be an arc or a directional sequence. Also $Z = Z(\mathbf{x}) = \{i: c_i(\mathbf{x}) = 0\}$ denotes the set of zero valued constraints at any point.

First order necessary conditions for $\mathbf{x}^*$ to minimize $\phi(\mathbf{x})$ are readily derived if the assumption is made that the vectors $\mathbf{a}_i^*$ $i \in Z^*$ are linearly independent. In this respect the result is weaker than that given in Chapter 14 but it does have some useful constructive aspects. It follows from (12.3.1) (with E empty) that

$$D_s\phi(\mathbf{x}) = \nu\mathbf{s}^{\mathrm{T}}\mathbf{g} - \sum_{i \notin Z} \sigma_i \mathbf{s}^{\mathrm{T}}\mathbf{a}_i + \sum_{i \in Z} (\mathbf{s}^{\mathrm{T}}\mathbf{a}_i)^- \tag{12.3.3}$$

where $\sigma_i = 1$ if $c_i < 0$ and $\sigma_i = 0$ if $c_i > 0$. Consider the proposition that

$$D_s\phi^* \triangleq D_s\phi(\mathbf{x}^*) \geqslant 0 \tag{12.3.4}$$

if and only if there exist multipliers λ_i^* $i \in Z^*$ such that

$$\nu\mathbf{g}^* - \sum_{i \notin Z^*} \sigma_i^* \mathbf{a}_i^* = \sum_{i \in Z^*} \lambda_i^* \mathbf{a}_i^* \tag{12.3.5}$$

$$0 \leqslant \lambda_i^* \leqslant 1 \qquad i \in Z^* \tag{12.3.6}$$

The 'if' result is straightforward since substituting (12.3.5) and (12.3.6) into (12.3.3) yields

$$D_s\phi^* = \sum_{i \in Z^*} ((\mathbf{s}^{\mathrm{T}}\mathbf{a}_i^*)^- + \lambda_i^* \mathbf{s}^{\mathrm{T}}\mathbf{a}_i^*) \geqslant 0 \tag{12.3.7}$$

For the converse result, it is assumed that there are no multipliers $\boldsymbol{\lambda}_i^*$ which satisfy both (12.3.5) and (12.3.6), and a direction $\mathbf{s}$ is constructed for which $D_s\phi^* < 0$, thus contradicting (12.3.4). Denote by $\mathbf{A}^*$ the matrix with columns $\mathbf{a}_i^*$ $i \in Z^*$. If (12.3.5) cannot be satisfied then following the construction in (9.1.4) there exist vectors $\boldsymbol{\lambda}$ and $\boldsymbol{\mu} \neq \mathbf{0}$ such that both $\mathbf{A}^{*\mathrm{T}}\boldsymbol{\mu} = \mathbf{0}$ and $\bar{\mathbf{g}}^* = \mathbf{A}^*\boldsymbol{\lambda} + \boldsymbol{\mu}$ where

$$\bar{\mathbf{g}}^* = \nu\mathbf{g}^* - \sum_{i \notin Z^*} \sigma_i^* \mathbf{a}_i^*. \tag{12.3.8}$$

Then $\mathbf{s} = -\boldsymbol{\mu}$ gives rise to

$$D_s\phi^* = \mathbf{s}^{\mathrm{T}}\bar{\mathbf{g}}^* = -\boldsymbol{\mu}^{\mathrm{T}}\boldsymbol{\mu} < 0,$$

and hence $\mathbf{s}$ is the required descent direction. Alternatively (12.3.5) may be true but (12.3.6) false for some $p \in Z^*$. Consider the vector $\mathbf{s} = \mathbf{A}^{*+\mathrm{T}}\mathbf{e}_p$ where $\mathbf{A}^+ = (\mathbf{A}^{\mathrm{T}}\mathbf{A})^{-1}\mathbf{A}^{\mathrm{T}}$ exists by virtue of the independence assumption. It follows that $\mathbf{s}^{\mathrm{T}}\mathbf{a}_p^* = 1$ and $\mathbf{s}^{\mathrm{T}}\mathbf{a}_i^* = 0$, $i \in Z^*$, $i \neq p$. Then

$$D_s\phi^* = \sum_{i \in Z^*} ((\mathbf{s}^{\mathrm{T}}\mathbf{a}_i^*)^- + \lambda_i^* \mathbf{s}^{\mathrm{T}}\mathbf{a}_i^*) = \lambda_p^*.$$

If (12.3.6) is false because $\lambda_p^* < 0$, then this construction provides a descent direction. Finally if (12.3.6) is false because $\lambda_p^* > 1$ then the direction $\mathbf{s} = -\mathbf{A}^{*+\mathrm{T}}e_p$ gives $D_s\phi^* = 1 - \lambda_p^*$ and again $\mathbf{s}$ is a descent direction.

To summarize this result, a first order necessary condition for $\mathbf{x}^*$ to minimize $\phi(\mathbf{x})$ is that $D_s\phi^* \geqslant 0$. The above discussion shows that this condition is equivalent to (12.3.5) and (12.3.6). If multipliers $\lambda_i^* = \sigma_i^*$ are defined for $i \notin Z^*$

then it is possible to write these conditions as

$$\nu\mathbf{g}^* = \sum_{i\in I\cup E} \lambda_i^* \mathbf{a}_i^* \tag{12.3.9a}$$

$$\left.\begin{array}{l} 0 \leqslant \lambda_i^* \leqslant 1 \\ c_i^* > 0 \Rightarrow \lambda_i^* = 0 \\ c_i^* < 0 \Rightarrow \lambda_i^* = 1 \end{array}\right\} \quad i\in I \tag{12.3.9b}$$

$$\left.\begin{array}{l} -1 \leqslant \lambda_i^* \leqslant 1 \\ c_i^* \neq 0 \Rightarrow \lambda_i^* = -\operatorname{sign} c_i^* \end{array}\right\} \quad i\in E \tag{12.3.9c}$$

In fact (12.3.9) gives first order conditions in the general case: the conditions for $i\in E$ are readily deduced in a similar way, or by using $|c_i| = c_i^+ + c_i^- = (-c_i)^- + c_i^-$ to include the E terms with the I terms. The differences between this system and the KT conditions (9.1.16) for the equivalent nonlinear programming problem (7.1.1) are not great. The parameter ν appears in (12.3.9) and not in (9.1.16). However, if (7.1.1) is scaled so that $f(\mathbf{x})$ is replaced by $\nu f(\mathbf{x})$ ($\nu > 0$) (this does not change local solutions of (7.1.1)) then the equation $\nu\mathbf{g}^* = \sum \mathbf{a}_i^* \lambda_i^*$ is also present in the KT conditions. Infeasible points of (7.1.1) are allowed in (12.3.9), but if $\mathbf{x}^*$ is infeasible for some constraint $c_i(\mathbf{x})$, then the corresponding multiplier must take the value $\lambda_i^* = -\operatorname{sign} c_i^*$. Thus the terms in (12.3.1) which are locally smooth at $\mathbf{x}^*$ can be written as

$$\bar{f}(\mathbf{x}) = \nu f(\mathbf{x}) - \sum_{i\notin Z^*} \lambda_i^* c_i(\mathbf{x}) \tag{12.3.10}$$

and in fact $\bar{\mathbf{g}}^*$ in (12.3.8) is the gradient vector of these terms. Moreover $\mathbf{x}^*$ equivalently solves the problem

$$\begin{array}{ll} \underset{\mathbf{x}}{\text{minimize}} & \bar{f}(\mathbf{x}) \\ \text{subject to} & c_i(\mathbf{x}) = 0 \qquad i\in Z^* \end{array} \tag{12.3.11}$$

and the multipliers λ_i^* $i\in Z^*$ are also the Lagrange multipliers of this problem.

The other significant difference between (9.1.16) and (12.3.9) is the presence of the bound $\lambda_i^* \leqslant 1$ for $i\in I$ in the latter. A simple interpretation of this is the following: assume that $\lambda_p^* > 1$ for some constraint $p\in Z^*$ in (12.3.11). Then following the construction above, the direction $\mathbf{s} = -\mathbf{A}^{*+\mathrm{T}}\mathbf{e}_p$ is infeasible with respect to constraint p, and has a directional derivative $D_s\phi^* = 1 - \lambda_p^*$, the $-\lambda_p^*$ term being the slope of the smooth part (12.3.10) and the 1 being the slope of penalty term c_p^-. Because $\lambda_p^* > 1$, the contribution of the penalty term does not have sufficient weight to dominate the smooth part and create a local minimum of $\phi(\mathbf{x})$ at $\mathbf{x}^*$. The situation is illustrated for a single constraint term in Figure 12.3.2. The cases where λ^* is outside $[0, 1]$ are illustrated in (ii) and (iii) and it is seen that the penalized function $\phi = f + c^-$ does not have a minimizer at $c^* = 0$, whereas (iv) shows that for λ^* in $(0, 1)$, adding in the penalty term does create a local minimizer. If it is desired to create a minimum in case

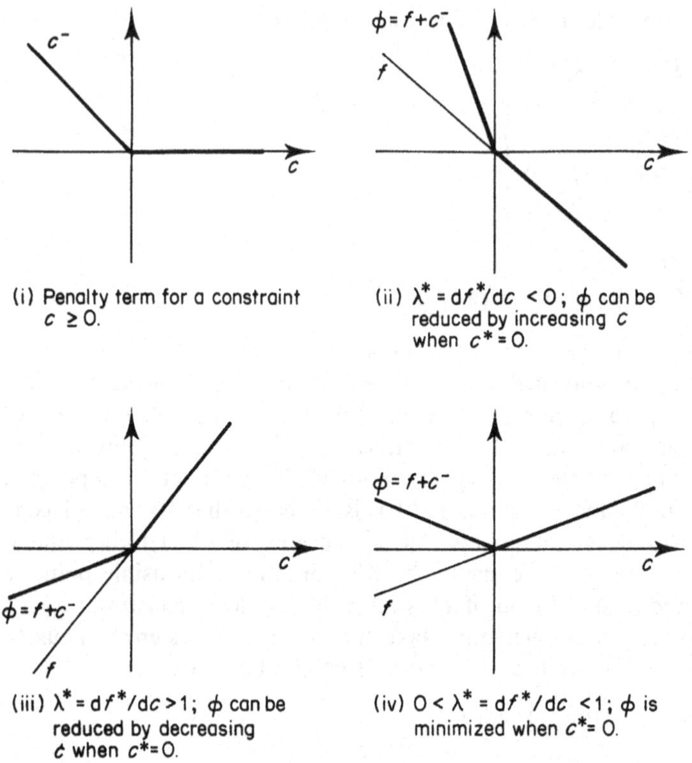

c^-
c
$\phi = f + c^-$
f
c
(i) Penalty term for a constraint $c \geq 0$.
(ii) $\lambda^* = df^*/dc < 0$; ϕ can be reduced by increasing c when $c^* = 0$.
c
$\phi = f + c^-$
f
$\phi = f + c^-$
c
f
(iii) $\lambda^* = df^*/dc > 1$; ϕ can be reduced by decreasing c when $c^* = 0$.
(iv) $0 < \lambda^* = df^*/dc < 1$; ϕ is minimized when $c^* = 0$.

Figure 12.3.2 Optimality conditions for an L_1 exact penalty function

(iii) where $\lambda^* > 1$, it is clearly necessary to increase the weight of the c^- term, or equivalently to scale down the function f by replacing it by νf where $0 < \nu < 1/\lambda^*$. This argument also enables the threshold value of ν for which $\phi(\mathbf{x})$ is minimized by $\mathbf{x}^*$ (where $\mathbf{x}^*, \boldsymbol{\lambda}^*$ is a KT point of (7.1.1)) to be stated. If

$$\nu \leqslant 1/\|\boldsymbol{\lambda}^*\|_\infty \tag{12.3.12}$$

then the first order conditions (12.3.9) and the KT conditions for the scaled problem are both satisfied, and both problems are solved by $\mathbf{x}^*$ to first order. If however $\nu > 1/\|\boldsymbol{\lambda}^*\|_\infty$ then for the scaled problem $\lambda_p^* > 1$ for some p and the penalty function can be decreased by leaving $\mathbf{x}^*$ in the direction $\mathbf{s}$ constructed as above. Note that the threshold value of ν depends only on first order information, and not second order information as in Sections 12.1 and 12.2.

The same approach can be used to obtain insight into second order conditions. Let the first order conditions (12.3.5) and (12.3.6) (for E empty) hold at $\mathbf{x}^*$ and in addition let $0 < \lambda_i^* < 1$ for all $i \in Z^*$ (essentially an extension of the concept of strict complementarity in Section 9.1). If $\mathbf{s}^T\mathbf{a}_i^* \neq 0$ for any $i \in Z^*$, then the inequality (12.3.7) is strict and it follows that all such directions are strict ascent directions ($D_s\phi^* > 0$), for which second order effects are not significant. However, for any $\mathbf{s}$ such that $\mathbf{s}^T\mathbf{a}_i^* = 0$ for all $i \in Z^*$, then $D_s\phi^* = 0$ from (12.3.7) and second order effects are important. For any such $\mathbf{s}$ an arc $\mathbf{x}(\theta)$ $\theta\in[0,\bar{\theta})$ can be constructed such that $\mathbf{x}(0)=\mathbf{x}^*$, $\dot{\mathbf{x}}(0)=\mathbf{s}$ and $c_i(\mathbf{x}(\theta))=0$ for all $i\in Z^*$ using the independence assumption, as in Lemma 9.2.2. Using the Lagrangian function (9.1.6), it follows from (12.3.1) and (12.3.9b) for sufficiently small θ that

$$\phi(\mathbf{x})=\mathscr{L}(\mathbf{x}(\theta),\boldsymbol{\lambda}^*)=\phi^*+\tfrac{1}{2}\boldsymbol{\delta}^{\mathrm{T}}\mathbf{W}^*\boldsymbol{\delta}+o(\boldsymbol{\delta}^{\mathrm{T}}\boldsymbol{\delta}) \qquad (12.3.13)$$

by using a Taylor expansion, where $\mathbf{W}^*=\nabla^2\mathscr{L}(\mathbf{x}^*,\boldsymbol{\lambda}^*)$ and $\boldsymbol{\delta}=\mathbf{x}(\theta)-\mathbf{x}^*$. By minimality of ϕ^* and taking the limit $\theta\downarrow 0$ it follows that

$$\mathbf{s}^{\mathrm{T}}\mathbf{W}^*\mathbf{s}\geqslant 0 \quad \forall\mathbf{s}\colon \mathbf{s}^{\mathrm{T}}\mathbf{a}_i^*=0 \qquad i\in Z^* \qquad (12.3.14)$$

and this is a *second order necessary condition* for $\mathbf{x}^*$ to minimize $\phi(\mathbf{x})$. It can be shown that second order sufficient conditions are obtained if the inequality (12.3.14) is strict (in addition to (12.3.5), (12.3.6) and strict complementarity). A more general derivation of all these results is presented in Chapter 14. As with the first order conditions, there is a close correlation between the second order conditions for nonlinear programming (cf. (9.3.8a)) and those for the L_1 penalty function. In fact if $\mathbf{x}^*$ is feasible and $\lambda^*<1$ is assumed then the conditions are equivalent. ($\lambda^*\leqslant 1$ is required for equivalence with (12.3.6) and $\lambda^*<1$ for strict complementarity: in the absence of this condition the result is not true—see the example below).

The above discussion on equivalence of optimality conditions enables statements to be made about the equivalence of solutions to the exact penalty function and the nonlinear programming problem. It is necessary to assume that $\mathbf{x}^*$ is feasible in the latter, in which case the above results can be summarised as follows. If $\nu\leqslant 1/\|\boldsymbol{\lambda}^*\|_\infty$ then minimizers to first order (that is KT points) are equivalent. If a regularity condition holds (e.g. linear independence of the vectors $\mathbf{a}_i^*$ $i\in\mathscr{A}^*$) and $\nu<1/\|\boldsymbol{\lambda}^*\|_\infty$ then minimizers to second order are equivalent. If in addition $\mathbf{x}^*$ satisfies the second order sufficient conditions, which are also equivalent, then $\mathbf{x}^*$ is a strict local solution to both problems. Some simple examples which illustrate non-equivalence can easily be given. For the problem

$$\text{minimize } 0 \text{ subject to } x^3+3x^2+3=0,$$

$x^*=0$ minimizes the L_1 penalty function but is infeasible in the nonlinear programming problem. In all the following cases $x^*=0$ solves the nonlinear programming problem but does not minimize the L_1 penalty function with $\nu=1$ for the reason given. For the problem

$$\text{minimize } x \text{ subject to } x^2\leqslant 0$$

x^* is not a KT point and linear independence does not hold. For the problem

$$\text{minimize } x^3 \text{ subject to } x^5\geqslant 0,$$

the curvature condition is not strict (x^* is a minimizer to second order). For the problem

$$\text{minimize } x-\tfrac{1}{2}x^2 \text{ subject to } 0\leqslant x\leqslant 1,$$

$\lambda^* = 1$ and hence the condition $\nu < 1/\|\boldsymbol{\lambda}^*\|_\infty$ is not strictly satisfied (x^* is a KT point, that is a minimizer to first order). Finally for the problem

$$\text{minimize } 2x - x^2 \text{ subject to } 0 \leqslant x \leqslant 1,$$

$\lambda^* = 2$ and the condition $\nu \leqslant 1/\|\boldsymbol{\lambda}^*\|_\infty$ is not satisfied and x^* is not even a minimizer to first order. However, given that $\mathbf{x}^*$ is feasible and a valid choice of ν is made, these exceptions only occur as limiting cases and it can be assumed that the nonlinear programming problem and the L_1 penalty function have equivalent solutions for all practical purposes.

Another useful set of first order conditions arises when $\phi(\mathbf{x})$ is minimized subject to simple bounds

$$\mathbf{l} \leqslant \mathbf{x} \leqslant \mathbf{u} \tag{12.3.15}$$

and is readily obtained by extending the argument above or as a special case of the result in Section 14.6. Assuming $\mathbf{l} < \mathbf{u}$ without loss of generality, a single multiplier π_i^* can be assigned to each pair of bounds in (12.3.15) and the necessary condition is that $\mathbf{x}^*$ is feasible and that there exist multipliers $\boldsymbol{\lambda}^*$ which satisfy (12.3.9b, c) and $\boldsymbol{\pi}^*$ such that

$$\nu\mathbf{g}^* = \sum_{i \in I \cup E} \lambda_i^* \mathbf{a}_i^* + \boldsymbol{\pi}^*$$

$$\left.\begin{aligned} &\pi_j^* \geqslant 0 \quad \text{if} \quad x_j^* = l_j \\ &\pi_j^* \leqslant 0 \quad \text{if} \quad x_j^* = u_j \\ &\pi_j^* = 0 \quad \text{otherwise} \end{aligned}\right\} \quad j = 1, 2, \ldots, n. \tag{12.3.16}$$

Second order conditions analgous to (12.3.14) are also readily obtained.

A particular case of the L_1 penalty function is the L_1LP *problem* in which the functions $f(\mathbf{x})$ and $c_i(\mathbf{x})$ in (12.3.1) are linear. It is of some importance as it provides a way of combining the phase 1 and phase 2 parts of an LP algorithm (cf. Section 8.4) in an efficient way. A convenient format which allows different relative weights to the constraints, and different constraint types is

$$\begin{aligned} &\text{minimize} \quad \mathbf{c}^T\mathbf{x} - \boldsymbol{\rho}^T(\mathbf{A}^T\mathbf{x} - \mathbf{b})^+ + \boldsymbol{\sigma}^T(\mathbf{A}^T\mathbf{x} - \mathbf{b})^- \\ &\text{subject to} \quad \mathbf{l} \leqslant \mathbf{x} \leqslant \mathbf{u} \end{aligned} \tag{12.3.17}$$

where $\mathbf{A} \in \mathbb{R}^{n \times m}$, $\boldsymbol{\rho} \leqslant \boldsymbol{\sigma}$ and $\mathbf{l} \leqslant \mathbf{u}$. For example for a unit weight one would set

$$\begin{aligned} &\rho_i = 0, \quad \sigma_i = 1 \quad \text{for a constraint} \quad \mathbf{a}_i^T\mathbf{x} \geqslant b_i \\ &\rho_i = -1, \quad \sigma_i = 0 \quad \text{for a constraint} \quad \mathbf{a}_i^T\mathbf{x} \leqslant b_i, \quad \text{and} \\ &\rho_i = -1, \quad \sigma_i = 1 \quad \text{for a constraint} \quad \mathbf{a}_i^T\mathbf{x} = b_i. \end{aligned}$$

An interesting result that (12.3.17) has a symmetric dual is given in Question 12.22. First order necessary conditions (which are also sufficient by convexity) are that if $\mathbf{x}$ solves (12.3.17), then there exist multipliers $\boldsymbol{\pi} \in \mathbb{R}^n$, $\boldsymbol{\lambda} \in \mathbb{R}^m$ such that

$$\begin{aligned} &\mathbf{c} = \mathbf{A}\boldsymbol{\lambda} + \boldsymbol{\pi} \\ &\mathbf{l} \leqslant \mathbf{x} \leqslant \mathbf{u} \qquad \text{primal feasibility} \end{aligned}$$

$$\begin{array}{ll} \boldsymbol{\rho} \leqslant \boldsymbol{\lambda} \leqslant \boldsymbol{\sigma} & \text{dual feasibility} \\ \left.\begin{array}{l} \pi_i > 0 \Rightarrow x_i = l_i \\ \pi_i < 0 \Rightarrow x_i = u_i \end{array}\right\} & \text{primal complementarity} \\ \left.\begin{array}{l} r_j > 0 \Rightarrow \lambda_j = \rho_j \\ r_j < 0 \Rightarrow \lambda_j = \sigma_j \end{array}\right\} & \text{dual complementarity} \end{array} \tag{12.3.18}$$

where r_j denotes $\mathbf{a}_j^{\mathrm{T}}\mathbf{x} - b_j$. These conditions are readily established by adding extra variables to convert (12.3.17) into an LP problem and this provides one way of solving the L_1LP problem. However an active set method for solving (12.3.17) can also be suggested along the lines of Sections 8.3 and 8.4 and can be expected to be more efficient. An active set $\mathscr{A}$ of n constraints (either from the bounds or the L_1 terms in (12.3.17)) determines the current point $\mathbf{x}^{(k)}$. The multipliers of the inactive L_1 terms are set so as to satisfy dual complementarity in (12.3.18). The smooth part of the L_1 function is then defined by $\bar{\mathbf{c}}^{\mathrm{T}}\mathbf{x} = (\mathbf{c} - \sum_{i \notin \mathscr{A}} \lambda_i \mathbf{a}_i)^{\mathrm{T}}\mathbf{x}$. As in (8.3.3), $\bar{\mathbf{c}}$ is used to define multipliers for the active constraints. If these multipliers are feasible in (12.3.18), then $\mathbf{x}^{(k)}$ is optimal, otherwise the most violated multiplier indicates which constraint to relax, as in (8.3.5). A line search for the best feasible point is carried out, and the new constraint which thus becomes active replaces the constraint being relaxed in the active set. A version of this algorithm which also gives guaranteed termination in the presence of degeneracy is suggested by Fletcher (1985a).

Another special case of the L_1 penalty function is the $\mathrm{L_1QP}$ problem. This can arise when combining the phase 1 and phase 2 parts of a QP algorithm, and also as a subproblem in the $\mathrm{SL_1QP}$ algorithm for nonlinear programming (see Section 12.4). It is again convenient to pose the problem in the form

$$\begin{array}{ll} \text{minimize} & \mathbf{g}^{\mathrm{T}}\mathbf{x} + \tfrac{1}{2}\mathbf{x}^{\mathrm{T}}\mathbf{G}\mathbf{x} - \boldsymbol{\rho}^{\mathrm{T}}(\mathbf{A}^{\mathrm{T}}\mathbf{x} - \mathbf{b})^{+} + \boldsymbol{\sigma}^{\mathrm{T}}(\mathbf{A}^{\mathrm{T}}\mathbf{x} - \mathbf{b})^{-} \\ \text{subject to} & \mathbf{l} \leqslant \mathbf{x} \leqslant \mathbf{u}. \end{array} \tag{12.3.19}$$

This problem can be transformed into a regular QP problem by adding extra variables, but as above it is more convenient to look for an active set method (e.g. Conn and Sinclair, 1975; Fletcher, 1985b). The active set comprises the currently active L_1 terms or bounds. At the current point $\mathbf{x}^{(k)}$ a line search is carried out towards the minimizer of the smooth part of the L_1 function for the current active set using any of the methods described in Section 10.1. The step is chosen to reach the last knot before the unconstrained minimizer (if such a knot exists: a knot is a point at which a derivative discontinuity occurs) or otherwise the unconstrained minimizer. If neither of these possibilities gives a feasible point, then a step to the nearest bound is taken. An index is added to the active set whenever a knot or bound is reached by the line search. If the minimizer of the smooth part of the L_1 function is reached by the line search, then the gradient of the smooth part $\bar{\mathbf{g}} = \mathbf{g} + \mathbf{G}\mathbf{x} - \sum_{i \notin \mathscr{A}} \lambda_i \mathbf{a}_i$ is used to determine multipliers for the active constraints, again using any of the methods of Section 10.1. Either the multipliers are feasible and the current point is optimal, or the most infeasible multiplier determines which constraint to relax, and a further line search is made. Fletcher (1985b) gives brief details of how the necessary matrix factors can be handled in this approach.

Another observation of interest is that the $\mathrm{L_1QP}$ problem can be written as a linear complementarity problem (cf. (10.6.4) for QP) in the form of the system

$$\begin{pmatrix}\boldsymbol{\pi}\\ \mathbf{r}\end{pmatrix}=\begin{bmatrix}\mathbf{G} & -\mathbf{A}\\ \mathbf{A}^{\mathrm{T}} & \mathbf{0}\end{bmatrix}\begin{pmatrix}\mathbf{x}\\ \boldsymbol{\lambda}\end{pmatrix}+\begin{pmatrix}\mathbf{g}\\ -\mathbf{b}\end{pmatrix}$$
$$\begin{pmatrix}\mathbf{l}\\ \boldsymbol{\rho}\end{pmatrix}\leqslant\begin{pmatrix}\mathbf{x}\\ \boldsymbol{\lambda}\end{pmatrix}\leqslant\begin{pmatrix}\mathbf{u}\\ \boldsymbol{\sigma}\end{pmatrix} \tag{12.3.20}$$

together with the complementarity conditions of (12.3.18). This can be solved by the methods of Section 10.6, suitably modified to allow for lower and upper bounds on the $(\mathbf{x}, \boldsymbol{\lambda})$ variables. The advantages and disadvantages of this approach to $\mathrm{L_1QP}$ are the same as those outlined in Section 10.6.

Finally some remarks about the practical choice of the parameter v are made when the L_1 penalty function is used, particularly in the general context of nonlinear programming. The smaller v is, the more f is damped out relative to the penalty terms in $\phi(\mathbf{x})$, and the less accurately is the solution located in the tangent plane of the active constraints (cf. Figure 12.1.2, $\sigma = 100$). Also when following a 'curved groove' along the surface $c_i(\mathbf{x}) = 0$ by a sequence of line searches, if v is decreased then the length of the correction which will reduce ϕ is decreased, and hence the total number of line searches is increased. Thus it is advantageous to keep v reasonably large, yet less than the threshold value $1/\|\boldsymbol{\lambda}^*\|_\infty$. Too small a value of v is indicated by the multipliers of the active constraints being uniformly small. Then it can be advantageous to increase v until these multipliers are within say about a power of 10 of their threshold value. On the other hand, a sequence $\phi^{(k)} \to -\infty$ may occur which may require a smaller value of v or a different starting point to be chosen. If the active multipliers differ widely in magnitude this may be an indication the constraints should be rescaled. Some algorithms allow this to be done automatically, in which case it can be better to use a format for the penalty function analogous to (12.3.17), that is

$$\phi(\mathbf{x}) = f(\mathbf{x}) - \boldsymbol{\rho}^{\mathrm{T}}(\mathbf{c}(\mathbf{x}))^{+} + \boldsymbol{\sigma}^{\mathrm{T}}(\mathbf{c}(\mathbf{x}))^{-} \tag{12.3.21}$$

in which each constraint function has its own weighting constant and their relative magnitude can be adjusted.

12.4 THE LAGRANGE–NEWTON METHOD (SQP)

A penalty function method is a somewhat indirect way of attempting to solve nonlinear constraint problems. A more direct and efficient approach is to iterate on the basis of certain approximations to the problem functions $f(\mathbf{x})$ and $\mathbf{c}(\mathbf{x})$, in particular by using linear approximations to the constraint functions $\mathbf{c}(\mathbf{x})$. This has to be done with some care to ensure rapid convergence properties close to the solution, and one particular method is seen to be fundamental in this respect. This method is most simply explained as being Newton's method applied to find the stationary point of the Lagrangian function (9.1.6), and hence might be referred to as the *Lagrange–Newton* method. The Lagrangian function is defined in terms of variables $\mathbf{x}$ and $\boldsymbol{\lambda}$, so a feature of the resulting methods is that a sequence of approximations $\mathbf{x}^{(k)}$, $\boldsymbol{\lambda}^{(k)}$ to both the solution vector $\mathbf{x}^*$ and the vector of optimum Lagrange multipliers $\boldsymbol{\lambda}^*$ is generated.

In the first instance, consider applying the method to the equality constraint problem (12.1.1) and define $\blacktriangledown$ as in (9.1.8) so that equation (9.1.7) is the stationary point (KT) condition at $\mathbf{x}^*, \boldsymbol{\lambda}^*$. As usual a Taylor series for $\blacktriangledown \mathscr{L}$ about $\mathbf{x}^{(k)}$, $\boldsymbol{\lambda}^{(k)}$ gives

$$\blacktriangledown \mathscr{L}(\mathbf{x}^{(k)} + \boldsymbol{\delta}\mathbf{x}, \boldsymbol{\lambda}^{(k)} + \boldsymbol{\delta\lambda}) = \blacktriangledown \mathscr{L}^{(k)} + [\blacktriangledown^2 \mathscr{L}^{(k)}]\begin{pmatrix} \boldsymbol{\delta}\mathbf{x} \\ \boldsymbol{\delta\lambda} \end{pmatrix} + \cdots \tag{12.4.1}$$

where $\blacktriangledown \mathscr{L}^{(k)} = \blacktriangledown \mathscr{L}(\mathbf{x}^{(k)}, \boldsymbol{\lambda}^{(k)})$, etc. Neglecting higher order terms and setting the left hand size to zero by virtue of (9.1.7) gives the iteration

$$[\blacktriangledown^2 \mathscr{L}^{(k)}]\begin{pmatrix} \boldsymbol{\delta}\mathbf{x} \\ \boldsymbol{\delta\lambda} \end{pmatrix} = -\blacktriangledown \mathscr{L}^{(k)}. \tag{12.4.2}$$

This is solved to give corrections $\boldsymbol{\delta}\mathbf{x}$ and $\boldsymbol{\delta\lambda}$ and is of course Newton's method for the stationary point problem. Formulae for $\blacktriangledown \mathscr{L}$ and $\blacktriangledown^2 \mathscr{L}$ are readily obtained from (9.1.6), giving the system

$$\begin{bmatrix} \mathbf{W}^{(k)} & -\mathbf{A}^{(k)} \\ -\mathbf{A}^{(k)\mathrm{T}} & \mathbf{0} \end{bmatrix}\begin{pmatrix} \boldsymbol{\delta}\mathbf{x} \\ \boldsymbol{\delta\lambda} \end{pmatrix} = \begin{pmatrix} -\mathbf{g}^{(k)} + \mathbf{A}^{(k)}\boldsymbol{\lambda}^{(k)} \\ \mathbf{c}^{(k)} \end{pmatrix}. \tag{12.4.3}$$

$\mathbf{A}^{(k)}$ is the Jacobian matrix of constraint normals evaluated at $\mathbf{x}^{(k)}$ and

$$\mathbf{W}^{(k)} = \nabla^2 f(\mathbf{x}^{(k)}) - \sum_i \lambda_i^{(k)} \nabla^2 c_i(\mathbf{x}^{(k)}) \tag{12.4.4}$$

is the Hessian matrix $\nabla_x^2 \mathscr{L}(\mathbf{x}^{(k)}, \boldsymbol{\lambda}^{(k)})$. In fact it is more convenient to write $\boldsymbol{\lambda}^{(k+1)} = \boldsymbol{\lambda}^{(k)} + \boldsymbol{\delta\lambda}$ and $\boldsymbol{\delta}^{(k)} = \boldsymbol{\delta}\mathbf{x}$, and to solve the equivalent system

$$\begin{bmatrix} \mathbf{W}^{(k)} & -\mathbf{A}^{(k)} \\ -\mathbf{A}^{(k)\mathrm{T}} & \mathbf{0} \end{bmatrix}\begin{pmatrix} \boldsymbol{\delta} \\ \boldsymbol{\lambda} \end{pmatrix} = \begin{pmatrix} -\mathbf{g}^{(k)} \\ \mathbf{c}^{(k)} \end{pmatrix} \tag{12.4.5}$$

to determine $\boldsymbol{\delta}^{(k)}$ and $\boldsymbol{\lambda}^{(k+1)}$. Then $\mathbf{x}^{(k+1)}$ is given by

$$\mathbf{x}^{(k+1)} = \mathbf{x}^{(k)} + \boldsymbol{\delta}^{(k)}. \tag{12.4.6}$$

The method requires initial approximations $\mathbf{x}^{(1)}$ and $\boldsymbol{\lambda}^{(1)}$, and uses (12.4.5) and (12.4.6) to generate the iterative sequence $\{\mathbf{x}^{(k)}, \boldsymbol{\lambda}^{(k)}\}$.

Just as with Newton's method in Section 3.1, it is possible to restate this method in terms of one in which the subproblem involves the minimization of a quadratic function. Consider the subproblem

$$\begin{aligned} &\underset{\boldsymbol{\delta}}{\text{minimize}} \quad q^{(k)}(\boldsymbol{\delta}) \\ &\text{subject to} \quad \mathbf{l}^{(k)}(\boldsymbol{\delta}) = \mathbf{0} \end{aligned} \tag{12.4.7}$$

where

$$q^{(k)}(\boldsymbol{\delta}) \triangleq \tfrac{1}{2}\boldsymbol{\delta}^{\mathrm{T}}\mathbf{W}^{(k)}\boldsymbol{\delta} + \mathbf{g}^{(k)\mathrm{T}}\boldsymbol{\delta} + f^{(k)} \tag{12.4.8}$$

and

$$\mathbf{l}^{(k)}(\boldsymbol{\delta}) \triangleq \mathbf{A}^{(k)\mathrm{T}}\boldsymbol{\delta} + \mathbf{c}^{(k)}. \tag{12.4.9}$$

From (10.2.1) and (10.2.2), first order conditions for (12.4.7) are given by equations (12.4.5) so that $\boldsymbol{\delta}^{(k)}$ is a stationary point of (12.4.7). If the Hessian matrix $\mathbf{W}^{(k)}$ is such that the reduced matrix $\mathbf{Z}^{(k)\mathrm{T}}\mathbf{W}^{(k)}\mathbf{Z}^{(k)}$ is positive definite (equivalently if $\boldsymbol{\delta}^{(k)}$ satisfies second order sufficient conditions for (12.4.7)), then $\boldsymbol{\delta}^{(k)}$ minimizes (12.4.7). (These conditions are ensured by continuity when $\mathbf{x}^{(k)}, \boldsymbol{\lambda}^{(k)}$ is sufficiently close to $\mathbf{x}^*, \boldsymbol{\lambda}^*$ and second order sufficient conditions hold at $\mathbf{x}^*, \boldsymbol{\lambda}^*$: see Question 12.17.) Thus the following iterative method is suggested, given initial estimates $\mathbf{x}^{(1)}$ and $\boldsymbol{\lambda}^{(1)}$.

For $k = 1, 2, \ldots$
(i) Solve (12.4.7) (or (12.4.11)) to determine $\boldsymbol{\delta}^{(k)}$ and let $\boldsymbol{\lambda}^{(k+1)}$ be the vector of Lagrange multipliers of the linear constraints. (12.4.10)
(ii) Set $\mathbf{x}^{(k+1)} = \mathbf{x}^{(k)} + \boldsymbol{\delta}^{(k)}$.

If a unique minimizer exists to (12.4.7) for all k then the iteration sequence so defined is identical to that given by the Lagrange–Newton method (12.4.5) and (12.4.6). In fact iteration (12.4.10) is preferable to (12.4.5) because the latter can cause convergence to a KT point of (12.1.1) which is not a minimizer. The situation is analogous to that in which Newton's method for unconstrained minimization is best interpreted as sequentially minimizing certain quadratic approximations $q^{(k)}(\boldsymbol{\delta})$ (Section 3.1).

The subproblem (12.4.7) bears a nice relationship to the original problem (12.1.1). The constraints in (12.4.7) are obtained by replacing the nonlinear constraints $\mathbf{c}(\mathbf{x}) = \mathbf{0}$ in (12.1.1) by their first order Taylor series approximation $\mathbf{l}^{(k)}(\boldsymbol{\delta}) = \mathbf{0}$ about $\mathbf{x}^{(k)}$, given by (12.4.9). Likewise the objective function $f(\mathbf{x})$ in (12.1.1) is replaced by the quadratic function $q^{(k)}(\boldsymbol{\delta})$ in (12.4.8). This is a second order Taylor series approximation about $\mathbf{x}^{(k)}$ but with the addition of constraint curvature terms in the Hessian. Including second order constraint terms in the subproblem is important in that otherwise second order convergence for nonlinear constraints would not be obtained. This is well illustrated by problem (12.1.5) in which the objective function is linear so that it is only the curvature of the constraint which causes a solution to exist. In this case the sequence determined by (12.4.10) is only well-defined if the constraint curvature terms are included.

This interpretation of (12.4.10) readily suggests a generalization for solving the nonlinear inequality constraint problem (12.1.2). Replacing $\mathbf{c}(\mathbf{x})$ by $\mathbf{l}^{(k)}(\boldsymbol{\delta})$ and $f(\mathbf{x})$ by $q^{(k)}(\boldsymbol{\delta})$ leads to the subproblem

$$\begin{aligned} &\underset{\boldsymbol{\delta}}{\text{minimize}} \quad q^{(k)}(\boldsymbol{\delta}) \\ &\text{subject to} \quad \mathbf{l}^{(k)}(\boldsymbol{\delta}) \geqslant \mathbf{0}. \end{aligned} \tag{12.4.11}$$

This can be used in an iterative framework like (12.4.10) in a similar way. Both (12.4.7) and (12.4.11) are *quadratic programming subproblems* and (12.4.10) has come to be known as the *sequential quadratic programming (SQP) method*. This type of method is first mentioned by Wilson (1963) as the basis of his SOLVER method (see Beale, 1967).

An alternative way of deriving (12.4.7) is to observe that second order sufficient conditions at $\mathbf{x}^*, \boldsymbol{\lambda}^*$ imply that $\mathbf{x}^*$ (that is $\boldsymbol{\delta}=\mathbf{0}$) solves the problem

$$\begin{aligned} &\underset{\boldsymbol{\delta}}{\text{minimize}} \quad \mathscr{L}(\mathbf{x}^* + \boldsymbol{\delta}, \boldsymbol{\lambda}^*) \\ &\text{subject to} \quad \mathbf{A}^{*\mathrm{T}}\boldsymbol{\delta} = \mathbf{0} \end{aligned} \tag{12.4.12}$$

because (9.3.5) and (9.3.4) ensure strictly positive curvature in the feasible region. Adding $0 = \boldsymbol{\lambda}^{*\mathrm{T}}\mathbf{A}^{*\mathrm{T}}\boldsymbol{\delta} = \mathbf{g}^{*\mathrm{T}}\boldsymbol{\delta}$ into the objective yields an equivalent problem

$$\begin{aligned} &\text{minimize} \quad \tfrac{1}{2}\boldsymbol{\delta}^{\mathrm{T}}\mathbf{W}^*\boldsymbol{\delta} + \mathbf{g}^{*\mathrm{T}}\boldsymbol{\delta} + f^* \\ &\text{subject to} \quad \mathbf{A}^{*\mathrm{T}}\boldsymbol{\delta} = \mathbf{0} \end{aligned} \tag{12.4.13}$$

which is also solved by $\boldsymbol{\delta}=\mathbf{0}$ and has Lagrange multipliers $\boldsymbol{\lambda}^*$. The constraints in (12.4.13) are linear approximations to $\mathbf{c}(\mathbf{x})=\mathbf{0}$ at $\mathbf{x}^*$ so by analogy, if $\mathbf{x}^{(k)}, \boldsymbol{\lambda}^{(k)}$ are approximations to $\mathbf{x}^*, \boldsymbol{\lambda}^*$, the solution of subproblem (12.4.7) is suggested. It follows clearly from this observation that the SQP method (12.4.10) has a fixed point property at $\mathbf{x}^*, \boldsymbol{\lambda}^*$. It also follows from this derivation that the Hessian $\mathbf{W}^{(k)}$ of $\mathscr{L}(\mathbf{x}, \boldsymbol{\lambda}^{(k)})$, and not $\mathbf{G}^{(k)}$, should be used in defining $q^{(k)}(\boldsymbol{\delta})$.

A numerical example which illustrates many of the features of the SQP method is given by the inequality constraint problem (9.1.15) from $\mathbf{x}^{(1)} = (\frac{1}{2}, 1)^{\mathrm{T}}$ and $\boldsymbol{\lambda}^{(1)} = \mathbf{0}$ (see Table 12.4.1). Since $\boldsymbol{\lambda}^{(1)} = \mathbf{0}$ no constraint curvature terms occur in $\mathbf{W}^{(1)}$, which is therefore the zero matrix since $f(\mathbf{x})$ is linear. Thus the initial subproblem is a linear programming problem and $\mathbf{x}^{(2)}$ is the vertex of the constraints linearized about $\mathbf{x}^{(1)}$. In fact, even though the constraint $c_1(\mathbf{x}) \geqslant 0$ is not active at the solution, the presence of its linearization is necessary to permit the first subproblem to be solvable. Moreover there exist different $\mathbf{x}^{(1)}$ for which the initial LP is unbounded, all of which illustrates that the well-behaved nature of (12.4.7) only holds necessarily in a neighbourhood of $\mathbf{x}^*, \boldsymbol{\lambda}^*$. In this case however the solution to the LP is well defined, and $\boldsymbol{\lambda}^{(2)} = (\frac{1}{3}, \frac{2}{3})^{\mathrm{T}}$ is the multiplier vector at its solution, indicating that both

Table 12.4.1 The SQP method applied to (9.1.15)

k	$x_1^{(k)}$	$x_2^{(k)}$	$\lambda_1^{(k)}$	$\lambda_2^{(k)}$	$c_1^{(k)}$	$c_2^{(k)}$
1	$\frac{1}{2}$	1	0	0	$\frac{3}{4}$	$-\frac{1}{4}$
2	$\frac{11}{12}$	$\frac{2}{3}$	$\frac{1}{3}$	$\frac{2}{3}$	-0.173611	-0.284722
3	0.747120	0.686252	0	0.730415	0.128064	-0.029130
4	0.708762	0.706789	0	0.706737	0.204445	-0.001893
5	0.707107	0.707108	0	0.707105	0.207108	$-0.28_{10}-5$

linearized constraints are active. Thus for the second iteration

$$\mathbf{W}^{(2)} = \mathbf{0} - \tfrac{1}{3}\begin{bmatrix} -2 & \\ & 0 \end{bmatrix} - \tfrac{2}{3}\begin{bmatrix} -2 & \\ & -2 \end{bmatrix} = \begin{bmatrix} 2 & \\ & \tfrac{4}{3} \end{bmatrix}$$

which is nicely positive definite. Solving the resulting QP problem causes $l_1^{(2)}(\boldsymbol{\delta}) \geqslant 0$ to become inactive so that $\lambda_1^{(3)} = 0$. Thus the correct active set is established, and the rapid convergence associated with Newton's method is observed on subsequent iterations. In comparing this with Table 12.2.1 say, it should be remembered that in the latter, each iteration requires an unconstrained minimization calculation and therefore many evaluations of the problem functions and their derivatives. In contrast, the SQP method only requires one evaluation of the problem functions and derivatives to determine the coefficients of the finite quadratic programming subproblem. Thus the SQP method, if it works, is generally considerably superior in terms of the number of function and derivative evaluations which are required.

An important feature of the method which Table 12.4.1 illustrates is that ultimately the convergence is second order. If second order sufficient conditions for the equality constraint problem (12.1.1) hold at $\mathbf{x}^*, \boldsymbol{\lambda}^*$, and if rank $\mathbf{A}^* = m$, then the Lagrangian matrix

$$\nabla^2 \mathscr{L}^* = \begin{bmatrix} \mathbf{W}^* & -\mathbf{A}^* \\ -\mathbf{A}^{*T} & \mathbf{0} \end{bmatrix} \tag{12.4.14}$$

is non-singular (see Question 12.4). The second order convergence of iteration (12.4.5) and (12.4.6) then follows by virtue of Theorem 6.2.1 applied to the system of $n+m$ equations $\nabla \mathscr{L}(\mathbf{x}, \boldsymbol{\lambda}) = \mathbf{0}$. This requires both $\mathbf{x}^{(k)}$ and $\boldsymbol{\lambda}^{(k)}$ to be sufficiently close to $\mathbf{x}^*$ and $\boldsymbol{\lambda}^*$ for some k. In fact the multiplier estimates $\boldsymbol{\lambda}^{(k)}$ play a relatively minor role, in that they only arise in the second order term involving $\mathbf{W}^{(k)}$, and this can be exploited to give a stronger result.

Theorem 12.4.1

If $\mathbf{x}^{(1)}$ is sufficiently close to $\mathbf{x}^*$, if the Lagrangian matrix $\begin{bmatrix} \mathbf{W}^{(1)} & -\mathbf{A}^{(1)} \\ -\mathbf{A}^{(1)T} & \mathbf{0} \end{bmatrix}$ is non-singular, and if second order sufficient conditions hold at $\mathbf{x}^*, \boldsymbol{\lambda}^*$ with rank $\mathbf{A}^* = m$, then the Lagrange–Newton iteration (12.4.5) *and* (12.4.6) *converges and the order is second order. If $\boldsymbol{\lambda}^{(1)}$ is such that* (12.4.7) *is solved uniquely by $\boldsymbol{\delta}^{(1)}$ then the same is true for the SQP method.*

Proof

Define errors $\mathbf{h}^{(k)} = \mathbf{x}^{(k)} - \mathbf{x}^*$ and $\boldsymbol{\Delta}^{(k)} = \boldsymbol{\lambda}^{(k)} - \boldsymbol{\lambda}^*$, and assume that f and the c_i are $\mathbb{C}^2$ and the elements of their Hessian matrices satisfy Lipschitz conditions, so that the Taylor series about $\mathbf{x}^{(k)}$

$$\mathbf{c}^* = \mathbf{c}^{(k)} - \mathbf{A}^{(k)T}\mathbf{h}^{(k)} + O(\|\mathbf{h}^{(k)}\|^2)$$

$$\mathbf{g}^* = \mathbf{g}^{(k)} - \nabla^2 f^{(k)}\mathbf{h}^{(k)} + O(\|\mathbf{h}^{(k)}\|^2)$$
$$\mathbf{a}_i^* = \mathbf{a}_i^{(k)} - \nabla^2 c_i^{(k)}\mathbf{h}^{(k)} + O(\|\mathbf{h}^{(k)}\|^2), \qquad i = 1, 2, \ldots, m,$$

are valid. It follows from (12.4.5) that $\mathbf{h}^{(k+1)}$, $\boldsymbol{\Delta}^{(k+1)}$ satisfy the equations

$$\begin{bmatrix} \mathbf{W}^{(k)} & -\mathbf{A}^{(k)} \\ -\mathbf{A}^{(k)\mathrm{T}} & \mathbf{0} \end{bmatrix}\begin{pmatrix} \mathbf{h}^{(k+1)} \\ \boldsymbol{\Delta}^{(k+1)} \end{pmatrix} = \begin{pmatrix} -\sum_i \Delta_i^{(k)}\nabla^2 c_i^{(k)}\mathbf{h}^{(k)} + O(\|\mathbf{h}^{(k)}\|^2) \\ O(\|\mathbf{h}^{(k)}\|^2) \end{pmatrix}$$
$$= \begin{pmatrix} O(\|\mathbf{h}^{(k)}\|^2) + O(\|\mathbf{h}^{(k)}\|\,\|\boldsymbol{\Delta}^{(k)}\|) \\ O(\|\mathbf{h}^{(k)}\|^2) \end{pmatrix}. \tag{12.4.15}$$

At $\mathbf{x}^*, \boldsymbol{\lambda}^*$ the Lagrangian matrix is non-singular (see Question 12.4) so for $\mathbf{x}^{(k)}$, $\boldsymbol{\lambda}^{(k)}$ in some neighbourhood of $\mathbf{x}^*, \boldsymbol{\lambda}^*$,

$$\begin{pmatrix} \mathbf{h}^{(k+1)} \\ \boldsymbol{\Delta}^{(k+1)} \end{pmatrix} = O(\|\mathbf{h}^{(k)}\|^2) + O(\|\mathbf{h}^{(k)}\|\,\|\boldsymbol{\Delta}^{(k)}\|)$$

and so there exists a constant $c > 0$ such that

$$\max(\|\mathbf{h}^{(k+1)}\|, \|\boldsymbol{\Delta}^{(k+1)}\|) \leqslant c\|\mathbf{h}^{(k)}\| \max(\|\mathbf{h}^{(k)}\|, \|\boldsymbol{\Delta}^{(k)}\|). \tag{12.4.16}$$

Thus, in a smaller neighbourhood, if $1 > c\max(\|\mathbf{h}^{(k)}\|, \|\boldsymbol{\Delta}^{(k)}\|) = \alpha$, say, then

$$\max(\|\mathbf{h}^{(k+1)}\|, \|\boldsymbol{\Delta}^{(k+1)}\|) \leqslant \alpha\|\mathbf{h}^{(k)}\| \leqslant \alpha \max(\|\mathbf{h}^{(k)}\|, \|\boldsymbol{\Delta}^{(k)}\|),$$

so the iteration converges and the order is seen to be quadratic from (12.4.16). Now let only $\mathbf{x}^{(1)}$ be in a neighbourhood of $\mathbf{x}^*$, so that $\mathbf{A}^{(1)}$ has full rank, and let $\boldsymbol{\lambda}^{(1)}$ be such that the Lagrangian matrix is non-singular. Then $\|\boldsymbol{\Delta}^{(1)}\| \geqslant \|\mathbf{h}^{(1)}\|$ and so as above there exists a constant, d say, such that

$$\max(\|\mathbf{h}^{(2)}\|, \|\boldsymbol{\Delta}^{(2)}\|) \leqslant d\|\mathbf{h}^{(1)}\|\,\|\boldsymbol{\Delta}^{(1)}\|.$$

If $\mathbf{x}^{(1)}$ is sufficiently close to $\mathbf{x}^*$ in that $\|\mathbf{h}^{(1)}\| < 1/(cd\|\boldsymbol{\Delta}^{(1)}\|)$ then $\max(\|\mathbf{h}^{(2)}\|, \|\boldsymbol{\Delta}^{(2)}\|) < 1/c$ and so $\mathbf{x}^{(2)}, \boldsymbol{\lambda}^{(2)}$ is in the neighbourhood for which convergence occurs.

The Lagrange–Newton method is equivalent to the SQP method if $\boldsymbol{\delta}^{(k)}$ in the solution to the former is the unique solution of (12.4.7) for all k. This is assured by continuity of second order sufficient conditions when $\boldsymbol{\lambda}^{(k)}$ is sufficiently close to $\boldsymbol{\lambda}^*$ (see Question 12.17) and this is true for $k \geqslant 2$ by the above discussion. However, it is necessary to make the *a priori* assumption that $\boldsymbol{\lambda}^{(1)}$ is suitably chosen so as to deduce the last part of the theorem. □

Essentially the minor role played by $\boldsymbol{\lambda}^{(k)}$ is illustrated by the fact that $\|\boldsymbol{\Delta}^{(k)}\|$ only occurs linearly on the right-hand side of (12.4.15), and it is this fact that is exploited in the theorem. For example, if $\mathbf{x}^{(1)} = \mathbf{x}^*$, then $\mathbf{x}^{(2)} = \mathbf{x}^*$ and $\boldsymbol{\lambda}^{(2)} = \boldsymbol{\lambda}^*$ irrespective of errors in $\boldsymbol{\lambda}^{(1)}$. This suggests that when using the method, it is more important to have $\mathbf{x}^{(1)}$ accurate than $\boldsymbol{\lambda}^{(1)}$. This is in contrast to the multiplier penalty function method of Section 12.2 where an inaccurate value of $\boldsymbol{\lambda}^{(1)}$ definitely limits the extent to which $\mathbf{x}(\boldsymbol{\lambda}^{(1)})$ agrees with $\mathbf{x}^*$, assuming $\boldsymbol{\sigma}$

is fixed. There is no difficulty in extending the theorem to handle inequality constraints and the details are sketched out in Question 12.18.

A possible disadvantage of the SQP method is the need to compute second derivatives to calculate the matrix $\mathbf{W}^{(k)}$. Variations of the method have been suggested in which updating formulae, analogous to those used in quasi-Newton methods, are used to revise a matrix $\mathbf{B}^{(k)}$ which approximates $\mathbf{W}^{(k)}$. Han (1976) suggests using the DFP formula (see Section 3.2) but with $\gamma^{(k)}$ being defined by

$$\gamma^{(k)} = \nabla \mathscr{L}(\mathbf{x}^{(k+1)}, \boldsymbol{\lambda}^{(k+1)}) - \nabla \mathscr{L}(\mathbf{x}^{(k)}, \boldsymbol{\lambda}^{(k+1)}) \tag{12.4.17}$$

and shows that the resulting algorithm is superlinearly convergent. Powell (1978a) prefers to use the BFGS formula on account of its success in solving unconstrained minimization problems. He also prefers to keep the matrix $\mathbf{B}^{(k)}$ positive definite so that the solution to the subproblem is always well defined. This is done by defining the vector

$$\boldsymbol{\eta}^{(k)} = \theta\gamma^{(k)} + (1-\theta)\mathbf{B}^{(k)}\boldsymbol{\delta}^{(k)}, \qquad 0 \leqslant \theta \leqslant 1 \tag{12.4.18}$$

($\gamma^{(k)}$ as in (12.4.17)) that is closest to $\gamma^{(k)}$ subject to the condition that

$$\boldsymbol{\delta}^{(k)\mathrm{T}}\boldsymbol{\eta}^{(k)} \geqslant 0.2\boldsymbol{\delta}^{(k)\mathrm{T}}\mathbf{B}^{(k)}\boldsymbol{\delta}^{(k)}.$$

$\boldsymbol{\eta}^{(k)}$ is then used in place of $\gamma^{(k)}$ in the updating formula. It might be thought that this device is somewhat artificial since the matrix $\mathbf{W}^*$ may not be positive definite, so that $\mathbf{B}^{(k)}$ may never be a close approximation to $\mathbf{W}^*$. However the projections of $\mathbf{B}^{(k)}$ and $\mathbf{W}^{(k)}$ into the tangent hyperplanes of the active constraints are likely to be close, and this is the important part of $\mathbf{B}^{(k)}$ insofar as the solution of the subproblem is concerned. Powell (1978b) is able to exploit this observation to prove superlinear convergence even when $\mathbf{W}^*$ is indefinite.

This quasi-Newton version of SQP has proved successful in practice on small and medium scale problems. There is, however, the difficulty that the matrix $\mathbf{B}^{(k)}$ is dense and the method does not therefore exploit any sparsity that is present in the matrix $\mathbf{W}^{(k)}$. This limits the size of problem that can be handled effectively. This point is considered further at the end of the section where it is shown that current research gives some hope for improvements.

These results show that the local properties of the SQP method are very satisfactory, so that the main difficulty which exists is the fact that the method may fail to converge remote from the solution, and that the solution to the subproblem ((12.4.7) or (12.4.11)) may not even exist (it may either be unbounded or infeasible). To induce global convergence requires some measure of goodness of $\mathbf{x}^{(k)}, \boldsymbol{\lambda}^{(k)}$ to be available which is minimized locally at the solution. One possibility is to introduce the error in the KT conditions

$$\phi(\mathbf{x}, \boldsymbol{\lambda}) = \|\mathbf{c}\|_2^2 + \|\mathbf{g} - \mathbf{A}\boldsymbol{\lambda}\|_2^2$$

where $\mathbf{c} = \mathbf{c}(\mathbf{x})$ etc., but this is not entirely satisfactory in that it does not give any bias to minimizing the objective function and can therefore cause convergence to any KT point to occur. A more suitable idea is to use an exact penalty function, that is a function $\phi(\mathbf{x})$, defined in terms of $f(\mathbf{x}), \mathbf{c}(\mathbf{x})$ and possibly derivatives thereof, which is minimized locally by $\mathbf{x}^*$. A popular choice has been the L_1 exact penalty function described in Section 12.3. Han (1977) suggests using the correction $\boldsymbol{\delta}$ defined by (12.4.5) (with $\mathbf{W}^{(k)}$ approximated by $\mathbf{B}^{(k)}$) as a direction of search, along which an inexact line search is carried out, using the L_1 exact penalty function as a criterion function. Han shows that if $\nu < 1/\|\boldsymbol{\lambda}^{(k+1)}\|_\infty$ then $\boldsymbol{\delta}$ is a descent direction for this function. He gives a global convergence result assuming that the matrices $\mathbf{B}^{(k)}$ are positive definite and $\mathbf{B}^{(k)}$ and $\mathbf{B}^{(k)-1}$ are bounded. He also assumes that the multipliers $\boldsymbol{\lambda}^{(k)}$ are bounded together with another assumption which excludes in inconsistent linearization which can occur when the vectors $\mathbf{a}_i^\infty i\in\mathscr{A}^\infty$ are dependent. Powell (1978a) also uses a line search with an L_1 exact penalty function, and changes the penalty parameters ($\boldsymbol{\sigma}$ and $\boldsymbol{\rho}$ in (12.3.2)) in order to obtain descent. Good practical results have been obtained with these algorithms on a number of standard test problems which substantiates to some extent the claim that they provide a robust practical realization of the SQP method. Table 12.4.2 compares the number of function and gradient evaluations required to solve Colville's (1968) first three problems and gives some idea of the relative improvement that is obtained. Even allowing for the fact that a QP calculation is required on each iteration, the improvement is still substantial. In fact if the QP software is able to use the fact that the active set on one iteration is likely to be a good estimate of that for the next iteration, then considerable savings are likely to be made in the time required for this part of the calculation.

Nonetheless difficulties do arise with the SQP method, not only when the subproblem is infeasible or unbounded, but also in a situation in which $\{\mathbf{x}^{(k)}\}$ approaches a limit point $\mathbf{x}^\infty$ (not necessarily a KT point) and the active constraint gradients $\mathbf{a}_i^\infty i\in\mathscr{A}^\infty$ are dependent. In this case the multipliers $\boldsymbol{\lambda}^{(k+1)}$ are not bounded, which can have serious implications. The requirement to get descent forces $\nu\downarrow 0$, and also the matrices $\mathbf{B}^{(k)}$ can become unbounded through the dependence of the updating formulae on $\boldsymbol{\lambda}^{(k+1)}$ (see (12.4.17)). Thus Chamberlain (1979) reports failure of Powell's algorithm, Powell (1977c) finds some difficulty in solving the Dembo 7 problems (as also do Coope and Fletcher (1980) with an augmented Lagrangian method) and Fletcher (1981) also gives two problems on which the SQP method with a L_1 line search fails. The situation is analogous to that in Section 6.2 in which the Newton–Raphson method with line search can fail when $\mathbf{A}^\infty$ is rank deficient (Powell, 1970b).

The most successful way of inducing global convergence for unconstrained

Table 12.4.2 Comparison of nonlinear programming techniques

Problem	Extrapolated barrier function	Multiplier penalty function	SQP method (Powell, 1978a)
TP1	177	47	6
TP2	245	172	17
TP3	123	73	3

minimization is arguably the restricted step or trust region approach or the use of a Levenberg–Marquardt parameter (Chapter 5). Both these ideas have been suggested in the context of SQP-like methods (in fact a step restriction does occur in Beale's (1967) description of the SOLVER method). Modifying $\mathbf{W}^{(k)}$ by adding a Marquardt–Levenberg term $\nu\mathbf{I}$ may help to ensure that (12.4.7) or (12.4.11) is not unbounded, but no longer has the effect of giving a bias towards steepest descent. In fact for a nonlinear equations problem ($m=n$) the correction $\boldsymbol{\delta}$ is unchanged, so the modification has no effect and therefore does not provide a guarantee of convergence. If a step length restriction $\|\boldsymbol{\delta}\| \leqslant h^{(k)}$ is added to (12.4.7) or (12.4.11) then the possibility of an unbounded correction is entirely removed. In this case the difficulty is that if $\mathbf{x}^{(k)}$ is infeasible and $h^{(k)}$ is sufficiently small then the resulting subproblem has no feasible point. Thus the idea in Section 5.1 of forcing convergence by reducing $h^{(k)}$ if necessary so as to give a descent step, is no longer valid. Therefore the straightforward generalization of the SQP method in this way is unsatisfactory. However, Fletcher (1981) shows how to use a step restriction (or trust region) so that all the above difficulties are removed. The method is easily explained: instead of substituting the Taylor series approximations (12.4.8) and (12.4.9) into the nonlinear programming problems ((7.1.1) in the general case), they are substituted directly into the L_1 exact penalty function (12.3.1), giving a piecewise quadratic approximating function $\psi^{(k)}(\boldsymbol{\delta})$ and hence a subproblem

$$\begin{aligned} \underset{\boldsymbol{\delta}}{\text{minimize}} \quad & \psi^{(k)}(\boldsymbol{\delta}) \triangleq \nu q^{(k)}(\boldsymbol{\delta}) + \sum_{i \in E} |l_i^{(k)}(\boldsymbol{\delta})| + \sum_{i \in I} (l_i^{(k)}(\boldsymbol{\delta}))^- \\ \text{subject to} \quad & \|\boldsymbol{\delta}\|_\infty \leqslant h^{(k)}. \end{aligned} \tag{12.4.19}$$

This subproblem is of a similar complexity to the QP subproblem (12.4.11), and in fact is an example of the $\mathrm{L_1QP}$ problem described in Section 12.3. Subproblem (12.4.19) differs from (12.4.11) in that there are no explicit constraints derived from the linear approximations $\mathbf{l}^{(k)}(\boldsymbol{\delta})$. Thus there are no difficulties with an infeasible subproblem. The use of a trust region guarantees boundedness of the subproblem: in practice the variables should be rescaled to ensure that the use of the L_∞ norm is realistic. The radius $h^{(k)}$ of the trust region is adjusted adaptively in a similar way to that described in Section 5.1. In fact this algorithm is a special case of the SNQP algorithm with trust region described in Section 14.5 in which the L_1 terms are represented by the composite function $h(\mathbf{l}^{(k)}(\boldsymbol{\delta}))$, and more details can be obtained by referring to (14.5.6). In the context of the L_1 exact penalty function, this algorithm is referred to as the *sequential L_1QP* (or $\mathrm{SL_1QP}$) method.

Globally the properties of $\mathrm{SL_1QP}$ are much superior to SQP. The multipliers $\boldsymbol{\lambda}^{(k+1)}$ are bounded by virtue of (12.3.16) and (12.3.9b, c) and this can be exploited as in Section 14.5 to prove global convergence to a stationary point, with no a priori assumptions on linear independence of the vectors $\mathbf{a}_i^{(k)}$. Preliminary results (Fletcher, 1981) are favourable and the method solves problems for which SQP fails. A quasi-Newton variant is possible which updates an approximation $\mathbf{B}^{(k)}$ to $\mathbf{W}^{(k)}$ as for SQP. Asymptotically the method is equivalent to SQP and it shares the same local convergence properties. Far from the solution however the methods differ and the trust region is seen to play an important part. Because the linearized constraint functions $l_i^{(k)}(\boldsymbol{\delta})$ appear only in the penalty term in (12.4.19), not all of these may be zeroed by the solution of the subproblem—even if they are equality constraints. Those that are zeroed can be interpreted as *locally active constraints.* The fact that this set is a usually subset of the active constraints in an SQP calculation gives some indication as to why dependent constraint gradients and unbounded multipliers do not give rise to difficulties in $\mathrm{SL_1QP}$, in contrast to SQP.

One adverse feature of using the L_1 exact penalty function (both in SQP and $\mathrm{SL_1QP}$) is the lack of smoothness caused by the L_1 terms. The derivative discontinuities give rise to 'grooves' in the surface and if they are curved and steep-sided they become difficult for an algorithm to follow. For example consider the equations

$$\begin{aligned} c_1(\mathbf{x}) &= \sigma(x_2 - x_1^2) \\ c_2(\mathbf{x}) &= 1 - x_1 \end{aligned} \tag{12.4.20}$$

derived from Rosenbrock's problem (1.2.2). The L_1 exact penalty function $\varphi(\mathbf{x}) = \|\mathbf{c}(\mathbf{x})\|_1$ has a groove along the parabola $x_2 = x_1^2$ and increasing σ increases the jump discontinuity in derivative across the groove. The performance of the $\mathrm{SL_1QP}$ method from the standard starting point $\mathbf{x}^{(1)} = (-1.2, 1)$ is given in Table 12.4.3. It is clearly seen that the performance deteriorates as the steepness factor σ increases. This occurs because the size of step along the linearized discontinuity that can reduce $\varphi(\mathbf{x})$ is progressively reduced, and this is also seen in the table. Another consequence of derivative discontinuities is the *Maratos effect* in which $\mathbf{x}^{(k)}, \boldsymbol{\lambda}^{(k)}$ may be arbitrarily close to $\mathbf{x}^*, \boldsymbol{\lambda}^*$ but a unit step of the SQP or $\mathrm{SL_1QP}$ method may fail to reduce the L_1 exact penalty function. The effect is illustrated in Figure 14.4.5 and is discussed in a more general setting in Section 14.4. It is shown there that the effect can be avoided by recalculating the step $\boldsymbol{\delta}^{(k)}$ after making a correction for second order errors that arise (the *second order correction* (SOC) step). Apart from an additional evaluation of the vector $\mathbf{c}(\mathbf{x})$, this can be done with negligible extra complication and cost.

Currently I regard $\mathrm{SL_1QP}$ as the most promising algorithm for nonlinear programming, particularly when second derivatives are available, and my

Table 12.4.3 Slow convergence caused by derivative discontinuities

σ	10	100	1000
Number of iterations	12	193	~2000
Typical trust region radius	0.25	0.008	0.0007

impression is that the SOC step adequately handles the difficulties caused by the non-smooth penalty function. However, some researchers view these difficulties in a more serious light and have considered using the SQP step in conjunction with a smooth exact penalty function. Schittkowski (1983) and Gill *et al.* (1986) consider using the augmented Lagrangian function $\varphi(\mathbf{x}, \boldsymbol{\lambda}, \boldsymbol{\sigma})$ described in Section 12.2. However, this is only an exact penalty function if $\boldsymbol{\lambda} = \boldsymbol{\lambda}^*$ is chosen, whereas an algorithm must necessarily work with an estimate $\boldsymbol{\lambda}^{(k)}$ of $\boldsymbol{\lambda}^*$. Any changes to $\boldsymbol{\lambda}^{(k)}$ redefine the penalty function, so it is not immediately clear that this can be used as a merit function in the same way say as the L_1 exact penalty function. Nonetheless the resulting method does seem capable of solving test problems effectively. However, Powell (1985c) shows that the algorithm runs into difficulties in the presence of near-linear dependence in the active constraint gradients. Boggs and Tolle (1984) and Powell and Yuan (1986) use the SQP method in conjunction with the smooth exact penalty function of Fletcher (1973) which is described in Section 12.6. These techniques are of some interest but it is as yet too early to assess what their impact will be.

Much research has recently been conducted into the quasi-Newton aspects of the SQP method, and this is summarized in the rest of this section. In particular there is now a much greater understanding of the extent to which $\mathbf{B}^{(k)}$ should approximate $\mathbf{W}^*$ in order to get rapid convergence. This theory is relevant only to the case of an equality constaint problem, so the rest of the section concentrates on this case. Assume therefore that (12.4.7) is used, with $\mathbf{B}^{(k)}$ replacing $\mathbf{W}^{(k)}$, so that the SQP subproblem essentially solves

$$\begin{bmatrix} \mathbf{B}^{(k)} & -\mathbf{A}^{(k)} \\ -\mathbf{A}^{(k)\mathrm{T}} & \mathbf{0} \end{bmatrix} \begin{pmatrix} \boldsymbol{\delta}^{(k)} \\ \boldsymbol{\lambda}^{(k+1)} \end{pmatrix} = \begin{pmatrix} -\mathbf{g}^{(k)} \\ \mathbf{c}^{(k)} \end{pmatrix} \tag{12.4.21}$$

using the methods of Sections 10.1 or 10.2. Since only asymptotic properties are being studied, it is assumed that the unit step given by (12.4.6) is taken, and $\boldsymbol{\lambda}^{(k+1)}$ is ignored, or may possibly be used to update $\mathbf{B}^{(k)}$. Let $\mathbf{Z} = \mathbf{Z}(\mathbf{x})$ be any null space matrix for $\mathbf{A}$ ($\mathbf{A}^{\mathrm{T}}\mathbf{Z} = \mathbf{0}$) such that $\mathbf{Z}(\mathbf{x})$ is $\mathbb{C}^1$ in a neighbourhood of $\mathbf{x}^*$. Then (12.4.21) can be written as

$$\begin{bmatrix} \mathbf{Z}^{(k)\mathrm{T}}\mathbf{B}^{(k)} \\ \mathbf{A}^{(k)\mathrm{T}} \end{bmatrix} \boldsymbol{\delta}^{(k)} = -\begin{pmatrix} \mathbf{Z}^{(k)\mathrm{T}}\mathbf{g}^{(k)} \\ \mathbf{c}^{(k)} \end{pmatrix}. \tag{12.4.22}$$

By virtue of the vector on the right, this can be regarded as a quasi-Newton method for the $n \times n$ system of nonlinear equations

$$\begin{pmatrix} \mathbf{Z}^{\mathrm{T}}\mathbf{g} \\ \mathbf{c} \end{pmatrix} = \mathbf{0} \tag{12.4.23}$$

($\mathbf{g} = \mathbf{g}(\mathbf{x})$ etc.) which is a multiplier-free realization of the KT conditions. Writing $\mathbf{g}^{\mathrm{T}}\mathbf{Z} = (\mathbf{g} - \mathbf{A}\boldsymbol{\lambda})^{\mathrm{T}}\mathbf{Z}$, operating with ∇, and noting that $(\mathbf{g} - \mathbf{A}\boldsymbol{\lambda})^* = \mathbf{0}$, it follows that

$$\nabla(\mathbf{g}^{\mathrm{T}}\mathbf{Z})^* = \mathbf{W}^*\mathbf{Z}^*. \tag{12.4.24}$$

Therefore the Dennis–Moré theorem (Theorem 6.2.3) gives a characterization of superlinear convergence for the SQP method as

$$\lim \frac{\left\| \left\{ \begin{bmatrix} \mathbf{Z}^{(k)\mathrm{T}}\mathbf{B}^{(k)} \\ \mathbf{A}^{(k)\mathrm{T}} \end{bmatrix} - \begin{bmatrix} \mathbf{Z}^{*\mathrm{T}}\mathbf{W}^{*} \\ \mathbf{A}^{*\mathrm{T}} \end{bmatrix} \right\} \boldsymbol{\delta}^{(k)} \right\|}{\| \boldsymbol{\delta}^{(k)} \|} = 0. \tag{12.4.25}$$

From the continuity of **A** and **Z**, this is more simply expressed as

$$\lim \frac{\| \mathbf{Z}^{(k)\mathrm{T}} (\mathbf{B}^{(k)} - \mathbf{W}^{*}) \boldsymbol{\delta}^{(k)} \|}{\| \boldsymbol{\delta}^{(k)} \|} = 0. \tag{12.4.26}$$

This is essentially the result of Boggs, Tolle and Wang (1982), although this simple derivation is similar to that of Tapia (1984) who gives other references. This result implies for example that it is sufficient for superlinear convergence to have $\mathbf{Z}^{(k)\mathrm{T}}(\mathbf{B}^{(k)} - \mathbf{W}^{*}) \to \mathbf{0}$ and that it is not necessary that $\mathbf{B}^{(k)} \to \mathbf{W}^{*}$.

Equation (12.4.21) can also be rearranged in a different way using the matrices **Y** and **Z** of Section 10.1 (e.g. (10.1.20)) and expression (10.1.15) giving

$$\boldsymbol{\delta}^{(k)} = -\mathbf{Z}^{(k)}(\mathbf{Z}^{(k)\mathrm{T}}\mathbf{B}^{(k)}\mathbf{Z}^{(k)})^{-1}(\mathbf{Z}^{(k)\mathrm{T}}\mathbf{g}^{(k)} - \mathbf{Z}^{(k)\mathrm{T}}\mathbf{B}^{(k)}\mathbf{Y}^{(k)}\mathbf{c}^{(k)}) - \mathbf{Y}^{(k)}\mathbf{c}^{(k)} \tag{12.4.27}$$

This shows the need for the reduced matrices $\mathbf{Z}^{(k)\mathrm{T}}\mathbf{B}^{(k)}\mathbf{Z}^{(k)}$ and $\mathbf{Z}^{(k)\mathrm{T}}\mathbf{B}^{(k)}\mathbf{Y}^{(k)}$ but not $\mathbf{Y}^{(k)\mathrm{T}}\mathbf{B}^{(k)}\mathbf{Y}^{(k)}$, and this reinforces the result in (12.4.26). However, if $\mathbf{x}^{(k)}$ is a feasible point, so that $\mathbf{c}^{(k)} = \mathbf{0}$, it follows from (12.4.27) that only the $(n-m) \times (n-m)$ *reduced approximate Hessian matrix*,

$$\mathbf{M}^{(k)} = \mathbf{Z}^{(k)\mathrm{T}}\mathbf{B}^{(k)}\mathbf{Z}^{(k)} \tag{12.4.28}$$

say, is required to compute $\boldsymbol{\delta}^{(k)}$. When quasi-Newton algorithms for linear constraints are considered, it has already been observed (e.g. (11.1.11)) that this amount of information is sufficient. Several recent publications (e.g. Womersley, 1981; Coleman and Conn, 1982a,b; Nocedal and Overton, 1985; Womersley and Fletcher, 1986) have considered using only $\mathbf{M}^{(k)}$ in the context of nonlinear constraints, essentially by setting $\mathbf{Z}^{(k)}\mathbf{B}^{(k)}\mathbf{Y}^{(k)} = \mathbf{0}$ in (12.4.27). It is particularly attractive to do this if n is large, **A** is sparse and $n - m \ll n$ because then it can provide a substantial reduction in the storage capacity required. Powell (1978b) gives the following argument to show that not a great deal is lost in neglecting the term involving $\mathbf{Z}^{(k)}\mathbf{B}^{(k)}\mathbf{Y}^{(k)}$ in (12.4.27). Let $\mathbf{M}^{(k)}$ be close to $\mathbf{Z}^{*\mathrm{T}}\mathbf{W}^{*}\mathbf{Z}^{*}$ and let $\mathbf{x}^{(k)}$ be an arbitrary point in a neighbourhood of $\mathbf{x}^{*}$. On iteration k the term $-\mathbf{Y}^{(k)}\mathbf{c}^{(k)}$ is a Newton step on to the constraints which causes the constraints to be satisfied to second order, whilst the term in the range of $\mathbf{Z}^{(k)}$ does not affect this property since $\mathbf{A}^{\mathrm{T}}\mathbf{Z} = \mathbf{0}$. Therefore on iteration $k+1$ the vector $\mathbf{c}^{(k+1)}$ is negligible and hence the step $\boldsymbol{\delta}^{(k+1)}$ is arbitrarily close to the step given by the quasi-Newton SQP method (12.4.27). Thus the iterates can exhibit *2-step superlinear convergence* in which $\|\mathbf{x}^{(k+2)} - \mathbf{x}^{*}\| / \|\mathbf{x}^{(k)} - \mathbf{x}^{*}\| \to 0$ and this is satisfactory for all practical purposes. Powell (1978b) shows that a suitable definition of closeness in the approximate reduced Hessian matrix is

$$\|(\mathbf{M}^{(k)} - \mathbf{Z}^{(k)\mathrm{T}}\mathbf{W}^*\mathbf{Z}^{(k)})\mathbf{V}^{(k)\mathrm{T}}\boldsymbol{\delta}^{(k)}\| / \|\boldsymbol{\delta}^{(k)}\| \to 0, \tag{12.4.29}$$

and that this is sufficient to ensure 2-step superlinear convergence.

The effect of the term—$\mathbf{Y}^{(k)}\mathbf{c}^{(k)}$ is very similar to that of the SOC step referred to earlier in this section (see also (12.5.3)), and Coleman and Conn (1982a, b) suggest an alternative algorithm in which the SOC step is calculated. They define $\boldsymbol{\delta}^{(k)}$ by $\boldsymbol{\delta}^{(k)} = \mathbf{h}^{(k)} + \mathbf{v}^{(k)}$ where

$$\begin{aligned} \mathbf{h}^{(k)} &= -\mathbf{Z}^{(k)}\mathbf{M}^{(k)-1}\mathbf{Z}^{(k)\mathrm{T}}\mathbf{g}^{(k)} \\ \mathbf{v}^{(k)} &= -\mathbf{Y}^{(k)}\mathbf{c}(\mathbf{x}^{(k)} + \mathbf{h}^{(k)}) \end{aligned} \tag{12.4.30}$$

are the 'horizontal' step and 'vertical' (SOC) step respectively. The use of $\mathbf{c}(\mathbf{x}^{(k)} + \mathbf{h}^{(k)})$ in (12.4.30) in place of $\mathbf{c}^{(k)}$ (cf. (12.4.27) with $\mathbf{Z}^{(k)}\mathbf{B}^{(k)}\mathbf{Y}^{(k)} = \mathbf{0}$) gives the Coleman–Conn iteration both a 2-step superlinear convergence property as above, and a 1-step superlinear convergence property involving the errors at $\mathbf{x}^{(k)} + \mathbf{h}^{(k)}$ (Byrd, 1984), assuming that (12.4.29) holds.

Algorithms based only on using $\mathbf{M}^{(k)}$ also need to update $\mathbf{M}^{(k)}$ after each iteration based on changes in first derivatives. Because $\mathbf{Z}^{*\mathrm{T}}\mathbf{W}^*\mathbf{Z}^*$ is positive definite if second order sufficient conditions hold at $\mathbf{x}^*$, it is quite natural to attempt to impose this condition on $\mathbf{M}^{(k)}$ for all k. If $\mathbf{Z}^{(k)}\mathbf{y}^{(k)}$ defines the component of the step $\boldsymbol{\delta}^{(k)}$ in the range of $\mathbf{Z}^{(k)}$ (see (10.1.11)), the discussion of Section 11.1 ((11.1.4) f.f.) suggests that it is appropriate to use the BFGS method to update $\mathbf{M}^{(k)}$, replacing $\boldsymbol{\gamma}^{(k)}$ by $\mathbf{Z}^{(k+1)\mathrm{T}}\mathbf{g}^{(k+1)} - \mathbf{Z}^{(k)\mathrm{T}}\mathbf{g}^{(k)}$ and $\boldsymbol{\delta}^{(k)}$ by $\mathbf{y}^{(k)}$. Some other formulae which are asymptotically equivalent to this are given by Nocedal and Overton (1985). They also point out that if the component of $\boldsymbol{\delta}^{(k)}$ in the range of $\mathbf{Y}^{(k)}$ is substantial, then there may be difficulties in using this update. For example the scalar that replaces $\boldsymbol{\gamma}^{(k)\mathrm{T}}\boldsymbol{\delta}^{(k)}$ in the BFGS formula may not be positive, leading to loss of positive definiteness. Thus Nocedal and Overton recommend that $\mathbf{M}^{(k+1)} = \mathbf{M}^{(k)}$ is set under certain circumstances, and the BFGS formula is not always used. Another pitfall to consider is that in practice $\mathbf{Z}^{(k)}$ may not be defined in a continuous manner (e.g. because of a different pivotal choice in variable elimination, or a switch of sign in Householder QR, or a change in the number of active constraints if an inequality problem is being solved). In this case the following procedure can be used. Given only $\mathbf{M}^{(k)}$, the best available estimate of $\mathbf{B}^{(k)}$ is the matrix $\bar{\mathbf{B}}^{(k)} = \mathbf{V}^{(k)}\mathbf{M}^{(k)}\mathbf{V}^{(k)\mathrm{T}}$ ($\mathbf{V}$ as in (10.1.20)). $\bar{\mathbf{B}}^{(k)}$ can then be reduced using a matrix $\mathbf{Z}^{(k+1)}$ calculated differently from $\mathbf{Z}^{(k)}$ by $\bar{\mathbf{M}}^{(k)} = \mathbf{Z}^{(k+1)\mathrm{T}}\bar{\mathbf{B}}^{(k)}\mathbf{Z}^{(k+1)}$. Clearly only the matrix $\mathbf{V}^{(k)\mathrm{T}}\mathbf{Z}^{(k+1)}$ is required to calculate $\bar{\mathbf{M}}^{(k)}$ from $\mathbf{M}^{(k)}$. $\bar{\mathbf{M}}^{(k)}$ is the best estimate that can be constructed from $\mathbf{M}^{(k)}$, consistent with the available information. In the inequality case, if the dimension $n - m$ increases then $\bar{\mathbf{M}}^{(k)}$ is singular and an additional change is needed to ensure positive definiteness. One possible method is that given in (11.2.4) and there are other obvious ideas.

The use of the matrix $\mathbf{M}^{(k)}$ in algorithms for inequality constraints is mentioned here only briefly. One idea is to use an active set approach as in Section 11.3 in which constraints are added to the active set if they become active in the line search, and are deleted if Lagrange multiplier estimates so indicate (Womersley, 1981; Coleman and Conn, 1982a, b; Womersley and Fletcher, 1986). Recently we have been developing an algorithm in which the active constraints are determined by solving an L_1LP subproblem (with trust region), which is also used to force global convergence (Fletcher and Sainz de la Maza, 1987).

12.5. NONLINEAR ELIMINATION AND FEASIBLE DIRECTION METHODS

An apparently attractive approach to the nonlinear programming problem is to try to produce direct methods which generalize the ideas in Chapters 10 and 11 and avoid the use of penalty functions. These methods attempt to maintain feasibility by searching from one feasible point to another along feasible arcs (lines in Chapters 10 and 11) and hence are referred to as *feasible direction methods.* They date back at least as far as Rosen (1960, 1961) and Zoutendijk (1960). Inequality constraints are handled by an active constraint strategy as in Sections 10.3 and 11.2 but attention is initially directed towards methods suitable for solving the equality problem (12.1.1). The presence of nonlinear constraints does not allow a line search to maintain feasibility so a simplistic explanation of what is done is that at any feasible point $\mathbf{x}^{(k)}$ a search direction $\mathbf{s}^{(k)}$ in the tangent plane is calculated and a feasible arc is then obtained by projecting any point on the resulting line into a corresponding point in the feasible region (see Figure 12.5.1). A search along the resulting feasible arc is then made to reduce $f(\mathbf{x})$. Many methods of this type have been suggested. Another possible approach to nonlinear programming is to use variables to eliminate constraints, if necessary by solving a nonlinear system of equations at each iteration. It is shown below that this is a special case of a *nonlinear generalized elimination method.* This turns out to be equivalent to the idea of a feasible direction method and a number of common methods are shown to be special cases of elimination. The projection step in the feasible direction method is then equivalent to the solution of the nonlinear system of equations in the elimination method.

It is shown in Section 10.1 that many methods for linear constraints are

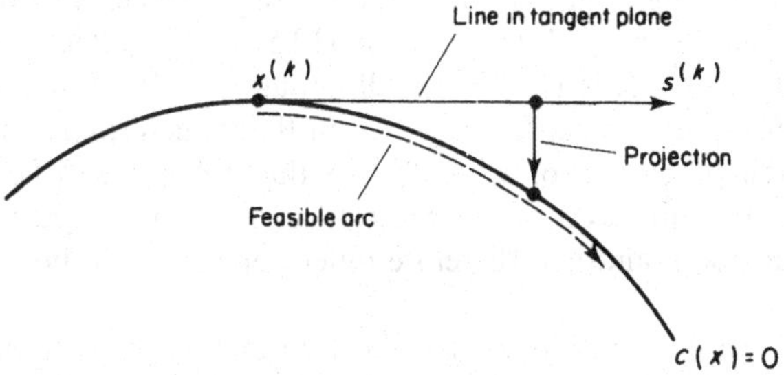

Figure 12.5.1 Feasible direction search

generalized elimination methods and correspond to different ways of choosing $\mathbf{V}$ in (10.1.20). Another way of regarding this construction is that a linear transformation to new variables $\mathbf{y}$ is made, defined by

$$\mathbf{y} = \begin{pmatrix} \mathbf{y}_1 \\ \mathbf{y}_2 \end{pmatrix} = [\mathbf{A}:\mathbf{V}]^{\mathrm{T}}\mathbf{x} - \begin{pmatrix} \mathbf{b} \\ \mathbf{0} \end{pmatrix}$$

where the partitions are $\mathbf{y}_1 \in \mathbb{R}^m$ and $\mathbf{y}_2 \in \mathbb{R}^{n-m}$. Since $[\mathbf{A}:\mathbf{V}]$ is non-singular the transformation is one to one. Then the method is derived by keeping $\mathbf{y}_1 = \mathbf{0}$, thus satisfying the constraints, and minimizing with respect to the remaining variables $\mathbf{y}_2$ (corresponding to $\mathbf{y}$ in (10.1.10)). A similar approach is possible when solving nonlinear constraint problems. It is assumed that $\mathbf{x}^{(k)}$ is a feasible point at which $\mathbf{A}^{(k)}$ has full rank. It is then appropriate to consider an analogous nonlinear transformation $\mathbf{x} \leftrightarrow \mathbf{y}$ defined by

$$\mathbf{y} = \begin{pmatrix} \mathbf{y}_1 \\ \mathbf{y}_2 \end{pmatrix} = \begin{pmatrix} \mathbf{c}(\mathbf{x}^{(k)} + \boldsymbol{\delta}) \\ \mathbf{V}^{\mathrm{T}}\boldsymbol{\delta} \end{pmatrix}. \tag{12.5.1}$$

$\mathbf{V}$ is such that $[\mathbf{A}^{(k)}:\mathbf{V}]$ is non-singular in which case the transformation is one-to-one in some neighbourhood of $\mathbf{x}^{(k)}$; in fact as (12.5.1) indicates, new $\mathbf{y}$ variables are defined on each iteration to help keep the transformation well defined. If $\mathbf{y}_2$ is any value of the free variables, the corresponding $\boldsymbol{\delta}$ is required which solves (12.5.1) with $\mathbf{y}_1 = \mathbf{0}$. This can be calculated by the Newton–Raphson iteration (Section 6.2) in an inner (r) iteration. The Jacobian matrix of the transformation (12.5.1) is $[\mathbf{A}:\mathbf{V}]$ and a suitable initial approximation is

$$\boldsymbol{\delta}^{(1)} = \mathbf{Z}\mathbf{y}_2. \tag{12.5.2}$$

The sequence of iterates is defined for $r \geqslant 1$ by

$$\begin{aligned} \boldsymbol{\delta}^{(r+1)} &= \boldsymbol{\delta}^{(r)} - [\mathbf{A}:\mathbf{V}]^{-\mathrm{T}}\begin{pmatrix} \mathbf{c}^{(r)} \\ \mathbf{0} \end{pmatrix} \\ &= \boldsymbol{\delta}^{(r)} - \mathbf{Y}\mathbf{c}^{(r)} \end{aligned} \tag{12.5.3}$$

where $\mathbf{A}$ (and hence $\mathbf{Y}$) and $\mathbf{c}^{(r)}$ are evaluated at $\mathbf{x}^{(k)} + \boldsymbol{\delta}^{(r)}$. When the r-iteration is deemed to have converged, $\boldsymbol{\delta}^{(r+1)}$ becomes the required vector $\boldsymbol{\delta}$. Both $\boldsymbol{\delta}$ and $\mathbf{x}(= \mathbf{x}^{(k)} + \boldsymbol{\delta})$ can be regarded as functions of the free variables $\mathbf{y}_2$. In practice it is more efficient in (12.5.3) to evaluate $\mathbf{A}$ and hence $\mathbf{Y}$ at $\mathbf{x}^{(k)}$, although the r-iteration no longer has second order convergence. Convergence of the r-iteration can be proved to occur if (12.5.2) is a sufficiently good initial approximation, which is true for small enough $\mathbf{y}_2$. In terms of Figure 12.5.1, $\boldsymbol{\delta}^{(1)}$ in (12.5.2) can be regarded as a step in the tangent plane and the iteration (12.5.3) is the projection of $\mathbf{x}^{(k)} + \boldsymbol{\delta}^{(1)}$ so that $\mathbf{x}^{(k)} + \boldsymbol{\delta}$ satisfies the nonlinear constraints. The direction of the projection is seen from (12.5.3) to lie in the column space of $\mathbf{Y}$ and will therefore differ depending on how $\mathbf{Y}$ is defined in (10.1.20).

The idea can be modified to give a search method by defining $\mathbf{p}^{(k)} \in \mathbb{R}^{n-m}$ as a search direction in the space of the free variables and taking $\mathbf{y}_2 = \alpha\mathbf{p}^{(k)}$ as a step in this direction. Then $\boldsymbol{\delta}(=\boldsymbol{\delta}(\alpha))$ is calculated from (12.5.2) and (12.5.3) and defines a feasible arc $\mathbf{x}(\alpha)=\mathbf{x}^{(k)}+\boldsymbol{\delta}(\alpha)$. The direction of the arc at $\mathbf{x}^{(k)}$ is the feasible direction $\mathbf{s}^{(k)}=\mathbf{Z}\mathbf{p}^{(k)}$ (see Figure 12.5.1). It is then possible to choose $\alpha^{(k)}$ to optimize $f(\mathbf{x}^{(k)}+\boldsymbol{\delta})$ by a line search along this feasible arc, assuming the arc exists and can be calculated for sufficiently large α, which may not always be the case.

One method of this type is the *GRG method* of Abadie and Carpentier (1969) which is equivalent to the direct elimination of variables. Thus $\mathbf{x}$ is partitioned into $\mathbf{x}_1 \in \mathbb{R}^m$ and $\mathbf{x}_2 \in \mathbb{R}^{n-m}$ and the variables $\mathbf{x}_2$ become the free variables $\mathbf{y}_2$. The matrix $\mathbf{Y}$ is $[\mathbf{A}_1^{-1}:\mathbf{0}]^{\mathrm{T}}$ (see (10.1.22)) and $\mathbf{Z}$ is $[-\mathbf{A}_2\mathbf{A}_1^{-1}:\mathbf{I}]^{\mathrm{T}}$. How $\mathbf{p}^{(k)}$ is chosen is described below; otherwise the method follows the above scheme. One way of writing the projection iteration (12.5.2) and (12.5.3) is to have $\{\mathbf{x}^{(k+1,r)}\}$ as a sequence of estimates of $\mathbf{x}^{(k+1)}$ defined by

$$\begin{aligned}\mathbf{x}^{(k+1,1)} &= \mathbf{x}^{(k)}+\alpha^{(k)}\mathbf{Z}\mathbf{p}^{(k)}\\ \mathbf{x}_1^{(k+1,r+1)} &= \mathbf{x}_1^{(k+1,r)}-\mathbf{A}_1^{-\mathrm{T}}\mathbf{c}^{(k+1,r)}, \qquad r=1,2,\ldots,\end{aligned} \tag{12.5.4}$$

whilst $\mathbf{x}_2^{(k+1)}=\mathbf{x}_2^{(k+1,1)}$ stays fixed, which shows essentially that the $\mathbf{x}_1$ variables are being chosen so as to eliminate the constraints. A summary of recent developments in the use of GRG methods is given by Lasdon (1985). Another feasible direction method is the *gradient projection method* of Rosen (1961) in which the feasible direction $\mathbf{s}^{(k)}$ is the projection of the negative gradient vector into the tangent plane. Rosen achieves this by representing the projection matrix directly, but a more stable approach is to use the orthogonal projection method of Section 10.1 to define $\mathbf{Y}$ and $\mathbf{Z}$, and then to proceed as described in this section. A feasible direction method which generalizes the Wolfe (1963a) reduced gradient method is also possible in a similar way.

The next step is to consider how the direction $\mathbf{s}^{(k)}(=\mathbf{Z}\mathbf{p}^{(k)})$ of the feasible arc might be calculated. The objective function can be regarded as defining a function $f_y(\mathbf{y})=f_x(\mathbf{x})$ where $\mathbf{x}$ and $\mathbf{y}$ are related by (12.5.1). As in Section 3.3, the chain rule shows that first derivatives are related by

$$\mathbf{g}_x=[\mathbf{A}:\mathbf{V}]\begin{pmatrix}\mathbf{g}_{y_1}\\ \mathbf{g}_{y_2}\end{pmatrix} \tag{12.5.5}$$

so that using (10.1.20), $\mathbf{g}_{y_2}=\mathbf{Z}^{\mathrm{T}}\mathbf{g}_x$ can again be interpreted as a reduced gradient vector. Thus early methods (Abadie and Carpentier, 1969; Rosen, 1961) choose $\mathbf{y}_2=\alpha\mathbf{p}^{(k)}$ where $\mathbf{p}^{(k)}=-\mathbf{Z}^{\mathrm{T}}\mathbf{g}^{(k)}$ is the reduced steepest descent direction, and the direction of the feasible arc is $\mathbf{s}^{(k)}=-\mathbf{Z}\mathbf{Z}^{\mathrm{T}}\mathbf{g}^{(k)}$. It is possible to improve these methods with little complication by using conjugate gradients (see Section 4.1) and the resulting methods have been used effectively on some problems. Nonetheless ideas of curvature do not figure strongly in these methods and better algorithms can be expected. The analogue of Newton's method (Section 3.1) requires second derivatives of f with respect to $\mathbf{x}$ or $\mathbf{y}$. Differentiating in (12.5.5) yields

$$\mathbf{G}_x=\sum_{i=1}^{m}\nabla^2 c_i\,\partial f/\partial y_i+[\mathbf{A}:\mathbf{V}]\mathbf{G}_y[\mathbf{A}:\mathbf{V}]^{\mathrm{T}} \tag{12.5.6}$$

so that the reduced Hessian at $\mathbf{x}^{(k)}$ (that is $\mathbf{y}_2 = \mathbf{0}$) is defined by

$$\nabla^2_{y_2} f_y(\mathbf{0}) = \mathbf{Z}^{\mathrm{T}}\mathbf{W}^{(k)}\mathbf{Z} \tag{12.5.7}$$

where $\mathbf{W}^{(k)} = \mathbf{G}^{(k)} - \sum \lambda_i^{(k)} \nabla^2 c_i^{(k)}$ with $\boldsymbol{\lambda}^{(k)} = \mathbf{Y}^{\mathrm{T}}\mathbf{g}^{(k)}$. Thus the Hessian of the Lagrangian function duly occurs along with what can be regarded a first order estimate of Lagrange multipliers (see (11.1.18)). The basic Newton's method therefore defines a correction $\mathbf{y}_2^{(k)}$ by solving the system

$$[\mathbf{Z}^{\mathrm{T}}\mathbf{W}^{(k)}\mathbf{Z}]\mathbf{y}_2 = -\mathbf{Z}^{\mathrm{T}}\mathbf{g}^{(k)} \tag{12.5.8}$$

which is then used with (12.5.2) and (12.5.3). Use of this formula to define a search direction, or its use in a modified Newton method (Section 3.1) are all possible extensions of this technique. It should be noted that the matrices $\mathbf{Z}$ and $\mathbf{Y}$ which arise above should strictly be written as $\mathbf{Z}^{(k)}$ and $\mathbf{Y}^{(k)}$ since they are determined by $\mathbf{A}^{(k)}$, which is no longer constant as it is in Chapter 10.

An illustration of this Newton method to solve problem (12.1.5) from the feasible point $\mathbf{x}^{(1)} = (0.8, 0.6)^{\mathrm{T}}$ is given in Table 12.5.1. Orthogonal factorizations $\mathbf{Z}^{\mathrm{T}}\mathbf{A} = \mathbf{0}$, $\mathbf{Z}^{\mathrm{T}}\mathbf{Z} = \mathbf{I}$, and $\mathbf{Y} = \mathbf{A}^{+}$ are used in the method. The tabulated values are related to (12.5.3) by $\mathbf{x}^{(k+1,r)} = \mathbf{x}^{(k)} + \boldsymbol{\delta}^{(r)}$. The convergence of the inner iteration (12.5.3) at a very rapid linear rate can be seen, although it is not quadratic since $\mathbf{Y}$ is evaluated at $\mathbf{x}^{(k)}$ and not $\mathbf{x}^{(k+1,r)}$. The rapid convergence of the outer iteration which uses (12.5.8) can also be observed. As in the previous section, an extension to quasi-Newton methods is also possible, in which an approximation $\mathbf{M}^{(k)}$ to the matrix $\mathbf{Z}^{(k)\mathrm{T}}\mathbf{W}^{(k)}\mathbf{Z}^{(k)}$ is updated using the BFGS formula say, involving differences in reduced gradient vectors

$$\boldsymbol{\gamma}^{(k)} = \mathbf{Z}^{(k+1)\mathrm{T}}\mathbf{g}^{(k+1)} - \mathbf{Z}^{(k)\mathrm{T}}\mathbf{g}^{(k)} \tag{12.5.9}$$

and differences in reduced coordinates $\boldsymbol{\delta}^{(k)} = \mathbf{y}_2^{(k)} - \mathbf{0} = \mathbf{y}_2^{(k)}$.

Another second order feasible direction method is summarized by Sargent (1974). This is based on using a nonlinear version of (12.4.5) in which the constraint linearization $\mathbf{c}^{(k)} + \mathbf{A}^{(k)\mathrm{T}}\boldsymbol{\delta} = \mathbf{0}$ is replaced by $\mathbf{c}(\mathbf{x}^{(k)} + \boldsymbol{\delta}) = \mathbf{0}$ as in (12.5.1). To allow the possibility of changing the length of the correction, a parameter

Table 12.5.1 Newton's method for the reduced nonlinear programming problem

k	r	$x_1^{(k,r)}$	$x_2^{(k,r)}$	$\lambda^{(k)}$	$y_2^{(k)}$	$c^{(r)}$
1	—	0.8	0.6	0.7	−0.142857	
2	1	0.714286	0.714286			−0.020408
	2	0.706122	0.708163			−0.000104
	3	0.706081	0.708132			−0.000001
	4	0.706080	0.708132			$-1_{10}-8$
2	—			0.707106	0.001451	
3	1	0.707108	0.707108			−0.000002
	2	0.707107	0.707107			$-4_{10}-13$
3	—			0.707107	0	

α is introduced, giving the nonlinear transformation

$$\begin{aligned} \mathbf{W}^{(k)}\boldsymbol{\delta} &= \alpha(\mathbf{A}^{(k)}\boldsymbol{\lambda} - \mathbf{g}^{(k)}) \\ \mathbf{c}(\mathbf{x}^{(k)} + \boldsymbol{\delta}) &= \mathbf{0} \end{aligned} \tag{12.5.10}$$

which is solved to give $\boldsymbol{\delta}^{(k)}$ and $\boldsymbol{\lambda}^{(k+1)}$. This system can also be solved in an inner (r) iteration by the Newton–Raphson method in the form

$$\begin{aligned} \mathbf{W}^{(k)}\boldsymbol{\delta}^{(k,r+1)} &= \alpha^{(k)}(\mathbf{A}^{(k)}\boldsymbol{\lambda}^{(k+1,r+1)} - \mathbf{g}^{(k)}) \\ \mathbf{c}(\mathbf{x}^{(k)} + \boldsymbol{\delta}^{(k,r)}) + \mathbf{A}^{\mathrm{T}}(\boldsymbol{\delta}^{(k,r+1)} - \boldsymbol{\delta}^{(k,r)}) &= \mathbf{0}. \end{aligned} \tag{12.5.11}$$

Setting $\mathbf{A} = \mathbf{A}^{(k)}$ at the expense of the quadratic order of convergence enables the linear system (12.5.11) to be solved for $\boldsymbol{\delta}^{(k,r+1)}$ and $\boldsymbol{\lambda}^{(k+1,r+1)}$ by any of the usual techniques described in Sections 10.1 and 10.2. Initial values $\boldsymbol{\delta}^{(k,1)}$ and $\boldsymbol{\lambda}^{(k+1,1)}$ for the r-iteration are obtained by solving the linearization of (12.5.10) in a similar way. When the r-iteration is deemed to have converged, $\mathbf{x}^{(k+1)}$ is set to $\mathbf{x}^{(k)} + \boldsymbol{\delta}^{(k,r+1)}$ and $\boldsymbol{\lambda}^{(k+1)}$ to $\boldsymbol{\lambda}^{(k+1,r+1)}$. To make the algorithm more robust, a line search is included, along the arc defined by solving (12.5.10) for different values of α, in order to reduce $f(\mathbf{x}^{(k)} + \boldsymbol{\delta})$ sufficiently. To ensure convergence, some sort of bias towards the (reduced) steepest descent direction must be included. This can be done for example by adding a Levenberg–Marquardt term $v\mathbf{I}$ to $\mathbf{W}^{(k)}$. Alternatively, a quasi-Newton approach is possible, and use of Powell's modification of the BFGS formula (see (12.4.18)) would be a good way of ensuring that $\mathbf{W}^{(k)}$ stays positive definite.

Convergence proofs for feasible direction methods have been given, but there are some hidden pitfalls, as Sargent (1974) points out. He considers the possibility of adding a penalty term, but then the distinction between direct methods and penalty function methods using low accuracy minimization becomes indistinct. As I see it, the main difficulty of feasible direction methods is the requirement to converge the inner r-iteration, because the basic Newton–Raphson method is not guaranteed to converge, and more elaborate alternatives would unduly complicate the overall method. Although the basic Newton–Raphson method may converge for small enough α, this may place an undue restriction on the length of step. Also there is a correlation between the convergence of the Newton–Raphson method and the rank of the Jacobian matrix, and it is not attractive to have to make implicit assumptions about the latter.

Additional complications arise when attention is switched from equality problems to consider the inequality constraint problem (12.1.2). The most obvious approach is to use an active set strategy as in Section 11.2. The general idea of keeping $c_i(\mathbf{x}) = 0$, $i \in \mathscr{A}$, and of systematically adding and dropping constraints remains the same, and estimates of Lagrange multipliers and precautions against zigzagging need also to be made in a similar way. There is however an additional difficulty which arises when the correction $\mathbf{Z}\mathbf{y}_2$ in (12.5.2) causes $\mathbf{x}^{(k+1,1)}$ to violate constraints not in the active set. Ideally the analogue of the linear constraint algorithm (11.2.2), step (e), would be to iterate the search along the feasible arc with respect to changes in α in an attempt to find the largest value of $\alpha = \bar{\alpha}^{(k)}$ which gives a feasible point. The constraint which is then limiting an increase in α can then be added into the active set, if appropriate. This process was originally used by Rosen (1961) but it adds yet another level of iteration into the method, which can therefore become very inefficient. Furthermore it is often possible to judge directly which constraint will become active, and this constraint can be zeroed with the other active constraints by extending the dimension of the r-iteration (12.5.3). However, there are then difficulties in writing down a strategy which covers all possibilities. For example $\mathbf{x}^{(k+1,1)}$ may violate so many constraints that it is not clear which ones to include in the r-iteration. Also constraints violated by $\mathbf{x}^{(k+1,1)}$ might not be active at all on the feasible arc. It is possible to introduce additional *ad hoc* rules to cover these cases but the resulting algorithm becomes cumbersome and unappealing.

In some ways it might be better to disregard the active set strategy and think in terms of a more complicated nonlinear transformation to replace (12.5.1) or (12.5.10). Sargent (1974) suggests

$$\begin{aligned} \mathbf{W}^{(k)}\boldsymbol{\delta} &= \alpha(\mathbf{A}^{(k)}\boldsymbol{\lambda} - \mathbf{g}^{(k)}) \\ \mathbf{c}(\mathbf{x}^{(k)} + \boldsymbol{\delta}) &\geqslant \mathbf{0}, \qquad \boldsymbol{\lambda} \geqslant \mathbf{0} \\ \boldsymbol{\lambda}^{\mathrm{T}}\mathbf{c}(\mathbf{x}^{(k)} + \boldsymbol{\delta}) &= 0 \end{aligned}$$

as the appropriate analogue of (12.5.10) for an inequality constraint problem. This can be solved by an inner iteration involving a sequence of quadratic programming subproblems. This is a very elegant solution to the difficulties caused by inequality constraints, although it does not help in solving the fundamental difficulty which is to force convergence of the inner iteration.

12.6 OTHER METHODS

Many other methods have been suggested for nonlinear programming, and a few of these are reviewed in this section. Although these methods are perhaps currently not as popular as those described previously, they often exhibit considerable ingenuity and contain interesting ideas which are worthy of note. An interesting penalty function for the equality problem (12.1.1) arises from the observation that the function

$$\phi(\mathbf{x}, f) = \left\| \begin{matrix} f(\mathbf{x}) - f \\ c_1(\mathbf{x}) \\ c_2(\mathbf{x}) \\ \vdots \\ c_m(\mathbf{x}) \end{matrix} \right\|_p \tag{12.6.1}$$

is minimized by $\mathbf{x}^*$ if the control parameter $f = f^*$. Thus a sequential penalty function method can be envisaged in which a sequence of estimates $\{f^{(k)}\} \to f^*$ is generated so that the minimizers $\mathbf{x}(f^{(k)}) \to \mathbf{x}^*$. Morrison (1968) suggests a method with $p=2$ in which the sum of squares function $[\phi(\mathbf{x}, f)]^2$ is minimized sequentially, but there is a potential difficulty that if $m<n$ then $\nabla^2\phi(\mathbf{x}^*, f)$ tends to a singular matrix as $f\to f^*$. This can slow down the rate of convergence of conventional algorithms. Gill and Murray (1976a) introduce two new features, one of which is a special purpose method for solving ill-posed least squares problems. They also give a new formula for updating the control parameter $f^{(k)}$ which has second order convergence and controls cancellation errors.

Another penalty function which relates the ideas in (12.6.1) to those contained in multiplier penalty functions (Section 12.2) and L_∞ exact penalty functions (Section 14.3) is given by Bandler and Charalambous (1972). The function

$$\begin{aligned} \phi(\mathbf{x}, p, \boldsymbol{\alpha}, f) &= \left(\sum_i f_i^p\right)^{1/p}, \qquad 0\leqslant i\leqslant m \\ f_0 &= f(\mathbf{x})-f \\ f_i &= f(\mathbf{x})-f-\alpha_i c_i(\mathbf{x}), \qquad i\geqslant 1 \end{aligned} \tag{12.6.2}$$

is defined, where $f<f^*$ and $f_i>0\ \forall i\geqslant 0$. There are many ways to force the minimizer $\mathbf{x}(p, \boldsymbol{\alpha}, f)$ of this function to converge to $\mathbf{x}^*$. In a neighbourhood of $\mathbf{x}^*$, and in the limit $p\to\infty$, the function becomes equivalent to the exact L_∞ penalty function. Thus one possible mode of application is a Polya-type algorithm in which f and $\boldsymbol{\alpha}$ are fixed, and a control sequence $\{p^{(k)}\}\to\infty$ is used. Another possibility is to fix p and $\boldsymbol{\alpha}$ and to choose a sequence of control parameters $f^{(k)}\to f^*$ as in (12.6.1). Finally p and f can remain fixed, in which case a sequence $\boldsymbol{\alpha}^{(k)}\to\boldsymbol{\lambda}^*/(m+1)$ of scaled Lagrange multiplier estimates can be used as control parameters. Charalambous (1977) shows how a suitable updating formula for the $\boldsymbol{\alpha}^{(k)}$ parameters can be determined.

Another attractive idea is to attempt to define an *exact penalty function* $\phi(\mathbf{x})$ which is minimized locally by the solution $\mathbf{x}^*$ to (12.1.1) or (12.1.2). A simple way to do this is described in Sections 12.3 and 14.3 and gives rise to a non-differentiable function. This requires special purpose techniques to solve the unconstrained minimization problem, but is a promising approach, and one particular algorithm is described in Section 14.5. It is also possible to determine a *smooth* exact penalty function. The key idea here (Fletcher, 1973) is to use an augmented Lagrangian function in which $\boldsymbol{\lambda}$ is not an independent vector but is a function $\boldsymbol{\lambda}(\mathbf{x})$ determined by a finite calculation. In particular the choice $\boldsymbol{\lambda}(\mathbf{x})=\mathbf{A}^{+\mathrm{T}}\mathbf{g}|_{\mathbf{x}}$, obtained by solving the over-determined least squares problem $\mathbf{A}\boldsymbol{\lambda}=\mathbf{g}$, gives rise to the penalty function

$$\phi(\mathbf{x})=f(\mathbf{x})-\boldsymbol{\lambda}(\mathbf{x})^{\mathrm{T}}\mathbf{c}(\mathbf{x})+\mathbf{c}(\mathbf{x})^{\mathrm{T}}\mathbf{S}\mathbf{c}(\mathbf{x}). \tag{12.6.3}$$

An obvious choice for $\mathbf{S}$ is $\frac{1}{2}\sigma\mathbf{I}$, in which case the penalty function is exact for any value of σ above a certain threshold. In fact the choice $\mathbf{S}=\sigma\mathbf{A}^+\mathbf{A}^{+\mathrm{T}}$ turns out to be more significant, since it is then possible to rearrange $\phi(\mathbf{x})$ to give

$$\phi(\mathbf{x})=f(\mathbf{x})-\boldsymbol{\pi}(\mathbf{x})^{\mathrm{T}}\mathbf{c}(\mathbf{x}) \tag{12.6.4}$$

where

$$\boldsymbol{\pi}(\mathbf{x})=\mathbf{A}^+(\mathbf{g}-\sigma\mathbf{A}^{+\mathrm{T}}\mathbf{c})|_{\mathbf{x}}$$

is the Lagrange multiplier vector of the subproblem

$$\begin{aligned} &\underset{\delta}{\text{minimize}} \quad \tfrac{1}{2}\sigma\boldsymbol{\delta}^T\boldsymbol{\delta} + \mathbf{g}^T\boldsymbol{\delta} \\ &\text{subject to} \quad \mathbf{A}^T\boldsymbol{\delta} + \mathbf{c} = \mathbf{0} \end{aligned} \tag{12.6.5}$$

where $\mathbf{g}$, $\mathbf{A}$, and $\mathbf{c}$ are evaluated at $\mathbf{x}$. In fact the solution $\boldsymbol{\delta}(\mathbf{x})$ of this system is discarded, but the multipliers $\boldsymbol{\pi}(\mathbf{x})$ are substituted into (12.6.4) to define $\phi(\mathbf{x})$. The constraints in (12.6.5) are linearizations of the nonlinear constraints about $\mathbf{x}$ (see (12.4.9)) so a generalization to solve inequality constraints is immediate. The resulting function $\phi(\mathbf{x})$ can be minimized by any suitable smooth unconstrained minimization technique. The main difficulty with this approach is that the calculation of $\boldsymbol{\lambda}(\mathbf{x})$ or $\boldsymbol{\pi}(\mathbf{x})$ requires first derivatives of f and $\mathbf{c}$. Thus $\nabla\phi$ requires second derivatives of f and $\mathbf{c}$, which are often not available. However, if these derivatives are available, then an $O(\|\mathbf{h}\|)$ estimate of $\nabla^2\phi(\mathbf{x}^*)$ can be made, which ensures second order convergence, and hence a Newton-like minimization method is very practicable in these circumstances. An interesting development of these ideas is given by di Pillo and Grippo (1979) who propose a smooth exact penalty function defined on the vector $(\mathbf{x}, \boldsymbol{\lambda})$.

Another idea which occurs frequently is to make linearizations of the nonlinear functions which arise in the problem, and then solve LP or QP subproblems. Some early methods such as the cutting plane method and MAP (see Wolfe 1967) and Beale (1967) respectively) make linear Taylor series approximations like (7.1.3) about the current iterate $\mathbf{x}^{(k)}$ to both the objective and constraint functions, and the resulting LP (possibly including some step-length restriction) is solved to determine $\mathbf{x}^{(k+1)}$. This may be satisfactory when the solution is at a vertex of the feasible region, but in general does not model the effect of curvature. Thus the convergence properties suffer and there can be numerical difficulties. The techniques of Section 12.4 are therefore recommended, since they avoid all these difficulties albeit at the expense of solving a QP subproblem. Another technique which uses linear approximations is *separable programming* (see Beale (1970) for instance) in which the objective and constraint functions are separable. A separable function is one which can be written as

$$f(\mathbf{x}) = \sum_i f_i(z_i) \tag{12.6.6}$$

where the scalar quantities z_i are either linear functions of $\mathbf{x}$ or other separable functions. Each separable function is replaced by a piecewise linear approximation whose knots are at fixed values of the variable z_i. It is then possible to reduce the problem to something like a linear programming problem, but with differences in the rules for changing the basis. The accuracy of the solution clearly depends on that of the piecewise linear approximations. In fact separable programming is most useful when a rough estimate of the solution is satisfactory and when the problem is substantially linear and sparse, such as in business or economic models.

Another type of algorithm which solves QP subproblems is given by Murray (1969). This is motivated by trying to avoid the ill-conditioning associated with

$\sigma \to \infty$ in the penalty function (12.1.3). A simple interpretation of the method arises from the fact that the minimizer $\mathbf{x}(\sigma)$ of (12.1.3) equivalently solves the constrained problem

$$\begin{aligned} &\underset{\mathbf{x}}{\text{minimize}} \quad f(\mathbf{x}) \\ &\text{subject to} \quad \mathbf{c}(\mathbf{x}) = \mathbf{c}(\mathbf{x}(\sigma)). \end{aligned} \tag{12.6.7}$$

A sequence of values $\sigma^{(k)} \to \infty$ is chosen although this is done as the algorithm proceeds. Murray (1969) does not give details but suggests that the rate of increase should depend on how well the solution is approximated. The quadratic approximation (12.4.8) to $f(\mathbf{x})$ and the linear approximation (12.4.9) to $\mathbf{c}(\mathbf{x})$ are made, and using the asymptotic result (12.1.10) the subproblem

$$\begin{aligned} &\underset{\boldsymbol{\delta}}{\text{minimize}} \quad q^{(k)}(\boldsymbol{\delta}) \\ &\text{subject to} \quad \mathbf{l}^{(k)}(\boldsymbol{\delta}) = \mathbf{c}^{(k)}\sigma^{(k)}/\sigma^{(k+1)} \end{aligned} \tag{12.6.8}$$

is defined. This is solved to determine a search direction along which $\phi(\mathbf{x}, \sigma^{(k+1)})$ is minimized. The line search offsets the apparent disadvantage that the 'best' choice $\sigma^{(k+1)} = \infty$ is not usually made. The method can be improved by introducing estimates of Lagrange multipliers and such a method is described by Biggs (1975). This method has been used widely in practice and has performed well, but it turns out to be similar to the methods of Section 12.4 and a detailed discussion of the method will not be given here. A convergence proof for the method is given by Biggs (1978) but seems to rely strongly on the requirement of uniform independence of the columns of the Jacobian matrix which is an unrealistic assumption.

QUESTIONS FOR CHAPTER 12

12.1. Show that if the penalty function (12.1.3) is applied to a quadratic programming problem with equality constraints, then $\phi(\mathbf{x}, \sigma)$ is a quadratic function of $\mathbf{x}$. State the resulting Hessian matrix $\nabla_x^2 \phi$.

12.2. Consider using the penalty function (12.1.3) to solve the problem: $\min -x_1x_2x_3$ subject to $72 - x_1 - 2x_2 - 2x_3 = 0$. Verify that the explicit expression for $\mathbf{x}(\sigma)$ given by $x_2 = x_3 = 24/(1 + \sqrt{(1 - 8/\sigma)})$, $x_1 = 2x_2$, satisfies $\nabla\phi(\mathbf{x}(\sigma), \sigma) = \mathbf{0}$. Verify also that $\mathbf{x}(\sigma) \to \mathbf{x}^*$ as $\sigma \to \infty$. Find $\mathbf{x}(\sigma)$ when $\sigma = 9$ and verify that $\nabla^2\phi(\mathbf{x}(9), 9)$ is positive definite.

12.3. Consider the problem (12.1.5). For the penalty function (12.1.3) show that the elements of the minimizer $\mathbf{x}(\sigma)$ satisfy the equations $x_1 = x_2$ and $2x_1^3 - x_1 - 1/(2\sigma) = 0$. Show that as $\sigma \to \infty$, $x_1 = \frac{1}{2}\sqrt{2} + a/\sigma + O(1/\sigma^2)$, and find a. Consider now the problem: $\min -x_1 - x_2$ subject to $1 - x_1^2 - x_2^2 \geqslant 0$ and $x_2 - x_1^2 \geqslant 0$. Show how the penalty function may be modified to solve this problem. For what values of σ do the minimizers of the two penalty functions agree?

12.4. Show that if second order conditions (9.3.5) hold, and if rank $\mathbf{A}^* = m$, then the Lagrangian matrix $\begin{bmatrix} \mathbf{W}^* & -\mathbf{A}^* \\ -\mathbf{A}^{*T} & \mathbf{0} \end{bmatrix}$ is non-singular. Let $\begin{bmatrix} \mathbf{W}^* & -\mathbf{A}^* \\ -\mathbf{A}^{*T} & \mathbf{0} \end{bmatrix}\begin{pmatrix} \mathbf{s} \\ \mathbf{t} \end{pmatrix} = \mathbf{0}$. If $\mathbf{s} = \mathbf{0}$, show that rank $\mathbf{A}^* = m$ implies $\mathbf{t} = \mathbf{0}$. If $\mathbf{s} \neq \mathbf{0}$ show that $\mathbf{s}$ satisfies (9.3.4) and $\mathbf{s}^T\mathbf{W}^*\mathbf{s} = \mathbf{0}$ which contradicts (9.3.5). Hence conclude that the Lagrangian matrix is non-singular.

12.5. Let $\mathbf{c}^{(k)} \to \mathbf{c}^*$, $\mathbf{A}^{(k)} \to \mathbf{A}^*$($\mathbf{A}$ is $n \times m, n > m$) and let rank $\mathbf{A}^* = m$. Use the Rayleigh quotient result ($\mathbf{u}^T\mathbf{M}\mathbf{u} \geqslant \lambda_n \mathbf{u}^T\mathbf{u}\ \forall \mathbf{u}$) and the definition of a singular value σ_i of $\mathbf{A}$ (square root of an eigenvalue λ_i of $\mathbf{A}^T\mathbf{A}$) to show that if $0 < \beta < \sigma_n$ then $\|\mathbf{A}^{(k)}\mathbf{c}^{(k)}\|_2 \geqslant \beta \|\mathbf{c}^{(k)}\|_2$ for all k sufficiently large. Hence establish (12.1.22).

12.6. For the inverse barrier function (12.1.24) show that estimates of Lagrange multipliers analogous to (12.1.8) are given by $\lambda_i^{(k)} = r^{(k)}/c_i^{(k)2}$. Hence show that $\lambda_i^{(k)} \to 0$ for an inactive constraint and that any limit point $\mathbf{x}^*, \lambda^*$ is a KT point. Show that $\mathbf{h}^{(k)}$ has asymptotic behaviour $O(r^{(k)^{1/2}})$ which can be used in an acceleration technique (see SUMT of Fiacco and McCormick, 1968). Show also that $o(r)$ estimates of f^* can be made in a similar way to (12.1.12).

12.7. For the logarithmic barrier function (12.1.25) show that estimates of Lagrange multipliers are given by $\lambda_i^{(k)} = r^{(k)}/c_i^{(k)}$, that $\mathbf{x}^*, \lambda^*$ is a KT point, that the asymptotic behaviour of $\mathbf{h}^{(k)}$ is $O(r)$, and that $o(r)$ estimates of f^* can be made.

12.8. Investigate the Hessian matrices of the barrier functions (12.1.24) and (12.1.25) at $\mathbf{x}(r^{(k)})$ and demonstrate ill-conditioning as $r^{(k)} \to 0$ if the number of active constraints is between 1 and $n-1$. What happens when $\lambda_i^* = 0$ for an active constraint?

12.9. When the penalty function (12.1.23) is applied to the problem

$$\begin{aligned} \text{minimize} \quad & -x_1 - x_2 + x_3 \\ \text{subject to} \quad & 0 \leqslant x_3 \leqslant 1 \\ & x_1^3 + x_3 \leqslant 1 \\ & x_1^2 + x_2^2 + x_3^2 \leqslant 1 \end{aligned}$$

the following data are obtained. Use this to estimate the optimum solution and multipliers, together with the active constraints, and give some indication of the accuracy which is achieved.

k	$\sigma^{(k)}$	$x_1(\sigma^{(k)})$	$x_2(\sigma^{(k)})$	$x_3(\sigma^{(k)})$
1	1	0.834379	0.834379	−0.454846
2	10	0.728324	0.728324	−0.087920
3	100	0.709557	0.709557	−0.009864
4	1000	0.707356	0.707356	−0.001017

12.10. Consider the problem

$$\begin{aligned}
\underset{\mathbf{x}}{\text{minimize}} \quad & \exp(x_1x_2x_3x_4x_5) \\
\text{subject to} \quad & x_1^2+x_2^2+x_3^2+x_4^2+x_5^2=10 \\
& x_2x_3=5x_4x_5 \\
& x_1^3+x_2^3=-1
\end{aligned}$$

given by Powell (1969). Investigate how accurately a local solution $\mathbf{x}^*$ and multipliers $\boldsymbol{\lambda}^*$ can be obtained, using the penalty function (12.1.3) and the sequence $\{\sigma^{(k)}\}=0.01,\ 0.1,\ 1,\ 10,\ 100$. Use a quasi-Newton subroutine which requires no derivatives. A suitable initial approximation to $\mathbf{x}(\sigma^{(1)})$ is the vector $(-2,2,2,-1,-1)^{\mathrm{T}}$, and it is sufficient to obtain each element of $\mathbf{x}(\sigma^{(k)})$ correct to three decimal places.

12.11. Consider using the multiplier penalty function (12.2.3) to solve the problem in Question 12.2. Assume that the controlling parameters $\lambda^{(1)}=0$ and $\sigma=9$ are chosen initially, in which case the minimizer $\mathbf{x}(0,9)$ is the same as the minimizer $\mathbf{x}(9)$ which has been determined in Question 12.2. Write down the new parameter $\lambda^{(2)}$ which would be given by using formulae (12.2.13) and (12.2.15). Which formula gives the better estimate of λ^*?

12.12. Use the Sherman–Morrison formula (Question 3.13) to show that if $\mathbf{S}$ is non-singular, then

$$\begin{bmatrix} \mathbf{W}+\mathbf{A}\mathbf{S}\mathbf{A}^{\mathrm{T}} & -\mathbf{A} \\ -\mathbf{A}^{\mathrm{T}} & \mathbf{0} \end{bmatrix}^{-1} = \begin{bmatrix} \mathbf{W} & -\mathbf{A} \\ -\mathbf{A}^{\mathrm{T}} & \mathbf{0} \end{bmatrix} - \begin{bmatrix} \mathbf{0} & \mathbf{0} \\ \mathbf{0} & \mathbf{S} \end{bmatrix}.$$

Hence use (10.2.6) to show that, in the notation of (12.2.6),

$$(\mathbf{A}^{\mathrm{T}}\mathbf{W}_{\sigma}^{-1}\mathbf{A})^{-1}=(\mathbf{A}^{\mathrm{T}}\mathbf{W}^{-1}\mathbf{A})^{-1}+\mathbf{S}.$$

Let $\boldsymbol{\sigma}_1$ be fixed so that $\mathbf{W}_{\sigma_1}$ is positive definite. Deduce that

$$(\mathbf{A}^{\mathrm{T}}\mathbf{W}_{\sigma}^{-1}\mathbf{A})^{-1}=\mathbf{S}+(\mathbf{A}^{\mathrm{T}}\mathbf{W}_{\sigma_1}^{-1}\mathbf{A})^{-1}-\mathbf{S}_1=\mathbf{S}+O(1)$$

which is (12.2.14).

12.13. Consider any iteration function

$$\boldsymbol{\lambda}^{(k+1)}=\boldsymbol{\phi}(\boldsymbol{\lambda}^{(k)})\triangleq\boldsymbol{\lambda}^{(k)}-\mathbf{M}^{(k)}\mathbf{c}^{(k)}$$

analogous to (12.2.13) or (12.2.15) for use with a multiplier penalty function. Show that

$$\nabla\boldsymbol{\phi}^{\mathrm{T}}(\boldsymbol{\lambda}^*)=\mathbf{I}-(\mathbf{A}^{*\mathrm{T}}\mathbf{W}_{\sigma}^{*-1}\mathbf{A}^*)\mathbf{M}^{*\mathrm{T}}.$$

Hence use the result of Question 12.12 and Theorem 6.2.2 to show that the Powell–Hestenes formula (12.2.15) converges at a linear rate, which can be made arbitrarily rapid by making $\mathbf{S}$ sufficiently large.

12.14. Show that the vector $\mathbf{x}(\boldsymbol{\lambda})$ which minimizes (12.2.1) equivalently solves the problem

$$\begin{aligned}&\underset{\mathbf{x}}{\text{minimize}} \quad f(\mathbf{x})\\&\text{subject to} \quad \mathbf{c}(\mathbf{x}) = \mathbf{c}(\mathbf{x}(\boldsymbol{\lambda})).\end{aligned}$$

If $\mathbf{x}(\boldsymbol{\lambda})$ minimizes (12.2.18) show that the equivalent problem is

$$\begin{aligned}&\underset{\mathbf{x}}{\text{minimize}} \quad f(\mathbf{x})\\&\text{subject to} \quad c_i(\mathbf{x}) \geqslant \min(c_i(\mathbf{x}(\boldsymbol{\lambda})), \theta_i), \qquad i = 1, 2, \ldots, m\end{aligned}$$

(Fletcher, 1975). These problems are neighbouring problems to (12.1.1) and (12.1.2) respectively.

12.15. Consider using the penalty function

$$\phi(\mathbf{x}, \boldsymbol{\lambda}, \boldsymbol{\sigma}) = f(\mathbf{x}) + \sum_i \sigma_i \exp(-\lambda_i c_i(\mathbf{x})/\sigma_i)$$

where $\sigma_i > 0$, to solve problem (12.1.1). Assume that a local solution $\mathbf{x}^*$ of this problem satisfies the second order sufficient conditions (9.3.5) and (9.3.4). Show that if control parameters $\boldsymbol{\lambda} = \boldsymbol{\lambda}^*$ and any $\boldsymbol{\sigma}$ are chosen, then $\mathbf{x}^*$ is a stationary point of $\phi(\mathbf{x}, \boldsymbol{\lambda}^*, \boldsymbol{\sigma})$. Show also that if, for all i, both $\lambda_i^* \neq 0$ and σ_i is sufficiently small, then $\mathbf{x}^*$ is a local minimizer of $\phi(\mathbf{x}, \boldsymbol{\lambda}^*, \boldsymbol{\sigma})$. (You may use without proof the lemma that if $\mathbf{A}$ is symmetric, if $\mathbf{D}$ is diagonal with sufficiently large positive elements, and if $\mathbf{v}^{\mathrm{T}}\mathbf{A}\mathbf{v} > 0$ for all $\mathbf{v}(\neq \mathbf{0})$ such that $\mathbf{B}^{\mathrm{T}}\mathbf{v} = \mathbf{0}$, then $\mathbf{A} + \mathbf{B}\mathbf{D}\mathbf{B}^{\mathrm{T}}$ is positive definite.)

What can be said about the case when some $\lambda_i^* = 0$? To what extent is this penalty function comparable to the Powell–Hestenes function?

If $\mathbf{x}(\boldsymbol{\lambda})$ is the unique global minimizer of $\phi(\mathbf{x}, \boldsymbol{\lambda}, \boldsymbol{\sigma})$ for all $\boldsymbol{\lambda}$, where $\boldsymbol{\sigma}$ is constant, and if $\mathbf{x}^* = \mathbf{x}(\boldsymbol{\lambda}^*)$, show that $\boldsymbol{\lambda}^*$ maximizes the function $\psi(\boldsymbol{\lambda}) \triangleq \phi(\mathbf{x}(\boldsymbol{\lambda}), \boldsymbol{\lambda}, \boldsymbol{\sigma})$. What are the practical implications of this result?

12.16. Solve the problem in Question 12.19 by generalized elimination of variables using the Newton method as shown in Table 12.5.1. Compare the sequence of estimates and the amount of work required with that in Question 12.19.

12.17. Assume that second order sufficient conditions (9.3.5) and (9.3.4) hold at $\mathbf{x}^*, \boldsymbol{\lambda}^*$. Let $(\mathbf{x}^{(k)}, \boldsymbol{\lambda}^{(k)}) \to (\mathbf{x}^*, \boldsymbol{\lambda}^*)$ and let a vector $\mathbf{s}^{(k)}(\|\mathbf{s}^{(k)}\|_2 = 1)$ exist such that $\mathbf{A}^{(k)\mathrm{T}}\mathbf{s}^{(k)} = \mathbf{0}$ but $\mathbf{s}^{(k)\mathrm{T}}\mathbf{W}^{(k)}\mathbf{s}^{(k)} \leqslant 0 \ \forall k$. Show by continuity that (9.3.5) and (9.3.4) are contradicted. Hence conclude that the subproblem (12.4.7) is well determined for $\mathbf{x}^{(k)}, \boldsymbol{\lambda}^{(k)}$ sufficiently close to $\mathbf{x}^*, \boldsymbol{\lambda}^*$.

Equivalently one can argue as follows. Second order sufficient conditions at $\mathbf{x}^*, \boldsymbol{\lambda}^*$ are equivalent to $\mathbf{Z}^{*\mathrm{T}}\mathbf{W}^*\mathbf{Z}^*$ being positive definite (Question 11.1). Define $\mathbf{Z}^*$ by (10.1.22) (direct elimination) assuming without loss of generality that $\mathbf{A}_1^*$ is non-singular. Then $\mathbf{Z}$ is a continuous function of $\mathbf{A}$ in a neighbourhood of $\mathbf{x}^*$. Thus for $\mathbf{x}^{(k)}, \boldsymbol{\lambda}^{(k)}$ sufficiently close to $\mathbf{x}^*, \boldsymbol{\lambda}^*$, $\mathbf{Z}^{(k)\mathrm{T}}\mathbf{W}^{(k)}\mathbf{Z}^{(k)}$ is positive definite also (continuity of eigenvalues) and hence second order sufficient conditions hold for (12.4.7).

12.18. Consider generalizing Theorem 12.4.1 to apply to the inequality constraint problem (12.1.2). If strict complementarity holds at $\mathbf{x}^*, \boldsymbol{\lambda}^*$, show by virtue of the implicit function theorem that a non-singular Lagrangian matrix (12.4.14) implies that the solution of (12.4.5) changes smoothly in some neighbourhood of $\mathbf{x}^*, \boldsymbol{\lambda}^*$. Hence show that the solution of the equality constraint problem also solves the inequality problem, so that constraints $c_i(\mathbf{x})$, $i \notin \mathscr{A}^*$, can be ignored. If strict complementarity does not hold, assume that the vectors $\mathbf{a}_i^*$, $i \in \mathscr{A}^*$, are independent and that the second order sufficient conditions of Theorem 9.3.2 hold. Deduce that the Lagrangian matrix (12.3.14) is non-singular for any subset of active constraints for which $\mathscr{A}_+^* \subseteq \mathscr{A} \subseteq \mathscr{A}^*$. Show that there is a neighbourhood of $\mathbf{x}^*, \boldsymbol{\lambda}^*$ in which the active constraints obtained from solving (12.4.11) satisfy this condition, so that a result like (12.4.16) holds for any such $\mathscr{A}$, and hence for $\mathscr{A}^*$. The rest of the theorem then follows.

12.19. Consider solving the problem: $\min x_1 + x_2$ subject to $x_2 \geqslant x_1^2$ by the Lagrange–Newton method (12.4.10) from $\mathbf{x}^{(1)} = \mathbf{0}$. Why does the method fail if $\lambda^{(1)} = 0$? Verify that rapid convergence occurs if $\lambda^{(1)} = 1$ is chosen.

12.20. The nonlinear L_1 norm minimization problem is

$$\text{minimize } f(\mathbf{x}) \triangleq \sum_{i=1}^{m} |r_i(\mathbf{x})| \qquad \mathbf{x} \in \mathbb{R}^n \tag{1}$$

where the functions $r_i(\mathbf{x})$ are given. A solution $\mathbf{x}^*$ of this problem can be found by applying unconstrained minimization techniques to the smooth function

$$\phi(\mathbf{x}, \varepsilon) = \sum_{i=1}^{m} (r_i^2(\mathbf{x}) + \varepsilon)^{1/2} \qquad \varepsilon > 0. \tag{2}$$

(El-Attar *et al.*, 1979). In this approach, a sequence of values $\varepsilon^{(k)} \to 0$ is taken, and for each $\varepsilon^{(k)}$, the minimizer $\mathbf{x}(\varepsilon^{(k)})$ of $\phi(\mathbf{x}, \varepsilon^{(k)})$ is determined.

(i) If global minimizers are calculated, that is

$$\phi(\mathbf{x}(\varepsilon^{(k)}), \varepsilon^{(k)}) \leqslant \phi(\mathbf{x}, \varepsilon^{(k)}) \quad \forall \mathbf{x},$$

and if a limit point $\mathbf{x}(\varepsilon^{(k)}) \to \mathbf{x}^*$ exists, then prove that $\mathbf{x}^*$ solves (1) (hint: relate $\phi(\mathbf{x}, \varepsilon)$ to $f(\mathbf{x})$ as $\varepsilon \to 0$).

(ii) At a solution $\mathbf{x}^*$ of (1) it can be shown that there exist multipliers λ_i^* $i = 1, 2, \ldots, m$ such that

$$\left.\begin{aligned} &\textstyle\sum \lambda_i^* \nabla r_i^* = \mathbf{0} \\ &-1 \leqslant \lambda_i^* \leqslant 1 \\ &\lambda_i^* = \operatorname{sign}(r_i^*) \quad \text{if} \quad r_i^* \neq 0 \end{aligned}\right\} \tag{3}$$

(r_i^* denotes $r_i(\mathbf{x}^*)$). Use the fact that $\nabla \phi(\mathbf{x}(\varepsilon^{(k)}), \varepsilon^{(k)}) = \mathbf{0}$ to determine certain multiplier estimates $\lambda_i^{(k)}$ $i = 1, 2, \ldots, m$, associated with $\mathbf{x}(\varepsilon^{(k)})$ which satisfy the condition

$$\textstyle\sum \lambda_i^{(k)} \nabla r_i(\mathbf{x}^{(k)}) = \mathbf{0}.$$

Assuming that $\boldsymbol{\lambda}^{(k)} \to \boldsymbol{\lambda}^*$ as $k \to \infty$, show that $\mathbf{x}^*, \boldsymbol{\lambda}^*$ satisfies (3).

(iii) If $r_i^* = 0$ how does $r_i(\mathbf{x}(\varepsilon^{(k)}))$ approach zero as a function of $\varepsilon^{(k)}$?
(iv) Solve the problem in which $n = 4$, $m = 11$ and

$$r_i(\mathbf{x}) = \frac{x_1(t_i^2 + x_2 t_i)}{t_i^2 + x_3 t_i + x_4} - d_i$$

where the data values are given by

d_i	t_i	d_i	t_i	d_i	t_i	d_i	t_i
0.1957	4	0.1600	0.5	0.0456	0.125	0.0235	0.0714
0.1947	2	0.0844	0.25	0.0342	0.1	0.0246	0.0625
0.1735	1	0.0627	0.167	0.0323	0.0823		

Use a quasi-Newton subroutine to minimize $\phi(\mathbf{x}, \varepsilon)$. Solve each problem to an accuracy of 10^{-8} in ϕ. Use the sequence $\varepsilon^{(k)} = 10^{-2k}$ $k = 1, 2, \ldots, 8$. An initial estimate of $\mathbf{x}^{(1)}$ is $(0.25, 0.4, 0.4, 0.4)^T$. Comment on the accuracy of your results and the extent to which the properties in (ii) and (iii) above are evident.

12.21. The least area of a right-angled triangle which contains a circle of unit radius can be obtained by solving the nonlinear programming problem

$$\begin{aligned} &\text{minimize} \quad x_1 x_2 \qquad \mathbf{x} \in \mathbb{R}^4 \\ &\text{subject to} \quad \frac{(x_1 x_3 + x_2 x_4)^2}{(x_1^2 + x_2^2)} - x_3^2 - x_4^2 + 1 = 0 \\ &\qquad x_1 \geqslant x_3 + 1 \\ &\qquad x_2 \geqslant x_4 + 1 \\ &\qquad x_3 \geqslant x_4 \\ &\qquad x_4 \geqslant 1. \end{aligned}$$

Attempt to solve this problem using the penalty function (12.1.23), with a sequence $\sigma^{(k)} = 10^k$, $k = 1, 2, \ldots, 6$. Use any quasi-Newton method to solve the unconstrained minimization problems which arise. Solve each problem to a tolerance of 5×10^{-6} in ϕ and take $(1, 1, \ldots, 1)^T$ as the initial estimate for minimizing $\phi(\mathbf{x}, \sigma^{(1)})$. Indicate your best estimate of the solution $(\mathbf{x}^*, \boldsymbol{\lambda}^*, f^*)$ and the active constraints $\mathscr{A}^*$. Say whether $\mathbf{x}^*, \boldsymbol{\lambda}^*$ is a KT point, and if so estimate the relative accuracy to which your estimate of $\mathbf{x}^*, \boldsymbol{\lambda}^*$ satisfies KT conditions.

12.22. By adding extra variables show how the L_1LP problem (12.3.7) can be transformed to an LP problem with equality constraints and general bounds. By taking the dual of the LP problem, show that the L_1LP problem has a symmetric dual L_1LP problem

$$\begin{aligned} &\underset{\boldsymbol{\lambda} \in \mathbb{R}^m}{\text{minimize}} \quad \mathbf{b}^T\boldsymbol{\lambda} + \mathbf{l}^T(\mathbf{A}\boldsymbol{\lambda} - \mathbf{c})^- - \mathbf{u}^T(\mathbf{A}\boldsymbol{\lambda} - \mathbf{c})^+ \\ &\text{subject to} \quad \boldsymbol{\rho} \leqslant \boldsymbol{\lambda} \leqslant \boldsymbol{\sigma}. \end{aligned}$$

Chapter 13

Other Optimization Problems

13.1 INTEGER PROGRAMMING

In this chapter some other types of problem are studied which have the feature that they can be reduced to or related to the solution of standard problems described in other chapters. *Integer programming* is the study of optimization problems in which some of the variables are required to take integer values. The most obvious example of this is the number of men, machines, or components, etc., required in some system or schedule. It also extends to cover such things as transformer settings, steel stock sizes, etc., which occur in a fixed ordered set of discrete values, but which may be neither integers, nor equally spaced. Most combinatorial problems can also be formulated as integer programming problems; this often requires the introduction of a *zero–one variable* which takes either of the values 0 or 1 only. Zero–one variables are also required when the model is dependent on the outcome of some decision on whether or not to take a certain course of action. Certain more complicated kinds of condition can also be handled by the methods of integer programming through the introduction of *special ordered sets*; see Beale (1978) for instance. There are many special types of integer programming problems but this section covers the fairly general category of *mixed integer programming* (both integer and continuous variables allowed, and any continuous objective and constraint functions as in (7.1.1)). The special case is also considered of *mixed integer LP* in which some integer variables occur in what is otherwise an LP problem. The *branch and bound method* is described for solving mixed integer programming problems, together with additional special features which apply to mixed integer LP. There are many special cases of integer programming problems and it is the case that the branch and bound method is not the most suitable in every case. Nonetheless this method is the one method which has a claim to be of wide generality and yet reasonably efficient. It solves a sequence of subproblems which are continuous problems of type (7.1.1) and is therefore well related to other material in this book. Other methods of less general applicability but which can be useful in certain special cases are reviewed by Beale (1978).

Integer programming problems are usually much more complicated and expensive to solve than the corresponding continuous problem on account of the discrete nature of the variables and the combinatorial number of feasible solutions which thus exists. A tongue-in-cheek possibility is the *Dantzig two-phase method.* The first phase is try to convince the user that he or she does not wish to solve an integer programming problem at all! Otherwise the continuous problem is solved and the minimizer is rounded off to the nearest integer. This may seem flippant but in fact is how many integer programming problems are treated in practice. It is important to realize that this approach can fail badly. There is of course no guarantee that a good solution can be obtained in this way, even by examining all integer points in some neighbourhood of the continuous solution. Even worse is the likelihood that the rounded-off points will be infeasible with respect to the continuous constraints in the problem (for example problem (13.1.5) below), so that the solution has no value. Nonetheless the idea of rounding up or down does have an important part to play when used in a systematic manner, and indeed is one main feature of the branch and bound method.

To develop the branch and bound method therefore, the problem is to find the solution $\mathbf{x}^*$ of the problem

$$\begin{aligned} P_I\text{:minimize} \quad & f(\mathbf{x}) \\ \text{subject to} \quad & \mathbf{x}\in R, \quad x_i \text{ integer } \forall i\in I \end{aligned} \tag{13.1.1}$$

where I is the set of integer variables and R is the (closed) feasible region of the continuous problem

$$P\text{:minimize } f(\mathbf{x}) \qquad \text{subject to } \mathbf{x}\in R, \tag{13.1.2}$$

which could be (7.1.1) for example. Let the minimizer $\mathbf{x}'$ of P exist: if it is feasible in P_I then it solves P_I. If not, then there exists an $i\in I$ for which x'_i is not an integer. In this case two problems can be defined by *branching* on variable x_i in problem P, giving

$$\begin{aligned} P^-\text{:minimize} \quad & f(\mathbf{x}) \\ \text{subject to} \quad & \mathbf{x}\in R, \quad x_i \leqslant [x'_i] \end{aligned} \tag{13.1.3}$$

and

$$\begin{aligned} P^+\text{:minimize} \quad & f(\mathbf{x}) \\ \text{subject to} \quad & \mathbf{x}\in R, \quad x_i \geqslant [x'_i]+1, \end{aligned} \tag{13.1.4}$$

where $[x]$ means the largest integer not greater than x. Also define P_I^- as problem (13.1.3) together with the integrality constraints x_i, $i\in I$, integer, and P_I^+ likewise. Two observations which follow are that $\mathbf{x}^*$ is feasible in either P^- or P^+ but not both, in which case it solves P_I^- or P_I^+. Also any feasible point in P_I^- or P_I^+ is feasible in P_I.

It is usually possible to repeat the branching process by branching on P^-

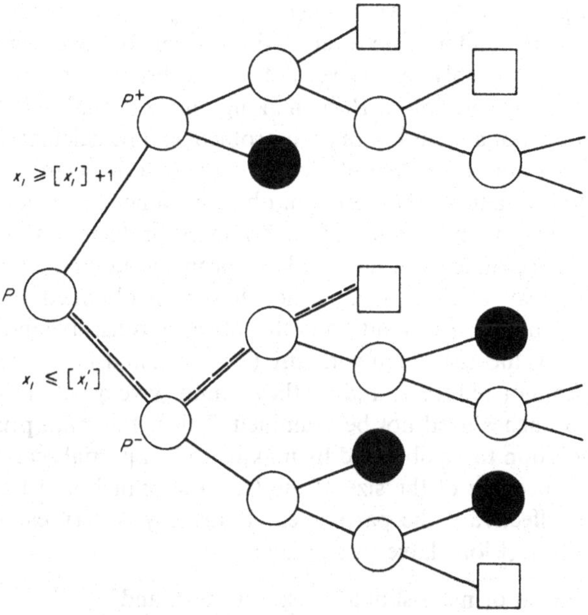

Figure 13.1.1 Tree structure in the branch and bound method

and P^+, and again on the resulting problems, so as to generate a *tree* structure (Figure 13.1.1). Each *node* corresponds to a continuous optimization problem, the *root* is the problem P, and the nodes are connected by the *branches* defined above. There are two special cases where no branching is possible at any given node: one is where the solution to the corresponding problem is integer feasible (□ in tree) and the other where the problem has no feasible point. (• in tree). Otherwise each node is a *parent problem* (○ in tree) and gives rise to two branched problems. If the feasible region is bounded then the tree is finite and each *path* through the tree ends in a □ or a •. Assuming that the solution $\mathbf{x}^*$ of P_I exists, then it is feasible along just one path through the tree (the broken line, say, in Figure 13.1.1) and is feasible in, and therefore solves, the □ problem which must exist at the end of the path. The solution of every □ problem is feasible in P_I and so the required solution vector $\mathbf{x}^*$ is that solution of a □ problem which has least objective function value. Thus the aim of the branch and bound method is to seek out □ nodes by exploring the tree in order to find the one with the least value. Often in the tree there are ○ nodes whose solution violates more than one integrality constraint, so in fact the tree is not uniquely defined until the choice of branching variable is defined. One anomalous (although unlikely) possibility is that $\mathbf{x}^*$ may be well defined, but the continuous problem P may be unbounded ($f(\mathbf{x}) \rightarrow -\infty, \mathbf{x} \in R$). This possibility is easily excluded in most practical cases by adding bounds to the variables so as to make R bounded.

To find the type of each node in the tree requires the solution of a continuous optimization problem. Also the number of nodes grows exponentially with the number of variables and may not even be finite. Therefore to examine the whole tree is usually impossibly expensive and so the branch and bound method attempts to find the solution of P_I by making only a partial search of the tree. Assuming that a global solution of each problem can be calculated, adding the branching constraint makes the solution of the parent problem infeasible in the branched problem, and so the optimum objective function value of the parent problem is a lower bound for that of the branched problem. Assume that some but not all of the problems in the tree have been solved, and let f_i be the best optimum objective function value which has been obtained by solving the problem at a □ node i. The solution of this problem is feasible in P_I so $f^* \leqslant f_i$. Now if f_j is the value associated with any ○ node j, and $f_j \geqslant f_i$, then branches from node j cannot reduce f_j and so they cannot give a better value than f_i. Hence these branches need not be examined. This is the main principle which enables the solution to be obtained by making only a partial search of the tree.

Nonetheless because of the size of the tree, the branch and bound method only becomes effective if the partial search strategy is very carefully chosen. Two important decisions have to be made:

(a) which problem or node should be solved next, and
(b) on which variable should a branch be made.

The following would seem to be an up-to-date assessment of the state of the art (see Beale (1978), for example, for more detailed references). An early method favoured the use of a *stack* (a last-in-first-out list or data structure). This algorithm keeps a stack of unsolved problems, each with a lower bound L on its minimum objective function value. This bound can be the value of the parent problem, although in linear programming problems a better bound can be obtained as described below. From those □ nodes which have been explored, the current best integer feasible solution $\hat{\mathbf{x}}$ with value $\hat{f}$ is also stored. Initially $\hat{f} = \infty$, the continuous problem is put in the stack with $L = -\infty$, and the algorithm proceeds as follows.

(i) If no problem is in the stack, finish with $\hat{\mathbf{x}}, \hat{f}$ as $\mathbf{x}^*, f^*$; otherwise take the top problem from the stack
(ii) If $L \geqslant \hat{f}$ reject the problem and go to (i).
(iii) Try to solve the problem: if no feasible point exists reject the problem and go to (i).
(iv) Let the solution be $\mathbf{x}'$ with value f': if $f' \geqslant \hat{f}$ reject the problem and go to (i).
(v) If $\mathbf{x}'$ is integer feasible then set $\hat{\mathbf{x}} = \mathbf{x}'$, $\hat{f} = f'$ and go to (i).
(vi) Select an integer variable i such that $[x_i'] < x_i'$, create two new problems by branching on x_i, place these on the stack with lower bound $L = f'$ (or a tighter lower bound derived from f'), and go to (i).

The rationale for these steps follows from the above discussion: in step (iv) any branched problems must have $f \geqslant f' \geqslant \hat{f} \geqslant f^*$ so cannot improve the optimum,

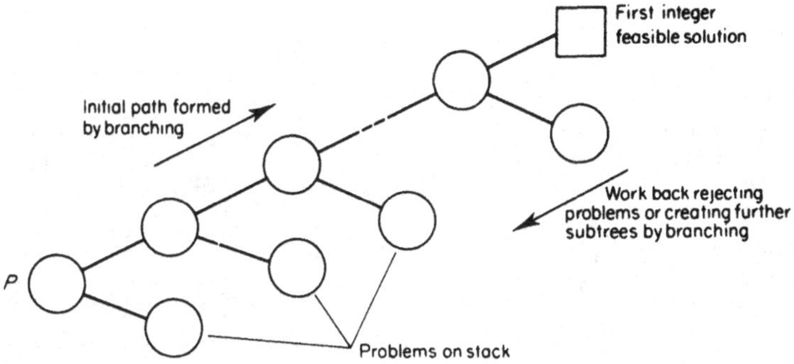

Figure 13.1.2 Progress of the stack method

so the node is rejected. Likewise in step (ii) where $f \geqslant L \geqslant \hat{f} \geqslant f^*$. Step (iii) corresponds to finding a $\bullet$ node and step (iv) to a $\square$ node. Step (vi) is the branching operation, and step (i) corresponds to having examined or excluded from consideration all nodes in the tree, in which case the solution at the best $\square$ node gives $\mathbf{x}^*$. The effect of this stack method is to follow a single path deep into the tree to a $\square$ node. Then the algorithm works back, either rejecting problems or creating further sub-trees by branching and perhaps updating the best integer feasible solution $\hat{\mathbf{x}}$ and $\hat{f}$. This is illustrated in Figure 13.1.2.

So far the description of the algorithm is incomplete in that rules are required for choosing the branching variable if more than one possibility exists, and for deciding which branched problem goes on the stack first. To make this decision requires estimates e_i^+ and e_i^- (>0) of the increase in f when adding the branching constraints $x_i \geqslant [x_i'] + 1$ and $x_i \leqslant [x_i']$ respectively. These estimates can be made using information about first derivatives (and also second derivatives in other than LP problems) in a way which is described below. Then the rule is to choose i to solve $\max_i \max(e_i^+, e_i^-)$ and place the branched problem corresponding to $\max(e_i^+, e_i^-)$ on the stack first. Essentially the worst case is placed on the stack first, and so its solution is deferred until later. The motivation for this rule is hopefully that, being the worst case, this problem will have a high value of f and so can be rejected at a later stage with little or no branching. A numerical example of the stack method to the integer LP problem

$$\begin{aligned} &\text{minimize} && x_1 + 10x_2 \\ &\text{subject to} && 66x_1 + 14x_2 \geqslant 1430 \\ & && -82x_1 + 28x_2 \geqslant 1306, \quad x_1, x_2 \text{ integer} \end{aligned} \tag{13.1.5}$$

is illustrated in Figure 13.1.3. Some shading on the grid shows the effect of adding extra constraints to the feasible region, and numbered points indicate solutions to the subproblems. The solution tree is shown, and the typical progress described in Figure 13.1.2 can be observed. After branching on problem 7, the stack contains problems 9, 8, 6, 4, 2 in top to bottom order. Problem 9 gives rise to an integer feasible solution, which enables the remaining problems to

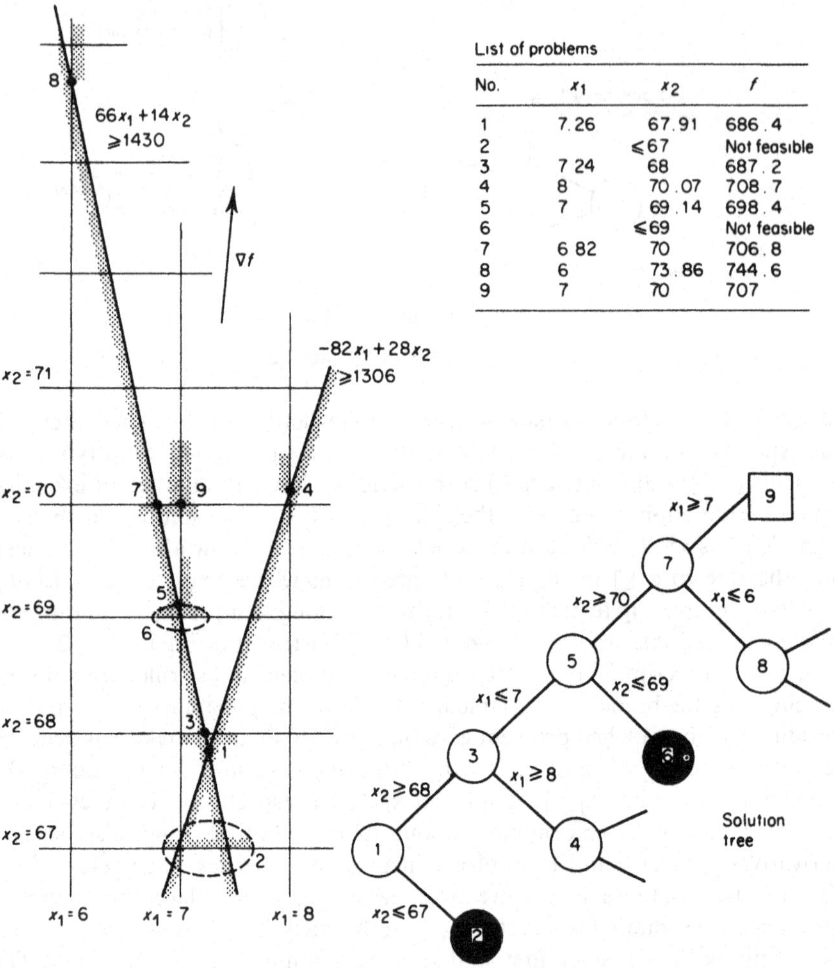

List of problems

No.	x_1	x_2	f
1	7.26	67.91	686.4
2		≤67	Not feasible
3	7.24	68	687.2
4	8	70.07	708.7
5	7	69.14	698.4
6		≤69	Not feasible
7	6.82	70	706.8
8	6	73.86	744.6
9	7	70	707

Figure 13.1.3 Solution of (13.1.5) by the branch and bound method

be rejected in turn without further branching. Notice that it may of course be necessary to branch on the same variable more than once in the tree.

In practice it has become apparent that the stack method contains some good features but can also be inefficient. The idea of following a path deep into the tree in order to find an integer feasible solution works well in that it rapidly determines a good value of $\hat{f}$ which can be used to reject other branches. This feature also gives a feasible solution if time does not permit the search to be completed. Furthermore it is very efficient in the amount of storage required to keep details of unsolved pending problems in the tree. However, the process of back-tracking from a □ node to neighbouring nodes in the tree can be unrewarding. Once a path can be followed no further it is preferable to select the best of all pending nodes to explore next, rather than that on top of the stack. To do this, it is convenient to associate an estimate E of its solution value with each unsolved problem which is created by branching. E is conveniently defined by $f' + e_i^+$ and $f' + e_i^-$ respectively, where f' is the value of the parent problem and e_i^+, e_i^- are the estimates referred to above. A path starting at the pending node with least E is then followed through the tree. Rules of this type are reviewed by Beale (1978) and Breu and Burdet (1974). Another useful idea is to have the user supply a cut-off value, above which solutions are of no interest. This can be used initially in place of $\hat{f} = \infty$ and may prevent deep exploration of unfavourable paths. A further disadvantage of the stack method is that the 'worst case' rule for choosing the branching variable is often inadequate. Present algorithms prefer to branch on the 'most important variable', that is the one which is likely to be most sensitive in determining the solution. This can be determined from user assigned 'priorities' on the variables (often zero–one variables should be given high priority). To choose between variables of equal priority the variable which solves $\max_i \min(e_i^-, e_i^+)$ is selected, that is the one for which the least estimate of the increase in f is largest. This quantifies the definition of the most important variable. Other possible selection rules are discussed by Breu and Burdet (1974).

The branch and bound strategies depend critically on making good estimates e_i^+ and e_i^- of the increases in f caused by adding either of the branching constraints. Such estimates are readily made for an integer LP problem and are most easily described in terms of the active set method (Section 8.3). Let $\mathbf{x}'$ be the solution of a parent node, value f'. Then the directions of feasible edges at $\mathbf{x}'$ are vectors $\mathbf{s}_q$ where $\mathbf{A}^{-T} = [\mathbf{s}_1, \mathbf{s}_2, \ldots, \mathbf{s}_n]$. The slope of f along $\mathbf{s}_q$ is given by the multiplier $\lambda_q (\geqslant 0)$. Consider finding e_i^+. A search along $\mathbf{s}_q$ increases x_i if $(\mathbf{s}_q)_i > 0$ at a rate $(\mathbf{s}_q)_i$, and increases f at a rate λ_q. Thus for small changes Δx_i, if a constraint $x_i \geqslant x_i' + \Delta x_i$ is added to the problem, then the increase Δf in the optimal solution value f' is given by $\Delta f = p_i^+ \Delta x_i$ where

$$p_i^+ = \min_{q:(\mathbf{s}_q)_i > 0} \frac{\lambda_q}{(\mathbf{s}_q)_i}. \tag{13.1.6}$$

For larger changes, previously inactive constraints may become active along the corresponding edge, and so further increases in f occur on restoring feasibility. Thus in general a lower bound $\Delta f \geqslant p_i^+ \Delta x_i$ on the increase in f is obtained. Thus if the fractional part of x_i' is $\phi_i (= x_i' - [x_i'])$ then the required estimate of the change in f is

$$e_i^+ = p_i^+ (1 - \phi_i). \tag{13.1.7}$$

It may be that no element $(\mathbf{s}_q)_i > 0$ exists, corresponding to the fact that no feasible direction exists which increases x_i. Then the resulting branched problem is infeasible (a $\bullet$ node). It can be convenient to define $e_i^+ = p_i^+ = \infty$ in this case. A similar development can be used when decreasing x_i. Consider a small change $\Delta x_i (> 0)$, and add the constraint $x_i \leqslant x_i' - \Delta x_i$ to the problem. Then

$\Delta f = p_i^- \Delta x_i$ where

$$p_i^- = \min_{q:(\mathbf{s}_q)_i<0} \frac{\lambda_q}{-(\mathbf{s}_q)_i}. \tag{13.1.8}$$

In general $\Delta f \geqslant p_i^- \Delta x_i$, and the required estimate is

$$e_i^- = p_i^- \phi_i. \tag{13.1.9}$$

The estimates (13.1.7) and (13.1.9) are used as described above in choosing the branching variable and estimating E for each unsolved problem created by branching. In fact since for integer LP these estimates give strict lower bounds on Δf, it follows that use of $L = E$ gives a tighter lower bound than the value of the parent problem $L = f'$. Other ways in which the structure of an LP problem can be used advantageously are given by Beale (1968). Often the dual formulation is recommended for solving the LP problem since constraints can be added whilst retaining feasibility. However, it seems to me that the standard form given in Question 8.8 would also be very suitable, using a cost function like (8.4.5) to restore feasibility. When solving linear constraint problems in which the objective is nonlinear and when the solution $\mathbf{x}'$ to any subproblem lies at a vertex, then similar estimates to the above can be made. These estimates are strict bounds if the function is convex. If the solution is not at a vertex then second order effects have to be introduced into (13.1.6) or (13.1.8) along the directions of zero slope which exist to avoid the trivial case $p_i^+ = p_i^- = 0$. I do not know whether this possibility has been investigated in practice although the information to do it is available. Similar estimates can also be made in integer problems with nonlinear constraints.

An example in which these estimates are calculated is given by problem (13.1.5) whose solution is described in Figure 13.1.3. Consider possible branches at the solution of the root problem (1). Feasible edges are defined by

$$\mathbf{A}^{-1} = \frac{1}{2996}\begin{bmatrix} 28 & 82 \\ -14 & 66 \end{bmatrix} = \begin{bmatrix} \mathbf{s}_1^T \\ \mathbf{s}_2^T \end{bmatrix}$$

and

$$\boldsymbol{\lambda} = \mathbf{A}^{-1}\mathbf{c} = \frac{1}{2996}\begin{pmatrix} 848 \\ 646 \end{pmatrix}.$$

Consider branching on x_2: searching along either $\mathbf{s}_1$ or $\mathbf{s}_2$ increases x_2 so it can be predicted that the branch $x_2 \leqslant 67$ is infeasible and $p_2^- = e_2^- = \infty$. For increasing x_2,

$$p_2^+ = \min\left(\frac{848}{82}, \frac{646}{66}\right) = \frac{646}{66}$$

and $e_2^+ = 0.09 \times 646/66 \approx 0.9$ which agrees with the observed increase in value in going from problem 1 to 3, since no other constraints become active. For branching on x_1,

$$p_1^+ = 848/28 \approx 30, \qquad p_1^- = 646/14 \approx 46$$
$$e_1^+ = 0.74 \times 30 \approx 22, \qquad e_1^- \approx 0.26 \times 46 \approx 12$$

which again agrees with observed changes in value. These estimates also illustrate the different rules for choosing branching variables. The 'worst case' rule in the stack method is achieved by $e_2^- = \infty$ and gives rise to a branch on x_2. The 'most important variable' rule is solved by $e_1^- = 12$ and causes a branch on x_1. In fact this is a much better branch since it yields problems 4 and 5 directly, and this supports the current preference for this type of rule.

13.2 GEOMETRIC PROGRAMMING

Another important type of problem which can also be transformed to the solution of a more simple problem is known as the *geometric programming problem.* This is a nonlinear programming problem in which the problem functions are constructed from *terms* of the form

$$t_i = c_i \prod_{j=1}^{n} x_j^{a_{ij}}, \tag{13.2.1}$$

where the a_{ij} are real and may be negative or fractional for instance. The coefficients c_i and t_i however must be positive, and a sum of such terms is referred to as a *posynomial.* The problem

$$\begin{aligned} &\underset{\mathbf{x}}{\text{minimize}} \quad g_0(\mathbf{x}) \\ &\text{subject to} \quad g_k(\mathbf{x}) \leqslant 1, \quad k = 1, 2, \ldots, m \\ &\qquad\qquad\quad \mathbf{x} > \mathbf{0} \end{aligned} \tag{13.2.2}$$

in which the functions $g_k(\mathbf{x})$ are posynomials is a convenient standard form for a *geometric programming problem.* The terms which occur within $g_0, g_1, \ldots, g_m$ are numbered consecutively from 1 to p, say, and sets J_k which collect the indices of terms in each g_k are defined by

$$J_k = \{r_k, r_k + 1, \ldots, s_k\},$$

so that (13.2.3)

$$r_0 = 1, \qquad r_{k+1} = s_k + 1, \qquad \text{and} \qquad s_m = p.$$

Thus the posynomials $g_k(\mathbf{x})$ are defined by

$$g_k = \sum_{i \in J_k} t_i, \qquad k = 0, 1, \ldots, m. \tag{13.2.4}$$

Geometric programming has met with a number of applications in engineering (see, for instance, Duffin, Peterson, and Zener (1967), Dembo and Avriel (1978), and Bradley and Clyne (1976), amongst many). There could well be scope for even more applications (either directly or by making suitable posynomial approximations) if the technique were more widely understood and appreciated.

The geometric programming problem (13.2.2) has nonlinear constraints but a transformation exists which causes considerable simplification. In some cases the solution can be found merely by solving a certain linear system, and in general it is possible to solve a more simple but equivalent minimization problem in which only linear constraints arise. In this problem the relative contribution of the various terms t_i is determined, and a subsequent calculation is then made to find the optimum variables $\mathbf{x}^*$. This transformation was originally developed as an application of the arithmetic–geometric mean inequality (see Duffin, Peterson, and Zener, 1967), hence the name geometric programming. However a more straightforward approach is to develop the transformation as an example of the duality theory of Section 9.5. First of all the transformation of variables

$$x_j = \exp(y_j), \quad j = 1, 2, \ldots, n \tag{13.2.5}$$

is made, which removes the constraints $x_j > 0$ and allows (13.2.1) to be written

$$t_i = c_i \exp(\textstyle\sum a_{ij} y_i). \tag{13.2.6}$$

In addition (13.2.2) is written equivalently as

$$\begin{aligned} &\underset{\mathbf{y}}{\text{minimize}} && \log g_0(\mathbf{y}) \\ &\text{subject to} && \log g_k(\mathbf{y}) \leqslant 0, \quad k = 1, 2, \ldots, m. \end{aligned} \tag{13.2.7}$$

The main result then follows by an application of the duality transformation to (13.2.7).

So that this transformation is applicable it is first of all important to establish that each function $\log g_k(\mathbf{y})$ is a convex function on $\mathbb{R}^n$. To do this let $\mathbf{y}_0, \mathbf{y}_1 \in \mathbb{R}^n$ and $\theta \in (0, 1)$. If $u_i = [t_i(\mathbf{y}_0)]^{1-\theta}$ and $v_i = [t_i(\mathbf{y}_1)]^{\theta}$ are defined, then it follows from (13.2.6) that

$$u_i v_i = t_i(\mathbf{y}_\theta)$$

where $\mathbf{y}_\theta = (1-\theta)\mathbf{y}_0 + \theta \mathbf{y}_1$. If $\mathbf{u}$ collects the elements $u_i, i \in J_k$, $\mathbf{v}$ similarly, then $\mathbf{u}^T\mathbf{v} = g_k(\mathbf{y}_\theta)$. If $p = 1/(1-\theta)$, $q = 1/\theta$ are defined, then Hölder's inequality

$$\mathbf{u}^T\mathbf{v} \leqslant \|\mathbf{u}\|_p \|\mathbf{v}\|_q \tag{13.2.8}$$

is valid (since $u_i > 0$, $v_i > 0$, and $p^{-1} + q^{-1} = 1$), where $\|\mathbf{u}\|_p \triangleq (\sum u_i^p)^{1/p}$. Taking logarithms of both sides in (13.2.8) establishes that

$$\log g_k(\mathbf{y}_\theta) \leqslant (1-\theta) \log g_k(\mathbf{y}_0) + \theta \log g_k(\mathbf{y}_1)$$

and so $\log g_k(\mathbf{y})$ satisfies the definition (9.4.3) of a convex function.

It follows from this result that (13.2.7) is a convex programming problem (see

(9.4.8)). Thus the dual problem (9.5.1) becomes

$$\begin{aligned} &\underset{\mathbf{y},\boldsymbol{\lambda}}{\text{maximize}} \quad \mathscr{L}(\mathbf{y},\boldsymbol{\lambda}) \triangleq \log g_0(\mathbf{y}) + \sum_{k=1}^{m} \lambda_k \log g_k(\mathbf{y}) \\ &\text{subject to} \quad \nabla_y \mathscr{L}(\mathbf{y},\boldsymbol{\lambda}) = \mathbf{0}, \quad \boldsymbol{\lambda} \geqslant \mathbf{0}. \end{aligned} \tag{13.2.9}$$

The Lagrangian can be written more simply by defining $\lambda_0 = 1$ so that

$$\mathscr{L}(\mathbf{y},\boldsymbol{\lambda}) = \sum_{k=0}^{m} \lambda_k \log g_k(\mathbf{y}). \tag{13.2.10}$$

As it stands, (13.2.9) has nonlinear constraints so it is important to show that it can be simplified by a further change of variable. Write the equality constraints as

$$\frac{\partial \mathscr{L}}{\partial y_j} = 0 = \sum_{k=0}^{m} \lambda_k \sum_{i \in J_k} \frac{a_{ij} t_i}{g_k}, \quad j = 1, 2, \ldots, n. \tag{13.2.11}$$

These equations suggest defining new non-negative variables δ_i by

$$\delta_i = \frac{\lambda_k t_i}{g_k}, \quad \forall i \in J_k, \quad k = 0, 1, \ldots, m, \tag{13.2.12}$$

which correspond to the terms (13.2.1), suitably weighted. Then (13.2.11) becomes

$$\sum_{i=1}^{p} \delta_i a_{ij} = 0, \quad j = 1, 2, \ldots, n, \tag{13.2.13}$$

and these equations are known as *orthogonality* constraints. Thus the equality constraint in (13.2.9) is equivalent to a set of (under-determined) linear equations in terms of the δ_i variables. An equivalent form of (13.2.4) is also implied by (13.2.12), that is

$$\sum_{i \in J_k} \delta_j = \lambda_k, \quad k = 0, 1, \ldots, m, \tag{13.2.14}$$

and is known as a *normalization* constraint. Furthermore it is possible to simplify the dual objective function (13.2.10), because (assuming $0 \log 0 = 0$)

$$\begin{aligned} \lambda_k \log g_k &= \lambda_k \log \lambda_k + \lambda_k \log (g_k/\lambda_k) \\ \text{by (13.2.14)} \quad &= \lambda_k \log \lambda_k + \sum_{i \in J_k} \delta_i \log (g_k/\lambda_k) \\ \text{by (13.2.12)} \quad &= \lambda_k \log \lambda_k + \sum_{i \in J_k} \delta_i \log (t_i/\delta_i) \\ \text{by (13.2.6)} \quad &= \lambda_k \log \lambda_k + \sum_{i \in J_k} \delta_i \log (c_i/\delta_i) + \sum_{i \in J_k} \delta_i \sum_{j=1}^{n} a_{ij} y_j \\ \text{by (13.2.13)} \quad &= \lambda_k \log \lambda_k + \sum_{i \in J_k} \delta_i \log (c_i/\delta_i). \end{aligned} \tag{13.2.15}$$

Thus the weighted terms δ_i and the Lagrange multipliers λ_k satisfy the following *dual geometric programming problem*

$$\begin{aligned}
&\underset{\boldsymbol{\delta},\boldsymbol{\lambda}}{\text{maximize}} \quad v(\boldsymbol{\delta},\boldsymbol{\lambda}) \triangleq \sum_{i=1}^{p} \delta_i \log(c_i/\delta_i) + \sum_{k=1}^{m} \lambda_k \log \lambda_k \\
&\text{subject to} \quad \sum_{i=1}^{p} \delta_i a_{ij} = 0 \qquad j = 1,2,\ldots,n \\
&\qquad\qquad\quad \sum_{i \in J_k} \delta_i = \lambda_k \qquad k = 0,1,\ldots,m \\
&\qquad\qquad\quad \delta_i \geqslant 0 \qquad i = 1,2,\ldots,p \\
&\qquad\qquad\quad \lambda_k \geqslant 0 \qquad k = 1,2,\ldots,m
\end{aligned} \tag{13.2.16}$$

where $\lambda_0 = 1$.

This then is the resulting transformed problem which is more easily solved than (13.2.2). The problem has all linear constraints and a nonlinear objective function so can be solved by the methods of Chapter 11. The number of degrees of freedom left when the constraints have been eliminated is known as the *degree of difficulty* of the problem, and is $p - n - 1$ if all the primal constraints are active. In some cases the degree of difficulty is known to be zero, in which case (13.2.16) reduces just to the solution of the orthogonality and normalization equations. Usually however it is necessary to solve (13.2.16) by a suitable numerical method. In fact (13.2.16) has other advantageous features. The dual objective function $v(\boldsymbol{\delta}, \boldsymbol{\lambda})$ is *separable* into one variable terms so that the Hessian matrix is diagonal and is readily calculated. If the λ_k variables are eliminated from (13.2.14) then the resulting function (of $\boldsymbol{\delta}$ only) is concave (see Duffin, Peterson, and Zener, 1967), so it follows from Theorem 9.4.1 that any solution of (13.2.16) is a global solution. A more thorough study of how best to solve (13.2.16) by using a linearly constrained Newton method of the type described in Chapter 11 is given by Dembo (1979). Amongst the factors which he takes into account are the need to eliminate in a stable fashion as described in Chapter 10 here. He also considers how best to take sparsity of the exponent matrix $[a_{ij}]$ into account, and extends (13.2.2) to include lower and upper bound constraints

$$0 \leqslant l_i \leqslant x_i \leqslant u_i.$$

(These can be written as posynomial constraints $u_i^{-1} x_i \leqslant 1$ and $l_i x_i^{-1} \leqslant 1$ but it is more efficient to treat them separately.) A particularly interesting feature is that when a posynomial constraint $g_k \leqslant 1$ is inactive at the solution then a whole block of variables δ_i, $i \in J_k$, and λ_k are zero. Thus Dembo considers modifications to the active set strategy which allow the addition or removal of such blocks of indices to or from the active set.

Once the dual solution $\boldsymbol{\delta}^*, \boldsymbol{\lambda}^*$ has been found, then the primal function value is given by $\log g_0^* = v(\boldsymbol{\delta}^*, \boldsymbol{\lambda}^*)$ (see Theorem 9.5.1). It is also important to be able to recover the optimum variables $\mathbf{y}^*$ and hence $\mathbf{x}^*$ in the primal problem. In principle it is possible to do this by solving the nonlinear equations (13.2.4) and (13.2.6) for the active primal constraints for which $\lambda_k^* > 0$ and hence $g_k^* = 1$.

However, a more simple possibility is to determine the optimum multipliers $w_j^*, j = 1, 2, \ldots, n$, of the orthogonality constraints in (13.2.16). These multipliers are available directly if (13.2.16) is solved by an active set method such as in Chapter 11.2. Then

$$w_j^* = y_j^* = \log x_j^* \tag{13.2.17}$$

determines the primal solution. To justify (13.2.17), the first order conditions (9.1.3) applied to (13.2.16) give

$$\begin{aligned} \log(\delta_i/c_i) + 1 &= \mu_k + \sum_j a_{ij} w_j, \qquad i \in J_k \\ -\log \lambda_k - 1 &= -\mu_k \end{aligned} \tag{13.2.18}$$

where w_j and μ_k are the multipliers of the orthogonality and normalization constraints respectively and k indexes the active primal constraints. Eliminating μ_k gives

$$\log \frac{\delta_i}{c_i \lambda_k} = \sum_j a_{ij} w_j$$

and since $g_k^* = 1$, it follows from (13.2.12) and (13.2.6) that

$$\sum a_{ij} y_j = \sum a_{ij} w_j.$$

Since the exponent matrix $[a_{ij}]$ has more rows than columns and may be assumed to have full rank (see below), the result (13.2.17) follows. There is no loss of generality in assuming that $[a_{ij}]$ has full rank. If not, then from (13.2.6) it is possible to eliminate one or more y_j variable in favour of the remaining y_j variables, so as to give a matrix $[a_{ij}]$ of full rank.

Finally a simple example of the geometric programming transformation is given. Consider the geometric programming problem in two variables with one constraint

$$\begin{aligned} &\underset{\mathbf{x}}{\text{minimize}} \quad g_0(\mathbf{x}) \triangleq \frac{2}{x_1 x_2^{1/2}} + x_1 x_2 \\ &\text{subject to} \quad g_1(\mathbf{x}) \triangleq \tfrac{1}{2} x_1 + x_2 \leqslant 1, \qquad x_1, x_2 > 0. \end{aligned} \tag{13.2.19}$$

There are four terms, so the orthogonality constraints are

$$\begin{aligned} -\delta_1 + \delta_2 + \delta_3 \qquad &= 0 \\ -\tfrac{1}{2}\delta_1 + \delta_2 \qquad + \delta_4 &= 0 \end{aligned}$$

and the normalization constraints are

$$\begin{aligned} \delta_1 + \delta_2 \qquad &= 1 \\ \delta_3 + \delta_4 &= \lambda_1. \end{aligned}$$

Assuming that the primal constraint is active, the δ_i variables can be eliminated in terms of $\lambda_1 (= \lambda)$ giving

$$\boldsymbol{\delta}(\lambda) = \frac{(2\lambda + 4, -2\lambda + 3, 3\lambda - 1, 4\lambda + 1)^{\mathrm{T}}}{7}. \tag{13.2.20}$$

This illustrates that there is one degree of difficulty ($=4-2-1$) and therefore a maximization problem in one variable (λ here) must be solved. Maximizing the dual objective function $v(\boldsymbol{\delta}(\lambda), \lambda)$ by a line search in λ gives $\lambda^* = 1.0247$ and $v^* = 1.10764512$. Also from (13.2.20) it follows that

$$\boldsymbol{\delta}^* = (0.86420, 0.13580, 0.72841, 0.29630)^{\mathrm{T}}.$$

Thus the dual geometric programming problem has been solved, and immediately $g_0^* = \exp v^* = 3.02722126$. Substituting in (13.2.18) for $k=1$ and $J_1 = \{3, 4\}$ gives $\mu_1^* = 1.02441$ and hence

$$\begin{pmatrix} 0.35184 \\ -1.24078 \end{pmatrix} = \begin{bmatrix} 1 & 0 \\ 0 & 1 \end{bmatrix} \begin{pmatrix} w_1^* \\ w_2^* \end{pmatrix}.$$

Hence by (13.2.17) the vector of variables which solves the primal is $\mathbf{x}^* = (1.4217, 0.28915)^{\mathrm{T}}$. In fact the primal problem is fairly simple and it is also possible to obtain a solution to this directly by eliminating x_2 and carrying out a line search. This calculation confirms the solution given by the geometric programming transformation.

13.3 NETWORK PROGRAMMING

Many optimization problems model physical situations involving flow in a *network*. These include fluid flow, charge flow (current) in an electrical network, traffic flow in a transportation network, data flow in a communications network, and so on. In such networks the conservation of flow conditions give rise to a set of linear constraints known as *network constraints* which have a particular algebraic structure. These constraints can be manipulated by techniques which are not only very elegant but also highly efficient, thus enabling very large problems to be solved having perhaps as many as 10^6 variables. Most of the section is concerned with the structure, representation and manipulation of network constraints, together with their use in linear programming and more complex optimization problems.

The appropriate mathematical concept for modelling networks is a *graph*, and a few simple concepts from graph theory are informally introduced. A graph consists of a finite collection of *nodes* and *arcs*, each arc representing a connection between two different nodes. The arcs are said to be *directed* if a sense of orientation of each arc is specified. A *path* is a sequence of adjacent arcs connecting any two nodes, together with an orientation (the orientation of arcs within the path need not coincide with the orientation of the path but any arc can only be traversed once in a given path). A *cycle* is a non-trivial path connecting one node to itself. A graph is *connected* if a path exists between any two nodes. A *tree* is a connected graph having no cycles. A graph G_1 is a *subgraph* of a graph G if both the nodes and arcs of G_1 are subsets of the nodes and arcs of G. A tree T is a *spanning tree* of a graph G if T is a subgraph of G having the same set of nodes. This section considers the case of a network which

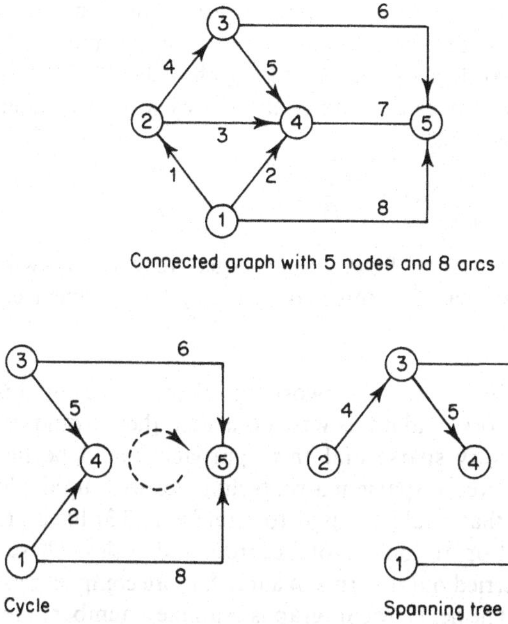

Figure 13.3.1 Graph theory concepts

is a connected graph having m nodes and n directed arcs (each being suitably numbered). These concepts are illustrated in Figure 13.3.1.

An important application of networks is to *network flow problems* in which it is required to find the flow, x_j say, in each arc j, measured in the direction of arc j, and subject to conservation of flow at the nodes. (By *flow* is meant the amount of substance (fluid, charge, traffic, etc.) transported per unit time.) It is also assumed that there exists an *external flow* b_i at each node i, with the convention that $b_i > 0$ if the flow is into the node i (a *source*), and $b_i < 0$ if the flow is out of node i (a *sink*). The total external flow $\sum b_i$ is necessarily assumed to be zero. An algebraic representation of this system is conveniently obtained by defining the $m \times n$ *node-arc incidence matrix* $\mathbf{A}$ by

$$a_{ij} = \begin{cases} +1 & \text{if arc } j \text{ is directed away from node } i \\ -1 & \text{if arc } j \text{ is directed towards node } i \\ 0 & \text{otherwise} \end{cases} \tag{13.3.1}$$

For the network illustrated in Figure 13.3.1 the node–arc incidence matrix is

$$\mathbf{A} = \begin{bmatrix} 1 & 1 & 0 & 0 & 0 & 0 & 0 & 1 \\ -1 & 0 & 1 & 1 & 0 & 0 & 0 & 0 \\ 0 & 0 & 0 & -1 & 1 & 1 & 0 & 0 \\ 0 & -1 & -1 & 0 & -1 & 0 & 1 & 0 \\ 0 & 0 & 0 & 0 & 0 & -1 & -1 & -1 \end{bmatrix}. \tag{13.3.2}$$

Rows of **A** correspond to nodes, columns to arcs. The non-zero elements in any row i give the arcs connected to node i and their orientation. Any column j has only two non-zero elements, one is $+1$ and the other -1 (the ± 1 *property*) and these give the two nodes that are connected by arc j. Consider row i: the total flow entering node i is

$$b_i + \sum_{j:a_{ij}=-1} x_j - \sum_{j:a_{ij}=1} x_j = b_i - \sum_j a_{ij}x_j.$$

By virtue of conservation of flow, this is equal to zero, showing that the conservation constraint can be expressed by the system of linear equations

$$\mathbf{Ax} = \mathbf{b}. \tag{13.3.3}$$

It is of course possible to solve network problems by using this algebraic representation of network, and if this were done then there would be little more to say. In fact **A** is very sparse and this approach would be inefficient for problems of any size, even if sparse matrix techniques were used. However, the algebraic techniques that could be used to handle (13.3.3) have graph theory analogues in terms of operations involving trees and cycles. These operations are most efficiently carried out in terms of a much more compact representation of the original graph. The next few paragraphs examine a number of relationships that exist between the linear algebraic representation and the graph theory representation.

First of all, observe that if all the equations in (13.3.3) are summed, the equation $\mathbf{0}^T\mathbf{x} = 0$ is obtained (by virtue of $\sum b_i = 0$ and the ± 1 property). Hence the equations are linearly dependent and rank $\mathbf{A} \leqslant m-1$. In fact it is shown in Theorem 13.3.1 below that the connectedness assumption ensures that rank $\mathbf{A} = m-1$ and the linear dependence can be removed by deleting any one equation. (The situation that the graph is not connected is explored in Question 13.6.) In terms of the graph, this is equivalent to deleting any node, which by analogy with an electrical network could be called the *earth node*. Columns of **A** corresponding to arcs connected to the earth node would then only have one non-zero element after this deletion. Alternatively the redundancy can be removed by postulating the existence of an earth node which is not in the original graph, and connecting this node to any node in the graph by adding an extra arc. This adds an extra column having only one non-zero element to the matrix **A**. In terms of the example in Figure 13.3.1 *either* node 5 say could be regarded as the earth node, thus deleting row 5 of **A** in (13.3.2), which becomes a 4×8 matrix, *or* an extra arc (9) could be directed from a separate earth node to node 5, so that an extra column $-\mathbf{e}_5$ is added to **A**, giving a 5×9 matrix. Clearly the overall effect of these two approaches is much the same. Subsequently in this section it is assumed that the graph is connected, has m nodes, and contains one or more arcs that are connected to an earth node (regarded as a single node) outwith the graph. For such a graph, a spanning tree is said to be *rooted at earth* if it includes one arc (only) connected to the earth node. These concepts are illustrated in Figure 13.3.2.

In both LP, QP, and other optimization problems, a system of linear equations

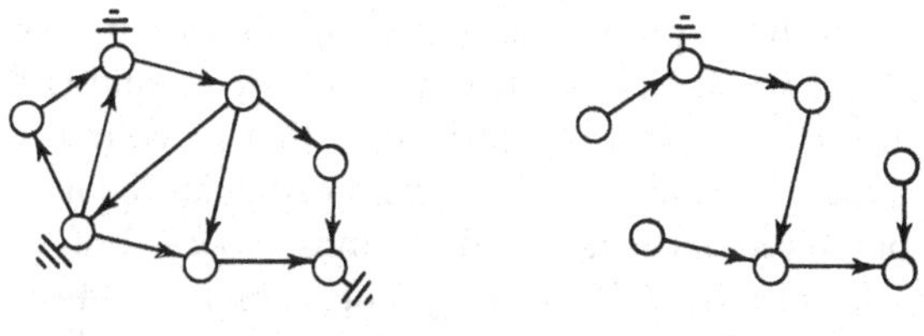

Figure 13.3.2 The earth node

like (13.3.3) is used to eliminate (temporarily) a subset of variables from the problem (e.g. (8.2.4) and (10.1.3)). In LP terminology there exists an index set B (the basis) having m elements such that the vectors $\mathbf{a}_j$ $j \in B$ are linearly independent ($\mathbf{a}_j = \mathbf{A}\mathbf{e}_j$). A remarkable result connects this concept with that of a spanning tree in the above graph. Define B as *quasi-triangular* if there exist $\mathbf{P}, \mathbf{Q}$ and $\mathbf{U}$ such that

$$\mathbf{P}^T\mathbf{A}\mathbf{Q} = \mathbf{U} \tag{13.3.4}$$

where $\mathbf{U}$ is upper triangular, $\mathbf{P}$ is a permutation matrix (both $m \times m$) and $\mathbf{Q}$ is an $n \times m$ matrix whose columns are the vectors $\mathbf{e}_j$ $j \in B$ in any order.

Theorem 13.3.1

For the graph defined in the previous paragraph

(i) $\exists$ *a basis* ($\Rightarrow$rank $\mathbf{A} = m$),

(ii) *to every basis there corresponds a spanning tree rooted at earth, and vice versa*

(iii) *every basis is quasi-triangular.*

Proof

Any spanning tree rooted at earth can be constructed by the following algorithm in which $N^{(k)}$ denotes a subset of the nodes, starting with $N^{(0)}$ as the earth node.

For $k = 1, 2, \ldots, m$ do
(a) Find any arc, α_k say, connecting any node in $N^{(k-1)}$ to a node, n_k say, not in $N^{(k-1)}$. (13.3.5)
(b) Set $N^{(k)} = N^{(k-1)} \cup n_k$.

Existence of an arc in (a) is assured by the connectedness assumption. Incidentally this algorithm shows that a rooted spanning tree has exactly m arcs. Now let $\mathbf{P}$ have columns $\mathbf{p}_k = \mathbf{e}_{n_k} \in \mathbb{R}^m$ and $\mathbf{Q}$ have columns $\mathbf{q}_k = \mathbf{e}_{\alpha_k} \in \mathbb{R}^n$, and define $\mathbf{U}$ by (13.3.4). Then $\mathbf{e}_{n_k}^T\mathbf{A}\mathbf{e}_{\alpha_k} = \pm 1$ since arc α_k is connected to node n_k. Also for $k > 1$ it follows that $\mathbf{e}_{n_l}^T\mathbf{A}\mathbf{e}_{\alpha_k} = \mp 1$ since arc α_k is also connected to another node n_l, and $l < k$ since $n_l \in N^{(k-1)}$. Other elements in column k of $\mathbf{U}$ are necessarily zero by the ± 1 property so $\mathbf{U}$ is upper triangular and non-singular. This construction proves (i) and $\Leftarrow$ in (ii).

Now let a basis B be given. The following algorithm constructs matrices $\mathbf{A}^{(k)} = \mathbf{P}^{(k)T}\mathbf{A}\mathbf{Q}^{(k)}$ for $k = 1, 2, \ldots, n$ where $\mathbf{P}^{(k)}, \mathbf{A}^{(k)} \in \mathbb{R}^{m \times m}$ and $\mathbf{Q}^{(k)} \in \mathbb{R}^{n \times m}$, starting from $\mathbf{P}^{(1)} = \mathbf{I}$ and $\mathbf{Q}^{(1)}$ being any matrix such that its columns are the vectors $\mathbf{e}_j\ j \in B$ in any order. The matrices $\mathbf{A}^{(k)}$ are non-singular, being row and column permutations of the vectors $\mathbf{a}_j\ j \in B$, and columns 1 to $k-1$ of $\mathbf{A}^{(k)}$ form an upper triangular matrix. Clearly this property holds for $k = 1$. For general k, sum rows k to m of $\mathbf{A}^{(k)}$: the resulting vector must be non-zero by linear independence of $\mathbf{a}_j\ j \in B$. Hence $\exists$ a column of $\mathbf{A}^{(k)}$ which has only a single non-zero element in rows k to m, by the ± 1 property. Let this element occur in position i, j: clearly $i \geqslant k$, also $j \geqslant k$ by the upper triangular property of rows 1 to $k-1$ of $\mathbf{A}^{(k)}$. Finally interchange columns i and k of $\mathbf{P}^{(k)}$ to get $\mathbf{P}^{(k+1)}$ and columns j and k of $\mathbf{Q}^{(k)}$ to get $\mathbf{Q}^{(k+1)}$. Clearly $\mathbf{A}^{(k+1)}$ satisfies the appropriate properties, so by induction $\mathbf{A}^{(m)}$ is upper triangular and non-singular. $\mathbf{P}^{(m)}$, $\mathbf{Q}^{(m)}$ and $\mathbf{A}^{(m)}$ can be identified with $\mathbf{P}, \mathbf{Q}$ and $\mathbf{U}$ respectively in (13.3.4). Hence any basis is quasi-triangular. It is easily seen that using $\mathbf{P}$ and $\mathbf{Q}$ to determine $\mathbf{n}$ and $\boldsymbol{\alpha}$ respectively, enables a rooted spanning tree to be constructed in the order given by algorithm (13.3.5). This proves (iii) and $\Rightarrow$ in (ii) and completes the proof. $\square$

There are a number of other interesting linear algebraic properties of the basis B which have an equivalent graph theoretic description in terms of the corresponding rooted spanning tree, T say. Let $\mathbf{A}_B = \mathbf{A}\mathbf{Q}$ be the basis matrix and consider row j of $\mathbf{A}_B^{-1}$. If arc α_j is removed from T, then two subtrees T_1 and T_2 are formed, one of which (T_1 say) is rooted at earth. Then row j has elements

$$[\mathbf{A}_B^{-1}]_{jk} = \begin{cases} s & \text{if } n_k \in T_2 \\ 0 & \text{if } n_k \in T_1 \end{cases} \tag{13.3.6}$$

where n_k refers to the node corresponding to row k (see (13.3.5)) and

$$s = \begin{cases} -1 & \text{if } a_j \text{ is directed away from the root in } T \\ 1 & \text{otherwise.} \end{cases}$$

To prove this result, the ordering in (13.3.5) can be chosen without loss of generality so that rows 1 to $k-1$ of $\mathbf{A}_B$ correspond to nodes in T_1 and rows k to m correspond to nodes in T_2. Then $\mathbf{A}_B$ is block upper triangular which proves (13.3.6) for $n_k \in T_1$. The result for $n_k \in T_2$ is proved inductively by considering the product of row j of $\mathbf{A}_B^{-1}$ with columns $\mathbf{A}_B$ in the upper triangular ordering, starting from column k.

Another interesting result concerns column k of $\mathbf{A}_B^{-1}$. Define $P(n_k)$ as the direct path from node n_k to earth on tree T. Then column k of $\mathbf{A}_B^{-1}$ has elements

$$[\mathbf{A}_B^{-1}]_{jk} = \begin{cases} 1 & \text{if arc } \alpha_j \in P(n_k) \text{ and its direction} \\ & \text{coincides with that of } P(n_k) \\ -1 & \text{if arc } \alpha_j \in P(n_k) \text{ and its direction} \\ & \text{is opposite to that of } P(n_k) \\ 0 & \text{otherwise } (\alpha_j \notin P(n_k)). \end{cases} \tag{13.3.7}$$

This result is readily proved from (13.3.6). An extension of this result is important in what follows. Define $P(n_k, n_l)$ as the unique path from node n_k to node n_l, both in tree T. Then the vector $\mathbf{y} = \mathbf{A}_B^{-1}(\mathbf{e}_k - \mathbf{e}_l)$ has elements

$$y_j = \begin{cases} 1 & \text{if arc } \alpha_j \in P(n_k, n_l) \text{ and its direction} \\ & \text{coincides with that of } P(n_k, n_l) \\ -1 & \text{if arc } \alpha_j \in P(n_k, n_l) \text{ and its direction} \\ & \text{is opposite to that of } P(n_k, n_l) \\ 0 & \alpha_j \notin P(n_k, n_l) \end{cases} \tag{13.3.8}$$

This is a direct consequence of (13.3.7).

Networks are of particular interest when they arise in optimization problems, and first of all the possibility of linear programming with networks is considered. If the problem is such that there is a cost per unit flow c_j associated with each arc j, then the objective function to be minimized is the total cost $\sum c_j x_j$ summed over all arcs, which is linear in $\mathbf{x}$. Often there are upper bounds on the flow in an arc (known as *capacities*) and also lower bounds which are usually zero. Taking these features together with the network equations $\mathbf{Ax} = \mathbf{b}$, a suitable standard form for the *linear network flow problem* is

$$\begin{aligned} &\text{minimize} \quad \mathbf{c}^{\mathrm{T}}\mathbf{x} \\ &\text{subject to} \quad \mathbf{Ax} = \mathbf{b}, \quad \mathbf{l} \leqslant \mathbf{x} \leqslant \mathbf{u}. \end{aligned} \tag{13.3.9}$$

This modifies the standard form (8.1.1) to include upper bounds on $\mathbf{x}$. Application of the simplex method to this problem is very much as described in Section 8.2, and the changes needed to accommodate the upper bounds are described in Question 8.8. In this section it is shown how each of the steps in the simplex method can be interpreted in graph theory terms. The outcome is again very revealing from the point of view of both insight and efficient computation.

Assume for the moment that a basic feasible solution is known. This is a vector of flows $\mathbf{x}$ which is feasible in (13.3.9) and basic, in that there exists a current rooted spanning tree T, and $x_j = l_j$ or u_j for arcs j that are not in this spanning tree (i.e. $j \in N$). Now consider the computation of $\boldsymbol{\pi} = \mathbf{A}_B^{-\mathrm{T}}\mathbf{c}_B$ in (8.2.7). From (13.3.4), this calculation is essentially a forward substitution in the order given by (13.3.5) and can be written

$$\pi_{n_k} = s c_{\alpha_k} + \pi_{n_l} \qquad k = 1, 2, \ldots, m \tag{13.3.10}$$

where n_l is the node in $N^{(k-1)}$ connected by arc α_k to node n_k and s is -1 or $+1$ according to whether or not the arc is directed towards node n_k. Note that when $k = 1$, then n_l refers to the earth node (n_0 say) and $\pi_{n_0} = 0$ is arbitrarily defined. To interpret this result it is instructive to use the concept of classical applied mathematics that the *potential difference* between two points is the work or energy required to move a unit amount of a substance from one point to the other. Taking a time derivative, potential difference is equivalently the power (energy/unit time) required per unit of flow (substance/unit time), which can be expressed as

$$\text{power} = \text{potential difference} \times \text{flow}. \tag{13.3.11}$$

If the objective function in linear network flow is regarded as the total power required to cause the given flow to occur, then the power required for a single arc is c_jx_j, and from (13.3.11) it follows that the coefficient c_j can be interpreted as the potential difference along arc j in the graph. Hence from (13.3.10) the elements of the vector $\boldsymbol{\pi}$ can be interpreted as the *potentials* at each node in the graph, and $\pi_{n_0} = 0$ is the assignment of an arbitrary zero potential to the earth node.

The next step is to consider calculating the vector of reduced costs $\hat{\mathbf{c}}_N = \mathbf{c}_N - \mathbf{A}_N^T\boldsymbol{\pi}$ in (8.2.6). Let $j \in N$ be a nonbasic arc which goes from node p to node q, where p and q are necessarily distinct nodes in T. Then (8.2.6) becomes

$$\hat{c}_j = c_j - \mathbf{a}_j^T\boldsymbol{\pi} = c_j - \pi_p + \pi_q \qquad j \in N. \tag{13.3.12}$$

In general the reduced cost $\hat{c}_j$ measures the marginal cost of increasing x_j subject to maintaining $\mathbf{Ax} = \mathbf{b}$, so in this case $\hat{c}_j$ can be interpreted as the marginal cost (c_j) of increasing the flow in arc j *less* the marginal cost $(\pi_p - \pi_q)$ of removing that flow from T. In other words, inserting arc j into T creates a *cycle* and $\hat{c}_j$ is the unit cost of putting flow into this cycle. This is illustrated in Figure 13.3.3 for the network and spanning tree from Figure 13.3.2. Thus the optimality test in the simplex method can be interpreted as follows. If $\hat{c}_j < 0$ and $x_j = l_j$ then the total cost can be reduced by putting flow into the cycle in the direction $p \to q$ and if $\hat{c}_j > 0$ and $x_j = u_j$ then the total cost can be reduced by removing flow from this cycle. If neither of these possibilities holds for all $j \in N$, then the flow pattern is optimal. If not, then the simplex method dictates that the largest modulus allowable $\hat{c}_j$ should indicate which arc is to become basic. In practice, although this computation is very simple, it nonetheless accounts for a substantial proportion of the total computing time. Thus it can be advantageous to search only a subset of the arcs to determine the arc which becomes basic (e.g. Grigoriadis, 1982).

If the current b.f.s. is not optimal then the simplex method goes on to the ratio test (8.2.11) which determines the first basic variable to reach its bound. In terms of a graph theoretic description this is readily seen to occur when one of the arcs in the cycle reaches either its upper or lower bound as the flow round the cycle is changed. A calculation based on this interpretation is the most efficient way of implementing the ratio test in practice. This is also

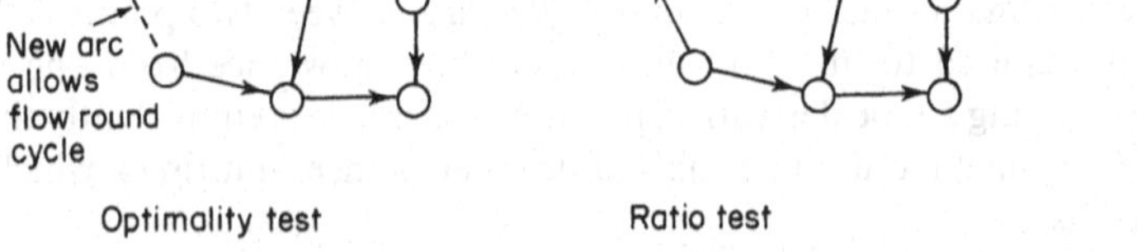

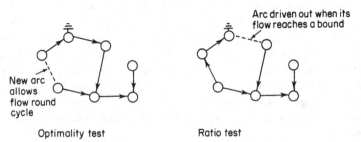

Figure 13.3.3 Simplex method in terms of spanning trees

illustrated in Figure 13.3.3. This interpretation can also be derived from a linear algebraic viewpoint. If arc j is being introduced, then the vector $\mathbf{A}_B^{-1}\mathbf{a}_j$ is required in (8.2.10) and this has the form given by (13.3.8). Thus $\mathbf{A}_B^{-1}\mathbf{a}_j$ represents the path in T connecting node p to node q. Together with arc j this path gives the cycle considered above. Thus the description can also be deduced from (13.3.8).

Finally (assuming that arc j does not move to become basic at its opposite bound) it is necessary to update the current representation of the basis, that is the rooted spanning tree T, when an arc in T is exchanged for one not in T (Figure 13.3.3). Clearly the most suitable representation for T is some form of list structure and various methods have been proposed (see for example the books of Jensen and Barnes (1980) and Kennington and Helgason (1980)). To give an idea of what is possible, only one such method is described here. This is the basis of the RNET code which Grigoriadis (1982) claims to be superior on the basis of a wide selection of computational experience. In RNET the vector of lower bounds $\mathbf{l}=\mathbf{0}$ is taken and it is assumed that all the data is integer valued. The vectors required by RNET are listed in Table 13.3.1. An explanation of how some of these vectors are used in representing the list structure for T is now given. The nodes in T are classified according to their *depth* (number of arcs away from the earth node in T): this information is stored in the vector D. The nodes are ordered according to a *depth-first* criterion: that is starting from the earth node a path of increasing depth is followed until a node is reached with no successor. Then the path is backtracked, until any node which has a different successor is located, and a path of increasing depth from this node is again followed. The process is repeated until all nodes are accounted for. The order in which the nodes are scanned is illustrated in Figure 13.3.4.

Table 13.3.1 Vectors required in the RNET list structure

Identifier	Domain → Range	Meaning
From*	$(j\in\mathscr{A})\rightarrow(i\in\mathscr{N})$	'From' node of arc j
To*	$(j\in\mathscr{A})\rightarrow(i\in\mathscr{N})$	'To' node of arc j
C*	$(j\in\mathscr{A})\rightarrow\{\text{integers}\}$	Unit cost of arc j
H*	$(j\in\mathscr{A})\rightarrow\{\text{integers}\}$	Upper bound for arc j
Γ	$(j\in\mathscr{A})\rightarrow\{0,1\}$	$1\Rightarrow$Nonbasic arc at upper bound
F	$(i\in\mathscr{N})\rightarrow(j\in\mathscr{N})$	Father of node i in T
D	$(i\in\mathscr{N})\rightarrow\{1,\ldots,m-1\}$	Depth of node i from root
P	$(i\in\mathscr{N})\rightarrow(j\in\mathscr{N})$	Successor of node i in depth-first sequence
ARC	$(i\in\mathscr{N})\rightarrow(j\in\mathscr{A})$	Arc number of arc $(F(i),i)$
B*	$(i\in\mathscr{N})\rightarrow\{\text{integers}\}$	External flow to node i
Δ	$(i\in\mathscr{N})\rightarrow\{0,1\}$	Flow direction: $1\Rightarrow$from $F(i)$
X	$(i\in\mathscr{N})\rightarrow\{\text{integers}\}$	Directed flow on arc i
U	$(i\in\mathscr{N})\rightarrow\{\text{integers}\}$	Node potentials π_i

*Means that the vector is problem data.
$\mathscr{A}=1,\ldots,n$ (arcs) $\mathscr{N}=0,\ldots,m$ (nodes)

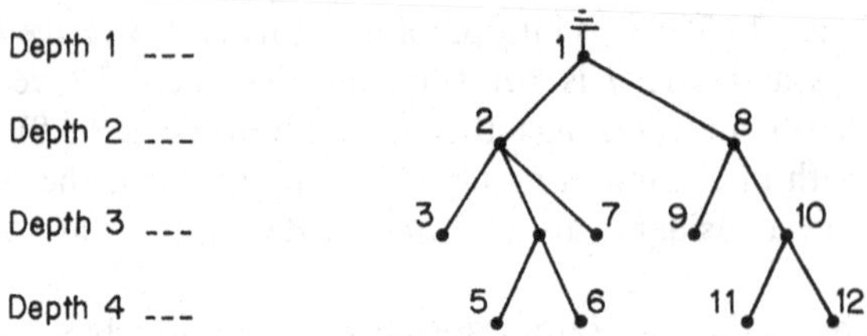

Figure 13.3.4 Depth-first ordering in rooted spanning tree

The successors of each node in this ordering are given in the vector P. Backtracking along a path is possible by using the array F which gives the *father* of node i (that is the next node in the direction of earth). Finding the arc j to enter the tree is straightforward and just involves scanning through the list of arcs (possibly a subset of them) and computing the reduced costs. Assume that this arc goes from node p to node q. The next step is to locate the cycle round which the flow will be changed. This is done by following the paths from p to earth and q to earth until a common node c is identified. Then $p \to c \to q \to p$ is the required cycle. To find c, the deepest path (from p say) is first followed until a node is reached having the same depth as q. Then both paths are followed to earth until the common node is located. Whilst making this search, the arc k to leave the tree can also be identified. The cycle is traversed again to update the flows. Finally the list structure must be updated to account for the change in T. Deleting arc k splits T into two subtrees T_1 and T_2, one of which is rooted at earth (T_1 say). The subtree T_2 is reconnected to T_1 using arc j. Note that only the nodes in T_2 must be reordered to get a valid list structure. This is a fairly intricate process to describe and the reader is referred to Grigoriadis (1982). Whilst updating this information, the node potentials for T_2 can also be adjusted.

To initialize the simplex method a basic feasible flow must be determined. The analogue of the artificial variables method of Section 8.4 is to add *artificial arcs* to the problem. If node i has an external flow $b_i > 0$ then an artificial arc is directed from node i to earth, and if $b_i < 0$ then from earth to node i. Artificial arcs have bounds $l_j = 0$ and $u_j = \infty$. In a two-phase method, phase I has an initial flow obtained by directing the external flows to or from earth along the artificial arcs. There are unit costs for the artificial arcs and zero costs for the remaining arcs. Solving the phase I problem has the effect of driving the flow off the artificial arcs. Having obtained feasibility, phase II then reverts to the original cost function. Artificial arcs can be removed once they become nonbasic. Other methods described in Section 8.4 can also be used, such as that based on (8.4.5) or the use of an L_1 exact penalty function. Indeed the latter is used in the RNET code referred to above.

This discussion of the simplex method for linear network flow finishes with a few miscellaneous comments. If the cost bounds and external flows which define the problem are all integer, then it follows from the unit triangular nature of the basis matrix that the flows and node potentials take integer values for any basic feasible solution. Thus there is no need to consider integer linear network flow as a more complex calculation. Dual problems can be derived in which the node potentials become the variables and the arc flows can be interpreted as multipliers. Degeneracy remains a problem and can in principle be tackled by any of the methods of Section 8.6, although the literature is not very forthcoming on this point. This is despite the fact that degeneracy is very common in network problems: Grigoriadis (1982) cites a problem with 3000 nodes and 35000 arcs which requires 11231 pivots, 8538 of which are degenerate. This proportion of degenerate pivots is typical of a wide range of problems. However, the integral nature of network problems at least removes the difficulty over the interaction of degeneracy and round-off error. Finally network computer codes can be remarkably effective: for instance a FORTRAN implementation of RNET solves the problem with 35000 variables referred to above on an IBM 370/168 computer in only 18.6 seconds!

Special cases of the linear network flow problem include the *transportation problem* in which nodes are partitioned into either supply nodes ($b_i > 0$) or demand nodes ($b_i < 0$) with no nodes having $b_i = 0$. Arcs link every supply node with every demand node, and there are no other arcs. The *assignment problem* is a further specialization in which additionally $|b_i| = 1$ for all i. Problems of finding the maximal flow or the shortest path in a network are also special cases. Often there are special purpose methods for solving these problems which take account of the structure and are more efficient than the network simplex method. For linear network flow there is an alternative to the simplex method, known as the *out-of-kilter algorithm*. Neither method is currently regarded as being uniformly superior. The ideas involved above in deriving the simplex network flow algorithm are also applicable in more general situations. These include the following: networks with gains or losses in the arcs, in fact any system $\mathbf{Ax} = \mathbf{b}$ in which the columns of $\mathbf{A}$ have at most two non-zero entries. Also included are multicommodity flow problems in which several substances share arcs: this essentially gives rise to block triangular basis matrices. Finally problems in which network constraints $\mathbf{Ax} = \mathbf{b}$ are augmented by other linear equations $\mathbf{Cx} = \mathbf{d}$ can also be solved, essentially by using the constraints $\mathbf{Ax} = \mathbf{b}$ to eliminate variables, by making use of the triangular property of any non-singular submatrix in $\mathbf{A}$. All the developments in this paragraph are reviewed in the book of Kennington and Helgason (1980).

Finally the possibilities are considered for solving nonlinear programming problems on networks, including the case of quadratic programming. If it is assumed that there is no interaction between arcs, then the objective function becomes *separable*, that is $f(\mathbf{x}) = \sum f_j(x_j)$, summed over all arcs. This implies that the Hessian $\mathbf{G} = \nabla^2 f$ is diagonal. A case of some interest is an electrical (DC) network in which a minimum power solution is sought. This case differs from linear network flow in that the power is assumed to be proportional to the *square* of flow (flow = current) and the constant of proportionality is known as resistance. Thus from (13.3.11) the potential difference is the product of resistance × current which gives Ohm's law. DC networks therefore give rise to QP problems. The first order conditions for such a network are known as Kirchhoff's laws.

It is instructive to relate QP on networks to the generalized elimination methods described in Section 10.1 in which a null space matrix is constructed (in this case a matrix $\mathbf{Z}$ such that $\mathbf{AZ}=\mathbf{0}$). This can again be interpreted in graph theoretic terms. For any cycle in the network, define an n-vector $\mathbf{z}$ by

$$z_j = \begin{cases} +1 & \text{if arc } j \text{ is in the cycle in the same direction as the cycle} \\ -1 & \text{if arc } j \text{ is in the cycle in the opposite direction} \\ 0 & \text{if arc } j \text{ is not in the cycle} \end{cases} \tag{13.3.13}$$

It is easily shown that $\mathbf{Az}=\mathbf{0}$ for any such vector $\mathbf{z}$, using (13.3.1). Consider node i: if it is not in the cycle then $a_{ij}z_j = 0\ \forall j$ and hence $\mathbf{e}_i^T\mathbf{Az}=0$. If node i is in the cycle then the path enters node i on one arc, p say, and leaves on another, q say. If $z_p = +1$ then arc p enters node i so $a_{ip} = -1$. If $z_p = -1$ then arc p leaves node i and $a_{ip} = +1$, so in both cases $a_{ip}z_p = -1$. Likewise if $z_q = +1$ then arc q leaves node i so $a_{iq} = +1$, if $z_q = -1$ then arc q enters node i so $a_{iq} = -1$, so in both cases $a_{iq}z_q = +1$. Thus $a_{ip}z_p + a_{iq}z_q = 0$. If the path passes through node i a number of times, then this result must be true for each pair of arcs entering and leaving node i. Other arcs j have either $a_{ij}=0$ or $z_j=0$. Hence $\mathbf{e}_i^T\mathbf{Az}=0$ if node i is on the cycle and so $\mathbf{Az}=\mathbf{0}$ (QED). Thus a valid null space matrix is obtained by taking $\mathbf{Z}$ to be any matrix whose columns are a subset of maximum rank of all such vectors $\mathbf{z}$, and $\mathbf{Z}$ is therefore seen to be associated with an independent set of cycles for the network. In the direct elimination method (10.1.22) $\mathbf{Z}$ is constructed by $\mathbf{Z}^T = [-\mathbf{A}_N^T\mathbf{A}_B^{-T}:\mathbf{I}]$ and by virtue of the discussion above on the ratio test, this too has a simple interpretation. Given a basis (rooted spanning tree T) then connecting any nonbasic arc into T in turn gives a cycle. The set of all such cycles for fixed T is the set that would be computed by direct elimination.

The simple structure of $\mathbf{Z}$ together with the diagonal form of $\mathbf{G}$ enables very effective methods for QP on networks to be determined. Likewise for nonlinear programming, obvious possibilities include Newton's method for linear constraint problems (11.1.9), a quasi-Newton method based on updating a diagonal approximation to the full Hessian, or an SQP method such as (11.2.7). A recent reference is Dembo and Klincewicz (1981). An alternative approach to separable nonlinear programming on networks is to use piecewise linear approximations to the cost functions and convert the problem into what is essentially a linear network flow problem (e.g. Kamesam and Meyer, 1984).

QUESTIONS FOR CHAPTER 13

13.1. Solve graphically the LP

$$\begin{aligned} \text{minimize} \quad & -x_1 - 2x_2 \\ \text{subject to} \quad & -2x_1 + 2x_2 \leqslant 3 \\ & 2x_1 + 2x_2 \leqslant 9, \end{aligned}$$

for the cases (i) no integer restriction, (ii) x_1 integer, (iii) x_1, x_2 integer. Show how case (iii) would be solved by the branch and bound method, determining problem lower bounds by $L = f' + e_i^+$ or $L = f' + e_i^-$. Give the part of the solution tree which is explored. Show also that although the method solves the problem finitely, the tree contains two infinite paths. Show that if the nearest integer is used as a criterion for which problem to explore first after branching, then it is possible that one of these infinite paths may be followed, depending on how ties are broken, so that the algorithm does not converge. Notice however that if the best pending node is selected after *every* branch then the algorithm does converge. If bounds $\mathbf{x} \geqslant \mathbf{0}$ are added to the problem, enumerate the whole tree: there are ten ○ nodes, six ● nodes and five □ nodes in all.

13.2. Solve the integer programming problem

$$\begin{array}{ll} \text{minimize} & x_1^4 + x_2^4 + 16(x_1 x_2 + (4 + x_2)^2) \\ \text{subject to} & x_1, x_2 \text{ integer} \end{array}$$

by the branch and bound method. In doing this, problems in which $x_i \leqslant n_i$ or $x_i \geqslant n_i + 1$ arise. Treat these as equations, and solve the resulting problem by a line search in the remaining variable (or by using a quasi-Newton subroutine with the ith row and column of $\mathbf{H}^{(1)}$ zeroed). Calculate Lagrange multipliers at the resulting solution and verify that the inequality problem is solved. Use the distance to the nearest integer as the criterion for which branch to select. Give the part of the solution tree which is explored.

13.3. Consider the geometric programming problem

$$\begin{array}{ll} \text{minimize} & x_1^{-1} x_2^{-1} x_3^{-1} \\ \text{subject to} & x_1 + 2x_2 + 2x_3 \leqslant 72, \qquad \mathbf{x} > \mathbf{0} \end{array}$$

derived from Rosenbrock's parcel problem (see Question 9.3). Show that the problem has zero degree of difficulty and hence solve the equations (13.2.13) and (13.2.14) to solve (13.2.16). Show immediately that the solution value is 1/3456. Find the multipliers w_j^*, $j = 1, 2, 3$, of the orthogonality constraints and hence determine the optimum variables from (13.2.17).

13.4. Consider the geometric programming problem

$$\begin{array}{ll} \text{minimize} & 40x_1 x_2 + 20x_2 x_3 \\ \text{subject to} & \tfrac{1}{5} x_1^{-1} x_2^{-1/2} + \tfrac{3}{5} x_2^{-1} x_3^{-2/3} \leqslant 1, \qquad \mathbf{x} > \mathbf{0}. \end{array}$$

Show that the problem has zero degree of difficulty and can be solved in a similar way to the previous problem.

13.5. Consider the geometric programming problem

$$\begin{array}{ll} \text{minimize} & 40x_1^{-1} x_2^{-1/2} x_3^{-1} + 20x_1 x_3 + 40x_1 x_2 x_3 \\ \text{subject to} & \tfrac{1}{3} x_1^{-2} x_2^{-2} + \tfrac{4}{3} x_2^{1/2} x_3^{-1} \leqslant 1, \qquad \mathbf{x} > \mathbf{0}. \end{array}$$

Show that the problem has one degree of difficulty. Hence express $\boldsymbol{\delta}$ in terms of the parameter $\lambda (= \lambda_1)$ as in (13.2.20), and solve the problem

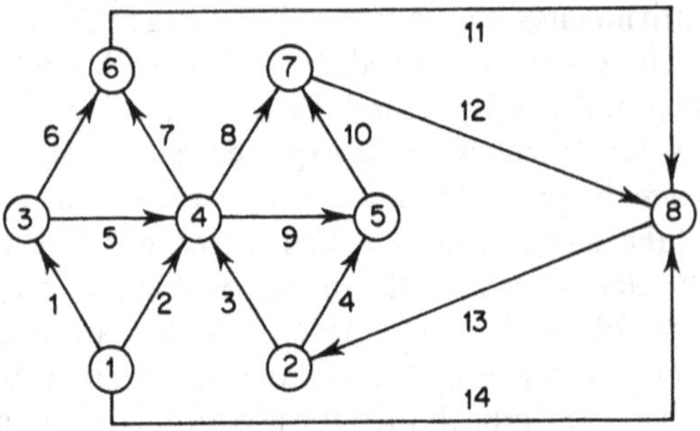

Figure 13.3.5 Network for Question 13.7

numerically by maximizing the function $v(\boldsymbol{\delta}(\lambda), \lambda)$ by a line search in λ. Then solve for the variables using (13.2.17). This and the previous problem are given by Duffin, Peterson, and Zener (1967) where many more examples of geometric programming problems can be found.

13.6. Consider a graph with no earth node which is formed from k connected subgraphs. Show that the matrix $\mathbf{A}$ defined in (13.1.1) has rank $m-k$.

13.7. Consider the linear network flow problem defined by the network shown in Figure 13.3.5. The bounds and costs for the arcs are given by $\mathbf{l}=\mathbf{0}$,

$$\mathbf{u}=(2,4,3,3,2,2,3,4,3,5,2,3,1,2)^{\mathrm{T}}$$

and

$$\mathbf{c}=(2,3,2,1,4,4,3,1,3,2,-5,1,3,-1)^{\mathrm{T}},$$

and the external flows on the nodes by

$$\mathbf{b}=(4,3,0,0,0,-2,-6,1)^{\mathrm{T}}.$$

Solve the problem using a subroutine for linear programming. Use the example to verify some of the graph theoretic properties of networks.

13.8. Show for a linear network flow problem that any basis matrix $\mathbf{A}_B=\mathbf{AQ}$ and the corresponding upper triangular matrix $\mathbf{U}$ in (13.3.4) have the same amount of sparsity, that is $2m-1$ non-zero elements. Consider LU factors of the basis matrix, that is

$$\mathbf{PA}_B=\mathbf{PAQ}=\mathbf{LU}$$

that might for example be computed by the Fletcher–Matthews method (Section 8.5). Construct an example of a linear network flow problem which shows that the total number of non-zeros in the LU factors can be greater than $2m-1$.

Chapter 14
Non-Smooth Optimization

14.1 INTRODUCTION

In Part 1 on unconstrained optimization an early assumption is to exclude all problems for which the objective function $f(\mathbf{x})$ is not smooth. Yet practical problems sometimes arise (*non-smooth optimization* (NSO) or *non-differentiable optimization* problems) which do not meet this requirement; this chapter studies progress which has been made in the practical solution of such problems. Examples of NSO problems occur when solving nonlinear equations $c_i(\mathbf{x}) = 0$, $i = 1, 2, \ldots, m$ (see Section 6.1) by minimizing $\|\mathbf{c}(\mathbf{x})\|_1$ or $\|\mathbf{c}(\mathbf{x})\|_\infty$. This arises either when solving simultaneous equations exactly ($m = n$) or when finding best solutions to over-determined systems ($m > n$) as in data fitting applications. Another similar problem is that of finding a feasible point of a system of nonlinear inequalities $c_i(\mathbf{x}) \leqslant 0$, $i = 1, 2, \ldots, m$, by minimizing $\|\mathbf{c}(\mathbf{x})^+\|_1$ or $\|\mathbf{c}(\mathbf{x})^+\|_\infty$ where $c_i^+ = \max(c_i, 0)$. A generalization of these problems arises when the equations or inequalities are the constraints in a nonlinear programming problem (Chapter 12). Then a possible approach is to use an exact penalty function and minimize functions like $\nu f(\mathbf{x}) + \|\mathbf{c}(\mathbf{x})\|$ or $\nu f(\mathbf{x}) + \|\mathbf{c}(\mathbf{x})^+\|$, in particular using the L_1 norm. This idea is attracting much research interest and is expanded upon in Section 14.3. Yet another type of problem is to minimize the *max function* $\max_i c_i(\mathbf{x})$ where the max is taken over some finite set. This includes many examples from electrical engineering including microwave network design and digital filter design (Charalambous, 1979). In fact almost all these examples can be considered in a more generalized sort of way as special cases of a certain *composite function* and a major portion of this chapter is devoted to studying this type of function. In particular the term *composite NSO* is used to describe this type of problem.

Another common source of NSO problems arises when using the decomposition principle. For example the LP problem

$$\begin{aligned} &\underset{\mathbf{x},\mathbf{y}}{\text{minimize}} \quad \mathbf{c}^{\mathrm{T}}\mathbf{x} + \mathbf{d}^{\mathrm{T}}\mathbf{y} \\ &\text{subject to} \quad \mathbf{A}\mathbf{x} + \mathbf{B}\mathbf{y} \leqslant \mathbf{b} \end{aligned}$$

can be written as the convex NSO problem

$$\underset{\mathbf{x}}{\text{minimize}}\, f(\mathbf{x}) \triangleq \mathbf{c}^{\mathrm{T}}\mathbf{x} + \min_{\mathbf{y}}\, [\mathbf{d}^{\mathrm{T}}\mathbf{y} : \mathbf{B}\mathbf{y} \leqslant \mathbf{b} - \mathbf{A}\mathbf{x}].$$

Application of NSO methods to column generation problems and to a variety of scheduling problems is described by Marsten (1975) and the application to network scheduling problems is described by Fisher, Northup, and Shapiro (1975). Both these papers appear in the Mathematical Programming Study 3 (Balinski and Wolfe, 1975) which is a valuable reference. Another good source is the book by Lemarechal and Mifflin (1978) and their test problems 3 and 4 illustrate further applications.

Insofar as I understand them, all these applications could in principle be formulated as max functions and hence as composite functions (see below). However in practice this would be too complicated or require too much storage. Thus a different situation can arise in which the only information available at any point $\mathbf{x}$ is $f(\mathbf{x})$ and a vector $\mathbf{g}$. Usually $\mathbf{g}$ is ∇f, but if f is non-differentiable at $\mathbf{x}$ then $\mathbf{g}$ is an element of the subdifferential ∂f (see Section 14.2) for convex problems, or the generalized gradient in non-convex problems. Since less information about $f(\mathbf{x})$ is available, this type of application is more difficult than the composite function application and is referred to here as *basic NSO*. However, both types of application have some structure in common.

In view of this common structure, algorithms for all types of NSO problem are discussed together in Section 14.4. For composite NSO it is shown that an algorithm with second order convergence can be obtained by making linearizations of the individual functions over which the composition is taken, together with an additional (smooth) quadratic approximation which includes curvature information from the functions in the composition. Variants of this algorithm have been used in a number of practical applications. A potential difficulty for this algorithm (and indeed for any NSO algorithm), known as the *Maratos effect* is described, in which the expected superlinear convergence may not be obtained. It is shown how this arises, and that it can be circumvented in a simple way by adding a *second order correction step* to the algorithm. This algorithm can also be made globally convergent by incorporating the idea of a trust region and details of this result are given in Section 14.5. The algorithm is applicable to both max function and L_1 and L_∞ approximation problems and to nonlinear programming applications of non-smooth exact penalty functions. Algorithms for basic NSO have not progressed as far because of the difficulties caused by the limited availability of information. One type of method—the *bundle method*—tries to accumulate information locally about the subdifferential of the objective function. Alternatively linearization methods have also been tried and the possibility of introducing second order information into these algorithms is being considered. In fact there is currently much research interest in NSO algorithms of all kinds and further developments can be expected.

A prerequisite for describing NSO problems and algorithms is a study of optimality conditions for non-smooth functions. This can be done at various levels of generality. I have tried to make the approach here as simple and readable as possible (although this is not easy in view of the inherent difficulty of the material), whilst trying to cover all practical applications. The chapter is largely self-contained and does not rely on any key results from outside sources. One requirement is the extension of the material in Section 9.4 on convex functions to include the non-smooth case. The concept of a subdifferential is introduced and the resulting definitions of directional derivatives lead to a simple statement of first order conditions. More general applications can be represented by the *composite function*

$$\phi(\mathbf{x}) \triangleq f(\mathbf{x}) + h(\mathbf{c}(\mathbf{x})) \tag{14.1.1}$$

and the problem of minimizing such a function is referred to as *composite NSO*. Here $f(\mathbf{x})$ ($\mathbb{R}^n \to \mathbb{R}^1$) and $\mathbf{c}(\mathbf{x})$ ($\mathbb{R}^n \to \mathbb{R}^m$) are smooth functions ($\mathbb{C}^1$), and $h(\mathbf{c})$ ($\mathbb{R}^m \to \mathbb{R}^1$) is convex but non-smooth ($\mathbb{C}^0$). Note that it would be possible more simply to express $\phi(\mathbf{x}) = h(\mathbf{c}(\mathbf{x}))$ and to regard (14.1.1) merely as a special case of this. I have not done this because in penalty function applications it is much more convenient to use (14.1.1). Other applications in which $f(\mathbf{x})$ is not present are readily handled by setting $f(\mathbf{x}) = 0$. Important special cases of $h(\mathbf{c})$ are set out in (14.1.3) below. Optimality conditions for (14.1.1) can be obtained as a straightforward extension of those for convex functions. This material is set out in Section 14.2 and both first and second order conditions are studied. An important aspect of this theory is to give a regularity condition under which the second order conditions are almost necessary and sufficient. This involves the idea of *structure functionals*. It is shown that these functionals enable an equivalence to be drawn in the neighbourhood of a local solution between a composite NSO problem and a nonlinear programming problem whose constraints involve the structure functionals.

There are NSO problems which do not fit into the category of composite NSO (for example $\phi(\mathbf{x}) = \max(0, \min(x_1, x_2))$), and which are not even well modelled locally by (14.1.1). A practical example is that of *censored* L_1 *approximation* (Womersley, 1984a), but few examples of this type occur in practice. Wider classes of function have been introduced which cover these cases, for example the *locally Lipschitz functions* of Clarke (1975), but substantial complication in the analysis is introduced. Also these classes do not directly suggest algorithms as the function (14.1.1) does. Furthermore the resulting first order conditions permit descent directions to occur and so are of little practical use (Womersley, 1982). Hence these classes are not studied any further here. Finally the problem of *constrained NSO* can be considered. First order conditions for smooth constraint functions are given by Watson (1978) and for constrained L_∞ approximation by Andreassen and Watson (1976). First and second order conditions for a single non-smooth constraint function involving a norm are given by Fletcher and Watson (1980). In Section 14.6 these results are extended to cover a composite objective function and a single composite constraint

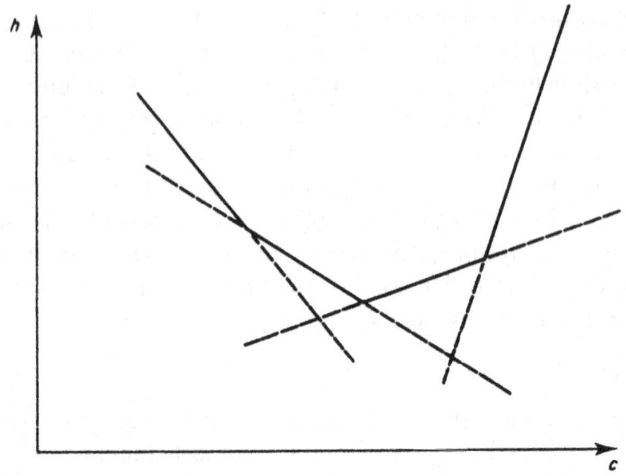

Figure 14.1.1 Polyhedral convex function

function (*constrained composite NSO*), and this is the most general case that needs to be considered. A regularity condition involving the use of structure functionals is also given.

In most cases (but not all—for example the exact penalty function $\phi = \nu f + \|\mathbf{c}\|_2$) a further specialization can be made in which $h(\mathbf{c})$ is restricted to be a *polyhedral convex function.* In this case the graph of $h(\mathbf{c})$ is made up of a finite number of supporting hyperplanes $\mathbf{c}^T\mathbf{h}_i + b_i$, and $h(\mathbf{c})$ is thus defined by

$$h(\mathbf{c}) \triangleq \max_i (\mathbf{c}^T\mathbf{h}_i + b_i) \tag{14.1.2}$$

where the vectors $\mathbf{h}_i$ and scalars b_i are given. Thus the polyhedral convex function is a max function in terms of linear combinations of the elements of $\mathbf{c}$. A simple illustration is given in Figure 14.1.1. Most interest lies in five special cases of (14.1.2), in all of which $b_i = 0$ for all i. These functions are set out below, together with the vectors $\mathbf{h}_i$ which define them given as columns of a matrix $\mathbf{H}$.

Case I	$h(\mathbf{c}) = \max_i c_i$	$\mathbf{H} = \mathbf{I} \quad (m \times m)$	
Case II	$h(\mathbf{c}) = \|\mathbf{c}^+\|_\infty$	$\mathbf{H} = [\mathbf{I}:\mathbf{0}] \quad (m \times (m+1))$	
Case III	$h(\mathbf{c}) = \|\mathbf{c}\|_\infty$	$\mathbf{H} = [\mathbf{I}: -\mathbf{I}] \quad (m \times 2m)$	
Case IV	$h(\mathbf{c}) = \|\mathbf{c}^+\|_1$	columns of $\mathbf{H}$ are all possible combinations of 1 and 0 $\quad (m \times 2^m)$	(14.1.3)
Case V	$h(\mathbf{c}) = \|\mathbf{c}\|_1$	columns of $\mathbf{H}$ are all possible combinations of 1 and -1 $\quad (m \times 2^m)$	

For example with $m = 2$ in case (V) the matrix $\mathbf{H}$ is $\mathbf{H} = \begin{bmatrix} 1 & 1 & -1 & -1 \\ 1 & -1 & 1 & -1 \end{bmatrix}$.
Contours of these functions for $m = 2$ are illustrated in Figure 14.1.2. The broken lines indicate the surfaces (*grooves*) along which the different linear pieces join up and along which the derivative is discontinuous. The term *piece* is used to

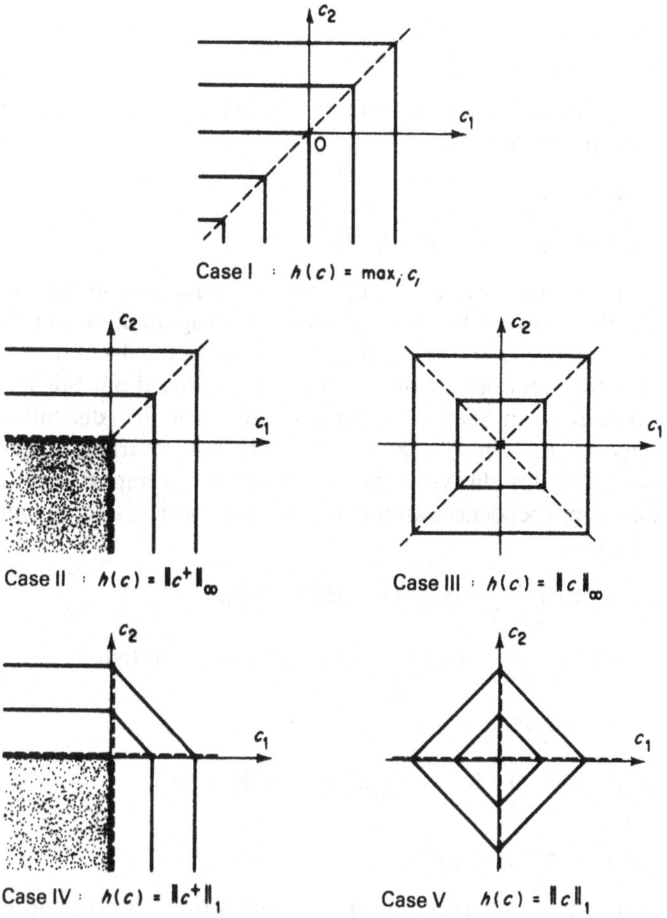

Figure 14.1.2 Contours of polyhedral convex functions

denote that part of the graph of a max function in which one particular function achieves the maximum value, for example the graph illustrated in case (II) of Figure 14.1.2 is made up of three linear pieces and the active equations are $h(\mathbf{c}) = c_2$ for $c_1 \leqslant c_2$ and $c_2 \geqslant 0$, $h(\mathbf{c}) = c_1$ for $c_2 \leqslant c_1$ and $c_1 \geqslant 0$, and $h(c) = 0$ for $c_1 \leqslant 0$ and $c_2 \leqslant 0$. The subdifferential $\partial h(\mathbf{c})$ for a polyhedral convex function has the simple form expressed in Lemma 14.2.2 as the convex hull of the gradients of the active pieces. The polyhedral nature of $h(\mathbf{c})$ has important consequences in regard to second order conditions, and this is considered later in Section 14.2.

In fact it is possible to derive optimality results for the composite function $\phi(\mathbf{x})$ in (14.1.1) when $h(\mathbf{c})$ is the polyhedral function (14.1.2) without using notions of convexity at all. Clearly $\phi(\mathbf{x})$ can be written

$$\phi(\mathbf{x}) = \max_i \, (f(\mathbf{x}) + \mathbf{c}(\mathbf{x})^{\mathrm{T}}\mathbf{h}_i + b_i) \tag{14.1.4}$$

which is equivalent to

$$\phi(\mathbf{x}) = \min v\colon v \geqslant f(\mathbf{x}) + \mathbf{c}(\mathbf{x})^{\mathrm{T}}\mathbf{h}_i + b_i \qquad \forall i. \tag{14.1.5}$$

Therefore $\mathbf{x}^*$ minimizes $\phi(\mathbf{x})$ locally iff $\mathbf{x}^*, v^*$ is a local solution of the nonlinear programming problem

$$\begin{aligned} &\underset{\mathbf{x},v}{\text{minimize}} \quad v \\ &\text{subject to} \quad v - f(\mathbf{x}) - \mathbf{c}(\mathbf{x})^{\mathrm{T}}\mathbf{h}_i \geqslant b_i \qquad \forall i. \end{aligned} \tag{14.1.6}$$

Thus first and second order optimality conditions for (14.1.4) can be obtained by applying the equivalent results in nonlinear programming to (14.1.6), that is Theorem 9.1.1, 9.3.1, and 9.3.2. It turns out however that this approach is less general and somewhat clumsy, so it is only sketched out briefly here, and the derivation given in Section 14.2 is preferred. The first difficulty concerns the regularity assumption $\mathscr{F}' = F'$. It is possible (but not trivial—see Question 14.1) to prove that this always holds, and therefore it is important not to make an unnecessary independence assumption. The appropriate Lagrangian function for Theorem 9.1.1 is

$$\mathscr{L}(\mathbf{x}, v, \boldsymbol{\mu}) = v - \sum_i \mu_i (v - f(\mathbf{x}) - \mathbf{c}(\mathbf{x})^{\mathrm{T}}\mathbf{h}_i - b_i) \tag{14.1.7}$$

and it is implied at $\mathbf{x}^*, v^*$ that multipliers $\boldsymbol{\mu}^*$ exist such that

$$\begin{aligned} &\frac{\partial}{\partial v}\mathscr{L}(\mathbf{x}^*, v^*, \boldsymbol{\mu}^*) = 0 \quad \text{or} \quad \sum \mu_i^* = 1 \\ &\nabla_x \mathscr{L}(\mathbf{x}^*, v^*, \boldsymbol{\mu}^*) = \mathbf{0} \quad \text{or} \quad \sum \mu_i^*(\mathbf{g}^* + \mathbf{A}^*\mathbf{h}_i) = \mathbf{0} \\ &\boldsymbol{\mu}^* \geqslant \mathbf{0} \\ &\mu_i^* > 0 \Rightarrow v^* = f^* + \mathbf{c}^{*\mathrm{T}}\mathbf{h}_i + b_i. \end{aligned} \tag{14.1.8}$$

As in previous chapters the notation $\mathbf{g} = \nabla f$ and $\mathbf{A} = \nabla \mathbf{c}^{\mathrm{T}}$ is used and $f^* = f(\mathbf{x}^*)$, etc. If $\boldsymbol{\lambda} = \mathbf{H}\boldsymbol{\mu}$ is written, then the existence of a vector $\boldsymbol{\mu}^*$ in the above is equivalent to the existence of a vector $\boldsymbol{\lambda}^*$ in the set

$$\partial h^* = \underset{i \in \mathscr{A}^*}{\operatorname{conv}} \mathbf{h}_i \tag{14.1.9}$$

such that $\mathbf{g}^* + \mathbf{A}^*\boldsymbol{\lambda}^* = \mathbf{0}$. In fact ∂h^* is the subdifferential of $h(\mathbf{c}^*)$ as shown in Lemma 14.2.2. $\mathscr{A}^*$ is the set of active constraints or equivalently the set of indices at which the max in (14.1.4) is attained, that is

$$\mathscr{A}^* = \{i\colon \mathbf{c}^{*\mathrm{T}}\mathbf{h}_i + b_i = h(\mathbf{c}^*)\}. \tag{14.1.10}$$

It is possible to use (14.1.9) to obtain equivalent but more convenient expressions for $\partial h(\mathbf{c})$ in cases (I) to (V) in (14.1.3) above as follows:

$$\partial \max_i c_i = \{\boldsymbol{\lambda}\colon \sum \lambda_i = 1,\ \boldsymbol{\lambda} \geqslant \mathbf{0},\ c_i < \max_i c_i \Rightarrow \lambda_i = 0\} \tag{14.1.11}$$

$$\begin{aligned} \partial \|\mathbf{c}^+\|_\infty = \{\boldsymbol{\lambda}\colon \sum \lambda_i = 1,\ \boldsymbol{\lambda} \geqslant \mathbf{0},\ c_i < \|\mathbf{c}^+\|_\infty \Rightarrow \lambda_i = 0 \\ \mathbf{c}^+ \neq \mathbf{0} \Rightarrow \sum \lambda_i = 1\} \end{aligned} \tag{14.1.12}$$

$$\begin{aligned}\partial\|\mathbf{c}\|_\infty = \{\boldsymbol{\lambda}\colon \mathbf{c}\neq\mathbf{0} \Rightarrow \textstyle\sum|\lambda_i| = 1 \\ |c_i| < \|\mathbf{c}\|_\infty \Rightarrow \lambda_i = 0 \\ |c_i| = \|\mathbf{c}\|_\infty \Rightarrow \lambda_i c_i \geqslant 0 \\ \mathbf{c}=\mathbf{0} \Rightarrow \textstyle\sum|\lambda_i| \leqslant 1\}\end{aligned} \quad (14.1.13)$$

$$\begin{aligned}\partial\|\mathbf{c}^+\|_1 = \{\boldsymbol{\lambda}\colon 0\leqslant\lambda_i\leqslant 1,\ c_i>0 \Rightarrow \lambda_i = 1 \\ c_i<0 \Rightarrow \lambda_i = 0\}\end{aligned} \quad (14.1.14)$$

$$\partial\|\mathbf{c}\|_1 = \{\boldsymbol{\lambda}\colon |\lambda_i|\leqslant 1,\ c_i\neq 0 \Rightarrow \lambda_i = \operatorname{sign} c_i\}. \quad (14.1.15)$$

The equivalence between these sets and those defined by $\operatorname{conv}_{i\in\mathscr{A}}\mathbf{h}_i$ is left as an exercise (Question 14.3). Expressions (14.1.12) to (14.1.15) can also be derived as special cases of (14.3.7) or (14.3.8).

It is also possible to apply the development of (9.3.9) onwards to problem (14.1.6) to obtain the second order conditions of Theorem 14.2.2 and 14.2.3, and also the regularity condition (14.2.31) (see Lemma 14.2.7). To do this an index $p\in\mathscr{A}_+^*$ is used to eliminate v (or s_{n+1}—see Question 14.1). The analysis is not entirely straightforward so again the more direct and more general approach of Section 14.2 is preferred.

The parameters $\boldsymbol{\lambda}^*\in\partial h^*$ which exist at the solution to an NSO problem are closely related to Lagrange multipliers and indeed they can be given a simple interpretation similar to (9.1.10). Consider a perturbed problem in which $\mathbf{c}(\mathbf{x})$ is replaced by $\mathbf{c}(\mathbf{x})+\boldsymbol{\varepsilon}$ giving a function $\phi_\varepsilon(\mathbf{x})$ and assume that the minimizer $\mathbf{x}(\boldsymbol{\varepsilon})$ is such that the same constraints are active in (14.1.6). It follows that the right-hand side of each active constraint is perturbed by an amount $\boldsymbol{\varepsilon}^{\mathrm{T}}\mathbf{h}_i$. Since μ_i^* is the multiplier of the ith constraint and using (9.1.10) it follows that the change to v is $\Delta v = \sum_i \boldsymbol{\varepsilon}^{\mathrm{T}}\mathbf{h}_i\mu_i^* = \boldsymbol{\varepsilon}^{\mathrm{T}}\boldsymbol{\lambda}^*$ and hence that

$$\frac{\mathrm{d}\phi^*}{\mathrm{d}\varepsilon_i} = \lambda_i^*. \quad (14.1.16)$$

Thus λ_i^* measures the first order rate of change of the optimum function value consequent on perturbations to c_i. This result can also be used to illustrate the need for the condition $\boldsymbol{\lambda}\geqslant\mathbf{0}$ in (14.1.11). For example if the first order conditions arising from (14.1.8) ($\exists\boldsymbol{\lambda}^*\in\partial h^*$ such that $\mathbf{g}^*+\mathbf{A}^*\boldsymbol{\lambda}^*=\mathbf{0}$) are satisfied except that $\lambda_i^*<0$ for some i, then it is possible to show that $\mathbf{x}^*$ is not optimal. Consider a perturbation $\boldsymbol{\varepsilon}$ with $\varepsilon_i>0$ and $\varepsilon_j=0, j\neq i$. Existence of the solution $\mathbf{x}(\boldsymbol{\varepsilon})$ follows under a suitable independence assumption using the implicit function theorem. At $\mathbf{x}(\boldsymbol{\varepsilon})$ the max in (14.1.2) is achieved by all $j\in\mathscr{A}^*$ if $\boldsymbol{\varepsilon}$ is small enough, so for all $j\in\mathscr{A}^*, j\neq i$,

$$c_j(\mathbf{x}(\boldsymbol{\varepsilon})) = c_i(\mathbf{x}(\boldsymbol{\varepsilon})) + \varepsilon_i.$$

Thus $c_i(\mathbf{x}(\boldsymbol{\varepsilon})) < c_j(\mathbf{x}(\boldsymbol{\varepsilon}))$ and hence $\phi_\varepsilon(\mathbf{x}(\boldsymbol{\varepsilon})) = \phi_0(\mathbf{x}(\boldsymbol{\varepsilon}))$ where ϕ_0 refers to the unperturbed function. It follows from (14.1.16) that $\mathrm{d}\phi_0^*/\mathrm{d}\varepsilon_i = \lambda_i^* < 0$ and so $\mathbf{x}^*$ is not a local solution. The situation is illustrated in Figure 14.1.3. Clearly increasing c_1 by ε does not reduced $\phi(\mathbf{x})$ in case (i) when $\boldsymbol{\lambda}\geqslant\mathbf{0}$ but does reduce $\phi(\mathbf{x})$ in case (ii). Another example of this is given in Question 14.12.

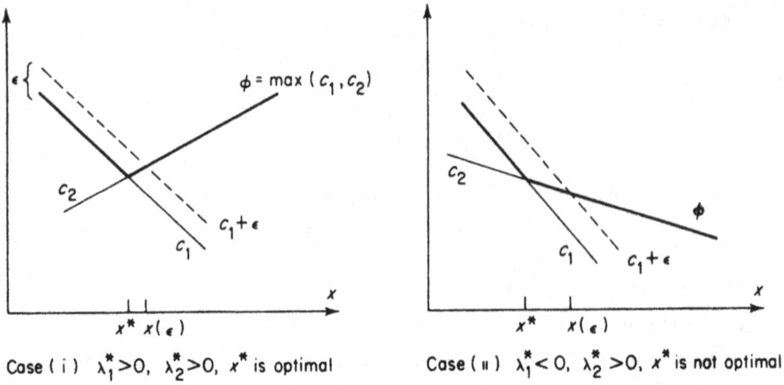

Figure 14.1.3. Interpretation of multipliers in NSO

The requirement that the active constraints in (14.1.6) remain the same is important and is in the nature of an independence assumption. This usually holds for case (I) and (II) problems in (14.1.3) and for case (III) problems when $\mathbf{c}^* \neq \mathbf{0}$. However for the L_1 norm functions of cases (IV) and (V) the assumption is not likely to hold. An alternative interpretation of Lagrange multipliers suitable for the L_1 case is given in Section 12.3. In Section 14.2 it is shown how all these cases, and also some others not involving polyhedral $h(\mathbf{c})$, can be handled within a unified framework by using a regularity assumption based on the concept of structure functionals.

14.2 OPTIMALITY CONDITIONS

In this section a number of results are proved leading to optimality conditions for the composite functions described in the previous section. Since these functions are defined in terms of non-smooth convex functions it is important to study the latter case first. Such a function is illustrated in Figure 14.2.1 and it can be seen that the non-differentiability at $\mathbf{x}^*$ allows the possibility of a number of supporting hyperplanes, in contrast to Figure 9.4.2. (Supporting hyperplanes to the epigraph of a convex function, which is a convex set, always exist; see Hadley, 1961.) Given a convex function $f(\mathbf{x})$ defined on a convex set $K \subset \mathbb{R}^n$, and given $\mathbf{x} \in$interior (K), then for each supporting hyperplane at $\mathbf{x}$ it follows that an inequality of the form

$$f(\mathbf{x}+\boldsymbol{\delta}) \geqslant f(\mathbf{x}) + \boldsymbol{\delta}^T\mathbf{g} \qquad \forall \mathbf{x}+\boldsymbol{\delta} \in K \tag{14.2.1}$$

holds, and $\mathbf{g}$ is a normal vector of the supporting hyperplane. Such a vector $\mathbf{g}$ is referred to as a *subgradient* at $\mathbf{x}$ and (14.2.1) is known as the *subgradient inequality* and is the generalization of (9.4.4) to non-smooth functions. The set of all subgradients at $\mathbf{x}$ is known as the *subdifferential* at $\mathbf{x}$ and is defined by

$$\partial f(\mathbf{x}) \triangleq \{\mathbf{g}\colon f(\mathbf{x}+\boldsymbol{\delta}) \geqslant f(\mathbf{x}) + \boldsymbol{\delta}^T\mathbf{g} \qquad \forall \mathbf{x}+\boldsymbol{\delta} \in K\}. \tag{14.2.2}$$

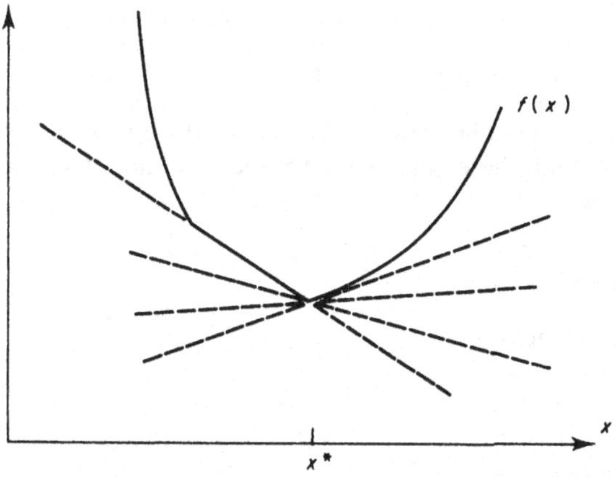

Figure 14.2.1 Supporting hyperplanes to a non-smooth convex function

The notation $\partial f^{(k)} = \partial f(\mathbf{x}^{(k)})$ analogous to $f^{(k)} = f(\mathbf{x}^{(k)})$, etc., is used. It is easy to see that $\partial f(\mathbf{x})$ is a closed convex set, which by Lemma 14.2.1 below is also bounded and therefore compact. In fact a more general result can be established.

Lemma 14.2.1

$\partial f(\mathbf{x})$ is bounded for all $\mathbf{x} \in B \subset$ interior (K) where B is compact.

Proof

If the lemma is false, $\exists$ a sequence $\mathbf{g}^{(k)} \in \partial f(\mathbf{x}^{(k)})$, $\mathbf{x}^{(k)} \in B$, such that $\|\mathbf{g}^{(k)}\|_2 \to \infty$. By compactness $\exists$ a subsequence $\mathbf{x}^{(k)} \to \mathbf{x}'$. Define $\boldsymbol{\delta}^{(k)} = \mathbf{g}^{(k)}/\|\mathbf{g}^{(k)}\|_2^2$. Then $\mathbf{x}^{(k)} + \boldsymbol{\delta}^{(k)} \in K$ for k sufficiently large, so by (14.2.1)

$$f(\mathbf{x}^{(k)} + \boldsymbol{\delta}^{(k)}) \geqslant f^{(k)} + \mathbf{g}^{(k)\mathrm{T}} \boldsymbol{\delta}^{(k)} = f^{(k)} + 1.$$

But in the limit, $f^{(k)} \to f'$, $\boldsymbol{\delta}^{(k)} \to \mathbf{0}$, and so $f(\mathbf{x}^{(k)} + \boldsymbol{\delta}^{(k)}) \to f'$, which is a contradiction, so the lemma is true. □

If f is differentiable at $\mathbf{x}$, then

$$f(\mathbf{x} + \boldsymbol{\delta}) = f(\mathbf{x}) + \boldsymbol{\delta}^{\mathrm{T}} \nabla f(\mathbf{x}) + o(\|\boldsymbol{\delta}\|)$$

and subtracting this from (14.2.1) gives

$$\boldsymbol{\delta}^{\mathrm{T}}(\mathbf{g} - \nabla f(\mathbf{x})) \leqslant o(\|\boldsymbol{\delta}\|).$$

Choosing $\boldsymbol{\delta} = \theta(\mathbf{g} - \nabla f(\mathbf{x}))$, $\theta \downarrow 0$, shows that $\mathbf{g} = \nabla f$. Hence in this case $\partial f(\mathbf{x})$ is the single vector $\nabla f(\mathbf{x})$.

In Section 14.1 the particular class of polyhedral convex functions $h(\mathbf{c})$, $\mathbb{R}^m \to \mathbb{R}^1$, is defined by

$$h(\mathbf{c}) = \max_i \mathbf{c}^{\mathrm{T}} \mathbf{h}_i + b_i \tag{14.2.3}$$

where $\mathbf{h}_i$ are the columns of a given finite matrix $\mathbf{H}$. Defining

$$\mathscr{A} = \mathscr{A}(\mathbf{c}) \triangleq \{i: \mathbf{c}^T\mathbf{h}_i + b_i = h(\mathbf{c})\} \tag{14.2.4}$$

as the set of supporting planes which are active at $\mathbf{c}$ and so attain the maximum, then it is clear that these planes determine the subdifferential $\partial h(\mathbf{c})$. This is proved as follows.

Lemma 14.2.2

$$\partial h(\mathbf{c}) = \underset{i\in\mathscr{A}(\mathbf{c})}{\operatorname{conv}} \mathbf{h}_i \tag{14.2.5}$$

Proof

$\partial h(\mathbf{c})$ is defined by

$$\partial h(\mathbf{c}) = \{\boldsymbol{\lambda}: h(\mathbf{c}+\boldsymbol{\delta}) \geqslant h(\mathbf{c}) + \boldsymbol{\delta}^T\boldsymbol{\lambda} \quad \forall \boldsymbol{\delta}\}. \tag{14.2.6}$$

Let $\boldsymbol{\lambda} = \mathbf{H}\boldsymbol{\mu} \in (14.2.5)$ where $\mu_i \geqslant 0$, $\sum \mu_i = 1$. Then for all $\boldsymbol{\delta}$,

$$\begin{aligned} h(\mathbf{c}) + \boldsymbol{\delta}^T\boldsymbol{\lambda} &= \max_i (\mathbf{c}^T\mathbf{h}_i + b_i) + \sum_{i\in\mathscr{A}} \boldsymbol{\delta}^T\mathbf{h}_i\mu_i \\ &\leqslant \max_{i\in\mathscr{A}} (\mathbf{c}^T\mathbf{h}_i + b_i) + \max_{i\in\mathscr{A}} \boldsymbol{\delta}^T\mathbf{h}_i \\ &= \max_{i\in\mathscr{A}} ((\mathbf{c}+\boldsymbol{\delta})^T\mathbf{h}_i + b_i) \leqslant h(\mathbf{c}+\boldsymbol{\delta}). \end{aligned}$$

Thus $\boldsymbol{\lambda} \in (14.2.6)$. Now let $\boldsymbol{\lambda} \in (14.2.6)$ and assume $\boldsymbol{\lambda} \notin (14.2.5)$. Then by Lemma 14.2.3 below, $\exists \mathbf{s} \neq \mathbf{0}$ such that $\mathbf{s}^T\boldsymbol{\lambda} > \mathbf{s}^T\boldsymbol{\mu}$ $\forall \boldsymbol{\mu} \in (14.2.5)$. Taking $\boldsymbol{\delta} = \alpha\mathbf{s}$, and since $\mathbf{h}_i \in (14.2.5)$ $\forall i \in \mathscr{A}$,

$$\begin{aligned} h(\mathbf{c}) + \boldsymbol{\delta}^T\boldsymbol{\lambda} &= \max_i (\mathbf{c}^T\mathbf{h}_i + b_i) + \alpha\mathbf{s}^T\boldsymbol{\lambda} \\ &> \mathbf{c}^T\mathbf{h}_i + b_i + \alpha\mathbf{s}^T\mathbf{h}_i \qquad \forall i \in \mathscr{A} \\ &= \max_{i\in\mathscr{A}} ((\mathbf{c}+\alpha\mathbf{s})^T\mathbf{h}_i + b_i) \\ &\geqslant \max_i ((\mathbf{c}+\alpha\mathbf{s})^T\mathbf{h}_i + b_i) = h(\mathbf{c}+\boldsymbol{\delta}) \end{aligned}$$

for α sufficiently small, since the max is then achieved on a subset of $\mathscr{A}$. Thus (14.2.6) is contradicted, proving $\boldsymbol{\lambda} \in (14.2.5)$. Hence the definitions of $\partial h(\mathbf{c})$ in (14.2.5) and (14.2.6) are equivalent. □

Examples of this result for a number of particular cases of $h(\mathbf{c})$ and $\mathbf{H}$ are given in more detail in Section 14.1. For convex functions involving a norm, an alternative description of $\partial h(\mathbf{c})$ is provided by (14.3.7) and (14.3.8). The above lemma makes use of the following important result.

Lemma 14.2.3 (Separating hyperplane lemma for convex sets)

If K is a closed convex set and $\boldsymbol{\lambda} \notin K$ then there exists a hyperplane which separates $\boldsymbol{\lambda}$ and K (see Figure 14.2.2).

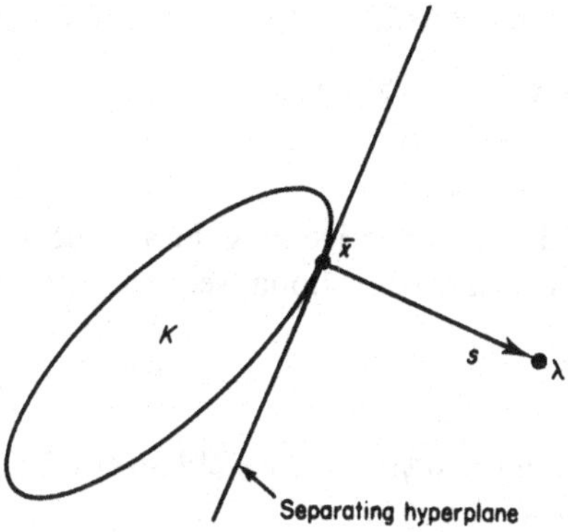

Figure 14.2.2 Existence of a separating hyperplane

Proof

Let $\mathbf{x}_0 \in K$. Then the set $\{\mathbf{x}: \|\mathbf{x}-\boldsymbol{\lambda}\|_2 \leqslant \|\mathbf{x}_0-\boldsymbol{\lambda}\|_2\}$ is bounded and so there exists a minimizer, $\bar{\mathbf{x}}$ say, to the problem: $\min \|\mathbf{x}-\boldsymbol{\lambda}\|_2$, $\mathbf{x} \in K$. Then for any $\mathbf{x} \in K$, $\theta \in [0, 1]$,

$$\|(1-\theta)\bar{\mathbf{x}} + \theta\mathbf{x} - \boldsymbol{\lambda}\|_2^2 \geqslant \|\bar{\mathbf{x}} - \boldsymbol{\lambda}\|_2^2$$

and in the limit $\theta \downarrow 0$ it follows that

$$(\mathbf{x}-\bar{\mathbf{x}})^{\mathrm{T}}(\boldsymbol{\lambda}-\bar{\mathbf{x}}) \leqslant 0 \qquad \forall \mathbf{x} \in K.$$

Thus the vector $\mathbf{s} = \boldsymbol{\lambda} - \bar{\mathbf{x}} \neq \mathbf{0}$ satisfies both $\mathbf{s}^{\mathrm{T}}(\boldsymbol{\lambda}-\bar{\mathbf{x}}) > 0$ and $\mathbf{s}^{\mathrm{T}}(\mathbf{x}-\bar{\mathbf{x}}) \leqslant 0$ $\forall \mathbf{x} \in K$ and hence

$$\mathbf{s}^{\mathrm{T}}\boldsymbol{\lambda} > \mathbf{s}^{\mathrm{T}}\mathbf{x} \qquad \forall \mathbf{x} \in K. \tag{14.2.7}$$

The hyperplane $\mathbf{s}^{\mathrm{T}}(\mathbf{x}-\bar{\mathbf{x}}) = 0$ thus separates K and $\boldsymbol{\lambda}$ as illustrated in Figure 14.2.2 □

At any point $\mathbf{x}'$ at which ∇f does not exist, it nonetheless happens that the directional derivative of the convex function in any direction is well defined. It is again assumed that $f(\mathbf{x})$ is defined on a convex set $K \subset \mathbb{R}^n$ and $\mathbf{x}' \in$ interior (K). The following preliminary result is required.

Lemma 14.2.4

Let $\mathbf{x}^{(k)} \to \mathbf{x}'$ and $\mathbf{g}^{(k)} \in \partial f^{(k)}$. Then any accumulation point of $\{\mathbf{g}^{(k)}\}$ is in $\partial f'$.

Proof

For any $\mathbf{y} \in K$, (14.2.1) can be written as

$$f(\mathbf{y}) \geqslant f^{(k)} + (\mathbf{y} - \mathbf{x}^{(k)})^{\mathrm{T}}\mathbf{g}^{(k)}.$$

Taking any subsequence for which $\mathbf{g}^{(k)} \to \mathbf{g}'$ it follows that

$$f(\mathbf{y}) \geqslant f' + (\mathbf{y} - \mathbf{x}')^{\mathrm{T}}\mathbf{g}' \qquad \forall \mathbf{y} \in K,$$

that is $\mathbf{g}' \in \partial f'$. □

A result for the directional derivative at $\mathbf{x}'$ in a direction $\mathbf{s}$ can now be given in a quite general way for any directional sequence (see Section 9.2).

Lemma 14.2.5

If $\mathbf{x}^{(k)} \to \mathbf{x}'$ *is any directional sequence with* $\delta^{(k)} > 0$ *such that* $\delta^{(k)} \to 0$ *and* $\mathbf{s}^{(k)} \to \mathbf{s}$ *in* (9.2.1) (*and allowing* $\mathbf{s} = \mathbf{0}$), *then*

$$\lim_{k \to \infty} \frac{f^{(k)} - f'}{\delta^{(k)}} = \max_{\mathbf{g} \in \partial f'} \mathbf{s}^{\mathrm{T}}\mathbf{g}. \qquad (14.2.8)$$

Proof

If $\mathbf{g}^{(k)} \in \partial f^{(k)}$ then it follows from (14.2.1) that for k sufficiently large

$$f' \geqslant f^{(k)} - \delta^{(k)}\mathbf{s}^{(k)\mathrm{T}}\mathbf{g}^{(k)}$$

and

$$f^{(k)} \geqslant f' + \delta^{(k)}\mathbf{s}^{(k)\mathrm{T}}\mathbf{g} \qquad \forall \mathbf{g} \in \partial f'$$

both hold, and hence

$$\mathbf{s}^{(k)\mathrm{T}}\mathbf{g}^{(k)} \geqslant \frac{f^{(k)} - f'}{\delta^{(k)}} \geqslant \max_{\mathbf{g} \in \partial f'} \mathbf{s}^{\mathrm{T}}\mathbf{g}. \qquad (14.2.9)$$

Since $\partial f^{(k)}$ is bounded in a neighbourhood of $\mathbf{x}'$ (Lemma 14.2.1), there exists a subsequence for which $\mathbf{g}^{(k)} \to \mathbf{g}'$, and $\mathbf{g}' \in \partial f'$ by Lemma 14.2.4. If (14.2.8) is not true then (14.2.9) gives a contradiction in the limit of such a subsequence. □

Thus (14.2.8) shows that the directional derivative is determined by an extreme supporting hyperplane whose subgradient gives the greatest slope: see Figure 14.2.1 for example.

It is possible to deduce from this result that if $\mathbf{x}^*$ is a local minimizer of $f(\mathbf{x})$, then $f^{(k)} \geqslant f^*$ for all k sufficiently large, and hence from (14.2.8) that

$$\max_{\mathbf{g} \in \partial f^*} \mathbf{s}^{\mathrm{T}}\mathbf{g} \geqslant 0 \qquad \forall \mathbf{s} : \|\mathbf{s}\| = 1. \qquad (14.2.10)$$

Thus a *first order necessary condition* for a local minimum is that the directional derivative is non-negative in all directions. This can be stated alternatively as

$$\mathbf{0} \in \partial f^* \qquad (14.2.11)$$

which generalizes the condition $\mathbf{g}^* = \mathbf{0}$ for smooth functions. Clearly (14.2.11) implies (14.2.10). If $\mathbf{0} \notin \partial f'$ then by Lemma 14.2.3 (with $\boldsymbol{\lambda} = \mathbf{0}$, $K = \partial f'$)

there exists a vector $\mathbf{s}=-\bar{\mathbf{g}}/\|\bar{\mathbf{g}}\|_2$ for which $\mathbf{s}^T\mathbf{g}<0\ \forall\ \mathbf{g}\in\partial f'$, where $\bar{\mathbf{g}}$ is the vector which minimizes $\|\mathbf{g}\|_2\ \forall\ \mathbf{g}\in\partial f'$. Applying this result at $\mathbf{x}^*$ shows that (14.2.10) and (14.2.11) are equivalent. It is now immediate from (14.2.1) that either (14.2.10) or (14.2.11) is also a sufficient condition for a global minimizer at $\mathbf{x}^*$. In fact the vector $\mathbf{s}=-\bar{\mathbf{g}}/\|\bar{\mathbf{g}}\|_2$ defined above is the *steepest descent vector* at $\mathbf{x}'$. Assuming $\mathbf{0}\notin\partial f'$ then from (14.2.8) the direction of least slope is defined by

$$\min_{\|\mathbf{s}\|_2=1}\max_{\mathbf{g}\in\partial f'}\mathbf{s}^T\mathbf{g}=\max_{\mathbf{g}\in\partial f'}\min_{\|\mathbf{s}\|_2=1}\mathbf{s}^T\mathbf{g}$$
$$=\max_{\mathbf{g}\in\partial f'}-\|\mathbf{g}\|_2=-\|\bar{\mathbf{g}}\|_2 \tag{14.2.12}$$

and hence the least slope is attained when $\mathbf{s}=-\bar{\mathbf{g}}/\|\bar{\mathbf{g}}\|_2$. The justification of interchanging the min and max operations is explored in Question 14.7.

The main aim in introducing the above development is to apply it to composite functions of the form

$$\phi(\mathbf{x})=f(\mathbf{x})+h(\mathbf{c}(\mathbf{x})) \tag{14.2.13}$$

where $f(\mathbf{x})$ ($\mathbb{R}^n\to\mathbb{R}^1$) and $\mathbf{c}(\mathbf{x})$ ($\mathbb{R}^n\to\mathbb{R}^m$) are smooth ($\mathbb{C}^1$) functions and $h(\mathbf{c})$ ($\mathbb{R}^m\to\mathbb{R}^1$) is convex but non-smooth ($\mathbb{C}^0$). Let $\mathbf{x}^{(k)}\to\mathbf{x}'$ be a directional sequence with $\delta^{(k)}\downarrow 0$ and $\mathbf{s}^{(k)}\to\mathbf{s}$ in (9.2.1). By Taylor series

$$f^{(k)}=f'+\delta^{(k)}\mathbf{g}'^T\mathbf{s}^{(k)}+o(\delta^{(k)})$$

so $(f^{(k)}-f')/\delta^{(k)}\to\mathbf{g}'^T\mathbf{s}$. Likewise

$$\mathbf{c}^{(k)}=\mathbf{c}'+\delta^{(k)}\mathbf{A}'^T\mathbf{s}^{(k)}+o(\delta^{(k)})$$

so $\mathbf{c}^{(k)}\to\mathbf{c}'$ is a directional sequence in $\mathbb{R}^m$ with $(\mathbf{c}^{(k)}-\mathbf{c}')/\delta^{(k)}\to\mathbf{A}'^T\mathbf{s}$. Thus by applying Lemma 14.2.5 to $h(\mathbf{c})$, it follows that

$$\lim_{k\to\infty}\frac{\phi^{(k)}-\phi'}{\delta^{(k)}}=\max_{\boldsymbol{\lambda}\in\partial h'}\mathbf{s}^T(\mathbf{g}'+\mathbf{A}'\boldsymbol{\lambda}) \tag{14.2.14}$$

and this gives the directional derivative at $\mathbf{x}'$ in the direction $\mathbf{s}$ for the function $\phi(\mathbf{x})$ in (14.2.13). Another more general result about directional derivatives is proved in Lemma 14.5.1 and its corollary.

It can be deduced from (14.2.14) that if $\mathbf{x}^*$ is a local minimizer of $\phi(\mathbf{x})$, then $\phi^{(k)}\geqslant\phi^*$ for all k sufficiently large, and hence

$$\max_{\boldsymbol{\lambda}\in\partial h^*}\mathbf{s}^T(\mathbf{g}^*+\mathbf{A}^*\boldsymbol{\lambda})\geqslant 0 \qquad \forall\mathbf{s}:\|\mathbf{s}\|=1. \tag{14.2.15}$$

This is a first order necessary condition for a local minimizer which like (14.2.10) can be interpreted as a non-negative directional derivative in all directions. Again the result can be stated alternatively as

$$\mathbf{0}\in\partial\phi(\mathbf{x}^*)\triangleq\{\boldsymbol{\gamma}:\ \boldsymbol{\gamma}=\mathbf{g}+\mathbf{A}\boldsymbol{\lambda}\quad\forall\boldsymbol{\lambda}\in\partial h\}_{\mathbf{x}=\mathbf{x}^*}. \tag{14.2.16}$$

The set $\partial\phi^*$ thus defined, although convex and compact, is not the subdifferential because ϕ may not be a convex function, but it is convenient to use the same notation. (It is in fact the *generalized gradient* of Clarke (1975).) Its definition in (14.2.16) is in the nature of a *generalized chain rule*, by analogy with the expression $\nabla\phi = \mathbf{g} + \mathbf{A}\nabla h$ for smooth functions. The equivalence of (14.2.15) and (14.2.16) is again a consequence of the existence of a separating hyperplane (Lemma 14.2.3). Yet another way to state this condition is to introduce the Lagrangian function

$$\mathscr{L}(\mathbf{x}, \boldsymbol{\lambda}) = f(\mathbf{x}) + \boldsymbol{\lambda}^{\mathrm{T}}\mathbf{c}(\mathbf{x}) \tag{14.2.17}$$

Then an equivalent statement of (14.2.16) is as follows.

Theorem 14.2.1 (First order necessary conditions)

If $\mathbf{x}^*$ *minimizes* $\phi(\mathbf{x})$ *in* (14.2.3) *then there exists a vector* $\boldsymbol{\lambda}^* \in \partial h^*$ *such that*

$$\nabla\mathscr{L}(\mathbf{x}^*, \boldsymbol{\lambda}^*) = \mathbf{g}^* + \mathbf{A}^*\boldsymbol{\lambda}^* = \mathbf{0}. \tag{14.2.18}$$

Proof

Immediate since $\partial\phi^*$ is the set of vectors $\nabla\mathscr{L}(\mathbf{x}^*, \boldsymbol{\lambda})$ for all $\boldsymbol{\lambda} \in \partial h^*$. □

This form illustrates the close relationship between the vector $\boldsymbol{\lambda}^* \in \partial h^*$ and the Lagrange multipliers in Theorem 9.1.1. These conditions are illustrated by Questions 14.8, 14.9, 14.10, and 14.12. In general, since $\phi(\mathbf{x})$ may not be convex, the conditions of Theorem 14.2.1 are not sufficient.

In view of this last observation it is important to consider second order conditions for $\mathbf{x}^*$ to be a minimizer of the composite function (14.2.13). These conditions again exhibit a close relationship with the results in Section 9.3. The approach is based on that taken by Fletcher and Watson (1980). In considering second order conditions it is necessary to restrict the possible directions to those having zero directional derivative so that second order effects become important. This is illustrated by the contours in Figure 14.2.3. $\mathbf{x}^*$ satisfies first order conditions but there are directions of zero slope along the derivative discontinuity $c_1(\mathbf{x}) = 0$ (broken line), so that first order conditions are not sufficient. However sufficient second order conditions can be derived which imply that $\mathbf{x}^*$ is a local minimizer—see Question 14.9. In general let $\boldsymbol{\lambda}^*$ be any vector which exists in Theorem 14.2.1 and consider the set

$$X = \{\mathbf{x}\colon h(\mathbf{c}(\mathbf{x})) = h(\mathbf{c}(\mathbf{x}^*)) + (\mathbf{c}(\mathbf{x}) - \mathbf{c}(\mathbf{x}^*))^{\mathrm{T}}\boldsymbol{\lambda}^*\}. \tag{14.2.19}$$

Define $\mathscr{G}^*$ as the set of normalized feasible directions with respect to X at $\mathbf{x}^*$. (That is $\mathbf{s} \in \mathscr{G}^*$ implies that there exists a directional sequence $\mathbf{x}^{(k)} \to \mathbf{x}^*$, feasible in (14.2.19), such that $\mathbf{s}^{(k)} \to \mathbf{s}$ in (9.2.1) with $\sigma = 1$.) $\mathscr{G}^*$ can be regarded as the set of stationary directions for $h(\mathbf{c}(\mathbf{x}))$, having linearized $h(\mathbf{c})$ but not $\mathbf{c}(\mathbf{x})$. It is possible to show that these directions are closely related to the set G^* of normalized directions of zero slope, that is

$$G^* \triangleq \left\{\mathbf{s}\colon \max_{\boldsymbol{\lambda} \in \partial h^*} \mathbf{s}^{\mathrm{T}}(\mathbf{g}^* + \mathbf{A}^*\boldsymbol{\lambda}) = 0,\ \|\mathbf{s}\|_2 = 1\right\}. \tag{14.2.20}$$

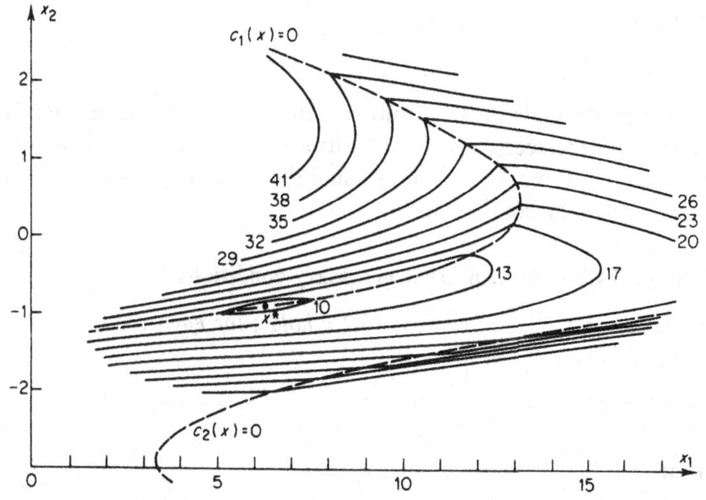

Figure 14.2.3 Contours for the scaled L_1 Freudenstein and Roth problem, Question 14.9

G^* can be regarded as the set of stationary directions for $h(\mathbf{c}(\mathbf{x}))$ having linearized both $h(\mathbf{c})$ and $\mathbf{c}(\mathbf{x})$. The extent to which $\mathscr{G}^*$ and G^* correspond is important and it is first shown that $\mathscr{G}^*$ is a subset of G^*.

Lemma 14.2.6

$\mathscr{G}^* \subset G^*$.

Proof

Let $\mathbf{s} \in \mathscr{G}^*$ so a directional sequence in X exists with $\mathbf{s}^{(k)} \to \mathbf{s}$, $\|\mathbf{s}\|_2 = 1$. Using (14.2.14), (14.2.13), (14.2.19), and a Taylor series it follows that

$$\begin{aligned}
\max_{\boldsymbol{\lambda} \in \partial h^*} \mathbf{s}^{\mathrm{T}}(\mathbf{g}^* + \mathbf{A}^*\boldsymbol{\lambda}) &= \lim_{k \to \infty} \frac{\phi^{(k)} - \phi^*}{\delta^{(k)}} \\
&= \lim_{k \to \infty} \frac{f^{(k)} - f^* + h(\mathbf{c}^{(k)}) - h(\mathbf{c}^*)}{\delta^{(k)}} \\
&= \lim_{k \to \infty} \frac{f^{(k)} - f^* + (\mathbf{c}^{(k)} - \mathbf{c}^*)^{\mathrm{T}}\boldsymbol{\lambda}^*}{\delta^{(k)}} \\
&= \mathbf{s}^{\mathrm{T}}(\mathbf{g}^* + \mathbf{A}^*\boldsymbol{\lambda}^*).
\end{aligned}$$

It then follows from (14.2.18) that $\mathbf{s} \in G^*$. □

To give a general result going the other way is not always possible although it can be done in important special cases. Further discussion of this is given later in this section: at present the regularity assumption

$$\mathscr{G}^* = G^* \tag{14.2.21}$$

is made (which depends on $\boldsymbol{\lambda}^*$ if more than one such vector exists). It is now possible to state the second order conditions. In doing this it is assumed that f and $\mathbf{c}$ are $\mathbb{C}^2$ functions. Note that as usual the regularity assumption is needed only in the necessary conditions.

Theorem 14.2.2 (Second order necessary conditions)

If $\mathbf{x}^*$ *minimizes* $\phi(\mathbf{x})$ *then Theorem* 14.2.1 *holds; for each vector* $\boldsymbol{\lambda}^*$ *which thus exists, if* (14.2.21) *holds, then*

$$\mathbf{s}^{\mathrm{T}}\nabla^2\mathscr{L}(\mathbf{x}^*, \boldsymbol{\lambda}^*)\mathbf{s} \geqslant 0 \qquad \forall \mathbf{s}\in G^*. \tag{14.2.22}$$

Proof

For any $\mathbf{s}\in G^*$, $\mathbf{s}\in\mathscr{G}^*$ by (14.2.21) and hence $\exists$ a directional sequence $\mathbf{x}^{(k)}\to\mathbf{x}^*$, feasible in (14.2.19). A Taylor expansion of $\mathscr{L}(\mathbf{x}, \boldsymbol{\lambda}^*)$ about $\mathbf{x}^*$ yields

$$\mathscr{L}(\mathbf{x}^{(k)}, \boldsymbol{\lambda}^*) = f^* + \mathbf{c}^{*\mathrm{T}}\boldsymbol{\lambda}^* + \mathbf{e}^{(k)\mathrm{T}}\nabla\mathscr{L}(\mathbf{x}^*, \boldsymbol{\lambda}^*) + \tfrac{1}{2}\mathbf{e}^{(k)\mathrm{T}}\nabla^2\mathscr{L}(\mathbf{x}^*, \boldsymbol{\lambda}^*)\mathbf{e}^{(k)} + o(\|\mathbf{e}^{(k)}\|^2)$$

where $\mathbf{e}^{(k)} = \mathbf{x}^{(k)} - \mathbf{x}^*$, and using (14.2.18) and (9.2.1),

$$\mathscr{L}(\mathbf{x}^{(k)}, \boldsymbol{\lambda}^*) = f^* + \mathbf{c}^{*\mathrm{T}}\boldsymbol{\lambda}^* + \tfrac{1}{2}\delta^{(k)^2}\mathbf{s}^{(k)\mathrm{T}}\nabla^2\mathscr{L}(\mathbf{x}^*, \boldsymbol{\lambda}^*)\mathbf{s}^{(k)} + o(\delta^{(k)^2}). \tag{14.2.23}$$

Since $\mathbf{x}^{(k)}$ is feasible in (14.2.19) it follows from (14.2.13) that

$$\phi^{(k)} = \phi^* + \tfrac{1}{2}\delta^{(k)^2}\mathbf{s}^{(k)\mathrm{T}}\nabla^2\mathscr{L}(\mathbf{x}^*, \boldsymbol{\lambda}^*)\mathbf{s}^{(k)} + o(\delta^{(k)^2}).$$

Since $\mathbf{x}^*$ is a local minimizer, $\phi^{(k)} \geqslant \phi^*$ for all k sufficiently large, so dividing by $\tfrac{1}{2}\delta^{(k)^2}$ and taking the limit yields (14.2.22). $\square$

For sufficient conditions, it is firstly observed that if the directional derivative is positive in all directions, that is

$$\max_{\boldsymbol{\lambda}\in\partial h^*} \mathbf{s}^{\mathrm{T}}(\mathbf{g}^* + \mathbf{A}^*\boldsymbol{\lambda}) > 0 \qquad \forall\, \mathbf{s}: \|\mathbf{s}\| = 1,$$

or equivalently G^* is empty, then the conditions of Theorem 14.2.1 imply that $\mathbf{x}^*$ is a strict and isolated local minimizer (an example is given in Question 14.12). This result is in fact a special case of Theorem 14.2.3 below. When G^* is non-empty, second order effects come into play and this is expressed in the following.

Theorem 14.2.3 (Second order sufficient conditions)

If there exists $\boldsymbol{\lambda}^*\in\partial h^*$ *such that* (14.2.18) *holds, and if*

$$\mathbf{s}^{\mathrm{T}}\nabla^2\mathscr{L}(\mathbf{x}^*, \boldsymbol{\lambda}^*)\mathbf{s} > 0 \qquad \forall\, \mathbf{s}\in G^* \tag{14.2.24}$$

then $\mathbf{x}^*$ *is a strict and isolated local minimizer of* $\phi(\mathbf{x})$.

Proof

Assume that $\mathbf{x}^*$ is not strict, so that $\exists$ a sequence and hence a directional sequence $\mathbf{x}^{(k)} \to \mathbf{x}^*$ such that $\phi^{(k)} \leqslant \phi^*$. By (14.2.18),

$$0 \leqslant \max_{\boldsymbol{\lambda} \in \partial h^*} \mathbf{s}^{\mathrm{T}}(\mathbf{g}^* + \mathbf{A}^*\boldsymbol{\lambda}) = \mu$$

say. If $\mu > 0$ then $\lim(\phi^{(k)} - \phi^*)/\delta^{(k)} = \mu$ (see (14.2.14)) which contradicts $\phi^{(k)} \leqslant \phi^*$. Thus $\mu = 0$ and hence $\mathbf{s} \in G^*$. Now from (14.2.17) and (14.2.13)

$$\begin{aligned}\mathscr{L}(\mathbf{x}^{(k)}, \boldsymbol{\lambda}^*) - \mathscr{L}(\mathbf{x}^*, \boldsymbol{\lambda}^*) &= f^{(k)} + \mathbf{c}^{(k)\mathrm{T}}\boldsymbol{\lambda}^* - f^* - \mathbf{c}^{*\mathrm{T}}\boldsymbol{\lambda}^* \\ &= \phi^{(k)} - \phi^* - (h(\mathbf{c}^{(k)}) - h(\mathbf{c}^*) - (\mathbf{c}^{(k)} - \mathbf{c}^*)^{\mathrm{T}}\boldsymbol{\lambda}^*) \\ &\leqslant \phi^{(k)} - \phi^*\end{aligned}$$

using the subgradient inequality. Hence from (14.2.23)

$$0 \geqslant \phi^{(k)} - \phi^* \geqslant \tfrac{1}{2}\delta^{(k)^2}\mathbf{s}^{(k)\mathrm{T}}\nabla^2\mathscr{L}(\mathbf{x}^*, \boldsymbol{\lambda}^*)\mathbf{s}^{(k)} + o(\delta^{(k)^2}).$$

Dividing by $\tfrac{1}{2}\delta^{(k)^2}$ and taking the limit contradicts (14.2.24) and establishes that $\mathbf{x}^*$ is a strict minimizer.

Now assume that $\mathbf{x}^*$ is not isolated so that $\exists$ a sequence of local solutions $\mathbf{x}^{(k)} \to \mathbf{x}^*$. A directional subsequence can therefore be extracted for which $\mathbf{s}^{(k)} \to \mathbf{s}$ in (9.2.1) with $\|\mathbf{s}\|_2 = 1$, and for which

$$\exists\, \boldsymbol{\lambda}^{(k)} \in \partial h^{(k)}, \qquad \mathbf{g}^{(k)} + \mathbf{A}^{(k)}\boldsymbol{\lambda}^{(k)} = \mathbf{0}, \qquad \boldsymbol{\lambda}^{(k)} \to \boldsymbol{\lambda}^*$$

by virtue of Theorem 14.2.1 at $\mathbf{x}^{(k)}$, and Lemma 14.2.4. By taking Taylor expansions about $\mathbf{x}^{(k)}$ and using the subgradient inequality

$$\begin{aligned}\phi^* &= \phi(\mathbf{x}^{(k)} - \boldsymbol{\delta}^{(k)}) + h(\mathbf{c}(\mathbf{x}^{(k)} - \boldsymbol{\delta}^{(k)})) \\ &\geqslant \phi^{(k)} - \boldsymbol{\delta}^{(k)\mathrm{T}}(\mathbf{g}^{(k)} + \mathbf{A}^{(k)}\boldsymbol{\lambda}^{(k)}) + o(\|\boldsymbol{\delta}^{(k)}\|) \\ &= \phi^{(k)} + o(\|\boldsymbol{\delta}^{(k)}\|)\end{aligned}$$

by $\mathbf{g}^{(k)} + \mathbf{A}^{(k)}\boldsymbol{\lambda}^{(k)} = \mathbf{0}$. Define μ as in the first part of the proof: this result provides a contradiction if $\mu > 0$. Hence $\mu = 0$ and $\mathbf{s} \in G^*$. Taylor expansions for $\mathbf{g}^{(k)}$ and $\mathbf{A}^{(k)}$ and $\boldsymbol{\lambda}^{(k)} \to \boldsymbol{\lambda}^*$ give

$$\mathbf{g}^{(k)} + \mathbf{A}^{(k)}\boldsymbol{\lambda}^{(k)} = \mathbf{g}^* + \mathbf{A}^*\boldsymbol{\lambda}^* + \nabla^2\mathscr{L}(\mathbf{x}^*, \boldsymbol{\lambda}^*)\boldsymbol{\delta}^{(k)} + o(\|\boldsymbol{\delta}^{(k)}\|).$$

It follows that $\boldsymbol{\delta}^{(k)\mathrm{T}}\nabla^2\mathscr{L}(\mathbf{x}^*, \boldsymbol{\lambda}^*)\boldsymbol{\delta}^{(k)} + o(\|\boldsymbol{\delta}^{(k)}\|^2) = 0$ and hence, in the limit, $\mathbf{s}^{\mathrm{T}}\nabla^2\mathscr{L}(\mathbf{x}^*, \boldsymbol{\lambda}^*)\mathbf{s} = 0$ which contradicts (14.2.24). Thus $\mathbf{x}^*$ is also an isolated minimizer. □

These conditions (with G^* non-empty) are illustrated by Questions 14.9 and 14.10.

The second order conditions for nonlinear programming given in Section 9.3 are almost necessary and sufficient because the regularity assumption is mild and there is a gap only in the case of zero curvature. The same is *not* true here and it is important to realize that there are well-behaved cases in which $\mathscr{G}^* \neq G^*$. These cases arise when the linearization of $h(\mathbf{c})$ in (14.2.19) is not valid; for example when $h(\mathbf{c})$ is smooth but nonlinear. A property which enables further progress to be made is the following. If there exists an open neighbourhood Ω of $\mathbf{c}^*$ such that

$$h(\mathbf{c}) = h(\mathbf{c}^*) + \max_{\boldsymbol{\lambda}\in\partial h^*} (\mathbf{c} - \mathbf{c}^*)^{\mathrm{T}}\boldsymbol{\lambda} \tag{14.2.29}$$

for all $\mathbf{c}\in\Omega$, then $h(\mathbf{c})$ is referred to as being *locally linear* at $\mathbf{c}^*$. This property is clearly true for any polyhedral convex function but also enables certain exact penalty function problems with smooth norms to be handled, as described in Lemma 14.3.1. Another construction is also required. Denote the dimension of $\partial h(\mathbf{c}^*)$ by l^* ($l^* \leqslant m$) and let the matrix $\mathbf{D}^*\in\mathbb{R}^{m\times l^*}$ have columns $\mathbf{d}_i^*$ $i = 1, 2, \ldots, l^*$ which provide a basis for $\partial h(\mathbf{c}^*) - \boldsymbol{\lambda}^*$. That is to say, $\partial h(\mathbf{c}^*)$ can be expressed as

$$\partial h(\mathbf{c}^*) = \{\boldsymbol{\lambda}\colon \boldsymbol{\lambda} = \boldsymbol{\lambda}^* + \mathbf{D}^*\mathbf{u},\ \mathbf{u}\in U^* \subset \mathbb{R}^{l^*}\} \tag{14.2.30}$$

where U^*, which is the image of $\partial h(\mathbf{c}^*) - \boldsymbol{\lambda}^*$ is also convex and compact. The idea is essentially that contained in Osborne's (1985) introduction of *structure functionals* for polyhedral convex functions, taken up more generally by Womersley (1984b). Note that $\boldsymbol{\lambda}^*$ is essentially an arbitrary vector in $\partial h(\mathbf{c}^*)$: the choice of the vector $\boldsymbol{\lambda}^*$ as defined in (14.2.18) is made for reasons of convenience. A mild independence assumption now enables (14.2.21) to be established when $h(\mathbf{c})$ is locally linear at $\mathbf{c}^*$.

Lemma 14.2.7 (Sufficient conditions for regularity)

If $\mathbf{x}^*$ *satisfies first order conditions* (14.2.18), *if* $h(\mathbf{c})$ *is locally linear at* $\mathbf{c}^*$, *and if*

$$\operatorname{rank}(\mathbf{A}^*\mathbf{D}^*) = l^* \tag{14.2.31}$$

then $\mathscr{G}^* = G^*$.

Proof

Because $G^* \supseteq \mathscr{G}^*$ it is sufficient to prove the reverse inclusion. Let $\mathbf{s}\in G^*$ and define

$$\partial h_s^* = \{\boldsymbol{\lambda}\in\partial h^*\colon \mathbf{s}^{\mathrm{T}}(\mathbf{g}^* + \mathbf{A}^*\boldsymbol{\lambda}) = 0\}$$

observing that ∂h_s^* depends on $\mathbf{s}$. It follows from (14.2.18) that

$$\mathbf{s}^{\mathrm{T}}\mathbf{A}^*(\boldsymbol{\lambda} - \boldsymbol{\lambda}^*) = 0 \qquad \forall\,\boldsymbol{\lambda}\in\partial h_s^*$$

and from (14.2.20) that

$$G^* = \left\{\mathbf{s}\colon \|\mathbf{s}\|_2 = 1,\quad \max_{\boldsymbol{\lambda}\in\partial h^*} \mathbf{s}^{\mathrm{T}}\mathbf{A}^*(\boldsymbol{\lambda} - \boldsymbol{\lambda}^*) = 0\right\} \tag{14.2.32}$$

and hence that

$$\mathbf{s}^{\mathrm{T}}\mathbf{A}^*(\boldsymbol{\lambda} - \boldsymbol{\lambda}^*) < 0 \qquad \forall\,\boldsymbol{\lambda}\in\partial h^*\backslash\partial h_s^*.$$

Let the dimension of ∂h_s^* be l_s^* ($l_s^* \leqslant l^*$) and assume without loss of generality that the vectors $\mathbf{d}_i^*$ $i = 1, 2, \ldots, l_s^*$ form a basis for $\partial h_s^* - \boldsymbol{\lambda}^*$. Hence

$$\mathbf{s}^T\mathbf{A}^*\mathbf{d}_i^* \begin{cases} = 0 & i = 1, 2, \ldots, l_s^* \\ < 0 & i = l_s^* + 1, \ldots, l^*. \end{cases}$$

If $l_s^* = n$ then it follows that $\mathbf{s}^T\mathbf{A}^*\mathbf{D}^* = \mathbf{0}^T$ and hence from the rank assumption it follows that $\mathbf{s} = \mathbf{0}$, which contradicts $\mathbf{s} \in G^*$. For $l_s^* < n$ it is possible to construct a smooth arc $\mathbf{x}(\theta)$ $\theta \in [0, \bar{\theta})$ for which $\mathbf{x}(0) = \mathbf{x}^*$ and $\dot{\mathbf{x}}(0) = \mathbf{s}$ in the manner of Lemma 9.2.2. (The appropriate nonlinear equation in (9.2.3) is $r_i(\mathbf{x}, \theta) = (\mathbf{c}(\mathbf{x}) - \mathbf{c}^*)^T\mathbf{d}_i^* - \theta\mathbf{s}^T\mathbf{A}^*\mathbf{d}_i^* = 0$ and the Jacobian matrix is $[\mathbf{A}^*\mathbf{D}^*:\mathbf{B}]$ where $\mathbf{B}$ is $nx(n - l^*)$.) It therefore follows that

$$(\mathbf{c}(\mathbf{x}(\theta)) - \mathbf{c}^*)^T\mathbf{d}_i^* = \theta\mathbf{s}^T\mathbf{A}^*\mathbf{d}_i \begin{cases} = 0 & i = 1, 2, \ldots, l_s^* \\ < 0 & i = l_s^* + 1, \ldots, l^* \end{cases}$$

and hence that

$$(\mathbf{c}(\mathbf{x}(\theta)) - \mathbf{c}^*)^T(\boldsymbol{\lambda} - \boldsymbol{\lambda}^*) \begin{cases} = 0 & \boldsymbol{\lambda} \in \partial h_s^* \\ < 0 & \boldsymbol{\lambda} \in \partial h^* \backslash \partial h_s^* \end{cases}$$

or equivalently that

$$\max_{\boldsymbol{\lambda} \in \partial h^*} (\mathbf{c}(\mathbf{x}(\theta)) - \mathbf{c}^*)^T(\boldsymbol{\lambda} - \boldsymbol{\lambda}^*) = 0.$$

Finally using (14.2.29) it follows that there exists a neighbourhood of $\mathbf{c}^*$ such that

$$h(\mathbf{c}(\mathbf{x}(\theta))) = h(\mathbf{c}^*) + (\mathbf{c}(\mathbf{x}(\theta)) - \mathbf{c}^*)^T\boldsymbol{\lambda}^*$$

and taking any sequence $\theta^{(k)} \downarrow 0$ gives a directional sequence that is feasible in (14.2.19), so that $\mathbf{s} \in \mathscr{G}^*$. □

Of course these are not the only sufficient conditions for $\mathscr{G}^* = G^*$, for example the same result holds if $\mathbf{c}(\mathbf{x})$ is affine (in place of (14.2.31)) or if it can be shown that G^* is empty. However, it does have the advantage that the rank condition (14.2.31) is readily checked when $\mathbf{c}(\mathbf{x})$ is nonlinear. Basis vectors for the set ∂h^* are readily determined in all practical cases, including the functions $h(\mathbf{c})$ listed in (14.1.3). For the max function (I), if $\mathscr{A}^* = \{i : c_i^* = h^*\}$ is the index set over which the max is attained, and $q \in \mathscr{A}^*$ is arbitrary, then $\mathbf{e}_j - \mathbf{e}_q$ $j \in \mathscr{A}^* \backslash q$ is a suitable basis set for the vectors $\mathbf{d}_i^*$. For the function $h(\mathbf{c}) = \|\mathbf{c}^+\|_\infty$ (II), if $h^* > 0$ then the same basis set as for (I) is valid. If $h^* = 0$ then $\mathscr{A}^*$ is the same and $\mathbf{e}_j^*$ $j \in \mathscr{A}^*$ is a basis set. For $h(\mathbf{c}) = \|\mathbf{c}\|_\infty$ (III) define $\mathscr{A}^* = \{j : |c_j^*| = h^*\}$. If $h^* > 0$ then a basis set is $\operatorname{sign}(c_j^*)\mathbf{e}_j - \operatorname{sign}(c_q^*)\mathbf{e}_q$ $j \in \mathscr{A}^* \backslash q$ where $q \in \mathscr{A}^*$ is arbitrary. If $h^* = 0$ (which implies $\mathbf{c}^* = \mathbf{0}$) then $\mathbf{d}_j^* = \mathbf{e}_j^*$ $j = 1, 2, \ldots, m$ is suitable.

In the case of functions based on the L_1 norm (IV and V) it is appropriate to define $\mathscr{A}^* = \{i : c_i^* = 0\}$. Then a suitable basis in both cases is the set of vectors $\mathbf{e}_i$ $i \in \mathscr{A}^*$. Thus the rank assumption (14.2.31) is equivalent to the assumption that the vectors $\mathbf{a}_i^*$ $i \in \mathscr{A}^*$ are linearly independent which is the standard assumption for regularity in nonlinear programming (Lemma 9.2.2). For the general polyhedral convex function (14.1.2), if $\mathscr{A}^*$ is defined by (14.1.10), then a basis set is any basis for the space spanned by the vectors $\mathbf{h}_j - \mathbf{h}_q$ for all $j \in \mathscr{A}^* - q$ where $q \in \mathscr{A}^*$ is arbitrary. A different example of some interest is $h(\mathbf{c}) = \|\mathbf{c}\|_2$ for which if $h(\mathbf{c}^*) = 0$ then $\mathbf{d}_j = \mathbf{e}_j$ $j = 1, 2, \ldots, m$ is a suitable basis for $\partial h^* - \boldsymbol{\lambda}^*$. Generalizations of this result are given in Lemma 14.3.1.

Another direction set of some interest is the column null space of the matrix $\mathbf{A}^*\mathbf{D}^*$,

$$\mathscr{N}^* = \{\mathbf{s}\colon \|\mathbf{s}\|_2 = 1, \mathbf{s}^T\mathbf{A}^*\mathbf{D}^* = \mathbf{0}^T\} \tag{14.2.33}$$

It follows easily from (14.2.32) and (14.2.30) that $\mathscr{N}^* \subseteq G^*$. In the case that $\boldsymbol{\lambda}^*$ is in the relative interior of ∂h^* (equivalently $\mathbf{0} \in \operatorname{int} U^*$ in (14.2.30)) then *strict complementarity* is said to hold (cf. Section 9.1) and $G^* = \mathscr{N}^*$ can be proved. Clearly if $\mathbf{s} \in G^*$ then

$$\mathbf{s}^T\mathbf{A}^*\mathbf{D}^*\mathbf{u} \leqslant 0 \qquad \forall \mathbf{u} \in U^*$$

and because $\mathbf{0} \in \operatorname{int} U^*$ it follows that $\mathbf{s}^T\mathbf{A}^*\mathbf{D}^* = \mathbf{0}^T$. Thus the reverse inclusion is true which proves the equivalence. The relationship between $\mathscr{N}^*$ and G^* is analogous to that between (9.3.4) and (9.3.11) for nonlinear programming.

A few special cases can be summarized. If $h(\mathbf{c})$ is smooth at $\mathbf{c}^*$, then $l^* = 0$ and $\mathscr{N}^* = G^* = \mathbb{R}^n$ whereas usually $\mathscr{G}^* = \varnothing$ (in absence of any locally linear behaviour). If $(\phi^*, \mathbf{x}^*)$ is a regular vertex of the epigraph of $\phi(\mathbf{x})$, then $l^* = n$. If strict complementarity also holds then $\mathscr{N}^* = G^* = \mathscr{G}^* = \varnothing$, the second order conditions (14.2.24) are vacuous and first order information is sufficient to ensure a strict local minimizer. It can also be the case that $(\phi^*, \mathbf{x}^*)$ is a degenerate vertex of the epigraph and $l^* > n$. However, if rank $\mathbf{A}^*\mathbf{D}^* = n$ and strict complementarity holds for some $\boldsymbol{\lambda}^* \in \partial h^*$ then it follows that $\mathscr{N}^* = G^* = \mathscr{G}^* = \varnothing$ as above and again first order information is sufficient.

These results enable a very illuminating comparison to be drawn between the composite nonsmooth optimization problem

$$\text{minimize } f(\mathbf{x}) + h(\mathbf{c}(\mathbf{x})) \tag{14.2.34}$$

and the nonlinear programming problem

$$\begin{aligned} &\text{minimize} \quad f(\mathbf{x}) + \mathbf{c}(\mathbf{x})^T\boldsymbol{\lambda}^* \\ &\text{subject to} \quad \mathbf{D}^{*T}(\mathbf{c}^* - \mathbf{c}(\mathbf{x})) = \mathbf{0}, \end{aligned} \tag{14.2.35}$$

as expressed by the following theorem. In this theorem, multipliers of the constraints in (14.2.35) are denoted by $\mathbf{u}^* \in \mathbb{R}^{l^*}$.

Theorem 14.2.4

If $h(\mathbf{c})$ *is locally linear about* $\mathbf{c}^*$ *then*

(i) if $\mathbf{x}^*$ *is a local minimizer of* (14.2.34) *then* $\mathbf{x}^*$ *is both a local solution and a KT point of* (14.2.35),

(ii) *if* $\mathbf{x}^*$ *satisfies second order sufficient conditions for* (14.2.34) *then* $\mathbf{x}^*$ *satisfies second order sufficient conditions for* (14.2.35).

(iii) *If* $\mathbf{x}^*, \mathbf{u}^*$ ($\mathbf{u}^* = \mathbf{0}$) *satisfies second order sufficient conditions for* (14.2.35) *and* $\boldsymbol{\lambda}^*$ *is in the relative interior of* $\partial h(\mathbf{c}^*)$ *then* $\mathbf{x}^*$ *satisfies second order sufficient conditions for* (14.2.34),

(iv) *the condition rank* $(\mathbf{A}^*\mathbf{D}^*) = l^*$ *is sufficient for regularity* ($\mathscr{G}^* = G^*$) *in both cases.*

Proof

In part (i) it follows from Theorem 12.2.1 that $\exists\ \boldsymbol{\lambda}^* \in \partial h(\mathbf{c}^*)$ such that $\mathbf{g}^* + \mathbf{A}^*\boldsymbol{\lambda}^* = \mathbf{0}$. Then (14.2.30) implies that $\mathbf{x}^*$ is a KT point for (14.2.35) with multiplier vector $\mathbf{u}^* = \mathbf{0}$. Let $\mathbf{x}$ be feasible in (14.2.35) so that $\mathbf{D}^{*\mathrm{T}}(\mathbf{c}^* - \mathbf{c}(\mathbf{x})) = \mathbf{0}$, and close enough to $\mathbf{x}^*$ so that (14.2.29) is valid for $\mathbf{c} = \mathbf{c}(\mathbf{x})$. It follows from these that $h(\mathbf{c}(\mathbf{x})) = h(\mathbf{c}^*) + \boldsymbol{\lambda}^{*\mathrm{T}}(\mathbf{c}(\mathbf{x}) - \mathbf{c}^*)$. Because $\mathbf{x}^*$ is a local minimizer of (14.2.34), it follows that $f(\mathbf{x}) + \boldsymbol{\lambda}^{*\mathrm{T}}\mathbf{c}(\mathbf{x}) \geqslant f^* + \boldsymbol{\lambda}^{*\mathrm{T}}\mathbf{c}^*$ and hence $\mathbf{x}^*$ is a local solution of (14.2.35).

For part (ii), second order sufficient conditions for (14.2.34) in Theorem 14.2.3 involve the matrix $\mathbf{W}^* = \nabla^2 f_i^* + \sum \lambda_i^* \nabla^2 c_i^*$. Those for (14.2.35) involve the matrix $\mathbf{W}^* = \nabla^2 f_i^* + \sum \lambda_i^* \nabla^2 c_i^* + \sum u_j^* \nabla^2 (\mathbf{D}^{*\mathrm{T}}\mathbf{c})_j$ which is the same matrix since $\mathbf{u}^* = \mathbf{0}$. Theorem 14.2.3 involves the direction set G^* in (14.2.20). Theorem 9.3 involves the direction set given in (9.3.11) which turns out to be the set $\mathscr{N}^*$ in (14.2.33). Part (ii) follows because $\mathscr{N}^* \subset G^*$ as pointed out after (14.2.33).

For part (iii) it follows that $\mathbf{g}^* + \mathbf{A}^*\boldsymbol{\lambda}^* = \mathbf{0}$, and by assumption $\boldsymbol{\lambda}^* \in \partial h(\mathbf{c}^*)$, so $\mathbf{x}^*$ satisfies first order conditions for (14.2.34). Moreover $\boldsymbol{\lambda}^* \in \text{rel int } \partial h(\mathbf{c}^*)$ (i.e. $\mathbf{0} \in \text{int } U^*$) implies $\mathscr{N}^* = G^*$. Hence second order sufficient conditions hold for (14.2.34) by following a reverse argument to that in the previous paragraph. For part (iv) the rank condition is that in (14.2.31) which is the same assumption that is needed to prove (9.3.12) (cf. Lemma 9.2.2). □

The implications of this result are that if $\boldsymbol{\lambda}^* \in \partial h(\mathbf{c}^*)$ then KT points for (14.2.34) and (14.2.35) are equivalent. Moreover if $h(\mathbf{c})$ is locally linear about $\mathbf{c}^*$ and the rank assumption (14.2.31) holds then regularity holds and the second order conditions are 'almost' necessary and sufficient. Thus, except in limiting cases, problems (14.2.34) and (14.2.35) have identical solutions. The objective function in (14.2.35) is essentially the "smooth" part of (14.2.34), and the $\mathbf{c}(\mathbf{x})^{\mathrm{T}}\boldsymbol{\lambda}^*$ term shifts the multipliers to $\mathbf{u}^* = \mathbf{0}$. The constraints in (14.2.35) are essentially the *structure functionals* given by Osborne (1985) in the polyhedral case and enable the nonsmooth part of (14.2.34) to be replaced by a system of nonlinear equations. It will be seen in the derivation of algorithms in Section 14.4 that this replacement allows us to analyse very readily the asymptotic behaviour of iteration sequences for the SNQP method. It also reinforces, but in a different way, the observation in Section 14.1 that the composite NSO problem is equivalent to a nonlinear programming problem ((14.1.6) in that case).

14.3 EXACT PENALTY FUNCTIONS

One of the most important applications of NSO is in the area of nonlinear programming through the use of an exact penalty function. For the special case of an L_1 exact penalty function, a simple presentation is given in Section 12.3. In this section a quite general treatment of the exact penalty function is given, involving an arbitrary norm. To simplify the presentation, two basic types of nonlinear programming problem are considered, the equality constraint problem

$$\begin{aligned} &\underset{\mathbf{x}}{\text{minimize}} \quad f(\mathbf{x}) \\ &\text{subject to} \quad \mathbf{c}(\mathbf{x}) = \mathbf{0} \end{aligned} \tag{14.3.1}$$

for which the corresponding exact penalty function problem is

$$\underset{\mathbf{x}}{\text{minimize}}\ \phi(\mathbf{x}) \triangleq \nu f(\mathbf{x}) + \|\mathbf{c}(\mathbf{x})\|, \tag{14.3.2}$$

and the inequality constraint problem

$$\begin{aligned} &\underset{\mathbf{x}}{\text{minimize}} \quad f(\mathbf{x}) \\ &\text{subject to} \quad \mathbf{c}(\mathbf{x}) \leqslant \mathbf{0} \end{aligned} \tag{14.3.3}$$

for which the corresponding exact penalty function problem is

$$\underset{\mathbf{x}}{\text{minimize}}\ \phi(\mathbf{x}) \triangleq \nu f(\mathbf{x}) + \|\mathbf{c}(\mathbf{x})^+\| \tag{14.3.4}$$

where c_i^+ denotes $\max(c_i, 0)$. There is no difficulty in generalizing further to the mixed problem (7.1.1) and this can be done within the framework of (14.3.3) by replacing $c_i = 0$ by $c_i \leqslant 0$ and $-c_i \leqslant 0$. Note that in (14.3.3) the inequality is in the reverse direction from that considered in Chapter 9. This is equivalent to a change of sign in the constraint residuals $\mathbf{c}(\mathbf{x})$ which must be kept in mind. These penalty functions are exact in the sense of Section 12.3, so that usually (14.3.1) and (14.3.2) (or (14.3.3) and (14.3.4)) are equivalent in that for sufficiently small ν a local solution of (14.3.1) is a local minimizer of (14.3.2), and *vice versa*. This result is made precise in Theorems 14.3.1 and 14.3.2 below. Practical considerations in choosing ν are discussed in Section 12.3. The conditions under which equivalence holds are usually satisfied in practice but there are examples where the problems are not equivalent. Examples where $\mathbf{x}^*$ solves (14.3.3) and not (14.3.4) are given at the end of Theorem 14.3.1. However, these examples are either pathological or limiting cases and need not greatly concern the user. The alternatively possibility is that a local minimizer $\mathbf{x}^*$ of (14.3.2) may not be feasible in (14.3.1), even though the latter may have a solution. This is the same situation as that illustrated in Figure 6.2.1 and described in Section 6.2: it is an inevitable consequence of the use of any penalty approach as a means of inducing global convergence. To circumvent the difficulty is a global minimization problem and hence generally impracticable. Corresponding advantages are that *best* solutions can be determined when no feasible point exists in (14.3.1) and that the difficulty of finding an initial feasible point is avoided. In practice the most likely unfavourable situation which arises when applying an exact penalty function is that a sequence $\{\mathbf{x}^{(k)}\}$ is calculated such that $\phi^{(k)} \to -\infty$, which is an indication that the calculation should be repeated with a smaller value of ν. It may be necessary to pre-scale the $c_i(\mathbf{x})$ to be of comparable magnitude so that the use of a single penalty parameter ν is reasonable. In this case it can be better to rewrite (14.3.2) for example as

$$\phi(\mathbf{x}) = f(\mathbf{x}) + \|\mathbf{S}\mathbf{c}(\mathbf{x})\|, \tag{14.3.5}$$

where $\mathbf{S} = \operatorname{diag} \sigma_i$ is a diagonal matrix of weights σ_i on each of the constraint functions. There is no difficulty in reorganizing the theory to account for this change.

The main attraction in using (14.3.2) or (14.3.4) is that if holds out the possibility of avoiding the sequential nature of the penalty functions in Sections 12.1 and 12.2 so that only a single unconstrained minimization calculation is required. Unfortunately (even if $\|\cdot\|_2$ is used) (14.3.2) and (14.3.4) are not smooth ($\mathbb{C}^1$) functions so that the many effective techniques for smooth minimization described in Part 1 cannot adequately be used. The study of algorithms for non-smooth problems is a relatively recent development and is described in some detail in Section 14.4 and some of the algorithms can be used here. Another approach (Han, 1977; Coleman and Conn, 1982a, b; Mayne, 1980) is to use an algorithm for nonlinear programming as a means of generating a direction of search, and to use the exact penalty function as the criterion function to be minimized (approximately) in the line search. This approach often works well in practice but unfortunately can fail (see Fletcher, 1981) and I think it is important to take into account the non-smooth nature of the penalty function when choosing the direction in which to search. The discussion of algorithms in Sections 14.4 and 4.5 leads to a globally convergent algorithm for composite NSO problems (Fletcher, 1982a) which works well when applied to minimize exact penalty functions (Fletcher 1981), and it is this type of algorithm which I currently favour. Use of the $\|\cdot\|_1$ is the most convenient for the reasons given in Sections 12.3 and 12.4 where more details are given.

The theory for exact penalty functions can be set out in a very concise and general way. The definition in (14.3.2) allows the use of any norm and in (14.3.4) any *monotonic* norm ($|\mathbf{x}| \leqslant |\mathbf{y}| \Rightarrow \|\mathbf{x}\| \leqslant \|\mathbf{y}\|$). This latter condition ensures that $\|\mathbf{c}^+\|$ is a convex function (see Question 14.2) and includes all L_p and scaled L_p norms for $1 \leqslant p \leqslant \infty$. Thus the results of this section are not just restricted to polyhedral norms. It is convenient to introduce the concept of a *dual norm*

$$\|\mathbf{u}\|_D \triangleq \max_{\|\mathbf{v}\| \leqslant 1} \mathbf{u}^{\mathrm{T}}\mathbf{v}. \tag{14.3.6}$$

The dual of $\|\cdot\|_1$ is $\|\cdot\|_\infty$, and vice versa, and the $\|\cdot\|_2$ is self-dual. Expressions for the subdifferential of the functions $\|\mathbf{c}\|$ and $\|\mathbf{c}^+\|$ are given by

$$\partial\|\mathbf{c}\| = \{\boldsymbol{\lambda}: \boldsymbol{\lambda}^{\mathrm{T}}\mathbf{c} = \|\mathbf{c}\|, \|\boldsymbol{\lambda}\|_D \leqslant 1\} \tag{14.3.7}$$

and

$$\partial\|\mathbf{c}^+\| = \{\boldsymbol{\lambda}: \boldsymbol{\lambda}^T\mathbf{c} = \|\mathbf{c}^+\|, \boldsymbol{\lambda} \geqslant \mathbf{0}, \|\boldsymbol{\lambda}\|_D \leqslant 1\}. \tag{14.3.8}$$

The proof of these expressions is sketched out in some detail in Questions 14.4 and 14.5. Expressions (14.1.12) to (14.1.15) are special cases of (14.3.7) and (14.3.8). The main results giving the equivalence between local solutions of (14.3.1) and (14.3.2) (or between (14.3.3) and (14.3.4)) can now be given. In fact the latter result only is given; the former case is similar (but easier). A point to bear in mind is that in these theorems $\boldsymbol{\lambda}^*$ refers to a multiplier vector for the constrained problem (14.3.3) and $v\boldsymbol{\lambda}^*$ is the equivalent multiplier vector for the exact penalty function problem (14.3.4).

Theorem 14.3.1

If $v < 1/\|\boldsymbol{\lambda}^*\|_D$ *and* $\mathbf{c}^{*+} = \mathbf{0}$ *then the second order sufficient conditions at* $\mathbf{x}^*$ *for problems* (14.3.3) *and* (14.3.4) *are equivalent. Therefore if they hold, the fact that* $\mathbf{x}^*$ *solves* (14.3.3) *implies that* $\mathbf{x}^*$ *solves* (14.3.4), *and vice versa.*

Proof

Second order conditions are given for problem (14.3.3) in Theorem 9.3.2 and for problem (14.3.4) in Theorem 14.2.3. The first order requirements are that $\mathbf{g}^* + \mathbf{A}^*\boldsymbol{\lambda}^* = \mathbf{0}$, $\boldsymbol{\lambda}^* = \mathbf{0}$, $\boldsymbol{\lambda}^{*T}\mathbf{c}^* = \|\mathbf{c}^{*+}\| = 0$ (with a suitable sign change), and by $v\mathbf{g}^* + \mathbf{A}^*(v\boldsymbol{\lambda}^*) = \mathbf{0}$, $v\boldsymbol{\lambda}^* \in \partial h^*$ respectively, which are clearly equivalent from (14.3.8) if $v < 1/\|\boldsymbol{\lambda}^*\|_D$. Next the sets G^* defined in (9.3.11) (with $\|\mathbf{s}\|_2 = 1$) and (14.2.20) are shown to be equivalent. For convenience refer to these as G_9^* and G_{14}^* respectively. Let $\mathbf{s} \in G_{14}^*$. From (14.2.20) and the above it follows that

$$\mathbf{s}^T\mathbf{A}^*(\boldsymbol{\lambda} - v\boldsymbol{\lambda}^*) \leqslant 0 \qquad \forall\, \boldsymbol{\lambda} \in \partial\|\mathbf{c}^{*+}\|. \tag{14.3.9}$$

Let $\mathscr{A}^*$ denote active constraints at $\mathbf{x}^*$ in (14.3.3). For $i \in \mathscr{A}^*$ and small ε it follows using $\|v\boldsymbol{\lambda}^*\|_D < 1$ that $\boldsymbol{\lambda} = v\boldsymbol{\lambda}^* + \varepsilon\mathbf{e}_i \in \partial\|\mathbf{c}^{*+}\|$ either if $\lambda_i^* = 0$ and $\varepsilon > 0$ or if $\lambda_i^* > 0$ and $\pm\varepsilon > 0$. Hence from (14.3.9),

$$\left.\begin{aligned} \mathbf{s}^T\mathbf{a}_i^* &\leqslant 0 \quad \text{if } \lambda_i^* = 0, \\ \mathbf{s}^T\mathbf{a}_i^* &= 0 \quad \text{if } \lambda_i^* > 0, \end{aligned}\right\} \quad i \in \mathscr{A}^*. \tag{14.3.10}$$

Thus $\mathbf{s} \in G_9^*$. Conversely let $\mathbf{s} \in G_9^*$. It follows from (14.3.10) that $\mathbf{s}^T\mathbf{a}_i^*\lambda_i^* = 0$ which also implies that $\mathbf{s}^T\mathbf{g}^* = 0$ from the first order conditions. Let $\boldsymbol{\lambda} \in \partial h^*$; then $\boldsymbol{\lambda}^T\mathbf{c}^* = 0$ and $\boldsymbol{\lambda} \geqslant \mathbf{0}$ imply that if $c_i^* < 0$ then $\lambda_i = 0$ so constraints $i \notin \mathscr{A}^*$ can be ignored. Otherwise $\mathbf{s}^T\mathbf{a}_i^*\lambda_i \leqslant 0$ follows from (14.3.10) and hence $\max_{\boldsymbol{\lambda} \in \partial h^*} \mathbf{s}^T(\mathbf{g}^* + \mathbf{A}^*\boldsymbol{\lambda}) = 0$. Thus $\mathbf{s} \in G_{14}^*$ and the equivalence of G_9^* and G_{14}^* is shown. Finally conditions (9.3.15) and (14.2.24) are clearly equivalent so the equivalence of the second order conditions is demonstrated. □

The proof of this theorem is very similar to that of Theorem 14.2.4, and indeed the results are very similar, showing as they do the equivalence between a composite NSO problem and a nonlinear programming problem. Similar theorems relating the solutions of (14.3.3) and (14.3.4) are given by Charalambous (1979) and Han and Mangasarian (1979).

The requirement that second order conditions hold and that $v < 1/\|\boldsymbol{\lambda}^*\|_D$ in Theorem 14.3.1 cannot easily be relaxed, as the following simple examples show. In all of these, $x^* = 0$ solves (14.3.3) but does not solve (14.3.4) using $\|\cdot\|_1$ for the reasons given. For the problem: min x subject to $x^2 \leqslant 0$, x^* is not a KT point. For the problem: min x^3 subject to $x^5 \geqslant 0$, x^* is a KT point and $\lambda^* = 0$ but the curvature condition (9.3.14) is not strict. For the problem: min $x - \frac{1}{2}x^2$ subject to $0 \leqslant x \leqslant 1$, x^* is a KT point with $\lambda^* = 1$. Then if $v = 1$ the condition $v < 1/\|\boldsymbol{\lambda}^*\|_D$ is not strict for L_1 norm. The last example illustrates another fact. The proof of Theorem 14.3.1 shows that $G_9^* \subset G_{14}^*$ without requiring that $v < 1/\|\boldsymbol{\lambda}^*\|_D$. (In the last example G_9^* is empty whilst G_{14}^* is the direction $\mathbf{s} = -1$.) Thus if $\mathbf{x}^*$ satisfies second order sufficient conditions for (14.3.4) it follows that both $v \leqslant 1/\|\boldsymbol{\lambda}^*\|_D$ and $\mathbf{x}^*$ satisfies second order sufficient conditions for (14.3.3). Finally if $v > 1/\|\boldsymbol{\lambda}^*\|_D$ then a result going the other way can be proved.

Theorem 14.3.2

If first order conditions $\mathbf{g}^* + \mathbf{A}^*\boldsymbol{\lambda}^* = \mathbf{0}$ *hold for* (14.3.3) *and if* $v > 1/\|\boldsymbol{\lambda}^*\|_D$ *then* $\mathbf{x}^*$ *is not a local minimizer of* (14.3.4).

Proof

Since $\|v\boldsymbol{\lambda}^*\|_D > 1$, the vector $v\boldsymbol{\lambda}^*$ is not in $\partial\|\mathbf{c}^{*+}\|$ and so the vector $v\mathbf{g}^* + \mathbf{A}^*(v\boldsymbol{\lambda}^*) = \mathbf{0}$ is not in $\partial\phi^*$. Hence by (14.2.6) $\mathbf{x}^*$ is not a local minimizer of $\phi(\mathbf{x})$. □

Another important result concerns the regularity assumption in (14.2.21). If $\mathbf{x}^*$, $\boldsymbol{\lambda}^*$ satisfy first order conditions for either (14.3.1) or (14.3.3) (including feasibility) then under mild assumptions (14.2.21) holds, even though the norm may not be polyhedral. Again the result is proved only in the more difficult case.

Lemma 14.3.1

If $\mathbf{x}^*, \boldsymbol{\lambda}^*$ *is a KT point for* (14.3.3), *if the vectors* $\mathbf{a}_i^*$, $i \in \mathscr{A}^*$, *are linearly independent, and if* $\|v\boldsymbol{\lambda}^*\|_D < 1$, *then* $\mathscr{G}^* = G^*$.

Proof

Let $\mathbf{s} \in G^*$. Then as in the proof of Theorem 14.3.1, (14.3.10) holds. By the independence assumption there exists an arc $\mathbf{x}(\theta)$ with $\mathbf{x}(0) = \mathbf{x}^*$ and $\dot{\mathbf{x}}(0) = \mathbf{s}$, for which

$$\begin{aligned} c_i(\mathbf{x}(\theta)) &= 0 \quad \text{if } \lambda_i^* > 0, \\ c_i(\mathbf{x}(\theta)) &\leqslant 0 \quad \text{if } \lambda_i^* = 0, \end{aligned} \qquad i \in \mathscr{A}^*$$

for $\theta \geqslant 0$ and sufficiently small (see the proof of Lemma 9.2.2). It follows that $\|\mathbf{c}(\mathbf{x}(\theta))^+\| = \mathbf{c}(\mathbf{x}(\theta))^{\mathrm{T}}\boldsymbol{\lambda}^* = 0$ so $\mathbf{x}(\theta)$ is feasible in (14.2.19) and hence $\mathbf{s} \in \mathscr{G}^*$.

An alternative and more general proof of this lemma can also be given. If $h(\mathbf{c}) = \|\mathbf{c}\|$ and $\mathbf{c}^* = \mathbf{0}$ then $\partial h^* = \{\boldsymbol{\lambda} : \|\boldsymbol{\lambda}\|_D \leqslant 1\}$ and hence a basis for $\partial h^* - \boldsymbol{\lambda}^*$ is the set of columns $\mathbf{e}_j$ $j = 1, 2, \ldots, m$. If $h(\mathbf{c}) = \|\mathbf{c}^+\|$ and $\mathbf{c}^{*+} = \mathbf{0}$ then from (14.3.8) it follows that $\partial h^* = \{\boldsymbol{\lambda} : c_i^* < 0 \Rightarrow \lambda_i = 0, \boldsymbol{\lambda} \geqslant 0, \|\boldsymbol{\lambda}\|_D \leqslant 1\}$. In both cases therefore a basis for $\partial h^* - \boldsymbol{\lambda}^*$ is the set of vectors $\mathbf{e}_i$ $i \in \mathscr{A}^*$. Thus independence of the vectors $\mathbf{a}_i^*$ $i \in \mathscr{A}^*$ implies that the rank condition (14.2.31) holds. The positive homogeneity of the functions $\|\cdot\|$ or $\|\cdot^+\|$ implies that $h(\mathbf{c})$ is locally linear about $\mathbf{c}^*$ when $\mathbf{c}^* = \mathbf{0}$ or $\mathbf{c}^{*+} = \mathbf{0}$. Also the KT conditions and $\|\nu\boldsymbol{\lambda}^*\|_D \leqslant 1$ imply that $\mathbf{x}^*$ satisfies first order conditions (14.2.18). Thus the assumptions of Lemma 14.2.7 are valid and it therefore follows that $\mathscr{G}^* = G^*$. □

To summarize all these results, if $\mathbf{x}^*$ is a feasible point of the nonlinear programming problem and if $\nu > 0$ is sufficiently small, then except in limiting cases, $\mathbf{x}^*$ solves the nonlinear programming problem if and only if $\mathbf{x}^*$ minimizes the corresponding exact penalty function (locally speaking).

In practice there are considerable advantages in using the L_1 norm to define (14.3.2) or (14.3.4). More details of this particular case, including an alternative derivation of the optimality conditions without using subgradients, is given in Section 12.3.

14.4 ALGORITHMS

This section reviews progress in the development of algorithms for many different classes of NSO problem. It happens that similar ideas have been tried in different situations and it is convenient to discuss them in a unified way. At present there is a considerable amount of interest in these developments and it is not possible to say yet what the best approaches are. This is particularly true when curvature estimates are updated by quasi-Newton like schemes. However some important common features are emerging which this review tries to bring out.

Many methods are line search methods in which on each iteration a direction of search $\mathbf{s}^{(k)}$ is determined from some model situation, and $\mathbf{x}^{(k+1)} = \mathbf{x}^{(k)} + \alpha^{(k)}\mathbf{s}^{(k)}$ is obtained by choosing $\alpha^{(k)}$ to minimize approximately the objective function $\phi(\mathbf{x}^{(k)} + \alpha\mathbf{s}^{(k)})$ along the line (see Section 2.3). A typical line search algorithm (various have been suggested) operates under similar principles to those described in Section 2.6 and uses a combination of sectioning and interpolation, but there are some new features. One is that since $\phi(\mathbf{x})$ may contain a polyhedral component $h(\mathbf{c}(\mathbf{x}))$, the interpolating function must also have the same type of structure. The simplest possibility is to interpolate a one variable function of the form

$$\psi(\alpha) = q(\alpha) + h(\mathbf{l}(\alpha)) \quad (14.4.1)$$

where $q(\alpha)$ is quadratic and the functions $\mathbf{l}(\alpha)$ are linear. For composite NSO applications q and $\mathbf{l}$ can be estimated from information about f and $\mathbf{c}$, and $\alpha^{(k)}$ is then determined by minimizing (14.4.1). For basic NSO, only the values of ϕ and $d\phi/d\alpha$ are known at any point so it must be assumed that $\psi(\alpha)$ has a more simple structure, for example the max of just two linear functions $l_i(\alpha)$. Many other possibilities exist. Another new feature concerns what acceptability tests to use, analogous to (2.5.1), (2.5.2), (2.5.4), and (2.5.6) for smooth functions. If the line search minimum is non-smooth (see Figure 14.4.1) then it is not appropriate to try to make the directional derivative small, as (2.5.6) does, since such a point may not exist. Many line searches use the Wolfe–Powell conditions (2.5.1) and (2.5.4). For this choice the range of acceptable α-values is the interval $[a, b]$ in Figure 14.4.1, and this should be compared with Figure 2.5.2. It can be seen that this line search has the effect that the acceptable point *always overshoots the minimizing value of* α, and in fact this may occur by a substantial amount, so as to considerably slow down the rate of convergence of the algorithm in which it is embedded. I have found it valuable to use a different test (Fletcher, 1981) that the line search is terminated when the predicted reduction based on (14.4.1) is sufficiently small, and in particular when the predicted reduction on any subinterval is no greater than 0.1 times the total reduction in $\phi(\mathbf{x})$ so far achieved in the search. This condition has been found to work well in ensuring that $\alpha^{(k)}$ is close to the minimizing value of α.

It is now possible to examine algorithms for NSO problems in many variables. In basic NSO only a limited amount of information is available at any given point $\mathbf{x}$, namely $\phi(\mathbf{x})$ and one element $\mathbf{g} \in \partial\phi(\mathbf{x})$ (usually $\nabla\phi(\mathbf{x})$ since $\phi(\mathbf{x})$ is almost everywhere differentiable). This makes the basic NSO problem more difficult than composite NSO in which values of f and $\mathbf{c}$ (and their derivatives)

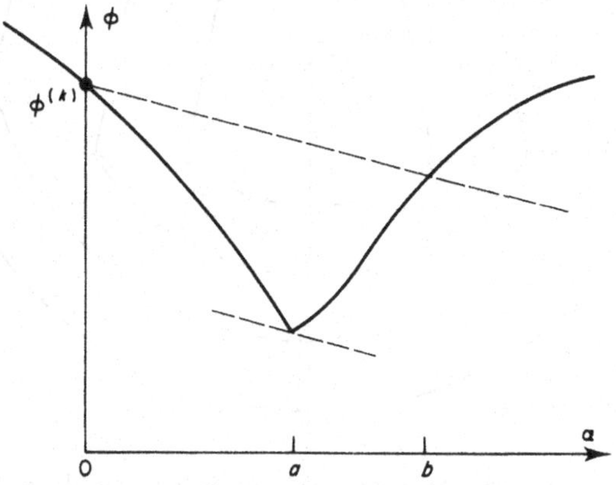

Figure 14.4.1 Line search for non-smooth functions

are given in (14.1.1). The simplest method for basic NSO is an analogue of the steepest descent method in which $\mathbf{s}^{(k)} = -\mathbf{g}^{(k)}$ is used as a search direction; this is known as *subgradient optimization*. Because of its simplicity the method has received much attention both in theory and in practice (see the references given in Lemarechal and Mifflin (1978)), but it is at best linearly convergent, as illustrated by the results for smooth functions (see Section 2.3). In fact the situation is worse in the non-smooth case because the convergence result (Theorem 2.4.1) no longer holds. In fact examples of non-convergence are easily constructed. Assume that the line search is exact and that the subgradient obtained at $\mathbf{x}^{(k+1)}$ corresponds to the piece which is active for $\alpha \geqslant \alpha^{(k)}$. Then the example

$$\begin{aligned} \phi(\mathbf{x}) &= \max_{i=1,2,3} c_i(\mathbf{x}) \\ c_1(\mathbf{x}) &= -5x_1 + x_2 \\ c_2(\mathbf{x}) &= x_1^2 + x_2^2 + 4x_2 \\ c_3(\mathbf{x}) &= 5x_1 + x_2 \end{aligned} \tag{14.4.2}$$

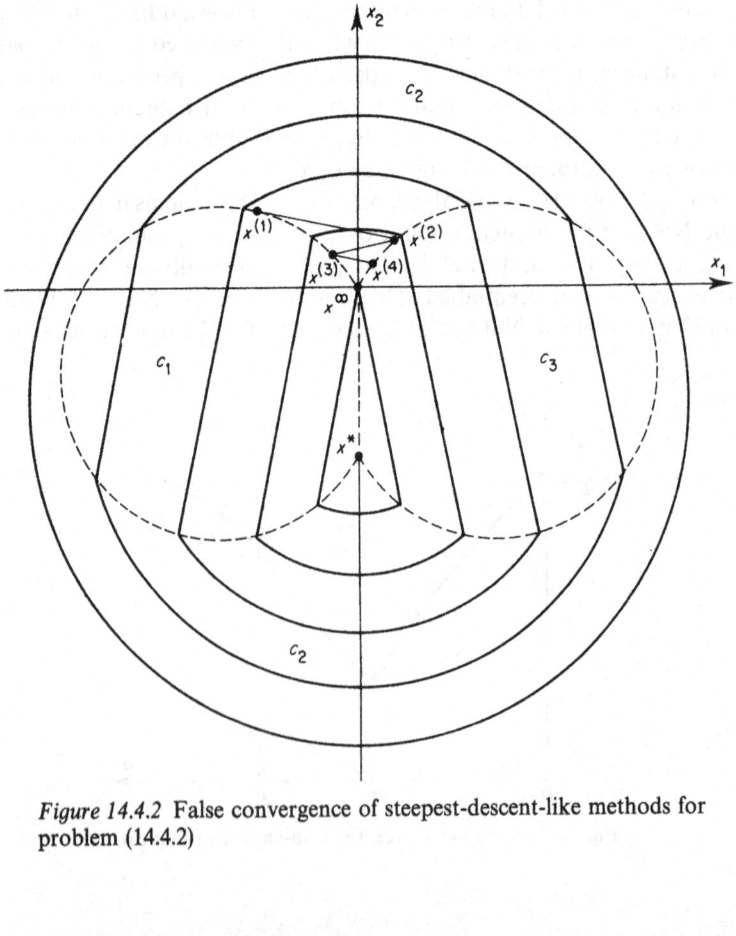

Figure 14.4.2 False convergence of steepest-descent-like methods for problem (14.4.2)

due to Demyanov and Malozemov (1971) illustrates convergence from the initial point $\mathbf{x}^{(1)}$ (see Figure 14.4.2) to the non-optimal point $\mathbf{x}^{\infty} = \mathbf{0}$. The solution is at $\mathbf{x}^* = (0, -3)^{\mathrm{T}}$. At $\mathbf{x}^{(1)}, \mathbf{x}^{(3)}, \ldots$ the subgradient corresponding to c_1 is used and at $\mathbf{x}^{(2)}, \mathbf{x}^{(4)}, \ldots$ the subgradient corresponding to c_3. It is easily seen from Figure 14.4.2 that the sequence $\{\mathbf{x}^{(k)}\}$ oscillates between the two curved surfaces of non-differentiability and converges to $\mathbf{x}^{\infty}$ which is not optimal. In fact it is not necessary for the surfaces of non-differentiability to be curved, and a similar polyhedral (piecewise linear) example can easily be constructed for which the algorithm also fails.

It can be argued that a closer analogue to the steepest descent method at a point of non-differentiability is to search along the the steepest descent direction $\mathbf{s}^{(k)} = -\bar{\mathbf{g}}^{(k)}$ where $\bar{\mathbf{g}}^{(k)}$ minimizes $\|\mathbf{g}\|_2$ for $\mathbf{g} \in \partial\phi^{(k)}$. This interpretation is justified for convex functions in (14.2.12) and a similar result holds when the composite function $\phi(\mathbf{x})$ is in use. Let $\partial\phi^{(k)}$ be defined by the convex hull of its extreme points, $\partial\phi^{(k)} = \operatorname{conv} \mathbf{g}_i^{(k)}$, say. (For composite functions $\mathbf{g}_i^{(k)} = \mathbf{g}^{(k)} + \mathbf{A}^{(k)}\mathbf{h}_i$, $i \in \mathscr{A}^{(k)}$, see (14.2.16).) Then $\bar{\mathbf{g}}^{(k)}$ is defined by solving the problem

$$\begin{aligned} \underset{\mathbf{g},\boldsymbol{\mu}}{\text{minimize}} \quad & \mathbf{g}^{\mathrm{T}}\mathbf{g} \\ \text{subject to} \quad & \mathbf{g} = \sum_i \mu_i \mathbf{g}_i^{(k)}, \\ & \sum_i \mu_i = 1, \qquad \boldsymbol{\mu} \geqslant \mathbf{0}, \end{aligned} \tag{14.4.3}$$

and $\mathbf{s}^{(k)} = -\bar{\mathbf{g}}^{(k)}$. This problem is similar to a least distance QP problem and is readily solved by the methods of Section 10.5. The resulting method terminates finitely when $\phi(\mathbf{x})$ is polyhedral (piecewise linear) and exact line searches are used. This latter condition ensures that the $\mathbf{x}^{(k)}$ are on the surfaces of non-differentiability for which $\partial\phi^{(k)}$ has more than one element. The algorithm is idealized however in that exact line searches are not generally possible in practice. Also the whole subdifferential $\partial\phi^{(k)}$ is not usually available, and even if it were the non-exact line search would cause $\partial\phi^{(k)} = \nabla\phi^{(k)}$ to hold and the method would revert to subgradient optimization.

The spirit of this type of method however is preserved in *bundle* methods (Lemarechal, 1978). A bundle method is a line search method which solves subproblem (14.4.3) to define $\mathbf{s}^{(k)}$, except that the vectors $\mathbf{g}_i^{(k)}$ are elements of a 'bundle' B rather than the extreme points of $\partial\phi^{(k)}$. Initially B is set to $B = \mathbf{g}^{(1)} \in \partial\phi^{(1)}$, and in the simplest form of the algorithm, subgradients $\mathbf{g}^{(2)}$, $\mathbf{g}^{(3)}, \ldots$ are added to B on successive iterations. The method continues in this way until $\mathbf{0} \in B$. Then the bundle B is reset, for instance to the current $\mathbf{g}^{(k)}$, and the iteration is continued. With careful manipulation of B (see for example the *conjugate subgradient method* of Wolfe (1975)) a convergence result for the algorithm can be proved and a suitable termination test obtained. However Wolfe's method does not terminate finitely at the solution for polyhedral $\phi(\mathbf{x})$. An example (due to Powell) has the pentagonal contours illustrated in Figure 14.4.3. At $\mathbf{x}^{(1)}$, B is $\mathbf{g}_1$, and at $\mathbf{x}^{(2)}$, B is $\{\mathbf{g}_1, \mathbf{g}_3\}$. At $\mathbf{x}^{(3)}$, B is $\{\mathbf{g}_1, \mathbf{g}_3, \mathbf{g}_4\}$ and $\mathbf{0} \in B$ so the process is restarted with $B = \mathbf{g}_3$ (or $B = \mathbf{g}_4$ with the same

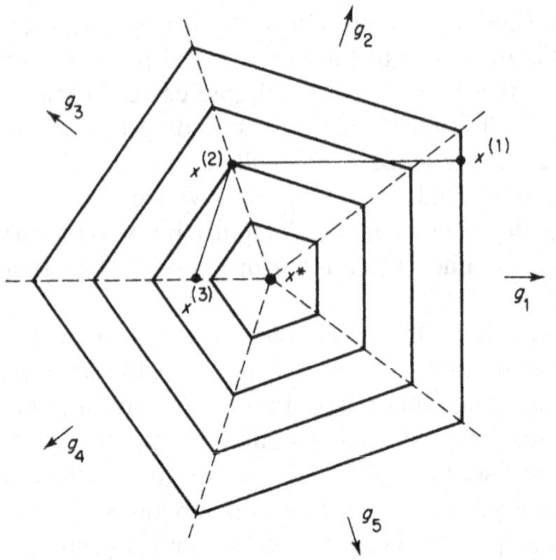

Figure 14.4.3 Non-termination of a bundle method

conclusions). The whole sequence is repeated and the points $\mathbf{x}^{(k)}$ cycle round the pentagon, approaching the centre but without ever terminating there.

This non-termination can be corrected by including more elements in B when restarting (for example both $\mathbf{g}_3$ and $\mathbf{g}_4$) or alternatively deleting old elements like $\mathbf{g}_1$. This can be regarded as providing a better estimate of $\partial\phi^{(k)}$ by B. However, it is by no means certain that this would give a better method: in fact the idealized steepest descent method itself can fail to converge. Problem (14.4.2) again illustrates this; this time the search directions are tangential to the surfaces of non-differentiability and oscillatory convergence to $\mathbf{x}^\infty$ occurs. This example illustrates the need to have information about the subdifferential at $\partial\phi^\infty$ (which is lacking in this example) to avoid the false convergence at $\mathbf{x}^\infty$. Generalizations of bundle methods have therefore been suggested which attempt to enlarge $\partial\phi^{(k)}$ so as to contain subgradient information at neighbouring points. One possibility is to use the ε-subdifferential.

$$\partial_\varepsilon\phi(\mathbf{x}) = \{\mathbf{g}\colon \phi(\mathbf{x}+\boldsymbol{\delta}) \geqslant \phi(\mathbf{x}) + \mathbf{g}^{\mathrm{T}}\boldsymbol{\delta} - \varepsilon \quad \forall\,\boldsymbol{\delta}\}$$

which contains $\partial\phi(\mathbf{x})$. Bundle methods in which B approximates this set are discussed by Lemarechal (1978) and are currently being researched.

A quite different type of method for basic NSO is to use the information $\mathbf{x}^{(k)}$, $\phi^{(k)}$, $\mathbf{g}^{(k)}$ obtained at any point to define the linear function

$$\phi^{(k)} + (\mathbf{x} - \mathbf{x}^{(k)})^{\mathrm{T}}\mathbf{g}^{(k)} \tag{14.4.4}$$

and to use these linearizations in an attempt to model $\phi(\mathbf{x})$. If $\phi(\mathbf{x})$ is convex then this function is a supporting hyperplane and this fact is exploited in the *cutting plane method*. The linearizations are used to define a model polyhedral convex function and the minimizer of this function determines the next iterate.

Specifically the linear program

$$\begin{aligned}&\underset{\mathbf{x},v}{\text{minimize}} \quad v \\ &\text{subject to} \quad v \geqslant \phi^{(i)} + (\mathbf{x}-\mathbf{x}^{(i)})^{\mathrm{T}}\mathbf{g}^{(i)}, \qquad i=1,2,\ldots,k\end{aligned} \tag{14.4.5}$$

is solved to determine $\mathbf{x}^{(k+1)}$. Then the linearization determined by $\mathbf{x}^{(k+1)}$, $\phi^{(k+1)}$, and $\mathbf{g}^{(k+1)}$ is added to the set and the process is repeated. On the early iterations a step restriction $\|\mathbf{x}-\mathbf{x}^{(k)}\|_\infty \leqslant h$ is required to ensure that (14.4.5) is not unbounded. A line search can also be added to the method. A similar method can be used on non-convex problems if old linearizations are deleted in a systematic way.

More sophisticated algorithms for basic NSO are discussed at the end of this section and attention is now turned to algorithms for minimizing the composite function (14.1.1) (which includes max functions, etc., when $h(\mathbf{c})$ is the polyhedral function (14.1.2)—see the special cases in (14.1.3)). Linear approximations have also been used widely and the simplest method is to replace $\mathbf{c}(\mathbf{x})$ by the first order Taylor series approximation

$$\mathbf{c}(\mathbf{x}^{(k)}+\boldsymbol{\delta}) \approx \mathbf{l}^{(k)}(\boldsymbol{\delta}) = \mathbf{c}^{(k)} + \mathbf{A}^{(k)\mathrm{T}}\boldsymbol{\delta} \tag{14.4.6}$$

and $f(\mathbf{x})$ (if present) by

$$f(\mathbf{x}^{(k)}+\boldsymbol{\delta}) \approx f^{(k)} + \mathbf{g}^{(k)\mathrm{T}}\boldsymbol{\delta}, \tag{14.4.7}$$

and to substitute these approximations into (14.1.6) which is an equivalent form of the original problem. The linear program

$$\begin{aligned}&\underset{\boldsymbol{\delta},v}{\text{minimize}} \quad v \\ &\text{subject to} \quad v - (\mathbf{g}^{(k)} + \mathbf{A}^{(k)}\mathbf{h}_i)^{\mathrm{T}}\boldsymbol{\delta} \geqslant f^{(k)} + \mathbf{c}^{(k)\mathrm{T}}\mathbf{h}_i + b_i \qquad \forall i\end{aligned} \tag{14.4.8}$$

is solved to determine $\boldsymbol{\delta}^{(k)}$ and hence $\mathbf{x}^{(k+1)} = \mathbf{x}^{(k)} + \boldsymbol{\delta}^{(k)}$. The only difference between this and (14.4.5) is that here there is sufficient information available at $\mathbf{x}^{(k)}$ to determine linear approximations to all the pieces, whereas in basic NSO this information has to be accumulated on a sequence of iterations. When applied to $\min \|\mathbf{c}(\mathbf{x})\|_\infty$ or $\min \|\mathbf{c}(\mathbf{x})\|_1$ the iteration based on (14.4.8) can be considered as a *Gauss–Newton method* by analogy with the same method for solving $\min \|\mathbf{c}(\mathbf{x})\|_2^2$ (see Section 6.1) which is based on the same linear approximations. An early study of this type of method is by Osborne and Watson (1969) and a more elaborate recent method is given by Charalambous and Conn (1978). As with Gauss–Newton methods, convergence is not guaranteed but this can likewise be rectified by going to a restricted step type of method. In this case $\|\mathbf{x}-\mathbf{x}^{(k)}\|_\infty \leqslant h^{(k)}$ is a suitable choice since it preserves a linear programming subproblem. This approach has been investigated by Madsen (1975).

Linear approximation methods are most successful when the linearizations of the active pieces at $\mathbf{x}^*$ fully determine $\mathbf{x}^*$ (that is to say when the first order conditions are sufficient, which occurs when the set G^* in (14.2.20) corresponding to directions of zero slope is empty and curvature effects are negligible). A necessary condition for this is that there are $n+1$ (or more) active pieces at $\mathbf{x}^*$. In these circumstances the order of convergence of the method based on (14.4.8) is second order (to prove this, show that the method is ultimately equivalent to the Newton–Raphson method for solving the active constraints in (14.4.8) or the structure functional constraints in (14.2.35)). This situation is most likely to occur in problems such as over-determined L_1 or L_∞ data fitting, where these methods can be very successful.

In general however it is not possible to exclude curvature effects, and when these are significant then methods based only on first order information converge slowly and become unreliable. It is therefore important to consider how second order information can be used to obtain second order convergence. Now it has been shown in Theorem 14.2.4 that the composite NSO problem is equivalent to the nonlinear programming problem (14.2.35). It is also known from Theorem 12.4.1 that the SQP method converges at second order for a nonlinear programming problem. Thus by writing down an NSO subproblem which is locally equivalent to the SQP subproblem, a method which converges at second order for the composite NSO problem can be given. The quadratic function

$$q^{(k)}(\boldsymbol{\delta}) = \tfrac{1}{2}\boldsymbol{\delta}^{\mathrm{T}}\mathbf{W}^{(k)}\boldsymbol{\delta} + \mathbf{g}^{(k)\mathrm{T}}\boldsymbol{\delta} + f^{(k)} \tag{14.4.9}$$

is again needed, and the SQP method for (14.2.35) with iterates $(\mathbf{x}^{(k)}, \mathbf{u}^{(k)})$ would involve the subproblem

$$\begin{aligned} &\underset{\boldsymbol{\delta}}{\text{minimize}} \quad q^{(k)}(\boldsymbol{\delta}) + \boldsymbol{\delta}^{\mathrm{T}}\mathbf{A}^{(k)}\boldsymbol{\lambda}^* \\ &\text{subject to} \quad \mathbf{D}^{*\mathrm{T}}(\mathbf{c}^* - \mathbf{l}^{(k)}(\boldsymbol{\delta})) = \mathbf{0}. \end{aligned} \tag{14.4.10}$$

(Here the result is used that $\mathbf{W}^{(k)} = \nabla^2 f_i^{(k)} + \sum \lambda_i^{(k)} \nabla^2 c_i^{(k)} = \nabla^2 f_i^{(k)} + \sum \lambda_i^* \nabla^2 c_i^{(k)} + \sum u_j^{(k)} \nabla^2 (\mathbf{D}^{*\mathrm{T}}\mathbf{c})_j^{(k)}$ where $\boldsymbol{\lambda}^{(k)} = \boldsymbol{\lambda}^* + \mathbf{D}^*\mathbf{u}^{(k)}$.) It is shown below under mild assumptions that this is equivalent to iterating with $(\mathbf{x}^{(k)}, \boldsymbol{\lambda}^{(k)})$ using the NSO subproblem

$$\underset{\boldsymbol{\delta}}{\text{minimize}} \quad \psi^{(k)}(\boldsymbol{\delta}) \triangleq q^{(k)}(\boldsymbol{\delta}) + h(\mathbf{l}^{(k)}(\boldsymbol{\delta})) \tag{14.4.11}$$

The kth iteration is simply to find the minimizer, $\boldsymbol{\delta}^{(k)}$ say, of (14.4.11) and set $\mathbf{x}^{(k+1)} = \mathbf{x}^{(k)} + \boldsymbol{\delta}^{(k)}$, whilst $\boldsymbol{\lambda}^{(k+1)}$ is set as the multipliers of this subproblem. This method is referred to here as the *sequential non-smooth quadratic programming (SNQP) method*. The first reference I have to such a method is that it is mentioned briefly by Pshenichnyi (1978) in the context of minimizing max functions, but it was first given in this general form in the previous edition of this book, where it was referred to as the *QL method*. The $\mathrm{SL_1QP}$ method in Section 12.4 is another example of such a method applied to minimizing non-smooth L_1 problems. The SNQP method can be seen to have a nice interpretation. The function $\psi^{(k)}(\boldsymbol{\delta})$ in (14.4.11) approximates the composite function $\phi(\mathbf{x})$ local to $\mathbf{x}^{(k)}$. This approximation is constructed by replacing $\mathbf{c}(\mathbf{x})$ in (14.1.1) or (14.2.34) by its linear approximation $\mathbf{l}^{(k)}(\boldsymbol{\delta})$ and $f(\mathbf{x})$ by the quadratic approximation $q^{(k)}(\boldsymbol{\delta})$. This quadratic has the Hessian matrix $\mathbf{W}^{(k)}$ which accounts for curvature

Table 14.4.1 Application of the SNQP method to problem (12.3.2)

k	1	2	3	4	5	6
$\mathbf{x}^{(k)}$	2	1.25	0.804310	0.712691	0.707133	0.707107
	0	0.5	0.801724	0.712981	0.707127	0.707107
$\lambda^{(k)}$	1	0.625	0.622844	0.692599	0.706968	0.707107
$\phi^{(k)}$	1	−0.9375	−1.316358	−1.409402	−1.414194	−1.414214
$\boldsymbol{\delta}^{(k)}$	−0.75	−0.445690	−0.091619	−0.005558	−0.000026	$2_{10}-10$
	0.5	0.301724	−0.088744	−0.005854	−0.000020	$-9_{10}-10$

in the functions $f(\mathbf{x})$ and $c_i(\mathbf{x})$. This is exactly the same way as the SQP method handles the equivalent nonlinear programming problem (14.2.35). Moreover in Section 12.4 the SQP method is itself shown to be equivalent to the Newton–Raphson method applied to the first order conditions that arise in the method of Lagrange multipliers. Thus the SNQP method can be regarded as a generalization of the Newton–Raphson method that is appropriate for composite NSO problems.

An illustration of the SNQP method applied to solve problem (12.3.2) with $\nu = 1$ is given. Initial approximations $\mathbf{x}^{(1)} = (2, 0)^T$ and $\lambda^{(1)} = 1$ are taken and the progress of the iterations is given in Table 14.4.1. It can be seen that the convergence is second order with $\|\boldsymbol{\delta}^{(k+1)}\|_2/\|\boldsymbol{\delta}^{(k)}\|_2^2 \approx 0.5$. The contours of the approximating function $\psi^{(k)}$ for $k = 1$ and $k = 2$ are illustrated in Figure 14.4.4. Each function has quadratic pieces with a discontinuous derivative on a linear surface (partly dotted) which is the linearization (through (14.4.6)) of the unit circle on which the discontinuity in Figure 12.3.1 occurs.

It is possible to under mild conditions to prove that the SNQP method conveges locally at second order in all cases of any real interest. Two preliminary lemmas are required and the main result then follows.

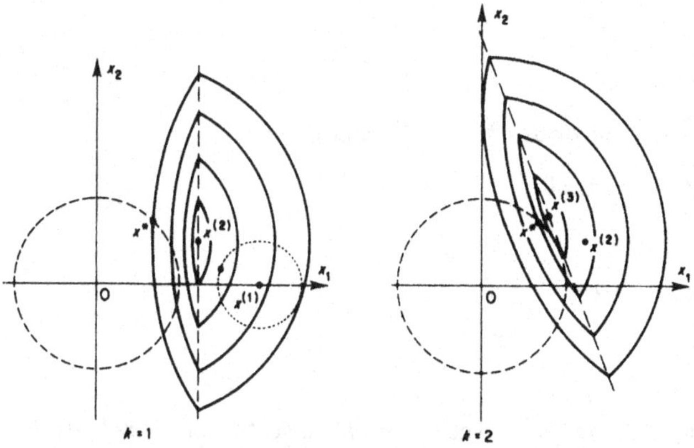

Figure 14.4.4 Contours of the approximating function in the SNQP method

Lemma 14.4.1

If $h(\mathbf{c})$ *is locally linear about* $\mathbf{c}^*$ *and if* $\mathbf{c}'\in\Omega$ *then*

$$\partial h(\mathbf{c}') \subseteq \partial h(\mathbf{c}^*). \tag{14.4.12}$$

Proof

Define $\mathbf{c}_\theta = \mathbf{c}^* + \theta(\mathbf{c}' - \mathbf{c}^*)$ where $\theta \geqslant 0$ and $\mathbf{c}_\theta \in \Omega$. Using the locally linear property (14.2.29)

$$\begin{aligned} h(\mathbf{c}_\theta) &= h(\mathbf{c}^*) + \max_{\boldsymbol{\lambda}\in\partial h^*} \theta(\mathbf{c}' - \mathbf{c}^*)^{\mathrm{T}}\boldsymbol{\lambda} \\ &= h(\mathbf{c}^*) + \theta \max_{\boldsymbol{\lambda}\in\partial h^*} (\mathbf{c}' - \mathbf{c}^*)^{\mathrm{T}}\boldsymbol{\lambda} \\ &= h(\mathbf{c}^*) + \theta(h(\mathbf{c}') - h(\mathbf{c}^*)) \end{aligned} \tag{14.4.13}$$

Consider $\boldsymbol{\lambda} \in \partial h(\mathbf{c}')$. The subgradient inequality gives

$$h(\mathbf{c}) \geqslant h(\mathbf{c}') + (\mathbf{c} - \mathbf{c}')^{\mathrm{T}}\boldsymbol{\lambda} \qquad \forall\, \mathbf{c} \tag{14.4.14}$$

so

$$h(\mathbf{c}_\theta) \geqslant h(\mathbf{c}') + (\mathbf{c}_\theta - \mathbf{c}')^{\mathrm{T}}\boldsymbol{\lambda}.$$

Substituting for $\mathbf{c}_\theta$ and using (14.4.13) gives

$$(1-\theta)(h(\mathbf{c}^*) - h(\mathbf{c}')) \geqslant (1-\theta)(\mathbf{c}^* - \mathbf{c}')^{\mathrm{T}}\boldsymbol{\lambda}.$$

Since Ω is open, $\exists$ points $\mathbf{c}_\theta \in \Omega$ with both $\theta < 1$ and $\theta > 1$, which implies that

$$h(\mathbf{c}^*) - h(\mathbf{c}') = (\mathbf{c}^* - \mathbf{c}')^{\mathrm{T}}\boldsymbol{\lambda}. \tag{14.4.15}$$

It follows from (14.4.14) that

$$h(\mathbf{c}) \geqslant h(\mathbf{c}^*) + (\mathbf{c} - \mathbf{c}^*)^{\mathrm{T}}\boldsymbol{\lambda} \qquad \forall\, \mathbf{c} \tag{14.4.16}$$

and hence $\boldsymbol{\lambda} \in \partial h(\mathbf{c}^*)$. □

Lemma 14.4.2

If $h(\mathbf{c})$ *is locally linear at* $\mathbf{c}^*$, *if* $\mathbf{c}'\in\Omega$, *and if*

$$\mathbf{D}^{*\mathrm{T}}(\mathbf{c}^* - \mathbf{c}') = \mathbf{0} \tag{14.4.17}$$

then

$$\partial h(\mathbf{c}') = \partial h(\mathbf{c}^*). \tag{14.4.18}$$

Proof

Let $\boldsymbol{\lambda}'$ denote an arbitrary vector in $\partial h(\mathbf{c}')$, so that $\boldsymbol{\lambda}' \in \partial h(\mathbf{c}^*)$ and

$$h(\mathbf{c}^*) - h(\mathbf{c}') = (\mathbf{c}^* - \mathbf{c}')^{\mathrm{T}}\boldsymbol{\lambda}' \tag{14.4.19}$$

both hold, from Lemma 14.4.1 and (14.4.15). In view of Lemma 14.4.1, only the reverse inclusion needs to be proved, so consider $\boldsymbol{\lambda} \in \partial h(\mathbf{c}^*)$. From (14.2.30) $\exists$ a vector $\mathbf{v}$ such that $\boldsymbol{\lambda} - \boldsymbol{\lambda}' = \mathbf{D}^*\mathbf{v}$ and hence from (14.4.17)

$$\mathbf{c}^{*\mathrm{T}}(\boldsymbol{\lambda} - \boldsymbol{\lambda}') = \mathbf{c}'^{\mathrm{T}}(\boldsymbol{\lambda} - \boldsymbol{\lambda}'). \tag{14.4.20}$$

The subgradient inequality (14.4.16) holds and it follows from (14.4.19) and (14.4.20) that (14.4.14) holds, and hence $\boldsymbol{\lambda} \in \partial h(\mathbf{c}')$. □

Theorem 14.4.1 (Second order convergence of the SNQP method)

Let $\mathbf{x}^*, \boldsymbol{\lambda}^*$ *satisfy the second order sufficient conditions of Theorem* 14.2.3, *and assume that* $h(\mathbf{c})$ *is locally linear at* $\mathbf{c}^*$, *and that both the rank assumption* (14.2.31) *and strict complementarity* $\mathbf{0} \in \mathrm{int}\ U^*$ *hold. In a neighbourhood of* $(\mathbf{x}^*, \boldsymbol{\lambda}^*)$ *the SNQP subproblem* (14.4.11) *has a locally unique minimizer. Assuming that this point is located when solving the subproblem, then the SNQP method is equivalent to the SQP method applied to* (14.2.35) *based on the subproblem* (14.4.10). *The iterates in the SNQP method converge to* $(\mathbf{x}^*, \boldsymbol{\lambda}^*)$ *and the order of convergence is quadratic.*

Proof

Consider the system of equations

$$\begin{bmatrix} \mathbf{W} & \mathbf{A}\mathbf{D}^* \\ (\mathbf{A}\mathbf{D}^*)^{\mathrm{T}} & \mathbf{0} \end{bmatrix} \begin{pmatrix} \boldsymbol{\delta} \\ \mathbf{u}^+ \end{pmatrix} = \begin{pmatrix} -\mathbf{g} - \mathbf{A}\boldsymbol{\lambda}^* \\ \mathbf{D}^{*\mathrm{T}}(\mathbf{c}^* - \mathbf{c}) \end{pmatrix} \tag{14.4.21}$$

where $\mathbf{c} = \mathbf{c}(\mathbf{x})$, $\mathbf{A} = \mathbf{A}(\mathbf{x})$, etc. and $\mathbf{W} = \mathbf{W}(\mathbf{x}, \boldsymbol{\lambda})$, and where $(\mathbf{x}, \boldsymbol{\lambda})$ is in some neighbourhood of $(\mathbf{x}^*, \boldsymbol{\lambda}^*)$. From second order sufficient conditions and strict complementarity it follows that

$$\mathbf{s}^{\mathrm{T}}\mathbf{W}^*\mathbf{s} > 0 \qquad \forall \mathbf{s} \in \mathcal{N}^*, \tag{14.4.22}$$

and hence by continuity that

$$\mathbf{s}^{\mathrm{T}}\mathbf{W}\mathbf{s} > 0 \qquad \forall \mathbf{s} \in \mathcal{N} = \{\mathbf{s}: \|\mathbf{s}\|_2 = 1,\ \mathbf{s}^{\mathrm{T}}\mathbf{A}\mathbf{D}^* = \mathbf{0}^{\mathrm{T}}\} \tag{14.4.23}$$

(see Question 12.17 for example). It follows from (14.4.22) and the rank assumption (14.2.31) that the matrix

$$\begin{bmatrix} \mathbf{W}^* & \mathbf{A}^*\mathbf{D}^* \\ (\mathbf{A}^*\mathbf{D}^*)^{\mathrm{T}} & \mathbf{0} \end{bmatrix} \tag{14.4.24}$$

is non-singular (see Question 12.4). Hence by the implicit function theorem there exists a neighbourhood of $(\mathbf{x}^*, \boldsymbol{\lambda}^*)$ such that for any $(\mathbf{x}, \boldsymbol{\lambda})$ in this neighbourhood, there is a uniquely defined continuous solution $(\boldsymbol{\delta}, \mathbf{u}^+)$ to (14.4.21), and at $(\mathbf{x}^*, \boldsymbol{\lambda}^*)$ the solution satisfies $\boldsymbol{\delta} = \mathbf{0}$ and $\mathbf{u}^+ = \mathbf{0}$. Equations (14.4.21) are the first order conditions for $(\boldsymbol{\delta}, \mathbf{u}^+)$ to solve the SQP subproblem (14.4.10) as defined for an iterate $(\mathbf{x}, \boldsymbol{\lambda})$. If $\boldsymbol{\lambda}^+ = \boldsymbol{\lambda}^* + \mathbf{D}^*\mathbf{u}^+$, the first equation in (14.4.21) is also the Lagrangian first order condition for $(\boldsymbol{\delta}, \boldsymbol{\lambda}^+)$ in the SNQP subproblem (14.4.11) as defined for an iterate $(\mathbf{x}, \boldsymbol{\lambda})$. Because $\mathbf{0} \in \mathrm{int}\ U^*$ it follows by continuity that $\mathbf{u}^+ \in U^*$ and hence $\boldsymbol{\lambda}^+ \in \partial h(\mathbf{c}^*)$. From the second condition in (14.4.21), continuity of $\boldsymbol{\delta}$ and Lemma 14.4.2, it follows that $\boldsymbol{\lambda}^+ \in \partial h(\mathbf{c} + \mathbf{A}^T\boldsymbol{\delta})$, and hence $(\boldsymbol{\delta}, \boldsymbol{\lambda}^+)$ satisfies first order conditions for $(\boldsymbol{\delta}, \boldsymbol{\lambda}^+)$ to solve the SNQP problem. Second order sufficient conditions for both problems involve (14.4.23) which is known to hold. It follows that a unique local minimizer exists for both subproblems, and in fact it is easily seen to be a global minimizer for the SQP subproblem. Thus if $\boldsymbol{\lambda} = \boldsymbol{\lambda}^* + \mathbf{D}^*\mathbf{u}$ and $\boldsymbol{\lambda}^+ = \boldsymbol{\lambda}^* + \mathbf{D}^*\mathbf{u}^+$, then calculating $(\boldsymbol{\delta}, \mathbf{u}^+)$ from an iterate $(\mathbf{x}, \mathbf{u})$ by solving the SQP subproblem is equivalent to calculating $(\boldsymbol{\delta}, \boldsymbol{\lambda}^+)$ from an iterate $(\mathbf{x}, \boldsymbol{\lambda})$ by solving the SNQP subproblem, assuming that the locally unique solution is located when solving the latter (see Question 14.16). Finally consider applying the SNQP method from an iterate $(\mathbf{x}^{(1)}, \boldsymbol{\lambda}^{(1)})$ sufficiently close to $(\mathbf{x}^*, \boldsymbol{\lambda}^*)$. Then as above, we can express $\boldsymbol{\lambda}^{(2)} = \boldsymbol{\lambda}^* + \mathbf{D}^*\mathbf{u}^{(2)}$. It then follows that for $k \geqslant 2$ the iterates $(\mathbf{x}^{(k)}, \boldsymbol{\lambda}^{(k)})$ in the SNQP method are equivalent to the iterates $(\mathbf{x}^{(k)}, \mathbf{u}^{(k)})$ in the SQP method, assuming that $\boldsymbol{\lambda}^{(k)} = \boldsymbol{\lambda}^* + \mathbf{D}^*\mathbf{u}^{(k)}$. Convergence and second order convergence follow from Theorem 12.4.1 □

The assumptions in Theorem 14.4.1 are mostly mild ones, the only significant restriction being that $h(\mathbf{c})$ must be locally linear at $\mathbf{c}^*$. As has been pointed out in Sections 14.2 and 14.3, this includes not only the polyhedral convex functions of Sections 14.1 but also the exact penalty function problems for smooth norms in Section 14.3. This covers all convex composite problems that are of any real interest. A final remark about order of convergence is that when $h(\mathbf{c}) = \|\mathbf{c}\|$ is a smooth norm (for example $\|\cdot\|_2$) and $\mathbf{c}^* \neq \mathbf{0}$ then $\mathscr{G}^* \neq G^*$ and the above development is not appropriate. In this case the most appropriate second order sufficient conditions are those for a smooth problem that $\nabla\phi^* = \mathbf{0}$ and $\nabla^2\phi^*$ is positive definite. However the SNQP method does not reduce to Newton's method so it is an open question as to whether second order convergence can be deduced.

These results on second order convergence make it clear that the SNQP method is attractive as a starting point for developing a general purpose algorithm for composite NSO problems. However the basic method itself is not robust and can fail to converge (when $m = 0$ it is just the basic Newton's method—see Section 3.1). An obvious precaution to prevent divergence is to require that the sequence $\{\phi^{(k)}\}$ is non-increasing. One possibility is to use the correction from solving (14.4.11) as a direction of search $\mathbf{s}^{(k)}$ along which to approximately minimize the objective function $\phi(\mathbf{x}^{(k)} + \alpha\mathbf{s}^{(k)})$. Both the basic method and the line search modification can fail in another way however, in that remote from the solution the function $\psi^{(k)}(\boldsymbol{\delta})$ may not have a minimizer. This is analogous to the case that $\mathbf{G}^{(k)}$ is indefinite in smooth unconstrained minimization. An effective modification in the latter case is the restricted step or trust region approach (Chapter 5) and it is fruitful to consider modifying the SNQP method in this way. Since (14.4.11) contains no side conditions the incorporation of a step restriction causes no difficulty. This is illustrated in Figure 14.4.4 ($k = 1$) in which the dotted circle with centre $\mathbf{x}^{(1)}$ is a possible trust region. Clearly there is no difficulty in minimizing $\psi^{(1)}(\boldsymbol{\delta})$ within this region: the solution is the point on the periphery of the circle. Contrast this situation with Section 12.4 where a trust region cannot be used in conjunction with the SQP method because there may not exist feasible points in the resulting problem. Fletcher (1982a) describes a prototype restricted step method for solving composite NSO problems which is globally convergent without the need to assume that vectors $\mathbf{a}_i^\infty$ are independent or multipliers $\lambda^{(k)}$ are bounded. This algorithm is described in more detail in Section 14.5 where the global convergence property is justified. Good practical experience with the method in L_1 exact penalty function applications is reported by Fletcher (1981), including the solution of test problems on which other methods fail.

There are two main computational costs with the SNQP method. One is that associated with solving the subproblem (14.4.11) and this of course depends on the form of the composite function $h(\cdot)$. When h is a polyhedral convex function the subproblem is equivalent to a QP problem and so the complexity of the subproblem is roughly the same as that for QP. In the L_1 case the subproblem becomes the L_1QP problem described in more detail in Section 12.3. The other main cost is the need to evaluate the problem functions f and $\mathbf{c}$ and their first and second derivatives at each iteration. The need to provide particularly second derivatives may be a disadvantage, but there is no difficulty in approximating the second derivative matrix $\mathbf{W}^{(k)}$ for instance by finite differences or by using a quasi-Newton update. One possibility is to use the modified BFGS formula given by Powell (1978a) as described in Section 12.4, and this is quite straightforward. Likewise the possibility of approximating the *reduced* second derivative matrix also suggests itself but it is by no means clear how to use this matrix to the best advantage in a general SNQP context. However, Womersley (1981) gives a method applicable to the particular case of max function.

There is one adverse feature of the SNQP method, and hence of any other superlinearly convergent algorithm for composite NSO which is not present with similar algorithms for smooth optimization. In the latter case when $\mathbf{x}^{(k)}$ is close to $\mathbf{x}^*$, the unit step of Newton's method reduces the objective function, so no line search or trust region restriction is brought into play to force the objective function to decrease. Thus the second order convergence, which depends on taking the unit step, is unaffected. The same situation does *not* hold for composite NSO as Maratos (1978) observes (see also Mayne (1980), Chamberlain *et al.* (1982) and Question 14.13). In particular there are well behaved composite NSO problems in which $(\mathbf{x}^{(k)}, \lambda^{(k)})$ can be arbitrarily close to $(\mathbf{x}^*, \lambda^*)$ and the unit step of the SNQP method can fail to reduce $\phi(\mathbf{x})$. The *Maratos effect* as it might be called, thus causes the unit step to be rejected and can obviate the possibility of obtaining second order convergence. More precisely for an iteration of the SNQP method, define the *actual reduction*

$$\Delta\phi^{(k)} = \phi(\mathbf{x}^{(k)}) - \phi(\mathbf{x}^{(k)} + \boldsymbol{\delta}^{(k)}), \tag{14.4.25}$$

the *predicted reduction*

$$\Delta\psi^{(k)} = \psi^{(k)}(\mathbf{0}) - \psi^{(k)}(\boldsymbol{\delta}^{(k)}) = \phi(\mathbf{x}^{(k)}) - \psi^{(k)}(\boldsymbol{\delta}^{(k)}) \tag{14.4.26}$$

and the ratio

$$r^{(k)} = \Delta\phi^{(k)}/\Delta\psi^{(k)}. \tag{14.4.27}$$

Also let

$$\varepsilon^{(k)} = \max(\|\mathbf{x}^{(k)} - \mathbf{x}^*\|, \|\boldsymbol{\lambda}^{(k)} - \boldsymbol{\lambda}^*\|). \tag{14.4.28}$$

Then for smooth unconstrained optimization (e.g. Theorems 3.1.1 and 5.2.2) it is possible to get both $\varepsilon^{(k+1)} = O(\varepsilon^{(k)2})$ and $r^{(k)} = 1 + O(\varepsilon^{(k)})$, whereas for composite NSO these results are inconsistent, and indeed $\Delta\phi^{(k)} < 0$ and hence $r^{(k)} < 0$ can occur which corresponds to the Maratos, effect. The reason for this is illustrated in Figure 14.4.5 in the non-smooth case. If $\mathbf{x}^{(k)}$ is close to (or on) a groove, the predicted first order changes to ϕ are small (or zero). Thus the predicted second order changes to ϕ can be significant. However, the functions $c_i(\mathbf{x})$ are only approximated linearly in calculating $\mathbf{x}^{(k+1)}$ and hence second order errors arise in $\phi(\mathbf{x}^{(k+1)})$ which can dominate the second order changes in the predicted reduction. This effect is due to the presence of derivative discontinuities in $\phi(\mathbf{x})$. These discontinuities also can occasionally cause algorithms for SNQP to converge slowly (for example see Table 12.4.3 for $\mathrm{SL_1QP}$). Both effects are most likely to be observed when the second order errors are large, that is when the grooves are significantly nonlinear, such as for example when the penalty term in an exact penalty function is relatively large. Thus if robust algorithms for NSO are to be developed, it is important to consider how these effects can be overcome.

The above observations indicate that a modification which corrects the second order errors in c_i is appropriate. This can be done (Fletcher, 1982b) by making a single 'second order correction' (SOC) step, which corrects the 'basic step' obtained by solving (14.4.11). Similar ideas occur in Coleman and Conn (1982a, b) and indirectly in Chamberlain *et al.* (1982). The idea can be simply explained as follows. The basic step $\boldsymbol{\delta}$ in the SNQP method is obtained by solving the

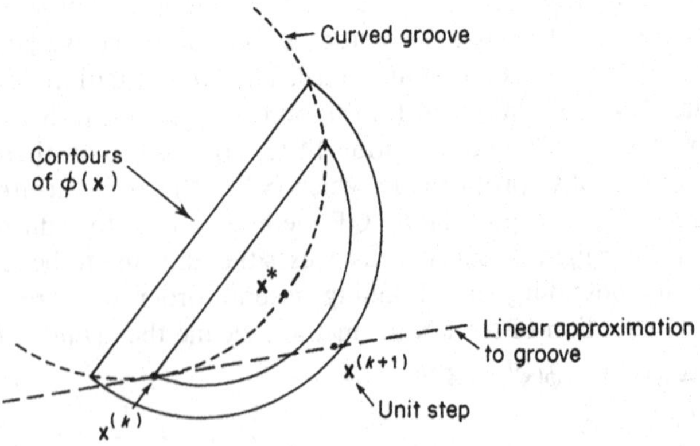

Figure 14.4.5 Maratos effect

subproblem

$$\underset{\boldsymbol{\delta}}{\text{minimize}} \quad q^{(k)}(\boldsymbol{\delta}) + h([c_i^{(k)} + \mathbf{a}_i^{(k)\mathrm{T}}\boldsymbol{\delta}]). \tag{14.4.31}$$

where $[\cdot]$ denotes the vector having the given elements. To account satisfactorily for second order changes it would be preferable to solve

$$\underset{\boldsymbol{\delta}}{\text{minimize}} \quad q^{(k)}(\boldsymbol{\delta}) + h([c_i^{(k)} + \mathbf{a}_i^{(k)\mathrm{T}}\boldsymbol{\delta} + \tfrac{1}{2}\boldsymbol{\delta}^{\mathrm{T}}\mathbf{G}_i^{(k)}\boldsymbol{\delta}]) \tag{14.4.32}$$

where $\mathbf{G}_i = \nabla^2 c_i$, but this subproblem is intractable. However, if the solution to (14.4.31) is $\hat{\boldsymbol{\delta}}$, and we define $\hat{\mathbf{x}} = \mathbf{x}^{(k)} + \hat{\boldsymbol{\delta}}$ and evaluate $\hat{\mathbf{c}} = \mathbf{c}(\hat{\mathbf{x}})$, then to second order

$$\hat{c}_i = c_i^{(k)} + \mathbf{a}_i^{(k)\mathrm{T}}\hat{\boldsymbol{\delta}} + \tfrac{1}{2}\hat{\boldsymbol{\delta}}^{\mathrm{T}}\mathbf{G}_i^{(k)}\hat{\boldsymbol{\delta}}. \tag{14.4.33}$$

Equation (14.4.33) can be rearranged to provide an expression for $\frac{1}{2}\hat{\boldsymbol{\delta}}^{\mathrm{T}}\mathbf{G}_i^{(k)}\hat{\boldsymbol{\delta}}$ which can be used to estimate the second order term in (14.4.32). Thus the SOC subproblem

$$\underset{\boldsymbol{\delta}}{\text{minimize}} \quad q^{(k)}(\boldsymbol{\delta}) + h([\hat{c}_i - \mathbf{a}_i^{(k)\mathrm{T}}\hat{\boldsymbol{\delta}} + \mathbf{a}_i^{(k)\mathrm{T}}\boldsymbol{\delta}]) \tag{14.4.34}$$

is suggested, and the step $\boldsymbol{\delta}^{(k)}$ that is used in the resulting algorithm to calculate $\mathbf{x}^{(k+1)} = \mathbf{x}^{(k)} + \boldsymbol{\delta}^{(k)}$ is that obtained by solving (14.4.34). Note that the only difference between (14.4.34) and (14.4.31) is a change to the constant term in the argument of $h(\cdot)$. For polyhedral problems this merely has the effect of shifting the linear approximation of the curved groove so that $\mathbf{x}^{(k+1)}$ is closer to the groove, as illustrated in Figure 14.4.6. This figure also shows that the SOC step is related to the projection step computed in feasible direction methods for nonlinear programming (cf. Figure 12.5.1) and in particular to that given by (12.5.11) for $\alpha^{(k)} = 1$.

The SOC modification has a number of practical advantages. It uses the

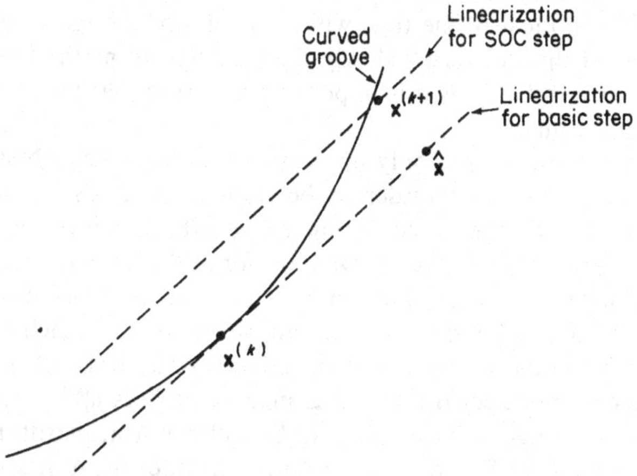

Figure 14.4.6 The second order correction step

same software for calculating both the basic and SOC step and so is easy to code. No extra derivative matrices need be calculated to define the SOC step and only one additional evaluation of the active constraint residuals $\mathbf{c}(\hat{\mathbf{x}})$ is required. Moreover, and of great importance, only a change to the constant values $\mathbf{c}^{(k)}$ in the basic step subproblem is made. Assuming that the software for the subproblem enables parametric changes to be made efficiently, it is likely that the same active set and same matrix factorizations can be used to solve the SOC subproblem directly. In these circumstances the SOC step can be computed at negligible cost. The above method represents the simplest way to use an SOC step: Fletcher (1982b) also suggests a more complicated algorithm in which the basic step is accepted for $\boldsymbol{\delta}^{(k)}$ in certain circumstances, and the SOC step is only used if it is judged to be necessary. An SOC step can also be used effectively when the SQP method is used in conjunction with a non-smooth penalty function.

Other second order methods for NSO have also been suggested, in particular the use of a *hybrid method* (or *2-stage method*). Stage 1 is the Gauss–Newton trust region method (14.4.8 ff.) which converges globally so is potentially reliable, but often converges only at first order which can be slow. Stage 2 is Newton's method applied to the first order conditions. This converges locally at second order but does not have any global properties. A hybrid method is one which switches between these methods and aims to combine their best features. To apply Newton's method requires a knowledge of the active pieces in the NSO problem (essentially a set of structure functionals as in (14.2.35) is required) and this knowledge can also be gained from the progress of the Gauss–Newton iteration. Hybrid methods are given by Watson (1979) for the minimax problem and by McLean and Watson (1980) for the L_1 problem. Hybrid methods using a quasi-Newton method for stage 2 are given by Hald and Madsen (1981, 1985). A recent method for general non-smooth problems, using the Coleman–Conn method for stage 2, is given by Fletcher and Sainz de la Maza (1987). Hybrid methods can work well but I anticipate one disadvantage. If the NSO problem is such that neighbourhood of the solution in which Newton's method works well is small, then most of the time will be taken up in stage 1, using the first order method. If this converges slowly then the hybrid method will not solve the problem effectively. Thus it is important to ensure that the stage 2 method is used to maximum effect.

The above techniques are only appropriate for composite NSO problems, and it is also important to consider applications to basic NSO in which second order information has been used. Womersley (1978) describes a method which requires $\phi^{(i)}$, $\nabla\phi^{(i)}$, and $\nabla^2\phi^{(i)}$ to be available for any $\mathbf{x}^{(i)}$, $i = 1, 2, \ldots, k$, and each such set of information is assumed to arise from an active piece of $\phi(\mathbf{x})$. This information is used to build up a linear approximation for each piece valid at $\mathbf{x}^{(k)}$, and also to give the matrix $\mathbf{W}^{(k)}$. This is then used as in (14.4.10). Information about pieces may be rejected if the Lagrange multipliers $\mu_i^{(k+1)}$ determine that the piece is no longer active. Also a Levenberg–Marquardt modification (Section 5.2) is made to $\mathbf{W}^{(k)}$ to ensure positive definiteness. A difficulty with the algorithm is that repeated approximations to the same piece tend to be collected, which cause degeneracy in the QP solver close to the solution. A modification is proposed in which an extra item of information is supplied at each $\mathbf{x}^{(i)}$ which is an integer *label* of the active piece. Such information can be supplied in most basic NSO applications. This labelling enables a single most recent approximation to each piece to be maintained and circumvents the degeneracy problem. Again this approach is considerably more effective than the subgradient or cutting plane methods as numerical evidence given by Womersley suggests (~ 20 iterations as against ~ 300 on a typical problem). A quasi-Newton version of the algorithm is described in Womersley (1981).

14.5 A GLOBALLY CONVERGENT PROTOTYPE ALGORITHM

The aim of this section is to show that the SNQP method based on solving the subproblem (14.4.11) can be incorporated readily with the idea of a step restriction which is known to give good numerical results in smooth unconstrained optimization (Chapter 5). Subproblem (14.4.11) contains curvature information and potentially allows second order convergence to occur, whereas the step length restriction is shown to ensure global convergence. The resulting method (Fletcher, 1981, 1982a) is applicable to the solution of all composite NSO problems, including exact penalty functions, best L_1 and L_∞ approximation, etc., but excluding basic NSO problems. The term 'prototype algorithm' indicates that the algorithm is presented in a simple format as a means of making its convergence properties clear. It admits of the algorithm being modified to improve its practical performance whilst not detracting from these properties. The motivation for using a step restriction is that it defines a *trust region* for the correction $\boldsymbol{\delta}$ by

$$\|\boldsymbol{\delta}\| \leqslant h^{(k)} \tag{14.5.1}$$

in which the Taylor series approximations (14.4.6) and (14.4.9) are assumed to be adequate. The norm in (14.5.1) is arbitrary but either the $\|\cdot\|_\infty$ or the $\|\cdot\|_2$ is the most likely choice, especially the former since (14.5.2) below can then be solved by QP-like methods (see (10.3.5) and Section 12.3). On each iteration the subproblem is to minimize (14.4.11) subject to this restriction, that is

$$\begin{aligned} &\underset{\boldsymbol{\delta}}{\text{minimize}} \quad \psi^{(k)}(\boldsymbol{\delta}) \\ &\text{subject to} \quad \|\boldsymbol{\delta}\| \leqslant h^{(k)}. \end{aligned} \tag{14.5.2}$$

The radius $h^{(k)}$ of the trust region is adjusted adaptively to be as large as possible subject to adequate agreement between $\phi(\mathbf{x}^{(k)} + \boldsymbol{\delta})$ and $\psi^{(k)}(\boldsymbol{\delta})$ being maintained. This can be quantified by defining the *actual reduction*

$$\Delta\phi^{(k)} = \phi(\mathbf{x}^{(k)}) - \phi(\mathbf{x}^{(k)} + \boldsymbol{\delta}^{(k)}) \tag{14.5.3}$$

and the *predicted reduction*

$$\Delta\psi^{(k)} = \phi(\mathbf{x}^{(k)}) - \psi^{(k)}(\boldsymbol{\delta}^{(k)}). \tag{14.5.4}$$

Then the ratio

$$r^{(k)} = \frac{\Delta\phi^{(k)}}{\Delta\psi^{(k)}} \tag{14.5.5}$$

measures the extent to which ϕ and $\psi^{(k)}$ agree local to $\mathbf{x}^{(k)}$. The rules for changing $h^{(k)}$ in the model algorithm are those given in Section 5.1 and are not elaborated on further, except to emphasize that in practice the rule for reducing $h^{(k)}$ can be more elaborate, based perhaps on some sort of interpolation (Fletcher 1981).

The kth iteration of the prototype algorithm is as follows.

(i) Given $\mathbf{x}^{(k)}$, $\boldsymbol{\lambda}^{(k)}$, and $h^{(k)}$, calculate $f^{(k)}$, $\mathbf{g}^{(k)}$, $\mathbf{c}^{(k)}$, $\mathbf{A}^{(k)}$, and $\mathbf{W}^{(k)}$ which determine $\phi^{(k)}$ and $\psi^{(k)}(\boldsymbol{\delta})$.
(ii) Find a global solution $\boldsymbol{\delta}^{(k)}$ to (14.5.2).
(iii) Evaluate $\phi(\mathbf{x}^{(k)} + \boldsymbol{\delta}^{(k)})$ and calculate $\Delta\phi^{(k)}$, $\Delta\psi^{(k)}$, and $r^{(k)}$.
(iv) If $r^{(k)} < 0.25$ set $h^{(k+1)} = \|\boldsymbol{\delta}^{(k)}\|/4$,
if $r^{(k)} > 0.75$ and $\|\boldsymbol{\delta}^{(k)}\| = h^{(k)}$ set $h^{(k+1)} = 2h^{(k)}$, (14.5.6)
otherwise set $h^{(k+1)} = h^{(k)}$.
(v) If $r^{(k)} \leqslant 0$ set $\mathbf{x}^{(k+1)} = \mathbf{x}^{(k)}$, $\boldsymbol{\lambda}^{(k+1)} = \boldsymbol{\lambda}^{(k)}$,
else $\mathbf{x}^{(k+1)} = \mathbf{x}^{(k)} + \boldsymbol{\delta}^{(k)}$, $\boldsymbol{\lambda}^{(k+1)} =$ multipliers from (14.5.2).

The parameters 0.25, 0.75, etc., which arise are arbitrary and are not very sensitive but the values given here are typical. The solution of constrained NSO problems like (14.5.2) is considered in more detail in Section 14.6 but it is a straightforward extension of the unconstrained case and multipliers $\boldsymbol{\lambda}^{(k+1)} \in \partial h(\mathbf{l}^{(k)}(\boldsymbol{\delta}^{(k)}))$ exist at the solution in an analogous way to Theorem 14.2.1. It is these multipliers that are used in step (v).

In proving global convergence, a result is used relating the directional derivative (14.2.14) of the composite function (14.2.13) and the difference quotient between two points $\mathbf{x}^{(k)}$ and $\mathbf{x}^{(k)} + \varepsilon^{(k)}\mathbf{s}$ in a common direction $\mathbf{s}$ ($\neq \mathbf{0}$), as both points approach a fixed point $\mathbf{x}'$. The result is a special case of one due to Clarke (1975) but is proved here directly for completeness.

Lemma 14.5.1

Let S be the set of all sequences $\mathbf{x}^{(k)} \to \mathbf{x}'$, $\varepsilon^{(k)} \to 0$, and let $\mathbf{c}^{(k)} \triangleq \mathbf{c}(\mathbf{x}^{(k)})$, etc., and $\mathbf{c}_{\varepsilon}^{(k)} \triangleq \mathbf{c}(\mathbf{x}^{(k)} + \varepsilon^{(k)}\mathbf{s})$; then

$$\limsup_{S} \frac{h(\mathbf{c}_{\varepsilon}^{(k)}) - h(\mathbf{c}^{(k)})}{\varepsilon^{(k)}} = \max_{\boldsymbol{\lambda} \in \partial h'} \mathbf{s}^{\mathrm{T}}\mathbf{A}'\boldsymbol{\lambda} \tag{14.5.7}$$

in the sense that the difference quotient is bounded above and the sup of all accumulation points of the quotient taken over all sequences in S is the directional derivative $\max_{\boldsymbol{\lambda} \in \partial h'} \mathbf{s}^{\mathrm{T}}\mathbf{A}'\boldsymbol{\lambda}$.

Proof

By the integral form of the Taylor series

$$\mathbf{c}_{\varepsilon}^{(k)} = \mathbf{c}^{(k)} + \varepsilon^{(k)} \int_0^1 [\mathbf{A}(\mathbf{x}^{(k)} + \theta\varepsilon^{(k)}\mathbf{s})]^{\mathrm{T}}\mathbf{s}\, \mathrm{d}\theta = \mathbf{c}^{(k)} + \varepsilon^{(k)}\mathbf{d}^{(k)} \tag{14.5.8}$$

say, where $\mathbf{d}^{(k)} \to \mathbf{A}'^{\mathrm{T}}\mathbf{s}$ as $k \to \infty$. Let $\boldsymbol{\lambda}_\varepsilon^{(k)} \in \partial h_\varepsilon^{(k)} \triangleq \partial h(\mathbf{c}_\varepsilon^{(k)})$. Using the subgradient inequality and (14.5.8),

$$h(\mathbf{c}^{(k)}) \geqslant h(\mathbf{c}_\varepsilon^{(k)}) + (\mathbf{c}^{(k)} - \mathbf{c}_\varepsilon^{(k)})^{\mathrm{T}}\boldsymbol{\lambda}_\varepsilon^{(k)} = h(\mathbf{c}_\varepsilon^{(k)}) - \varepsilon^{(k)}\mathbf{d}^{(k)\mathrm{T}}\boldsymbol{\lambda}_\varepsilon^{(k)}$$

or

$$\mathbf{d}^{(k)\mathrm{T}}\boldsymbol{\lambda}_\varepsilon^{(k)} \geqslant \frac{h(\mathbf{c}_\varepsilon^{(k)}) - h(\mathbf{c}^{(k)})}{\varepsilon^{(k)}}. \tag{14.5.9}$$

Since $\partial h_\varepsilon^{(k)}$ is bounded in a neighbourhood of $\mathbf{x}'$ (Lemma 14.2.1) the difference quotient is bounded above. Now consider a subsequence for which the difference quotient accumulates, and let

$$\lim \frac{h(\mathbf{c}_\varepsilon^{(k)}) - h(\mathbf{c}^{(k)})}{\varepsilon^{(k)}} > \max_{\boldsymbol{\lambda} \in \partial h'} \mathbf{s}^{\mathrm{T}}\mathbf{A}'\boldsymbol{\lambda}. \tag{14.5.10}$$

Since $\boldsymbol{\lambda}_\varepsilon^{(k)}$ is bounded there exists a thinner subsequence for which $\boldsymbol{\lambda}_\varepsilon^{(k)} \to \boldsymbol{\lambda}'$ and since $\mathbf{c}_\varepsilon^{(k)} \to \mathbf{c}'$ it follows by Lemma 14.2.4 that $\boldsymbol{\lambda}' \in \partial h'$. Thus from (14.5.9),

$$\lim \frac{h(\mathbf{c}_\varepsilon^{(k)}) - h(\mathbf{c}^{(k)})}{\varepsilon^{(k)}} \leqslant \mathbf{s}^{\mathrm{T}}\mathbf{A}'\boldsymbol{\lambda}' \leqslant \max_{\boldsymbol{\lambda} \in \partial h'} \mathbf{s}^{\mathrm{T}}\mathbf{A}'\boldsymbol{\lambda} \tag{14.5.11}$$

which contradicts (14.5.10), so that the reverse inequality ($\leqslant$) in (14.5.10) is true. Finally by taking $\mathbf{x}^{(k)} = \mathbf{x}'$ and $\varepsilon^{(k)} \downarrow 0$, the discussion before (14.2.14) shows that there is a sequence in S which attains equality with the directional derivative. Thus (14.5.7) is established. □

Corollary

Define $\phi(\mathbf{x})$ *by* (14.2.13) *where* $f \in \mathbb{C}^1$. *Then*

$$\limsup_{S} \frac{\phi(\mathbf{x}^{(k)} + \varepsilon^{(k)}\mathbf{s}) - \phi(\mathbf{x}^{(k)})}{\varepsilon^{(k)}} = \max_{\boldsymbol{\lambda} \in \partial h'} \mathbf{s}^{\mathrm{T}}(\mathbf{g}' + \mathbf{A}'\boldsymbol{\lambda}). \tag{14.5.12}$$

Proof

The result follows by using an analogous Taylor series for $f(\mathbf{x})$ as in the proof of the lemma. □

It is now possible to state the main result of this section.

Theorem 14.5.1 (Global convergence of prototype algorithm)

Let $\mathbf{x}^{(k)} \in B \subset \mathbb{R}^n \,\forall k$ *where* B *is bounded and let* f, $\mathbf{c}$ *be* $\mathbb{C}^2$ *functions whose second derivative matrices are bounded on* B. *Then there exists an accumulation point* $\mathbf{x}^\infty$ *of algorithm* (14.5.6) *at which the first order conditions of Theorem* 14.2.1 *hold, that is equivalently*

$$\max_{\boldsymbol{\lambda} \in \partial h^\infty} \mathbf{s}^{\mathrm{T}}(\mathbf{g}^\infty + \mathbf{A}^\infty\boldsymbol{\lambda}) \geqslant 0 \quad \forall \mathbf{s}. \tag{14.5.13}$$

Proof

There exists a convergent subsequence $\mathbf{x}^{(k)} \to \mathbf{x}^\infty$ for which either

(i) $r^{(k)} < 0.25, h^{(k+1)} \to 0$, and hence $\|\boldsymbol{\delta}^{(k)}\| \to 0$ or
(ii) $r^{(k)} \geqslant 0.25$ and $\inf h^{(k)} > 0$.

In either case (14.5.13) is shown to hold. In case (i) let $\exists$ a descent direction $\mathbf{s}$ ($\|\mathbf{s}\| = 1$) at $\mathbf{x}^\infty$, that is

$$\max_{\boldsymbol{\lambda} \in \partial h^\infty} \mathbf{s}^T(\mathbf{g}^\infty + \mathbf{A}^\infty \boldsymbol{\lambda}) = -d, \qquad d > 0. \tag{14.5.14}$$

By Taylor series

$$\begin{aligned} f(\mathbf{x}^{(k)} + \varepsilon^{(k)}\mathbf{s}) &= f^{(k)} + \varepsilon^{(k)}\mathbf{s}^T\mathbf{g}^{(k)} + o(\varepsilon^{(k)}) \\ &= q^{(k)}(\varepsilon^{(k)}\mathbf{s}) + o(\varepsilon^{(k)}) \end{aligned} \tag{14.5.15}$$

by (14.4.9), since $\boldsymbol{\lambda}^{(k)}$ is bounded by Lemma 14.2.1 and $\nabla^2 f$, $\nabla^2 c_i$ are bounded by assumption. Likewise by (14.4.6),

$$\mathbf{c}(\mathbf{x}^{(k)} + \varepsilon^{(k)}\mathbf{s}) = \mathbf{l}^{(k)}(\varepsilon^{(k)}\mathbf{s}) + o(\varepsilon^{(k)}) \tag{14.5.16}$$

and hence by (14.1.2), the boundedness of ∂h, and (14.4.11), it follows that

$$\begin{aligned} \phi(\mathbf{x}^{(k)} + \varepsilon^{(k)}\mathbf{s}) &= q^{(k)}(\varepsilon^{(k)}\mathbf{s}) + h(\mathbf{l}^{(k)}(\varepsilon^{(k)}\mathbf{s}) + o(\varepsilon^{(k)})) + o(\varepsilon^{(k)}) \\ &= q^{(k)}(\varepsilon^{(k)}\mathbf{s}) + h(\mathbf{l}^{(k)}(\varepsilon^{(k)}\mathbf{s})) + o(\varepsilon^{(k)}) \\ &= \psi^{(k)}(\varepsilon^{(k)}\mathbf{s}) + o(\varepsilon^{(k)}). \end{aligned} \tag{14.5.17}$$

Writing $\varepsilon^{(k)} = \|\boldsymbol{\delta}^{(k)}\|$ and considering a step along $\mathbf{s}$ in the subproblem, it follows by the optimality of $\boldsymbol{\delta}^{(k)}$ that

$$\begin{aligned} \Delta\psi^{(k)} &\geqslant \phi^{(k)} - \psi^{(k)}(\varepsilon^{(k)}\mathbf{s}) \\ &= \phi^{(k)} - \phi(\mathbf{x}^{(k)} + \varepsilon^{(k)}\mathbf{s}) + o(\varepsilon^{(k)}) \\ &\geqslant \varepsilon^{(k)}(d + o(1)) + o(\varepsilon^{(k)}) = d\varepsilon^{(k)} + o(\varepsilon^{(k)}) \end{aligned} \tag{14.5.18}$$

by the corollary to Lemma 14.5.1 and (14.5.14). But (14.5.17) implies that

$$\Delta\phi^{(k)} = \Delta\psi^{(k)} + o(\varepsilon^{(k)})$$

and hence $r^{(k)} = \Delta\phi^{(k)}/\Delta\psi^{(k)} = 1 + o(\varepsilon^{(k)})/\Delta\psi^{(k)} = 1 + o(1)$ from (14.5.18) since $d > 0$, which contradicts $r^{(k)} < 0.25$. Thus $d \leqslant 0$ for all $\mathbf{s}$ and hence (14.5.13) holds at $\mathbf{x}^\infty$.

In case (ii) the argument of case (ii) of Theorem 5.1.1 is largely followed (see also Fletcher, 1982a). The only extra step is to deduce from $\mathbf{x}^{(k)} \in B$ that $\partial h^{(k)}$ is uniformly bounded for all k, so that the parameters $\boldsymbol{\lambda}^{(k)}$ are bounded. Thus a thinner subsequence can be chosen such that $\boldsymbol{\lambda}^{(k)} \to \boldsymbol{\lambda}^\infty$ and hence $\mathbf{W}^{(k)} \to \mathbf{W}^\infty$. Then functions $q^\infty(\boldsymbol{\delta})$, $\mathbf{l}^\infty(\boldsymbol{\delta})$, and $\psi^\infty(\boldsymbol{\delta})$ are defined and it is concluded as in Theorem 5.1.1 that $\boldsymbol{\delta} = \mathbf{0}$ minimizes $\psi^\infty(\boldsymbol{\delta})$. It follows that the first order conditions (14.5.13) hold at $\mathbf{x}^\infty$. In case (ii) it is also possible to conclude that second order conditions hold. □

Note that the existence of a bounded region B which the theorem requires is implied if any level set $\{\mathbf{x}:\phi(\mathbf{x}) \leqslant \phi^{(k)}\}$ is bounded. Also the theorem assumes that the sequence $\{\mathbf{x}^{(k)}\}$ is infinite; if not then $\Delta\psi^{(k)} = 0$ for some k, the iteration terminates, and first order conditions are satisfied.

One point to emphasize about the theorem is that there are no hidden assumptions that certain vectors $\mathbf{a}_i^{\infty}$ are linearly independent or that the multipliers $\boldsymbol{\lambda}^{(k)}$ are bounded. Methods for NSO or nonlinear programming can often be proved to be convergent under such assumptions, yet can fail in practice. Thus it is important that this theorem avoids such assumptions. Another point is that $\mathbf{W}^{(k)}$ does not need to be defined as in (14.4.10) but can be any bounded matrix. Thus the theorem indicates that a corresponding quasi-Newton method, using for example $\mathbf{B}^{(k)}$ as defined in (12.3.18) in place of $\mathbf{W}^{(k)}$, can only fail if $\mathbf{B}^{(k)}$ becomes unbounded (a weaker condition on $\mathbf{B}^{(k)}$ for convergence is given by Yuan (1985a)). A weakness in the theorem requiring further thought is that it is assumed that global minima are calculated in the subproblems. This is guaranteed by convexity when $\mathbf{B}^{(k)}$ is positive definite. However, when exact second derivatives are used, $\mathbf{W}^{(k)}$ may be indefinite and it is then unrealistic to guarantee that global minima are calculated. Finally the theorem is seen to subsume the result of Madsen (1975) for the first order method like (14.4.8) in which essentially $\mathbf{W}^{(k)} = \mathbf{0}$.

It is important to consider the effect of the trust region modification on the SNQP algorithm. Clearly if the trust region bound is inactive for sufficiently large k then the satisfactory order of convergence properties of the unmodified algorithm carry through. However, Yuan (1984) gives examples which show that the trust region bound is not inactive for sufficiently large k, and the order is first order. This clearly points to the need to modify the algorithm to retain both global and second order convergence properties. As in Section 14.4 the use of an SOC step to follow the basic step suggests itself, and an appropriate subproblem for calculating the SOC step is to add a trust region bound to (14.4.34) giving

$$\begin{aligned} &\underset{\boldsymbol{\delta}}{\text{minimize}} \quad q^{(k)}(\boldsymbol{\delta}) + h([\hat{c}_i - \mathbf{a}_i^{(k)\mathrm{T}}\hat{\boldsymbol{\delta}} + \mathbf{a}_i^{(k)\mathrm{T}}\boldsymbol{\delta}]) \\ &\text{subject to} \quad \|\boldsymbol{\delta}\| \leqslant h^{(k)}. \end{aligned} \tag{14.5.19}$$

Of course if the trust region bound is inactive for sufficiently large k this algorithm retains the satisfactory order of convergence properties of the SOC algorithm. Yuan (1985b) extends this result by showing that if $\boldsymbol{\lambda}^{(k)} \to \boldsymbol{\lambda}^*$ then the trust region bound does become inactive for sufficiently large k.

In comparison with smooth unconstrained optimization (compare Theorem 5.1.2) these results are nonetheless not all that one might hope for, and it is desirable to say something about second order conditions at a limit point. To do this requires more attention to be paid to the choice of multiplier estimates $\boldsymbol{\lambda}^{(k)}$. The current rule in algorithm (14.5.6) is that if the kth iteration is successful in reducing $\phi(\mathbf{x})$, then $\boldsymbol{\lambda}^{(k+1)}$ is chosen as the multiplier vector from the kth subproblem, else $\boldsymbol{\lambda}^{(k+1)} = \boldsymbol{\lambda}^{(k)}$ is set. This rule is advantageous in that it requires no extra computation, and the multipliers are bounded a priori from Lemma 14.2.1, which is important for global convergence. (It has been proposed that $\boldsymbol{\lambda}^{(k)}$ should be selected as the least squares solution of the system $\mathbf{g}^{(k)} + \mathbf{A}^{(k)}\boldsymbol{\lambda} = \mathbf{0}$ from (14.2.18) but this is disadvantageous on both counts. It is possible to ensure boundedness by solving a constrained least squares problem, but at the expense of a further increase in complexity.) However, it is shown by Fletcher (1982b) that the current rule may cause algorithm (14.5.6) (with or without SOC) to stick a point which, although it satisfies first order conditions (Theorem 14.2.1), fails to satisfy second order necessary conditions. Thus a second order descent direction exists at the point under consideration. This difficulty can be circumvented by the equally simple rule that $\boldsymbol{\lambda}^{(k+1)}$ is always set to the multipliers of the kth subproblem, irrespective of the change in $\phi(\mathbf{x})$. This rule would seem to be undesirable when the step $\boldsymbol{\delta}^{(k)}$ arising from the subproblem is large but unsuccessful. One way to compromise between these two possibilities is to devise a rule which has the property that if

$$h^{(k+1)} < \min_{j<k} h^{(j)} \tag{14.5.20}$$

(that is if $h^{(k+1)}$ is less than any previous value) then $\boldsymbol{\lambda}^{(k+1)}$ is always set to the multipliers from the kth subproblem. The best choice of rule is currently under consideration.

14.6 CONSTRAINED NON-SMOOTH OPTIMIZATION

In Section 14.1 it is argued that the majority of NSO problems fall into the category of *composite NSO* problems. However, cases do arise in which it is required to minimize composite functions subject to constraints on the variables (for example, problem (14.5.2)) and such problems are the subject of this section. A suitable general format (*constrained composite NSO*) is

$$\begin{aligned} &\underset{\mathbf{x}\in\mathbb{R}^n}{\text{minimize}} \quad \phi(\mathbf{x}) \triangleq f(\mathbf{x}) + h(\mathbf{c}(\mathbf{x})) \\ &\text{subject to} \quad t(\mathbf{r}(\mathbf{x})) \leqslant 0 \end{aligned} \tag{14.6.1}$$

where the objective function is the composite function (14.1.1). There is a single inequality constraint involving a composite function with $\mathbf{r}(\mathbf{x})$ ($\mathbb{R}^n \to \mathbb{R}^p$) being smooth ($\mathbb{C}^1$) and $t(\mathbf{r})$ ($\mathbb{R}^p \to \mathbb{R}$) being convex but non-smooth ($\mathbb{C}^0$). This formulation covers constraints such as $\|\mathbf{x}\|_\infty \leqslant \rho$ which appear in restricted step methods, or a general system of inequality constraints $\mathbf{r}(\mathbf{x}) \leqslant \mathbf{0}$ through the composition $t(\mathbf{r}) = \max_i r_i$. Equality constraints $r_i(\mathbf{x}) = 0$ can be included by writing them as $r_i(\mathbf{x}) \leqslant 0$ and $-r_i(\mathbf{x}) \leqslant 0$. It also includes constraints like $\|\mathbf{r}(\mathbf{x})\|_1 \leqslant \rho$ which are not conveniently handled by (7.1.1) say. Thus (14.6.1) represents a very wide range of practical optimization problems.

In practice, those applications in which $\mathbf{r}(\mathbf{x})$ is nonlinear can be handled satisfactorily by using an exact penalty function and including penalty terms derived from $\mathbf{r}(\mathbf{x})$ in the objective function. Thus most practical interest in (14.6.1) arises when $\mathbf{r}(\mathbf{x})$ is affine and $t(\mathbf{r})$ is polyhedral. A suitable type of method for solving such problems is to modify the SNQP method described in Sections 14.4 and 14.5. Thus a sequence of NQP subproblems is solved, only differing from (14.5.2) in the addition of the polyhedral constraint $t(\mathbf{r}(\mathbf{x})) \leqslant 0$. These modified subproblems can be solved by QP-like techniques modelled on those mentioned at the end of Section 12.4. The global and local convergence properties of the modified SNQP method are therefore likely to be analogous to those for the SNQP method itself (Section 14.5) although to my knowledge these have not been investigated.

The main aim of this section is to give a comprehensive presentation of local optimality conditions for (14.6.1) which reduce to those for constrained smooth optimization (Section 9.2) when $h(\mathbf{c}(\mathbf{x}))$ is not present ($m = 0$), or to those for unconstrained non-smooth optimization (Section 14.2) when $t(\mathbf{r}(\mathbf{x}))$ is not present ($p = 0$). This can be done without losing any of the essential details of the simpler cases. Thus there appear in what follows analogues of feasible direction sets, constraint qualification, an independence assumption and a separating hyperplane lemma, leading to a statement of first order necessary conditions. Both necessary and sufficient second order conditions are also set out together with a study of the regular case in which these conditions are almost necessary *and* sufficient. Finally some examples of the various conditions in action are given. The approach is based on that of Fletcher and Watson (1980) but with some extra features to handle the more general form of the constraint in (14.6.1).

Firstly the concept of feasible directions for the constraint $t(\mathbf{r}(\mathbf{x})) \leqslant 0$ is considered, using the concept of a directional sequence defined in (9.2.1) Thus at a feasible point $\mathbf{x}'$ the set of feasible directions

$$\mathscr{F}' = \{\mathbf{s} \colon \mathbf{s} \neq \mathbf{0}, \exists \{\mathbf{x}^{(k)}\}, t(\mathbf{r}(\mathbf{x}^{(k)})) \leqslant 0, \mathbf{x}^{(k)} \to \mathbf{x}', \mathbf{s}^{(k)} \to \mathbf{s}, \delta^{(k)} \downarrow 0\} \tag{14.6.2}$$

can be defined. This is related to the set

$$F' = \{\mathbf{s} \colon \mathbf{s} \neq \mathbf{0}, t' = 0 \Rightarrow \max_{\mathbf{u} \in \partial t'} \mathbf{s}^{\mathrm{T}} \mathbf{R}' \mathbf{u} \leqslant 0\} \tag{14.6.3}$$

where $\mathbf{R} = \nabla \mathbf{r}^{\mathrm{T}}$, which can be regarded as the set of feasible directions for the linearized constraint at $\mathbf{x}'$. It is very convenient if the sets $\mathscr{F}'$ and F' are the same so it is important to consider the extent to which this is true. The mechanism of linearization is seen in the following lemma which establishes an inclusion.

Lemma 14.6.1

$$F' \supseteq \mathscr{F}'$$

Proof

Let $\mathbf{s} \in \mathscr{F}'$ so that there exists a directional sequence $\mathbf{x}^{(k)} \to \mathbf{x}'$ such that $\mathbf{s}^{(k)} \to \mathbf{s}$. A Taylor series about $\mathbf{x}'$ gives

$$\mathbf{r}^{(k)} = \mathbf{r}' + \delta^{(k)} \mathbf{R}'^{\mathrm{T}} \mathbf{s}^{(k)} + o(\delta^{(k)})$$

so that $\mathbf{r}^{(k)} \to \mathbf{r}'$ is a directional sequence in $\mathbb{R}^p$ with $(\mathbf{r}^{(k)} - \mathbf{r}')/\delta^{(k)} \to \mathbf{R}'^{\mathrm{T}}\mathbf{s}$. Thus by Lemma 14.2.5 applied to $t(\mathbf{r})$

$$\max_{\mathbf{u}\in\partial t'} \mathbf{s}^{\mathrm{T}}\mathbf{R}'\mathbf{u} = \lim \frac{t^{(k)} - t'}{\delta^{(k)}} \leqslant 0$$

if $t' = 0$ by feasibility (t' denotes $t(\mathbf{r}(\mathbf{x}'))$ etc.). Thus $\mathbf{s}\in F'$. $\square$

Unfortunately a result going the other way is not true, e.g. $t(\mathbf{r}(\mathbf{x})) = \max(x_2 - x_1^3, -x_2)$ as illustrated in Figure 9.2.1. However, this is an unlikely situation and usually the *constraint qualification* $\mathscr{F}' = F'$ can be assumed to hold. Indeed this result can be guaranteed under mild conditions as the following lemma shows. This uses the notation analogous to that in (14.2.30) that if $\partial t'$ has dimension q' and $\mathbf{u}'\in\partial t'$ is arbitrary, then there exists a matrix $\mathbf{F}'\in\mathbb{R}^{p\times q'}$ with columns $\mathbf{f}_i'$ $i = 1, 2, \ldots, q'$ which provide a basis for $\partial t' - \mathbf{u}'$, in that

$$\partial t' = \{\mathbf{u}\colon \mathbf{u} = \mathbf{u}' + \mathbf{F}'\mathbf{v},\ \mathbf{v}\in V' \subset \mathbb{R}^{q'}\}. \tag{14.6.4}$$

Lemma 14.6.2

Sufficient conditions for $\mathscr{F}' = F'$ at a feasible point $\mathbf{x}'$ are

(i) $t' < 0$

or if $t' = 0$ and the function $t(\mathbf{r})$ is locally linear about $\mathbf{r}'$, then either

(ii) rank $(\mathbf{R}'[\mathbf{u}'\!:\!\mathbf{F}']) = q' + 1$, or

(iii) *the function $\mathbf{r}(\mathbf{x})$ is affine.*

Proof

If $t' < 0$ then $\mathscr{F}' = \mathbb{R}^n\backslash\mathbf{0}$ and the result is trivial, so subsequently assume that $t' = 0$. Because $\mathscr{F}' \subseteq F'$ it is only necessary to establish the reverse inclusion, so let $\mathbf{s}\in F'$. If $\max_{\mathbf{u}\in\partial t'} \mathbf{s}^{\mathrm{T}}\mathbf{R}'\mathbf{u} < 0$ then take $\mathbf{x}^{(k)} = \mathbf{x}' + \theta^{(k)}\mathbf{s}$ for any sequence $\theta^{(k)} \downarrow 0$ and it follows that $t^{(k)} \leqslant 0$ for sufficiently large k and hence $\mathbf{s}\in\mathscr{F}'$ (if not $\exists$ a subsequence for which $t^{(k)} > 0$: using a Taylor expansion and Lemma 14.2.5 it follows that $\max_{\mathbf{u}\in\partial t'} \mathbf{s}^{\mathrm{T}}\mathbf{R}'\mathbf{u} \geqslant 0$ which is a contradiction). If $\max_{\mathbf{u}\in\partial t'} \mathbf{s}^{\mathrm{T}}\mathbf{R}'\mathbf{u} = 0$, let $\mathbf{u}'\in\partial t'$ be any vector for which $\mathbf{s}^{\mathrm{T}}\mathbf{R}'\mathbf{u}' = 0$. Without loss of generality $\mathbf{u}'$ can be regarded as the arbitrary vector in (14.6.4) above. Also define

$$\partial t_s' = \{\mathbf{u}\in\partial t'\colon \mathbf{s}^{\mathrm{T}}\mathbf{R}'\mathbf{u}' = 0\},$$

let the dimension of $\partial t_s'$ be q_s' ($q_s' < q'$), and without loss of generality let $\mathbf{f}_i'$ $i = 1, \ldots, q_s'$ be a basis for $\partial t_s' - \mathbf{u}'$. Hence

$$\mathbf{s}^{\mathrm{T}}\mathbf{R}'\mathbf{f}_i' \begin{cases} = 0 & i = 1, 2, \ldots, q_s' \\ < 0 & i = q_s' + 1, \ldots, q'. \end{cases}$$

If $q_s' + 1 = n$ it follows that $\mathbf{s}^{\mathrm{T}}\mathbf{R}'[\mathbf{u}'\!:\!\mathbf{F}'] = \mathbf{0}^{\mathrm{T}}$ and hence that $\mathbf{s} = \mathbf{0}$ by virtue of the rank assumption, which contradicts $\mathbf{s}\in F'$. If $q_s' + 1 < n$ it is possible following the construction of Lemma 14.2.7 (with $\mathbf{D}^*$ replaced by $[\mathbf{u}'\!:\!\mathbf{F}']$) to determine a smooth arc $\mathbf{x}(\theta)$ $\theta \in [0, \bar{\theta})$ for which $\mathbf{x}(0) = \mathbf{x}'$, $\dot{\mathbf{x}}(0) = \mathbf{s}$, and

$$(\mathbf{r}(\mathbf{x}(\theta)) - \mathbf{r}')^{\mathrm{T}}\mathbf{u}' = \theta \mathbf{s}^{\mathrm{T}}\mathbf{R}'\mathbf{u}' = 0$$

$$(\mathbf{r}(\mathbf{x}(\theta)) - \mathbf{r}')^{\mathrm{T}}\mathbf{f}'_i = \theta \mathbf{s}^{\mathrm{T}}\mathbf{R}'\mathbf{f}'_i \begin{cases} = 0 & i = 1, 2, \ldots, q'_s, \\ < 0 & i = q'_s + 1, \ldots, q'. \end{cases}$$

It follows that

$$\max_{\mathbf{u} \in \partial t'} (\mathbf{r}(\mathbf{x}(\theta)) - \mathbf{r}')^{\mathrm{T}}\mathbf{u} = 0. \tag{14.6.5}$$

Then using the definition (14.2.29) of a locally linear function it follows that there exists a neighbourhood of $\mathbf{r}'$ such that $t(\mathbf{r}(\mathbf{x}(\theta))) = 0$, and taking any sequence $\theta^{(k)} \downarrow 0$ gives a directional sequence with $\mathbf{s} \in \mathscr{F}'$.

Finally in case (iii), if $\mathbf{r}(\mathbf{x})$ is affine, it follows that the ray $\mathbf{x}(\theta) = \mathbf{x}' + \theta\mathbf{s}$ has $\mathbf{r}(\mathbf{x}(\theta)) = \mathbf{r}' + \theta \mathbf{R}'^{\mathrm{T}}\mathbf{s}$. Thus it is possible to deduce (14.6.5) in the above directly from $\max_{\mathbf{u} \in \partial t'} \mathbf{s}^{\mathrm{T}}\mathbf{R}'\mathbf{u} = 0$, leading again to the conclusion that $\mathbf{s} \in \mathscr{F}'$. □

Note that this result is somewhat stronger than that of Fletcher and Watson (1980) in which the assumption rank $\mathbf{R}' = p$ restricts the range of $\mathbf{r}(\mathbf{x})$ to have dimension $p \leqslant n$. Also Lemma 14.6.2 allows a constraint like $\|\mathbf{r}(\mathbf{x})\|_1 \leqslant \rho$ which cannot usually be handled effectively by using Lemma 9.2.2 and a polyhedral expansion of $\|\mathbf{r}\|_1$, because of the degenerate nature of the resulting constraint set.

In moving on to discuss necessary conditions at a local solution of (14.6.1) it is convenient to define the set of *descent directions* at $\mathbf{x}'$

$$\mathscr{D}(\mathbf{x}') = \mathscr{D}' = \left\{ \mathbf{s}\colon \max_{\boldsymbol{\lambda} \in \partial h'} \mathbf{s}^{\mathrm{T}}(\mathbf{g}' + \mathbf{A}'\boldsymbol{\lambda}) < 0 \right\}.$$

Then the most basic necessary condition is the following.

Lemma 14.6.3

If $\mathbf{x}^*$ *is a local minimizer then* $\mathscr{F}^* \cap \mathscr{D}^* = \emptyset$ (no feasible descent directions).

Proof

Let $\mathbf{s} \in \mathscr{F}^*$ so that there exists a feasible directional sequence $\mathbf{x}^{(k)} \to \mathbf{x}^*$ with $\mathbf{s}^{(k)} \to \mathbf{s}$. By Taylor expansion about $\mathbf{x}^*$

$$f^{(k)} = f^* + \delta^{(k)}\mathbf{g}^{*\mathrm{T}}\mathbf{s}^{(k)} + o(\delta^{(k)})$$
$$\mathbf{c}^{(k)} = \mathbf{c}^* + \delta^{(k)}\mathbf{A}^{*\mathrm{T}}\mathbf{s}^{(k)} + o(\delta^{(k)}).$$

By local optimality, $\phi^{(k)} \geqslant \phi^*$ for sufficiently large k and hence

$$0 \leqslant \frac{\phi^{(k)} - \phi^*}{\delta^{(k)}} = \frac{f^{(k)} - f^*}{\delta^{(k)}} + \frac{h(\mathbf{c}^{(k)}) - h(\mathbf{c}^*)}{\delta^{(k)}}.$$

Taking the limit as $k \to \infty$, and using Lemma 14.2.5 and the fact that $\mathbf{c}^{(k)} \to \mathbf{c}^*$

is a directional sequence with direction $\mathbf{A}^{*\mathrm{T}}\mathbf{s}$, it follows that

$$0 \leqslant \max_{\lambda \in \partial h'} \mathbf{s}^{\mathrm{T}}(\mathbf{g}^* + \mathbf{A}^*\boldsymbol{\lambda}).$$

This contradicts $\mathbf{s} \in \mathscr{D}^*$ and so proves the Lemma. □

For the reasons set out in Section 9.2 it is not possible to proceed further without making a regularity assumption

$$\mathscr{F}^* \cap \mathscr{D}^* = F^* \cap \mathscr{D}^* \tag{14.6.6}$$

However this is a very weak assumption and is certainly implied by the mild sufficient conditions for constraint qualification ($\mathscr{F}^* = F^*$) of Lemma 14.6.2. With this assumption the necessary condition of Lemma 14.6.3 becomes $F^* \cap \mathscr{D}^* = \varnothing$. It is now possible to relate this condition to the existence of multipliers in Theorem 14.6.1 below. A preliminary lemma is required which extends the concept of a separating hyperplane.

Lemma 14.6.4

If, in $\mathbb{R}^n$, C is a closed convex cone, B is a non-empty compact convex set and $B \cap C = \varnothing$, then there exists hyperplane $\mathbf{s}^{\mathrm{T}}\mathbf{x} = 0$ which separates B and C.

Proof

By construction. For each point $\mathbf{b} \in B$ the closest point in C can be determined as in Lemma 9.2.5. Because B is a compact there exist points $\hat{\mathbf{b}} \in B$ and $\hat{\mathbf{g}} \in C$ which minimize $\|\mathbf{b} - \mathbf{g}\|_2$ over all $\mathbf{b} \in B$ and $\mathbf{g} \in C$. Hence for any $\mathbf{b} \in B$

$$\|\hat{\mathbf{b}} - \hat{\mathbf{g}} + \theta(\mathbf{b} - \hat{\mathbf{b}})\|_2^2 \geqslant \|\hat{\mathbf{b}} - \hat{\mathbf{g}}\|_2^2 \qquad \theta \in [0, 1]$$

and taking the limit $\theta \downarrow 0$ it follows that

$$(\mathbf{b} - \hat{\mathbf{b}})^{\mathrm{T}}(\hat{\mathbf{g}} - \hat{\mathbf{b}}) \leqslant 0.$$

But by Lemma 9.2.5, if $\mathbf{s} = \hat{\mathbf{g}} - \hat{\mathbf{b}}$ then $\mathbf{s}^{\mathrm{T}}\mathbf{x} \geqslant 0$ for all $\mathbf{x} \in C$ and $\mathbf{s}^{\mathrm{T}}\hat{\mathbf{b}} < 0$. It follows that $\mathbf{s}^{\mathrm{T}}\mathbf{b} < 0$ for all $\mathbf{b} \in B$ thus establishing the lemma. □

In stating the first main result an appropriate Lagrangian function

$$\mathscr{L}(\mathbf{x}, \boldsymbol{\lambda}, \mathbf{u}, \pi) = f(\mathbf{x}) + \boldsymbol{\lambda}^{\mathrm{T}}\mathbf{c}(\mathbf{x}) + \pi\mathbf{u}^{\mathrm{T}}\mathbf{r}(\mathbf{x}) \tag{14.6.7}$$

is introduced.

Theorem 14.6.1 (First order necessary conditions)

If $\mathbf{x}^*$ locally solves (14.6.1) *and the regularity condition* (14.6.6) *holds, then there exist multipliers $\boldsymbol{\lambda}^* \in \partial h^*$, $\mathbf{u}^* \in \partial t^*$ and $\pi^* \geqslant 0$ such that*

$t^* \leqslant 0$ (feasibility)

$\pi^* t^* = 0$ (complementarity)

and

$$\begin{aligned} \mathbf{0} &= \nabla \mathscr{L}(\mathbf{x}^*, \boldsymbol{\lambda}^*, \mathbf{u}^*, \pi^*) \\ &= \mathbf{g}^* + \mathbf{A}^*\boldsymbol{\lambda}^* + \pi^*\mathbf{R}^*\mathbf{u}^* \end{aligned} \tag{14.6.8}$$

Proof

The result essentially utilizes a general form of Farkas' lemma that $F^* \cap \mathscr{D}^* = \varnothing$ iff the conditions of the theorem hold. Clearly if the latter hold and $\mathbf{s} \in F^*$ then

$$t^* < 0 \Rightarrow \pi^* = 0 \Rightarrow \mathbf{g}^* + \mathbf{A}^*\boldsymbol{\lambda}^* = \mathbf{0}$$

and

$$t^* = 0 \Rightarrow \mathbf{s}^{\mathrm{T}}\mathbf{R}^*\mathbf{u}^* \leqslant 0 \Rightarrow \mathbf{s}^{\mathrm{T}}(\mathbf{g}^* + \mathbf{A}^*\boldsymbol{\lambda}^*) \geqslant 0$$

from (14.6.3) and (14.6.8). In both cases $\mathbf{s} \in \mathscr{D}^*$ is contradicted so that $F^* \cap \mathscr{D}^* = \varnothing$. Conversely if the conditions of the theorem do not hold then a direction $\mathbf{s} \in F^* \cap \mathscr{D}^*$ can be constructed as follows. The conditions of the theorem are equivalent to the statement that the closed convex cone

$$\begin{aligned} C = \{\mathbf{y}\colon & t^* < 0 \Rightarrow \mathbf{y} = \mathbf{0} \\ & t^* = 0 \Rightarrow \mathbf{y} = -\pi\mathbf{R}^*\mathbf{u} \quad \forall \pi \geqslant 0, \quad \mathbf{u} \in \partial t^*\}, \end{aligned}$$

which is convex by convexity of ∂t^*, and the compact convex set

$$B = \{\mathbf{b}\colon \mathbf{b} = \mathbf{g}^* + \mathbf{A}^*\boldsymbol{\lambda} \quad \forall \boldsymbol{\lambda} \in \partial h^*\}$$

have a common point (from (14.6.8)). Therefore if there is no common point it follows from Lemma 14.6.4 that there exists a direction $\mathbf{s}$ such that $\max_{\boldsymbol{\lambda} \in \partial h^*} \mathbf{s}^{\mathrm{T}}(\mathbf{g}^* + \mathbf{A}^*\boldsymbol{\lambda}) < 0$, that is $\mathbf{s} \in \mathscr{D}^*$, and $t^* = 0 \Rightarrow \max_{\mathbf{u} \in \partial t^*} \mathbf{s}^{\mathrm{T}}\mathbf{R}^*\mathbf{u} \leqslant 0$, that is $\mathbf{s} \in F^*$. The theorem is then a consequence of Lemma 14.6.3 and assumption (14.6.6).

□

It is also possible to analyse the effect of second order changes within the framework of (14.6.1) if the additional assumption is made that $\mathbf{c}(\mathbf{x})$ and $\mathbf{r}(\mathbf{x})$ are $\mathbb{C}^2$ functions (but $h(\mathbf{c})$ and $t(\mathbf{r})$ remain $\mathbb{C}^0$ and convex). As in Sections 9.2 and 14.2 the first step is to define certain sets of feasible directions related to directions of zero slope, so that second order effects become important. In general let $\mathbf{x}^*, \boldsymbol{\lambda}^*, \mathbf{u}^*, \pi^*$ satisfy the first order conditions of Theorem 14.6.1 and consider the set

$$\begin{aligned} X = \{\mathbf{x}\colon\ & \mathbf{h}(\mathbf{c}(\mathbf{x})) = h^* + (\mathbf{c}(\mathbf{x}) - \mathbf{c}^*)^{\mathrm{T}}\boldsymbol{\lambda}^*, \\ & t(\mathbf{r}(\mathbf{x})) \leqslant 0, \\ & \pi^*(\mathbf{r}(\mathbf{x}) - \mathbf{r}^*)^{\mathrm{T}}\mathbf{u}^* = 0\}. \end{aligned} \tag{14.6.9}$$

Define $\mathscr{G}^*$ as the set of normalized feasible directions with respect to X at $\mathbf{x}^*$, that is

$$\mathscr{G}^* = \{\mathbf{s}\colon\ \|\mathbf{s}\|_2 = 1, \quad \exists\,\{\mathbf{x}^{(k)}\}, \mathbf{x}^{(k)} \in X, \mathbf{x}^{(k)} \to \mathbf{x}^*, \mathbf{s}^{(k)} \to \mathbf{s}, \delta^{(k)} \downarrow 0\}. \tag{14.6.10}$$

This set is closely related to the set

$$G^* = \left\{ \mathbf{s}: \|\mathbf{s}\|_2 = 1, \quad \mathbf{s} \in F^*, \quad \max_{\boldsymbol{\lambda} \in \partial h^*} \mathbf{s}^T(\mathbf{g}^* + \mathbf{A}^*\boldsymbol{\lambda}) = 0, \quad \pi^* \max_{\mathbf{u} \in \partial t^*} \mathbf{s}^T\mathbf{R}^*\mathbf{u} = 0 \right\} \tag{14.6.11}$$

which can be interpreted as the set of linearized feasible directions that have zero slope with regard to both $\phi(\mathbf{x})$ and (if $\pi^* > 0$) $t(\mathbf{r}(\mathbf{x}))$. The extent to which $\mathscr{G}^*$ and G^* correspond is important and it is shown by linearizing the functions which define $\mathscr{G}^*$ that $\mathscr{G}^*$ is a subset of G^*.

Lemma 14.6.5

$$\mathscr{G}^* \subseteq G^*$$

Proof

Let $\mathbf{s} \in \mathscr{G}^*$ which implies $\mathbf{s} \in \mathscr{F}^*$ and hence $\mathbf{s} \in F^*$. As in Lemma 14.2.6 it follows that $\max_{\boldsymbol{\lambda} \in \partial h^*} \mathbf{s}^T(\mathbf{g}^* + \mathbf{A}^*\boldsymbol{\lambda}) = 0$. By a similar argument, if $\pi^* > 0$ (and $t^* = 0$) it follows that

$$0 = \lim \frac{(\mathbf{r}^{(k)} - \mathbf{r}^*)^T\mathbf{u}^*}{\delta^{(k)}} = \mathbf{s}^T\mathbf{R}^*\mathbf{u}^*$$

and because $\mathbf{s} \in F^*$ it follows from (14.6.3) that $\pi^* \max_{\mathbf{u} \in \partial t^*} \mathbf{s}^T\mathbf{R}^*\mathbf{u} = 0$. Therefore $\mathbf{s} \in G^*$. $\square$

As in Section 14.2 it is not generally possible to derive the reverse inclusion, but there are important special cases where this can be done, associated with locally linear functions. Further discussion of this is given later in this section: at present the regularity assumption

$$\mathscr{G}^* = G^* \tag{14.6.12}$$

is made (which depends on $\boldsymbol{\lambda}^*, \mathbf{u}^*, \pi^*$ if these elements are not unique). It is now possible to state and prove the second order conditions: as usual the regularity assumption is needed only in the necessary conditions.

Theorem 14.6.2 (Second order necessary conditions)

If $\mathbf{x}^*$ *locally solves* (14.6.1) *and* (14.6.6) *holds then Theorem* 14.6.1 *is valid. For each triple* $\boldsymbol{\lambda}^*, \mathbf{u}^*, \pi^*$ *which thus exists, if* (14.6.12) *holds, then*

$$\mathbf{s}^T\nabla^2\mathscr{L}(\mathbf{x}^*, \boldsymbol{\lambda}^*, \mathbf{u}^*, \pi^*)\mathbf{s} \geq 0 \quad \forall \mathbf{s} \in G^*. \tag{14.6.13}$$

Proof

Let $\mathbf{s} \in G^*$. Then $\mathbf{s} \in \mathscr{G}^*$ and $\exists$ a directional sequence feasible in (14.6.9). A Taylor expansion of $\mathscr{L}(\mathbf{x}, \boldsymbol{\lambda}^*, \mathbf{u}^*, \pi^*)$ about $\mathbf{x}^*$ yields

$$\begin{aligned}\mathscr{L}(\mathbf{x}^{(k)}, \boldsymbol{\lambda}^*, \mathbf{u}^*, \pi^*) &= \mathscr{L}(\mathbf{x}^*, \boldsymbol{\lambda}^*, \mathbf{u}^*, \pi^*) + \mathbf{e}^{(k)T}\nabla\mathscr{L}(\mathbf{x}^*, \boldsymbol{\lambda}^*, \mathbf{u}^*, \pi^*) \\ &\quad + \tfrac{1}{2}\mathbf{e}^{(k)T}\nabla^2\mathscr{L}(\mathbf{x}^*, \boldsymbol{\lambda}^*, \mathbf{u}^*, \pi^*)\mathbf{e}^{(k)} + o(\|\mathbf{e}^{(k)}\|^2)\end{aligned} \tag{14.6.14}$$

where $\mathbf{e}^{(k)} = \mathbf{x}^{(k)} - \mathbf{x}^*$. The definition of $\mathscr{L}$ yields

$$\begin{aligned}
&\mathscr{L}(\mathbf{x}^{(k)}, \boldsymbol{\lambda}^*, \mathbf{u}^*, \pi^*) - \mathscr{L}(\mathbf{x}^*, \boldsymbol{\lambda}^*, \mathbf{u}^*, \pi^*) \\
&\quad = f^{(k)} - f^* + (\mathbf{c}^{(k)} - \mathbf{c}^*)^{\mathrm{T}}\boldsymbol{\lambda}^* + \pi^*(\mathbf{r}^{(k)} - r^*)^{\mathrm{T}}\mathbf{u}^* \qquad (14.6.15)\\
&\quad = f^{(k)} - f^* + h(\mathbf{c}^{(k)}) - h^* = \phi^{(k)} - \phi^*
\end{aligned}$$

by feasibility in (14.6.9). Substituting $\mathbf{e}^{(k)} = \delta^{(k)}\mathbf{s}^{(k)}$ and the local minimality of ϕ^* then gives

$$0 \leqslant \phi^{(k)} - \phi^* = \tfrac{1}{2}\delta^{(k)^2}\mathbf{s}^{(k)\mathrm{T}}\nabla^2\mathscr{L}(\mathbf{x}^*, \boldsymbol{\lambda}^*, \mathbf{u}^*, \pi^*)\mathbf{s}^{(k)} + o(\delta^{(k)^2}).$$

Finally dividing by $\frac{1}{2}\delta^{(k)^2}$ and taking the limit establishes (14.6.13). □

Theorem 14.6.3 (Second order sufficient conditions)

If at $\mathbf{x}^*, t^* \leqslant 0$ *and there exist* $\boldsymbol{\lambda}^* \in \partial h^*$, $\mathbf{u}^* \in \partial t^*$ *and* $\pi^* \geqslant 0$ *such that* $\pi^* t^* = 0$ *and* (14.6.8) *holds (first order conditions), and if*

$$\mathbf{s}^{\mathrm{T}}\nabla^2\mathscr{L}(\mathbf{x}^*, \boldsymbol{\lambda}^*, \mathbf{u}^*, \pi^*)\mathbf{s} > 0 \qquad \forall \mathbf{s} \in G^* \qquad (14.6.16)$$

then $\mathbf{x}^*$ *is a strict local solution of* (14.6.1).

Proof

Assume the contrary that ∃ a feasible sequence and hence a feasible direction sequence $\mathbf{x}^{(k)} \to \mathbf{x}^*$, $\mathbf{s}^{(k)} \to \mathbf{s}$, $\|\mathbf{s}\|_2 = 1$ such that $\phi^{(k)} \leqslant \phi^*$. Now $\mathbf{s} \in \mathscr{F}^* \Rightarrow \mathbf{s} \in F^*$, so using (14.6.8),

$$t^* < 0 \Rightarrow \pi^* = 0 \Rightarrow \mathbf{g}^* + \mathbf{A}^*\boldsymbol{\lambda}^* = \mathbf{0}$$

and from $\mathbf{s} \in F^*$

$$t^* = 0 \Rightarrow \max_{\mathbf{u} \in \partial t^*} \mathbf{s}^{\mathrm{T}}\mathbf{R}^*\mathbf{u} \leqslant 0$$

so in both cases it follows that

$$0 \leqslant \max_{\boldsymbol{\lambda} \in \partial h^*} \mathbf{s}^{\mathrm{T}}(\mathbf{g}^* + \mathbf{A}^*\boldsymbol{\lambda}) = \mu \quad \text{say.}$$

If $\mu > 0$ then from Lemma 14.2.5, $\lim(\phi^{(k)} - \phi^*)/\delta^{(k)} = \mu$ which contradicts $\phi^{(k)} \leqslant \phi^*$. Thus $\mu = 0$. Let $\mathbf{s}^{\mathrm{T}}(\mathbf{g}^* + \mathbf{A}^*\boldsymbol{\lambda}^*) < 0$; then the first order conditions imply $\pi^* > 0$, $t^* = 0$ and $\mathbf{s}^{\mathrm{T}}\mathbf{R}^*\mathbf{u}^* > 0$ which contradicts $\mathbf{s} \in F^*$. Thus $\mathbf{s}^{\mathrm{T}}(\mathbf{g}^* + \mathbf{A}^*\boldsymbol{\lambda}^*) = 0$ and hence $\pi^*\mathbf{s}^{\mathrm{T}}\mathbf{R}^*\mathbf{u}^* = 0$. It follows from $\mathbf{s} \in F^*$ that $\pi^* \max_{\mathbf{u} \in \partial t^*} \mathbf{s}^{\mathrm{T}}\mathbf{R}^*\mathbf{u}^* = 0$ and hence $\mathbf{s} \in G^*$. Now from (14.6.15)

$$\begin{aligned}
&\mathscr{L}(\mathbf{x}^{(k)}, \boldsymbol{\lambda}^*, \mathbf{u}^*, \pi^*) - \mathscr{L}(\mathbf{x}^*, \boldsymbol{\lambda}^*, \mathbf{u}^*, \pi^*) \\
&\quad = \phi^{(k)} - \phi^* - (h^{(k)} - h^* - (\mathbf{c}^{(k)} - \mathbf{c}^*)^{\mathrm{T}}\boldsymbol{\lambda}^*) \\
&\qquad + \pi^*(t^{(k)} - t^* - (t^{(k)} - t^* - (\mathbf{r}^{(k)} - \mathbf{r}^*)^{\mathrm{T}}\mathbf{u}^*)) \\
&\quad \leqslant \phi^{(k)} - \phi^* + \pi^*(t^{(k)} - t^*) \leqslant \phi^{(k)} - \phi^*
\end{aligned}$$

using the subgradient inequality and then feasibility. Hence from (14.6.14) and (14.6.8)

$$0 \geqslant \phi^{(k)} - \phi^* \geqslant \tfrac{1}{2}\delta^{(k)^2}\mathbf{s}^{(k)\mathrm{T}}\nabla^2\mathscr{L}(\mathbf{x}^*, \boldsymbol{\lambda}^*, \mathbf{u}^*, \pi^*)\mathbf{s}^{(k)} + o(\delta^{(k)^2}).$$

Dividing by $\frac{1}{2}\delta^{(k)^2}$ and taking the limit contradicts (14.6.16) and establishes the theorem. □

Corollary

If the directional derivative $\max_{\boldsymbol{\lambda}\in\partial h^*}\mathbf{s}^{\mathrm{T}}(\mathbf{g}^* + \mathbf{A}^*\boldsymbol{\lambda})$ *is positive for all feasible directions in* F^*, *or equivalently if* G^* *is empty, then the first order conditions are sufficient to imply that* $\mathbf{x}^*$ *is a strict local solution of* (14.6.1).

Proof

Immediate from the statement of Theorem 14.6.2. □

The remarks after Theorem 14.2.3 about when there is near equivalence in the second order conditions apply equally here. The conditions are 'almost' necessary and sufficient only if it can be shown that $\mathscr{G}^* = G^*$ and this result is only true for particular types of composite function having the locally linear property. Effectively either $h(\mathbf{c})$ and $t(\mathbf{r})$ are restricted to being polyhedral, or $h(\mathbf{c})$ can be the penalty term in certain exact penalty function applications (Lemma 14.3.1). A mild independence assumption then enables $\mathscr{G}^* = G^*$ to be established using the constructions of (14.2.30) and (14.6.4) as follows.

Lemma 14.6.6 (Sufficient conditions for regularity)

If $\mathbf{x}^*$ *satisfies the first order conditions of Theorem* 14.6.1, *if* $h(\mathbf{c})$ *and* $t(\mathbf{r})$ *are* locally linear at $\mathbf{c}^*$ *and* $\mathbf{r}^*$ *and if*

$$\operatorname{rank}([\mathbf{A}^*\mathbf{D}^*:\mathbf{R}^*\mathbf{u}^*:\mathbf{R}^*\mathbf{F}^*]) = l^* + q^* + 1 \tag{14.6.17}$$

then $\mathscr{G}^* = G^*$. *Alternatively the rank assumption may be replaced by an assumption that the functions* $\mathbf{c}(\mathbf{x})$ *and* $\mathbf{r}(\mathbf{x})$ *are affine.*

Proof

Because $G^* \supseteq \mathscr{G}^*$ it is sufficient to establish the reverse inclusion. Let $\mathbf{s}\in G^*$. If $t^* < 0$ or if $t^* = 0$, $\pi^* = 0$ and $\max_{\mathbf{u}\in\partial t^*}\mathbf{s}^{\mathrm{T}}\mathbf{R}^*\mathbf{u} < 0$ then an arc and hence a directional sequence is constructed as in Lemma 14.2.7. This sequence is easily shown to be feasible ($t^{(k)} \leqslant 0$): in the latter case the argument of Lemma 14.6.2 is used. Thus $\mathbf{s}\in\mathscr{G}^*$. If $\pi^* = 0$ and $\max_{\mathbf{u}\in\partial t^*}\mathbf{s}^{\mathrm{T}}\mathbf{R}^*\mathbf{u} = 0$ then without loss of generality $\mathbf{u}^*$ can be taken to attain the max, so that $\mathbf{s}^{\mathrm{T}}\mathbf{R}^*\mathbf{u}^* = 0$. This enables this case and the general case $\pi^* > 0$ to be treated together, as follows. Define

$$\partial h_s^* = \{\boldsymbol{\lambda}\in\partial h^* : \mathbf{s}^{\mathrm{T}}(\mathbf{g}^* + \mathbf{A}^*\boldsymbol{\lambda}) = 0\}$$
$$\partial t_s^* = \{\mathbf{u}\in\partial t^* : \mathbf{s}^{\mathrm{T}}\mathbf{R}^*\mathbf{u} = 0\}$$

which depend on s. Now $\mathbf{s} \in G^*$ and the first order conditions imply that

$$\mathbf{s}^T(\mathbf{g}^* + \mathbf{A}^*\boldsymbol{\lambda}^*) = \pi^*\mathbf{s}^T\mathbf{R}^*\mathbf{u}^* = 0$$

so that

$$\begin{aligned} &\mathbf{s}^T\mathbf{A}^*(\boldsymbol{\lambda} - \boldsymbol{\lambda}^*) = 0 \qquad \forall \boldsymbol{\lambda} \in \partial h_s^* \\ &\mathbf{s}^T\mathbf{R}^*\mathbf{u} = 0 \qquad \forall \mathbf{u} \in \partial t_s^* \end{aligned}$$

and from (14.6.11)

$$\mathbf{s}^T\mathbf{A}^*(\boldsymbol{\lambda} - \boldsymbol{\lambda}^*) < 0 \quad \forall \boldsymbol{\lambda} \in \partial h^* \backslash \partial h_s^*$$

$$\mathbf{s}^T\mathbf{R}^*\mathbf{u} < 0 \quad \forall \mathbf{u} \in \partial t^* \backslash \partial t_s^*$$

Let the dimensions of ∂h_s^* and ∂t_s^* be l_s^* and q_s^* respectively and let $\mathbf{u}^*$ be the arbitrary vector $\mathbf{u}'$ in (14.6.4). Assume without loss of generality that the vectors $\mathbf{d}_i^*$ $i = 1, \ldots, l_s^*$ and $\mathbf{f}_i^*$ $i = 1, \ldots, q_s^*$ form bases for $\partial h_s^* - \boldsymbol{\lambda}^*$ and $\partial t_s^* - \mathbf{u}^*$ respectively. Then

$$\mathbf{s}^T\mathbf{A}^*\mathbf{d}_i^* \begin{cases} = 0 & i = 1, 2, \ldots, l_s^* \\ < 0 & i = l_s^* + 1, \ldots, l^* \end{cases}$$

$$\mathbf{s}^T\mathbf{R}^*\mathbf{u}^* = 0$$

$$\mathbf{s}^T\mathbf{R}^*\mathbf{f}_i^* \begin{cases} = 0 & i = 1, 2, \ldots, q_s^* \\ < 0 & i = q_s^* + 1, \ldots, q^* \end{cases}$$

If $l_s^* + q_s^* + 1 = n$ then it follows from the rank assumption that $\mathbf{s} = \mathbf{0}$ which contradicts $\mathbf{s} \in G^*$. For $l_s^* + q_s^* + 1 < n$ it is possible as in Lemma 14.2.7 to construct a smooth arc $\mathbf{x}(\theta)$ $\theta \in [0, \bar{\theta})$ for which $\mathbf{x}(0) = \mathbf{x}^*$, $\dot{\mathbf{x}}(0) = \mathbf{s}$ and

$$\begin{aligned} &(\mathbf{c}(\mathbf{x}(\theta)) - \mathbf{c}^*)^T\mathbf{d}_i^* = \theta\mathbf{s}^T\mathbf{A}^*\mathbf{d}_i^* \qquad i = 1, 2, \ldots, l^* \\ &(\mathbf{r}(\mathbf{x}(\theta)) - \mathbf{r}^*)^T\mathbf{u}^* = \theta\mathbf{s}^T\mathbf{R}^*\mathbf{u}^* \\ &(\mathbf{r}(\mathbf{x}(\theta)) - \mathbf{r}^*)^T\mathbf{f}_i^* = \theta\mathbf{s}^T\mathbf{R}^*\mathbf{f}_i^* \qquad i = 1, 2, \ldots, q^*. \end{aligned} \tag{14.6.18}$$

Consequently in a similar way to Lemma 14.2.7

$$\max_{\boldsymbol{\lambda} \in \partial h^*} (\mathbf{c}(\mathbf{x}(\theta)) - \mathbf{c}^*)^T(\boldsymbol{\lambda} - \boldsymbol{\lambda}^*) = 0 \tag{14.6.19}$$

$$\max_{\mathbf{u} \in \partial t^*} (\mathbf{r}(\mathbf{x}(\theta) - \mathbf{r}^*)^T\mathbf{u} = 0 \tag{14.6.20}$$

and hence from the local linearity assumption

$$\begin{aligned} &h(\mathbf{c}(\mathbf{x}(\theta))) = h^* + (\mathbf{c}(\mathbf{x}(\theta)) - \mathbf{c}^*)^T\boldsymbol{\lambda}^* \\ &t(\mathbf{r}(\mathbf{x}(\theta))) = 0. \end{aligned}$$

Adding to these the equation

$$\pi^*(\mathbf{r}(\mathbf{x}(\theta)) - \mathbf{r}^*)^T\mathbf{u}^* = 0 \tag{14.6.21}$$

from (14.6.18), it follows that $\mathbf{x}(\theta) \in X$ in (14.6.9) and hence by taking any sequence $\theta^{(k)} \downarrow 0$, $\mathbf{s}$ is shown to be a feasible direction in $\mathcal{G}^*$.

Alternatively if the assumption is made that $\mathbf{c}(\mathbf{x})$ and $\mathbf{r}(\mathbf{x})$ are affine in place of (14.6.17), then the arc $\mathbf{x}(\theta) = \mathbf{x}^* + \theta\mathbf{s}$ has

$$\mathbf{c}(\mathbf{x}(\theta)) = \mathbf{c}^* + \mathbf{A}^{*\mathrm{T}}\mathbf{s}$$
$$\mathbf{r}(\mathbf{x}(\theta)) = \mathbf{r}^* + \mathbf{R}^{*\mathrm{T}}\mathbf{s}$$

and it is possible to deduce (14.6.19), (14.6.20) and (14.6.21) directly from the equations

$$\max_{\boldsymbol{\lambda}\in\partial h^*} \mathbf{s}^{\mathrm{T}}\mathbf{A}^*(\boldsymbol{\lambda} - \boldsymbol{\lambda}^*) = 0$$

$$\max_{\mathbf{u}\in\partial t^*} \mathbf{s}^{\mathrm{T}}\mathbf{R}^*\mathbf{u} = 0$$

$$\mathbf{s}^{\mathrm{T}}\mathbf{R}^*\mathbf{u}^* = 0$$

which arise earlier in the proof. □

This section finishes with a number of remarks and examples to clarify the way in which the conditions operate. It is intrinsic in the development of the conditions that $t(\mathbf{r})$ has the property of local linearity. An example for which this is not the case is a constraint like $\|\mathbf{r}(\mathbf{x})\|_2 \leqslant \rho$ for $\rho > 0$, in which $t(\mathbf{r}) = \|\mathbf{r}\|_2$ is locally smooth and nonlinear. Such a constraint may be brought within the scope of Lemma 14.6.6 by replacing it by $|\bar{r}(\mathbf{x})| \leqslant \rho$ where $\bar{r}(\mathbf{x}) = \|\mathbf{r}(\mathbf{x})\|_2$. Thus the problem is essentially transformed to one in which $t(\bar{r})$ is a polyhedral function on $\mathbb{R}^1$. ∂t^* contains just the single element $u^* = \operatorname{sign}(\bar{r}(\mathbf{x}^*))$ and if $\pi^* > 0$ the third condition in (14.6.9) simply reduces to the requirement that $\bar{r}(\mathbf{x}) = \bar{r}(\mathbf{x}^*)$.

An important application of the conditions is the use of the single constraint $t(\mathbf{r}(\mathbf{x})) \leqslant 0$ to represent a set of inequality constraints $\mathbf{r}(\mathbf{x}) \leqslant \mathbf{0}$ using the polyhedral convex function $t(\mathbf{r}) = \max_i r_i$. The first order conditions (14.6.8) involve the term $\pi^*\mathbf{R}^*\mathbf{u}^*$ and the elements $\pi^* u_i^*$ become the Lagrange multipliers of the individual constraints $r_i(x) \leqslant 0$. If $\mathscr{A}^*$ indexes the active constraints then the definition of ∂t^* shows that $\mathbf{u}^*$ satisfies $\mathbf{u}^* \geqslant \mathbf{0}$, $\sum_{i\in\mathscr{A}^*} u_i^* = 1$ and $u_i^* = 0 \; i \notin \mathscr{A}^*$ which shows that $\pi^* u_i^*$ satisfies the usual complementarity conditions for inequality constraints, and π^* is the L_1 norm of the Lagrange multiplier vector $[\pi^* u_i^*]$. The set X in (14.6.9) involves the conditions $\max_i r_i(\mathbf{x}) \leqslant 0$ and $\pi^*(\mathbf{r}(\mathbf{x}) - \mathbf{r}^*)^{\mathrm{T}}\mathbf{u}^* = 0$ which is equivalent to $\sum_{i\in\mathscr{A}^*_+} \pi^* u_i^* r_i(\mathbf{x}) = 0$ where $\mathscr{A}^*_+$ denotes active constraints with positive multipliers. This is again equivalent to $r_i(\mathbf{x}) = 0 \; i \in \mathscr{A}^*_+$ which can be recognized as the condition for defining $\mathscr{G}^*$ in Section 9.3. Likewise the condition $\max_{\mathbf{u}\in\partial t^*} \mathbf{s}^{\mathrm{T}}\mathbf{R}^*\mathbf{u} \leqslant 0$ in F^* in (14.6.3) is equivalent to $\mathbf{s}^{\mathrm{T}}\mathbf{R}^*\mathbf{e}_i \leqslant 0$ for $i \in \mathscr{A}^*$ and the condition $\pi^* \max_{\mathbf{u}\in\partial^*} \mathbf{s}^{\mathrm{T}}\mathbf{R}^*\mathbf{u} = 0$ is equivalent to $\mathbf{s}^{\mathrm{T}}\mathbf{R}^*\mathbf{e}_i = 0$ for $i \in \mathscr{A}^*_+$ so that these conditions which define G^* can be identified with the equivalent conditions in (9.3.11) in Section 9.3.

Finally a numerical example of the first and second order conditions is given in which the constraint involves the L_1 norm. Consider the problem

$$\begin{array}{ll} \text{minimize} & \|\mathbf{x}\|_\infty \\ \text{subject to} & \|\mathbf{r}(\mathbf{x})\|_1 \leqslant 0.5 \end{array} \qquad (14.6.22)$$

where $\mathbf{r}:\mathbb{R}^3 \to \mathbb{R}^4$ is defined by

$$\begin{aligned} r_1 &= x_1^2 + x_2^2 - 1 \\ r_2 &= x_1 x_2 - 0.5 \\ r_3 &= x_1 + \tfrac{1}{2}(x_2^2 - 1) \\ r_4 &= -x_1 + x_3^2 + 0.54. \end{aligned} \tag{14.6.23}$$

This problem is an example of (14.6.1) with $f(\mathbf{x}) = 0$, $\mathbf{c}(\mathbf{x}) = \mathbf{x}$, $h(\mathbf{c}) = \|\mathbf{c}\|_\infty$ and $t(\mathbf{r}) = \|\mathbf{r}\|_1 - 0.5$. At $\mathbf{x}^* = (0.6, 0.8, 0)^T$, $\mathbf{r}^* = (0, -0.02, 0.42, -0.06)^T$ and the first order conditions are satisfied by $\mathbf{A}^* = \mathbf{I}$, $\pi^* = \frac{5}{7}$ and

$$\boldsymbol{\lambda}^* = \begin{pmatrix} 0 \\ 1 \\ 0 \end{pmatrix} \quad \mathbf{R}^* = \begin{bmatrix} 1.2 & 0.8 & 1 & -1 \\ 1.6 & 0.6 & 0.8 & 0 \\ 0 & 0 & 0 & 0 \end{bmatrix} \quad \mathbf{u}^* = \begin{pmatrix} -1 \\ -1 \\ 1 \\ -1 \end{pmatrix}$$

The set $G^* = \{\mathbf{s}: \|\mathbf{s}\|_2 = 1, s_2 = 0, s_1 \leqslant 0\}$, the dimensions of ∂h^* and ∂t^* are $l^* = 0$ and $t^* = 1$ respectively, and the basis for $\partial t^* - \mathbf{u}^*$ is $\mathbf{F}^* = [1 \quad 0 \quad 0 \quad 0]^T$. Thus the matrix $[\mathbf{R}^*\mathbf{u}^* \vdots \mathbf{R}^*\mathbf{F}^*]$ has rank 2, and since $h(\mathbf{c})$ and $t(\mathbf{r})$ are polyhedral it follows from Lemma 14.6.6 that the regularity assumption $\mathscr{G}^* = G^*$ holds. In fact the set $\mathscr{G}^*$ can be obtained explicitly as follows. The first condition that defines X in (14.6.9) is $\|\mathbf{x}\|_\infty = x_2$ which is locally vacuous. When $\pi^* > 0$ the remaining conditions imply that $t(\mathbf{r}(\mathbf{x})) = 0$, and the last condition $(\mathbf{r} - \mathbf{r}^*)^T\mathbf{u}^* = 0 = \mathbf{r}^T\mathbf{u}^* - 0.5$ implies that $\operatorname{sign} r_i = u_i^*$ and in particular that $r_1 \leqslant 1$. After some manipulation using (14.6.23) the condition $(\mathbf{r} - \mathbf{r}^*)^T\mathbf{u}^*$ yields

$$d_1^2 + \tfrac{1}{2}d_2^2 + 1.4d_2 + d_3^2 = 0$$

where $d_i = x_i - x_i^*$ $i = 1, 2, 3$. Thus $d_2 = O(d_1^2) + O(d_3^2) \leqslant 0$ and feasible directions are seen to have $s_1 = s_3 = 0$. In addition the condition $r_1 \leqslant 0$ imposes $d_1 \leqslant 0$ (for otherwise $d_1 > 0$ and $d_2 = O(d_1^2)$ gives a contradiction) so the extra condition $s_1 \leqslant 0$ is implied and hence $\mathscr{G}^* = G^*$. Because $d_2 < 0$ if d_1 or d_2 is non-zero it can be seen that $\mathbf{x}^*$ is not a minimizing point. However, this result can also be deduced from the Lagrangian condition (14.6.13) because

$$\mathbf{s}^T\nabla^2\mathscr{L}^*\mathbf{s} = -\tfrac{5}{7}\mathbf{s}^T \begin{bmatrix} 2 & 1 & 0 \\ 1 & 1 & 0 \\ 0 & 0 & 2 \end{bmatrix} \mathbf{s} < 0$$

for both $\mathbf{s} = \mathbf{e}_1$ and $\mathbf{s} = \mathbf{e}_3$, that is for all $\mathbf{s} \in G^*$. This shows that $\mathbf{x}^*$ fails to satisfy the second order necessary conditions, and hence $\mathbf{x}^*$ is not a solution since $\mathscr{F}^* = F^*$ and $\mathscr{G}^* = G^*$.

In fact the solution to the problem is given by $x_1^* = x_2^* = (1 + \sqrt{6})/5 \doteq 0.689898$ and $x_3^* = \sqrt{(x_1^* - 0.54)} \doteq 0.387167$. The multipliers $\boldsymbol{\lambda}^* = (0.436700, 0.563299)^T$, $\mathbf{u}^* = (-1, -1, 1, 0)^T$ and $\pi^* = 0.408247$ satisfy the first order conditions. Because $\boldsymbol{\lambda}^*$ and $\mathbf{u}^*$ are in the relative interiors of ∂h^* and ∂t^*, because $l^* + q^* + 1 = 3$, and because the rank condition (14.6.17) is satisfied, it follows that $\mathscr{G}^* = G^* = \varnothing$ and the first order conditions are sufficient to show that $\mathbf{x}^*$ is a solution. Another example which is a variation on this theme is given in Question 14.15.

QUESTIONS FOR CHAPTER 14

14.1. For the nonlinear program (14.1.6) prove that $\mathscr{F}' = F'$ (defined in Section 9.2). Let $\bar{\mathbf{s}} = (\mathbf{s}^{\mathrm{T}}, s_{n+1})^{\mathrm{T}}$ be a feasible direction, and use the linearized constraint equations to show that

$$s_{n+1} \geqslant \max_{i \in \mathscr{A}'} \mathbf{s}^{\mathrm{T}}(\mathbf{g}' + \mathbf{A}'\mathbf{h}_i). \tag{a}$$

Let $\delta^{(k)} \downarrow 0$ be any sequence, and define $\mathbf{x}^{(k)} = \mathbf{x}' + \delta^{(k)}\mathbf{s}$. If equality holds in (a), and if $v^{(k)} = \phi(\mathbf{x}^{(k)})$, show that $\mathbf{x}^{(k)}, v^{(k)}$ gives a feasible directional sequence in (14.1.6) for sufficiently large k. If strict inequality holds in (a), and if $v^{(k)} = v^* + \delta^{(k)} s_{n+1}$, again show that $\mathbf{x}^{(k)}$, $v^{(k)}$ is a feasible directional sequence.

14.2. Prove that $\|\mathbf{c}^+\|$ is a convex function of $\mathbf{c}$ when the norm is monotonic ($|\mathbf{x}| \leqslant |\mathbf{y}| \Rightarrow \|\mathbf{x}\| \leqslant \|\mathbf{y}\|$). Deduce that $\mathbf{c}_\theta^+ \leqslant (1-\theta)\mathbf{c}_0^+ + \theta\mathbf{c}_1^+$ as an intermediate stage.

14.3. Establish the equivalence between each of (14.1.11) to (14.1.15) and the general expression $\partial h(\mathbf{c}) = \mathrm{conv}_{i \in \mathscr{A}} \mathbf{h}_i$. In (14.1.12) let $\lambda_i = \mu_i$, $i \leqslant m$, and μ_{m+1} acts as the slack variable for $\sum_i \lambda_i \leqslant 1$. In (14.1.13) define $\lambda_i = \mu_i - \mu_{m+i}$ or $\mu_i = \max(\lambda_i, 0)$, $\mu_{m+i} = \max(-\lambda_i, 0)$. In (14.1.14) use the fact that the cube $0 \leqslant \lambda_i \leqslant 1$ has extreme points which are all combinations of 1 and 0, and similarly for (14.1.15).

14.4. Establish the equivalence between the subdifferential expression

$$\partial\|\mathbf{c}\| = \{\boldsymbol{\lambda}\colon \|\mathbf{c}+\mathbf{h}\| \geqslant \|\mathbf{c}\| + \boldsymbol{\lambda}^{\mathrm{T}}\mathbf{h} \quad \forall\, \mathbf{h}\} \tag{b}$$

and (14.3.7). Use the generalized Cauchy inequality $\mathbf{a}^{\mathrm{T}}\mathbf{b} \leqslant \|\mathbf{a}\|\,\|\mathbf{b}\|_D$ on $(\mathbf{c}+\mathbf{h})^{\mathrm{T}}\boldsymbol{\lambda}$ to show that $\boldsymbol{\lambda} \in$(14.3.7) implies $\boldsymbol{\lambda} \in$(b). If $\boldsymbol{\lambda} \in$(b), use the triangle inequality to show that $\mathbf{h}^{\mathrm{T}}\boldsymbol{\lambda} \leqslant \|\mathbf{h}\|$ $\forall\, \mathbf{h}$ and (14.3.4) to show $\|\boldsymbol{\lambda}\|_D \leqslant 1$. Hence $\boldsymbol{\lambda}^{\mathrm{T}}\mathbf{c} \leqslant \|\mathbf{c}\|$. Then with $\mathbf{h} = -\mathbf{c}$ in (b) show that $\boldsymbol{\lambda}^{\mathrm{T}}\mathbf{c} \geqslant \|\mathbf{c}\|$ and hence $\boldsymbol{\lambda} \in$(14.3.7).

14.5. If the norm is monotonic (see Question 14.2) establish the equivalence between

$$\partial\|\mathbf{c}^+\| = \{\boldsymbol{\lambda}\colon \|(\mathbf{c}+\mathbf{h})^+\| \geqslant \|\mathbf{c}^+\| + \boldsymbol{\lambda}^{\mathrm{T}}\mathbf{h} \quad \forall\, \mathbf{h}\} \tag{c}$$

and (14.3.8). The proof is similar to that in Question 14.4. In the first part also use the fact that $\boldsymbol{\lambda} \geqslant \mathbf{0}$ implies $(\mathbf{c}+\mathbf{h})^{+\mathrm{T}}\boldsymbol{\lambda} \geqslant (\mathbf{c}+\mathbf{h})^{\mathrm{T}}\boldsymbol{\lambda}$. In the second part also use the monotonic norm property to establish $\|\mathbf{h}^+\| \geqslant \mathbf{h}^{\mathrm{T}}\boldsymbol{\lambda}$ $\forall\, \mathbf{h}$. Then $\mathbf{h} = -\mathbf{e}_i$ yields $\lambda_i \geqslant 0$. Use $\mathbf{h}^+ \leqslant |\mathbf{h}|$ to show $\|\boldsymbol{\lambda}\|_D \leqslant 1$. Then proceed much as in Question 14.4.

14.6. Show that the subdifferential in (14.2.2) is a closed convex set.

14.7. Justify equation (14.2.12) when $\mathbf{0}\notin\partial f'$. Show by straightforward arguments that

$$\min_{\|\mathbf{s}\|_2=1}\max_{\mathbf{g}\in\partial f'}\mathbf{s}^T\mathbf{g}\geqslant\max_{\mathbf{g}\in\partial f'}\min_{\|\mathbf{s}\|_2=1}\mathbf{s}^T\mathbf{g}=-\|\bar{\mathbf{g}}\|_2.$$

Then use the separating hyperplane result (Lemma 14.2.3) to show that equality is achieved when $\mathbf{s}=-\bar{\mathbf{g}}/\|\bar{\mathbf{g}}\|_2$. Show by considering $f=|\mathbf{x}|$ that the result is not true when $\mathbf{0}\in\partial f'$. The result can be generalized to include the case $\mathbf{0}\in\partial f'$ by writing $\|\mathbf{s}\|_2\leqslant 1$ in place of $\|\mathbf{s}\|_2=1$.

14.8. Consider the Freudenstein and Roth equations

$$c_1(\mathbf{x})=x_1-x_2^3+5x_2^2-\ 2x_2-13$$
$$c_2(\mathbf{x})=x_1+x_2^3+\ x_2^2-14x_2-29$$

(see Chapter 6) and consider minimizing $\|\mathbf{c}(\mathbf{x})\|_p$ for $1\leqslant p\leqslant\infty$. A local solution in all cases is $\mathbf{x}^*=(11.4128,-0.8968)^T$. Find the sets $\partial\|\mathbf{c}^*\|_p$ (that is ∂h^*) for $p=1,2,\infty$ and the vector $\boldsymbol{\lambda}^*$ which satisfies Theorem 14.2.1. For $p=1$ the local solution is not unique and any $x_1^*\in[6.4638, 11.4128]$ gives a solution. Find $\partial\|\mathbf{c}^*\|_1$ when $x_1^*=6.4638$.

14.9. Consider minimizing $\|\mathbf{c}\|_1$ when the equation $c_1(\mathbf{x})$ in Question 14.8 is scaled by multiplying the right-hand side by 2. Show that $\mathbf{x}^*=(6.4638, -0.8968)^T$ satsifies the second order sufficient conditions of Theorem 14.2.3 and find $\boldsymbol{\lambda}^*$. Find a trajectory which is feasible in (14.2.19) (see Figure 14.2.3) and hence find the sets $\mathscr{G}^*$ and G^* and verify that they are non-empty and equal.

14.10. Consider minimizing $\|\mathbf{c}\|_\infty$ for the system which results on adding an equation $c_3(\mathbf{x})=\alpha x_1$ to those in Question 14.8. If $\alpha=0.4336$ show that $\mathbf{x}^*=(11.4128,-0.8968)^T$ satisfies the second order sufficient conditions of Theorem 14.2.3 and find $\boldsymbol{\lambda}^*$. Find a trajectory which is feasible in (14.2.19) and hence find the sets $\mathscr{G}^*$ and G^* and verify that they are non-empty and equal.

14.11. For the problems defined in Questions 14.9 and 14.10 derive the dimension l^* and the basis $\mathbf{D}^*$ for the set $\partial h^*-\boldsymbol{\lambda}^*$ in (14.2.30) and show that the rank assumption (14.2.31) is satisfied in both cases.

14.12. Consider the problem defined in (14.4.2). At $\mathbf{x}^\infty$ find the multipliers $\boldsymbol{\lambda}^\infty$ for which $\mathbf{g}^\infty+\mathbf{A}^\infty\boldsymbol{\lambda}^\infty=\mathbf{0}$ and show that $\lambda_2^\infty<0$ which implies that $\phi(\mathbf{x})$ can be reduced by making the inequality $v\geqslant c_2(\mathbf{x})$ in (14.1.6) inactive (see end of Section 14.1). Find the multipliers $\boldsymbol{\lambda}^*$ at $\mathbf{x}^*$ and show that the conditions of Theorem 14.2.1 hold. Apply the method of (14.4.8) from $\mathbf{x}^{(1)}=(0,-4)^T$ and verify that second order convergence is obtained. Why does this happen in the absence of curvature information?

14.13. In problem (12.3.2) let $\mathbf{x}^{(1)}$ lie on the unit circle arbitrarily close to $\mathbf{x}^*$, and let $\boldsymbol{\lambda}^{(1)}=\boldsymbol{\lambda}^*$. Show that there exists a range of values of v with $v<1/\lambda^*$ for which the unit step determined by solving (14.4.11) fails to reduce $\phi(\mathbf{x})$ (the Maratos effect).

14.14. Consider the unconstrained NSO problem: $\min f(\mathbf{x}) + \sum_{i=1}^{m} |r_i(\mathbf{x})|$, where $\mathbf{r}(\mathbf{x}) = \mathbf{A}^{\mathrm{T}}\mathbf{x} + \mathbf{b}$ $(\mathbf{x} \in \mathbb{R}^n)$ and $f(\mathbf{x})$ is convex. By introducing variables

$$r_i^+ = \max(r_i, 0), \qquad r_i^- = \max(-r_i, 0),$$

show that $r_i^+ - r_i \geq 0$, $r_i^- + r_i \geq 0$, and $r_i^+ + r_i^- = |r_i|$. Hence show that the unconstrained problem can be restated as

$$\begin{aligned} \underset{\mathbf{x},\mathbf{r}^+,\mathbf{r}^-}{\text{minimize}} \quad & f(\mathbf{x}) + \mathbf{e}^{\mathrm{T}}(\mathbf{r}^+ + \mathbf{r}^-) \\ \text{subject to} \quad & \mathbf{r}^+ - \mathbf{A}^{\mathrm{T}}\mathbf{x} - \mathbf{b} \geq \mathbf{0} \\ & \mathbf{r}^- + \mathbf{A}^{\mathrm{T}}\mathbf{x} + \mathbf{b} \geq \mathbf{0}, \qquad \mathbf{r}^+ \geq \mathbf{0}, \mathbf{r}^- \geq \mathbf{0}, \end{aligned}$$

where $\mathbf{e} = (1, 1, \ldots, 1)^{\mathrm{T}}$. Show that this is a convex programming problem. Write down the dual of this problem, denoting the multipliers of the constraints by $\boldsymbol{\lambda}^+, \boldsymbol{\lambda}^-, \boldsymbol{\mu}^+, \boldsymbol{\mu}^-$ respectively. By eliminating $\boldsymbol{\mu}^+$ and $\boldsymbol{\mu}^-$, and writing $\boldsymbol{\lambda}^+ - \boldsymbol{\lambda}^- = \boldsymbol{\lambda}$, show that the dual can be restated more simply as

$$\begin{aligned} \underset{\mathbf{x},\boldsymbol{\lambda}}{\text{maximize}} \quad & f(\mathbf{x}) + \boldsymbol{\lambda}^{\mathrm{T}}(\mathbf{A}^{\mathrm{T}}\mathbf{x} + \mathbf{b}) \\ \text{subject to} \quad & \mathbf{g}(\mathbf{x}) + \mathbf{A}\boldsymbol{\lambda} = \mathbf{0}, \qquad -\mathbf{e} \leq \boldsymbol{\lambda} \leq \mathbf{e}, \end{aligned}$$

where $\mathbf{g} = \nabla_x f$ (see also Watson, 1978).

14.15. Show that the point $\mathbf{x}^* = (0.6, 0.8, 0)$ satisfies first but not second order necessary conditions for the problem

$$\begin{aligned} \text{minimize} \quad & \|\mathbf{r}(\mathbf{x})\|_1 \\ \text{subject to} \quad & \|\mathbf{x}\|_\infty \leq 0.8 \end{aligned}$$

where $\mathbf{r}(\mathbf{x})$ is defined by (14.6.23). Find the solution of the problem.

14.16. Apropos of Theorem 14.4.1, construct a problem satisfying the conditions of the theorem such that the SNQP subproblem as defined at $(\mathbf{x}^*, \boldsymbol{\lambda}^*)$ does not have $\boldsymbol{\delta} = \mathbf{0}$ and $\boldsymbol{\lambda}^+ = \boldsymbol{\lambda}^*$ as its global minimizer. (Hint: choose $\mathbf{W}^*$ to be indefinite.)

References

Abadie, J. and Carpentier, J. (1969). Generalization of the Wolfe reduced gradient method to the case of nonlinear constraints, in *Optimization* (Ed. R. Fletcher), Academic Press, London.

Akaike, H. (1959). On a successive transformation of probability distribution and its application to the analysis of the optimum gradient method, *Ann. Inst. Stat. Math. Tokyo*, **11**, 1–16.

Al-Baali, M. (1985). Descent property and global convergence of the Fletcher–Reeves method with inexact line search, *IMA J. Num. Anal.*, **5**, 121–124.

Al-Baali, M. and Fletcher, R. (1985). Variational methods for nonlinear least squares, *J. Oper. Res. Soc.*, **36**, 405–421.

Al-Baali, M. and Fletcher, R. (1986). An efficient line search for nonlinear least squares, *J. Opt. Theo. Applns.*, **48**, 359–378.

Andreassen, D. O. and Watson, G. (1976). Nonlinear Chebyshev approximation subject to constraints, *J. Approx. Theo.*, **18**, 241–250.

Apostol, T. M. (1957). *Mathematical Analysis*, Addison-Wesley, Reading, Mass.

Appelgren, L. (1971). Integer programming methods for a vessel scheduling problem, *Transportation Science*, **5**, 64–78.

Balinski, M. L. and Gomory, R. E. (1963). A mutual primal-dual simplex method, in *Recent Advances in Mathematical Programming* (Eds R. L. Graves and P. Wolfe), McGraw-Hill, New York.

Balinski, M. L. and Wolfe, P. (Eds) (1975). *Nondifferentiable Optimization*, Mathematical Programming Study 3, North-Holland, Amsterdam.

Bandler, J. W. and Charalambous, C. (1972). Practical least p-th optimization of networks, *IEEE Trans. Microwave Theo. Tech., (1972 Symposium Issue)*, **20**, 834–840.

Barnes, J. G. P. (1965). An algorithm for solving nonlinear equations based on the secant method, *Computer J.*, **8**, 66–72.

Barrodale, I. (1970). On computing best L_1 approximations, in *Approximation Theory* (Ed. A. Talbot), Academic Press, London.

Barrodale, I. and Roberts, F. D. K. (1973). An improved algorithm for discrete l_1 linear approximation, *SIAM J. Num. Anal.*, **10**, 839–848.

Bartels, R. H. (1971). A stabilization of the simplex method, *Numer. Math.*, **16**, 414–434.

Beale, E. M. L. (1959). On quadratic programming, *Naval Res. Log. Quart.*, **6**, 227–244.

Beale, E. M. L. (1967). Numerical methods, in *Nonlinear Programming* (Ed. J. Abadie), North-Holland, Amsterdam.

Beale, E. M. L. (1968). *Mathematical Programming in Practice*, Pitman, London.

Beale, E. M. L. (1970). Advanced algorithmic features for general mathematical programming systems, in *Integer and Nonlinear Programming* (Ed. J. Abadie), North-Holland, Amsterdam.

Beale, E. M. L. (1972). A derivation of conjugate gradients, in *Numerical Methods for Nonlinear Optimization* (Ed. F. A. Lootsma), Academic Press, London.

Beale, E. M. L. (1978). Integer programming, in *The State of the Art in Numerical Analysis* (Ed. D. A. H. Jacobs), Academic Press, London.

Bennett, J. M. (1965). Triangular factors of modified matrices, *Numer. Math.*, **7**, 217–221.

Benveniste, R. (1979). A quadratic programming algorithm using conjugate search directions, *Math. Prog.*, **16**, 63–80.

Biggs, M. C. (1973). A note on minimization algorithms which make use of non-quadratic properties of the objective function, *J. Inst Math. Applns.*, **12**, 337–338.

Biggs, M. C. (1975). Constrained minimization using recursive quadratic programming: some alternative subproblem formulations, in *Towards Global Optimization* (Eds L. C. W. Dixon and G. P. Szego), North-Holland, Amsterdam.

Biggs, M. C. (1978). On the convergence of some constrained minimization algorithms based on recursive quadratic programming, *J. Inst. Maths. Applns.*, **21**, 67–81.

Björck, A. (1985). Stability analysis of the method of semi-normal-equations for linear least squares problems, Linköping Univ. Report LiTH-MAT-R-1985–08.

Boggs, P. T. and Tolle, J. W. (1984). A family of descent functions for constrained optimization, *SIAM J. Num. Anal.*, **21**, 1146–1161.

Boggs, P. T., Tolle, J. W. and Wang, P. (1982). On the local convergence of quasi-Newton methods for constrained optimization, *SIAM J. Control Optim.*, **20**, 161–171.

Bradley, J. and Clyne, H. M. (1976). Applications of geometric programming to building design problems, in *Optimization in Action* (Ed. L. C. W. Dixon), Academic Press, London.

Branin, F. H. and Hoo, S. K. (1972). A method for finding multiple extrema of a function of n variables, in *Numerical Methods for Nonlinear Optimization* (Ed. F. A. Lootsma), Academic Press, London.

Brent, R. P. (1973a). *Algorithms for Minimization without Derivatives*, Prentice-Hall Inc., Englewood Cliffs, N. J.

Brent, R. P. (1973b). Some efficient algorithms for solving systems of nonlinear equations, *SIAM J. Num. Anal.*, **10**, 327–344.

Breu, R. and Burdet, C.-A. (1974). Branch and bound experiments in zero-one programming, in *Approaches to Integer Programming* (Ed. M. L. Balinski), Mathematical Programming Study 2, North-Holland, Amsterdam.

Brodlie, K. W. (1975). A new direction set method for unconstrained minimization without evaluating derivatives, *J. Inst. Maths. Applns.*, **15**, 385–396.

Brodlie, K. W. (1977). Unconstrained minimization, in *The State of the Art in Numerical Analysis* (Ed. D. A. H. Jacobs), Academic Press, London.

Brodlie, K. W., Gourlay, A. R. and Greenstadt, J. L. (1973). Rank-one and rank-two corrections to positive definite matrices expressed in product form, *J. Inst. Maths. Applns.*, **11**, 73–82.

Brown, K. M. and Dennis, J. E. (1971). A new algorithm for nonlinear least squares curve fitting, in *Mathematical Software* (Ed. J. R. Rice), Academic Press, New York.

Broyden, C. G. (1965). A class of methods for solving nonlinear simultaneous equations, *Maths. Comp.*, **19**, 577–593.

Broyden, C. G. (1967). Quasi-Newton methods and their application to function minimization, *Maths. Comp.*, **21**, 368–381.

Broyden, C. G. (1970). The convergence of a class of double rank minimization algorithms, parts I and II, *J. Inst. Maths. Applns.*, **6**, 76–90 and 222–231.

Broyden, C. G. (1975). *Basic Matrices*, Macmillan, London.

Broyden, C. G., Dennis, J. E. and Moré, J. J. (1973). On the local and superlinear convergence of quasi-Newton methods, *J. Inst. Maths. Applns.*, **12**, 223–245.

Buckley, A. (1975). Constrained minimization using Powell's conjugacy approach, AERE Harwell report CSS 22.

Bunch, J. R. and Parlett, B. N. (1971). Direct methods for solving symmetric indefinite systems of linear equations, *SIAM J. Num. Anal.*, **8**, 639–655.

Buys, J. D. (1972). Dual algorithms for constrained optimization problems, Ph.D. thesis, Univ. of Leiden.

Byrd, R. H. (1984). On the convergence of constrained optimization methods with accurate Hessian information on a subspace, Univ. of Colorado at Boulder, Dept. of Comp. Sci. Report CU-CS-270-84.

Byrd, R. H. and Shultz, G. A. (1982). A practical class of globally convergent active set strategies for linearly constrained optimization, Univ. of Colorado at Boulder, Dept. of Comp. Sci. Report CU-CS-238-82.

Carroll, C. W. (1961). The created response surface technique for optimizing nonlinear restrained systems, *Operations Res.*, **9**, 169–84.

Chamberlain, R. M. (1979). Some examples of cycling in variable metric methods for constrained minimization, *Math. Prog.*, **16**, 378–383.

Chamberlain, R. M., Lemarechal, C., Pedersen, H. C. and Powell, M. J. D. (1982). The watchdog technique for forcing convergence in algorithms for constrained optimization, in *Algorithms for Constrained Minimization of Smooth Nonlinear Functions* (Eds A. G. Buckley and J.-L. Goffin), Mathematical Programming Study 16, North-Holland, Amsterdam.

Charalambous, C. (1977). Nonlinear least p-th optimization and nonlinear programming, *Math. Prog.*, **12**, 195–225.

Charalambous, C. (1979). Acceleration of the least p-th algorithm for minimax optimization with engineering applications, *Math. Prog.*, **17**, 270–297.

Charalambous, C. and Conn, A. R. (1978). An efficient method to solve the minimax problem directly, *SIAM J. Num. Anal.*, **15**, 162–187.

Charnes, A. (1952). Optimality and degeneracy in linear programming, *Econometrica*, **20**, 160–170.

Clarke, F. H. (1975). Generalized gradients and applications, *Trans. Amer. Math. Soc.*, **205**, 247–262.

Coleman, T. F. and Conn, A. R. (1982a). Nonlinear programming via an exact penalty function: asymptotic analysis, *Math. Prog.*, **24**, 123–136.

Coleman, T. F. and Conn, A. R. (1982b). Nonlinear programming via an exact penalty function: global analysis, *Math. Prog.*, **24**, 137–161.

Colville, A. R. (1968). A comparative study on nonlinear programming codes, IBM NY Scientific Center Report 320–2949.

Concus, P. (1967). Numerical solution of the minimal surface equation, *Maths. Comp.*, **21**, 340–350.

Conn, A. R. (1973). Constrained optimization using a nondifferentiable penalty function, *SIAM J. Num. Anal.*, **13**, 145–154.

Conn, A. R. (1979). An efficient second order method to solve the (constrained) minimax problem, Univ. of Waterloo Dept. of Combinatorics and Optimization Res. Report CORR-79-5.

Conn, A. R. and Sinclair, J. W. (1975). Quadratic programming via a non-differentiable penalty function, Univ. of Waterloo Dept. of Combinatorics and Optimization Report CORR 75/15.

Coope, I. D. (1976). Conjugate direction algorithms for unconstrained optimization, Ph.D. thesis, Univ. of Leeds Dept. of Mathematics.

Coope, I. D. and Fletcher, R. (1980). Some numerical experience with a globally convergent algorithm for nonlinearly constrained optimization, *J. Opt. Theo. Applns.*, **32**, 1–16.

Cottle, R. W. and Dantzig, G. B. (1968). Complementary pivot theory of mathematical programming, *J. Linear Algebra Applns.*, **1**, 103–125.

Courant, R. (1943). Variational methods for the solution of problems of equilibrium and vibration, *Bull. Amer. Math. Soc.*, **49**, 1–23.

Curtis, A. R., Powell, M. J. D. and Reid, J. K. (1974). On the estimation of sparse Jacobian matrices, *J. Inst. Maths. Applns.*, **13**, 117–120.

Curry, H. (1944). The method of steepest descent for nonlinear minimization problems, *Quart. Appl. Math.*, **2**, 258–261.

Dantzig, G. B. (1963). *Linear Programming and Extensions*, Princeton University Press, Princeton, N. J.

Dantzig, G. B., Orden, A. and Wolfe, P. (1955). The generalized simplex method for minimizing a linear form under linear inequality restraints, *Pacific J. Maths.*, **5**, 183–195.

Dantzig, G. B. and Wolfe, P. (1960). A decomposition principle for linear programs, *Operations Res.*, **8**, 101–111.

Davidon, W. C. (1959). Variable metric method for minimization, AEC Res. and Dev. Report ANL-5990 (revised).

Davidon, W. C. (1968). Variance algorithms for minimization, *Computer J.* **10**, 406–410.

Davidon, W. C. (1975). Optimally conditioned optimization algorithms without line searches, *Math. Prog.*, **9**, 1–30.

Davidon, W. C. (1980). Conic approximations and collinear scaling for optimizers, *SIAM J. Num. Anal.*, **17**, 268–281.

Dembo, R. S. (1976). A set of geometric programming test problems and their solutions, *Math. Prog.*, **10**, 192–213.

Dembo, R. S. (1979). Second order algorithms for the posynomial geometric programming dual, Part I: Analysis, *Math. Prog.*, **17**, 156–175.

Dembo, R. S. and Avriel, M. (1978). Optimal design of a membrane separation process using geometric programming, *Math. Prog.*, **15**, 12–25.

Dembo, R. S. and Klincewicz, J. G. (1981). A scaled reduced gradient algorithm for network flow problems with convex separable costs, in *Network Models and Applications* (Ed. D. Klingman and J. M. Mulvey), Mathematical Programming Study 15, North-Holland Amsterdam.

Demyanov, V. F. and Malozemov, V. N. (1971). The theory of nonlinear minimax problems, *Uspekhi Matematcheski Nauk*, **26**, 53–104.

Dennis, J. E. (1977). Nonlinear least squares and equations, in *The State of the Art in Numerical Analysis* (Ed. D. A. H. Jacobs), Academic Press, London.

Dennis, J. E., Gay, D. M. and Welsch, R. E. (1981). An adaptive nonlinear least-squares algorithm, *ACM Trans. Math. Software*, **7**, 348–368

Di Pillo, G. and Grippo, L. (1979). A new class of augmented Lagrangians in nonlinear programming, *SIAM J. Control Optim.*, **17**, 618–628.

Dixon, L. C. W. (1972). Quasi-Newton algorithms generate identical points, *Math. Prog.*, **2**, 383–387.

Dixon, L. C. W. (1973). Conjugate directions without line searches, *J. Inst. Maths. Applns.*, **11**, 317–328.

Dixon, L. C. W. (ed.) (1976). *Optimization in Action*, Academic Press, London.

Dixon, L. C. W. and Szego, G. P. (eds) (1975). *Towards Global Optimization*, North-Holland, Amsterdam.

Duffin, R. J., Peterson, E. L. and Zener, C. (1967). *Geometric Programming—Theory and Application*, John Wiley, New York.

El-Attar, R. A., Vidyasagar, M. and Dutta, S. R. K. (1979). An algorithm for l_1-norm minimization with application to nonlinear l_1 approximation, *SIAM J. Num. Anal.*, **16**, 70–86.

Eriksson, J. (1980). A note on solution of large sparse maximum entropy problems with linear equality constraints, *Math. Prog.*, **18**, 146–154.

Fiacco, A. V. and McCormick, G. P. (1968). *Nonlinear Programming*, John Wiley, New York.

Fisher, M. L., Northup, W. D. and Shapiro, J. F. (1975). Using duality to solve discrete optimization problems: theory and computational experience, in *Nondifferentiable Optimization* (Eds M. L. Balinski and P. Wolfe), Mathematical Programming Study 3, North-Holland, Amsterdam.

Fletcher, R. (1965). Function minimization without evaluating derivatives—a reivew, *Computer J.*, **8**, 33–41.

Fletcher, R. (1969). A technique for orthogonalization, *J. Inst. Maths. Applns.*, **5**, 162–116.

Fletcher, R. (1970a). A new approach to variable metric algorithms, *Computer J.*, **13**, 317–322.

Fletcher, R. (1970b). The calculation of feasible points for linearly constrained optimization problems, AERE Harwell Report AERE-R6354.

Fletcher, R. (1971a). A modified Marquardt subroutine for nonlinear least squares, AERE Harwell Report AERE-R6799.

Fletcher, R. (1971b). A general quadratic programming algorithm, *J. Inst. Maths. Applns.*, **7**, 76–91.

Fletcher, R. (1972a). Conjugate direction methods, in *Numerical Methods for Unconstrained Optimization* (Ed. W. Murray), Academic Press, London.

Fletcher, R. (1972b). An algorithm for solving linearly constrained optimization problems, *Math. Prog.*, **2**, 133–165.

Fletcher, R. (1972c). Minimizing general functions subject to linear constraints, in *Numerical Methods for Nonlinear Optimization* (Ed. F. A. Lootsma), Academic Press, London.

Fletcher, R. (1973). An exact penalty function for nonlinear programming with inequalities, *Math. Prog.*, **5**, 129–150.

Fletcher, R. (1975). An ideal penalty function for constrained optimization, *J. Inst. Maths. Applns.*, **15**, 319–342, and in *Nonlinear Programming 2* (Eds O. L. Mangasarian, R. R. Meyer and S. M. Robinson), Academic Press, London (1975).

Fletcher, R. (1978). On Newton's method for minimization, in *Proc. IX Int. Symp. on Math. Programming* (Ed. A. Prekopa), Akademiai Kiado, Budapest.

Fletcher, R. (1981). Numerical experiments with an L_1 exact penalty function method, in *Nonlinear Programming 4* (Eds O. L. Mangasarian, R. R. Meyer and S. M. Robinson), Academic Press, New York.

Fletcher, R. (1982a). A model algorithm for composite nondifferentiable optimization problems, in *Nondifferential and Variational Techniques in Optimization* (Eds D. C. Sorensen and R. J.-B. Wets), Mathematical Programming Study 17, North-Holland, Amsterdam.

Fletcher, R. (1982b). Second order corrections for nondifferentiable optimization, in *Numerical Analysis, Dundee 1981* (Ed. G. A. Watson), Lecture Notes in Mathematics 912, Springer-Verlag, Berlin.

Fletcher, R. (1985a). Degeneracy in the presence of round-off errors, Univ. of Dundee Dept. of Math. Sci. Report NA/79.

Fletcher, R. (1985b). An l_1 penalty method for nonlinear constraints, in *Numerical Optimization 1984* (Eds P. T. Boggs, R. H. Byrd and R. B. Schnabel), SIAM Publications, Philadelphia.

Fletcher, R. and Freeman, T. L. (1977). A modified Newton method for minimization, *J. Opt. Theo. Applns.*, **23**, 357–372.

Fletcher, R. and Jackson, M. P. (1974). Minimization of a quadratic function of many variables subject only to lower and upper bounds, *J. Inst. Maths. Applns.*, **14**, 159–174.

Fletcher, R. and Matthews, S. P. J. (1984). Stable modification of explicit LU factors for simplex updates, *Math. Prog.*, **30**, 267–284.

Fletcher, R. and Matthews, S. P. J. (1985). A stable algorithm for updating triangular factors under a rank one change, *Maths. Comp.*, **45**, 471–485.

Fletcher, R. and McCann, A. P. (1969). Acceleration techniques for nonlinear programming, in *Optimization* (Ed. R. Fletcher), Academic Press, London.

Fletcher, R. and Powell, M. J. D. (1963). A rapidly convergent descent method for minimization, *Computer. J.*, **6**, 163–168.

Fletcher, R. and Powell, M. J. D. (1974). On the modification of LDL^T factorizations, *Maths. Comp.*, **29**, 1067–1087.

Fletcher, R. and Reeves, C. M. (1964). Function minimization by conjugate gradients, *Computer J.*, **7**, 149–154.

Fletcher, R. and Sainz de la Maza, E. (1987). Nonlinear programming and nonsmooth optimization by successive linear programming, Univ. of Dundee, Dept. of Math. Sci. Report NA/100.

Fletcher, R. and Sinclair, J. W. (1981). Degenerate values for Broyden methods, *J. Opt. Theo. Applns.*, **33**, 311–324.

Fletcher, R. and Watson, G. A. (1980). First and second order conditions for a class of nondifferentiable optimization problems, *Math. Prog.*, **18**, 291–307; abridged from a Univ. of Dundee Dept. of Mathematics Report NA/28 (1978).

Fletcher, R. and Xu, C. (1985). Hybrid methods for nonlinear least squares, Univ. of Dundee Dept. of Math. Sci. Report NA/92, (to appear in *IMA J. Num. Anal.*).

Forrest, J. J. H. and Tomlin, J. A. (1972). Updated triangular factors of the basis to maintain sparsity in the product form simplex method, *Math. Prog.*, **2**, 263–278.

Frisch, K. R. (1955). The logarithmic potential method of convex programming, Oslo Univ. Inst. of Economics Memorandum, May 1955.

Gacs, P. and Lovasz, L. (1979). Khachian's algorithm for linear programming, Stanford Univ. Dept. of Comp. Sci. Report CS 750.

Gay, D. M. (1979). Some convergence properties of Broyden's method, *SIAM J. Num. Anal.*, **16**, 623–630.

Gay, D. M. (1985). A variant of Karmarkar's linear programming algorithm for problems in standard form, AT&T Bell Labs, Num. Anal. Manuscript 85–10.

Gentleman, W. M. (1973). Least squares computations by Givens' transformations without square roots, *J. Inst. Maths. Applns.*, **12**, 329–336.

Gerber, R. R. and Luk, F. T. (1980). A generalized Broyden's method for solving simultaneous linear equations, Cornell Univ. Dept. of Comp. Sci. Report TR-80-438.

Gill, P. E. and Murray, W. (1972). Quasi-Newton methods for unconstrained optimization, *J. Inst. Maths. Applns.*, **9**, 91–108.

Gill, P. E. and Murray, W. (1973). A numerically stable form of the simplex algorithm, *J. Linear Algebra Applns.*, **7**, 99–138.

Gill, P. E. and Murray, W. (1974a). Newton type methods for linearly constrained optimization, in *Numerical Methods for Constrained Optimization* (Eds P. E. Gill and W. Murray), Academic Press, London.

Gill, P. E. and Murray, W. (1974b). Methods for large-scale linearly constrained problems, in *Numerical Methods for Constrained Optimization* (Eds P. E. Gill and W. Murray), Academic Press, London.

Gill, P. E. and Murray, W. (1974c). Quasi-Newton methods for linearly constrained optimization, in *Numerical Methods for Constrained Optimization* (Eds P. E. Gill and W. Murray), Academic Press, London.

Gill, P. E. and Murray, W. (1976a). Nonlinear least squares and nonlinearly constrained optimization, in *Numerical Analysis, Dundee 1975* (Ed. G. A. Watson), Lecture Notes in Mathematics 506, Springer-Verlag, Berlin.

Gill, P. E. and Murray, W. (1976b). Minimization of a nonlinear function subject to bounds on the variables, NPL Report NAC 72.

Gill, P. E. and Murray, W. (1978a). Modification of matrix factorizations after a rank-one change, in *The State of the Art in Numerical Analysis* (Ed. D. A. H. Jacobs), Academic Press, London.

Gill, P. E. and Murray, W. (1978b). Numerically stable methods for quadratic programming, *Math. Prog.*, **14**, 349–372.

Gill, P. E., Murray, W. and Picken, S. M. (1972). The implementation of two modified Newton algorithms for unconstrained optimization, NPL Report NAC24.

Gill, P. E., Murray, W. and Pitfield, R. A. (1972). The implementation of two revised quasi-Newton algorithms for unconstrained optimization, NPL Report NAC11.

Gill, P. E., Gould, N. I. M., Murray, W., Saunders, M. A. and Wright, M. H. (1984a). Weighted Gram–Schmidt method for convex quadratic programming, *Math. Prog.*, **30**, 176–195.

Gill, P. E., Murray, W., Saunders, M. A. and Wright, M. H. (1984b). Sparse matrix methods in optimization, *SIAM J. Sci. Stat. Comp.*, **5**, 562–589.

Gill, P. E., Murray, W., Saunders, M. A., Tomlin, J. A. and Wright, M. H. (1985). On projected Newton barrier methods for linear programming and an equivalence to Karmarkar's projective method, Stanford Univ. Systems Optimization Lab. Report SOL 85–11.

Gill, P. E., Murray, W., Saunders, M. A. and Wright, M. H. (1986). Some theoretical properties of an augmented Lagrangian merit function, Stanford Univ. Systems Optimization Lab. Report SOL 86–6.

Goldfarb, D. (1969). Extension of Davidon's variable metric method to maximization under linear inequality and equality constraints, *SIAM J. Appl. Math.*, **17**, 739–764.

Goldfarb, D. (1970). A family of variable metric methods derived by variational means, *Maths. Comp.*, **24**, 23–26.

Goldfarb, D. (1972). Extensions of Newton's method and simplex methods for solving quadratic programs, in *Numerical Methods for Nonlinear Optimization* (Ed. F. A. Lootsma), Academic Press, London.

Goldfarb, D. (1980). Curvilinear path steplength algorithms for minimization which use directions of negative curvature, *Math. Prog.*, **18**, 31–40.

Goldfarb, D. and Idnani, A. (1981). Dual and primal-dual methods for solving strictly convex quadratic programs, in *IIMAS Workshop in Numerical Analysis (1981) Mexico* (Ed. J. P. Hennart), Lecture Notes in Mathematics, Springer-Verlag, Berlin.

Goldfarb, D. and Idnani, A. (1983). A numerically stable dual method for solving strictly convex quadratic programs, *Math. Prog.*, **27**, 1–33.

Goldfarb, D. and Reid, J. K. (1977). A practicable steepest-edge algorithm, *Math. Prog.*, **12**, 361–371.

Goldfeld, S. M., Quandt, R. E. and Trotter, H. F. (1966). Maximisation by quadratic hill-climbing, *Econometrica*, **34**, 541–551.

Goldstein, A. A. (1965). On steepest descent, *SIAM J. Control*, **3**, 147–151.

Goldstein, A. A. and Price, J. F. (1967). An effective algorithm for minimization, *Numer. Math.*, **10**, 184–189.

Grandinetti, L. (1979). Factorization versus nonfactorization in quasi-Newtonian algorithms for differentiable optimization, in *Methods for Operations Research*, Hain Verlay.

Graves, G. W. and Brown, G. G. (1979). Computational implications of degeneracy in large scale mathematical programming, X Symposium on Mathematical Programming, Montreal, August 1979.

Greenstadt, J. L. (1970). Variations of variable metric methods, *Maths. Comp.*, **24**, 1–22.

Grigoriadis, M. (1982). Minimum-cost network flows, Part I: an implementation of the network simplex method, Rutgers Univ. Lab. for Comp. Sci. Report LCSR-TR-37.

Griewank, A. and Toint, Ph. L. (1984). Numerical experiments with partially separable optimization problems, in *Numerical Analysis, Dundee 1983* (Ed. D. F. Griffiths), Lecture Notes in Mathematics 1066, Springer-Verlag, Berlin.

Hadley, G. (1961). *Linear Algebra*, Addison-Wesley, Reading, Mass.

Hadley, G. (1962). *Linear Programming*, Addison-Wesley, Reading, Mass.

Hald, J. and Madsen, K. (1981). Combined LP and quasi-Newton methods for minimax optimization, *Math. Prog.*, **20**, 49–62.

Hald, J. and Madsen, K. (1985). Combined LP and quasi-Newton methods for nonlinear l_1 optimization, *SIAM J. Num. Anal.*, **22**, 68–80.

Han, S. P. (1976). Superlinearly convergent variable metric algorithms for general nonlinear programming problems, *Math. Prog.*, **11**, 263–282.

Han, S. P. (1977). A globally convergent method for nonlinear programming, *J. Opt. Theo. Applns.*, **22**, 297–309.

Han, S. P. and Mangasarian, O. L. (1979). Exact penalty functions in nonlinear programming, *Math. Prog.*, **17**, 251–269.

Hardy, G. H. (1960). *A Course of Pure Mathematics* (10th edn), Cambridge Univ. Press, Cambridge, England.

Hebden, M. D. (1973). An algorithm for minimization using exact second derivatives, AERE Harwell Report TP515.

Hestenes, M. R. (1969). Multiplier and gradient methods, *J. Opt. Theo. Applns.*, **4**, 303–320, and in *Computing Methods in Optimization Problems, 2* (Eds L. A. Zadeh, L. W. Neustadt and A. V. Balakrishnan), Academic Press, New York (1969).

Hestenes, M. R. and Stiefel, E. (1952). Methods of conjugate gradients for solving linear systems, *J. Res. N.B.S.*, **49**, 409–436.

Holt, J. N. and Fletcher, R. (1979). An algorithm for constrained nonlinear least squares, *J. Inst. Maths. Applns.*, **23**, 449–464.

Hoshino, S. (1972). A formulation of variable metric methods, *J. Inst. Maths. Applns.*, **10**, 394–403.

Huang, H. Y. (1970). Unified approach to quadratically convergent algorithms for function minimization, *J. Opt. Theo. Applns.*, **5**, 405–423.

Jensen, P. A. and Barnes, J. W. (1980). *Network Flow Programming*, John Wiley, New York.

Kamesam, P. V. and Meyer, R. R. (1984). Multipoint methods for separable nonlinear networks, in *Mathematical Programming at Oberwolfach* (Eds B. Korte and K. Ritter), Mathematical Programming Study 22, North-Holland, Amsterdam.

Karmarkar, N. (1984). A new polynomial-time algorithm for linear programming, *Combinatorics*, **4**, 373–395.

Kennington, J. L. and Helgason, R. V. (1980). *Algorithms for Network Programming*, John Wiley, New York.

Kershaw, D. S. (1978). The incomplete Cholesky-conjugate gradient method for the iterative solution of systems of linear equations, *J. Comp. Phys.*, **26**, 43–65.

Khachiyan, L. G. (1979). A polynomial algorithm in linear programming, *Doklady Akad. Nauk. USSR*, **244**, 1093–1096, translated as *Soviet Mathematics Doklady*, **20**, 191–194.

Kiefer, J. (1957). Optimal sequential search and approximation methods under minimum regularity conditions, *SIAM J. Appl. Math.*, **5**, 105–136.

Klee, V. and Minty, G. J. (1971). How good is the simplex algorithm?, in *Inequalities III* (Ed. O. Shisha), Academic Press.

Kuhn, H. W. and Tucker, A. W. (1951). Nonlinear programming, in *Proceedings of the Second Berkeley Symposium on Mathematical Statistics and Probability* (Ed. J. Neyman), University of California Press.

Lancaster, P. (1969). *Theory of Matrices*, Academic Press, New York.

Lasdon, L. S. (1985). Nonlinear programming algorithms—applications, software and comparisons, in *Numerical Optimization 1984* (Eds P. T. Boggs, R. H. Byrd and R. B. Schnabel), SIAM Publications, Philadelphia.

Lemarechal, C. (1978). Bundle methods in nonsmooth optimization, in *Nonsmooth Optimization* (Eds C. Lemarechal and R. Mifflin), IIASA Proceedings 3, Pergamon, Oxford.

Lemarechal, C. and Mifflin, R. (eds) (1978). *Nonsmooth Optimization*, IIASA Proceedings 3, Pergamon, Oxford.

Lemke, C. E. (1965). Bimatrix equilibrium points and mathematical programming, *Management Sci.*, **11**, 681–689.

Levenberg, K. (1944). A method for the solution of certain nonlinear problems in least squares, *Quart. Appl. Math.*, **2**, 164–168.

Lustig, I. J. (1985). A practical approach to Karmarkar's algorithm, Stanford Univ. Systems Optimization Lab. Report SOL 85-5.

Madsen, K. (1975). An algorithm for minimax solution of overdetermined systems of nonlinear equations, *J. Inst. Maths. Applns.*, **16**, 321–328.

Maratos, N. (1978). Exact penalty function algorithms for finite dimensional and control optimization problems, Ph.D. thesis, Univ. of London.

Marquardt, D. W. (1963). An algorithm for least squares estimation of nonlinear parameters, *SIAM J.*, **11**, 431–441.

Marsten, R. E. (1975). The use of the boxstep method in discrete optimization, in *Nondifferentiable Optimization* (Eds M. L. Balinski and P. Wolfe), Mathematical Programming Study 3, North-Holland, Amsterdam.

Mayne, D. Q. (1980). On the use of exact penalty functions to determine step length in optimization algorithms, in *Numerical Analysis. Dundee 1979* (Ed. G. A. Watson), Lecture Notes in Mathematics 773, Springer-Verlag, Berlin.

McCormick, G. P. (1977). A modification of Armijo's step-size rule for negative curvature, *Math. Prog.*, **13**, 111–115.

McCormick. G. P. and Pearson, J. D. (1969). Variable metric methods and unconstrained optimization, in *Optimization* (Ed. R. Fletcher), Academic Press, London.

McLean, R. A. and Watson, G. A. (1980). Numerical methods for nonlinear discrete L_1 approximation problems, in *Numerical Methods of Approximation Theory* (Eds L. Collatz, G. Meinardus and H. Warner), ISNM 52, Birkhauser-Verlag, Basle.

Moré, J. J. (1978). The Levenberg–Marquardt algorithm: implementation and theory, in *Numerical Analysis*, Dundee 1977 (Ed. G. A. Watson), Lecture Notes in Mathematics 630, Springer-Verlag, Berlin.

Moré, J. J. and Sorensen, D. C. (1982). Newton's method, Argonne Nat. Lab. Report ANL-82-8.

Morrison, D. D. (1968). Optimization by least squares, *SIAM J. Num. Anal.*, **5**, 83–88.

Murray, W. (1969). An algorithm for constrained minimization, in *Optimization* (Ed. R. Fletcher), Academic Press, London.

Murray, W. (1971). An algorithm for finding a local minimum of an indefinite quadratic program, NPL Report NAC 1.

Murray, W. (1972). Second derivative methods, in *Numerical Methods for Unconstrained Optimization* (Ed. W. Murray), Academic Press, London.

Murtagh, B. A. and Sargent, R. W. H. (1969). A constrained minimization method with quadratic convergence, in *Optimization* (Ed. R. Fletcher), Academic Press, London.

Nelder, J. A. and Mead, R. (1965). A simplex method for function minimization, *Computer J.*, **7**, 308–313.

Nocedal, J. and Overton, M. L. (1985). Projected Hessian updating algorithms for nonlinearly constrained optimization, *SIAM J. Num. Anal.*, **22**, 821–850.

Oren, S. S. (1974). On the selection of parameters in self-scaling variable metric algorithms, *Math. Prog.*, **7**, 351–367.

Osborne, M. R. (1972). Topics in optimization, Stanford Univ. Dept. of Comp. Sci. Report STAN-CS-72-279.

Osborne, M. R. (1985). *Finite Algorithms in Optimization and Data Analysis*, John Wiley, Chichester.

Osborne, M. R. and Watson, G. A. (1969). An algorithm for minimax approximation in the nonlinear case, *Computer J.*, **12**, 63–68.

Pietrzykowski, T. (1969). An exact potential method for constrained maxima, *SIAM J. Num. Anal.*, **6**, 217–238.

Polak, E. (1971). *Computational Methods in Optimization: A Unified Approach*, Academic Press, New York.

Powell, M. J. D. (1964). An efficient method for finding the minimum of a function of several variables without calculating derivatives, *Computer J.*, **7**, 155–162.

Powell, M. J. D. (1965). A method for minimizing a sum of squares of nonlinear functions without calculating derivatives, *Computer J.*, **11**, 302–304.

Powell, M. J. D. (1969). A method for nonlinear constraints in minimization problems, in *Optimization* (Ed. R. Fletcher), Academic Press, London.

Powell, M. J. D. (1970a). A new algorithm for unconstrained optimization, in *Nonlinear Programming* (Eds J. B. Rosen, O. L. Mangasarian and K. Ritter), Academic Press, New York.

Powell, M. J. D. (1970b). A hybrid method for nonlinear equations, in *Numerical Methods for Nonlinear Algebraic Equations* (Ed. P. Rabinowitz), Gordon and Breach, London.

Powell, M. J. D. (1971). On the convergence of the variable metric algorithm, *J. Inst. Maths. Applns.*, **7**, 21–36.

Powell, M. J. D. (1972a). Unconstrained minimization and extensions for constraints, AERE Harwell Report TP495.

Powell, M. J. D. (1972b). Some properties of the variable metric algorithm, in *Numerical Methods for Nonlinear Optimization* (Ed. F. A. Lootsma), Academic Press, London.

Powell, M. J. D. (1972c). Quadratic termination properties of minimization algorithms, I and II, *J. Inst. Maths. Applns.*, **10**, 333–342 and 343–357.

Powell, M. J. D. (1972d). Unconstrained minimization algorithms without computation of derivatives, AERE Harwell Report TP483.

Powell, M. J. D. (1972e). Problems related to unconstrained optimization, in *Numerical Methods for Unconstrained Optimization* (Ed. W. Murray), Academic Press, London.

Powell, M. J. D. (1973). On search directions for minimization algorithms, *Math. Prog.*, **4**, 193–201.

Powell, M. J. D. (1975a). A view of minimization algorithms that do not require derivatives, *ACM Trans. Math. Software*, **1**, 97–107.

Powell, M. J. D. (1975b). Convergence properties of a class of minimization algorithms, in *Nonlinear Programming 2* (Eds O. L. Mangasarian, R. R. Meyer and S. M. Robinson, Academic Press, New York.

Powell, M. J. D. (1976). Some global convergence properties of a variable metric algorithm for minimization without exact line searches, in *SIAM-AMS Proceedings, Vol. IX* (Eds R. W. Cottle and C. E. Lemke), SIAM Publications, Philadelphia.

Powell, M. J. D. (1977a). Quadratic termination properties of Davidon's new variable metric algorithm, *Math. Prog.*, **12**, 141–147.

Powell, M. J. D. (1977b). Restart procedures for the conjugate gradient method, *Math. Prog.*, **12**, 241–254.

Powell, M. J. D. (1977c). Constrained optimization by a variable metric method, Cambridge Univ. DAMTP Report 77/NA6.

Powell, M. J. D. (1978a). A fast algorithm for nonlinearly constrained optimization calculations, in *Numerical Analysis, Dundee 1977* (Ed. G. A. Watson), Lecture Notes in Mathematics 630, Springer-Verlag, Berlin.

Powell, M. J. D. (1978b). The convergence of variable metric methods for nonlinearly constrained optimization calculations, in *Nonlinear Programming 3* (Eds O. L. Mangasarian, R. R. Meyer and S. M. Robinson), Academic Press, New York.

Powell, M. J. D. (1985a). On the quadratic programming algorithm of Goldfarb and Idnani, in *Mathematical Programming Essays in Honor of George B. Dantzig, Part II* (Ed. R. W. Cottle), Mathematical Programming Study 25, North-Holland, Amsterdam.

Powell, M. J. D. (1985b). On error growth in the Bartels–Golub and Fletcher–Matthews algorithms for updating matrix factorizations, Cambridge Univ. DAMTP Report 85/NA10.

Powell, M. J. D. (1985c). The performance of two subroutines for constrained optimization on some difficult test problems, in *Numerical Optimization 1984* (Eds P. T. Boggs, R. H. Byrd and R. B. Schnabel), SIAM Publications, Philadelphia.

Powell, M. J. D. (1985d). How bad are the BFGS and DFP methods when the objective function is quadratic?, Univ. of Cambridge DAMTP Report 85/NA4.

Powell, M. J. D. (1985e). Updating conjugate directions by the BFGS formula, Univ. of Cambridge DAMTP Report 85/NA11.

Powell, M. J. D. and Yuan, Y. (1986). A recursive quadratic programming algorithm that uses differentiable penalty functions, *Math. Prog.*, **35**, 265–278.

Pshenichnyi, B. N. (1978). Nonsmooth optimization and nonlinear programming, in *Nonsmooth Optimization* (Eds C. Lemarechal and R. Mifflin), IIASA Proceedings 3, Pergamon, Oxford.

Rall, L. B. (1969). *Computational Solution of Nonlinear Operator Equations*, John Wiley, New York.

Reid, J. K. (1971a). On the method of conjugate gradients for the solution of large sparse systems of equations, in *Large Sparse Sets of Linear Equations* (Ed. J. K. Reid), Academic Press, London.

Reid, J. K. (ed.) (1971b). *Large Sparse Sets of Linear Equations*, Academic Press, London.

Reid, J. K. (1975). A sparsity-exploiting version of the Bartels–Golub decomposition for linear programming bases, AERE Harwell Report CSS20.

Rhead, D. (1974). On a new class of algorithms for function minimization without evaluating derivatives, Univ. of Nottingham, Dept. of Mathematics Report.

Robinson, S. M. (1982). Generalised equations and their solutions, Part II: applications to nonlinear programming, in *Optimality and Stability in Mathematical Programming* (Ed. M. Guignard), Mathematical Programming Study 19, North-Holland, Amsterdam.

Rockafeller, R. T. (1974). Augmented Lagrange multiplier functions and duality in non-convex programming, *SIAM J. Control*, **12**, 268–285.

Rosen, J. B. (1960). The gradient projection method for nonlinear programming, Part I: Linear constraints, *J. SIAM*, **8**, 181–217.

Rosen, J. B. (1961). The gradient projection method for nonlinear programming, Part II: Nonlinear constraints, *J. SIAM*, **9**, 514–532.

Sargent, R. W. H. (1974). Reduced gradient and projection methods for nonlinear programming, in *Numerical Methods for Constrained Optimization* (Eds P. E. Gill and W. Murray), Academic Press, London.

Saunders, M. A. (1972). Product form of the Cholesky factorization for large-scale linear Programming, Stanford Univ. Report STAN-CS-72-301.

Schubert, L. K. (1970). Modification of a quasi-Newton method for nonlinear equations with a sparse Jacobian, *Maths. Comp.*, **24**, 27–30.

Schittkowski, K. (1983). On the convergence of a sequential quadratic programming method with an augmented Lagrangian line search function, *Math. Operationsforschung u. Stat., Ser. Optimization*, **14**, 197–216.

Schittkowski, K. and Stoer, J. (1979). A factorization method for the solution of constrained linear least-squares problems allowing data changes, *Numer. Math.*, **31**, 431–463.

Shanno, D. F. (1970). Conditioning of quasi-Newton methods for function minimization, *Maths. Comp.*, **24**, 647–656.

Shanno, D. F. and Phua, K.-H. (1978). Matrix conditioning and nonlinear optimization, *Math. Prog.*, **14**, 149–160.

Sherman, A. H. (1978). On Newton iterative methods for the solution of systems of nonlinear equations, *SIAM J. Num. Anal.*, **15**, 755–771.

Sinclair, J. E. and Fletcher, R. (1974). A new method of saddle point location for the calculation of defect migration energies, *J. Phys. C: Solid State Phys.*, **7**, 864–870.

Sinclair, J. W. (1979). On quasi-Newton methods, Ph.D. thesis, Univ. of Dundee Dept. of Mathematics.

Smith, C. S. (1962). The automatic computation of maximum likelihood estimates, N.C.B. Sci. Dept. Report SC846/MR/40.

Spendley, W., Hext, G. R. and Himsworth, F. R. (1962). Sequential application of simplex designs in optimization and evolutionary operation, *Technometrics*, **4**, 441–461.

Stewart, G. W. (1967). A modification of Davidon's minimization method to accept difference approximations to derivatives, *J. Ass. Comput. Mach.*, **14**, 72–83.

Stoer, J. (1975). On the convergence rate of imperfect minimization algorithms in Broyden's β-class, *Math. Prog.*, **9**, 313–335.

Swann, W. H. (1972). Direct search methods, in *Numerical Methods for Unconstrained Optimization* (Ed. W. Murray), Academic Press, London.

Swann, W. H. (1974). Constrained optimization by direct search, in *Numerical Methods for Constrained Optimization* (Eds P. E. Gill and W. Murray), Academic Press, London.

Tapia, R. (1984). On the characterization of Q-superlinear convergence of quasi-Newton methods for constrained optimization, Rice Univ. Dept. of Math. Sci. Report 84–2.

Toint, Ph.L. (1977). On sparse and symmetric updating subject to a linear equation, *Maths. Comp.*, **31**, 954–961.

Varga, R. S. (1962). *Matrix Iterative Analysis*, Prentice-Hall, Englewood Cliffs, N.J.

Walsh, G. R. (1975). *Methods of Optimization*, John Wiley, London.

Watson, G. A. (1978). A class of programming problems whose objective function contains a norm, *J. Approx. Theo.*, **23**, 401–411.

Watson, G. A. (1979). The minimax solution of an overdetermined system of nonlinear equations, *J. Inst. Maths. Applns.*, **23**, 167–180.

Wilkinson, J. H. (1965). *The Algebraic Eigenvalue Problem*, Oxford Univ. Press, Oxford.

Wilkinson, J. H. (1977). Some recent advances in numerical linear algebra, in *The State of the Art in Numerical Analysis* (Ed. D. A. H. Jacobs), Academic Press, London.

Wilson, R. B. (1963). A simplicial algorithm for concave programming, Ph.D. dissertation, Harvard Univ. Graduate School of Business Administration.

Winfield, D. (1973). Function minimization by interpolation in a data table, *J. Inst. Maths. Applns.*, **12**, 339–348.

Wolfe, P. (1959). The secant method for simultaneous nonlinear equations, *Comm. Ass. Comput. Mach.*, **2**, 12–13.

Wolfe, P. (1961). A duality theorem for nonlinear programming, *Quart. Appl. Math.*, **19**, 239–244.

Wolfe, P. (1963a). Methods of nonlinear programming, in *Recent Advances in Mathematical Programming* (Eds R. L. Graves and P. Wolfe), McGraw-Hill, New York.

Wolfe, P. (1963b). A technique for resolving degeneracy in linear programming, *J. SIAM*, **11**, 205–211.

Wolfe, P. (1965). The composite simplex algorithm, *SIAM Rev.*, **7**, 42–54.

Wolfe, P. (1967). Methods of nonlinear programming, in *Nonlinear Programming* (Ed. J. Abadie), North-Holland, Amsterdam.

Wolfe, P. (1968a). Another variable metric method, working paper.

Wolfe, P. (1968b). Convergence conditions for ascent methods, *SIAM Rev.*, **11**, 226–235.

Wolfe, P. (1972). On the convergence of gradient methods under constraint, *IBM J. Res. and Dev*, **16**, 407–411.

Wolfe, P. (1975). A method of conjugate subgradients, in *Nondifferentiable Optimization* (Eds M. L. Balinski and P. Wolfe), Mathematical Programming Study 3, North-Holland, Amsterdam.

Wolfe, P. (1980). The ellipsoid algorithm, *Optima* (Math. Prog. Soc. newsletter), Number 1.

Womersley, R. S. (1978). An approach to nondifferentiable optimization, M.Sc. thesis, Univ. of Dundee Dept. of Mathematics.

Womersley, R. S. (1981). Numerical methods for structured problems in nonsmooth optimization, Ph.D. thesis, Univ. of Dundee Dept. of Mathematics.

Womersley, R. S. (1982). Optimality conditions for piecewise smooth functions, in

Nondifferential and Variational Techniques in Optimization (Eds D. C. Sorensen and R. J.-B. Wets), Mathematical Programming Study 17, North-Holland, Amsterdam.
Womersley, R. S. (1984a). Censored discrete linear l_1 approximation, ANU Dept. of Math. Sci. Res. Report No. 8.
Womersley, R. S. (1984b). Minimizing nonsmooth composite functions, ANU Dept. of Math. Sci. Res. Report No. 13.
Womersley, R. S. and Fletcher, R. (1986). An algorithm for composite nonsmooth optimization problems, *J. Opt. Theo. Applns.*, **48**, 493–523.
Yuan, Y. (1984). An example of only linear convergence of trust region algorithms for nonsmooth optimization, *IMA J. Num. Anal.*, **4**, 327–335.
Yuan, Y. (1985a). Conditions for convergence of trust region algorithms for nonsmooth optimization, *Math. Prog.*, **31**, 220–228.
Yuan, Y. (1985b). On the superlinear convergence of a trust region algorithm for nonsmooth optimization, *Math. Prog.*, **31**, 269–285.
Zoutendijk, G. (1960). *Methods of Feasible Directions*, Elsevier, Amsterdam.

Subject Index